conversion factors

#109

	TO CONVERT		MULTIPLY BY ↓
	FROM	TO	
6.947×10^{-14}	erg/molecule	kcal/mole	1.439×10^{13}
4.336×10^{-2}	$\dfrac{\text{electron volts}}{\text{molecule}}$	kcal/mole	23.061
3.498×10^{2}	$\dfrac{\text{cm}^{-1}}{\text{molecule}}$	kcal/mole	2.859×10^{-3}
5.034×10^{15}	$\dfrac{\text{cm}^{-1}}{\text{molecule}}$	erg/molecule	1.986×10^{-16}
2.390×10^{-4}	kcal/mole	joules/mole	4.1840×10^{3}
↑ MULTIPLY BY	TO	FROM	
	TO CONVERT		

understanding chemistry

GEORGE C. PIMENTEL

University of California,
Berkeley

Petit salaud!
Caramba!
Canaille!
eat shit!

RICHARD D. SPRATLEY

University of
British Columbia

understanding
chemistry

HOLDEN–DAY, INC.

San Francisco
Cambridge
London
Amsterdam

THE COVER

The cover was designed by Steve Osborn. It takes for its setting a field of flowers, signifying the real world around us that we are trying to appreciate through understanding. Through the spectral sequence, the flowers are seen in higher and higher magnification, culminating in the chemist's atomic view of their inner form. The atomic representation, then, furnishes the inner field that we find supported by representations appropriate to the major unifying theories of chemistry, thermochemistry (front cover) and quantum mechanics (rear cover). It is fitting that we pass from the real world we live in to the atomic world we visualize through a spectral rainbow, since the spectrum of the hydrogen atom is the keystone of quantum mechanics, the basis for understanding the structure and behavior of atoms and molecules.

The composition, printing, and binding of this text were done by The Maple Press Company, York, Pennsylvania. The text typeface is Monotype News Gothic.

To our daughters
Jan and Tess, Chris and Amy
Sarah and Rohan

preface

Understanding Chemistry is designed for an introductory college course that includes students specializing in all branches of the physical, biological, and premedical sciences and students in the humanities and social sciences who have taken high school chemistry. No mathematics beyond elementary algebra is required. The material presented here has been used successfully for three years in the general chemistry course at Berkeley, for which no college mathematics is required. It presents, we believe, both a valid characterization of scientific activity and a firm foundation for further scientific study.

This text takes as its premise the idea that a modern introductory chemistry course must be structured around a meaningful introduction to chemical principles. Such an introduction

—is essential to a useful consideration of the vast accumulation of knowledge in this field. With the ever-accelerating growth of this knowledge, it would be bewildering, soon overwhelming, if presented without the coherence and meaning imparted by our unifying principles.

—is necessary to a valid picture of chemistry as it exists today. In all branches of chemistry, guiding principles are constantly at work, facilitating new applications and stimulating new discoveries.

—is optimum to a display of the interplay between observation and codification—between experiment and theory. This interplay defines scientific activity.

The course begins with a carefully structured discussion of the nature of scientific activity. The relationships between observations, unifying principles, and predictions are explored, using examples from the most basic ideas of chemistry. These simple, but fundamental, concepts not only reveal most clearly the *modus operandi* of scientific investigation, they also review the essential chemical topics needed for the course. Many problems are offered at the end of Chapter 1 to sharpen up these basic tools: the atomic theory, the mole concept, weight relations, gas laws, the Periodic Table, uncertainty, the electron-dot view of bonding.

The subject of equilibrium is entered through solubility, acid-base behavior, and electrochemical cells. These topics provide colorful and significant opportunities for laboratory experimentation and they build up a valuable fund of chemical facts and vocabulary. They lead to exploration of energy and randomness, factors that motivate and determine the course and the pace of chemical change.

The text then turns to the microscopic world of atoms and molecules conjured up by chemists to help them explain the macroscopic world around us. The simplest atom, H, and the simplest molecule, H_2^+, provide keys to our understanding of the structures of atoms and of molecules. This understanding brings coherence to the central subject of chemistry, chemical bonding.

The text concludes with a substantial amount of descriptive chemistry. With a firm base of macroscopic theory and microscopic understanding, the myriad of known chemical facts can be appreciated by example and trend. Because of its special significance to life processes, the chemistry of carbon compounds is afforded an entire chapter. Organic chemistry is treated in sufficient depth to reach biologically important molecules (proteins, DNA, and enzymes) and the energy effects in their reactions. The rich chemistry of the lower part of the Periodic Table (the "long rows") is generously typified, and the subjects of nuclear reactions and nuclear energy liberation are placed naturally in the context of actinide chemistry.

Each chapter is accompanied by an abundance of exercises and problems. They provide a guide to the important ideas associated with each chapter and demonstrate the applicability of these ideas to everyday problems of concern to society. As a study aid, answers are given in the back of the book to all odd-numbered problems. A huge volume of chemical information is efficiently compiled in eight appendices. The Periodic Table and Table of Atomic Weights are placed for convenient reference on the inside covers. The Subject Index is quite comprehensive and it will aid in the use of the book. It contains a complete listing of definitions to facilitate return to the introduction of a particular idea. It also lists the Tables so that the mass of information presented in illustrative tables can be readily tapped. The Subject Index is supported by a Formula Index to make accessible the large amount of descriptive material dispersed throughout the book.

Throughout this book, we have placed most emphasis on concept and significance—"*What does it mean?*" and "*Where does it lead?*" However, the lasting value of your study will probably not be based in the particular answers you choose. These conclusions will erode and change as time passes and as knowledge grows. What will last, and serve you well, will be the habit of asking these questions and the ability to judge what is a useful answer. That is what we call "critical thinking." It is sometimes—all too rarely—applied to human affairs. It is really what science is all about. Try it on for size—you won't be sorry.

Oh, by the way, science is lots of fun. Have some.

George C. Pimentel
Richard D. Spratley

January 1971

acknowledgments

We have many people to thank for their help in the development of the material in this book. Our faculty and graduate student colleagues who participated in the introductory chemistry instruction during the last three years contributed much to the content and pedagogy. Many freshmen also contributed through their conscientious efforts as our material was taking form. Our friends and families offered the continual encouragement and forbearance necessary to its completion. Judi Plitt was the copy editor; her attention to detail, her thoroughness, and her pleasant disposition and calm to the very last are very much appreciated. Our publisher, Holden-Day, Inc., and particularly its president, Frederick Murphy, has given us every possible support, including enthusiastic encouragement, reassuring confidence, and unlimited patience throughout the deliberate years that we have devoted to this book. Finally, we want to thank Susan Arbuckle for her tireless efforts, her accurate typing, her delicious candies, and her inexhaustible cheer.

contents

**understanding
chemistry**

one   science and chemistry

1-1   Prologue

The 50-ton telescope slowly swung 'round to the North and then, in its giant yoke, it tilted down toward the dark horizon. Deep in the observatory, the 120-inch beam of light was focussed into a one-quarter inch image of Mars, to the delight of two scientists bundled in heavy coats against the cold of the laboratory. With delicate mirror adjustments, they carefully directed the now-slender beam into the optical path of an infrared spectrometer. In this sensitive instrument, the faint light that had traversed 100 million miles was dispersed into an invisible infrared spectrum—not invisible, however, to a germanium semiconductor detector cooled to the temperature of liquid helium. The spectrometer grating slowly turned, sweeping the spectrum past the detector. As the "color" of the light striking the germanium changed, a miniscule voltage was generated, fluctuating in response to the changing light intensity. The electrical signal was amplified and amplified again, then represented by a pen record on a suitable graph.

The two scientists, the designers of the infrared spectrometer, are not astronomers tracking the orbit of the planets—neither are they physicists measuring the temperature of the Red Planet. They are chemists and their aim is the chemical analysis of the Martian atmosphere!

This is the face of modern chemistry. Methods that used to be classified as physics are fully employed by organic, inorganic, and physical chemists alike. Added to the chemists' traditional tools—balance and buret, flask and funnel—are such impressive techniques as gas chromatography, nuclear magnetic resonance spectroscopy, optical rotary dispersion, infrared spectroscopy, nuclear quadrupole coupling measurements, X-ray crystallography. The success of these revolutionary methods is almost incredible—over one and one-half million compounds have been synthesized, of which perhaps 900,000 were not known in nature prior to their preparation in the laboratory. New compounds are being discovered at the rate of 300 per day, day in, day out. The products of chemistry so much influence our everyday life that we hardly take notice—annually, thousands of tons of inorganic fertilizer multiply manyfold our food production; millions of gallons of refined organic liquids fuel our automobiles; drugs

1

such as atabrine, sulfathiozole, penicillin, and 5-fluorouracil give promise of paring from the list of threats to man's survival such ailments as malaria, poliomyelitis, leprosy, and soon, no doubt, even cancer.

By examining in this text this new face of chemistry, we shall try to learn about science, perhaps the most dominant human endeavor of our time. The fruits of science lie all around us—they affect our lives in countless ways, they simultaneously protect us against hazards of our environment and, in the hands of unwary politicians, threaten us with self-destruction. Our intimate and growing knowledge of the life process penetrates to the corners of personal philosophy. What is this powerful process called science? What force impels us in its quest? How will it reshape our existence in the next decade? In the next century? What are its capabilities and what are its limitations?

All of these questions are difficult to answer in a few words—some answers are subjective and each of us must find his own through familiarity with the ways of science. That will be our primary goal in this text—to familiarize you with science through your involvement in its activities, as exemplified in chemistry. As you see the state of chemistry today and how we came there, you will measure the power of science, its capabilities, and its bounds. Perhaps you will, as well, begin to share the excitement and personal involvement felt by the scientist and, at last, you will begin to sense and be readied for the impact science will have on our world around us. Then you will be ready to participate in this exploration and to direct its efforts to the benefit of mankind.

1-2   What is scientific activity?

(a) SCIENCE AND COMMUNICATION

Science is man's systematic study of himself and his environment. Its roots are, no doubt, simply man's will to survive in a threatening scene. This will to survive impels us to investigate nature and to organize what we learn to make it benefit and prolong our existence. This type of activity can be seen in other animals in a rudimentary form—when a family moves into a new home, their pet dog or cat will investigate his new abode with the same curiosity, interest, and nervousness as the children. In less than a week, you will see him put his study to work as he sleeps in the warmest spot in the house. A jungle animal will locate and return to the best hunting grounds and the reliable watering places. A bird will fly to the south as winter closes in. Such behavior in animals is often designated instinctive behavior and indeed it is. But we must recognize that our own curiosity, interest, and nervousness about the things around us are instinctive, too.

Nevertheless, there is a world of difference between the legendary curiosity of a cat and scientific activity as conducted by *Homo sapiens*. This activity is, of course, but one facet of man's mental capacity, far

—Courtesy of the Fisher Collection of Alchemical and Historical Pictures, Pittsburgh.

Figure 1-1 The beakers and flasks of the chemist of old (a) have been replaced or supplemented by intricate electronic instruments such as the infrared spectrometer shown in (b).

exceeding that of the next form of life on the intellectual ladder. Anthropologists tell us that this capacity developed about three million years ago over a period of two or three hundred thousand years, only an hour or so in an evolutionary day. This period, revealed by the simultaneous appearance of rapidly increasing cranial volume on the one hand and, on the other, of sophisticated tools, housing, art, pottery, and other artifacts of dawning civilization, apparently was triggered by the development of communication. Whether the inference is correct or not, certainly communication is the key to all scientific progress. It adds a cumulative dimension to man's investigation of his environment. It is not necessary for each of us to discover for himself that the atomic theory is magnificiently successful in explaining the properties of gases and the weights of chemical reactants. These ideas were developed some 150 years ago; experiments and interpretations were made available to all future generations in books. It is not necessary for a chemist in Novosibirsk to discover for himself that the xenon–oxygen compounds are excellent oxidizing agents but dangerously explosive. These properties were discovered in laboratories in Vancouver and Berkeley and the facts were shared with other workers at scientific meetings and in technical journals. The chemist in Novosibirsk can add these results to his own studies of the heat effects that accompany the formation of xenon trioxide. And he will, in turn, communicate his results back to Canada, to California, to all other chemists, and finally, to students of chemistry like yourself. Each of us gains height from our fellows—we

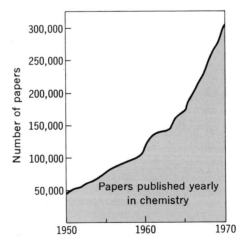

Figure 1-2 *Growth of the chemical literature. During the 20-year period 1950–1970, the annual production of research papers in chemistry increased sixfold. In fact, the chemical literature published during the 60's is as large as that published prior to 1960—throughout history. The growth of the literature is a good measure of the growth of activity in chemical research.*

are standing on the shoulders of our predecessors—all through the medium of communication. *A crucial part of scientific activity is communication,* oral and written, of observations and interpretations, of decisions and doubts, of our knowledge as it grows.

(b) SCIENCE AND CREATIVITY

Another aspect of man's investigation of his surroundings seems to differentiate him from other animals. He displays an imaginative dimension that permits him to develop abstract and intuitive interconnections. This dimension is called creativity. The richness of our science—indeed, the existence of all of our culture—can be attributed to man's creativity.

It is not easy to place an all-encompassing definition on creativity. We think of an artist vividly expressing in a painting his emotions, his reaction to the things around him, his relationships with people. A composer does the same, but with musical expression. What has a colorful pattern or a rhythmic melody to do with love, or loneliness, or patriotism, or relaxation? Each is an abstract and intuitive interconnection among human sensations and experience.

In the same way, a scientist makes interconnections among his observations and his experiences. He relates properties of systems that are not well understood to properties of systems that have been more thoroughly studied. He seeks hidden likenesses that give coherence to his knowledge of the surroundings. All levels of abstraction are involved in this search, but all involve the exercise of creativity.

Here are some simple likenesses that anyone can perceive.

—A piece of gold is bright and shiny—like a piece of silver.
—Gasoline can be poured—like water.
—Mercury is bright and shiny—like silver—and it can be poured—like water.

Such obvious relationships form the basis of the most rudimentary scientific activities—description and classification. They lead to class definitions: Gold and silver are easily distinguished by their different colors, but they are classified together as metals; gasoline and water have entirely different odors, but they are both called liquids; mercury is both a metal and a liquid. This classification process helps us develop certain expectations: Mercury should conduct electricity (since it is a metal) and boil when heated (since it is a liquid). This is the reward from this creative process. It arose from the imaginative coupling of parts of the environment that, though distinguishably different, are interestingly similar. That is the essence of creativity in science—recognizing likenesses in the presence of differences, "looking at one thing and seeing another."

Figure 1-3 Likenesses. As the scientist probes the behavior of his compounds, he strives to relate properties that are new or poorly understood to those of more familiar systems.

(c) SCIENCE AND EXPLANATION

But science is much more than just classification. It involves explanation and understanding. So what does it mean to explain? What does it mean to understand? Another set of likenesses will show.

 —Sea water conducts electricity—like a metal.
 —The sun emits light—like a white-hot tungsten wire.
 —A glowing gas in a fluorescent light fixture conducts electricity—like a metal—and emits light—like a white-hot tungsten wire.

 Sea water and metals have in common the property that they conduct electricity—they are both said to have "mobile charge carriers." The sun emits light as does a hot tungsten wire—we conclude that the sun is hot. A glowing gas conducts electricity, so it, too, must have mobile charge carriers. But touch the glass tube of a fluorescent light fixture—it is *not* hot like the tungsten wire or the sun! But wait! How do we know for sure that the sun really is hot? We can't reach out and touch it, after all. If a glowing gas can emit light without being hot, then how do we know that the sun is at a very high temperature, like a glowing tungsten wire, rather than glowing at a moderate temperature, like a fluorescent light bulb?

 Almost any scientist would be pleased to expound for a few minutes on the answer to that naive question. (Or is it naive?) He might explain that electrons are accelerated in a fluorescent tube by very high voltages that don't exist on the sun. (How does he know that they don't? By the way, what is an electron?) Besides, the spectral distribution of the light from the glowing tube is rather different from that of a hot tungsten wire and the sun's light looks more like that of the wire. (How did he learn that?) And there are nuclear processes occurring at the center of the sun that cause it to be hot. (Can he see into the center of the sun?) Yet, he might add, there is evidence that electrons are present in the sun's gaseous exterior (oh, so the sun *is* a gas, containing charge carriers, after all), but they arise by different mechanisms. (My, this fellow knows a lot about something 90 million miles away!)

Yes, we do know a lot about the sun, despite its distance. We have an elaborate and carefully considered set of likenesses that relate the sun's behavior to events that occur close at hand where they can be studied systematically. This set of likenesses constitutes a *model* of the behavior of the less readily accessible object of interest—the sun. They constitute an explanation and, equally, an understanding of the sun.

So we see that explanation and understanding are just the elaborate end result of seeking likenesses. They permit knowledge in one area of experience to be applied in another area of relative ignorance.

(d) SCIENCE AND MATHEMATICS

The most useful likenesses are those that have been made quantitative. Then the powerful methods of mathematics can be used to extract subtle implications of a model. The behavior of a gas provides a clear example.

Any serious student of New Year's Eve parties knows that as gas is blown into a balloon, the balloon gets bigger. The air inside presses against the balloon walls, stretching the rubber. When the balloon reaches full size, we have to blow harder and harder to push in more air. The air pressure in the balloon gets higher and higher. Let's study this important phenomenon in the laboratory.

Figure 1-4 shows a 2-liter bulb, instead of a balloon, and attached to it is a U-shaped tube containing mercury. In the right arm of the tube, there is a vacuum above the mercury, but any gas in the bulb can enter the left arm to be in contact with the mercury surface. Figure 1-4(a) shows us the system with no gas in the bulb. The mercury levels are at the same height. Now, in Figure 1-4(b), 0.32 gram of oxygen gas is introduced. The mercury levels change, lifting mercury in the right arm against the force of gravity. Evidently the oxygen gas presses down on the mercury in the left arm. In Figure 1-4(c), 0.64 gram of oxygen gas has been added. The left-hand mercury level is pressed down even further.

This U-shaped device, called a mercury manometer, is handy because it permits us to discuss this gas behavior quantitatively. The difference in heights of the right and left columns, coupled with the density of mercury, measures the gas pressure. It has long been expressed in the simple length units, "millimeters of mercury" or, more briefly, "mm Hg." A more recent, if less explanatory, term for "mm Hg" is "torr." In Figure 1-4(b), the mercury levels differ by 93 mm; the pressure is 93 torr. In Figure 1-4(c), with twice as much gas, the pressure is 186 torr, twice as big. More gas—more pressure. Why is that?

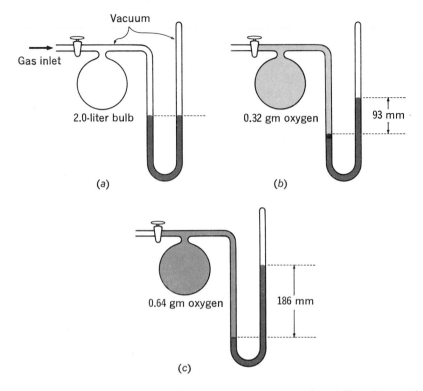

Figure 1-4 The measurement of gas pressure: At room temperature, 0.32 g of oxygen in a two-liter bulb exerts a pressure sufficient to support a 93 mm column of mercury. (In scientific circles, "room temperature" is defined to be 25°C or 298°K.) When the amount of oxygen is doubled, the pressure also doubles.

The "Why?" question asks for an explanation. We need a useful model— one that we understand—with which to "explain" the gas behavior. For example, a gas can be likened to a collection of particles that behave in their motion and collisions as do billiard balls. The dynamics of billiard ball collisions are intuitively familiar. By looking at a gas and "seeing" a collection of tiny billiard balls, we can "understand" the observation that increasing the amount of gas increases the pressure. The pressure is due to collisions between the particles and the walls, including the surface of the mercury in the U tube. More particles, more pressure! In a similar way, we can understand why expanding a gas into a larger volume decreases its pressure, whereas raising the temperature increases it. If the volume is increased, the tiny particles strike the walls less often as they move about. Since the pressure is attributed to the particles striking the wall, the pressure will be lowered. Raising the temperature, on the other hand, causes the particles to move more rapidly. They strike the walls more often and at higher speeds. Again attributing pressure to the particles colliding with the walls, we see that the pressure must rise.

We find the following agreements between the billiard ball model of a gas and our common experience with how gases behave:

Pressure increases as the amount of gas is increased.   (1-1)

Pressure decreases as the volume is increased.   (1-2)

Pressure increases as the temperature is increased.   (1-3)

This qualitative accord with common knowledge about gases encourages us that the particle model of gas behavior is a useful likeness. Can we make it quantitative? Let's try to apply to gases our quantitative knowledge of the kinetic energy and momentum held by a billiard ball of mass m and velocity v:

$$\text{kinetic energy} = \tfrac{1}{2}mv^2 \qquad (1\text{-}4)$$

$$\text{momentum} = mv \qquad (1\text{-}5)$$

If we assume that gaseous neon is a collection of particles that are all alike, then each particle should have the same mass, m_{Ne}, characteristic of neon particles. A cubical box of neon contains many neon particles, n_{Ne}. In this box, the particles are moving about with some average velocity, v_{Ne}.

In terms of these quantities, let's try to deduce the pressure that is exerted against one of the walls in the neon box. If the neon particles behave like billiard balls, the pressure is determined by how "hard" each particle hits the wall and how many particles hit the wall per second. How "hard" the particle hits the wall is determined by its momentum, $m \cdot v$. It's like making a goal-line stand in a football game: How hard our right tackle must hit the man opposite him to stop him cold is determined by his mass and how fast he charges, $m \cdot v$. If we want to stop him and reverse his movement, it takes $2(mv)$. If the offensive fullback and halfback come pounding in too, the pressure on our right tackle increases proportionately. The pressure, either on the wall of a bulb or on our right tackle, is the momentum transferred per second to a unit area. We need only express this momentum transfer in mathematical form.

$$\text{Pressure} = \left(\frac{\text{momentum transfer}}{\text{collision}}\right)\left(\frac{\text{collisions}}{\text{second}}\right)\left(\frac{1}{\text{wall area}}\right) \qquad (1\text{-}6)$$

If a neon particle approaches the wall perpendicular to its surface with momentum mv and leaves with the same momentum in the opposite

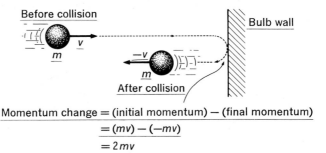

Before collision

Bulb wall

After collision

$$\text{Momentum change} = (\text{initial momentum}) - (\text{final momentum})$$
$$= (mv) - (-mv)$$
$$= 2mv$$

Figure 1-5 Momentum transfer per collision: After collision with the wall, the velocity is equal in magnitude to the velocity before collision, but opposite in sign (it is going in the opposite direction). Hence the momentum change is 2(mv).

direction, then the momentum transferred to the wall in this single collision is:

$$\frac{\text{momentum transfer}}{\text{collision}} = 2(mv) \tag{1-7}$$

At least that is what happens when billiard balls strike the table cushion or when an opposing lineman is driven back.

The second term in (1-6), collisions per second, depends upon the box dimension and the particle's velocity (as it bounces back and forth between the walls). We can assume, in a simple-minded way, that one third of the molecules are moving in the east-west direction, one way or the other. So $n/3$ molecules are bouncing back and forth, alternately hitting the east, then the west walls of the box. The east wall receives a collision each time one of the particles travels the box dimension, ℓ, and back, a distance of 2ℓ.

$$\frac{\text{collisions}}{\text{second}} = \frac{(\text{no. of particles bouncing back and forth})}{(\text{time for a particle to travel distance } 2\ell)}$$

$$\frac{\text{collisions}}{\text{second}} = \frac{(n/3)}{2\ell/v} = \left(\frac{n}{3}\right)\left(\frac{v}{2\ell}\right) = \frac{nv}{6\ell} \tag{1-8}$$

The area of a wall of the cubical box is just ℓ^2. So now we can substitute (1-7) and (1-8) into (1-6) and calculate the pressure.

$$\text{Pressure} = \left(\frac{\text{momentum transfer}}{\text{collision}}\right)\left(\frac{\text{collisions}}{\text{second}}\right)\left(\frac{1}{\text{wall area}}\right)$$

$$= (2mv) \cdot \left(\frac{nv}{6\ell}\right) \cdot \left(\frac{1}{\ell^2}\right) \tag{1-9}$$

$$= \frac{1}{3}\left(\frac{n}{\ell^3}\right)(mv^2)$$

The quantity ℓ^3 is just the volume V of the box and, by (1-4), mv^2 is two times the kinetic energy, KE.

$$P = \frac{1}{3}\left(\frac{n}{V}\right)(2 \text{ KE}) = \frac{2}{3} \cdot \frac{n \cdot \text{KE}}{V} \qquad (1\text{-}10)$$

In (1-10) we see a precise mathematical statement of the qualitative conclusions (1-1) and (1-2). Pressure is directly proportional to the number of particles and inversely proportional to the volume they occupy. This simple expression came from our billiard ball model. Since it agrees with the measured properties of real gases, it is useful.

If we multiply both sides of (1-10) by the volume, V, the equation appears in a more familiar form.

$$PV = n \cdot \tfrac{2}{3}(\text{KE}) \qquad (1\text{-}11)$$

On the left is the "PV product," pressure times volume. On the right is n, the number of particles, and their average kinetic energy. Neither equation (1-10) nor (1-11) seems to say anything about conclusion (1-3), the effect of temperature. It relates the PV product to the average kinetic energy, instead. Yet there is a connection: We must heat a gas—add energy—to raise its temperature. This added energy must increase the energy of motion held by the particles. If we can decide how temperature and kinetic energy are related, our expression (1-11) will be complete.

This brings us face-to-face with the actual meaning of temperature. Everyone knows several ways to measure the temperature of an object. The easiest way is to touch the object—but that is only qualitative and it can be painful. The color of a hot piece of metal is another clue—we specified earlier, a "white-hot" tungsten wire to indicate something hotter than "red-hot." A mercury thermometer is the familiar quantitative device—quantitative because it has a scale on it. But even then, some thermometers are calibrated in Fahrenheit degrees and some in Centigrade degrees. Besides, what does the density of mercury, as it changes with temperature, tell us about the kinetic energy of gas particles? There is no easy answer.

Perhaps it would be easier to use the gas itself as a thermometer. That is what scientists have done—they developed what is called the "absolute temperature scale." In terms of experimental operations, it means measuring the pressure-volume product for a fixed amount of gas (a fixed value of n). By the billiard ball model, that gives a number proportional to the average gas kinetic energy. With a suitable proportionality constant, c, let's call that number the temperature. We are now adding to our quantitative model of gas behavior.

We define an "absolute temperature," T, such that

$$\text{average kinetic energy} = c \cdot T \qquad (1\text{-}12)$$

Inserting (1-12) into (1-11), the gas behavior becomes

$$PV = n \cdot \tfrac{2}{3}(cT)$$

This can be rearranged into the familiar form known as "the perfect gas law."

$$\boxed{PV = nRT}$$
(1-13)

In this expression, R is the "suitable" proportionality constant mentioned earlier. The temperature, essentially defined by the equation (1-13), is called the absolute temperature and is measured in "degrees Kelvin." Experiment shows that it is simply related to Centigrade temperature, t (see Fig. 1-6).

$$T(°K) = t(°C) + 273.16$$
(1-14)

The advantage of using absolute temperature is that it simplifies our understanding of gas behavior. We see that temperature, so defined, directly measures the average kinetic energy of our hypothetical gas particles.

With this highly developed model, we can begin exploring gas behavior more widely. For example, suppose we compare two identical flasks, one containing neon and one containing argon. If these two flasks are immersed in a thermostatted water bath, they will come to the same temperature. Then let's add neon to the neon bulb until the two flasks have the same pressure. Now expression (1-13) permits us to calculate the number of particles in each container:

in the neon bulb $\qquad n_{Ne} = \dfrac{P_{Ne} \cdot V_{Ne}}{RT_{Ne}}$ (1-15)

in the argon bulb $\qquad n_{Ar} = \dfrac{P_{Ar} \cdot V_{Ar}}{RT_{Ar}}$ (1-16)

Since we have arranged that the pressures, temperatures, and volumes be equal,

$$P_{Ne} = P_{Ar}$$

$$V_{Ne} = V_{Ar}$$

$$T_{Ne} = T_{Ar}$$

we conclude that

$$n_{Ne} = n_{Ar}$$

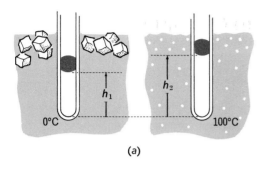

Temp. (°C)	P·V (relative to value at 0°C)
0	(1.00)
50	1.19
100	1.36
200	1.73

(a)

(b) Typical data

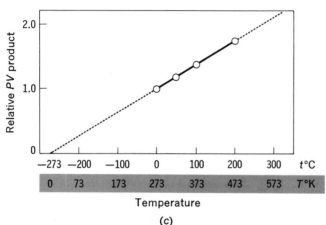

(c)

Figure 1-6   The absolute temperature scale: (a) A drop of mercury traps some air in a narrow tube. The height of the air column, h, is proportional to the volume of the trapped gas. When the temperature is raised from 0°C (ice bath) to 100°C (boiling water), the gas expands while the pressure, determined by the atmosphere above the mercury drop, remains constant. The PV product is thus proportional to the column height. (b) Some typical data, expressed relative to the height measured at 0°C. (c) A plot of the PV product and a determination of the absolute zero of temperature.

This gives us an easy way to obtain equal numbers of neon and argon particles (if they exist). It is a paraphrase of the famous hypothesis attributed to Amadeo Avogadro over one and a half centuries ago:

Avogadro's Hypothesis
"Equal volumes of gases contain equal numbers of particles (at the same temperature and pressure)."

This is only one of many successful applications of this quantitative (mathematical) billiard ball model. It contains useful implications about such different phenomena as gas diffusion and the velocity of sound. As the range of applicability has broadened, our confidence in the model has grown. To proclaim this confidence, the model is now called a theory—the *Kinetic Theory of Gases*. That is the game of science.

We see that from the most elementary act of description to the most abstract development of mathematical theory, scientific activity consists of discovering and exploiting likenesses. These likenesses give coherence to our observations of what goes on around us. We call this coherence "understanding." It permits us to develop expectations for the future from experience in the past. This means we can have a larger measure of safety and comfort in a sometimes threatening environment—we may even gain control of it, to remove the threats and mold it to man's benefit.

It is a simple game—but powerful—and pleasurable—and productive. Let's see how it goes in the corner of science called chemistry.

1-3   What do chemists talk about at lunch?

No one has ever seen the particles we have been mentioning. You would hardly realize that, however, to listen to a group of chemists. They speak of atoms and molecules as if they invented them (which they did). Seldom does anyone mention that this is only a model, devised and shaped to be similar to experience observed in beakers and flasks. Let's listen in on some "chemist talk" and see what the model is like.

(a) ATOMS: THE PEOPLE OF A LILLIPUTIAN WORLD

All the things around us—the solid earth, the liquid ocean, and the gaseous atmosphere—are imagined to be made up of particles called atoms. These atoms must be so small that we do not see them, but many, many properties of substances show that the stuff of our world behaves as if it is composed of these tiny particles.

An atom has a characteristic structure that determines its personality. It contains positive and negative charges that attract and repel each other by the familiar electrostatic forces. A positive charge rides on a particle called a proton. A negative charge rides on a particle called an electron. The proton and electron charges are exactly equal in magnitude, but opposite in sign. Their masses, however, are quite different. An electron weighs only 1/1836 as much as a proton.

An atom contains another kind of particle, a neutron. This fellow has no electrical charge, but he weighs almost the same amount as a proton. Protons and neutrons attract each other by another, mysterious kind of force called "nuclear force." As a result, all of the protons and neutrons bundle together tightly—despite the electrical repulsions among the protons—into a small cluster called the *nucleus* of the atom.

The featherweight electrons are too flighty to be held in the confines of the nucleus. They remain near to it because of its positive charge, but they do not surrender completely their freedom of movement.

There are over 100 different kinds of atoms—chemists have given them names like hydrogen, helium, lithium, and so on. But these are also the names of substances we find around us. How does one atom differ from another? How does the chemist know which kind of atom should have the same name as a particular substance? We'll answer the first question now and the second question later.

The behavior of an atom is determined by the number of protons in the nucleus. That number is, of course, an integer: one, two, three, . . . up to above 100. We'll denote this integer nuclear charge by the symbol Z. It determines the number of electrons liable to be found in the atom. The nucleus strongly attracts electrons (pulling them away from other atoms) as long as there are fewer than Z electrons near it. When the nucleus, with charge $+Z$, finds exactly Z electrons around itself, the net charge is $(+Z) + (-Z) = 0$. The atom is neutral.

All atoms with the same nuclear charge behave alike—they "have the same chemistry," a chemist would say. The mass of an atom is fixed by the sum of the number of protons and neutrons, but electrons couldn't care less about neutrons, so the neutrons in the nucleus are almost without

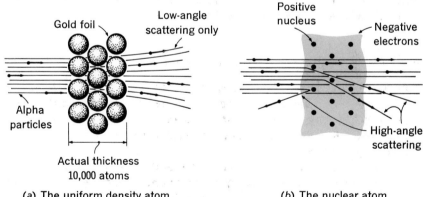

Gold foil

Low-angle scattering only

Positive nucleus

Negative electrons

Alpha particles

Actual thickness 10,000 atoms

High-angle scattering

(a) The uniform density atom   (b) The nuclear atom

Figure 1-7   The structure of the atom: During most of the nineteenth century, atoms were believed to be structureless, indivisible particles. By 1900, it was clear that atoms contained both negative and positive charges. Negative electrons were pictured to be stuck in a sphere of positive electricity of uniform density. In 1911, Ernest Rutherford passed alpha-particles (doubly charged helium atoms from radioactive decay of radium) through gold foil. He expected little scattering, as pictured in (a). The actual results (b) were quite surprising, as expressed in his own words: "It was almost incredible as if you fired a 15 inch shell at a piece of tissue paper and it came back and hit you. . . . when I made calculations, I saw that it was impossible to get anything of that order of magnitude unless. . . the greater part of the mass of the atom was concentrated in a minute nucleus. It was then that I had the idea of an atom with a minute massive center carrying a charge."

chemical influence. To take a simple case, a neutral atom that consists of two protons, two neutrons, and two electrons is called a helium atom. Its mass is four, determined by the two protons and two neutrons. Its behavior is determined by the two electrons that mate with the two protons. Another kind of neutral atom has two protons, one neutron, and two electrons. This atom is also called helium because its behavior is also determined by its two electrons. True, its mass is different—this atom has only three heavy particles. To distinguish the two kinds of helium, we designate them as helium-four and helium-three or, in shorthand, ^{4}He and ^{3}He. Such atoms with the same nuclear charge but different masses are called *isotopes*.

Table 1-1 lists a few kinds of atoms, their names, shorthand symbols, and their proton-neutron-electron makeup. At least that's what chemists say.

(b) MOLECULES: THE SOCIAL CLIQUES OF ATOMS

Few of the atoms we know go through life alone. Most of them are found in close company with other atoms. For example, our atmosphere contains oxygen atoms and nitrogen atoms but, if you go looking for an oxygen atom, you find two of them bound together. Nitrogen atoms, too, are found in pairs. If you look hard enough, particularly in the smog over Los Angeles, you'll occasionally find a nitrogen atom going around with an oxygen atom.

These aggregates of atoms are called *molecules*. The ones we have mentioned each contain two atoms—they are called *diatomic molecules*. There are more complicated molecules hovering in the air around us though. The molecules called water, carbon dioxide, ozone, and nitrogen dioxide each have three atoms—they are called *triatomic molecules*. A water molecule has two hydrogen atoms and one oxygen atom. Each carbon dioxide molecule has two oxygen atoms and one carbon atom. Ozone has three oxygen atoms held together, while a nitrogen dioxide molecule has one nitrogen atom bound to two oxygen atoms. Chemists tell the atomic makeup of these molecules with symbols—water is represented as H_2O, carbon dioxide as CO_2, ozone as O_3 and nitrogen dioxide as NO_2. The diatomic molecules mentioned earlier are symbolized N_2, O_2, and NO: nitrogen, oxygen, and nitric oxide, respectively.

(c) ELEMENTS AND COMPOUNDS

A chemist defines an *element* (or an elementary substance) to be a substance in which all atoms have the same nuclear charge. To prove that a substance is an element used to be rather complicated, but nowadays atoms can be weighed with a device called a *mass spectrometer*. Information about their nuclear charge is obtained by measuring the energy needed to remove an electron from an atom—the *ionization energy*. We'll say more

Table 1-1 A Few Kinds of Atoms and Their Isotopes

Name	Symbol	Number of Protons, Z	Number of Electrons	Number of Neutrons	Mass	Isotope Abundance (%)
hydrogen	^{1}H	1	1	0	1	99.984
	^{2}H or D*	1	1	1	2	0.016
helium	^{3}He	2	2	1	3	$1.34 \cdot 10^{-4}$
	^{4}He	2	2	2	4	100
lithium	^{6}Li	3	3	3	6	7.40
	^{7}Li	3	3	4	7	92.6
beryllium	^{9}Be	4	4	5	9	100
boron	^{10}B	5	5	5	10	18.83
	^{11}B	5	5	6	11	81.17
carbon	^{12}C	6	6	6	12	98.892
	^{13}C	6	6	7	13	1.108
nitrogen	^{14}N	7	7	7	14	99.64
	^{15}N	7	7	8	15	0.36
oxygen	^{16}O	8	8	8	16	99.76
	^{17}O	8	8	9	17	0.04
	^{18}O	8	8	10	18	0.20
fluorine	^{19}F	9	9	10	19	100
neon	^{20}Ne	10	10	10	20	90.0
	^{21}Ne	10	10	11	21	0.27
	^{22}Ne	10	10	12	22	9.73
sodium	^{23}Na	11	11	12	23	100
chlorine	^{35}Cl	17	17	18	35	75.4
	^{37}Cl	17	17	20	37	24.6
gold	^{197}Au	79	79	118	197	100
uranium	^{235}U	92	92	143	235	0.72
	^{238}U	92	92	146	238	99.28

*Among all the stable isotopes, only this one has a special name. ^{2}H is called deuterium and is symbolized by D. The radioactive hydrogen isotope, ^{3}H, also has a special name, tritium, symbolized T.

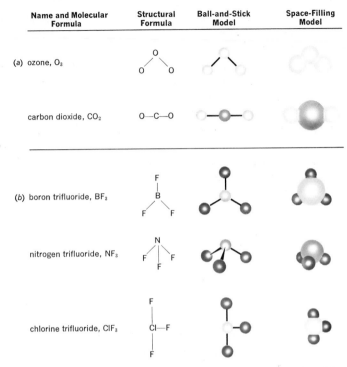

Name and Molecular Formula	Structural Formula	Ball-and-Stick Model	Space-Filling Model

(a) ozone, O_3

carbon dioxide, CO_2

(b) boron trifluoride, BF_3

nitrogen trifluoride, NF_3

chlorine trifluoride, ClF_3

Figure 1-8   *Different representations of molecules: (a) triatomic molecules; (b) tetratomic molecules.*

about these sophisticated measurements later, but for the moment, let's go along with the chemist in his definition of an element.

A *compound*, then, must contain two or more different kinds of atoms. By controlled chemical reactions, a compound can be separated into the elements of which it is composed. This process is known as chemical analysis and it can be carried out with great precision. A chemist would have no great difficulty showing that a sample of the compound nitric oxide contains the elements nitrogen and oxygen. He could then go on to show that 30.01 grams of nitric oxide (symbolized NO) contains exactly 14.01 grams of nitrogen (symbolized N_2) and 16.00 grams of oxygen (symbolized O_2).

Notice something about elements and compounds: Though an element contains only one kind of atom, it can be made up of either atoms or molecules. Helium gas is an element that we find to be made up of single atoms moving about independently; it is *monatomic*. The substances oxygen and ozone, O_2 and O_3, are both elementary substances (each contains only atoms of nuclear charge eight, oxygen atoms), but these substances exist as molecules. An element can be a gas, a liquid, or a solid. Some gaseous elements are monatomic, but some are made up of molecules. A compound can be a gas, a liquid, or a solid. Since a compound must contain more than one kind of atom, it can never be monatomic. Thus all gaseous compounds are made up of molecules.

Oxygen and ozone are the cause for one more note of interest. They are both elementary substances—each contains only oxygen atoms. But their molecules are different: Oxygen is symbolized O_2, meaning two oxygen atoms are bound together, and ozone is symbolized O_3, meaning three oxygen atoms are bound together. These two substances are given different names because they are different in behavior. We breathe oxygen to keep us alive whereas ozone "burns" the lungs. Ozone attacks rubber after a few seconds, but oxygen does not. Ozone bleaches dyes and kills microorganisms that exist quite comfortably in air. The stuff with the properties that we describe for ozone changes on heating into the stuff with the properties that we describe for oxygen. All of these properties account for the fact that elementary oxygen is said to have two forms: oxygen, O_2, and ozone, O_3.

(d)   WEIGHT RELATIONS IN CHEMICAL REACTIONS

We have listened to the language of chemists—they talk of atoms and molecules that they admit they've never viewed individually in the light of day. They talk of molecules of ozone changing into molecules of oxygen and of molecules of one kind combining with molecules of another kind to produce still other kinds of molecules. Why are chemists so confident with this model? The weight relations observed experimentally in chemical reactions show one of the reasons why.

We remarked that the smog constituent nitric oxide, symbolized NO, can be analyzed to show that 30.01 grams of it contain 14.01 grams of nitrogen and 16.00 grams of oxygen. If this amount of nitric oxide is mixed with the elementary substance ozone, O_3, the compound nitrogen dioxide, NO_2, is produced. The weight of nitrogen dioxide produced turns out to be 46.01 grams. Careful analysis of this product shows that it still contains the 14.01 grams of nitrogen, but now contains 32.00 grams of oxygen. What a simple world we live in—in nitric oxide, 14.01 grams of nitrogen combine with 16.00 grams of oxygen, whereas in nitrogen dioxide, 14.01 grams of nitrogen combine with precisely twice as much oxygen, 32.00 grams.

It was evidence exactly like this that caused chemists to invent the submicroscopic atomic model. How simply the formulas paraphrase our facts! The compound nitrogen dioxide, NO_2, has just twice as many oxygen atoms per nitrogen atom as the compound NO. Naturally NO_2 has twice the weight of oxygen per 14.01 grams of nitrogen as that found in NO. Of course, no one is proposing that one nitrogen atom weighs 14.01 grams or that one oxygen atom weighs 16.00 grams. All we're saying is that however many nitrogen atoms there are in 14.01 grams of the substance nitrogen, there is exactly the same number of oxygen atoms in 16.00 grams of the substance oxygen and twice as many in 32.00 grams of oxygen. In symbols, if there are N_0 nitrogen atoms in 14.01 grams of nitrogen, there are N_0 oxygen atoms in 16.00 grams of oxygen and $2N_0$ oxygen atoms in 32.00 grams. This is just grocery store arithmetic. A dozen medium grade A eggs weigh less than a dozen large grade A's, but both dozens contain 12 eggs. Two dozen large grade A's will, of course contain 24 eggs. The number N_0 might be called the "chemist's dozen."

We can, if we wish, tell our story about the ozone and nitric oxide reaction in shorthand notation.

nitric	reacts			nitrogen	
oxide	with	ozone	to form	dioxide	
NO	$+$	O_3	$\rightarrow$	NO_2	(1-17)

The expression (1-17) leaves something to be desired, though. In the shorthand notation, it is clear that there are four atoms of oxygen among the reactants (one in NO and three in O_3) and only two in NO_2. Ozone releases an oxygen atom to nitric oxide, but that leaves two more atoms, an oxygen molecule, left over. We'd better add this to (1-17) to keep track of all of the atoms:

$$NO + O_3 \rightarrow NO_2 + O_2 \tag{1-18}$$

Expression (1-18) now has all of the atoms in the reactants (NO and O_3) accounted for in the products (NO_2 and O_2). It is called a "balanced equation" for the reaction. This equation contains a lot of information useful to a chemist. It can be read in a number of ways:

One nitric oxide molecule reacts with one ozone molecule
to produce one nitrogen dioxide molecule and one oxygen molecule
$$(1\text{-}19)$$

or

one dozen nitric oxide molecules react with one dozen ozone molecules
to produce one dozen nitrogen dioxide molecules and one dozen oxygen molecules
$$(1\text{-}20)$$

or

N_0 nitric oxide molecules react with N_0 ozone molecules
to produce N_0 nitrogen dioxide molecules and N_0 oxygen molecules
$$(1\text{-}21)$$

or, from our experimental weighings,

30.01 grams of nitric oxide react with N_0 ozone molecules
to produce 46.01 grams of nitrogen dioxide and N_0 oxygen molecules
$$(1\text{-}22)$$

This last statement would be more useful if we specified the weights of the ozone and oxygen. Fortunately it isn't difficult to deduce these weights

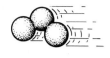

(a) Before collision   (b) Collision occurs

(c) After collision

Figure 1-9 *Visualization of a chemical reaction: The collision of a nitric oxide molecule,*
NO, with an ozone molecule, O_3, results in the formation of a nitrogen dioxide molecule,
NO_2, and an oxygen molecule, O_2.

from what we already know. Expression (1-18) says there is the same
number of oxygen atoms in an oxygen molecule as in an NO_2 molecule—two.
Reading (1-18) in the form (1-21), we can say that the N_0 molecules of NO_2
contain $2N_0$ atoms of oxygen atoms and the N_0 molecules of O_2 also contain
$2N_0$ atoms of oxygen. We have already decided by experiment that $2N_0$
oxygen atoms in NO_2 weigh 32.00 grams. So the N_0 molecules of O_2 must
also weigh 32.00 grams.

Exercise 1-1. Recognizing that N_0 molecules of ozone contain $3N_0$ atoms
of oxygen, deduce the weight of ozone that should appear in (1-22).

Now we can rewrite (1-18) and (1-22) to connect symbols and weights.

$$NO \; + \; O_3 \; \rightarrow \; NO_2 \; + \; O_2 \qquad\qquad (1\text{-}23)$$

$$30.01 \; + \; 48.00 \; \rightarrow \; 46.01 \; + \; 32.00$$
grams   grams   grams   grams

Thus the equation (1-23) tells us experimental quantities—the weights
involved in a reaction. If we wish to convert 30.01 grams of NO to NO_2,
(1-23) tells us how much ozone we'll need—48.00 grams. It also told us
earlier that we'd end up with something besides NO_2—32.00 grams of O_2
will be produced. All of this information came from keeping track of the
atoms in a reaction (i.e., we "balanced" the equation) and from experimental evidence about the relative weights of atoms.

Exercise 1-2. In an electric arc, nitrogen molecules and oxygen molecules
from air (N_2 and O_2) react to form nitric oxide (NO). Balance the equation
for the reaction between 32.00 grams of oxygen (N_0 molecules of oxygen)

and N_2 to form NO. This means deciding the correct coefficients x and y in the equation:

$$x\,N_2 + O_2 \rightarrow y\,NO$$

Exercise 1-3. Use the values of x and y to convince yourself that the 32.00 grams of O_2 will react with 28.02 grams of N_2. Then decide the weight of nitric oxide that will be produced. What weight of N_2 will react with 0.3200 grams of O_2?

(e) THE MOLE: A CHEMIST'S DOZEN

We have not yet said how big N_0 is. We started talking about 30.01 grams of nitric oxide, but we might just as well have selected one tenth as much or twice as much. It's like arguing about whether a dozen cookies should number 12 or 13. There is no basis for choice except convenience and convention—and we've decided on 12. It is only confusing to argue that a dozen should be 12 for eggs but 13 for cookies (the "baker's dozen")—it is an arbitrary choice and its usefulness depends upon everyone knowing what the choice is.

As they began recognizing that weight relations led to an atomic model, chemists began casting about for a convenient choice of N_0, their "dozen." Since atoms can't be seen, it isn't convenient to count them out, so a nice even number, like 10, or 12, or 10,000, isn't especially convenient. The situation is more like measuring out beans than eggs. No one would go to the store to buy 17,000 beans—you buy two pounds of them—something easy to measure.

Chemists decided on a convenient weight of a particular element to fix N_0. They selected 16.000 grams of oxygen—the number of atoms in that easily measured weight defines the number N_0. They call this number of atoms "one mole of atoms" just as we say that 12 eggs are "one dozen eggs." For a long time, chemists worked comfortably with the mole concept without knowing exactly the magnitude of N_0. (Do *you* know how many beans there are in a pound?) Nowadays there are many ways to measure N_0—the number of atoms in a mole (16.000 grams of oxygen). It is a whopping big number!

$$N_0 = 602{,}350{,}000{,}000{,}000{,}000{,}000{,}000$$

This number cries out for an abbreviation. It is, hence, always represented by a shorthand notation that tells how many factors of ten are needed to move the decimal point to the first digit. In this notation, we write

$$N_0 = 6.0235 \cdot 10^{23}$$
Avogadro's Number

(1-24)

Because Amadeo Avogadro was one of the first scientists to recognize the significance of N_0, this value is called "Avogadro's Number."*

Exercise 1-4. To get a feeling for the magnitude of Avogadro's Number, calculate how many moles of $\frac{1}{4}$-pound pieces of cheese would be needed to be clustered together to build the moon—which has a mass of $7.4 \cdot 10^{25}$ grams. (One pound is $454 = 4.54 \cdot 10^2$ grams.)

(f) THE ATOMIC WEIGHT SCALE

Armed with the atomic model, chemists began systematic studies of weight relations in chemical reactions interpreted in terms of balanced equations. Gradually, molecular formulas became known, and as they did, the weight relations led to a scale called the "Atomic Weight Scale." It lists, for each element, the weight of one mole of atoms of that element. This scale, shown on the inside back cover of this book, is given in every chemical handbook, and a large fraction of it is in the working memory of every chemist. Coupled with a balanced equation, this scale helps the chemist calculate in measurable amounts how much of something reacts with how much of something else to produce a desired amount of some product.

Exercise 1-5. Using the atomic weight scale, calculate how many moles of carbon atoms are contained in the Hope Diamond, which weighs $44\frac{1}{2}$ carats or 8.9 grams. (Diamond is one of the elementary forms of carbon. Graphite is another.)

(g) MOLECULAR WEIGHTS

Calculating the weight of a mole of molecules is an everyday use of the atomic weight scale. There is nothing to it, if you know the formula of a molecule. Chlorine has the molecular formula Cl_2—one mole of chlorine molecules contains two moles of chlorine atoms. The atomic weight of a mole of chlorine atoms is 35.45 grams. The molecular weight (MW) of a mole of chlorine molecules is $2 \cdot (35.45) = 70.90$ grams. The molecular formula for sulfuric acid is H_2SO_4—a mole of H_2SO_4 contains two moles of hydrogen atoms, one mole of sulfur atoms, and four moles of oxygen atoms.

$$\text{mol. wt. } H_2SO_4 = 2(\text{at. wt. H}) + (\text{at. wt. S}) + 4(\text{at. wt. O})$$

$$MW(H_2SO_4) = \quad 2(1.008) \quad + \quad (32.07) \quad + \quad 4(16.000)$$

$$= 98.09 \text{ grams/mole}$$

Ho hum. So what else is new?

---

* Recently the definition of a mole has been altered to put it in terms of mass spectrometric measurements, which can be made with high accuracy. The carbon-12 isotope is now assigned atomic weight 12.000 · · · · . This changes the entire scale by −0.0045%, an unimportant change from a chemist's point of view.

(a) A cubic foot of gas (any gas)

(b) A two-ounce shot of pure alcohol

(c) A four-inch crescent wrench

(d) A moon of quarter-pound pieces of cheese

(e)

Figure 1-10 One mole of some familiar substances. (a) A cubic foot of hot air (160°F) contains one mole of gases, mostly N_2 and O_2. (b) Pure ethanol, C_2H_5OH. (c) 56 grams of iron (steel is iron containing a tiny percentage of carbon). (d) A moon mole. (e) A mole.

Exercise 1-6 Show that the molecular weights of carbon dioxide, CO_2, and propane, C_3H_8, are almost identical but not perfectly so.

(h) MEASURING THE MOLECULAR WEIGHT OF A GAS

The perfect gas model relates the pressure-volume-temperature behavior of a gas to the number of gas particles. If absolute temperature is used, the "perfect gas law" results—expression (1-13):

$$PV = nRT$$

In (1-13), R was noncommittally referred to as a "suitable proportionality constant." Now we can say what is "suitable." The value of R depends upon the units selected for all of the other quantities in (1-13). Temperature is always given in degrees Kelvin. Pressure is usually expressed in atmospheres (1 atmosphere = 760 mm of Hg = 760 torr). Volume is sometimes conveniently specified in cubic centimeters (cc) and sometimes in liters. The quantity n is the number of particles, usually given in moles. Now we can calculate R from that simple experiment portrayed in Figure 1-4.

Into a bulb of measured volume 2.000 liters is placed 0.03200 gram of oxygen gas. The bulb is immersed in an ice bath to establish a uniform temperature of 0°C. The pressure is measured—it is 8.520 torr = 0.01121 atmosphere. Now we have measured all of the quantities in (1-13) except R.

$P = 0.01121$ atmosphere

$V = 2.000$ liters

$t = 0°C,$ so $T = 273.16 + t = 273.16°K$

$$n = \frac{0.03200 \text{ gram}}{\text{mol. wt. } O_2} = \frac{0.03200}{2(16.000)} = \frac{0.03200}{32.00} = 0.001000 \text{ mole}$$

$$R = \frac{PV}{nT} = \frac{(0.01121)(2.000)}{(0.001000)(273.16)} = 0.08205 \frac{\text{liter-atmosphere}}{\text{degree K mole}}$$

$$\boxed{\begin{array}{l} R = 0.08205 \dfrac{\ell\text{-atm}}{°\text{K mole}} \\[2mm] \text{or} \\[2mm] R = 82.05 \dfrac{\text{cc-atm}}{°\text{K mole}} \end{array}}$$

Exercise 1-7 Use R to show that the volume of one mole of a perfect gas at one atmosphere pressure and 0°C is 22.414 liters. Calculate the volume one mole would occupy at one atmosphere pressure and 25°C.

Notice that Exercise 1-7 made no reference to the composition of the gas. It is applicable to *any* perfect gas. And most gases behave according to the perfect gas law, (1-13), at moderate pressures—pressures below one atmosphere, as are normally used in the laboratory. That means gas measurements can be used to measure molecular weights. We know R now, so we can weigh a known volume of a substance, measure its pressure and temperature, and calculate how many moles we have. The weight divided by the number of moles is the molecular weight!

Here is a typical experiment. A student wishes to measure the molecular weight of a liquid sample known to contain only carbon and hydrogen. He finds the weight of the sample to be 1.408 grams to the nearest 3 milligrams. The liquid is transferred to a suitable apparatus and vaporized at 100°C within plus or minus $\frac{1}{2}$° (100°C $=$ 373.16°K). While the pressure is held constant at 1.000 $\pm$ 0.001 atmosphere, the volume is measured to be 612 $\pm$ 1 cc.

$$PV = nRT$$

or

$$n = \frac{PV}{RT} = \frac{(1.000)(612)}{(82.05)(373.2)}$$

$$n = 0.02000 \text{ mole}$$

$$\text{mol. wt.} = \frac{\text{grams}}{\text{moles}} = \frac{1.408}{0.02000} = 70.40 \text{ grams/mole}$$

If the compound contains five atoms of carbon per molecule, the atoms would contribute 5(12.01) $=$ 60.05 to the molecular weight. The rest would be hydrogen:

$$
\begin{array}{r}
70.40 \\
-60.05 \\
\hline
10.35
\end{array}
$$

Hence we can calculate the number of atoms of hydrogen needed to make up the experimental value

$$\frac{10.35}{1.008} = 10.27$$

We conclude with the molecular formula $C_5H_{10.3}$. The nearest integer formula is C_5H_{10}: The sample might be cyclopentane, one of several compounds with the formula C_5H_{10} (see Fig. 1-11).

Exercise 1-8. Show that the compound cannot contain six carbon atoms per molecule and that if it is assumed to contain four carbon atoms, a molecular formula C_4H_{22} is obtained, a formula that chemists reject on the basis of their understanding of chemical bonding.

Exercise 1-9. Cyclobutane, C_4H_8, has the same percentage of carbon as cyclopentane, C_5H_{10}. Calculate the volume that would be occupied by 1.408 grams of cyclobutane vaporized at 100°C and 1.000 atmosphere to show that it would be easily distinguishable from cyclopentane.

1-4 How do we know we're right?

The example just given raises one of the most fundamental questions of science. As we seek our hidden likenesses, build our models, and polish our theories, how do we know we're on the right track?

Consider the molecular weight determination. The experimental result turned out to be $C_5H_{10.3}$. Without the slightest twinge of conscience, we drew the conclusion, "The nearest integer formula is C_5H_{10}." Why didn't we conclude that the atomic weights of carbon and hydrogen are incorrect in the table? Or that the gas law doesn't hold? Or that the atomic theory is not useful, after all?

We don't draw any of these rather drastic conclusions for an obvious reason. There are some opportunities for error in the measurements that led to the molecular weight. The odds are that some minor experimental difficulty accounts for the discrepancy between $C_5H_{10.3}$ and C_5H_{10}. After all, every measurement has some limitation on its accuracy. On the other

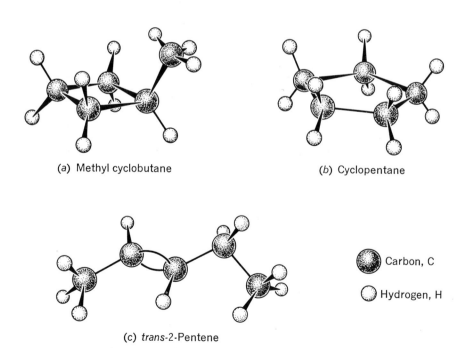

(a) Methyl cyclobutane

(b) Cyclopentane

(c) trans-2-Pentene

Carbon, C

Hydrogen, H

Figure 1-11 Ball-and-stick models of three different molecules, each with the same formula, C_5H_{10}. These are the molecules of different, known compounds.

hand, the atomic theory came from just such measurements as these. If the measurements are intrinsically uncertain, why are we so sure about the theory? We are face-to-face again with the basic activities of science. How do we build a theory? How do we choose between two competing theories? How certain can we be that a given theory is "really right"?

(a) BUILDING A THEORY: EXPLAINING OBSERVATIONS

Building a theory means paraphrasing what you know about a phenomenon of study in terms of a model phenomenon that is well understood. This is the result of the creative search for likenesses, as discussed earlier. Whether the model is satisfactory or not depends on whether it reproduces what we know about the phenomenon of study. Naturally, we mean that it reproduces what we know within the accuracy with which we have made our observations. If the model, or theory, or equation reproduces the observed behavior as accurately as we can measure the behavior, it is certainly good enough. There is no need for a model to reproduce what we know more accurately than we know it—in fact, there is no meaning to that suggestion.

(b) PREDICTIONS BY A THEORY

One of man's strongest aspirations has always been the desire to predict the future. Stated more scientifically, he wants to develop reliable and useful expectations. That is the real aim of a theory. It is satisfying to bring coherence to our existing knowledge of the environment. It becomes meaningful when it is applied to the problems of our continued existence.

There are two kinds of expectation that can be sought. The first might be called interpolation. It means using theory to predict what will happen in an impending event that is within the range of experience that led to the establishment of the theory. We might predict what will happen to the pressure in an automobile tire when it becomes warm from driving at high speed. The perfect gas law encompasses this situation and, hence, it gives a confident answer.

The other kind of prediction is more exciting—and more risky. We would like also to use a theory to predict outside and beyond the experience upon which the theory is based. The further we go beyond this experience, the less confidence we can have in the prediction. Again the perfect gas law provides an excellent example.

The perfect gas law states that the pressure-volume product will be constant for a fixed number of moles of gas at a constant temperature. If the pressure is raised, the volume will decrease, so that $P \cdot V$ remains the same. Table 1-2 shows typical data, in this case obtained by measuring the pressure and the volume, each to an accuracy of plus or minus one percent (designated $\pm 1\%$).

Table 1-2   Pressure-Volume Behavior for One
Mole of Ammonia at 25°C (1% Accuracy)

Pressure (atmospheres)	Volume (liters)	$P \cdot V(\ell \cdot atm)$
0.100	246	24.6
0.200	123	24.6
0.400	62.0	24.8
0.800	30.4	24.3
1.00	24.4	24.4
2.00	12.1	24.2
		Av. 24.5 $\pm$ 0.2

The perfect gas law and the simple kinetic theory (based upon billiard ball behavior) are in accord with the everyday experience expressed in Table 1-2. This simple kinetic theory is widely used because it reproduces a mountain of such experience. Within the one percent accuracy of the knowledge shown in Table 1-2, the theory and the perfect gas law are completely satisfactory models.

Now we might contemplate interpolation. With great confidence and no particular pride, we would answer the question, "What volume would be occupied by the one mole of ammonia at a pressure of 0.500 atmosphere and at 25°C?" Easy as pie—to one percent accuracy, $PV = 24.5$ liter-atmosphere. Therefore,

$$V = \frac{24.5 \; \ell \cdot atm}{P \quad atm} = \frac{24.5}{0.500} \; \ell = 49.0 \text{ liters}$$

That doesn't seem like much of a prediction—we do it so often that we take this type of interpolation for granted. Less obvious "interpolation" types of prediction are more satisfying, however. The pressure-volume behavior of ammonia and the kinetic theory that explains it imply that there will be associated behavior that seems different in kind: Acoustic behavior is one such. An acoustic disturbance is generated by a periodic increase and decrease of volume adjacent to a vibrating diaphragm. The perfect gas law implies a periodic, inverse pressure variation. Via collisions among molecules, these periodic pressure variations are propagated away from the diaphragm. Our collisional model of the gas allows us to expect this behavior and to calculate some of its details, such as the velocity of sound in ammonia.

Extrapolation is the exciting realm of prediction, however. It means applying our model outside the range of experience upon which the model was based. A number of possible extrapolations from the Table 1-2 data for ammonia are obvious. We could assume that the $PV = nRT$ behavior of ammonia applies to another gas, say, to that unknown substance whose molecular weight we measured to be 70.4 grams/mole. We do this with an

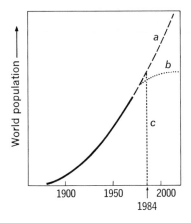

Figure 1-12 Extrapolation carries uncertainty: The solid curve represents the world's population growth up to the present. A continuation of this growth curve, depicted by the dashed curve (a), permits a prediction of the population in the year 2000 (too large). Curve (b) is a possible population growth that could not be predicted from previous experience: It might result from sensible, hence unlikely, moderation of population growth. Curve (c) also could not be accurately predicted: It could happen tomorrow if someone pushes the wrong button.

air of confidence because our experience is much broader than represented in Table 1-2. Many, many gases already have been found to follow the perfect gas law (to one percent accuracy). We don't feel bold in guessing that it will work again. Yet we are going outside of our experience—maybe we will be surprised. Sometimes we are.

A second extrapolation would be to assume that the behavior in Table 1-2 applies to much higher accuracy than our measurements. After all, the perfect gas law doesn't mention anything about $\pm 1\%$. It says that at constant temperature, the pressure-volume product is *constant*—period. The simple kinetic theory we've developed says much more—it says that the PV product would be constant to 0.1% or 0.01% or to whatever precision is possible. Should we believe this "prediction"? Well, that would be an extrapolation; hence we should retain some reservations—we must be prepared for surprises.

Table 1-3 shows some data for ammonia obtained with sufficiently refined experimental techniques to reduce the experimental uncertainty to $\pm 0.01\%$. These experimental data show that the perfect gas theory does not fit the real world of ammonia to this accuracy. The PV product drops as pressure rises. This sort of discrepancy proves to be common. If measurements are extended to sufficient accuracy, every gas shows some deviation from the $PV = nRT$ behavior. Our model of gas behavior is imperfect. With typical egocentricity, however, we say that the *gas* is imperfect! The ammonia is, of course, correct—the theory must be wrong. Yet we refer to the $PV = nRT$ model as "the perfect gas law" and when

Table 1-3   Pressure-Volume Behavior for One
Mole of Ammonia at 25°C (0.01% Accuracy)

Pressure (atmospheres)	Volume (liters)	$P \cdot V(\ell \cdot atm)$
0.10000	244.52	24.452
0.20000	122.19	24.438
0.40000	61.033	24.413
0.8000	30.453	24.362
1.0000	24.338	24.338
2.0000	12.107	24.214

a gas, such as ammonia, does not behave according to this simple theory, we attribute the discrepancy to "gas imperfections."

Having discovered these discrepancies and having become convinced that they cannot be ascribed to experimental difficulties, we are forced to question our theory. Should we look for a patch to place on the model to restore its agreement with the facts, even to 0.01% accuracy? Or should we abandon completely the billiard ball model even though it fits many kinds of gas behavior quite well? No, the usefulness of the model argues against abandonment. Let's hunt for a patch.

Two possible patches come to mind. They both bring into question tacit assumptions of the perfect gas model (which isn't so perfect, after all). When we applied the billiard ball energy-momentum equations, we assumed that the tiny billiard balls exert no force on each other at any time except at the instant of a collision. Perhaps there is some force of attraction between the ammonia molecules that increases as the molecules are held closer together. That would tend to lower the PV product. Perhaps we could figure out a modification to our mathematical gas law to include such an effect, small though it is, and restore agreement between theory and experiment.

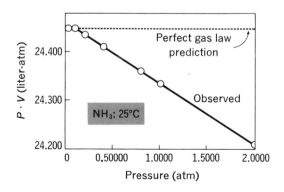

Figure 1-13   PV data for ammonia at 25°C plotted on a greatly expanded scale to show the regular decrease in PV with increasing pressure.

Another possibility is that ammonia gas does not consist solely of NH_3 molecules. Perhaps a fraction of them exist as double molecules, N_2H_6. These would be called "dimers" and, insofar as they exist, 15.0 grams of NH_3, the molecular weight, would contain less than a mole of molecules. This, too, would lower the PV product, even if both NH_3 and N_2H_6 act as perfect gases.

There—see how easy science is! A short time after discovering a discrepancy between theory and experiment, we have thought of two possible explanations. Now all we have to do is decide which is better.

(c) CHOOSING BETWEEN COMPETING THEORIES

The most interesting parts of science are those in which scientists disagree. The disagreement means there is more to be learned. (Please understand that even when all scientists agree, there is probably more to be learned, but it isn't yet obvious.) Nonscientists, however, tend to be nonplussed by such a situation. Many people have the impression that scientific activity is so "objective" and literal that there always ought to be a clear-cut right and wrong to everything. Hence all scientists ought to agree!

But science is not all that objective. It is, after all, a human activity and each person engaged in it sees the world through his own eyes. Even with the benefit of all earlier knowledge—his own experience coupled with that stored for him in libraries—he interprets that knowledge with certain ends in mind and certain specific interests. That will influence his attitude toward a theory, particularly while it is first taking form.

To illustrate this, let's consider the simple question "How does one choose between two competing theories?" Several criteria can be used. They don't always lead to the same conclusion. That's when the fun begins.

—*Simplicity and Ease of Use.* Scientists devise their theories and models to make it easier to understand and to use our knowledge. There is no point in working with a cumbersome theory if there is a sufficiently good explanation that is facile in application and easy to comprehend. The words "sufficiently good" are key words here. What is sufficiently good for one purpose may not do for another. All chemists know about the existence of gas imperfections, but the perfect gas law is used every day. In many experiments, there is no need to worry about discrepancies of one or two tenths of a percent because of other limitations of the experiment. For example, if the problem is to measure the molecular weight of an unknown compound, the degree of purity of the compound probably will limit the possible accuracy. To guarantee a purity in excess of 99% can be quite difficult when the identity of the compound is not known. Under any circumstances, the easier a theory is to apply, the greater will be its popularity, provided it fits the facts.

—*Accuracy of Fit.* The accuracy and detail with which a theory reproduces experience is another basis for choosing between two theories. No matter how useful a simple theory might be (such as the perfect gas law), it is unsatisfying if there is a real and known discrepancy with the facts. Ultimately, we prefer a theory that does fit what is known to one that is only in approximate agreement.

—*Scope.* Sometimes two theories are equally useful in a particular area, but one applies to other areas as well. The theory that interconnects apparently different types of phenomena is preferable because it gives coherence to our knowledge and simplifies our attempts to deal with it.

These three criteria all add up to one simple question—"How well does a theory fit what we already know?" They tell us that the business of a theory is to fit, as accurately as possible, as much of our existing knowledge as possible and to do so in a convenient, usable way.

Yet this simple formula is not sufficient for some. Some scientists would contend that the most important criterion has not yet been mentioned: ability to predict. Let's examine this potentiality to see if it helps us choose between two theories.

—*Predictive Power.* Prediction can mean either interpolation or extrapolation. The first is already encompassed by the criteria mentioned earlier. That is what we mean by "fit what we already know." So here we must concern ourselves with our ability to extrapolate. If a theory would reliably predict outside the range of the experience upon which it is based, it would allow the scientist to generate new knowledge in his armchair— an appealing prospect!

Despite its appeal, however, this criterion—extrapolation to new knowledge—is not at all helpful as we choose between two competing theories. To see this, imagine that theory A predicts a new, previously undiscovered phenomenon, X. Theory B can be either in agreement with A or in disagreement. If both theories A and B agree that X should occur, then we can't choose between them on the basis of this phenomenon. If B contradicts A, then we have a challenge. But now we have no basis for preference! Both theories made predictions: A says X *will* occur, B says X *will not* occur. If making predictions is what we look for, each theory answers our call. As long as phenomenon X is outside our knowledge, either theory could be right. No choice is possible.

Now what should we do? The answer is obvious. Someone could do an experiment—someone could go looking for phenomenon X. If suitable conditions can be found for a definitive test, then the test will answer which theory should be preferred. Let's suppose that a dozen scientists in labs all around the world address themselves to the discovery of phenomenon X. And suppose, five years later, all finally agree on the answer— X does not occur. Now we know that theory B is to be preferred—and we have a dozen scientists who are quite disgruntled with theory A. But at least now we can decide on a basis that we all understand—B fits what we know better than A because now phenomenon X (the nonexistence of it) is within our knowledge.

So we couldn't choose on the basis of extrapolation because both theories predict something about X. Only after we know the answer about

X can we use it as a basis for preference. But now we are no longer extrapolating. *Extrapolation beyond existing knowledge gives no basis at all for*
preference between competing theories.

Then why do some scientists argue for predictive capability? In a certain
sense, theory A, which activated 60 man-years of fruitless effort (12 scientists times five years each), deserves to be criticized for making that prediction about X! There are two reasons why this isn't the way scientists feel.

First, the exploration of contradictory predictions between two theories
is a powerful dimension in our search for better understanding of the
environment. Regardless of which theory, A or B, turns out to be the better,
the possible X phenomenon points to a basis for the choice, once X is
within the realm of experience. That will move us ahead. If A turns out
to be wrong, better to have done with it and concentrate on B.

Second, the stakes in extrapolation are sometimes quite high. If no one
can choose between theories A and B on the basis of existing knowledge,
then we have no basis for denying the possibility that A's prediction about
X might be correct. If it is correct, then the 60 man-years might have re-
sulted in the discovery of a new, previously undiscovered phenomenon.
That's the sort of advance any scientist would like to make at least once in
a lifetime. That is why the dozen disgruntled scientists will now turn to
some extrapolative prediction of theory B and start another five-year search
for some new, previously unheard-of phenomenon, Y, that it predicts. And
they may be discouraged, but they won't be disillusioned. The issue between theories A and B is settled, after all. And scientists know that in the
exploration of the unknown, not every step is an advance. Yet there is no
other way to advance than by taking steps.

(d) AN EXAMPLE: GAS IMPERFECTIONS

Table 1-4 presents pressure-volume data for nitrogen dioxide, NO_2, at
100°C. These can be contrasted to the ammonia data given in Tables 1-2
and 1-3. It seems that ammonia's troubles are small compared to those of
nitrogen dioxide. For NH_3, the PV product looked sensibly constant at the
1% level of measurement—we perceived the failure of the perfect gas
picture only with 0.01% accuracy. For NO_2 at 100°C, the PV product

Table 1-4 Pressure-Volume Behavior for One Mole of Nitrogen Dioxide at 100°C

Pressure (atmospheres)	Volume (liters)	$P \cdot V(\ell \cdot atm)$
0.100	243	24.3
0.200	120	24.0
0.400	59.4	23.7
0.800	29.0	23.2
1.00	22.9	22.9
2.00	10.9	21.8

changes by 7% over the same pressure range, 0.1 to 1 atmosphere. Now it seems more urgent to find a patch for our theory. Two possibilities were suggested for ammonia—let's explore them in more detail.

One way to explain the NH_3 (and NO_2) gas imperfections is to say that the molecules, if not too far apart, exert attractive forces on each other. These intermolecular attractions tend to "pull the molecules back" as they hit the container walls, so they lower the pressure a bit. When pressure is raised, the molecules are closer (on the average), so these effects become larger. We could take this into account in the kinetic theory by adding distance-dependent forces between molecules. Empirically, we can insert a small, pressure-dependent corrective term in the gas law:

perhaps $$PV = nRT(1 - bP) \tag{1-25}$$

The coefficient b is a measure of the molecular attractions. It is multiplied by P to show that these attractions become more significant as the molecules are crowded closer together.

The second possible explanation for gas imperfections is that dimers of ammonia form and, though both the monomer, NH_3, and the dimer, N_2H_6, behave as perfect gases, their mixture does not. We imagine some equilibrium mixture in which both monomer and dimer are present, but the mixture composition changes as pressure increases. The equilibria can be presented as balanced equations:

$$2\,NH_3 = N_2H_6 \tag{1-26}$$

$$2\,NO_2 = N_2O_4 \tag{1-27}$$

In fact, Le Chatelier's Principle tells us what would be the effect of such equilibria. Le Chatelier, 'way back in 1884, studied equilibria like those postulated in (1-26) or (1-27). He observed that when he tried to change some condition, such as temperature or pressure, the equilibrium shifted somewhat in the direction to reduce, in part, the applied change.

Le Chatelier's Principle
   For a system at equilibrium, when a condition is changed, the equilibrium will shift in the direction that will reduce, in part, the applied change.

This means that if the temperature is raised, the equilibrium shifts in the direction that causes absorption of heat, so the temperature rise is moderated. If the pressure is raised, the equilibrium shifts in the direction that reduces the expected pressure rise. Consequently, raising the pressure would cause both (1-26) and (1-27) to shift to the right, toward N_2H_6 and N_2O_4. That would reduce the number of molecules present (it takes two monomers to make one dimer) so n would be smaller in the perfect gas equation. Therefore, the PV product would drop as pressure rises.

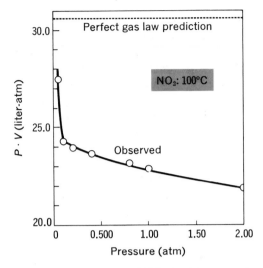

Figure 1-14 PV data for nitrogen dioxide at 100°C. At this temperature, the perfect gas law predicts a value of 30.6 liter-atm/mole. The observed PV product is not even close to that.

These two competing theories each explain qualitatively gas imperfections like those in Tables 1-3 and 1-4. They are about equally simple to apply and they both can restore quantitative agreement between the kinetic theory of gases and the actual gas behavior. In fact, chemists favor one explanation for ammonia (intermolecular attraction) and the other explanation for nitrogen dioxide (dimer formation). The nature of the intermolecular forces that cause the ammonia gas imperfections will be discussed in Section 17-4 (under "hydrogen bonds"). The equilibrium question opens the field of chemical thermodynamics, the subject of Chapters Seven to Ten. So we'll see later how chemists came to prefer the one theory in one case and the other theory in another. At this point, we'll explore only one more question in the competition between these two theories—their intrinsic predictions.

Consider the equilibrium theory for NO_2 and Le Chatelier's Principle first. The principle says that if the pressure is increased, the equilibrium will shift toward N_2O_4 because that reduces in part the pressure rise. Hence, as pressure is raised, the fraction of NO_2 in the dimer form ought to increase. This view carries with it, though, the prediction that Le Chatelier's prediction about temperature should hold as well. If temperature is raised, the equilibrium should shift in the direction that absorbs heat. This direction can be learned by carefully measuring the temperature as NO_2 expands. We find that expansion of the gas (which ought to decrease the fraction of NO_2 in the dimer form) causes a fall in temperature. Heat must be absorbed as NO_2 is formed from N_2O_4,

$$N_2O_4 + \text{some heat} = 2\,NO_2 \tag{1-28}$$

Now Le Chatelier's Principle tells us that as temperature is raised, more NO_2 will be formed from N_2O_4 because a shift in the equilibrium toward NO_2 absorbs heat. Conversely, lowering the temperature will shift the equilibrium toward N_2O_4 because that releases heat and moderates the temperature drop. Here are two predictions. At very high temperatures, we should lose all of the N_2O_4 and then, with only NO_2, the perfect gas law should apply reasonably well. A molecular weight measurement then should be meaningful and the answer should be $14.0 + 2(16.0) = 46.0$. At very low gas temperatures, we should lose all of the NO_2 and then, with only N_2O_4, the perfect gas law should again apply. Experiments show that these predictions are true. In fact, as the experiments are performed, there is visual evidence that the equilibrium shifts. At 100°C, the NO_2 gas is a deep orange, whereas at 0°C the color almost disappears. Chemists now know that the monomer, NO_2, absorbs blue light to give it an orange color whereas the dimer, N_2O_4, is colorless. Plainly the equilibrium theory is useful and applicable to nitrogen dioxide—it even explains the color changes that we observe with temperature.

Exercise 1-10. A weight 2.30 grams of nitrogen dioxide is placed in a 2.000-liter bulb and the pressure is measured at 0°C to be 0.309 atmosphere. Calculate the apparent molecular weight of the NO_2–N_2O_4 mixture at this low temperature. Now the temperature is raised to 100°C and the pressure is measured to be 0.728 atmosphere. What is the new apparent molecular weight?

Finally, consider the molecular attraction view of the ammonia gas behavior. What predictions can we find in this theory? There is one that fairly leaps out at us. The expression (1-25) says that the measured PV product will go down as pressure rises—because of the $b \cdot P$ term in the parentheses

$$PV = nRT(1 - bP)$$

The thought that immediately comes to mind is that something spectacular should happen if P is raised so much that $(1 - bP)$ approaches zero! The gas should disappear, it seems! From the magnitude of b, we can even predict the pressure needed, 194 atmospheres. That is something worth checking—perhaps we are on the verge of a new phenomenon—the disappearance of matter. Or perhaps our theory is about to join the multitudes of discarded theories out on the junkpile behind the laboratory. Off to the lab to perform the experiment. Table 1-5 shows the results.

Indeed we did discover something! All of a sudden, as the pressure rose from 9.8 to 9.9 atmospheres, the PV product dropped from a value only 5% different from the perfect gas expectation to a ridiculous value a factor

Table 1-5 Pressure-Volume Behavior for One Mole of Ammonia at 25°C and High Pressures

Pressure (atmospheres)	Volume (liters)	$P \cdot V(\ell \cdot$ atm$)$
2.00	12.1	24.2
4.00	5.98	23.9
8.00	2.92	23.4
9.80	2.36	23.1
9.90	0.0200	0.198
10.0	0.0200	0.200
15.0	0.0200	0.300
50.0	0.0200	1.00

of 100 lower. The "matter" didn't disappear, though; it assumed a volume of 0.0200 liter and then it stubbornly refused to compress further (to 1% accuracy remember!). This incompressible stuff flows like water—it is liquid ammonia! Our theory said that something would happen at high pressures, but long before we reached the expected catastrophe, the gas condensed to the liquid phase. By adding symbols to show the phase, gas or liquid (g or ℓ), we can write an equation for the change:

$$\text{At 25°C, 9.8 atm} \qquad NH_3(g) = NH_3(\ell) \qquad (1\text{-}29)$$

Naturally, we aren't going to await the Nobel Prize for this prediction by equation (1-25) and the resulting discovery in Table 1-5. Someone discovered a while ago that gases condense to liquids if compressed enough and that liquids don't follow the perfect gas law. We have reenacted this bit of history to show how an attempt to explain tiny gas imperfections at moderate pressures led us to postulate intermolecular attractions. This led us to expect a change of behavior when the attractions become a dominant part of the gas behavior. These considerations result in the seeds of understanding of why gases condense to liquids. That is the flow of science. That is what this book is all about.

1-5 Uncertainty in measurement—uncertainty in science

"The weight of the sample is 1.408 grams to the nearest 3 milligrams."
"The liquid was vaporized at a temperature of 100°C within plus or minus $\frac{1}{2}$°."
"The volume is measured to be 612 ± 1 cc."

In each of these statements, a quantitative measurement is specified in terms of two numbers. One number tells the magnitude of the quantity measured—"*the weight* . . . *is 1.408 grams*", "*at a temperature of 100°C*", "*the volume is* . . . *612 cc*". The second number indicates the degree of

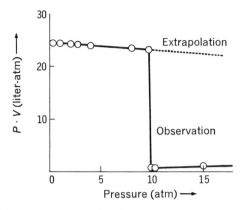

Figure 1-15 P · V data for ammonia at 25°C extrapolated to higher pressures. The dangers of extrapolation are clearly illustrated.

reliability—the uncertainty—*"to the nearest 3 milligrams"*, *"within plus or minus $\frac{1}{2}°$"*, *" $\pm 1\ cc$"*. **Every quantitative measurement involves some uncertainty.** The uncertainty should be specified because the significance of the measurement can't be assessed without some clue to its reliability.

A quantitative model or theory attempts to reproduce our quantitative knowledge of the environment. The only way to test a theory or to choose between competing theories is to compare to experimental facts. Hence, no theory can with certainty be more accurate than the measurements used to test it. If every bit of observational knowledge is uncertain to some extent, then *every theory involves some uncertainty.*

This confession of fallibility is the hallmark of science. We candidly admit that we can never be absolutely sure a theory is right. We are always ready for the possibility that an improvement in accuracy or explorations into new domains will reveal limitations and need for change in even the most time-tested theory. As our observational knowledge grows, our theories evolve, but they are always limited by the existing uncertainty in the relevant knowledge.

(a) MANIFESTATIONS OF UNCERTAINTY

If uncertainty is so important, we must know how to measure it. There are two avenues to follow. First, we can repeat the same measurement several times to see how much the result fluctuates. Second, we can find an entirely different experimental approach to the same quantity. We shall see that these two avenues give different information about uncertainty.

Consider our molecular weight determination in Section 1-3(h). The result, 70.4 grams/mole, is not in exact agreement with the calculated value, 70.13 grams/mole, derived from the assigned formula, C_5H_{10}. Is this discrepancy real? Could it be due to some carelessness on the part of the

Table 1-6 Repeated Measurements of Molecular Weight by the Gas Density Method

Student	Measured Molecular Weight (grams/mole)
A	70.4
B	71.3
C	70.2
D	70.7
E	69.9

student who made the measurement? Or could it be due to some system-atic difficulty in the technique, such as gas imperfection? In a first attempt to answer these crucial questions (Is it or isn't it C_5H_{10}?), let's repeat the measurement.

Table 1-6 lists the results obtained by four other students, each deter-mining the molecular weight of that same substance by measuring the gas volume. Look at the results! The situation seems worse than before! Instead of one discrepancy to worry about, we now have five different answers, none of which is exactly 70.13! But these different answers in duplicate experiments are only a manifestation of the uncertainty in any single measurement. The situation is really better because now we have a basis for estimating this uncertainty.

Before dealing with these new data, let's try our second attack on uncertainty—an entirely different experimental approach. A mass spec-trometer may provide an answer. In this device, electrons bombard mole-cules to give each one an electric charge. Then the molecules are acceler-ated with a known energy and the velocity is measured. Again we can use the billiard ball expression, $E = \frac{1}{2}mv^2$. We measure E and v, so m can be calculated.

Table 1-7 lists four measurements by this technique. Three of these measurements are below 70, but one agrees (at last) with the calculated value, 70.1 grams/mole. Most important, though, is the inescapable im-pression that the mass spectrometer tends to give lower values than the vapor density method. This is the second manifestation of uncertainty—different answers can be obtained from different kinds of measurement of the same quantity.

Where do we go from here?

Table 1-7 Repeated Measurements of Molecular Weight by a Mass Spectrometer

Experiment	Measured Molecular Weight (grams/mole)
1	69.9
2	70.1
3	69.9
4	69.7

(b) THE AVERAGE—A MEASURE OF CENTRAL TENDENCY

How shall we deal with this array of data? In Table 1-6, the values range from 69.9 to 71.3 grams/mole. Student C came closest to the "theoretical" result, 70.1, and student E's value agrees with two of the mass spectrometer results. It would be presumptuous and arbitrary, however, to select one of the numbers because we think it is right when we have four other measurements that disagree. A better idea would be to take a consensus of the five values—to find their "central tendency."

The simplest measure of central tendency is the "center value." In Table 1-6, it is 70.4. There are two lower values and two higher values. In Table 1-7, it is 69.9, the average of the center pair. These choices make sense if we assume that the differences among measurements are caused by little inconsistencies and peculiarities of technique that are equally likely to give too high or too low results. These inconsistencies give random errors and, among several measurements, one in the middle ought to be best if one is to be singled out.

A more sophisticated way to cope with random errors is to take the average. That, too, should cause the effects of random errors to be lessened. We merely add up all the values and divide by the number of measurements.

Vapor Density
$$
\begin{array}{r}
70.4 \\
71.3 \\
70.2 \\
70.7 \\
\underline{69.9} \\
352.5
\end{array}
\qquad (1\text{-}30)
$$

$$
\begin{array}{r}
70.5 \\
\overline{5)\,352.5}
\end{array}
$$

Exercise 1-11. Show that the average of the mass spectrometer results is 69.90 grams/mole.

Because of the tendency for random error effects to cancel, *the average of several measurements is always more significant than a single measurement.*

(c) AVERAGE DEVIATION—A MEASURE OF DISPERSION

In Table 1-6, we see that the values range from 69.9 to 71.3. This range measures, in a way, the spread or dispersion of this type of experiment.

This is useful when there are only two or three measurements. A more meaningful measure can be obtained by examining how much, on the average, a single measurement deviates from the average. For the vapor density method, the measurements range from 0.1 to 0.8. If we ignore algebraic signs, they average 0.4.

	Deviations from
Vapor Density	the Average
70.4 − 70.5 = −0.1	0.1
71.3 − 70.5 = +0.8	0.8
70.2 − 70.5 = −0.3	0.3
70.7 − 70.5 = +0.2	0.2
69.9 − 70.5 = −0.6	0.6
	2.0

(1-31)

$$\frac{0.4}{5)\overline{2.0}}$$

average deviation $= \pm 0.4$

Exercise 1-12. Show that the average deviation in the mass spectrometer data is much smaller, ± 0.1.

Now we are prepared to make a rather complete statement about the central tendency and the dispersion for these two types of experiment. The vapor density method gives an average molecular weight of 70.5 grams/mole and the uncertainty in any single measurement is probably ± 0.4 gram/mole. The mass spectrometer method gives an average of 69.9 grams/mole and the uncertainty in any single measurement is probably ± 0.1 gram/mole.

(d) UNCERTAINTY IN THE AVERAGE

The average deviation indicates how reliable a single measurement might be. Our five attempts with this method indicate that with a single measurement of 70.4, we can be reasonably confident that the average of many measurements would be in the range 70.4 − 0.4 to 70.4 + 0.4, or between 70.0 and 70.8.

However, we no longer need to restrict our attention to student A's measurement. The average of the five values must be more meaningful because of the cancelling of random errors. The uncertainty in the average, 70.5 grams/mole, must be less than ± 0.4.

Statistics confirm that expectation. Assuming only that errors are equally likely to give too high or too low a result, it can be shown that the uncertainty in the average of N measurements is smaller than the uncertainty in one measurement by a factor near $1/\sqrt{N}$. Hence uncertainties can be assigned to our average values:

vapor density (5 values)

$$\text{uncertainty} = \frac{\pm 0.4}{\sqrt{5}} = 0.18 \simeq 0.2$$

Exercise 1-13. Calculate the uncertainty in the average of the four mass spectrometer results.

Now we can summarize succinctly all of the information about our two types of experiment.

$$\text{vapor density: molecular weight} = 70.5 \pm 0.2 \qquad (1\text{-}32)$$

$$\text{mass spectrometer: molecular weight} = 69.90 \pm 0.05 \qquad (1\text{-}33)$$

(e) RANDOM VERSUS SYSTEMATIC ERRORS

Unfortunately, we don't seem to be getting a decisive value for the molecular weight. Our analysis gives rather convincing evidence that the two methods give different results and that neither of these is 70.13 grams/mole, the theoretical value. How can we be certain which value is correct?

The first thought that comes to mind is that the mass spectrometer probably gives the better estimate—its uncertainty is only one quarter that of the vapor density method. Surprisingly, this is not necessarily true. The uncertainty we quoted refers only to the variability caused by random error. But the two methods plainly give different values because of systematic differences of some kind. *The average deviation gives no clue to the magnitude of systematic errors.* A systematic error is one that always works in the same direction: It gives either too high a result all the time or one that is always too low. No matter how many measurements are averaged, these errors do not cancel out.

There are three probable systematic errors in the vapor density experiment, for example. If the compound is volatile, some may evaporate after the sample has been weighed, while it is being transferred to the volume-measuring apparatus. Then its gas volume will be a bit too small, so the calculated number of moles will be too small. Since n, the number of moles, is divided into the weight, a smaller n makes the molecular weight a bit too large.

A second systematic error will enter if the compound is not a perfect gas. If, like ammonia, the gas has a PV product below that of a perfect gas, we will calculate n below its actual value. Again, our molecular weight result will tend to be too high.

A third systematic error will be incurred if the unknown sample is not perfectly pure. This could give a result uniformly either too high or too low, depending upon the molecular weight of the impurity.

Recognizing the presence of these sinister errors is one problem; removing them, once recognized, is another. If, for some reason, vaporization is suspected, then better techniques for handling the weighed liquid can be devised. If there is evidence of impurity (such as discoloration), there are steps that might improve sample purity, even though the compound has unknown composition. It could be fractionally distilled, frozen and recrystallized, or dried if there might be water present.

The remaining possible error, gas imperfection, may be the most difficult both to detect and to correct. To detect this trouble, it is necessary to try another type of measurement that doesn't depend upon perfect gas behavior. The mass spectrometer will do fine.

This complex instrument, however, has its own catalog of possible systematic errors. We must measure voltages to learn the molecular energies and clock the molecular motion to measure their velocities. These measurements might involve quite small random errors, but large systematic errors. We cannot be sure but that the mass spectrometer consistently gives results a few percent too low. This kind of problem can be reduced, however, by calibration. With a substance of accurately known molecular weight, the instrument can be tested. Any discrepancy can be used as an additive correction term in normal use of the instrument.

(f)   PRECISION VERSUS ACCURACY

With two types of errors, random and systematic, we are faced with two kinds of uncertainty. To differentiate them, scientists use two terms: precision and accuracy.

> —**Precision** *is determined by the uncertainty associated with random errors.*
> —**Accuracy** *is determined by the uncertainty associated with systematic errors.*

Precision is readily learned by repeating a measurement many times. The average deviation specifies it quantitatively. Accuracy must be learned by repeating a measurement in different ways. It is difficult to estimate. Calibration, making the measurement under circumstances in which the answer can be confidently expected, sometimes gives a reliable estimate. Almost always, a scientist will try to learn and will state his precision. Often the best he can do about accuracy is to design the experiment carefully to avoid systematic errors, to calibrate the technique, and then to cross his fingers.

Figure 1-16 Precision versus accuracy. (a) There is a distribution of darts around the bull's eye just as there is always a distribution of results when a measurement is repeated many times. These random errors determine the precision associated with the measurement (or the skill of the dart-thrower). (b) The accuracy of the results has been affected by a systematic error introduced by the wind. If the wind is steady, it is a simple matter for the thrower to adjust to the new conditions. It is often very difficult, however, to assess the effects of hidden systematic errors when making scientific measurements.

(g) STATING UNCERTAINTY—SIGNIFICANT FIGURES

It isn't always convenient to keep repeating the precision and accuracy of a measured number. There is, however, a simple rule that is helpful to your colleagues. When a number is stated, pay deliberate attention to the number of digits you specify. Only those digits that have real significance should be listed. For example, our final value for the two molecular weight measurements were, by vapor density, 70.5 ± 0.2 and, by mass spectrometer, 69.90 ± 0.05. When we refer to the first of these, it would be meaningless and misleading to call this number 70.50. Three figures are all that are needed, 70.5, and all that are appropriate. The last digit, .5, may be off by ± 0.2. The mass spectrometric result, 69.90, needs four digits—four significant figures. We are reasonably confident of the third digit, .9. The fourth is the last digit that is still informative; it may be off by $\pm .05$.

Exercise 1-14. By calculating the percentage uncertainty in the two numbers, show that the quantity 0.02000 ± 0.00004 has the same number of significant figures (four) as the quantity $2.000 \cdot 10^{-2} \pm 0.004 \cdot 10^{-2}$.

Exercise 1-15. Rewrite each of the following numbers in exponential notation and with the correct number of significant figures, three.

$$220.41 \pm 1.2$$
$$219{,}958 \pm 3251$$
$$0.002201 \pm 0.0000098$$

It is a useful rule in calculations to carry one more digit than the uncertainty warrants. Then the final result can be shortened (rounded off) to the proper number of significant figures ("sig. figs.," as they are called in the trade).

1-6 Conclusion

This chapter has been about the nature of scientific activity. The examples that were used not only display the ways the scientific machinery works, but also they define the starting ground for the rest of this book. The atomic model will be elaborated and chemical bonding will be a major topic later on, but meanwhile, we will speak of atoms and molecules as though they are old friends. The reactions of molecules, their balanced equations, and the weight relations they imply will be assumed to be at our disposal as we explore more advanced topics. The same will be true for perfect gas behavior, compactly represented by the expression $PV = nRT$. If any of these tools of chemistry have become a bit rusty, the problems below will provide sufficient honing.

So, enough talk about science. It is time to become engaged in it. That is the best way to learn about chemistry, and also the most enjoyable. To enter this fascinating field, we'll begin talking about the ionization energy of an atom. This is the energy needed to remove one electron from the atom, or two electrons, or three, or more. We'll see that there is a wealth of information to be drawn from such data. Through this single measurement, we'll learn something about the structure of the atom, we'll see the Periodic Table appear before our eyes, and, later, we'll find a basis for understanding chemical bonding.

Problems

1. Suppose you were trying to design a chameleon that would be blue when asleep in his underground burrow but green when he is climbing trees and frolicking in the grass. What sort of ideas would come to mind if you were to think of (a) a girl's blush, (b) a piece of litmus paper, (c) a Venetian blind, (d) a suntan, (e) fluorescent fingernail polish?

2. A bulb of UF_6 gas at temperature T_1 is placed in thermal contact with a bulb of F_2 gas at temperature T_2 until the two gases reach the same temperature. (a) How does the average kinetic energy of the UF_6 molecules compare to that of the F_2 molecules?
(b) How does the average velocity of the UF_6 molecules compare to that of the F_2 molecules?
(c) How does the average momentum $(m \cdot v)$ of the UF_6 molecules compare to that of the F_2 molecules?

3. At noon, at the equator of Mars, the temperature reaches 295°K, but at midnight, the temperature drops to about 190°K. What are these temperatures, expressed in °C? Dry ice (solid CO_2) has a one atmosphere vapor pressure at −78°C. What is this temperature, expressed in °K?

4. Venus is a cloud-covered planet. Infrared light, which does not penetrate the clouds, shows that the cloud-top temperatures are near 235°K, but radar measurements show that the surface may be as hot as 700°K. What are these temperatures expressed in °C? Liquid mercury, Hg, freezes at -39°C and lead, Pb, melts at $+327$°C. What are these temperatures, expressed in °K?

5. Which of the following substances are elements and which are compounds?
(a) gaseous nitrogen, $N_2(g)$, at room temperature
(b) liquid nitrogen, $N_2(\ell)$, at 77°K, its normal boiling point
(c) ammonia gas, $NH_3(g)$
(d) "dry ice," solid carbon dioxide, $CO_2(s)$
(e) graphite, a solid form of carbon, C(graphite)
(f) diamond, another solid form of carbon, C(diamond)

6. Which of the following substances are elements and which are compounds?
(a) "flowers of sulfur," a solid, $S_8(s)$
(b) liquid sulfuric acid, $H_2SO_4(\ell)$
(c) white phosphorus, a solid, $P_4(s)$
(d) phosphine gas, $PH_3(g)$
(e) nitric acid, a liquid, $HNO_3(\ell)$
(f) solid sodium chloride, $NaCl(s)$
(g) metallic sodium, $Na(s)$
(h) metallic mercury, $Hg(\ell)$

7. How many moles of nitrogen atoms are present in one mole of each of the following compounds?
(a) ammonia, NH_3
(b) hydrazine, N_2H_4
(c) nitrogen, N_2
(d) nitrous acid, HONO ($= HNO_2$)
(e) nitric acid, $HONO_2$ ($= HNO_3$)
(f) urea,

$$\underset{\text{H}_2\text{NCNH}_2}{\overset{\overset{\displaystyle O}{\|}}{}}$$

(g) caffeine,

8. How many moles of oxygen atoms are present in one mole of each of the following compounds?
(a) water, H_2O
(b) hydrogen peroxide, H_2O_2
(c) oxygen, O_2
(d) acetic acid, CH_3COOH

(e) citric acid,

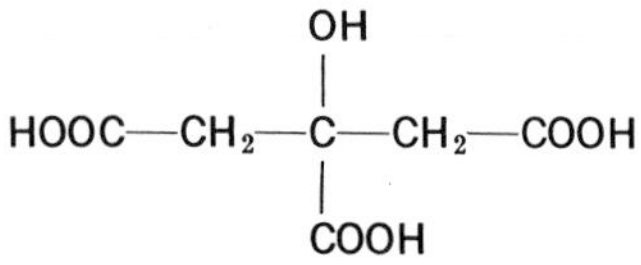

$$HOOC-CH_2-\overset{\overset{\displaystyle OH}{|}}{\underset{\underset{\displaystyle COOH}{|}}{C}}-CH_2-COOH$$

(f) cholesterol,

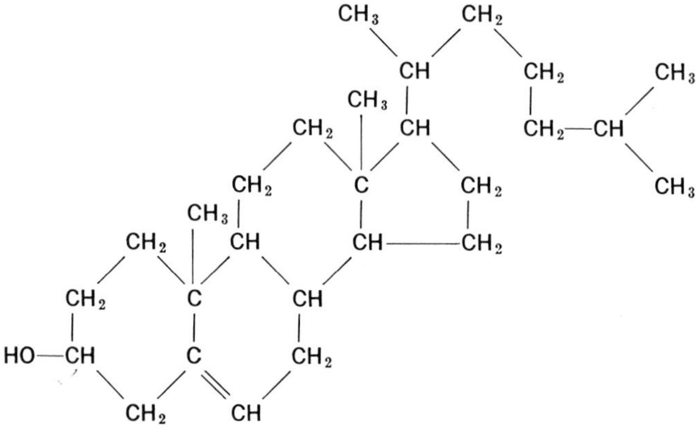

9. What is the molecular weight of each of the substances listed in Problem 7 (to the nearest 0.01 gram)?

10. What is the molecular weight of each of the substances listed in Problem 8 (to the nearest 0.01 gram)?

11. The weight of liquid mercury needed to fill a 10-milliliter graduated cylinder is 135.5 grams. How many Hg atoms are held in the graduated cylinder?

12. It takes 2.94 grams of helium to pump up a one-foot diameter balloon. How many helium atoms are held in the balloon? (one inch = 2.54 cm)

13. Solid ammonium nitrate, NH_4NO_3, decomposes on heating to 400°C to form nitrous oxide gas, N_2O, and water vapor, H_2O.
(a) Balance the equation for the reaction.
(b) Calculate the number of grams of water that will form from the decomposition of 0.100 mole of ammonium nitrate.

14. (a) Balance the equation for the reaction between gaseous carbon monoxide, CO, and gaseous oxygen, O_2, to form gaseous carbon dioxide, CO_2 (i.e., find x and y in the equation).

$$x\ CO(g) + O_2(g) = y\ CO_2(g)$$

(b) Suppose an experiment is performed in which 32.00 grams of oxygen react with carbon monoxide to form 88.02 grams of carbon dioxide. What is the weight of carbon monoxide that was consumed?

15. (a) The mineral hematite, which is mainly ferric oxide, Fe_2O_3, is reduced at red heat to the metal ("smelted") in steps, first to the "mixed oxide," Fe_3O_4, then to ferrous oxide, FeO, and finally to metallic iron, Fe. Carbon monoxide gas, CO, is

the reducing agent (produced from "coke," solid carbon). Balance the equations for the reactions by determining the missing coefficients.

$$Fe_2O_3(s) + \underline{\quad}CO(g) \longrightarrow \underline{\quad}Fe_3O_4(s) + \underline{\quad}CO_2(g)$$
$$Fe_3O_4(s) + \underline{\quad}CO(g) \longrightarrow \underline{\quad}FeO(s) + \underline{\quad}CO_2(g)$$
$$FeO(s) + \underline{\quad}CO(g) \longrightarrow \underline{\quad}Fe(s) + \underline{\quad}CO_2(g)$$

(b) In the smelting of iron (see part (a)), how many moles of Fe_3O_4 are produced from 1.00 kg of hematite, Fe_2O_3?

16. By X-ray studies, the crystal and molecular structure of Ni were determined for a monster-molecule $[Cr(NH_2CH_2CH_2NH_2)_3][Ni(CN)_5] \cdot \frac{3}{2} H_2O$, which has the interminable name, tris(ethylenediamine)chromium(III) pentacyanonickelate(II) sesquihydrate.
(a) How many moles of nitrogen atoms are contained in one mole of this compound?
(b) What is the molecular weight of this compound?

17. When solid ammonium sulfate, $(NH_4)_2SO_4$, is heated, gaseous ammonia, NH_3, and liquid sulfuric acid, H_2SO_4, are formed.
(a) Write the balanced equation for the reaction.
(b) How many grams of $(NH_4)_2SO_4$ will produce 0.300 mole of ammonia?

18. In Problem 17, what volume would be occupied by the ammonia at a temperature of 500°K and a pressure of 700 torr?

19. Carbon dioxide can be produced by the reaction
solutions to 19 & 23 are in the last page of class notebook

$$Na_2CO_3(s) + 2\, HCl(\ell) \longrightarrow H_2O(\ell) + CO_2(g) + 2\, NaCl(s)$$

How many grams of HCl are needed to produce 100 cc of dry CO_2 gas at 200°C and 450 torr pressure?
.112 g

20. In Problem 13, suppose the 0.100 mole of NH_4NO_3 is decomposed in a 5.0-liter bulb at 400°C.
(a) What will be the total pressure in the bulb?
(b) Now suppose the gaseous mixture is cooled to 20°C, at which temperature liquid water condenses (vapor pressure = 17.6 torr), but N_2O remains as a gas. What will be the total pressure?

21. Solid lithium hydride, LiH, is a convenient source of hydrogen for balloonists. It reacts with excess water according to the equation

$$LiH(s) + H_2O(\ell) \longrightarrow LiOH(aq) + H_2(g)$$

(a) How many grams of LiH will be needed to fill a balloon 10 feet in diameter (that is, 1.52 meters in radius) with dry hydrogen if the internal pressure is 800 torr and the temperature is 25°C?
(b) How many pounds would this be? (1 pound = 454 grams)

22. Repeat the calculation of Problem 21 for a balloon filled to a total pressure of 800 mm with water vapor present at its equilibrium vapor pressure, 24 torr at 25°C.

23. Solid calcium carbide, $CaC_2(s)$, reacts with water to produce a gas whose molecules each contain one hydrogen atom for each carbon atom. At 273°K and 2.00 atmospheres pressure, 1.50 liter of this gas weighs 3.49 grams. Determine the correct formula of the gas.

24. For the reactions involved in the smelting of iron (see Problem 15) calculate the volume of carbon monoxide required to convert 1.00 kg of hematite, Fe_2O_3, to metallic iron if the CO is at 600°C (873°K) and at 1.00 atm.

25. Plot the PV product at constant temperature versus P for one mole of oxygen gas and for one mole of carbon dioxide gas, using the data below. (Use an expanded PV scale in which the difference between 22.4 and 22.3 is 10 cm on the ordinate scale.)
(a) From your graph, determine the value of the PV product for one mole of a perfect gas at 0°C (i.e., the PV product at $P = 0$).
(b) Use your value of PV at $P = 0$ to calculate R, the gas constant.
(c) Determine, from your graph, the values of (b) in the imperfect gas equation, $PV = nRT(1 - bP)$ for each of these substances.
(d) Calculate, for each gas, the error that would be involved in calculating the volume of one mole at one atmosphere pressure if perfect gas behavior is assumed.

	O_2 (273.16°K)	CO_2 (273.16°K)
Pressure	$P \cdot V$(liter-atm)	$P \cdot V$(liter-atm)
1.0000	22.3939	22.2643
0.5000	22.4045	22.3397
0.2500	22.4096	22.3775
0.1667		22.3897

26. The salt mentioned in Problem 16 was initially characterized as a dihydrate, but the analytical data fit a sesquihydrate equally well. To what accuracy must a molecular weight determination be made to distinguish between the dihydrate ($\cdot 2 H_2O$) and the sesquihydrate ($\cdot \frac{3}{2} H_2O$)?

27. Two types of measurements of the concentration of a solution of HCl in water give different average values, even after many repeat experiments.

method A: HCl concentration = 2.19 ± 0.03 moles/liter
method B: HCl concentration = 1.95 ± 0.08 moles/liter

Answer each question A, B, or "can't tell."
(a) Which method has the larger random errors?
(b) Which method has the larger systematic errors?
(c) Which method has higher precision?
(d) Which method has higher accuracy?
(e) If the "best value" were taken to be the average of the two methods, what would be the number, expressed to the correct number of significant figures?

28. Four repeat molecular weight determinations give measurements of 150, 152, 169, and 157. Calculate the average and the average deviation from the average.

two ionization energy and the periodic table

There are over 100 elements and about one and a half million compounds, each with its own fascinating chemistry. Some compounds have similar formulas—CO_2, SO_2, NO_2, ClO_2—OF_2, SF_2, MgF_2, XeF_2—but sometimes similar formulas mean similar chemistry and sometimes not. For example, a good chemical stockroom would have twelve different bottles labeled C_5H_{10}, the formula we deduced in the molecular weight determination of Chapter One. Only one of the twelve would be called cyclopentane. Each of the others would have its own identifying name to indicate that it can be distinguished from cyclopentane, as well as from the other C_5H_{10} compounds. Yet three of the compounds have chemistry much like that of cyclopentane—the other eight are quite different.

Among the elements, each one displays distinctive properties, but some of them are so alike that they are classified together in families. Chemists have exploited these familial likenesses to organize the staggering volume of chemical information we have accumulated. Particularly useful in this organization are the similarities revealed by the ionization energies of the atoms. Ionization energies are not difficult to measure and they furnish a basis for grouping the elements in the marvelous array called the Periodic Table. That will be our entrée into chemistry.

2-1 Measurement of ionization energy

The atom contains a very small nucleus that carries most of the atomic mass. This nucleus also carries a positive, integer charge, $+Z$. A number of electrons, each with -1 charge, are moving around the nucleus. If there are Z electrons, then the net atomic charge is $(+Z) + (Z)(-1) = 0$. The atom with zero charge is said to be neutral.

The energy required to remove an electron from an atom is called the ionization energy: E_1 for the first electron removed, E_2 for the second, and E_m for the mth electron. Measurement of E_m entails transferring energy to the atom and detecting electrons when they are dislodged. For example, atoms can be bombarded with projectiles of known energy and any electrons released can be detected by their electrical conductivity. The

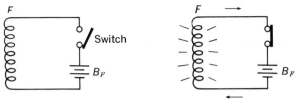

(a) Filament circuit

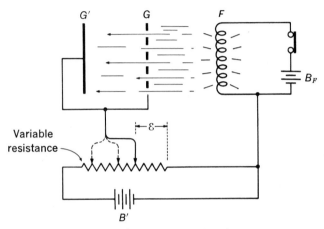

(b) Electron accelerating
circuit added

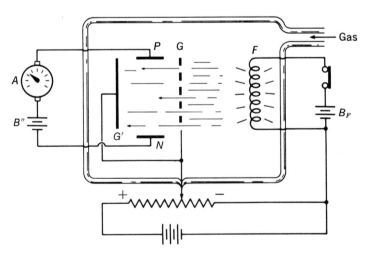

(c) Ionization detection
circuit added

Figure 2-1   *Build-up of apparatus for measuring ionization energies (see text for explanations of symbols).*

projectile energy can be increased slowly until conductivity just begins. That threshold energy is the ionization energy.

Figure 2-1 builds up a simple apparatus for measuring ionization energies. Electrons themselves make good projectiles. They can be vaporized out of a red-hot piece of tungsten and accelerated to a known energy with a measured voltage. Figure 2-1(a) shows the filament circuit. A small battery B_F causes current to flow through the tungsten filament F, heating it to red heat. In Figure 2-1(b), another battery B' places a positive charge $\mathcal{E}$ on the wire grid G and the plate G'. Electrons coming out of the filament are accelerated to the energy equivalent to voltage $\mathcal{E}$ as they reach the wire grid G. Some of the rapidly moving electrons go hurtling through the holes in the grid but then they merely coast along at constant velocity because there is no voltage difference between G and G'. The projectile electrons have a velocity, hence energy, determined by the slide wire setting, $\mathcal{E}$.

Figure 2-1(c) shows the complete device, now enclosed in a glass envelope so air can be removed and a test gas added. A third battery, B'', is connected to two more electrodes, P and N, through a sensitive current detector, microammeter A. The battery B'' places a positive charge on P and a negative charge on N.

Now suppose neon is placed in the cell. In the region between G and G', electron projectiles may strike neon atoms and, if $\mathcal{E}$ is sufficiently large, ionization of the neon atom may occur. This process removes electrons from neon atoms and leaves some of them with a positive charge. The dislodged electrons are attracted to the positive electrode P and the positive neon atoms are attracted to the negative electrode N. As some of these charges reach P and N, they cause current to flow through ammeter A.

Figure 2-2 shows the results of two experiments, one using neon gas and the second using sodium vapor obtained by heating sodium metal at a temperature near 500°C. In the graphical presentation, the horizontal axis, the *abscissa*, shows the voltage $\mathcal{E}$ as it is varied by moving the slide wire

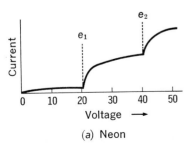

(a) Neon

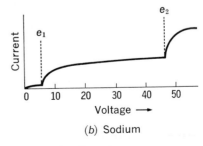

(b) Sodium

Figure 2-2   *Measurement of ionization energies of neon and sodium atoms.*

contact. The vertical axis, the *ordinate,* shows the current flowing through ammeter A. There is almost no current at all at low voltages. For neon, there is a spurt of current when the voltage reaches 21.6 volts. As the voltage $\mathcal{E}$ is raised, the current rises steadily, but then, at 41.1 volts, there is a second spurt of current. The first voltage, 21.6 volts corresponds to e_1, the energy needed to remove one electron from a neutral neon atom. The second voltage, 41.1 volts, corresponds to e_2, the energy needed to remove the second electron from a Ne^+ ion. The corresponding values for gaseous sodium atoms are $e_1 = 5.15$ volts and $e_2 = 47.3$ volts.

Chemists usually express energies in kilocalories per mole of reaction. Here's how the calculation goes. The bombarding electrons have energies, in ergs, equal to the accelerating voltage times the charge per electron.

$$e_1 = \frac{\text{energy per}}{\text{ionization}} = (\text{voltage}) \cdot (\text{electron charge})$$

$$= (21.6) \cdot (1.602 \cdot 10^{-19}) \text{ volt-coulomb}$$

$$= 3.46 \cdot 10^{-18} \text{ joule} = 3.46 \cdot 10^{-11} \text{ erg} \qquad (2\text{-}1)$$

The energy per mole of ionization reactions is obtained by multiplying by Avogadro's Number, $N_0 = 6.023 \cdot 10^{23}$. Then we can change units to kilocalories by multiplying by the conversion factor, $2.39 \cdot 10^{-11}$ kcal/erg.

$$E_1 = N_0 e_1 = (6.023 \cdot 10^{23})(3.46 \cdot 10^{-11}) = 2.08 \cdot 10^{13} \text{ ergs/mole} \qquad (2\text{-}2)$$

$$E_1 = (2.08 \cdot 10^{13})(2.39 \cdot 10^{-11}) = 497 \text{ kcal/mole} \qquad (2\text{-}3)$$

The steps (2-1), (2-2), and (2-3) are combined in a convenient factor in Appendix a. Here we see that to convert from electron volts to kcal/mole, we need only multiply by 23.061.

Exercise 2-1. Use the conversion table in Appendix a to calculate the values of E_2 for neon and E_1 and E_2 for sodium in kcal/mole.

Our experimental results can now be expressed in conventional form.

$$Ne(g) \longrightarrow Ne^+(g) + e^- \qquad E_1 = 497 \text{ kcal/mole} \qquad (2\text{-}4)$$

$$Ne^+(g) \longrightarrow Ne^{+2}(g) + e^- \qquad E_2 = 947 \text{ kcal/mole} \qquad (2\text{-}5)$$

$$Na(g) \longrightarrow Na^+(g) + e^- \qquad E_1 = 119 \text{ kcal/mole} \qquad (2\text{-}6)$$

$$Na^+(g) \longrightarrow Na^{+2}(g) + e^- \qquad E_2 = 1091 \text{ kcal/mole} \qquad (2\text{-}7)$$

Already we see interesting contrasts. It is much easier to remove one electron from a sodium atom than from a neon atom. On the other hand, it is much more difficult to remove the second electron from sodium, even more so than to remove the second electron from neon. It is information like this that we shall explore, searching for trends, similarities, and differences.

2-2 Trends: first ionization energies and chemical formulas

An enormous amount of research effort has been devoted to ionization energy measurement, a reflection of its importance. First ionization energies are known with reasonable accuracy for almost every one of the elements, even the least volatile and the most reactive. Let's examine the experimental numbers for the first 20 elements.

(a) E_1 VERSUS Z

Electrons are held near the nucleus because of electrostatic attraction. Hence one might expect ionization energy to rise as Z increases. That expectation is readily tested. Table 2-1 lists the first 20 elements and their measured values of E_1.

Look at the values of E_1 for the first five elements, H, He, Li, Be, B: 314, 567, 124, 215, 191. Immediately it is clear that E_1 is not simply correlated with Z. It is only after a bit of searching through Table 2-1 that one regular-

Table 2-1   First Ionization Energies for the First Twenty Elements

Element	Symbol	Z	E_1 (kcal/mole)
hydrogen	H	1	313.6
helium	He	2	567
lithium	Li	3	124
beryllium	Be	4	215
boron	B	5	191
carbon	C	6	260
nitrogen	N	7	335
oxygen	O	8	314
fluorine	F	9	402
neon	Ne	10	497
sodium	Na	11	119
magnesium	Mg	12	176
aluminum	Al	13	138
silicon	Si	14	188
phosphorus	P	15	242
sulfur	S	16	239
chlorine	Cl	17	300
argon	Ar	18	363
potassium	K	19	100
calcium	Ca	20	141

ity does stand out. Three times in this list, there is a precipitous change in E_1 between adjacent elements—between helium and lithium, between neon and sodium, and between argon and potassium. This is shown more vividly in a graphical presentation of these same data, as in Figure 2-3.

Just a glance at Figure 2-3 reveals an inescapable regularity. Between elements lithium and neon there is a gradual rise in E_1 as Z increases, but with a couple of small jogs at elements boron and oxygen. This contour is closely duplicated between elements sodium and argon. As Z rises from 3 to 10, we see the general rising trend we expected (except for the small jogs) and then it starts over again as Z rises from 11 to 18 (except for the same small jogs). This repetitious, or periodic, aspect of ionization energies is a start toward classifying the elements. It remains to be seen whether this periodicity appears in chemistry. We'll look first for periodicity in chemical formulas.

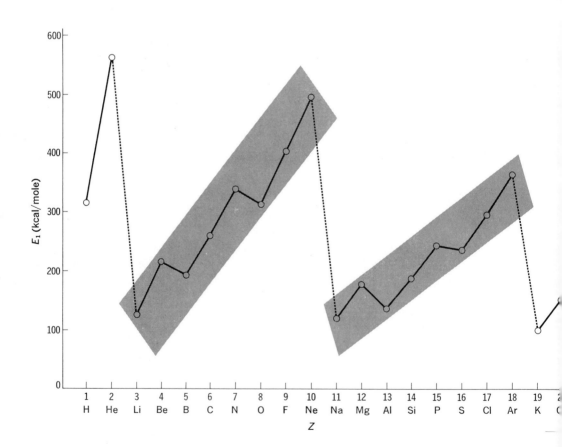

Figure 2-3   First ionization energy and nuclear charge: elements 1 to 20.

(b) CHEMICAL FORMULAS VERSUS Z

Chemical formulas always involve simple integer numbers of atoms—that is why we believe in atoms. But some formulas appear in the chemical stockroom and others do not. We find bottles labeled CH_4 and CCl_4, but none labeled NH_4 or NCl_4. These formulas, both those present and those missing, compile a lot of chemical experience and they give us a first opportunity to look for periodicity in chemistry. We'll concentrate on compounds containing chlorine, fluorine, and hydrogen because these elements combine to form stable compounds with most of the elements.

Table 2-2 lists the formulas of some stable Cl, F, and H compounds for the first 20 elements, those whose ionization energies are shown in Figure 2-3. All of the formulas are of the type MX_n. There are some notable vacancies. Three elements have no compounds listed: helium, neon, and argon. Because of this similarity, they might be classified together as a family (as they are). Referring back to Figure 2-3, these are the elements at the peaks of the sawtooth contours, $Z = 2$, 10, and 18. Equally notable is that, between lithium and neon, n in the formulas MX_n systematically increases from 1 to 4, then it decreases again to zero. This simple sequence repeats between sodium and argon.

Thus we see that the sequences identified by ionization energy trends (Fig. 2-3) are significant in chemistry. After the first two elements, there are two eight-element groups, or periods, with identical trends, both in E_1 and formulas MX_n. Let's add to Figure 2-3 to see if the rest of the elements cluster in sequences of eight.

(c) THE LONG PERIODS

Figure 2-4 extends Figure 2-3 to include elements with Z up to 56. This shows no clusters of eight! Instead, there are two groups of eighteen elements placed between the distinctive sawtooth spikes. The shading draws attention, however, to the fact that in each group the values of E_1 begin and end with the trends displayed twice among the first 20 elements. It seems as though ten elements have injected themselves in the middle of a group of eight. The plot thickens. What numerology accounts for these sequences—2–8–8–18–18? What model of an atom would parallel this intriguing regularity? These multiple ionization energies must hold clues, both to the structure of the atom and to the systematic chemistry shown in Table 2-2.

2-3   **Multiple ionization energies,** E_m

We already glanced at values of E_2, the energies needed to remove a second electron, for a neon atom and for a sodium atom. In each case, it is much harder to get off that second electron than it is to remove the first,

Table 2-2 *Some Known Compounds* of the First Twenty Elements with Cl, F, and* H

Element	H	He	Li	Be	B	C	N	O	F	Ne	Na	Mg	Al	Si	P	S	Cl	Ar	K	Ca
Z	1	2	3	4	5	6	7	8	9	10	11	12	13	14	15	16	17	18	19	20
Chloride	HCl	—	LiCl	BeCl₂	BCl₃	*CCl₄*	NCl₃	OCl₂	FCl	—	NaCl	MgCl₂	AlCl₃	*SiCl₄*	PCl₃	SCl₂	ClCl	—	KCl	CaCl₂
Fluoride	HF	—	LiF	BeF₂	BF₃	CF₄	NF₃	OF₂	FF	—	NaF	MgF₂	AlF₃	SiF₄	PF₃	SF₂	ClF	—	KF	CaF₂
Hydride	HH	—	LiH	—	—	CH₄	NH₃	*OH₂*†	FH†	—	NaH	MgH₂	—	SiH₄	PH₃	SH₂†	ClH†	—	KH	CaH₂
MXₙ n =	1	0	1	2	3	4	3	2	1	0	1	2	3	4	3	2	1	0	1	2

* Solids are listed in bold type, liquids in italics. All other compounds are gaseous at normal temperatures.

† Formula usually written in opposite order: H₂O, HF, H₂S, HCl.

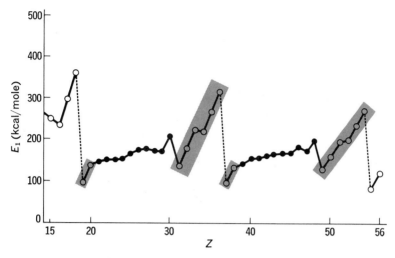

Figure 2-4 *First ionization energy and nuclear charge: elements 15 to 56.*

twice as hard for neon and nine times as hard for sodium. Here are similar data for a beryllium atom, which has only four electrons to its name.

$$\text{Be(g)} \longrightarrow \text{Be}^+(\text{g}) + \text{e}^- \qquad E_1 = 215 \text{ kcal/mole}$$

$$\text{Be}^+(\text{g}) \longrightarrow \text{Be}^{+2}(\text{g}) + \text{e}^- \qquad E_2 = 420 \text{ kcal/mole}$$

$$\text{Be}^{+2}(\text{g}) \longrightarrow \text{Be}^{+3}(\text{g}) + \text{e}^- \qquad E_3 = 3548 \text{ kcal/mole}$$

$$\text{Be}^{+3}(\text{g}) \longrightarrow \text{Be}^{+4}(\text{g}) + \text{e}^- \qquad E_4 = 5020 \text{ kcal/mole}$$

(2-8)

Again we see that it becomes harder to remove successive electrons as the charge on the ion increases. Electrostatics must play a role here and should help us understand these trends.

(a) IONIZATION ENERGY AND ELECTROSTATICS

Somehow, the electron we're trying to remove is moving around the nucleus, attracted by its positive nuclear charge, $+Z$. Of course, it is also repelled by the other electrons, so the force of attraction is the difference between the attraction to the nucleus and the repulsion by the electrons. Immediately it is clear why the second electron is harder to remove than the first. The electron pulled out of a neutral atom looks back to see the $+Z$ charge of the nucleus surrounded by $(Z - 1)$ electrons. The net charge is $+1$. The second electron is pulled away from a positive ion; it looks back to see the $+Z$ charge surrounded by $(Z - 2)$ electrons. The net charge is $+2$. The third electron would see a net charge of $+3$; the mth electron to go would see $+m$.

That gives us a basis for interpreting E_m. From simple electrostatics, the energy needed to pull a negative charge q_1 away from a positive charge q_2 is proportional to the magnitudes of q_1 and q_2 and inversely proportional to the distance between them, r_m.

$$E_m = \frac{q_1 q_2}{r_m} \qquad (2\text{-}9)$$

If we were to use the simple model suggested above, $-q_1$ would be the electron charge and q_2 would be a multiple of q_1, that is, the charge on the ion after the electron is gone. As the mth electron leaves, q_2 would be $+m \cdot q_1$. The distance r_m would be some average distance of the electron from the nucleus in the atom. Our expression (2-9) becomes

$$E_m = -m \cdot \frac{q_1^2}{r_m} \qquad (2\text{-}10)$$

Expression (2-10) indicates that E_m/m is more easily interpreted than E_m itself. Dividing through by m removes the simple electrostatic part, the least interesting component.

$$\frac{E_m}{m} = -\frac{q_1^2}{r_m} \qquad (2\text{-}11)$$

(b) E_m/m FOR THE BERYLLIUM ATOM

The values of E_m for beryllium (2-8) will test the usefulness of E_m/m. After dividing each energy by the appropriate integer, we obtain:

$$
\begin{array}{lll}
Be(g) \longrightarrow Be^+(g) + e^- & E_1/1 = 215 \text{ kcal/mole} & \\
Be^+(g) \longrightarrow Be^{+2}(g) + e^- & E_2/2 = 210 \text{ kcal/mole} & \\
Be^{+2}(g) \longrightarrow Be^{+3}(g) + e^- & E_3/3 = 1183 \text{ kcal/mole} & (2\text{-}12) \\
Be^{+3}(g) \longrightarrow Be^{+4}(g) + e^- & E_4/4 = 1255 \text{ kcal/mole} &
\end{array}
$$

Now it is immediately obvious that there are two kinds of electrons in beryllium. The first two, the easiest to remove, have almost identical values of E_m/m, about 210 kcal/mole. The last two, the hardest to remove, have much higher and, again, about equal values of E_m/m, about 1200 kcal/mole.

Looking back at (2-11), we see that one property of the atom, r_m, remains to account for the two types of electron. If the first two electrons have the same value of E_m/m, they must be about the same distance from the nucleus. If the third and fourth electrons require much higher values of E_m/m, they must move, on the average, much closer to the nucleus. In

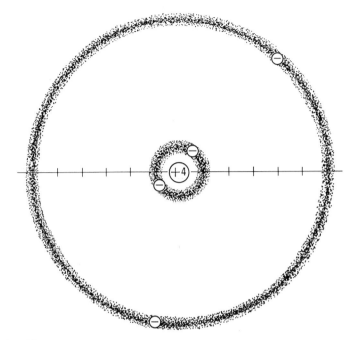

Figure 2-5 The location of electrons in beryllium as suggested by E_m/m.

fact, if $E_3/3$ and $E_4/4$ are six times as high as $E_2/2$, the third and fourth electrons must be at one sixth the radius.

Figure 2-5 shows the picture of the atom that is developing. There are two electrons held in tightly by the nucleus. We'll call them type 1. Then there are two more electrons held much more loosely—at a much larger radius. These are electrons of type 2. We are learning about the structure of the atom.

(c) E_m/m FOR THE FIRST TWELVE ELEMENTS

That was so informative, it is worth pursuit. Let's examine E_m/m for the first twelve elements. To cope with all the numbers involved, we'll need a graphical presentation again. Figure 2-6 shows the result.

Again the graphical plot is revealing. Look at the two type-1 electrons of beryllium. The shading indicates the unambiguous connection between these two electrons and a similar pair on every other atom except hydrogen, which has only one. The slope of the shaded area is, of course, more of the nuclear charge effect we expected earlier. As nuclear charge goes up, these electrons are pulled in closer and closer to the nucleus. Chemists

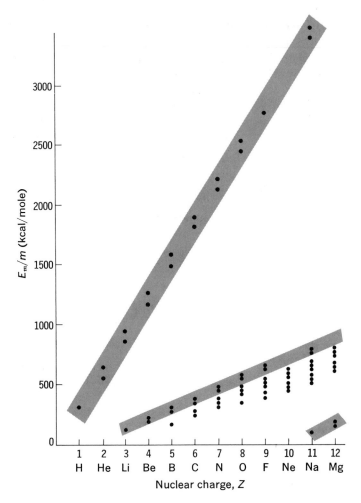

Figure 2-6   A graphical representation of E_m/m for the first twelve elements.

call the identifying number "type 1" the "quantum number" and designate it n. Figure 2-6 shows that an atom can accept one or two but no more
than two electrons with quantum number $n = 1$. If an atom, or ion,
possesses more than two electrons, the others must move at a much
greater distance from the nucleus.

From lithium to neon, there is a second type of electron identified—
there can be up to eight of these but no more. These are said to have
quantum number $n = 2$. Again as Z increases, from lithium to neon, these
quantum number 2 electrons become more and more difficult to remove
and they are pulled in closer and closer to the nucleus. Then, when the

atom has eight electrons with quantum number $n = 2$, it all starts over again at sodium and magnesium.

Figure 2-7 shows the size implications of the E_m/m values for lithium, fluorine, and sodium, all contrasted to beryllium. All radii are referred to that of the first electron of beryllium, so we don't have to worry about scale. If E_m/m is above E_1 of beryllium, the radius is proportionately smaller. The figure shows that the atoms shrink in size across the sequence lithium to neon. Then, however, sodium is large again, with its single quantum number 3 electron. The figure also makes it obvious why lithium and sodium are chemical look-alikes. Each atom has a single electron in its outermost region and the atoms are almost identical in size. The inner electrons differ in number, but they are held so tightly in either case that they are inaccessible. The outermost electrons, those most easily removed, determine the chemistry of an atom—they are called the *valence electrons*.

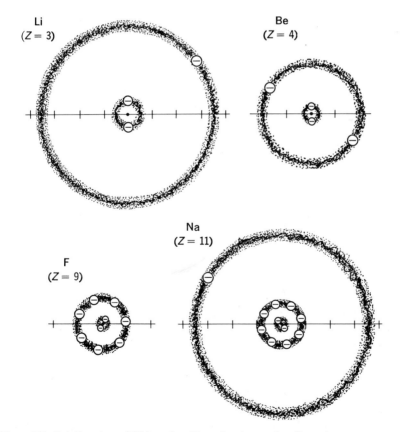

Figure 2-7 *Relative sizes of lithium, beryllium, fluorine, and sodium atoms.*

2-4 Valence electrons, inert gases, and chemical bonding

Our analysis of ionization energies showed that the elements helium, neon, and argon have three properties in common:

—high ionization energy
—no chemical compounds
—the next element in the Periodic Table has an easily removed electron that occupies a different region of space.

Each of these properties suggests that there is a special stability to the number of electrons possessed by an inert gas atom. This apparent special stability provides the basis for a simple theory of chemical bonding, one proposed many years ago by G. N. Lewis. Though extremely rudimentary, this bonding model has been quite successful and still it is used daily by practicing chemists. Lewis saw that the chemical formulas of many substances could be understood with the simple rule that atoms "strive" to achieve the specially stable inert gas electron distributions by sharing electrons, two at a time. Hydrogen gives us a clear example.

(a) THE HYDROGEN MOLECULE: H TRYING TO LOOK LIKE He

The neutral helium atom has two electrons of quantum number $n = 1$. Except for hydrogen, every other neutral atom also has two electrons of quantum number $n = 1$. We infer that a hydrogen atom could also accommodate two such electrons, and doesn't merely because it has only one electron to deal with.

It is easy to imagine how two hydrogen atoms, if brought together, would solve this frustrating situation. Each of the atoms has the "capability" to accommodate two electrons of quantum number $n = 1$, and between the two of them, they have two. All they need to do is share their pair of electrons and each atom can enjoy that specially stable feeling that helium has.

We can represent this electron-sharing pictorially by showing the electrons as dots.*

$$H \cdot + \cdot H \longrightarrow H : H \qquad\qquad (2\text{-}13)$$

The formula shows that each H atom in H_2 is near two electrons, so each has the electron population of the inert gas helium. The resulting molecule, H_2, should be more stable than are the two H atoms when apart. That means a chemical bond is formed to hold the molecule together. Furthermore, each H atom now has all the quantum number 1 electrons it can accommodate. That means there is no residual bonding capacity: We can-

---

* Electrons from different atoms will often be shown by different symbols (●, ○, ✕) to aid us in our bookkeeping. There is, however, never any way to tell one electron from another.

not expect more complicated molecules to be formed with additional hydrogen atoms. Indeed, no one has ever made H_3, H_4, H_5, or higher molecules. We see that electron-pair sharing fits the bonding habits of hydrogen very well.

(b) FLUORINE AND HYDROGEN FLUORIDE: F TRYING TO LOOK LIKE Ne

Fluorine, like the hydrogen atom, just precedes an inert gas in the Periodic Table. Like hydrogen, it can accommodate one more electron to gain the special stability that seems to go with being an inert gas, this time, neon. Just as two H atoms find their answer in electron-sharing, two fluorine atoms can couple, sharing a pair of electrons. Thus the two atoms bond together to form a stable molecule in which each atom has the neon electron population.

$$\overset{\circ\circ}{\underset{\circ\circ}{\circ}} \overset{\cdot\cdot}{F} \cdot \; + \; \cdot \overset{\cdot\cdot}{F} \underset{\cdot\cdot}{:} \longrightarrow \overset{\circ\circ}{\underset{\circ\circ}{\circ}} \overset{\cdot\cdot}{F} \overset{}{:} \overset{\cdot\cdot}{F} \underset{\cdot\cdot}{:} \qquad (2\text{-}14)$$

Notice that in the atom, the symbol F is shown surrounded by only seven electrons, though a neutral fluorine atom has nine. The reason is clear in Figure 2-7: The seven $n = 2$ electrons are out on the edges of the atom. These outermost electrons (the valence electrons) are the ones a neighboring atom would encounter. The two $n = 1$ electrons are held so tightly, so close to the nucleus, that they are quite ineffective in bonding. Hence these two are incorporated into the symbol for the nucleus. *In all electron-dot formulas only valence electrons are shown.*

As the two fluorine atoms come together to share two electrons, each nucleus "sees" eight electrons in its valence region. Each atom now has the electron population of the inert gas neon, so a stable molecule results. No additional electrons can be accommodated, so no additional bonds will be formed.

The formation of hydrogen fluoride combines the "desire" of the H atom to achieve the helium electron configuration and of the F atom to achieve the neon electron configuration.

$$H \cdot \; + \; \cdot \overset{\cdot\cdot}{F} \underset{\cdot\cdot}{:} \longrightarrow H \overset{\cdot\cdot}{:} \overset{}{F} \underset{\cdot\cdot}{:} \qquad (2\text{-}15)$$

The shared electron pair gives each atom an inert gas population and it leaves no residual need for electrons. The molecule HF should be a stable molecule.

Again chemical experience is neatly explained. On the chemical shelf we do find stable molecular substances with formula F_2, elemental fluorine, and HF, hydrogen fluoride.

(c) OXYGEN MOLECULES: O_2, H_2O AND H_2O_2

By analogy to fluorine, we might picture the formation of O_2 as follows:

$$: \overset{\circ}{\underset{\circ\circ}{O}} \cdot + \cdot \overset{\cdot\cdot}{\underset{\cdot}{O}} : \longrightarrow : \overset{\circ}{\underset{\circ\circ}{O}} : \overset{\cdot\cdot}{\underset{\cdot}{O}} : \qquad (2\text{-}16)$$

The molecular formula on the right suggests the formation of a bond (two electrons shared), but it also shows that neither oxygen atom has achieved the inert gas electron arrangement of neon. Hence the scheme indicates that another pair of electrons can be shared.

$$: \overset{\circ}{\underset{\circ\circ}{O}} \cdot + \cdot \overset{\cdot}{\underset{\cdot\cdot}{O}} : \longrightarrow \cdot \overset{\cdot}{\underset{\circ}{O}} : : \overset{\cdot}{\underset{\circ}{O}} \cdot \qquad (2\text{-}17)$$

Now the electron arrangement gives each oxygen atom eight valence electrons (all four of the electrons between the atoms are shared) as in neon, and it explains the existence of diatomic oxygen molecules, O_2.

Turning to the reaction between hydrogen and oxygen, we find further developments. If a hydrogen atom shares an electron pair with oxygen, the hydrogen atom achieves the helium inert gas arrangement, but oxygen does not achieve the neon arrangement

$$H \cdot + \cdot \overset{\cdot}{\underset{\cdot\cdot}{O}} : \longrightarrow H : \overset{\cdot}{\underset{\cdot\cdot}{O}} : \qquad (2\text{-}18)$$

The molecule OH is expected to be stable, but the oxygen atom has residual bonding capacity so it will be reactive. Unlike the oxygen molecule case, a single hydrogen atom cannot furnish a second electron to help out. Consequently OH can react with either another hydrogen atom or, indeed, with another OH molecule.

$$H \cdot + H : \overset{\cdot}{\underset{\cdot\cdot}{O}} : \longrightarrow H : \overset{\overset{\textstyle H}{\cdot\cdot}}{\underset{\cdot\cdot}{O}} : \qquad (2\text{-}19)$$

$$H : \overset{\cdot}{\underset{\cdot\cdot}{O}} : + H : \overset{\cdot}{\underset{\circ\circ}{O}} : \longrightarrow : \overset{\overset{\textstyle H}{\cdot\cdot}}{\underset{\circ\circ}{O}} : \overset{\cdot\cdot}{\underset{\circ\circ}{O}} : \qquad (2\text{-}20)$$
$$\underset{\textstyle H}{}$$

Both reactions (2-19) and (2-20) give molecules in which every atom has an inert gas electron arrangement with no residual bonding capacity in any atom. Both molecules are expected to be stable, and unreactive, H_2O and H_2O_2. These are water and hydrogen peroxide, both familiar substances.

Thus, the Lewis scheme tells us the bonding capacity of an atom. As an atom forms bonds with other atoms, it still has residual bonding capacity as long as it has not achieved an inert gas population. A molecule that includes an atom having residual bonding capacity will be stable but extremely reactive. A molecule in which every atom does have an inert gas arrangement will be stable and can probably be found on someone's store-room shelf. That is not to say that these stable molecules have no reactivity at all, though. Under some conditions, H_2O_2 will decompose to O_2 and H_2O; F_2 and H_2, when mixed, react explosively to form HF. Even so, molecules that have no residual bonding capacity generally have reactivities that can be moderated and controlled. So these are the molecules that can be stored and these are the ones that make useful starting materials as we put chemistry to work.

(d) LINE REPRESENTATIONS

In each of the compounds O_2, H_2O, and H_2O_2, we see that an oxygen atom needs to share two electron pairs with other atoms to reach the inert gas arrangement. To represent this bonding situation it isn't really necessary to show all the electrons. Chemists, trying to simplify life, habitually show only the two bonding pairs, each pair indicated by a line drawn between the two atoms. One such line between two atoms means that they share one pair of electrons and hence they are connected by a "single bond." Two such lines between atoms means that two pairs of electrons are shared, so the bond is called a "double bond." Figure 2-8 compares this simpler representation with the electron-dot picture. In each of the first three relatively unreactive compounds there are two lines connected to each oxygen atom. In the last example, OH, one line is shown connecting the oxygen atom to the hydrogen atom and a dot displays the fact that

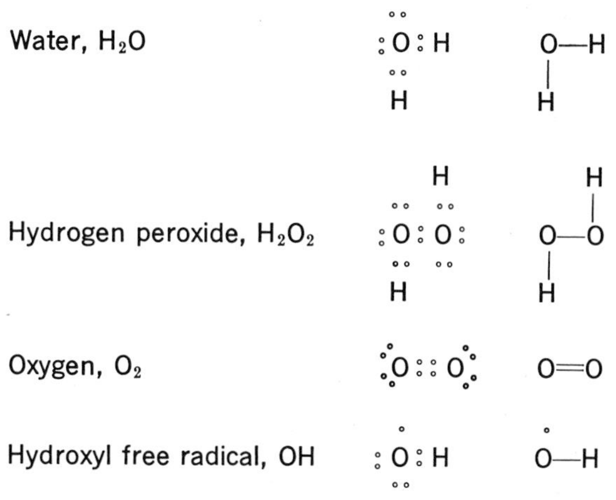

Figure 2-8 Electron dots and line representations of bonds to oxygen.

oxygen has the capacity to share a second electron pair. This dot is the seventh, "unpaired" electron in the electron-dot representation. It indicates residual bonding capacity and hence a tendency toward extreme reactivity. Such a molecule, one with an unshared valence electron, is called a "free radical."

(e) CARBON COMPOUNDS WITH HYDROGEN

These ideas are readily illustrated with the hydrogen compounds of carbon. If we consider first a single carbon atom, electron-dot formulas predict the *stability* of the molecules CH, CH_2, CH_3, and CH_4. The first three display residual bonding capacity for the carbon, so only the last, CH_4, should be unreactive.

CH $\overset{\circ}{\underset{\circ}{.C}} : H$ or $.\overset{\circ}{\underset{\circ}{C}}\!-\!H$ stable, but reactive   (2-21)

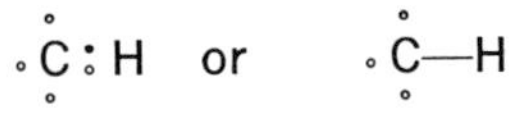

CH_2 $\overset{\circ}{.C} : H$ or $.\overset{\circ}{C}\!-\!H$ stable, but reactive   (2-22)

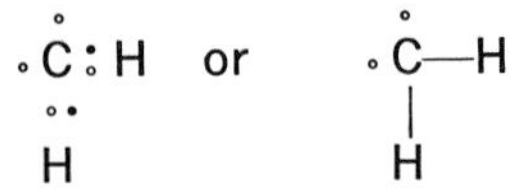

CH_3 $H : \overset{\circ}{C} : H$ or $H\!-\!\overset{\circ}{C}\!-\!H$ stable, but reactive   (2-23)

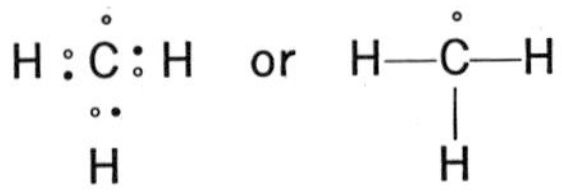

CH_4 $H : C : H$ or $H\!-\!C\!-\!H$ stable, unreactive   (2-24)

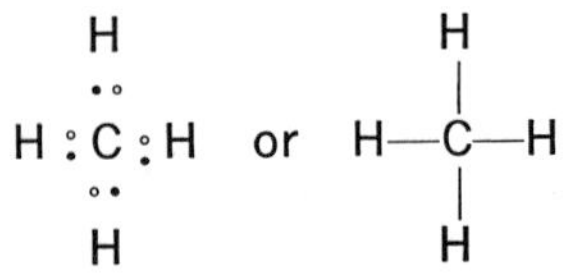

We recognize CH_4, of course, as methane, the household gas that is piped into our homes as a convenient fuel. The other three molecules, CH, CH_2 and CH_3, are all well known to chemists, but not to housewives. These molecules all have lifetimes of the order of microseconds at normal temperatures and pressures. Each of these species can, however, react with another identical species to form a molecule that does give an inert gas structure to each atom:

$$CH + CH \longrightarrow C_2H_2 \text{ (acetylene)}$$

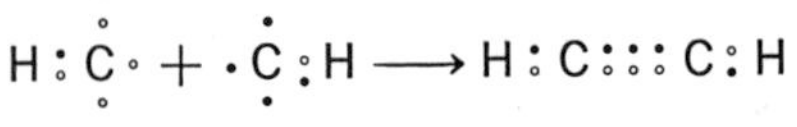

$$H : \overset{\circ}{\underset{\circ}{C}} . + . \overset{.}{C} : H \longrightarrow H : C ::: C : H \qquad (2\text{-}25a)$$

or

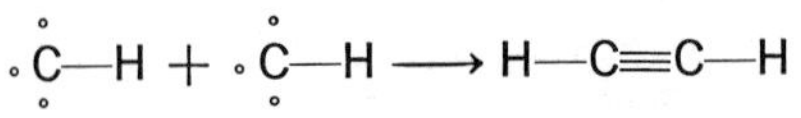

$$. \overset{\circ}{\underset{\circ}{C}}\!-\!H + . \overset{\circ}{\underset{\circ}{C}}\!-\!H \longrightarrow H\!-\!C\!\equiv\!C\!-\!H \qquad (2\text{-}25b)$$

$$CH_2 + CH_2 \longrightarrow C_2H_4 \text{ (ethylene)}$$

$$
\begin{array}{ccc}
\text{H} & \text{H} & \text{H}
\end{array}
$$

$$\text{H}:\overset{\circ}{\underset{\circ}{\text{C}}}\cdot + \cdot\overset{\bullet\bullet}{\underset{\bullet}{\text{C}}}:\text{H} \longrightarrow \quad :\overset{\bullet\bullet}{\text{C}}::\overset{\bullet}{\text{C}}: \qquad\qquad (2\text{-}26a)$$

$$
\begin{array}{ccc}
\text{H} & \text{H} & \text{H}
\end{array}
$$

or

$$\qquad\qquad (2\text{-}26b)$$

$$CH_3 + CH_3 \longrightarrow C_2H_6 \text{ (ethane)}$$

$$
\begin{array}{cc}
\text{H} & \text{H} \qquad\qquad \text{H H}
\end{array}
$$

$$\text{H}:\overset{\circ}{\underset{\circ}{\text{C}}}\cdot + \cdot\overset{\bullet\bullet}{\underset{\bullet}{\text{C}}}:\text{H} \longrightarrow \text{H}:\overset{\bullet\bullet}{\underset{\bullet\bullet}{\text{C}}}:\overset{\bullet\bullet}{\underset{\bullet\bullet}{\text{C}}}:\text{H} \qquad\qquad (2\text{-}27a)$$

$$
\begin{array}{cc}
\text{H} & \text{H} \qquad\qquad \text{H H}
\end{array}
$$

or

$$
\begin{array}{cc}
\text{H} & \text{H} \qquad\qquad \text{H H}
\end{array}
$$

$$\text{H}-\overset{|}{\underset{|}{\text{C}}}\cdot + \cdot\overset{|}{\underset{|}{\text{C}}}-\text{H} \longrightarrow \text{H}-\overset{|}{\underset{|}{\text{C}}}-\overset{|}{\underset{|}{\text{C}}}-\text{H} \qquad\qquad (2\text{-}27b)$$

$$
\begin{array}{cc}
\text{H} & \text{H} \qquad\qquad \text{H H}
\end{array}
$$

All the carbon–hydrogen compounds predicted in this simple model, C_2H_2, C_2H_4 and C_2H_6, are familiar compounds commonly handled in tank-car amounts. Each has four bond lines connected to each carbon atom.

Clearly the inert gas, electron population scheme is easy to apply and can predict lots of chemistry. Table 2-3 shows a variety of common molecular species, including some familiar ions, that have electron populations consistent with the Lewis electron-pair, inert gas structure model of bonding.

Exercise 2-2. Draw electron-dot formulas for the stable molecules that might be formed between a single nitrogen atom and one or more hydrogen atoms. Convince yourself that NH and NH_2 would be stable but reactive, NH_3 would be stable and unreactive, and NH_4 should not exist.

Exercise 2-3. Draw the electron-dot formula for the rocket fuel hydrazine, N_2H_4.

(f) TOO MANY STRUCTURES: "RESONANCE" SAVES THE DAY

Despite the many successes of the inert gas, electron-pair scheme for explaining chemical formulas, there are some difficulties. There are quite a few common substances whose existence is consistent with more than one acceptable inert gas arrangement. In every such case, any one of these

Table 2-3 *Electron-Dot Formulas and Inert Gas Structures of Some Stable Compounds and Ions*

Monofluoro-
methane
CH_3F

Difluoro-
methane
CH_2F_2

Trifluoromethane
CHF_3

Tetrafluoro-
methane
CF_4

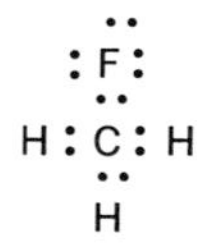

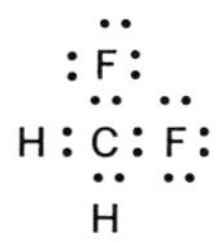

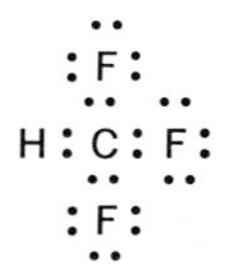

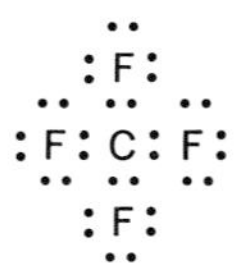

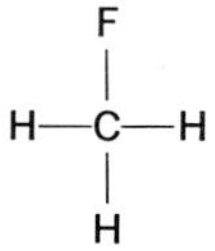

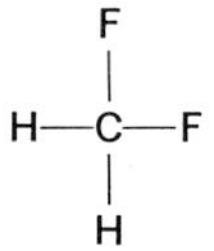

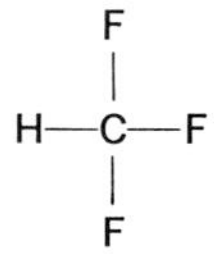

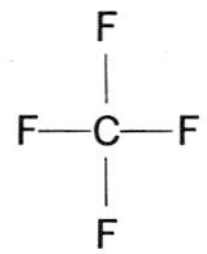

Hypochlorite
ion
ClO^-

Chlorite ion
ClO_2^-

Chlorate ion
ClO_3^-

Perchlorate ion
ClO_4^-

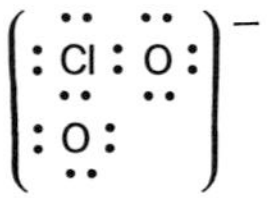

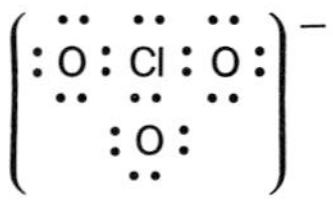

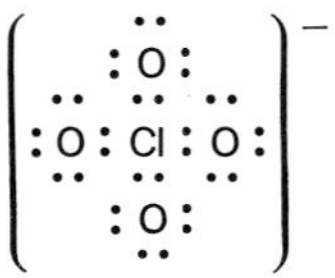

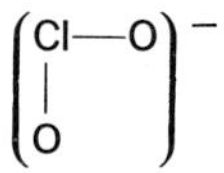

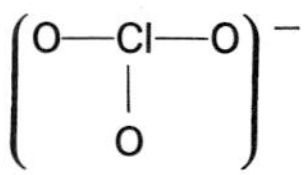

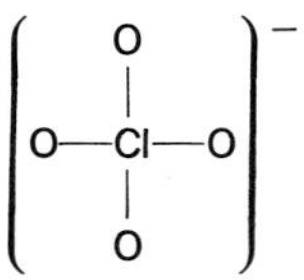

Table 2-4 Electron-Dot Formulas and Inert Gas Structures for Molecules That Involve Resonance

		Observed Structure

Ozone
O₃

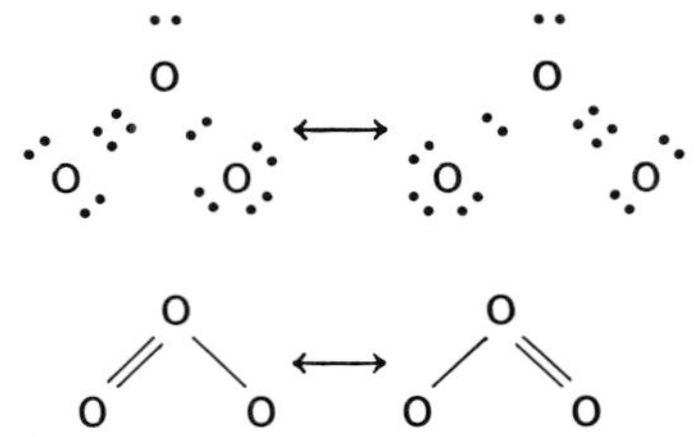

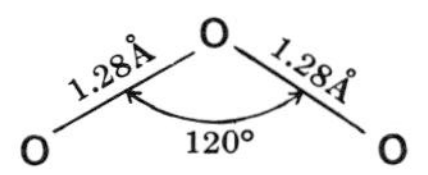

Sulfur trioxide
SO₃

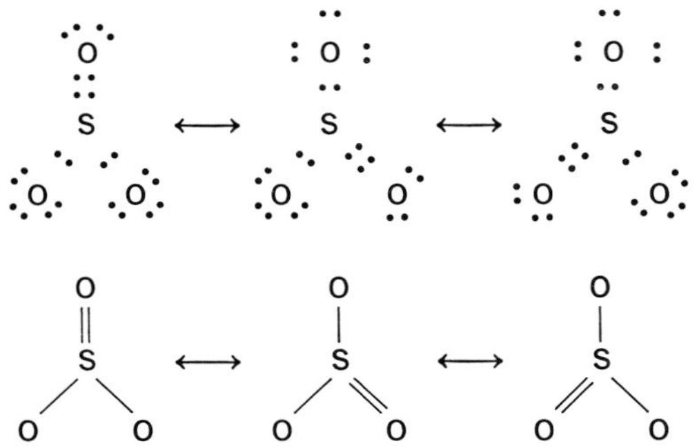

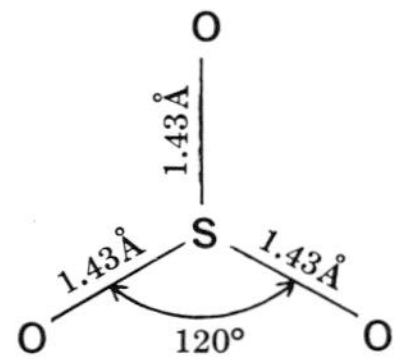

Nitrogen dioxide
NO₂

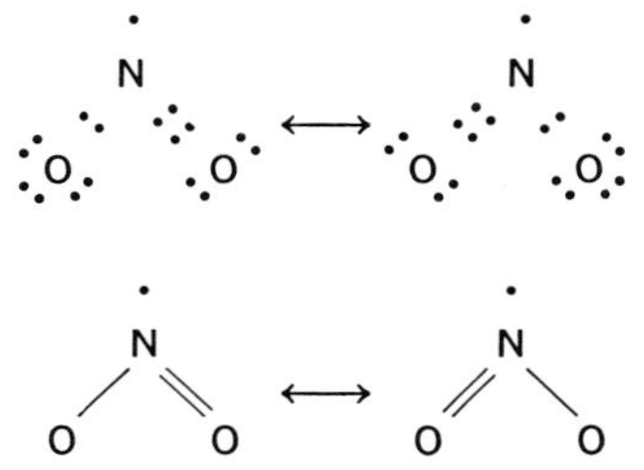

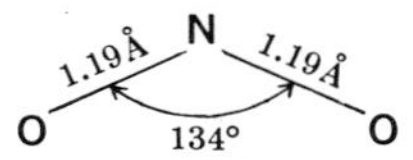

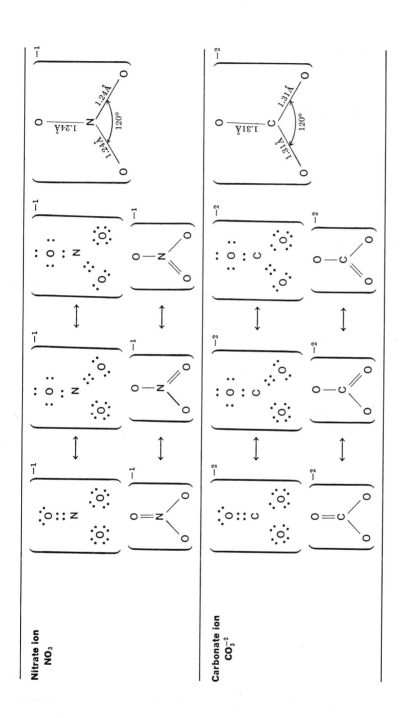

Nitrate ion
NO₃

Carbonate ion
CO₃⁻²

Table 2-4 Electron-Dot Formulas and Inert Gas Structures for Molecules That Involve Resonance (Continued)

Boron trifluoride
BF₃

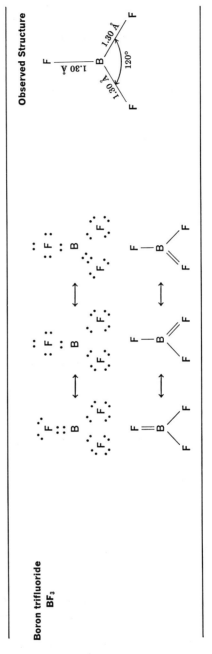

Observed Structure

structures implies that nonequivalent bonds could be formed, but they never are.

Sulfur dioxide, SO_2, is an example. Two electron arrangements can be drawn in which each atom has an inert gas arrangement nearby. Each structure implies one single bond and one double bond—one long bond and one short.

$$\overset{\circ\circ}{\underset{\circ\circ}{S}}\overset{\bullet\bullet}{\underset{\bullet\bullet}{O}}\text{:} \qquad \overset{\|}{\underset{}{S}}\text{—}O \qquad\qquad\qquad (2\text{-}28a)$$
$$\underset{\bullet\bullet}{\overset{\bullet\bullet}{\cdot}O\cdot}\qquad\qquad O$$

$$\overset{\circ\circ}{\underset{\circ\circ}{S}}\text{::}\overset{\bullet\bullet}{O}\overset{\bullet}{\cdot} \qquad S{=}O \qquad\qquad\qquad (2\text{-}28b)$$
$$\text{:}\underset{\bullet\bullet}{O}\text{:}\qquad\quad O$$

Sulfur dioxide, however, is known to have two exactly equal sulfur–oxygen bond lengths in the structure (2-29).

$$
\begin{array}{c}
\text{S}\\
\text{1.432 Å} \nearrow \quad \nwarrow \text{1.432 Å}\\
O \quad \overset{120°}{} \quad O
\end{array}
\qquad\qquad (2\text{-}29)
$$

This discrepancy was not hard for chemists to resolve. The two structures (2-28a) and (2-28b) were already confusing because who was to say which was preferable? Energy-wise, the electrons ought to be just as comfortable in one as in the other. The solution is obvious: Electrons are highly mobile, so they try to occupy *both* equivalent structures at the same time. Chemists introduce the term "resonance" to suggest that the electrons move back and forth between (2-28a) and (2-28b), as pictured in (2-30).

$$\overset{\circ\circ}{\underset{\circ\circ}{S}}\overset{\circ\circ}{\underset{\circ\circ}{O}}\text{:}\longleftrightarrow \overset{\circ\circ}{\underset{\circ\circ}{S}}\text{::}\overset{\circ}{O}\overset{\circ}{\cdot} \qquad\qquad (2\text{-}30)$$
$$\underset{\circ\,\circ}{\overset{\circ\circ}{O}}\qquad\quad \underset{\circ\circ}{\overset{\circ\circ}{\text{:}O\text{:}}}$$

$$S\text{—}O \longleftrightarrow S{=}O$$
$$\overset{\|}{O}\qquad\quad \overset{|}{O}$$

Now the two sulfur–oxygen bonds will be equal, averaged over the rapid electron movement back and forth between the two arrangements.* The

---

* This averaging over two (or more) hypothetical contributing structures is signified by the double-headed arrow in (2-30). That symbol should not be confused with the double arrow representation ($\rightleftarrows$) used to designate an equilibrium situation between real and distinguishable chemical species.

actual molecule will not be like either of our fictional "resonance structures," but rather, it will be like the superposition of the two. Instead of displaying one single and one double bond, the molecule will have two $1\frac{1}{2}$ order bonds.

Table 2-4 lists a number of examples that are in agreement with the known molecular structures *if* the resonance idea is used.

This rather awkward resonance idea is necessitated by a basic weakness in the electron-dot or simple electron-sharing view of bonding. The super-position of the two (or more) fictional resonance structures is necessary if we wish to save this simple scheme. Chemists have found electron-dot formulas, based on inert gas populations, sufficiently useful to tolerate the cumbersome resonance idea for almost four decades.

2-5 Chemical look-alikes—chemical families

This simple bonding scheme correlates nicely with the periodicities in the chemical formulas displayed in Table 2-2. If bonding capacity is deter-mined by inert gas populations, the chemistry of an element should be fixed by its position relative to the nearest inert gas in the Periodic Table. We can expect all of the elements in the column that just precedes that of the inert gases (chlorine, bromine, iodine, and astatine) to have chemis-try like fluorine. Each of the elements that are two positions preceding an inert gas should be like oxygen; these are the elements sulfur, selenium, tellurium, and polonium. The bonding scheme confirms the conclusion reached earlier by looking at regularities in E_m/m: Elements can be classi-fied into families. Elements within a family display similar chemistry. We can hope and expect that this similarity extends beyond the mere chemical formulas. This idea can be tested further by examining the chemistry of those elements that are ionization energy look-alikes of beryllium.

To decide which elements are in the beryllium family, we'll find all those elements with Z two higher than an inert gas. Table 2-5 shows the elements that fill the bill. For each of these elements, the first two electrons are much more easily removed than the third. For historical reasons, these elements are called the *alkaline earths*.

Table 2-5 *Ionization Energies and Relative Sizes of the Alkaline Earth Elements*

Element	Symbol	Z	E_1	$E_2/2$	$E_3/3$	r_1
beryllium	Be	4	215	210	1183	(1.00)
magnesium	Mg	12	176	173	616	1.22
calcium	Ca	20	141	137	394	1.51
strontium	Sr	38	131	127	335	1.61
barium	Ba	56	120	115	273	1.79

(a) THE ALKALINE EARTH ELEMENTS

All of the alkaline earth elements are metallic in the elemental state. Table 2-6 shows their physical properties. The volume per mole of atoms, the *molar volume*, furnishes a tangible measure of atomic size. In fact, if atoms were assumed to be cubical in shape (maybe they are!), their sizes would be proportional to the cube root of the volume per mole. The last column in Table 2-6 shows these cube roots, all divided by that of beryllium. The trend parallels qualitatively that shown in the last column of Table 2-5, based upon ionization energies.

Table 2-6 *Physical Properties of the Alkaline Earth Elements*

Element	Symbol	m.p. (°K)	b.p. (°K)	Heat of Fusion (kcal/ mole)	Molar Volume (cc)	$\left(\dfrac{\text{Molar Vol.}}{\text{Molar Vol. Be}}\right)^{1/3}$
beryllium	Be	1556	3243	2.3	4.9	(1.00)
magnesium	Mg	923	1380	2.2	14.1	1.42
calcium	Ca	1123	1513	2.2	25.9	1.75
strontium	Sr	1043	1657	2.2	33.7	1.91
barium	Ba	977	1911	4	39.2	2.00

The other physical properties shown, the melting points, boiling points, and heats of fusion,* vary relatively little within the family, except for beryllium. A glance at the chemistry of the metals reveals more important similarities, however. For example, beryllium metal reacts with chlorine to form a solid compound $BeCl_2$ and the metal reacts with oxygen to form a solid compound BeO. Experiment with the other alkaline earth elements shows that they all do the same.

$$Be(\text{metal}) + Cl_2(\text{gas}) \longrightarrow BeCl_2(\text{solid})$$
$$Mg(m) + Cl_2(g) \longrightarrow MgCl_2(s)$$
$$Ca(m) + Cl_2(g) \longrightarrow CaCl_2(s) \qquad (2\text{-}31)$$
$$Sr(m) + Cl_2(g) \longrightarrow SrCl_2(s)$$
$$Ba(m) + Cl_2(g) \longrightarrow BaCl_2(s)$$

$$Be(\text{metal}) + \tfrac{1}{2} O_2(\text{gas}) \longrightarrow BeO(\text{solid})$$
$$Mg(m) + \tfrac{1}{2} O_2(g) \longrightarrow MgO(s)$$
$$Ca(m) + \tfrac{1}{2} O_2(g) \longrightarrow CaO(s) \qquad (2\text{-}32)$$
$$Sr(m) + \tfrac{1}{2} O_2(g) \longrightarrow SrO(s)$$
$$Ba(m) + \tfrac{1}{2} O_2(g) \longrightarrow BaO(s)$$

* The heat of fusion is the energy required to melt one mole of the substance.

The reactions (2-31) and (2-32) are so alike that they can be written as generalized, family reactions in which the element is designated M:

$$M(m) + Cl_2(g) \longrightarrow MCl_2(s) \tag{2-31}$$

$$M(m) + \tfrac{1}{2} O_2(g) \longrightarrow MO(s) \tag{2-32}$$

This condensation of information is the benefit that comes from recognizing meaningful relationships.

(b) THE ALKALINE EARTH CHLORIDES

The similarity of the reactions (2-31) suggests that the products should also be alike. Table 2-7 shows the facts.

All of the chlorides are white, crystalline solids with very low electrical conductivities. They have high melting points and, with the exception of $BeCl_2$, high heats of fusion. These data show that the chloride crystals are quite stable. It takes a high temperature and lots of energy to break up the regular crystal arrangement to form the liquid. Yet all of these substances dissolve readily in water with quite high solubilities. Furthermore, heat is *released* when the crystal lattice is broken up to form a water solution—an *aqueous solution*. We must conclude that the aqueous solutions of the alkaline earth chlorides are about as stable as the solids themselves.

All of these chlorides dissolve in water to give solutions that readily conduct electrical current. Experiments devoted to the composition of these solutions reveal no evidence that molecules MCl_2 are present. The electrical conductivity shows what has happened. As the solid dissolves, mobile, charged species—called *ions*—M^{+2} and Cl^- are formed. The stability of the chloride solutions must be associated with the stability of these ions in their interaction with the solvent water molecules. To remind each other of the importance of these ion–water interactions, chemists symbolize M^{+2} in water as M^{+2}(aqueous) or M^{+2}(aq). The set of reactions need not be written out in detail—they all have the form:

$$MCl_2(s) + \text{excess water} \longrightarrow M^{+2}(aq) + 2\,Cl^-(aq) \tag{2-33}$$

When excess water is used to form a solution, it is usually omitted as a reactant, so (2-33) appears as

$$MCl_2(s) \longrightarrow M^{+2}(aq) + 2\,Cl^-(aq) \tag{2-34}$$

Substances that dissolve in water according to reactions like (2-33) are called *salts* and, because they form electrically conducting solutions, they are also called *electrolytes*.

Table 2-7 *Properties of the Alkaline Earth Chlorides*

Compound	Formula	Heat of Fusion (kcal/mole)	m.p. (°K)	b.p. (°K)	Molar Volume (cc)	Solubility in Water (298°K) (moles/liter)	Heat Released upon Solution in Water (kcal/mole)
beryllium chloride	$BeCl_2$	3.0	678	820	42.1	(very soluble)	51.1
magnesium chloride	$MgCl_2$	10.0	981	1691	41.1	5.7 M	37.1
calcium chloride	$CaCl_2$	6.8	1045	1873	44.2	6.7 M	19.8
strontium chloride	$SrCl_2$	4.1	1148	—	51.9	3.4 M	12.4
barium chloride	$BaCl_2$	5.4	1235	1940	54.0	1.7 M	3.2

(c) THE ALKALINE EARTH OXIDES

Further evidence of likeness among these elements is provided by the oxides produced in reactions (2-32). Each of them reacts with water to form a hydroxide according to the generalized reaction (2-35). The hydroxide dissolves in excess water to a much smaller extent than do the chlorides. Nevertheless, this solubility shows that the hydroxides are electrolytes and they can be likened to the chlorides with the molecular ion OH^-, hydroxide ion, substituting for Cl^-, the chloride ion.

$$MO(s) + H_2O(\ell) \longrightarrow M(OH)_2(s) \qquad\qquad (2\text{-}35)$$

$$M(OH)_2(s) \longrightarrow M^{+2}(aq) + 2\ OH^-(aq) \qquad\qquad (2\text{-}36)$$

Despite the fact that all of the alkaline earth oxides react as in (2-35) and (2-36), the hydroxide solubilities differ by much larger factors than we found among the chlorides. Table 2-8 shows these solubilities, expressed in moles/liter (M = moles/liter). For example, if one tenth of a mole of $Mg(OH)_2$ is stirred with one liter of water, only 0.05% of the solid dissolves. When the concentration of Mg^{+2} ion reaches $5 \cdot 10^{-4}\ M$, the solubility limit prevents any further dissolving. Since reaction (2-36) gives two OH^- ions for every Mg^{+2} ion, the OH^- concentration now is $2(5 \cdot 10^{-4}) = 1.0 \cdot 10^{-3}\ M$. In contrast, if 0.10 mole of $Ba(OH)_2$ is stirred with one liter of water, all of the $Ba(OH)_2$ dissolves to give a solution containing $0.10\ M\ Ba^{+2}$ ion and $0.20\ M\ OH^-$ ion, well below the solubility limit.

Plainly this behavior would permit a mixture of barium hydroxide and magnesium hydroxide to be reasonably separated merely by exposing the mixture to water. For example, if a solid mixture of 0.10 mole of $Ba(OH)_2$ and 0.10 mole of $Mg(OH)_2$ is stirred into one liter of water, all of the barium hydroxide dissolves, but a negligible fraction of the magnesium hydroxide enters the solution. This type of behavior is of great utility. It permits the chemist to prepare pure compounds from mixtures and, in the course of doing so, to determine the composition of the mixture. If he wishes merely to learn *what* is present, the determination of composition is called *qualitative analysis*. If he wishes also to find out *how much* is present, it is called *quantitative analysis*.

Table 2-8   *Aqueous Solubilities of Alkaline Earth Hydroxides* (298°K)

Compound	Solubility in Water (moles/liter)
beryllium hydroxide	$8 \cdot 10^{-6}$
magnesium hydroxide	$5 \cdot 10^{-4}$
calcium hydroxide	$1.7 \cdot 10^{-2}$
strontium hydroxide	$6.7 \cdot 10^{-2}$
barium hydroxide	$2.2 \cdot 10^{-1}$

Exercise 2-4. From the solubility of $Ba(OH)_2$, show that 3.4 grams of barium oxide will dissolve in 100 ml of water.

Exercise 2-5. A 2.00 gram sample of BaO mixed with MgO is stirred with 100 ml of water. The undissolved magnesium hydroxide is filtered, dried, and heated to form MgO in the reverse of reaction (2-35). The dried MgO weighs 0.30 gram. Show that the concentration of barium ion in solution is 0.111 mole/liter.

(d) PRECIPITATION REACTIONS

The partial solubility of two hydroxides contrasts with the behavior of a mixture of chlorides. Both $BaCl_2$ and $MgCl_2$ will dissolve if they are stirred with water, whereas only the $Ba(OH)_2$ dissolves in a $Ba(OH)_2$–$Mg(OH)_2$ mixture. That raises an interesting question: What would happen if a mixture of 0.10 mole of solid $Ba(OH)_2$ and 0.10 mole of solid $MgCl_2$ were stirred with one liter of water? We know that both of these solids are reasonably soluble in water, so they would both dissolve.

$$Ba(OH)_2(s) \longrightarrow Ba^{+2}(aq) + 2\ OH^-(aq) \tag{2-37}$$

$$MgCl_2(s) \longrightarrow Mg^{+2}(aq) + 2\ Cl^-(aq) \tag{2-38}$$

The 0.10 mole of $Ba(OH)_2$ would give a barium ion concentration of 0.10 M and a hydroxide ion concentration twice as large, 0.20 M. We will abbreviate "barium ion concentration" with the symbol $[Ba^{+2}]$. Hence these concentrations can be written, $[Ba^{+2}] = 0.10\ M$ and $[OH^-] = 0.20\ M$. Turning to the $MgCl_2$, it would give a Mg^{+2} concentration of 0.10 M and Cl^- concentration of 0.20 M, i.e., $[Mg^{+2}] = 0.10\ M$ and $[Cl^-] = 0.20\ M$. Here is a strange situation. We have a Mg^{+2} concentration of 0.10 M, from $MgCl_2$, in the same solution with an OH^- concentration of 0.20 M, from $Ba(OH)_2$. These far exceed the solubility limit for $Mg(OH)_2$, which we took from Table 2-8 to be $[Mg^{+2}] = 5 \cdot 10^{-4}\ M$ and $[OH^-] = 1.0 \cdot 10^{-3}\ M$. What will happen? Experiment shows that solid $Mg(OH)_2$ forms, lowering again the magnesium ion and hydroxide ion concentrations until the solubility limit is no longer exceeded.

$$Mg^{+2}(aq) + 2\ OH^-(aq) \longrightarrow Mg(OH)_2(s) \tag{2-39}$$

This formation of solid from a solution is called *precipitation*.

Exercise 2-6. Show that you could prepare solid magnesium hydroxide by mixing 500 ml of a 0.20 M solution of barium hydroxide with 500 ml of a 0.20 M solution of magnesium chloride. Calculate the concentrations of all ions in the solution immediately after mixing and predict what would happen.

(e) SULFATES AND CARBONATES

The hydroxide solubilities increase in the series Be to Ba. That is not always the trend, however. Table 2-9 shows the opposite trends for the compounds formed by the alkaline earth ions with the sulfate ion, SO_4^{-2}, and with the carbonate ion, CO_3^{-2}. Here we see that a $MgSO_4$–$BaSO_4$ mixture will partially dissolve in water, this time leaving the barium salt undissolved. Also, the mixing of a magnesium sulfate solution with a barium chloride solution would cause the precipitation reaction to form solid $BaSO_4$.

Table 2-9 Aqueous Solubilities (298°K) of Alkaline Earth Sulfates and Carbonates (Excluding Beryllium)

Positive Ion	Sulfate Solubility in Water (moles/liter)	Carbonate Solubility in Water (moles/liter)
Mg^{+2}	3.0	$1.1 \cdot 10^{-3}$
Ca^{+2}	$5 \ \cdot 10^{-3}$	$1.3 \cdot 10^{-4}$
Sr^{+2}	$7.6 \cdot 10^{-5}$	$6.7 \cdot 10^{-6}$
Ba^{+2}	$1.0 \cdot 10^{-5}$	$8.7 \cdot 10^{-6}$

Exercise 2-7. Solid sodium sulfate, Na_2SO_4, is an electrolyte with high solubility in water. It dissolves to give Na^+ ions and SO_4^{-2} ions. What would happen if a 0.008 M solution of Na_2SO_4 were mixed with an equal volume of solution containing both 0.010 M $CaCl_2$ and 0.008 M $BaCl_2$?

(f) SUMMARY ON ALKALINE EARTH CHEMISTRY

Two common characteristics of a chemical family, as portrayed in the Periodic Table, are displayed by the alkaline earths. First, we found great similarity in the reactions of the elements in the family. Second, there are significant and useful differences among them. We have already seen how these differences can be used in a qualitative way to separate compounds, to prepare a desired compound from others, and to determine what is present in a solution or a solid. They are even more powerful when used quantitatively, which opens the subject of chemical equilibrium, the subject of the next chapter. Before proceeding to that subject, however, let's return to ionization energies.

2-6 Ionization energies and the Periodic Table

Figures 2-3 and 2-4 show how a particular type of measurement, ionization energy, displays repetitious regularities among the elements. These periodic trends were used to segregate the elements into families. Then those elements with similar ionization energy behavior were found to be similar in chemistry, at least for the alkaline earths, the family we studied. This success with the alkaline earths is richly borne out among the other families identified by the ionization energy trends. They lead to the pictorial grouping of the elements called the *Periodic Table*. This grouping was first proposed in its modern form by a Russian chemist, Dimitri Mendeleev, in 1869. Mendeleev didn't have the ionization energies at his disposal since none had been measured (the electron hadn't been invented yet). So he devised the array on the basis of a variety of types of information—physical properties, chemical formulas, and chemical reactions. This Periodic Table is shown on the inside front cover of this book. The horizontal rows correspond to the periods identified in Figures 2-3 and 2-4. There are dramatic changes in the chemistry across a row. These changes can be understood in terms of ionization energies, as we shall see later.

The vertical columns in the Periodic Table identify groups of elements that have similar chemistry. These groups, or families, give us significant predictive power. Once we have measured properties of two or three elements is a given period, we can make reasonable estimates for the analogous properties of the other elements known to be in the same family. We will return to the Periodic Table often as we explore in chemistry. It is, and it deserves to be, engraved in the memory of every practicing chemist.

Problems

$$313 \; \text{kcal} \times \frac{1 \, ev}{23} \cdot \frac{mole}{kcal} = 3.6$$

1. Table 2-1 gives $E_1 = 313.6$ kcal/mole for the ionization energy of gaseous hydrogen atoms. Calculate in electron volts (ev) the accelerating voltage that would just be enough to cause ionization of an H atom. (1 ev/particle $= 23.061$ kcal/mole)

2. Molecules also display ionization when bombarded by electrons in an apparatus like that shown in Figure 2-1. For example, current just begins to flow with an accelerating voltage of 15.6 volts when N_2 is used and at 12.1 volts when O_2 is used. Calculate in kcal/mole the ionization energies of N_2 and O_2 (to give N_2^+ and O_2^+) and compare the values to those given in Table 2-1 for N atoms and O atoms.

3. For the first twelve elements (H to Mg), replot the values of E_1 shown in Figure 2-3 (and given in Appendix b) on a compressed energy scale that extends from 0 to 1800 kcal/mole. Now plot on this same graph the values of E_2. Explain why
(a) H atom does not appear in the E_2 plot.
(b) Li and Na appear at peaks in the E_2 plot and at valleys in the E_1 plot.

4. From the plot in Problem 3, predict which elements will have positive ions, M^+, that behave like inert gases, which will form only a monohydride, MH^+ (as does a halogen), and which will form a dihydride, MH_2^+ (as does the oxygen family).

5. (a) Calculate E_m/m for all five electrons of the boron atom (see Appendix b).
(b) Estimate the average radius of each successive electron relative to the average radius of the first electron removed.
(c) Draw a picture of the boron atom like those shown in Figure 2-7.

6. Repeat Problem 5 for all six electrons of the carbon atom.

7.   The five species, H, He^+, Li^{+2}, Be^{+3}, and B^{+4}, are particularly simple; each has only one electron per atom (type 1). Calculate E_m/m for each and estimate the average radius of each ion relative to that of H atom. Make a plot of r/r_H versus ionic charge for these one-electron ions.

8.   Repeat Problem 7 for the two-electron atoms, He, Li^+, Be^{+2}, B^{+3}, and C^{+4}. Estimate the average radii relative to that of He. Plot the values of r/r_{He} on the same graph made in Problem 7. Now add the values of r/r_{Ne} for the ions isoelectronic with neon: Na^+, Mg^{+2}, Al^{+3}, and Si^{+4}. Why do the slopes differ?

9.   For the eighteen elements hydrogen to argon use E_1 to estimate the average radius of each atom relative to that of lithium. Plot these ratios, r/r_{Li}, versus Z. Circle each point that refers to removal of a type-2 electron. Draw an $\times$ through each point that refers to a type-3 electron.

10.   Repeat Problem 9 for the eighteen elements of the first long row of the Periodic Table (K to Kr). Plot these ratios, r/r_{Li}, versus Z on the same graph used in Problem 9, placing K under H and Kr under Ar.

11.   From the values of E_1, E_2, and E_3, calculate the energies needed to form gaseous ferrous ions, $Fe^{+2}(g)$, and gaseous ferric ions, $Fe^{+3}(g)$, from gaseous iron atoms, Fe(g). $E_1 + E_2$ $E_1 + E_2 + E_3$

12.   The three ions, Na^+, Mg^{+2}, and Al^{+3}, are isoelectronic (same number of electrons). However, the energies *released* when these gaseous ions are dissolved in water differ greatly:

$$Na^+(g) \longrightarrow Na^+(aq) \quad \text{energy released is 97 kcal}$$

$$Mg^{+2}(g) \longrightarrow Mg^{+2}(aq) \quad \text{energy released is 460 kcal}$$

$$Al^{+3}(g) \longrightarrow Al^{+3}(aq) \quad \text{energy released is 1121 kcal}$$

(a) Calculate the energies *absorbed* to form the gaseous ions $Na^+(g)$, $Mg^{+2}(g)$, and $Al^{+3}(g)$ from the neutral, gaseous atoms Na(g), Mg(g), and Al(g).
(b) Calculate the ratio of energy released on aquation divided by energy absorbed to form the gaseous ion.

13.   Draw the electron-dot formulas and the line representations for hydrogen sulfide, H_2S, and dihydrogen disulfide, H_2S_2, the sulfur counterparts to water and hydrogen peroxide (H_2O and H_2O_2).

14.   Draw the electron-dot formulas and the line representations for ammonia, NH_3, phosphine, PH_3, and arsine, AsH_3.

15.   Whereas elemental oxygen has the formula O_2, the common form of elemental sulfur has the formula S_8. Draw an electron-dot formula and a line representation for S_8 that allow every sulfur to have the inert gas electron population of Ar.

16.   Draw the electron-dot formulas and line representations for carbon tetrafluoride, CF_4, carbon tetrachloride, CCl_4, fluoroform, HCF_3, and chloroform, $HCCl_3$.

17.   Draw the electron-dot formulas and line representations of ethylene, C_2H_4, tetrafluoroethylene, C_2F_4, and tetrachloroethylene, C_2Cl_4.

18.   Draw the electron-dot formulas and line representations of methanol, CH_3OH, ethanol, CH_3CH_2OH, and acetone, CH_3COCH_3.

Problems 85

$$MgCl_{2(s)} \rightarrow Mg^{+2}_{aq} + 2Cl^-_{aq}$$

$$\frac{2.5 \times 10^{-2}}{1 \times 10^{-1}} = .25 M$$

$$2 \times .25 M = .50$$

19. If 0.025 mole of magnesium chloride is added to 100 ml (0.100 ℓ) of water, all of the solid dissolves. Write the equation for this reaction and calculate the concentrations in moles/liter of magnesium ion, $Mg^{+2}(aq)$, and chloride ion, $Cl^-(aq)$.

.25M   .50M

20. If 0.432 gram of calcium oxide, CaO, is added to 450 ml (0.450 ℓ) of water, all of the solid dissolves. Write the equation for this reaction and calculate the concentrations in moles/liter of calcium ion, $Ca^{+2}(aq)$, and hydroxide ion, $OH^-(aq)$.

21. 100 ml of 0.10 M $CaCl_2$ is mixed with 100 ml 0.20 M $SrCl_2$. What are the concentrations of all ions in the resulting solution?

22. If 0.432 gram of magnesium oxide, MgO, is added to 450 ml of water, most of the solid remains undissolved. This undissolved solid is removed by filtration, dried, and weighed. From the solubility of $Mg(OH)_2$ given in Table 2-8, calculate how many grams of dried MgO are found.

23. Equal volumes of 0.20 M $MgSO_4$ (magnesium sulfate) and 0.20 M $BaCl_2$ (barium chloride) are mixed. A white precipitate forms. Write the equation for the precipitation reaction and calculate the concentrations of all ions remaining in solution.

24. A mixture of 0.0050 mole each of solid $MgCl_2$, $CaCl_2$, and $BaCl_2$ is dissolved in 250 ml of water. Now 250 ml of 0.040 M sodium hydroxide solution, NaOH (a soluble, strong electrolyte) is added and a white precipitate forms. This precipitate is removed by filtration. To the 500 ml of clear filtrate solution is added 500 ml of 0.010 M sodium sulfate solution, Na_2SO_4. Again a white precipitate appears. What is the composition of each precipitate? List the concentrations of all ions that remain in solution, in order of decreasing concentration.

$$\frac{1 \times 10^{-1}}{1 \times 10^{-1}}$$

21)

$$CaCl_2 \rightarrow Ca^{++} + 2Cl^-$$
$$\quad\quad 1M \quad\quad 2M$$

$$SrCl_2 \rightarrow Sr^- + 2Cl^-$$
$$\quad\quad 2m \quad\quad 4M$$

$6MCl^-$

C_2F_4

$FoCl$

C_2H_4

three a microscopic view of equilibrium

In Chapter Two, we found solubility trends to be quite useful. A solid of high solubility can be separated from one of low solubility. A solid of low solubility can be prepared by mixing two solutions of other appropriate, soluble compounds. In each case, a chemical reaction is initiated and allowed to run its course—to reach equilibrium. Then we take advantage of the new situation that exists—in these simple examples, we separate the solid from the liquid for the purpose of purification, or preparation.

That is the plot in most of man's use and control of chemistry. Substances are brought together that have some tendency to change (to react). The reaction proceeds, heading toward equilibrium. If and when equilibrium is attained, we take advantage of this new state. The trick is to understand equilibria sufficiently well so that we know what to mix and what we will have when changes cease—when equilibrium is attained.

So we will devote quite a bit of thought to equilibrium. It is helpful, first, to consider the equilibrium situation on the microscopic level in order to obtain a feeling for "what is going on." With that intuitive setting, we can profitably examine the equilibrium state for some particular reactions. Then we will have a background of experience with which to enter the exciting realm of chemical thermodynamics, the set of principles that tell us what factors determine the equilibrium state and why changes take place. Thus we shall see how chemists understand and control chemical reactions.

3-1 The equilibrium state

In mechanics, an equilibrium state is one of balance between opposing forces. It is recognized by an appearance of quiescence—nothing is changing as time passes. Let us see how much of this description we can transfer to a chemical system.

(a) SOME CHEMICAL SYSTEMS: WHICH ARE AT EQUILIBRIUM?

Figure 3-1 shows three systems for which a number of observations indicate a constancy of properties. The first is a bunsen burner flame. In a quiet

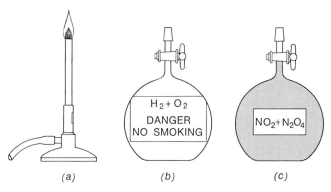

Figure 3-1 Systems with constant properties: Which are at equilibrium?

room, the flame appears quite immobile; a thermocouple placed at a particular place inside the flame or near the flame indicates a constant temperature. Yet everyone recognizes that a bunsen burner flame is not an equilibrium situation. The reason is that gas and air are continuously being added to the flame at the base of the burner, and hot carbon dioxide and water are ejected at the top of the flame. *Equilibrium can exist only in a closed system:* a system to which neither energy nor material substances are being continuously supplied or removed. The constant addition of reactants, and removal of both products and heat, means the flame is not a closed system, so it cannot be at equilibrium. A system that achieves constancy of properties in this way is called a *steady state.*

The second system in Figure 3-1 is a bottle containing a mixture of hydrogen and oxygen. Again, it is common knowledge that such a mixture is explosive; only the tiniest spark is needed to initiate violent proof that it is not at equilibrium, chemistry-wise. The reason the mixture just sits there, innocently looking at you, is that the explosive reaction $2\,H_2 + O_2 \longrightarrow 2\,H_2O$ is extremely slow under normal conditions of temperature and pressure. Once the reaction gets started, however, the hydrogen and oxygen molecules know what to do! Nevertheless, this gives us a second caveat to place on the use of the constancy of properties as a means of recognizing an equilibrium situation. *The rate of any possible change that might occur must be noticeably rapid.* Then the constancy of properties implies equilibrium.

The third system in Figure 3-1 is a bottle containing a mixture of NO_2 and N_2O_4. The bottle is colored because one of the constituents, NO_2, is colored. This system *is* at equilibrium with respect to the reaction

$$2\,NO_2(g) \rightleftharpoons N_2O_4(g) \tag{3-1}$$

How do we know? By three simple tests. *First,* we verify that it is a closed system. It is: The bottle is tightly stoppered and no heat is being transferred. *Secondly*, we measure properties that depend on the equilibrium and see that they do not vary over time. For example, as reaction (3-1) proceeds, the pressure drops (two molecules of NO_2 produce only one of N_2O_4) and the orange color of NO_2 is reduced in intensity. However, neither the pressure nor the color is changing. *Thirdly*, we make a small and momentary change in some condition that is known to affect the reaction in question. For example, we could first lower the temperature a few degrees and then restore it to its initial value. Immediately we would perceive a slight decrease in the orange color of the NO_2, and then a restoration to the original color. The reaction proceeds a bit to the right, producing more N_2O_4 and consuming a bit of NO_2, and then, as the temperature is raised again, the original color is restored. *This proves that the reaction in question proceeds rapidly and reversibly under our conditions of study.* Therefore we have an equilibrium system.

A system is at equilibrium when:

—*neither energy nor material substance is being added or removed (it is a closed system);*

—*the measurable bulk properties of the system are not changing with time;*

—*possible changes proceed at a measurable pace, as revealed by the response to a small disturbance that causes observable properties to change and then to return to their initial values when the disturbance is removed.*

(b) A CLOSER LOOK AT EQUILIBRIUM

Despite the generally placid appearance of chemical systems at equilibrium, scientists are convinced that things are not as quiet on the atomic level. In fact, it is found experimentally that atoms are constantly in motion. In the gas phase, molecules dash about, bouncing off the walls of the container. Molecules in liquids and solids are not this active, but they are, nevertheless, constantly jiggling and vibrating. Keeping these characteristics in mind, let's consider a very simple system—a closed vessel containing a little liquid water—and watch how it attains equilibrium.

When the system is at equilibrium, there will be some water vapor present along with the liquid water; the exact amount will depend only on the temperature. The gas-phase molecules will be dashing around, hitting both the sides of the flask *and* the surface of the liquid. Will they have a chance of sticking in the liquid, or will the original group of molecules that evaporate stay in the gas phase for as long as the conditions remain the same? This can be tested quite easily using the apparatus in Figure 3-2, which consists of two identical flasks joined by a stopcock. Each contains liquid water, but the right-hand flask contains some radioactive water, HTO. (The symbol T is used to represent tritium, the hydrogen isotope of mass

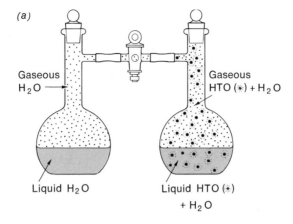

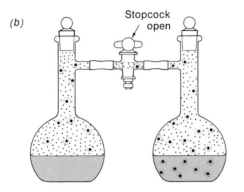

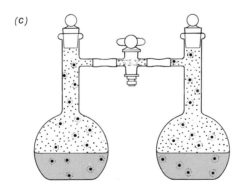

Figure 3-2 *The dynamic nature of equilibrium. (a) Both flasks contain* H_2O, *and the right-hand flask contains some radioactive HTO* (T $= {}^3_1$H). *(b) When the stopcock is opened, the vapors mix; though the pressure remains constant, gaseous HTO passes into the left-hand flask. (c) After a time, if liquid is removed from the left-hand flask, it too will exhibit radioactivity; again, while pressure remains constant, HTO condenses.*

number three, $^3_1$H.) Both flasks have the same temperature and water vapor pressure. When the stopcock is opened, the vapors mix, and soon equal numbers of radioactive molecules are present in the gas in each container. What about the liquid in the left-hand flask? If, after a time, some is removed and tested with a Geiger counter, it will be found that this flask contains some radioactive water. A plot, against time, of the radioactivity in the liquid phases in each flask would look like Figure 3-3. Thus, we must conclude that molecules are constantly entering and leaving the surface of the liquid. Equilibrium is not a static situation, but a dynamic one. For our liquid $\rightleftharpoons$ vapor equilibrium, we must have the condition

$$\text{rate of evaporation} = \text{rate of condensation}$$

or

$$\begin{bmatrix} \text{number of molecules} \\ \text{leaving surface of} \\ \text{liquid per second} \end{bmatrix} = \begin{bmatrix} \text{number of molecules} \\ \text{entering surface of} \\ \text{liquid per second} \end{bmatrix}$$

A simple calculation shows that in every second, there are about 10^{22} H_2O molecules striking each square centimeter of water surface at room temperature. Chemists believe that almost all of these molecules condense, removing them from the vapor phase. For the vapor pressure to remain constant, this same huge number of molecules must be evaporating every second from that same square centimeter of surface. Thus we see that the constancy of observable properties is actually achieved by a whirlwind of activity at the molecular level. **Equilibrium occurs,** not when all activity ends, but **when the rates of opposing processes become equal.**

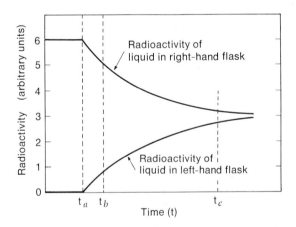

Figure 3-3   *The vapor–liquid equilibrium is dynamic. At t_a, when the stopcock is opened, all HTO radioactivity is in the right-hand flask. A short time later, at t_b, HTO is observable in the left-hand flask, though the vapor pressure remains constant. A long time later, at t_c, the HTO is almost uniformly distributed between the liquids in the two flasks, though the vapor pressure has not changed all the while.*

(c) THE EQUILIBRIUM CONSTANT IN A PARTICULAR REACTION

A chemist mixes hydrogen and iodine in the gas phase at 700°K. Reaction begins, as shown by a decreasing intensity of the color of the purple iodine gas; colorless hydrogen iodide is produced.

$$H_2(g) + I_2(g) = 2 HI(g) \qquad (3-2)$$

After a time, the iodine color stops decreasing—equilibrium has been reached. The remaining color shows that some reactants remain unconsumed. Now, one of the most important questions a chemist can ask about the system is, how far did reaction proceed? Were most of the reactants consumed, so that now the bulb contains mainly HI? Or was only a small percentage of the iodine converted?

One possible way to answer this question would be to examine the ratio

$$\frac{\text{concentration of products at equilibrium}}{\text{concentration of reactants at equilibrium}} \qquad (3-3)$$

This number would be large if reaction (3-2) proceeds almost to completion and small if it proceeds barely at all. In fact, we could look at the ratio for various mixtures to see how it depends on the relative concentrations. For this purpose, we might examine the quotient

$$\frac{(\text{conc. HI})}{(\text{conc. } H_2)(\text{conc. } I_2)} \qquad (3-4)$$

Table 3-1 shows, in the fourth column, the magnitude of ratio (3-4) for five different mixtures. We see that the number is high (hydrogen iodide is favored) and it depends upon the mixture composition.

Many decades ago, chemists learned empirically that for a reaction such as (3-2), the most informative way to represent the equilibrium situation is

Table 3-1   Concentration Relations for H_2–I_2 Mixtures at Equilibrium, $T = 698.6°K$

	Experimental Results		Calculated Ratios	
$[H_2]$	$[I_2]$	$[HI]$	$\dfrac{[HI]}{[H_2][I_2]}$	$\dfrac{[HI]^2}{[H_2][I_2]}$
(mole/liter)	(mole/liter)	(mole/liter)		
1.8313×10^{-3}	3.1292×10^{-3}	17.671×10^{-3}	$3.08 \times 10^{+3}$	54.5
2.9070	1.7069	16.482	3.32	54.6
4.5647	0.7378	13.544	4.02	54.4
0.4789	0.4789	3.531	15.4	54.4
1.1409	1.1409	8.410	6.47	54.4

to modify the ratio (3-3) by squaring the concentration of a constituent that appears in the balanced equation with a coefficient 2, cubing the concentration of a constituent with a coefficient 3, and so on. Thus, our quotient (3-4) would be better written as

$$\frac{(\text{conc. HI})^2}{(\text{conc. } H_2)(\text{conc. } I_2)} \qquad (3\text{-}5)$$

With the same equilibrium concentrations, Table 3-1 lists the values of quotient (3-5) in the last column. Now the ratios are always the same! Apparently we now have a constant number that characterizes equilibrium at the temperature of study. This number is called the equilibrium constant K:

$$K = \frac{(\text{conc. HI})^2}{(\text{conc. } H_2)(\text{conc. } I_2)}$$

or, as concentrations are usually symbolized by brackets, we write

$$K = \frac{[HI]^2}{[H_2][I_2]} \qquad (3\text{-}6)$$

(d) THE EQUILIBRIUM CONSTANT FOR ANY REACTION

In the last section we have merely reminisced, for a particular case, on the empirical discovery of the Equilibrium Law. For a generalized reaction between reactants $A, B, \ldots$ to give products $C, D, \ldots$, the balanced equation would be

$$w\,A + x\,B + \cdots \rightleftharpoons y\,C + z\,D + \cdots \qquad (3\text{-}7)$$

and the generalized concentration relationship would be

$$K = \frac{[C]^y[D]^z \ \cdots}{[A]^w[B]^x \ \cdots} \qquad \textbf{The Equilibrium Law} \qquad (3\text{-}8)$$

Experimentally we find that the concentration quotient (3-8) (including the integer exponents w, x, y, and z) turns out to be a constant at a given temperature. The constant is symbolized K and is called the equilibrium constant. Once we measure this number (experimentally), we can calculate the end result, that is, the equilibrium situation, for any set of starting conditions. That makes K an important and useful quantity, one we must understand.

We shall see later that the magnitude of the equilibrium constant K can be understood in thermodynamic terms. In this chapter we will examine the magnitude of K in mechanistic terms.

3-2 Equilibrium and reaction rate

The Equilibrium Law (3-8) can be derived for a given reaction merely by equating the rate of the reaction in the forward direction to its rate in the reverse direction. Though the derivation is simple, scientists with purer souls object to it. Both the argument and the objection are educational, so we will review them both, using the real reaction (3-2) as an example.

(a) THE RATE OF FORMATION OF PRODUCTS

We'll begin with a mixture of H_2 and I_2 and consider first the formation of HI. If we try to describe on a microscopic scale the events leading to formation of HI, it is natural to picture a hydrogen molecule colliding with an iodine molecule and, in an exchange of atoms, forming two HI molecules (see Fig. 3-4). Some of the factors that fix the rate at which this process will occur are straightforward. Clearly, the rate depends upon collisions between H_2 and I_2 molecules if our picture is correct. Therefore, the rate to the right is proportional to the concentration of each constituent that must participate in the collision, and in turn, the concentrations are determined by the pressures p_{H_2} and p_{I_2}.

$$H_2 + I_2 \xrightarrow{\text{(rate)}_R} \text{products} \tag{3-9}$$

$$\text{(rate)}_R \text{ is proportional to } (p_{H_2}) \times (p_{I_2}) \tag{3-10}$$

Since HI neither helps nor hinders the collisional process (3-9), its partial pressure should not affect (rate)$_R$, so it does not appear in (3-10). The proportionality constant, k_R, in (3-10) includes lots of information we chemists would like to decipher. It tells us what fraction of the H_2–I_2 collisions are effective, how important the collisional geometry is at the time of encounter, and how the rotation of the molecules and vibrations of the atoms in the molecules affect the atomic rearrangement. Furthermore, it contains the temperature dependence of the process pictured in Figure 3-4. In our ignorance of all these details, we merely write

$$\text{(rate)}_R = k_R(p_{H_2})(p_{I_2}) \tag{3-11}$$

(b) THE RATE OF FORMATION OF REACTANTS

As reaction (3-2) proceeds, hydrogen iodide accumulates. Then, occasionally, two hydrogen iodide molecules collide, and a fraction of these

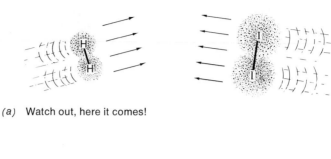

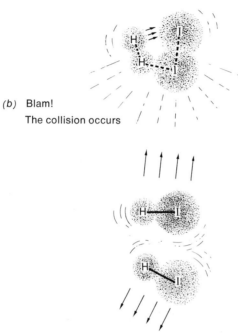

(a)   Watch out, here it comes!

(b)   Blam!
The collision occurs

(c)   Look what happened!

Figure 3-4   A reactive collision between a hydrogen molecule and an iodine molecule to form hydrogen iodide.

collisions disrupt the HI molecules to form H_2 and I_2. This process to the left occurs at a rate limited by the frequency of HI–HI collisions, as pictured in Figure 3-5:

$$\text{reactants} \xleftarrow{\text{(rate)}_L} \text{HI} + \text{HI} \tag{3-12}$$

Because two HI molecules are involved in the collision, raising the pressure of HI is doubly effective. There are not only more HI molecules to collide, but also more to collide with. Hence, the rate to the left is proportional

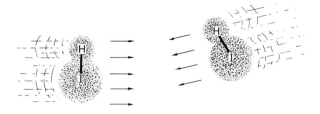

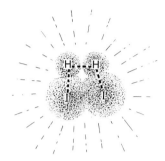

(a) Brace yourself!

(b) Ouch!

(c) Now look what happened!

Figure 3-5 Hydrogen iodide molecules can collide too, to give back hydrogen and iodine.

to the square of the HI pressure:

$$(\text{rate})_\text{L} \text{ is proportional to } (p_\text{HI})^2 \tag{3-13}$$

Once again, there is a fascinatingly interesting proportionality constant k_L that concerns the detailed dynamics of collisions such as those pictured in Figure 3-5. Knowing little about these dynamics, we merely write

$$(\text{rate})_\text{L} = k_\text{L}(p_\text{HI})^2 \tag{3-14}$$

(c) EQUILIBRIUM: A BALANCE BETWEEN OPPOSING REACTIONS

When H_2 and I_2 are first mixed, the rate of formation of HI, reaction (3-9), is rapid, and since no HI is present initially, the rate of reforming H_2 and I_2 by the reverse reaction (3-12) is zero. As the reaction proceeds, however, both H_2 and I_2 are used up, so $(rate)_R$ slows down. Meanwhile, the HI is accumulating, so $(rate)_L$ is increasing. As this process continues, $(rate)_R$ dropping and $(rate)_L$ rising, the two reaction rates approach each other and finally become equal. At this point the rate of formation of HI is exactly counterbalanced by an equal rate of loss of HI through the reverse reaction. Then, despite the continued reaction to right and left, a dynamic balance exists, and concentrations no longer change. We call this equilibrium. It is characterized as follows:

$$\text{At equilibrium} \qquad (rate)_R = (rate)_L \qquad\qquad (3\text{-}15)$$

Substituting (3-11) and (3-14) into (3-15),

$$k_R(p_{H_2})(p_{I_2}) = k_L(p_{HI})^2 \qquad\qquad (3\text{-}16)$$

Rearranging,

$$\frac{k_R}{k_L} = \frac{(p_{HI})^2}{(p_{H_2})(p_{I_2})} \qquad\qquad (3\text{-}17)$$

However, the ratio of two constants must also be a constant, so $k_R/k_L = K$. Comparing the pressure relationships in (3-17) to the concentration relationships in (3-6), we see that equating forward and reverse reaction rates leads to the Equilibrium Law

$$K = \frac{(p_{HI})^2}{(p_{H_2})(p_{I_2})} \qquad\qquad (3\text{-}18)$$

(d) WHY SOME LIKE IT AND SOME DON'T

This reaction rate derivation of the Equilibrium Law has great merit. It conveys a dynamic picture that every chemist believes exists. Although an equilibrium mixture seems static and quiescent from a macroscopic view, the chemist looks into it with his intuitive submicroscopic vision and sees lots of molecular reactions going on: reactants meeting to give products and, at the same time, products meeting to give reactants. He also sees that the static condition represents a state of balance between these opposing processes. This dynamic view of equilibrium, and the approach to equilibrium, is an extremely valuable one.

So what's the objection? If all chemists—including the purest of them—believe that equilibrium is dynamic and that it involves equality of opposing rates, why do some of them object to the reaction rate derivation of the Equilibrium Law? The answer is that the rate derivation requires a microscopic view of *how* the reaction proceeds, whereas the thermodynamic argument does not. In fact, one of the greatest strengths of thermodynamic deductions is that they are independent of a detailed understanding of chemical processes at the molecular level. We can debate—profitably, mind you—whether or not the H_2–I_2 reaction mechanism is as simple as shown in Figures 3-4 and 3-5. In fact, this worry merits some detailed consideration.

(e) BACK TO THE RATE DERIVATION—WHAT IF OUR MECHANISM IS WRONG?

Fortunately, it can be shown that the rate derivation of the Equilibrium Law is independent of the actual mechanism. The H_2–I_2 reaction is a fine case in point, since the mechanism has been challenged only recently. John H. Sullivan proposed that the reaction, which proceeds rapidly only at elevated temperatures, actually involves an equilibrium, thermal dissociation of iodine atoms and a series of subsequent reactions as follows:[*]

$$I_2(g) \rightleftharpoons 2\ I(g) \tag{3-19}$$

$$I(g) + H_2(g) \rightleftharpoons HI(g) + H(g) \tag{3-20}$$

$$H(g) + I_2(g) \rightleftharpoons HI(g) + I(g) \tag{3-21}$$

Sullivan further postulated that reaction (3-20) is much slower than reaction (3-21). This means that whenever (3-20) occurs (consuming an iodine atom), (3-21) immediately follows (replacing the iodine atom).

These deductions, which pertain to the rate of the reaction when no hydrogen iodide is present, can be applied to the equilibrium situation. Equilibrium is, of course, characterized by constant composition. Hence, we can conclude that the rate at which H_2 is being consumed must be equal to the rate at which it is being produced. All the arguments used in the derivation (3-15) to (3-18) can be applied to (3-20):

$$I(g) + H_2(g) \underset{k_{L,20}}{\overset{k_{R,20}}{\rightleftharpoons}} HI(g) + H(g) \tag{3-22}$$

where

rate of loss of $H_2 = k_{R,20}(p_I)(p_{H_2})$

---

[*] J. H. Sullivan, *Journal of Chemical Physics*, Vol. **46**, p. 73, 1967.

and

$$\text{rate of production of } H_2 = k_{L,20}(p_{HI})(p_H)$$

At equilibrium,

$$k_{R,20}(p_I)(p_{H_2}) = k_{L,20}(p_{HI})(p_H)$$

hence,

$$\frac{k_{R,20}}{k_{L,20}} = K_{20} = \frac{(p_{HI})(p_H)}{(p_{H_2})(p_I)} \tag{3-23}$$

However, if reaction (3-20) does not cause either net production or net consumption of H_2, then it will cause no net production or consumption of H atoms. This means that the constancy of composition at equilibrium *requires* that reaction (3-21), the only other source of H atoms, must *also* neither consume nor produce H atoms. Once again

$$H(g) + I_2(g) \underset{k_{L,21}}{\overset{k_{R,21}}{\rightleftharpoons}} HI(g) + I(g) \tag{3-24}$$

where

$$\text{rate of loss of } H = k_{R,21}(p_H)(p_{I_2})$$

and

$$\text{rate of production of } H = k_{L,21}(p_{HI})(p_I)$$

At equilibrium,

$$k_{R,21}(p_H)(p_{I_2}) = k_{L,21}(p_{HI})(p_I)$$

hence,

$$\frac{k_{R,21}}{k_{L,21}} = K_{21} = \frac{(p_I)(p_{HI})}{(p_H)(p_{I_2})} \tag{3-25}$$

We now have two equilibrium relationships among pressures: (3-23) and (3-25). Neither is easy to verify experimentally because each involves the partial pressures of hydrogen atoms and of iodine atoms. These two constituents, because of their reactivity, are present at very minute concentration, too low to measure. We can, however, combine the two expressions in a way that eliminates the undetectable intermediates. Multiplying (3-23) by (3-25) gives

$$K_{20} \times K_{21} = \frac{(p_{HI})(p_H)}{(p_{H_2})(p_I)} \times \frac{(p_I)(p_{HI})}{(p_H)(p_{I_2})}$$

By cancelling, and recognizing that the product of two constants is itself a constant, $K_{20} \times K_{21} = K$, and hence,

$$K = \frac{(p_{HI})^2}{(p_{H_2})(p_{I_2})}$$

This is exactly the result that has been observed experimentally and that was derived by assuming another kinetic mechanism.

This is not an accident. *Any possible reaction mechanism will lead to the same equilibrium relationship, the Equilibrium Law!* In fact, a thermodynamic argument can be framed that requires this to be so (the "Principle of Microscopic Reversibility").* So there is, after all, no objection to deriving the Equilibrium Law by equating the opposing rates, as long as it is realized that the correct outcome does not imply that the mechanism assumed must be the most important one occurring. We are on a one-way street. Since the same Equilibrium Law is obtained for any mechanism, we can derive this law by equating the rates of the opposing reactions for the most likely mechanism. Having obtained the correct Equilibrium Law relationship, we cannot, however, conclude that this gives any assurance that the mechanism is the one actually doing the work. All mechanisms lead to the same law.

3-3 Equilibrium: a state of dynamic balance

Since this mechanistic view of equilibrium is in agreement with the empirical Equilibrium Law, we are inclined to regard it as a valid model. In fact, it is generally accepted today that, at equilibrium in a chemical reaction, both reactants and products are present. The bulk properties of the system do not change because reactants are forming products at exactly the same rate at which the products are reforming the reactants. The position of equilibrium favors products if the forward rate is intrinsically rapid compared with the reverse rate. Even then, there is some concentration level of the reactants that is sufficiently small so that, finally, the forward and reverse reaction rates become equal. Equilibrium is established.

---

* The argument is based on the empirical impossibility of the perpetual motion machine that could be made by coupling two mechanisms with different Equilibrium Laws for the same reaction, one proceeding in the forward direction and the other proceeding in the reverse direction.

With that understanding of the nature of equilibrium, we can consider some specific equilibrium situations. We'll begin with equilibria in which two different phases are involved: solids and gases, solids and solutions, and liquids and gases. These are called *heterogeneous equilibria*. Then in the subsequent chapter, we will consider some examples of *homogeneous equilibria* in solutions.

Problems

1. Sugar is added to a glass of iced tea till no more will dissolve. With respect to solubility of sugar, does this glass of tea constitute a closed system in which solubility equilibrium exists?

2. Although heat is generated inside a spacecraft by instruments, a constant temperature can be maintained by radiation of energy to deep space. With respect to thermal control, is the spacecraft a closed system?

3. When you are driving at 60 miles per hour, the car engine temperature remains constant near 190°F. Is this an equilibrium temperature?

4. The solubility of air in water is determined by the rate at which the molecules in air strike the surface and dissolve, but in dynamic balance with the rate at which dissolved molecules move to the surface and escape from solution.
(a) Would the solubility of air in water differ if measured at Denver (altitude, 6000 feet) from the value measured at Berkeley (altitude, 100 feet)?
(b) A scuba diver who comes up too rapidly from depths below 100 feet is liable to suffer from a painful sensation caused by gas bubbles forming in his blood vessels (the "bends"). Explain the phenomenon in terms of equilibrium, keeping in mind your answer to (a).

5. The human cooling system depends upon evaporation of perspiration. Explain in terms of the dynamic nature of equilibrium why one's cooling system fails on a humid day.

6. Two Stanford Research Institute chemists* studied the high-temperature equilibrium in the "disproportionation" reaction by which gaseous niobium tetrachloride, $NbCl_4$, decomposes to form gaseous niobium trichloride, $NbCl_3$, and niobium pentachloride, $NbCl_5$.

$$2 NbCl_4(g) = NbCl_3(g) + NbCl_5(g)$$

At 1073°K, they found the following partial pressures at equilibrium (all given in atmospheres).

$p_4 = NbCl_4$	$p_3 = NbCl_3$	$p_5 = NbCl_5$
0.100	$6.71 \cdot 10^{-3}$	$6.71 \cdot 10^{-3}$
0.102	$5.00 \cdot 10^{-2}$	$9.00 \cdot 10^{-4}$
0.200	$1.28 \cdot 10^{-2}$	$1.28 \cdot 10^{-2}$
0.0100	$5.00 \cdot 10^{-3}$	$1.02 \cdot 10^{-4}$

* *Journal of Physical Chemistry*, Vol. **73**, p. 3054, 1969.

(a) Calculate the quotient p_3p_5/p_4 for each set of equilibrium conditions, its average, $\bar{Q}'$, its average deviation from the average $\bar{\delta}'$, and the ratio $\bar{\delta}'/\bar{Q}'$.

(b) Calculate the quotient p_3p_5/p_4^2, its average, $\bar{Q}$, its average deviation from the average $\bar{\delta}$, and the ratio $\bar{\delta}/\bar{Q}$.

(c) Use your best value of the equilibrium constant, K, to calculate the $NbCl_5$ partial pressure in an equilibrium mixture known to contain equal partial pressures of $NbCl_3$ and $NbCl_4$, each at $5.10 \cdot 10^{-3}$ atm. Estimate the uncertainty in your calculated values.

7. Hydrazine, N_2H_4, and nitrogen dioxide, NO_2, furnish a good rocket fuel. Write the Equilibrium Law expression that relates the pressures of reactants and products at equilibrium, presuming that the exhaust flame products are N_2 and H_2O.

$$2\,N_2H_4(g) + 2\,NO_2(g) = 3\,N_2(g) + 4\,H_2O(g)$$

8. An important source of ammonia for fertilizer manufacture is the Haber process, by which nitrogen and hydrogen are reacted at a high temperature (about $800°K$) and at a high pressure (about 1000 atm). What is the Equilibrium Law relation among the gas pressures?

$$N_2(g) + 3\,H_2(g) = 2\,NH_3(g)$$

9. Sulfuric acid is one of the most important commercial chemicals because it is used in the manufacture of fertilizers. We make H_2SO_4 by burning sulfur to sulfur dioxide, followed by the equilibrium oxidation of sulfur dioxide to sulfur trioxide. Then the SO_3 is dissolved in water to give H_2SO_4.

$$S_8(s) + 8\,O_2(g) = 8\,SO_2(g)$$

$$SO_2(g) + \tfrac{1}{2}\,O_2(g) = SO_3(g)$$

$$SO_3(g) + H_2O(\ell) = H_2SO_4(\ell)$$

What is the Equilibrium Law relation for the second reaction?

10. Aqueous ferrous ion, $Fe^{+2}(aq)$, reacts with stannic ion, $Sn^{+4}(aq)$, to form ferric ion, $Fe^{+3}(aq)$, and stannous ion, $Sn^{+2}(aq)$.

$$2\,Fe^{+2}(aq) + Sn^{+4}(aq) = 2\,Fe^{+3}(aq) + Sn^{+2}(aq)$$

At equilibrium, what is the relationship among the concentrations, according to the Equilibrium Law?

11. When ammonia is added to an aqueous cupric chloride solution, the sky-blue color of cupric ion, $Cu^{+2}(aq)$, changes to the midnight blue of the copper ammonia complex ion, $Cu(NH_3)_4^{+2}(aq)$. Write the equation for the reaction and the Equilibrium Law relationship among the concentrations of these species.

12. An important reaction in smog chemistry is the reaction between oxygen, O_2, and nitric oxide, NO, to form nitrogen dioxide, NO_2.

$$2\,NO(g) + O_2(g) = 2\,NO_2(g)$$

The forward reaction rate constant, k_f, has been measured at $298°K$ to be 11.2 $atm^{-2}\,sec^{-1}$, and the equilibrium constant for the reaction is $1.71 \cdot 10^{+12}$ atm^{-1}. From the fact that $K = k_f/k_r$, calculate the rate constant for the reverse reaction, k_r. What are the units of k_r?

four   heterogeneous equilibria

Heterogeneous reactions provide excellent opportunities to apply the Equilibrium Law. It gives us a quantitative answer to the important question, "When known amounts of substances (reactants) are mixed, how much reaction will have occurred to produce new substances (products) after equilibrium has been reached?"[*] We shall begin with some especially simple equilibria in which a solid or a liquid substance is in equilibrium with itself in a different phase—either the vapor phase or in solution. Then we can be more ambitious and proceed to discover more interesting reactions that involve several reactants and products.

4-1   Phase equilibria and molecular solutions

Perhaps the simplest of all equilibria involve the vaporization of a pure liquid or solid—the latter process is usually called sublimation. We will deal with each in turn.

(a) SOLID-VAPOR EQUILIBRIA

In the autumn, winter clothes brought out of storage sometimes smell of mothballs—pellets of the white crystalline solid naphthalene, $C_{10}H_8$. It is clear both to us and to the moths that some naphthalene vapor is in equilibrium with the solid at room temperature.

$$C_{10}H_8(s) \rightleftharpoons C_{10}H_8(g) \tag{4-1}$$

We can apply the Equilibrium Law to write down an equilibrium constant, K', for this reaction:

$$K' = \frac{[C_{10}H_8(g)]}{[C_{10}H_8(s)]} \tag{4-2}$$

Imagine an experiment in which some solid naphthalene is placed in a

---

[*] An equally important question is *not* answered: "When a reaction occurs, how *long* will it take to reach equilibrium?" This is the subject of Chapter Eleven, Chemical Kinetics.

1.00-liter bulb from which all the air is then removed (Fig. 4-1(a)). Initially the pressure in the bulb is zero. In a very short time naphthalene vapor fills the bulb at its equilibrium pressure (Fig. 4-1(b)). This can be measured with a very sensitive manometer. If the experiment is carried out at room temperature—defined for scientific purposes to be 25°C or 298°K—the measured pressure is 1.05×10^{-4} atmosphere.

If we wish to express the equilibrium amount of vapor in moles per liter (in expression (4-2)), the perfect gas law is a help.

$$PV = nRT \tag{4-3}$$

To find the concentration, we need only solve for the ratio

$$[C_{10}H_8(g)] = \left(\frac{n}{V}\right) (\text{moles/liter}) = \frac{P}{RT} \tag{4-4}$$

Substituting the appropriate values, we obtain:

$$[C_{10}H_8(g)] = \frac{(1.05 \times 10^{-4} \text{ atm})}{(0.082 \text{ liter-atm/mole °K})(298°K)}$$

$$= 4.30 \times 10^{-6} \text{ mole/liter} \tag{4-5}$$

The concentration of solid needed in equation (4-2) is a different story. Suppose we had a mothball of weight W. From the density, ρ, of solid naphthalene, 1.145 gms/cm³, we can calculate the volume, V, occupied by the solid. Then with the molecular weight (MW), 128.2 gms/mole, we can

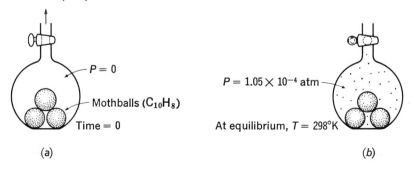

To vacuum pump

$P = 0$

Mothballs ($C_{10}H_8$)

Time $= 0$

(a)

$P = 1.05 \times 10^{-4}$ atm

At equilibrium, $T = 298°K$

(b)

Figure 4-1   A simple equilibrium: the vaporization of solid naphthalene.

calculate the number of moles, n, in our pellet and thence the molar concentration, n/V, in moles per liter.

$$V = \frac{W(\text{gms})}{\rho(\text{gms/cm}^3)} = \frac{W}{\rho}\,\text{cm}^3 = \frac{W}{1000\rho}\,\text{liters} \tag{4-6}$$

$$n = \frac{W(\text{gms})}{\text{MW(gms/mole)}} = \frac{W}{\text{MW}}\,\text{moles} \tag{4-7}$$

The concentration is the quotient of (4-6) and (4-7):

$$[\text{C}_{10}\text{H}_8(\text{s})] = \frac{n}{V} = \frac{W/\text{MW}}{W/1000\rho} = \frac{1000\rho}{\text{MW}}\frac{\text{moles}}{\text{liter}} \tag{4-8}$$

Notice that the concentration of the pure solid does not depend upon how much solid is placed in the flask. So the quantity $[\text{C}_{10}\text{H}_8(\text{s})]$ is a constant (it turns out to be 8.93 moles/liter), as long as any solid is present. This permits us to simplify the equilibrium expression (4-2) by multiplying both sides by this constant number, and to define a new equilibrium constant, K_c.

$$K_c = K' \cdot [\text{C}_{10}\text{H}_8(\text{s})] = \frac{[\text{C}_{10}\text{H}_8(\text{g})]}{[\text{C}_{10}\text{H}_8(\text{s})]} \cdot [\text{C}_{10}\text{H}_8(\text{s})]$$

$$K_c = [\text{C}_{10}\text{H}_8(\text{g})] \tag{4-9}$$

This is the way equilibrium expressions are always written for reactions involving solids—only variable concentrations appear; the concentrations of pure solids are embodied in the value of K_c.

By the way, the equilibrium expression (4-9) contains only one quantity, the gaseous naphthalene concentration. Since we measured the equilibrium vapor pressure and calculated the implied concentration, we now know the magnitude of K_c.

$$K_c = [\text{C}_{10}\text{H}_8(\text{g})] = 4.30 \cdot 10^{-6}\,\text{mole/liter}$$

In fact, K can be expressed directly in the dimensions of pressure, since that is what we measure and since concentration is simply proportional to it. This is usually done:

$$K_p = P_{\text{equilibrium}} = 1.05 \cdot 10^{-4}\,\text{atm} \tag{4-10}$$

The pressure of gas in equilibrium with a pure solid (or a pure liquid) is called the vapor pressure, $P_{\text{vap.}}$. The simple form of the Equilibrium Law, expression (4-10), tells us that a pure solid has one and only one equilibrium vapor pressure at a particular temperature. If the vapor above the solid is at a lower pressure than $K_p = P_{vap.}$, then the solid will vaporize until the pressure is raised to equal K_p. If the pressure of vapor above the solid exceeds K_p, then gas will condense, lowering the pressure, again until it reaches $1.05 \cdot 10^{-4}$ atm, the value of K_p.

Most of the solids around us have extremely small vapor pressures, usually immeasurably small. Table 4-1 lists a few, the smallest of which must be calculated from thermal data, using the methods of chemical thermodynamics.

Exercise 4-1. K_c for the sublimation of solid iodine at 298°K is 1.66×10^{-5} mole per liter. Show that the vapor pressure of iodine at this temperature is 0.308 torr (mm of Hg).

Exercise 4-2. Show that the naphthalene bulb contains $2.6 \cdot 10^{18}$ molecules at equilibrium, whereas there are less than a billion atoms of gaseous sodium in a liter in equilibrium with solid sodium at room temperature.

(b) LIQUID–VAPOR EQUILIBRIA

The vapor pressures of liquids at room temperature are generally much higher than those of solids, so the equilibrium constants K_p and K_c for vaporization are proportionately larger. The equilibrium expression is identical. Consider the vaporization of liquid water.

$$H_2O(\ell) \rightleftharpoons H_2O(g) \tag{4-11}$$

$$K' = \frac{[H_2O(g)]}{[H_2O(\ell)]} \tag{4-12}$$

Like a solid, a pure liquid has a constant concentration. For example, consider water for which expressions (4-6) and (4-7) are again applicable:

$$V = \frac{W}{1000\rho} \quad \text{and} \quad n = \frac{W}{MW}$$

Table 4-1 *Vapor Pressures of Some Pure Solids at 298°K*

Substance	$P_{vap.}$(atm)	$P_{vap.}$(torr)
iodine, $I_2(s)$	4.06×10^{-4}	3.08×10^{-1}
naphthalene, $C_{10}H_8(s)$	1.05×10^{-4}	7.99×10^{-2}
sulfur, $S_8(s)$	4.87×10^{-8}	3.69×10^{-5}
sodium, $Na(s)$	3.40×10^{-14}	2.58×10^{-11}
lead, $Pb(s)$	4.28×10^{-29}	3.25×10^{-26}

so that, again,

$$\frac{n}{V} = \frac{1000\rho}{MW} \frac{\text{moles}}{\text{liter}}$$

Substituting the appropriate values of density (ρ) and molecular weight (MW) for water,

$$\frac{n}{V} = \frac{(1000 \text{ cm}^3/\text{liter}) \cdot (1.00 \text{ gm}/\text{cm}^3)}{(18.02 \text{ gms}/\text{mole})}$$

$$[H_2O(\ell)] = \frac{n}{V} = 55.5 \frac{\text{moles}}{\text{liter}} \qquad (4\text{-}13)$$

If equation (4-12) is multiplied on both sides by the concentration of pure liquid water,

$$K_c = K' \cdot (55.5 \text{ moles}/\text{liter}) = [H_2O(g)] \qquad (4\text{-}14)$$

Similarly, K can be expressed in pressure units

$$K_p = P_{\text{vap.}} \qquad (4\text{-}15)$$

Thus, like a solid, a pure liquid has one and only one equilibrium vapor pressure at a particular temperature. If the vapor above the liquid is at a lower pressure than $K_p = P_{\text{vap.}}$, liquid will vaporize until the pressure rises to $P_{\text{vap.}}$. If the pressure exceeds $P_{\text{vap.}}$, gas will condense, lowering the pressure until it reaches $P_{\text{vap.}}$.

Because of the fluidity of a liquid, an additional phenomenon called boiling can occur. This merely consists of vapor formation within the body of the liquid (bubbles) and it can take place only when the *total* pressure on the liquid is below $P_{\text{vap.}}$. A bubble, which contains only the vapor, cannot exert a pressure larger than $P_{\text{vap.}}$. If the liquid has an inert gas pressing down on it at a higher pressure (for example, air presses down on a liquid in an open beaker) then any bubbles will collapse as rapidly as they form. Boiling will not occur. Under such conditions, vaporization can occur only at the surface of the liquid. However, if the temperature is raised until $P_{\text{vap.}}$ exceeds the total pressure on the liquid, bubbles can form and grow within the liquid; then boiling will occur.

The "normal boiling point" of a liquid is that temperature at which it boils under an applied pressure of exactly one atmosphere (760 torr). It can be made to boil, however, at any temperature merely by reducing the applied total pressure (vapor plus other gases) below the equilibrium vapor pressure at that temperature. This boiling phenomenon should not obscure

Table 4-2   Vapor Pressure of Some Pure Liquids at 298°K

Substance	$P_{vap.}$(atm)	$P_{vap.}$(torr)
glacial acetic acid, $CH_3COOH(\ell)$	0.00164	1.25
water, $H_2O(\ell)$	0.0313	23.8
ethanol, $CH_3CH_2OH(\ell)$	0.0782	59.4
methanol, $CH_3OH(\ell)$	0.173	131
chloroform, $CHCl_3(\ell)$	0.247	188
bromine, $Br_2(\ell)$	0.281	213.9
methyl chloride, $CH_3Cl(\ell)$	5.27	4000

the fact that a pure liquid has one and only one equilibrium vapor pressure at a particular temperature.

The vapor pressures of some pure substances are given in Table 4-2. We see that one of the liquids, methyl chloride, has a vapor pressure exceeding one atmosphere at room temperature. Hence it would boil violently if it were placed in an open beaker. Only if methyl chloride is kept in a closed container and if the vapor pressure reaches 5.27 atm will we find liquid present at room temperature. This is quite practical, however; some gases commonly used as household fuels (propane, butane) are stored as liquids in high-pressure cylinders.

Exercise 4-3.   Show that $K_c = 0.215$ mole/liter for the vaporization of methyl chloride at room temperature and a pressure of 5.27 atm. What would one liter of methyl chloride vapor weigh under these equilibrium conditions?

(c) MOLECULAR SOLUTIONS

As a final example of molecular species indulging in heterogeneous equilibria, we will discuss the formation of molecular solutions. If, for example, solid iodine is placed in contact with a liquid, molecules of I_2 leave the solid and enter the solution, retaining their molecular identities.

$$I_2(\text{solid}) \rightleftharpoons I_2(\text{solution}) \tag{4-16}$$

A solution in which the dissolved molecule retains its molecular identity without undergoing either dissociation or reaction is called a *molecular solution*. An equilibrium constant can be written down in which, as usual, the concentration of the pure solid is incorporated in the equilibrium constant.

$$K_c = I_2(\text{solution}) \tag{4-17}$$

The numerical value of K_c will depend on the particular liquid in which the iodine is dissolved. Solid iodine, for instance, dissolves very slightly in water and somewhat more readily in carbon tetrachloride, CCl_4. The equilibrium constants for these processes are known:

$$I_2(s) = I_2(aq) \qquad K_c = 1.34 \cdot 10^{-3} \text{ mole/liter} \qquad (4\text{-}18)$$

$$I_2(s) = I_2(CCl_4) \qquad K_c = 1.12 \cdot 10^{-2} \text{ mole/liter} \qquad (4\text{-}19)$$

The magnitudes of K_c (4-18 and 4-19) tell us that iodine is about ten times more soluble in carbon tetrachloride than in water.

Gases also dissolve to form molecular solutions. Air bubbled through fish tanks provides the occupants with dissolved oxygen essential to fishy life. If the oxygen pressure is P atm, we have

$$O_2(g, P \text{ atm}) \rightleftharpoons O_2(aq) \qquad (4\text{-}20)$$

and

$$K = \frac{[O_2(aq)]}{P_{O_2(g)}} \qquad (4\text{-}21)$$

The magnitude of K can be determined by a measurement of the solubility of oxygen in water under a known pressure of oxygen gas. Experiment shows that when $P_{O_2(g)} = 1$ atmosphere, the solubility, $[O_2(aq)]$ is $1.38 \cdot 10^{-3}$ mole/liter. Substituting into (4-21),

$$K = \frac{[O_2(aq)]}{P_{O_2(g)}} = \frac{1.38 \cdot 10^{-3} \text{ mole/liter}}{1 \quad \text{atm}} = 1.38 \cdot 10^{-3} \qquad (4\text{-}22)$$

With this value of K, we can calculate, for example, the solubility of oxygen in our fishbowl at 0.20 atm, the oxygen partial pressure in air at one atmosphere.

$$K = 1.38 \cdot 10^{-3} = \frac{[O_2(aq)]}{0.20}$$

$$O_2(aq) = \left[1.38 \cdot 10^{-3} \frac{\text{mole/liter}}{\text{atm}} \right] (0.20 \text{ atm})$$

$$= 2.76 \cdot 10^{-4} \text{ mole/liter} \qquad (4\text{-}23)$$

Table 4-3 collects some equilibrium constants for aqueous solutions of familiar gases. All of the solubilities are rather low and there are some interesting contrasts. Some of the data suggest that molecular size is not very influential: O_2 and CH_4 have about the same solubility, only about 20% larger than that of H_2 whereas the solubility of ethylene, C_2H_4, is higher and that of ethane, C_2H_6, is lower. Among the inert gases, however,

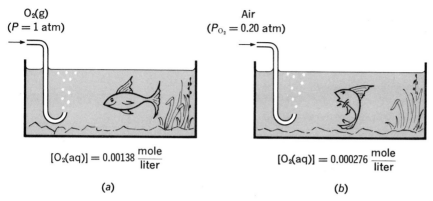

Figure 4-2   Gases form molecular solutions.

there is a regular trend: Solubility increases in the series Ne, Ar, Kr, Xe. These observations are interpretable in terms of the interactions between water molecules and the dissolved molecules (the solute), a subject we will take up again in Chapter Seventeen.

Exercise 4-4.   Calculate the equilibrium concentration of nitrogen in the blood of a deep-sea diver if he is breathing air (80% nitrogen) at a depth of 400 feet (32 feet of water = 1 atmosphere pressure; assume blood has the same solvent properties as water and that $T = 298°K$.)

4-2   The nature of solutions

Solutions are employed in countless ways. Automobile gasoline contains dissolved additives that increase combustion efficiency and prolong engine life; cough syrup, vinegar and martinis are solutions; the ocean is an immense solution with many components. A large percentage of chemical reactions carried out in the laboratory and in chemical industry occur in a solution medium.

Table 4-3   Equilibrium Constants for Aqueous Molecular Solutions of Some Simple Gases (mole/liter-atm; $T = 298°K$)

He	4.25×10^{-4}	H_2	8.34×10^{-4}
Ne	4.25×10^{-4}	O_2	1.38×10^{-3}
Ar	1.38×10^{-3}	N_2	7.23×10^{-4}
Kr	2.30×10^{-3}	CO	9.55×10^{-4}
Xe	4.51×10^{-3}	CH_4	1.36×10^{-3}
		C_2H_4	4.91×10^{-3}
		C_2H_6	1.72×10^{-4}

In Section 4-1(c) we studied some simple molecular solutions. In the remainder of this and in the following chapter we will focus attention upon aqueous solutions. Many of these are called *ionic solutions.* Perhaps we should begin by deciding just what we mean by this term, "ionic solution."

(a) IS THE SOLUTION IONIC OR MOLECULAR?

Neither sugar nor iodine aqueous solutions conduct electrical current. Experiments show the presence of neutral sugar molecules and of neutral iodine molecules in the water medium. These are molecular solutions. In Chapter One we noted that sea water conducts electricity. Sea water must contain mobile charge carriers. Then, in Chapter Two, we described the properties of alkaline earth halides in their water solutions, which also conduct electricity. There, the mobile charge carriers were identified as "ions." Sodium chloride, like the alkaline earth halides, dissolves in water to give ions. These are ionic solutions. In most dilute ionic solutions, there seem to be no neutral solute molecules present at all.

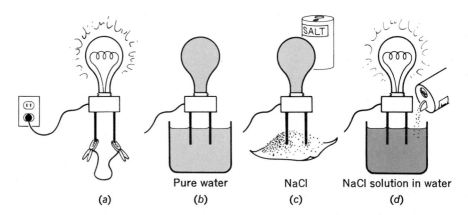

| | | Pure water | NaCl | NaCl solution in water |
| (a) | (b) | (c) | (d) |

Figure 4-3   Ionic solutions conduct electricity.

As mentioned earlier, substances that behave like sodium chloride are given the class name "salts." They are characterized by crystallinity in the solid state (like the familiar crystallinity of sodium chloride) and by their ability to form ionic solutions in water. Here are the reactions by which a number of familiar salts dissolve in water to give ionic solutions.

sodium chloride
$$NaCl(s) = Na^+(aq) + Cl^-(aq) \qquad (4\text{-}24)$$

calcium chloride
$$CaCl_2(s) = Ca^{+2}(aq) + 2\ Cl^-(aq) \qquad (4\text{-}25)$$

sodium hydroxide
$$NaOH(s) = Na^+(aq) + OH^-(aq) \qquad (4\text{-}26)$$

calcium hydroxide	$Ca(OH)_2(s) = Ca^{+2}(aq) + 2\ OH^-(aq)$	(4-27)
sodium nitrate	$NaNO_3(s) = Na^+(aq) + NO_3^-(aq)$	(4-28)
sodium carbonate	$Na_2CO_3(s) = 2\ Na^+(aq) + CO_3^{-2}(aq)$	(4-29)
ammonium chloride	$NH_4Cl(s) = NH_4^+(aq) + Cl^-(aq)$	(4-30)
ammonium sulfate	$(NH_4)_2SO_4(s) = 2\ NH_4^+(aq) + SO_4^{-2}(aq)$	(4-31)

The reactions (4-26) to (4-31) show that some ions are diatomic; hydroxide ion is an example. Furthermore, some ions, nitrate, carbonate, ammonium and sulfate ions, are polyatomic. In none of these cases, either, is there evidence for neutral solute molecules in dilute aqueous solutions. What is special about dilute solutions that causes them to contain ions rather than neutral molecules? To answer that, we'd better begin by defining "dilute."

(b) CONCENTRATING ON SOLUTIONS

There are countless ways to describe the amount of one substance (called the "solute") dissolved in another (called the "solvent"). Perhaps the simplest way to proceed is to state the amount of dissolved solute and the final volume of the solution—this is the approach we have used so far. As chemists, we care about the number of moles rather than the weight of substance so we use the unit *molarity*, M, which is defined as the number of moles of solute in a liter of *solution*. The simplicity of this description is illustrated in Figure 4-4(a) for 0.100 M sucrose (sugar) solution.

Life is more complicated when the solute dissociates—as salts do. When, as illustrated in Figure 4-4(b), 0.100 mole (11.1 gm) $CaCl_2$ is dissolved in enough water to make a liter of solution, we end up with 0.100 mole of $Ca^{+2}(aq)$ and 0.200 mole $Cl^-(aq)$ but *no* $CaCl_2$ at all. Nevertheless it is usual to describe a solution in terms of the initial amount of undissociated solute added. Often we can't do much better than this, for the solute may dissociate only partly, leaving us with a solution containing several different species with unknown concentrations. A solution such as the salt solution, then, is called a 0.100 M $CaCl_2$ solution, despite our conviction that all of the $CaCl_2$ has dissociated to ions.

(c) IS THE SOLUTION DILUTE?

If a small amount of solute is dissolved in a lot of solvent the solution is said to be *dilute*. As more and more solute is added, the solution becomes more *concentrated* until it may be difficult to distinguish between solute and

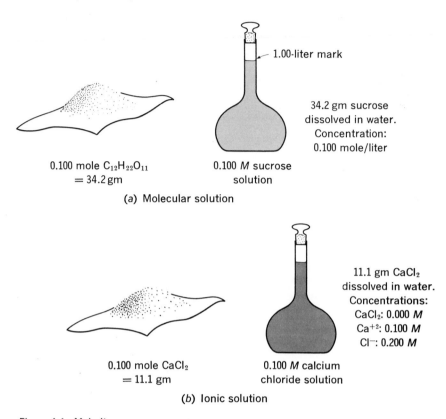

1.00-liter mark

34.2 gm sucrose
dissolved in water.
Concentration:
0.100 mole/liter

0.100 mole $C_{12}H_{22}O_{11}$
= 34.2 gm

0.100 M sucrose
solution

(a) Molecular solution

11.1 gm $CaCl_2$
dissolved in **water.**
Concentrations:
$CaCl_2$: 0.000 M
Ca^{+2}: 0.100 M
Cl^-: 0.200 M

0.100 mole $CaCl_2$
= 11.1 gm

0.100 M calcium
chloride solution

(b) Ionic solution

Figure 4-4   Molarity.

solvent. In general, the solvent is designated as that component present in excess. In the case of a gaseous or a liquid–liquid solution it is often a matter of choice or convenience to specify which is solute and which is solvent.

From our point of view in this book, a solution is dilute when the individual solute molecules or ions are almost entirely in an environment of solvent molecules. Then there cannot be much interaction between solute species, be they ionic or molecular. Of course, there will always be interaction between solute and solvent. We can get a feeling for when a solution is dilute by this criterion by calculating the ratio of solvent to solute species in 0.100 M calcium chloride. The total concentration of ions is the sum of the calcium ion concentration, 0.10 M, and the chloride ion concentration, 0.20 M:

$$\text{total ion concentration} = 0.10 + 0.20$$

$$[\text{ions}] = 0.30$$

We have already calculated the number of water molecules present in a liter of water (in Section 4-1(b)), 55.5 moles. Hence the molar concentration of water is 55.5 M.* Thus the ratio is

$$\frac{\text{number solvent molecules}}{\text{number of ions}} = \frac{\text{concentration solvent molecules}}{\text{concentration solute species}}$$

$$\frac{55.5}{0.30} = 185$$

There are over 180 solvent molecules for each and every calcium or chloride ion in tenth molar calcium chloride! We can consider 0.1 M solutions to be effectively dilute by our standards. The ratio in a 1.00 M solution, however, is around 18:1. Here the odds against interaction are not so high and such solutions do not usually qualify as dilute. In 6 M solutions, common for some familiar laboratory reagents, there are only 5 to 10 water molecules per ion, hardly enough to keep the ions well separated.

(d) WHAT'S SO IMPORTANT ABOUT WATER?

It is not uncommon that a compound ionizes when dissolved in water, but forms a molecular solution when dissolved in some other solvent. For example, hydrogen chloride gas, HCl, dissociates completely into H+(aq) ions and Cl−(aq) ions in water, yet 70% of the HCl molecules are undissociated in a 0.1 M solution in the solvent formic acid, HCOOH. Hydrochloric acid dissolved in carbon tetrachloride shows no conductivity at all—no dissociation occurs in this solvent. A substance that causes the formation of ions is called an ionizing solvent. For HCl, formic acid is called an ionizing solvent and carbon tetrachloride is not. One of the most useful and most abundant ionizing solvents is water.

In Chapter Two, we remarked that the forces that hold a salt crystal together must be very strong. This was illustrated by the characteristically high melting points of salts (that of NaCl is 1074°K), a measure of the effort needed to disrupt the solid structure. Nevertheless, many salts dissolve readily in water. The energy necessary to break the bonds in the solid and to pull the ions into solution must come from attractive forces between solvent molecules and the ions. How can we "explain" these strong attractive forces?

The simplest explanation concocted by chemists is based upon a simple electrostatic model. The behavior of pure water in an electric field shows that the positive and negative charges in a water molecule are not evenly distributed. The molecules act as though there is more negative charge at the oxygen end of the molecule than at the hydrogen end. The water

* In using 55.5 moles/liter as the water concentration in a solution, we assume a negligible volume to be occupied by the solute. In fact, the solute replaces roughly the amount of water equal to the solute concentration. That means that our 0.10 M CaCl$_2$ solution actually contains about 55.5 − 0.3 ≅ 55.2 M water instead of 55.5 M.

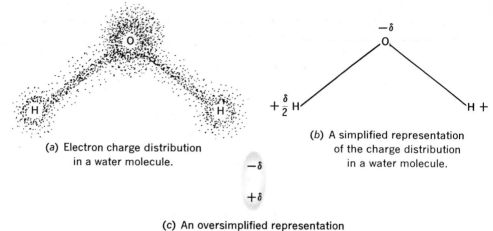

(a) Electron charge distribution in a water molecule.

(b) A simplified representation of the charge distribution in a water molecule.

(c) An oversimplified representation of the charge distribution in a water molecule.

Figure 4-5 The charge distribution in a water molecule.

molecule can be pictured, electrically speaking, as in Figure 4-5. Such a charge distribution, positive charge on one end and negative on the other, is called a *dipole*. A molecule with a dipole is said to have a *dipole moment*.

It is fairly obvious how the cigar-shaped charge representation of a water molecule (Fig. 4-5(c)) can give us a simple explanation for the stability of ions in water. Clearly a molecule with such an oversimplified charge distribution will have more than a casual interest in an electrically charged ion. We can picture the positive ends of the water dipoles enthusiastically orienting themselves close to a negatively charged chloride ion. The negative ends of other water dipoles will have similar fun nestling up to positively charged calcium ions. Electrostatic attractions in these special orientations readily account for the stability of ionic solutions in a dipolar solvent—the stability that permits them to dissolve salt crystals despite the intrinsic stability of the crystals.

We'll look into a more sophisticated view of this aquation process in Chapter Seventeen. The uncomplicated model of Figure 4-6 satisfies most

$Cl^-(aq)$ $Ca^{+2}(aq)$

Figure 4-6 Dipoles oriented around ions: a very simple representation of aqueous ions.

chemists, however. Without arguing the point, we can estimate roughly the magnitude of the energy effects involved in the interactions that must be present in an ionic solution.

In Chapter Two, we noted that most salts dissolve in water with only a small heat effect; sometimes a few kilocalories are absorbed, sometimes a few are evolved. To take a specific case, sodium chloride absorbs 1.3 kilocalories per mole when it is dissolved in an excess of water.

$$NaCl(solid) \longrightarrow Na^+(aq) + Cl^-(aq) \qquad \frac{heat}{absorbed} = 1.3 \frac{kcal}{mole}$$

$$(4\text{-}32)$$

In contrast, the energy needed to form gaseous sodium and chloride ions from the solid is over 140 times larger:

$$NaCl(solid) \longrightarrow Na^+(gas) + Cl^-(gas) \qquad \frac{heat}{absorbed} = 184 \frac{kcal}{mole}$$

$$(4\text{-}33)$$

Expression (4-33) gives our clue to the energy of interaction between ions and solvent. Imagine reaction (4-32) being carried out in two steps, the first of which is reaction (4-33) and the second of which is the dissolving of gaseous ions in water:

$$NaCl(solid) \xrightarrow[184\ kcal]{heat\ absorbed} Na^+(gas) + Cl^-(gas) \xrightarrow[x]{heat\ absorbed} Na^+(aq) + Cl^-(aq)$$

over-all heat effect = 1.3 kcal absorbed

$$(4\text{-}34)$$

It is natural to equate the over-all heat effect, 1.3 kcal absorbed, to the sum of the heat effects in the step-wise reaction that accomplished the same change.*

$$1.3\ kcal = 184 + x\ kcal$$

$$x = 1.3 - 184 = -183\ kcal\ absorbed$$

This procedure says that minus 183 kcal is absorbed as a mole of Na^+ and Cl^- ions dissolve in water. The "minus" merely indicates that energy is liberated instead of absorbed. Figure 4-7 shows in diagrammatic form that it takes lots of energy to form gaseous ions from solid sodium chloride, but we get it all back when the ions interact with water.

---

* Brace yourself. We just used chemical thermodynamics for the first time. We believe that the over-all heat effect equals the sum of the step-wise heat in (4-34) because we believe that energy is conserved. More later!

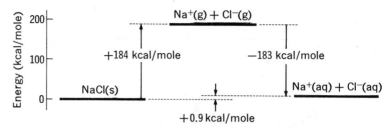

Figure 4-7 *Energy diagram for the dissociation of sodium chloride into ions in the gas phase and in aqueous solution.*

If we were to take the picture of Figure 4-6 literally, it shows four solvent molecules per ion, hence eight solvent–ion interactions per pair of ions. That would mean that the energy per interaction is, on the average, $183/8 \cong 20$–25 kcal. Such large energy effects imply that ions should be very stable in the water environment.

With this result, it is no longer surprising that water forms stable ionic solutions. Furthermore, we see that the ionization process is not merely the *dispersing* of ions in solution, but rather the reaction of solvent molecules with the atoms in the crystal lattice. In fact, these interactions often involve energies on the same order as those involved in the formation of chemical bonds.

Water, important as an ionizing solvent, is also our most abundant chemical. Life on earth is largely based on aqueous chemistry: Chemical life processes are carried out in aqueous solution (about two thirds of the body weight is water); food preparation and agriculture involve aqueous chemistry. Much of our planet has normal temperatures between the freezing and boiling points of aqueous solutions. Since both we and our environment are so aqueous, we will emphasize aqueous chemistry throughout our studies of solution equilibria. That is not to say that reactions in non-aqueous solvents are not of importance—they find their way into a host of experimental and industrial applications. Non-aqueous solvents will also be considered, a bit in this chapter, and again later in the book.

4-3   Sparingly soluble solids and ionic solutions

Table salt dissolves readily in water—it has a high solubility in water. Other salts have much lower solubilities in water. When a spatula full of silver chloride is added to a beaker of water, no visible amount of solid dissolves—AgCl is only sparingly soluble in water. When equilibrium is reached between solid AgCl and its aquated ions (this salt, also, is totally dissociated in aqueous solution) the solution is said to be *saturated*. The *solubility* of a substance is defined as the number of moles of solid dissolved per liter of saturated solution. Let's look now at some details about solubility equilibria in dilute solutions.

(a) THE SOLUBILITY PRODUCT

Silver chloride dissolves according to the reaction

$$AgCl(s) \rightleftharpoons Ag^+(aq) + Cl^-(aq) \tag{4-35}$$

The equilibrium expression for AgCl in contact with a saturated solution is

$$K' = \frac{[Ag^+(aq)][Cl^-(aq)]}{[AgCl\,(s)]} \tag{4-36}$$

By now it should be habitual to incorporate the constant concentration of the pure solid into the equilibrium constant.

$$K = K'[AgCl(s)] = [Ag^+(aq)][Cl^-(aq)] \tag{4-37}$$

The constant K appears in (4-37) as a *product*, or *ion product*, of two concentrations fixed by the solubility. Consequently such an equilibrium constant is called the *solubility product* and it is designated K_{sp}. In the equilibrium expression, the (aq) is usually dropped to simplify the appearance. Hence (4-37) is usually seen in the form

$$K_{sp} = [Ag^+][Cl^-] \tag{4-38}$$

One way to learn the magnitude of K_{sp} is to measure how much solid silver chloride dissolves in a liter of pure water. Careful experiments show that only 1.9 milligrams dissolve per liter. The number of moles of AgCl in a liter is then

$$moles\ AgCl = \frac{1.9 \cdot 10^{-3}\ gm}{MW} = \frac{1.9 \cdot 10^{-3}}{107.8 + 35.5} = 1.33 \cdot 10^{-5}\ mole \tag{4-39}$$

Since the silver chloride dissociates completely to ions in the liter of water, (4-39) gives the ion concentrations.

$$[Ag^+] = [Cl^-] = 1.33 \cdot 10^{-5}\ mole/liter \tag{4-40}$$

Substituting (4-40) into (4-38), we can calculate K_{sp}.

$$K_{sp} = (1.33 \cdot 10^{-5})(1.33 \cdot 10^{-5})$$

$$K_{sp} = 1.77 \cdot 10^{-10} \tag{4-41}$$

Exercise 4-5. Calculate the solubility product of lead sulfate, $PbSO_4$, from its measured solubility in pure water at 298°K, $1.26 \cdot 10^{-4}$ mole/liter.

Exercise 4-6. The solubility product for silver bromide, AgBr, is $5.30 \cdot 10^{-13}$. Reverse the calculation of the last exercise to show that the solubility of AgBr in water is $7.3 \cdot 10^{-7}$ mole/liter.

Salts of formula MX_2 are treated in exactly the same way, but they provide us with an extra twist or two. Consider a saturated solution of calcium fluoride:

$$CaF_2(s) = Ca^{+2}(aq) + 2\ F^-(aq) \tag{4-42}$$

The equilibrium expression for reaction (4-42) is

$$K_{sp} = [Ca^{+2}][F^-]^2 \tag{4-43}$$

Tables of solubility products list K_{sp} for CaF_2 as

$$K_{sp} = 3.98 \times 10^{-11}$$

From (4-42) we see that *two* moles of F^-(aq) are produced for every *one* of Ca^{+2}(aq). Hence, the concentration of F^- is double that of Ca^{+2}.

$$[F^-] = 2[Ca^{+2}] \tag{4-44}$$

If we substitute (4-44) into (4-43) we have

$$K_{sp} = [Ca^{+2}](2[Ca^{+2}])^2$$
$$= 4[Ca^{+2}]^3 \tag{4-45}$$

and, solving for $[Ca^{+2}]$,

$$[Ca^{+2}] = \sqrt[3]{K_{sp}/4}$$
$$= \sqrt[3]{3.98 \times 10^{-11}/4}$$
$$= 2.15 \times 10^{-4} \text{ mole/liter}$$

The fluoride concentration is just twice as large:

$$[F^-] = 2[Ca^{+2}] = 2(2.15 \cdot 10^{-4}) = 4.30 \cdot 10^{-4} \text{ mole/liter}$$

A word of caution is necessary here. Sometimes competing equilibria and effects due to interactions between ions cause the equilibrium situation to be more complicated. This is always the case in concentrated solu-

Table 4-4   Solubility Products of Salts at 298°K (moles/liter)

MX	$K_{sp} = [M^{+n}][X^{-n}]$	MX$_2$ or M$_2$X	$K_{sp} = (M^{+2})(X^-)^2$ or $(M^+)^2(X^{-2})$
AgBrO$_3$	$5.3 \cdot 10^{-5}$	Ag$_2$SO$_4$	$1.6 \cdot 10^{-5}$
AgIO$_3$	$3.0 \cdot 10^{-8}$	Ag$_2$C$_2$O$_4$	$5.0 \cdot 10^{-12}$
AgN$_3$	$2.9 \cdot 10^{-9}$	Ag$_2$CrO$_4$	$1.9 \cdot 10^{-12}$
AgCl	$1.8 \cdot 10^{-10}$	Ag$_2$S	$5.5 \cdot 10^{-51}$
AgBr	$5.3 \cdot 10^{-13}$		
AgI	$8.3 \cdot 10^{-17}$	BaF$_2$	$1.7 \cdot 10^{-6}$
		Ba(IO$_3$)$_2$	$1.3 \cdot 10^{-9}$
BaCO$_3$	$1.6 \cdot 10^{-9}$		
BaSO$_4$	$1.1 \cdot 10^{-10}$	Ca(OH)$_2$	$1.3 \cdot 10^{-6}$
BaCrO$_4$	$8.5 \cdot 10^{-11}$	CaF$_2$	$4.0 \cdot 10^{-11}$
		Cu(IO$_3$)$_2$	$7.4 \cdot 10^{-8}$
CaSO$_4$	$2.5 \cdot 10^{-5}$	Cu(OH)$_2$	$1.6 \cdot 10^{-19}$
CaCO$_3$	$4.8 \cdot 10^{-9}$		
CaC$_2$O$_4$	$2.3 \cdot 10^{-9}$	MgF$_2$	$6.6 \cdot 10^{-9}$
		Mg(OH)$_2$	$1.8 \cdot 10^{-11}$
CuCl	$3.2 \cdot 10^{-7}$		
CuBr	$5.9 \cdot 10^{-9}$	PbCl$_2$	$1.6 \cdot 10^{-5}$
CuI	$1.1 \cdot 10^{-12}$	PbI$_2$	$8.3 \cdot 10^{-9}$
CuS	$4 \cdot 10^{-38}$	Pb(IO$_3$)$_2$	$2.6 \cdot 10^{-13}$
FeS	$1 \cdot 10^{-19}$	SrF$_2$	$2.9 \cdot 10^{-9}$
MgC$_2$O$_4$	$8.6 \cdot 10^{-5}$	Zn(OH)$_2$	$4.5 \cdot 10^{-17}$
PbSO$_4$	$1.6 \cdot 10^{-8}$		
PbCrO$_4$	$2 \cdot 10^{-16}$	MX$_3$	$K_{sp} = (M^{+3})(X^-)^3$
PbS	$1 \cdot 10^{-29}$	Al(OH)$_3$	$5 \cdot 10^{-33}$
		Ce(IO$_3$)$_3$	$3.2 \cdot 10^{-10}$
SrCrO$_4$	$3.6 \cdot 10^{-5}$	Fe(OH)$_3$	$6 \cdot 10^{-38}$
SrSO$_4$	$7.6 \cdot 10^{-7}$	La(IO$_3$)$_3$	$6.3 \cdot 10^{-12}$
ZnS	$4.5 \cdot 10^{-24}$		

tions. Table 4-4 lists some solubility product constants applicable to salts dissolved in dilute aqueous solutions. A brief discussion of some of the complicating effects is found in Section 4-4.

Exercise 4-7. Calculate the solubility product of silver sulfate, Ag$_2$SO$_4$, from its solubility in pure water at 298°K, $1.58 \cdot 10^{-2}$ M. Compare your calculation to the value listed in Table 4-4. (m/ℓ)

(b) QUANTITATIVE SEPARATIONS BY PRECIPITATION

In Exercise 2-7, we considered qualitatively the question, "What would happen if 0.008 M Na$_2$SO$_4$ solution were mixed with an equal volume of solution containing both 0.010 M $CaCl_2$ and 0.008 M $BaCl_2$?" Both argument and experience led us to the conclusion that barium sulfate would precipitate. We might attack that question again now, with the aid of our more sophisticated understanding of solubility equilibria.

Immediately after mixing equal amounts of the two solutions, the volume is doubled, so all concentrations are halved. The ions present all will act independently, forgetting their parentage, so the initial ion concentrations will be as follows:

$$[SO_4^{-2}] = \frac{0.008}{2} = 0.0040 \ M$$

$$[Na^+] = 2[SO_4^{-2}] = 2(0.0040) = 0.008 \ M$$

$$[Ca^{+2}] = \frac{0.010}{2} = 0.0050 \ M \tag{4-46}$$

$$[Ba^{+2}] = \frac{0.008}{2} = 0.0040 \ M$$

$$[Cl^-] = 2[Ca^{+2}] + 2[Ba^{+2}] = 2(0.0050) + 2(0.0040) = 0.0180 \ M$$

Turning to Table 4-4, we find solubility products for both calcium sulfate, $CaSO_4$, and barium sulfate, $BaSO_4$. Sodium sulfate has such high solubility that its solubility product is not listed. We find:

$$K_{sp}(CaSO_4) = [Ca^{+2}][SO_4^{-2}] = 2.5 \cdot 10^{-5} \tag{4-47}$$

$$K_{sp}(BaSO_4) = [Ba^{+2}][SO_4^{-2}] = 1.1 \cdot 10^{-10} \tag{4-48}$$

Expression (4-47) says that in equilibrium with solid calcium sulfate, the product of $[Ca^{+2}]$ and $[SO_4^{-2}]$ equals $2.5 \cdot 10^{-5}$. The actual concentrations in our solution mixture, (4-46), can be substituted into (4-47) to see where we stand relative to equilibrium. The concentration product in the solution is:

$$[Ca^{+2}][SO_4^{-2}] = [0.0050] \ [0.0040] = 2.0 \cdot 10^{-5} \tag{4-49}$$

This concentration product, $2.0 \cdot 10^{-5}$, is smaller than K_{sp}. Hence, if solid $CaSO_4$ were present, more would dissolve, raising the concentrations until their product equals $2.5 \cdot 10^{-5}$. Of course there is no solid $CaSO_4$ present, and the calculation shows that we do not have a saturated solution. *No solid $CaSO_4$ will form.*

Contrast this situation with that for barium sulfate. The initial concentration product, just after mixing the solution is

$$[Ba^{+2}][SO_4^{-2}] = [0.0040][0.0040] = 1.6 \cdot 10^{-5} \tag{4-50}$$

The number $1.6 \cdot 10^{-5}$ exceeds the solubility product of $BaSO_4$, $1.1 \cdot 10^{-10}$, by a very large factor! Plainly, precipitation must occur to lower the two concentrations $[Ba^{+2}]$ and $[SO_4^{-2}]$ until their product is lowered to $1.1 \cdot 10^{-10}$. In our example, they both start out at $0.0040 \, M$ and, as solid $BaSO_4$ is formed, they are reduced at the same rate. When equilibrium is finally reached, the concentrations will still be equal and their product will equal K_{sp}:

$$[Ba^{+2}][SO_4^{-2}] = [Ba^{+2}][Ba^{+2}] = K_{sp}$$

$$[Ba^{+2}] = \sqrt{K_{sp}} = \sqrt{1.1 \cdot 10^{-10}}$$

$$[Ba^{+2}] = [SO_4^{-2}] = 1.05 \cdot 10^{-5} \qquad (4\text{-}51)$$

Our conclusion confirms the qualitative prediction of Exercise 2-7: $BaSO_4$ will precipitate and $CaSO_4$ will not. Now we can add, however, a quantitative statement about the process. After mixing the solutions and after equilibrium has been reached, the composition of the solution will be changed from the initial values, (4-46), to the following.

$$[SO_4^{-2}] = 0.000010 \, M$$

$$[Na^{+2}] = 0.008 \, M$$

$$[Ca^{+2}] = 0.0050 \, M \qquad (4\text{-}52)$$

$$[Ba^{+2}] = 0.000010 \, M$$

$$[Cl^-] = 0.0180 \, M$$

Furthermore, we see that 99.75% of the Ba^{+2} and SO_4^{-2} was precipitated. This shows quantitatively how completely we can separate Ca^{+2} and Ba^{+2} by this use of solubility equilibria.

(c) CONTROL OF SOLUBILITY WITH A COMMON ION

For some purposes, the separation brought about in the last section may not be sufficiently complete. Perhaps we need a solution containing calcium ion that is more completely free of barium ion than achieved above. Perhaps we are trying to determine quantitatively how much barium was present in the original solution to higher accuracy than 99.75%. In either case, we can achieve our goal simply by properly adjusting the sulfate ion concentration.

Consider the change that would occur if, in Exercise 2-7, the concentration of sodium sulfate were doubled. The concentrations (4-46) obtained immediately after mixing the solutions would be the same except for the SO_4^{-2} and Na^+ ion concentrations, each of which would be twice as big.

$$[SO_4^{-2}] = 0.0080 \; M$$

$$[Na^{+2}] = 0.016 \; M$$

$$[Ca^{+2}] = 0.0050 \; M \qquad\qquad (4\text{-}53)$$

$$[Ba^{+2}] = 0.0040 \; M$$

$$[Cl^-] = 0.0180 \; M$$

Precipitation of $BaSO_4$ would, of course, occur again, but as solid forms, the barium ion and sulfate ion concentrations would not be equal to each other. In fact, if all of the barium ion were to precipitate, reducing $[Ba^{+2}]$ to zero, there would remain 0.0040 M excess sulfate ion concentration left over. From the initial conditions (4-53) until equilibrium is reached, the sulfate ion concentration will always exceed the barium ion concentration by this amount, 0.0040 M:

$$[SO_4^{-2}] = [Ba^{+2}] + 0.0040 \qquad\qquad (4\text{-}54)$$

Since (4-54) is always true, it is true at equilibrium, so our solubility product condition at equilibrium becomes

$$K_{sp} = [Ba^{+2}][SO_4^{-2}] = [Ba^{+2}]([Ba^{+2}] + 0.0040) \qquad\qquad (4\text{-}55)$$

Expanding (4-55), we obtain the equation to be solved:

$$[Ba^{+2}]^2 + 0.0040[Ba^{+2}] - K_{sp} = 0 \qquad\qquad (4\text{-}56)$$

This is the familiar quadratic equation whose solution is readily calculated with a bit of algebra:

$$ax^2 + bx + c = 0 \qquad\qquad (4\text{-}57)$$

$$x = \frac{-b \pm \sqrt{b^2 - 4ac}}{2a} \qquad\qquad (4\text{-}58)$$

A mathematical purist can turn the crank on (4-58) with $x = [Ba^{+2}]$, $a = 1$, $b = 0.0040$, and $c = K_{sp} = 1.1 \cdot 10^{-10}$ to derive the answer, $[Ba^{+2}] = 2.75 \cdot 10^{-8} \; M$. The process is a bit laborious, however, and often unnecessarily so. The result of the last section showed that the barium ion concentration is extremely small. Perhaps we can get away with an approximate solution to (4-55) based upon the guess that $[Ba^{+2}]$ is negligibly small compared to 0.0040 M. Perhaps

$$[Ba^{+2}] + 0.0040 \cong 0.0040 \tag{4-59}$$

If (4-59) is a useful approximation, (4-55) becomes

$$K_{sp} = [Ba^{+2}]([Ba^{+2}] + 0.0040) \cong [Ba^{+2}](0.0040) \tag{4-60}$$

The new expression (4-60) can be solved in a trice.

$$[Ba^{+2}] \cong \frac{K_{sp}}{0.0040} = \frac{1.1 \cdot 10^{-10}}{0.0040} = 2.75 \cdot 10^{-8} \, M \tag{4-61}$$

We see immediately that (4-61) is the same result obtained by solving the quadratic equation directly. Approximation (4-59) was obviously a useful approximation, as is readily verified by comparing the magnitude of $[Ba^{+2}]$, now that we know it, to 0.0040 M.

$$[Ba^{+2}] + 0.0040 = 2.75 \cdot 10^{-8} + 0.0040 = 0.0040000275 \cong 0.0040$$

Now we can look back to the effect on the precipitation of barium as we doubled the Na_2SO_4 concentration from 0.0040 M to 0.0080 M. The barium ion concentration at equilibrium dropped from $1.0 \cdot 10^{-5} \, M$ to $2.75 \cdot 10^{-8}$, a factor of 360. This enormous effect gives us powerful control over equilibrium conditions.

There are two interesting asides. First, notice that the chemical reaction, precipitation of barium sulfate, was initiated by mixing a solution of barium chloride with a solution of sodium sulfate. We might write the equation for the reaction as follows:

$$Ba^{+2}(aq) + 2\,Cl^-(aq) + 2\,Na^+(aq) + SO_4^{-2}(aq)$$
$$= BaSO_4(s) + 2\,Cl^-(aq) + 2\,Na^+(aq) \tag{4-62}$$

This is, however, a strange looking equation for a reaction because some of the reactants appear also as products, $Cl^-(aq)$ and $Na^+(aq)$. This shows, of course, that they do not enter the reaction. Neither are their concentrations changed as the reaction proceeds: Neither Cl^- nor Na^+ is consumed or produced. We see that these constituents, though necessarily present, contribute nothing to the chemical change we are examining. Consequently, chemists refer to such species as "spectator ions" and omit them from the equation for the reaction. *The equation for a reaction that lists only those species that participate in the chemical change is called the* **net reaction.** The net reaction for (4-62) is

$$Ba^{+2}(aq) + SO_4^{-2}(aq) = BaSO_4(s) \tag{4-63}$$

The second aside has to do with conditions at equilibrium. Notice that the final equilibrium concentrations reached after mixing the $BaCl_2$ and Na_2SO_4 solutions are identical to those reached by adding solid $BaSO_4$ to a 0.0040 M Na_2SO_4 solution. In the latter case, as solid dissolves, the Ba^{+2} ion concentration rises to its equilibrium value, S (for solubility). At the same time, the SO_4^{-2} ion concentration rises by the same amount, adding to the SO_4^{-2} ions already present from the Na_2SO_4 solution. Sodium sulfate is said to be a substance with a "common ion," i.e., it furnishes an additional source of one of the ions formed by the dissolving substance.

$$[Ba^{+2}] = S$$

$$[SO_4^{-2}] = S + 0.0040 \tag{4-64}$$

Substituting (4-64) into the equilibrium expression, we obtain

$$K_{sp} = [Ba^{+2}][SO_4^{-2}] = (S)(S + 0.0040) \tag{4-65}$$

The expression (4-65) is exactly the same algebraic problem as (4-55), so we already know the answer, $S = [Ba^{+2}] = 2.75 \cdot 10^{-8} \, M$.

This result is always true: The conditions at equilibrium are the same whether they are reached by precipitation or by dissolving. In fact, the easiest way to solve most precipitation problems is to *assume first that solid precipitates until the constituent in shortest supply is completely consumed and then that some of the solid redissolves to reach equilibrium* in a common ion solution. The strong dependence of solubility upon the con-

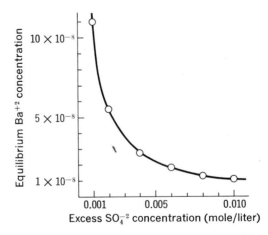

Figure 4-8 *The common ion effect. The solubility of $BaSO_4(s)$ is plotted against the concentration of sulfate ions added as the soluble salt, Na_2SO_4. On the vertical scale, measured in units of 10^{-8} mole/liter, the solubility of $BaSO_4$ in pure water would be 1050.*

centration of excess common ion is regularly used by chemists to control equilibrium conditions.

Exercise 4-8. What will be the sulfate ion concentration at equilibrium if solid $BaSO_4$ is added to a 0.040 M solution of $Ba(NO_3)_2$?

Of course, more complicated salts behave in the same way. Consider the solubility of calcium fluoride, $CaF_2(s)$, in 0.010 M sodium fluoride, NaF. The reaction is:

$$CaF_2(s) = Ca^{+2}(aq) + 2 F^-(aq) \qquad (4\text{-}66)$$

If we call the solubility S, then, since the only source of $[Ca^{+2}]$ is the reaction (4-66),

$$[Ca^{+2}] = S$$

The fluoride ion, however, comes from two sources, an amount 0.010 M from the common ion solution of NaF and $2S$ from reaction (4-66)

$$[F^-] = 2S + 0.010$$

The solubility product expression is, then

$$K_{sp} = [Ca^{+2}][F^-]^2 = (S)(2S + 0.010)^2 \qquad (4\text{-}67)$$

The equation (4-67) expands into a cubic, but before grappling with it, we should explore possible useful approximations. Can we get away with the assumption that $2S$ is negligible compared to 0.010?

Perhaps $\qquad [2S + 0.010] \cong 0.010 \qquad\qquad\qquad (4\text{-}68)$

so that

$$K_{sp} = (S)(2S + 0.010)^2 \cong (S)(1.0 \cdot 10^{-4})$$

$$S = \frac{K_{sp}}{1.0 \cdot 10^{-4}} = \frac{3.98 \cdot 10^{-11}}{1.0 \cdot 10^{-4}} = 3.98 \cdot 10^{-7} M \qquad (4\text{-}69)$$

Before accepting the conclusion that $[Ca^{+2}] = S = 3.98 \cdot 10^{-7}$, we must go back to approximation (4-68) to check its validity. Is it true that $[2S + 0.010]$ $= [2 \cdot 3.98 \cdot 10^{-7} + 0.010] \cong 0.010$? Plainly, the answer is "Yes"—$2 \cdot 3.98 \cdot 10^{-7}$ is indeed negligible compared to 0.010.

(d) APPROXIMATION TECHNIQUES

Sometimes such a simple approximation as (4-68) doesn't work. The clue is that the attempt to verify the approximation doesn't substantiate it. Then the quadratic formula (4-58) can be applied. Of course, the problem may be a cubic equation, or even a quartic—then analytical solutions can be quite difficult. Fortunately it is still possible to use approximations, but this time in several steps. The approach is called the "method of successive approximations." In this method, the first answer obtained, the one used to show that the first approximation was not adequate, is substituted back into the original equation and the calculation repeated. This gives a second approximation that is usually better than the first. It can be, in turn, substituted back for a third go-round. This process is repeated until two successive approximations give identical results within the desired accuracy.

Consider the solubility of silver sulfate, $Ag_2SO_4(s)$, in 0.050 M silver nitrate solution, $AgNO_3$.

$$Ag_2SO_4(s) = 2\,Ag^+(aq) + SO_4^{-2}(aq) \tag{4-70}$$

$$K_{sp} = 1.58 \cdot 10^{-5}$$

If we represent the sulfate ion concentration as S, the equilibrium concentrations will be

$$[SO_4^{-2}] = S$$

$$[Ag^+] = 2S + 0.050$$

$$K_{sp} = [Ag^+]^2[SO_4^{-2}] = (2S + 0.050)^2(S) \tag{4-71}$$

To simplify solution of the cubic expression (4-71), we guess that perhaps we can neglect $2S$ compared to 0.050:

$$\text{Perhaps}\quad 2S + 0.050 \cong 0.050 \tag{4-72}$$

Now (4-71) is easily solved:

$$K_{sp} = 1.58 \cdot 10^{-5} \cong (0.050)^2(S)$$

$$S = \frac{15.8 \cdot 10^{-6}}{25 \cdot 10^{-4}} = 0.0063\ M \tag{4-73}$$

The result (4-73) must be checked against the approximation (4-72):

$$2S + 0.050 = 2 \cdot 0.0063 + 0.050 = 0.062 \neq 0.050$$

Unfortunately, the result, 0.062, is 24% larger than the value we assumed, 0.050. We must conclude that the result (4-73) is not acceptable.

This is the point at which we begin a second approximation. Instead of assuming that $2S + 0.050 \cong 0.050$, let's assume the first rough calculation of S given by (4-73) provides a better estimate:

Perhaps $\quad 2S + 0.050 \cong 2(0.0063) + 0.050 = 0.062 \qquad (4\text{-}74)$

If so, then

$$K_{sp} = 1.58 \cdot 10^{-5} \cong (0.062)^2(S)$$

$$S = \frac{15.8 \cdot 10^{-6}}{38 \cdot 10^{-4}} = 0.0041 \qquad (4\text{-}75)$$

Again we must check our assumption (4-74):

$$2S + 0.050 = 2(0.0041) + 0.050 = 0.058$$

This second calculation, based on the assumption that $2S + 0.050 \cong 0.062$, gives a result with $2S + 0.050 = 0.058$. Now the result is too low, but only by 7%. Things are getting better. ~~fuck you~~
We might make one more approximation step:

Perhaps $\quad 2S + 0.050 \cong 2(0.0041) + 0.050 = 0.058 \qquad (4\text{-}76)$

If so, then

$$S = \frac{15.8 \cdot 10^{-6}}{(0.058)^2} = \frac{15.8 \cdot 10^{-6}}{33.6 \cdot 10^{-4}} = 0.0047 \qquad (4\text{-}77)$$

Checking our approximation,

$$2S + 0.050 = 2(0.0047) + 0.050 = 0.059 \cong 0.058$$

Now approximation (4-76) is corroborated quite well—within 2%. We conclude that we have calculated S to a high enough accuracy. The solubility of Ag_2SO_4 in 0.050 M $AgNO_3$ gives an equilibrium sulfate ion concentration of 0.0047 M.

Exercise 4-9. Use the calculated solubility (4-77) to make one more approximation to see how much further S will change.

(e) NON-AQUEOUS SOLVENTS

Many salts have high solubility in water and almost all salts are dissociated completely to ions as they dissolve. Relatively few liquids have this ionizing

Table 4-5 Physical Properties of H₂O, NH₃, and SO₂: Substances That Act as Ionizing Solvents

	H₂O	NH₃	SO₂
melting point (°K)	273	195	198
boiling point (°K)	373	240	263
heat of fusion (kcal/mole)	1.44	1.35	1.77
heat of vaporization (kcal/mole)	10.5	5.6	6.0
density of liquid (g/cc)	1.00	0.677	1.46
molar volume liquid (cc/mole)	18.0	25.2	43.6
vapor pressure at 273°K (atm)	0.0060	4.24	1.53
dieletric constant	80.1	26.7	15.6
dipole moment (Debye)	1.82	1.47	1.61

solvent action. On the other hand, there are legions of organic liquids—substances containing carbon—in which salts have extremely low solubility and, insofar as a salt does dissolve, in which no ions are formed. This includes such familiar liquids as carbon tetrachloride, CCl_4, hexane, C_6H_{14}, and benzene, C_6H_6. In turn, water is a very poor solvent for these organic liquids.

Two of the liquids that do act as ionizing solvents for salts are ammonia, NH_3, and sulfur dioxide, SO_2.* Ammonia, like water, causes rather complete dissociation into ions. In sulfur dioxide, salts are generally only partially dissociated; some neutral molecules dissolve. This difference from water is consistent with the fact that SO_2 is a good solvent for many organic liquids that do not dissolve appreciably in water.

Table 4-5 compares some of the physical properties of liquid water, ammonia, and sulfur dioxide. Both ammonia and sulfur dioxide boil at temperatures below the freezing point of water, so they must be either refrigerated or used under moderate pressures. Aside from this, the three substances are rather similar. In particular, notice that the molecular dipole moments are all about the same and rather large. In Section 4-2(d), the ionizing solvent action of water was attributed to orientation of its electrical dipoles around ions. Presumably this same model might be applied to the NH_3 and SO_2 solvent behavior.

Quantitatively, salt solubilities display some interesting contrasts. Table 4-6 lists the solubilities of some common salts. Both NaCl and KCl have rather small solubility in NH_3 and SO_2. Both NaBr and KBr have higher solubility in NH_3, nearer the water solubilities, but SO_2 again dissolves rather little of either salt. The iodides finally make the three solvents look comparable.

Turning to the silver salts, we see even more distinctive differences. Both

---

* Other familiar liquids that show some ionizing solvent properties are pure H_2SO_4, pure HF, pure acetic acid, pure formic acid, and dinitrogen tetroxide. The word "pure" is added merely to emphasize the complete absence of water.

ammonia and sulfur dioxide are better solvents for AgCl, AgBr, and AgI than is water. Furthermore, the trend in the halogen series is reversed. Notice that AgI has 1400 times *lower* solubility in water than AgCl whereas AgI has 150 times *higher* solubility in ammonia than AgCl.

Finally, observe that none of the five salts Na_2SO_4, $MgCl_2$, $MgBr_2$, $CaBr_2$, or $BaBr_2$ has high solubility in either liquid NH_3 or SO_2. Each of these salts possesses a doubly charged ion; apparently neither solvent accommodates highly charged ions as well as does water. The nitrates of calcium and barium contradict this generalization, but their solubilities seem to stand out as an exception.

Having glanced at the solubilities, it is surely not clear that the differences can be simply explained by referring to the molecular dipole moments. Though the dipole moment of SO_2 is between that of H_2O and NH_3, the solubilities of salts in SO_2 are often not intermediate. Furthermore, sulfur dioxide dissociates salts less completely than does ammonia. The orientation of simple charge dipoles, as pictured in Figure 4-6 for water, provides at best only a qualitative starting point in understanding the ion–solvent interactions.

Table 4-6 *Solubilities (moles/liter) of Salts in Water, Ammonia, and Sulfur Dioxide at 298°K Unless Noted*

Salt	Water	Ammonia	Sulfur Dioxide*
NaCl	6.1	0.35	"insoluble"
NaBr	8.8	9.1	0.0020
NaI	12.0	7.3	1.5
KCl	4.7	0.0036	0.0080
KBr	5.5	0.77	0.058
KI	8.7	7.4	3.6
AgCl	$1.3 \cdot 10^{-5}$	0.039	<0.0010
AgBr	$0.072 \cdot 10^{-5}$	0.213	0.0023
AgI	$0.00092 \cdot 10^{-5}$	6.0	0.0010
Na_2SO_4	1.13	0.0	"insoluble"
$MgCl_2$	5.7	—	0.00215
$MgBr_2$	2.7	0.00026*	0.0020
$CaBr_2$	6.3*	0.00052*	—
$BaBr_2$	3.5	0.00039	—
$NaNO_3$	8.6*	7.8	—
KNO_3	3.1	0.70	—
$Ca(NO_3)_2$	21	3.3	—
$Ba(NO_3)_2$	0.33	2.5	—

*Solubility at 273°K.

Exercise 4-10. Write the net reaction that will occur if equal volumes of 0.02 M AgCl in ammonia and 0.02 M KBr in ammonia are mixed.

Exercise 4-11. Show that the solubility product for KCl in ammonia is $1.3 \cdot 10^{-5}$.

4-4   Difficulties in solubility product calculations

Sometimes solubility product calculations don't match up very well with experiment. For example, the solubility product of AgCl, $K_{sp} = 1.77 \cdot 10^{-10}$, quite accurately reproduces the experimental solubility of AgCl in pure water, $1.33 \cdot 10^{-5} \, M$. In contrast, the solubility in 0.20 M NaCl is predicted to be $8.9 \cdot 10^{-10}$, but experiment shows that $5.3 \cdot 10^{-6}$ mole of AgCl dissolves in a liter of that common ion solution. This is 6000 times higher than calculated! Even more perplexing is the observation that the solubility of AgCl in 0.01 M NaNO$_3$—a solution containing *no* common ions—is $1.43 \cdot 10^{-5} \, M$, eight percent higher than in pure water.

The situation is reminiscent of our discussion of gas behavior in Chapter One. In that case, two proposals were explored to account for the observed deviations from perfect gas behavior. The very large deviations displayed by NO$_2$ (see Table 1-4) were interpreted to indicate that the gas contained other molecular species. The much smaller deviations of gaseous NH$_3$ (see Table 1-3) were attributed to interactions between the molecules that are not taken into account in the perfect gas model but which are not negligible with 0.01% accuracy. Surprisingly, these same two proposals are useful in explaining the solubility discrepancies described above.

(a) COMPLEX FORMATION

Consider first the solubility of AgCl in common ion solutions of NaCl. The dashed curve in Figure 4-9 shows the solubility derived from the simple solubility product calculation

$$K_{sp} = [Ag^+][Cl^-] \tag{4-78}$$

The solid curve shows the experimental solubilities measured in the laboratory. As the chloride ion concentration is raised from that observed in pure water, at first the solubility drops just as predicted from K_{sp}. At about $3 \cdot 10^{-4} \, M$ NaCl, however, the experimental solubility noticeably exceeds that expected from K_{sp}. If the NaCl concentration rises above about $3 \cdot 10^{-3}$, then the solubility actually begins to rise! Now the effect of adding more NaCl is to draw more and more AgCl into solution.

Perhaps our experience with gaseous NO$_2$ furnishes a guide. In Section 1-4(d), we saw that the rapid drop in the PV product with rising pressure could be explained by assuming an equilibrium involving a more complex

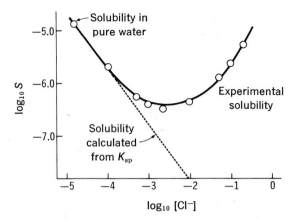

Figure 4-9 *Comparison of the measured and calculated solubility S of AgCl in NaCl solutions with chloride concentrations* [Cl$^-$].

molecule, N_2O_4. Could there be some ions in AgCl solution that are more complex than Ag$^+$(aq) and Cl$^-$(aq)? Chemists find that it is useful to postulate that there are. A saturated solution of AgCl(s) in 1.0 M NaCl is pictured to include a variety of "complex ions," $AgCl_2^-$, $AgCl_3^{-2}$, and $AgCl_4^{-3}$ and even some undissociated AgCl(aq), as well. If so, then the solubility behavior can be calculated only if these equilibria are considered (just as the PV product of NO_2 can be calculated only if the NO_2–N_2O_4 equilibrium is considered). So we must write down an array of chemical reactions, all in equilibrium at once and each having its own equilibrium constant.

$$AgCl(s) = Ag^+(aq) + Cl^-(aq) \qquad K_{sp} = K_0 = [Ag^+][Cl^-]$$

$$AgCl(s) = AgCl(aq) \qquad K_1 = [AgCl]$$

$$Cl^-(aq) + AgCl(s) = AgCl_2^-(aq) \qquad K_2 = \frac{[AgCl_2^-]}{[Cl^-]}$$

$$2\,Cl^-(aq) + AgCl(s) = AgCl_3^{-2}(aq) \qquad K_3 = \frac{[AgCl_3^{-2}]}{[Cl^-]^2}$$

$$3\,Cl^-(aq) + AgCl(s) = AgCl_4^{-3}(aq) \qquad K_4 = \frac{[AgCl_4^{-3}]}{[Cl^-]^3}$$

$$(4\text{-}79)$$

The numerical values of the constants K_0 to K_4 are selected as needed to fit as closely as possible the experimental curve shown in Figure 4-9. Table 4-7 shows the values so obtained, along with the similar constants for several other salts that show complex ion formation.

Table 4-7 *Equilibrium Constants for Formation of Complexes from Solid Salts**

$$(n - 1)X^- + MX(s) = MX_n^{-(n-1)} \qquad K_n = \frac{[MX_n^{-(n-1)}]}{[X^-]^{n-1}}$$

Salt (MX)	$K_{sp} = K_0$	K_1	K_2	K_3	K_4
			Equilibrium Constant		
AgCl	1.77×10^{-10}	2.00×10^{-7}	2.00×10^{-5}	2.00×10^{-5}	3.47×10^{-5}
AgBr	7.94×10^{-13}	1.10×10^{-8}	1.00×10^{-5}	7.08×10^{-5}	6.03×10^{-4}
AgI	4.47×10^{-17}	6.03×10^{-9}	3.98×10^{-6}	2.51×10^{-3}	1.10×10^{-2}
AgCN	1.20×10^{-16}	$(10^{-7})\dagger$	2.40×10^{-5}	4.79×10^{-6}	6.46×10^{-5}
CuCl	1.86×10^{-7}	$(10^{-5})\dagger$	7.59×10^{-2}	3.39×10^{-2}	
CuCN	3.24×10^{-20}	$(10^{-13})\dagger$	5.89×10^{-5}	4.37×10^{-1}	8.71×10^{-3}
TlCl	9.12×10^{-4}	7.08×10^{-4}	1.82×10^{-4}	2.00×10^{-5}	
TlBr	1.55×10^{-5}	3.31×10^{-5}	2.40×10^{-5}	7.94×10^{-6}	1.58×10^{-6}

* Data from J. N. Butler, *Ionic Equilibrium*, Addison-Wesley, 1964.
† Values in parentheses estimated.

The equilibrium expressions (4-79) are readily solved for the concentrations of the succeeding silver complexes:

$$[Ag^+] = K_0/[Cl^-] = K_{sp}/[Cl^-]$$
$$[AgCl(aq)] = K_1$$
$$[AgCl_2^-] = K_2[Cl^-] \qquad\qquad (4\text{-}80)$$
$$[AgCl_3^{-2}] = K_3[Cl^-]^2$$
$$[AgCl_4^{-3}] = K_4[Cl^-]^3$$

The sum of all of the silver-containing species, listed in (4-80), is the solubility.

$$\text{solubility} = \text{total } Ag^+ \text{ in solution}$$
$$= [Ag^+] + [AgCl(aq)] + [AgCl_2^-] + [AgCl_3^{-2}] + [AgCl_4^{-3}] \quad (4\text{-}81)$$

Substituting (4-80) into (4-81),

$$S = \frac{K_{sp}}{[Cl^-]} + K_1 + K_2[Cl^-] + K_3[Cl^-]^2 + K_4[Cl^-]^3 \qquad (4\text{-}82)$$

The total chloride in solution is only a bit more complicated:

$$\begin{aligned}
\frac{total\ Cl^-}{in\ solution} &= [Cl^-] + [AgCl(aq)] + 2[AgCl_2^-] + 3[AgCl_3^{-2}] \\
&\qquad\qquad + 4[AgCl_4^{-3}] \quad (4\text{-}83) \\
&= [Cl^-] + K_1 + 2K_2[Cl^-] + 3K_3[Cl^-]^2 + 4K_4[Cl^-]^3 \qquad (4\text{-}84)
\end{aligned}$$

If the NaCl concentration is quite small, all of the terms after the first one in (4-82) will be negligible.

Let's use the numerical values for the K_n's given in Table 4-7 to calculate S first in a solution with $[Cl^-] = 3 \cdot 10^{-4}\ M$ and then in a solution with $[Cl^-] = 3 \cdot 10^{-2}\ M$.

If

$$[Cl^-] = 3 \cdot 10^{-4}$$

$$\begin{aligned}
S &= \frac{1.77 \cdot 10^{-10}}{3 \cdot 10^{-4}} + 2.00 \cdot 10^{-7} + (2.00 \cdot 10^{-5})(3 \cdot 10^{-4}) \\
&\qquad + (2.00 \cdot 10^{-5})(3 \cdot 10^{-4})^2 + (3.47 \cdot 10^{-5})(3 \cdot 10^{-4})^3 \\
&= 5.90 \cdot 10^{-7} + 2.00 \cdot 10^{-7} + 0.06 \cdot 10^{-7} + 1.8 \cdot 10^{-12} + 9.3 \cdot 10^{-16}
\end{aligned}$$

$$S = 8.0 \cdot 10^{-7} \qquad\qquad\qquad\qquad (4\text{-}85)$$

If

$$[Cl^-] = 3 \cdot 10^{-2}$$

$$\begin{aligned}
S &= \frac{1.77 \cdot 10^{-10}}{3 \cdot 10^{-2}} + 2.00 \cdot 10^{-7} + (2.00 \cdot 10^{-5})(3 \cdot 10^{-2}) \\
&\qquad + (2.00 \cdot 10^{-5})(3 \cdot 10^{-2})^2 + (3.47 \cdot 10^{-5})(3 \cdot 10^{-2})^3 \\
&= 0.0590 \cdot 10^{-7} + 2.00 \cdot 10^{-7} + 6.00 \cdot 10^{-7} + 0.18 \cdot 10^{-7} \\
&\qquad\qquad\qquad\qquad\qquad\qquad\qquad + 0.0093 \cdot 10^{-7}
\end{aligned}$$

$$S = 8.2 \cdot 10^{-7} \qquad\qquad\qquad\qquad (4\text{-}86)$$

The results (4-85) and (4-86) show about the same solubility, in agreement with the experimental curve of Figure 4-9 (they'd better agree! The curve was used to determine the K's). However, inspection of the contributions to the solubilities shows great differences. In the $[Cl^-] = 3 \cdot 10^{-4}\ M$ solution, the $Ag^+(aq)$ ion concentration is $5.9 \cdot 10^{-7}\ M$, which accounts for 74% of the dissolved silver ion species. At $[Cl^-] = 3 \cdot 10^{-2}$, however, $[Ag^+] = 0.059 \cdot 10^{-7}$, now only 0.7% of the total. Now, the bulk of the dissolved silver ion is present either as neutral AgCl molecules (24%) or as the complex ion $AgCl_2^-$ (73%).

Exercise 4-12. Use expression (4-84) to show that there is negligible error involved in the approximation that the value of $[Cl^-]$ in a $3 \cdot 10^{-4}$ M NaCl solution saturated with AgCl is still equal to $3 \cdot 10^{-4}$ M even after all of the chloride-containing complex ions are formed.

(b) INTERACTIONS AMONG IONS IN SOLUTION

Now let's turn to the solubility discrepancies that are observed in sodium nitrate solutions. The facts are presented in Figure 4-10. The dashed line shows what K_{sp} has to say about the solubility (the more complicated calculation using all of the terms in (4-82) gives the same answer). Experiment shows, however, an easily noticeable increase in solubility at sodium nitrate concentrations of $1 \cdot 10^{-2}$ M. Even so, the discrepancies shown in Figure 4-10 are much smaller than those displayed in Figure 4-9 (notice the logarithmic scale in Fig. 4-9). Again we are reminded of the gas imperfection explanations—the large perfect gas discrepancies of NO_2 were explained in terms of an equilibrium, the smaller discrepancies of NH_3 were given a different explanation. The situation is similar here. The effect of sodium nitrate is not explained in terms of nitrate complex ions. Instead, interactions between ions in solution are considered to be the cause.

Harking back to our derivation of the Equilibrium Law in Chapter Three, we made the assumption that the only concentration factors of importance were those of the ions that needed to find lattice positions to make the crystal grow. Those ions were pictured as moving about freely in a water medium, uninfluenced by each other or by idle "spectator ions" that do not fit in the crystal lattice. However, electrically charged species exert significant force on each other and it may well be that ions do not move

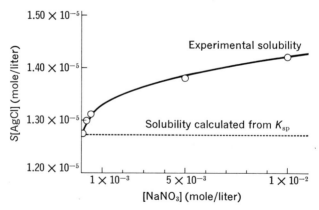

Figure 4-10   Comparison of the measured and calculated solubilites, S, of AgCl in NaNO₃ solutions.

freely and independently in solution, particularly as they get closer together in a concentrated solution. In fact, we might view the addition of more ions to be in effect a change of the solvent. The more charges there are in the solution, the more favorable it is as an environment for the ions of the salt that is dissolving. With this in mind, we could say that the solubility of a salt would probably be influenced somewhat by the concentration, C, and the charge, Z, of every ionic species present in the solution.

A quantitative model that attempts to account for coulombic (electrostatic) interactions between ions is based upon the quantity called *ionic strength*, I, defined as follows

$$I = \tfrac{1}{2}\Sigma C_i Z_i^2 \qquad\qquad (4\text{-}87)$$

in which

   Σ means "the sum of"

   C_i = concentration of the ith type of ion

   Z_i = charge on the ith type of ion

This model, called the Debye–Hückel theory of ionic solutions, would attribute to our AgCl–NaNO$_3$ solution an ionic strength equal to:

$$I = \tfrac{1}{2}([Ag^+] + [Cl^-] + [Na^+] + [NO_3^-])$$

By an elaborate consideration of interionic attractions, Debye and Hückel concluded that the logarithm of the salt solubility should depend upon the square root of ionic strength. Figure 4-11 shows the data of Figure 4-10

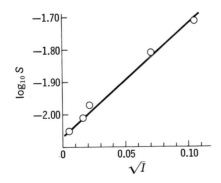

Figure 4-11   *Debye–Hückel relation between ionic strength, I, and solubility of AgCl(s). The data of Figure 4-10 have been replotted.*

replotted according to this expectation: The logarithm of solubility is plotted against $\sqrt{I}$. The reasonable agreement between the experimental points and a straight line shows the value of the model.

Now we are faced with the dilemma of an equilibrium constant that won't stay constant. To the extent that the solubility changes with ionic strength, we are robbed of the ability to calculate solubility from K_{sp}. The escape that is commonly used is to define an "effective concentration" or *activity*, a, such that a constant value of K can be used. We define a so that at any ionic strength,

$$K_{sp} = (a_{Ag^+})(a_{Cl^-}) \tag{4-88}$$

The activity is related to the molar concentration by a factor f called the *activity coefficient*.

$$a_{Ag^+} = f_+[Ag^+]$$
$$a_{Cl^-} = f_-[Cl^-] \tag{4-89}$$

and

$$K_{sp} = (f_+)(f_-) \cdot [Ag^+][Cl^-] \tag{4-90}$$

The trick is to find f_+ and f_-.

The Debye–Hückel theory comes to the rescue. It proposes that the *mean activity coefficient*, $f_{\pm} = \sqrt{f_+ f_-}$ can be calculated from I:

$$\log_{10} f_{\pm} = 0.509 Z_+ Z_- \sqrt{I} \tag{4-91}$$

in which Z_+ and Z_- are the charges on the positive and negative ions, respectively. A minus sign always results when the values of Z_+ and Z_- are substituted, since one is always a plus charge and the other is always negative. Thus, the Debye–Hückel theory tells us that the mean activity coefficient depends only on the charges on the ions and the ionic strength, and not on the particular ions involved. This is only approximately true, so more elaborate treatments have considered, as well as charge, the effect that ionic size might have.

Empirically we can measure solubilities and tabulate the implied values of the activity coefficients. We find that in quite dilute solutions, say, less than $10^{-3} M$, activity and molar concentration are nearly identical. At ion concentrations exceeding 0.01 M, equations (4-90) and (4-91) permit more accurate calculation of solubility than can be obtained neglecting ionic interactions. In anticipation that this will be done, when high accuracy is desired, chemists usually tabulate equilibrium constants after extrapolating their measurements to zero ionic strength. For that special situation,

(4-91) and (4-90) have the following values:

At $I = 0$

$$\log_{10} f_{\pm} = 0.509 \cdot (Z_+)(Z_-) \sqrt{0} = 0$$
$$f_{\pm} = 10^0 = 1 \tag{4-92}$$
$$K_{sp} = f_+ f_- [Ag^+] [Cl^-] = f_{\pm}^2 [Ag^+][Cl^-]$$
$$K_{sp} = [Ag^+][Cl^-]$$

We can see the magnitude of the effect at $I \neq 0$ by calculating the solubility of AgCl in 0.01 M NaNO$_3$ solution. If we call the solubility S, then

$$[Ag^+] = [Cl^-] = S$$

and

$$K_{sp} = (f_+)(f_-)S^2 = f_{\pm}^2 S^2$$

so that

$$S = \frac{\sqrt{K_{sp}}}{f_{\pm}} \tag{4-93}$$

To calculate $f_{\pm}$, we first need the ionic strength, I, defined by (4-87).

$$I = \tfrac{1}{2}([Ag^+] + [Cl^-] + [Na^+] + [Cl^-]) = \tfrac{1}{2}(S + S + 0.01 + 0.01)$$

Since the solubility is quite low, we can approximate,

$$I = \tfrac{1}{2}(0.01 + 0.01) = 0.01$$

Now $f_{\pm}$ can be estimated by (4-91):

$$\log_{10} f_{\pm} = (0.509)(+1)(-1)(\sqrt{0.01})$$
$$= -0.0509$$
$$f_{\pm} = 10^{-0.0509} = \frac{1}{1.12} = 0.89 \tag{4-94}$$

The mean activity coefficient (4-94) shows that S is increased by about 12% by 0.01 M NaNO$_3$:

$$S = \frac{\sqrt{K_{sp}}}{f_{\pm}} = \frac{\sqrt{1.77 \cdot 10^{-10}}}{0.89} = 1.49 \cdot 10^{-5} \ M$$

Exercise 4-13. Show that the Debye–Hückel theory predicts that the solubility of silver chloride in a solution containing 0.01 M NaNO$_3$ and 0.02 M KNO$_3$ is about 23% higher than in pure water.

Exercise 4-14. Show that the Debye–Hückel theory predicts that the solubility of BaSO$_4$ in a solution containing 0.01 M NaNO$_3$ is about 60% higher than in pure water.

(c) CONCLUSIONS

For many purposes, it is possible to ignore the effects of ionic strength, or to approximate them with the simple Debye–Hückel theory. We cannot be casual, however, when complexes form. An intelligent understanding of the chemistry of a solution depends crucially upon knowing the species that are present in the solution. The heterogeneous equilibria provided some simple and very useful examples with which to work. The principles we have derived will be applicable also to homogeneous equilibria, equilibrium reactions in which all constituents are in solution. There we shall see even more forcefully the lesson learned with AgCl solubility, the importance of recognizing all equilibria that are taking place in solution. So let's proceed to consider homogeneous dissociation equilibria in solution—in particular those of acids and bases.

Problems

Note: In all solubility problems, neglect both complex ion and ionic strength effects unless they are specifically mentioned.

1.   The vapor pressures of liquid water and of ice become equal at the freezing point (273.16°K) at 4.6 torr. What is the equilibrium constant, K_c (in moles/liter) for the vaporization of either liquid water or ice at 273.16°K?

2.   Butane, C$_4$H$_{10}$, is a good household fuel for rural use. Suppose a supply truck places two moles of liquid butane into an empty storage tank of 24.5 liters capacity at 298°K. What will be the pressure in the tank after filling? What will be the pressure in the tank the next morning if, during the night, snow falls to cover the tank? (The vapor pressure of liquid butane is 1830 torr at 298°K and 776 torr at 273°K.)

3.   Calcium chloride is a useful drying agent (desiccant) because an equilibrium between CaCl$_2$ and water vapor causes the formation of a new crystal that incorporates water molecules, calcium chloride dihydrate, CaCl$_2 \cdot 2$ H$_2$O.

$$\tfrac{1}{2}\,\text{CaCl}_2 \cdot 2\,\text{H}_2\text{O(s)} = \tfrac{1}{2}\,\text{CaCl}_2\text{(s)} + \text{H}_2\text{O(g)}$$

$$K = 8.0 \cdot 10^{-4} \text{ atm } (298°\text{K})$$

(a) Calculate the equilibrium vapor pressure in torr over a mixture of the two solids, CaCl$_2 \cdot 2$ H$_2$O and CaCl$_2$.

(b) Under a small bell jar at room temperature is placed beaker A containing one mole of liquid water, beaker B containing one mole of CaCl$_2$(s), and beaker C containing one mole of CaCl$_2 \cdot 2$ H$_2$O. What would be found under the bell jar several days later, when equilibrium has been attained?

4. Which of the pure liquids listed in Table 4-2 would boil if their containers were opened in a spacecraft in which the atmosphere is maintained at 10 torr of pure oxygen?

5. The diver mentioned in Exercise 4-4 quickly comes from a depth of 400 feet under water up to the surface. How many ml of nitrogen gas are released as bubbles in one liter of the diver's blood? (Assume $T = 298°K$.)

6. Water in equilibrium with solid thallium bromate, $TlBrO_3$, contains $2.0 \cdot 10^{-2}$ mole of bromate ion, BrO_3^-. What is the equilibrium constant (the solubility product, K_{sp}) for the solubility of $TlBrO_3$?

7. It takes 0.0200 gram of $BaF_2(s)$ to saturate 15.0 ml of water. What is the solubility product, K_{sp}?

8. The solubility product of lead sulfate is $1.6 \cdot 10^{-8}$. What are the concentrations of $Pb^{+2}(aq)$ and $SO_4^{-2}(aq)$ if one mole of $PbSO_4$ is added to one liter of water? $\quad x^2 = Ksp$

9. The solubility product of silver iodate, $AgIO_3$, is $3.0 \cdot 10^{-8}$. What are $[Ag^+]$ and $[IO_3^-]$ at equilibrium if excess $AgIO_3(s)$ is added to a volume of water?

$= 4x^3$

10. The solubility product of calcium hydroxide is $1.3 \cdot 10^{-6}$. What is the calcium $C(OH)_2 = Ca^{2+} + 2OH$ ion concentration in an aqueous solution saturated with $Ca(OH)_2(s)$?

$4x^3 = 1.3 \times 10^{-6}$

11. The solubility product of barium iodate, $Ba(IO_3)_2$, is $1.3 \cdot 10^{-9}$. What is $[Ba^{+2}]$ in $\sqrt[3]{} = .325 \times 10^{-9}$ an aqueous solution saturated with $Ba(IO_3)_2(s)$? $= 325 \times 10^{-12}$

12. To 1.0 liter of $0.0020 \, M \, CaSO_4$ solution is added 1.36 g (10.0 millimoles) of solid $CaSO_4$. At equilibrium, how much solid will remain undissolved? Express your $x = 688 \times 10^{-4}$ answer both in grams and in millimoles. Ba^{2+}

$SO_4^- = 2x$

13. Repeat Problem 9 if the excess $AgIO_3(s)$ is added to $0.10 \, M \, NaIO_3$ solution.

14. Repeat Problem 10 if excess $Ca(OH)_2(s)$ is added to $0.10 \, M \, NaOH$ solution. $1.3 \sqrt{6} = x \quad (2x+.1)^2 = 1.3 \times 10^{-7} = x$

15. Repeat Problem 9 if the excess $AgIO_3(s)$ is added to $0.00020 \, M \, NaIO_3$ solution. (Use successive approximations till your answer is reliable to $\pm 10\%$.)

16. Repeat Problem 10 if excess $Ca(OH)_2(s)$ is added to $0.020 \, M \, NaOH$ solution, (Use successive approximations till your answer is reliable to $\pm 10\%$.)

$CaSO_4$ .2 Moles

17. To 500 ml of $0.20 \, M \, CaCl_2$ is added 500 ml of $0.40 \, M \, Na_2SO_4$. A precipitate forms. Calculate the equilibrium ion concentrations in solution and the weight of the precipitate. $[Cl^-] = .2M \quad [Na] = .4$

18. The solubility of $PbCl_2(s)$ in pure water is 0.016 mole/liter. To 1.0 liter of $0.015 \, M \, AgNO_3$ is added 0.020 mole of solid $PbCl_2$. When equilibrium is reached, what will be the ion concentrations? What will be the amount (in moles) and composition of any solid that is left?

19. To the equilibrium mixture formed in Problem 17 is added 0.50 mole of solid Na_2CO_3. Calculate the new equilibrium ion concentrations and state the composition of the solid that is left.

20. The solubility product, K_{sp}, for barium sulfate is slightly larger than that of barium chromate: $BaSO_4$, $K_{sp} = 1.1 \cdot 10^{-10}$; $BaCrO_4$, $K_{sp} = 8.5 \cdot 10^{-11}$.
(a) Calculate the equilibrium concentrations of Ba^{+2} and SO_4^{-2} if excess solid $BaSO_4$ is added to pure water.
(b) Calculate the equilibrium concentrations of Ba^{+2} and CrO_4^{-2} if excess solid $BaCrO_4$ is added to pure water.

(c) Describe qualitatively what will happen to the equilibrium mixture obtained in (a) if excess solid $BaCrO_4$ is added to it.

(d) Calculate the new concentrations when equilibrium is reached in (c).

21. Use the molar volumes of ammonia and sulfur dioxide given in Table 4-5 to calculate the number of moles/liter in each of these pure liquids.

22. From the solubilities given in Table 4-6 and the results of Problem 21, calculate the number of solvent molecules per ion in saturated solutions of potassium iodide in water, in ammonia, and in sulfur dioxide. (Neglect the volume occupied by the ions themselves.)

23. Calculate the solubility of thallium bromide, TlBr, in $3.0 \cdot 10^{-2} \ M$ NaBr solution under the following conditions.

(a) Assume that the only important equilibrium is expressed by the solubility product, $K_{sp} = 1.55 \cdot 10^{-5}$.

(b) Assume that all of the equilibria expressed by the five equilibrium constants in Table 4-7 are important.

(c) Calculate the percentage error involved in (a).

24. Repeat Problem 23, considering the solubility of TlBr in $3.0 \cdot 10^{-1} \ M$ NaBr.

25. Excess solid silver bromide is added to 1.0 liter of water. Then 0.05 M NaBr solution is added slowly. How many ml are needed to make the concentration [Ag^+] just equal to [AgBr], the concentration of undissociated silver bromide.

26. In the text, the concentration of each species is calculated for the solubility of AgCl(s) in $3 \cdot 10^{-4}$ and $3 \cdot 10^{-2} \ M$ NaCl. Repeat these calculations for $1.0 \cdot 10^{-3} \ M$ NaCl and $1.0 \cdot 10^{-2} \ M$ NaCl. Plot the total concentration of dissolved solid and also the concentration of each of the species [Ag^+], [AgCl], [$AgCl_2$] and [$AgCl_2^{-2}$], using for the abscissa scale $\log_{10}$[Cl^-] and for ordinate, concentration $\cdot 10^{+7}$.

27. Considering the effect of ionic strength, estimate the solubility of silver chloride in a solution containing 0.020 M $Ba(NO_3)_2$. Compare this result to the solubility in pure water and to that calculated in Exercise 4-13. (Notice that the total concentration of ions in 0.020 M $Ba(NO_3)_2$ is the same as in Exercise 4-13.)

28. Calculate the mean activity coefficients in 0.10 M $NaNO_3$ solution; in 0.10 M $Ba(NO_3)_2$ solution; in 0.10 M $Al(NO_3)_3$ solution, (a) for AgCl, for which $Z_+ = +1$ and $Z_- = -1$; (b) for $PbCl_2$, for which $Z_+ = +2$ and $Z_- = -1$.

five dissociation equilibria: acids and bases

5-1 What is an acid, anyway?

There are hydrogen compounds with formulas analogous to the crystalline salts $NaCl$, $NaNO_3$ and Na_2SO_4. Two of these, HCl and HNO_3, are gases and the third, H_2SO_4, is a liquid, but they dissolve readily in water and, like the salts, give conducting solutions. Ions must be formed. An HCl solution, added to silver nitrate solution, precipitates $AgCl$—that is the characteristic chemistry of Cl^-. The H_2SO_4 solution, added to barium chloride solution, precipitates $BaSO_4$—the chemistry of SO_4^{-2}. It is natural to assume that these hydrogen-containing compounds are dissociated into ions in water, as are the salts.

$$NaCl(s) \longrightarrow Na^+(aq) + Cl^-(aq)$$
$$HCl(g) \longrightarrow H^+(aq) + Cl^-(aq)$$
(5-1)

$$NaNO_3(s) \longrightarrow Na^+(aq) + NO_3^-(aq)$$
$$HNO_3(g) \longrightarrow H^+(aq) + NO_3^-(aq)$$
(5-2)

$$Na_2SO_4(s) \longrightarrow 2\,Na^+(aq) + SO_4^{-2}(aq)$$
$$H_2SO_4(\ell) \longrightarrow 2\,H^+(aq) + SO_4^{-2}(aq)$$
(5-3)

According to reactions (5-1) to (5-3), the aqueous solutions of HCl, HNO_3, and H_2SO_4 contain a common ion, $H^+(aq)$. They should have chemical properties in common—the chemistry of this species, $H^+(aq)$. They do. Each solution

—has a sour taste—like vinegar, only more so;

—reacts with zinc to produce hydrogen gas; (5-4)

—causes the dye litmus to turn red.

Because of these likenesses, these compounds are given a class name; they are called *acids:* HCl—hydrochloric acid; HNO_3—nitric acid; H_2SO_4—

sulfuric acid. Any substance whose aqueous solutions display this array of properties (5-4) is said to be acidic.

This is the simple logic of classification. If the grouping proves to be useful, the class persists. This class, the acids, is as old as chemistry itself.

The longevity and growth of the class of compounds called acids has been paralleled by another type of compound epitomized by NaOH, KOH, and $Ca(OH)_2$.

$$NaOH(s) \longrightarrow Na^+(aq) + OH^-(aq)$$

$$KOH(s) \longrightarrow K^+(aq) + OH^-(aq) \tag{5-5}$$

$$Ca(OH)_2(s) \longrightarrow Ca^{+2}(aq) + 2\ OH^-(aq)$$

The aqueous solutions of these salts also contain a common ion, the hydroxide ion, $OH^-(aq)$. They should have chemical properties in common —the chemistry of this species, $OH^-(aq)$. They do. Each solution

—has a bitter taste;

—has a slippery feel;

—causes the dye litmus to turn blue; (5-6)

—reacts with an acid solution to diminish or remove its acidic properties.

Because of these likenesses, these compounds also are given a class name; they are called *bases*. Any substance whose aqueous solutions display this array of properties (5-6) is said to be basic.

(a) WHERE HAS ALL THE ACID GONE?

The most important property of a base is its ability to react with an acid solution to diminish the acidic properties. In fact, if a solution of NaOH is carefully added to a solution of HCl until the number of moles of NaOH and HCl are exactly equal, the solution has neither acidic nor basic properties! Where has all the acid gone? What happened to the species $H^+(aq)$, to which we attribute the acidic properties?

Plainly, the base species, $OH^-(aq)$, has reacted with $H^+(aq)$ to produce something else. The something else is water.

$$H^+(aq) + OH^-(aq) \longrightarrow H_2O(\ell) \tag{5-7}$$

Reaction (5-7) is called *"neutralization."* It reveals why acids and bases have such far-reaching importance. They are linked through our most important chemical, water.

(b) SELF-IONIZATION OF WATER

Reaction (5-7) is, of course, a sensible, law-abiding reaction, like any other. An acid and a base, when mixed, react until they reach equilibrium. Then they settle down to a dynamic balance between forward and reverse reactions. Let's take a look at the reverse reaction—we haven't mentioned it before.

$$H_2O(\ell) \longrightarrow H^+(aq) + OH^-(aq) \qquad\qquad (5\text{-}8)$$

This reaction has several rather strange implications. It says water is an acid (it forms $H^+(aq)$) and also a base (it forms $OH^-(aq)$). Yet pure water has neither the array of properties (5-4) nor (5-6). The reaction says that pure water contains ions. Yet the crude tests we used to detect ionic solutions (see Fig. 4-3) revealed no conductivity in pure water. Most important, reactions (5-7) and (5-8) imply an Equilibrium Law relationship that connects the concentrations of $H^+(aq)$ and $OH^-(aq)$ in *any* aqueous solution whether acidic or basic.

$$H_2O(\ell) = H^+(aq) + OH^-(aq) \qquad\qquad (5\text{-}9)$$

$$K' = \frac{[H^+][OH^-]}{[H_2O]} \qquad\qquad (5\text{-}10)$$

or, more conventionally,

$$\boxed{K_W = K'[H_2O] = [H^+][OH^-]} \qquad\qquad (5\text{-}11)$$

To a chemist, none of these implications is strange. "Yes," he would say, "water *is* both an acid and a base." Yes, water does contain ions; they can be detected with an extremely sensitive conductivity device. We just need an instrument that can record much smaller conductivities than can be sensed with the light-bulb set-up of Figure 4-3. The microammeter we used for measuring ionization energies (see Fig. 2-1) will do the trick easily. Such measurements show that pure water has electrical conductivity, but it is lower than that in 0.1 M KCl by a factor of 250,000. That tells us something about the equilibrium constants K and K'. They must be quite small. Indeed they are. Table 5-1 shows the magnitude of K at several temperatures. The value of K' can be obtained merely by dividing by 55.5, the concentration of water in water.

With the value of K at 298°K, we can calculate the ion concentrations to see why it is difficult to measure the electrical conductivity of water. The concentrations of $H^+(aq)$ and $OH^-(aq)$ will be equal if the only source is the

Table 5-1 *The Equilibrium Constant for Self-Ionization* of Water*

$$H_2O(\ell) = H^+(aq) + OH^-(aq)$$

$t(°C)$	$T(°K)$	K_w
0	273	$0.114 \cdot 10^{-14}$
10	283	0.295
20	293	0.676
25	298	$1.00 \cdot 10^{-14}$
30	303	1.47
40	313	2.71
50	323	5.30

* Also called "auto-ionization."

self-ionization of water, reaction (5-9). Hence we can calculate $[H^+]$:

$$[H^+][OH^-] = [H^+]^2 = 1.00 \cdot 10^{-14}$$

$$[H^+] = \sqrt{1.00 \cdot 10^{-14}}$$

$$[H^+] = [OH^-] = 1.00 \cdot 10^{-7} \, M \tag{5-12}$$

This is quite a low ion concentration, and it explains the rather low electrical conductivity of water. Nevertheless, the equilibrium constant reveals the ionizing properties of water, the solvent, on water, the solute.

Finally, we note the reciprocal relation between $[H^+]$ and $[OH^-]$.

$$[H^+] = \frac{K_w}{[OH^-]} \tag{5-13a}$$

and

$$[OH^-] = \frac{K_w}{[H^+]} \tag{5-13b}$$

If a base, a source of $OH^-(aq)$, is added to an acid solution, the concentration of OH^- rises. Since $[OH^-]$ appears in the denominator of (5-13a), $[H^+]$ must decrease. Hence the equilibrium reaction (5-9) explains why addition of $OH^-(aq)$ to an acid diminishes the acidic properties associated with a high concentration of $H^+(aq)$.

Exercise 5-1. Use the equilibrium expression (5-13) to show that the concentration of hydroxide ion is $1.0 \cdot 10^{-9} \, M$ in a solution containing $1.0 \cdot 10^{-5} \, M$ HCl.

Exercise 5-2. Show that as pure water is warmed from 25°C to 50°C, the concentration of $H^+(aq)$ rises from $1.0 \cdot 10^{-7} \, M$ to $2.3 \cdot 10^{-7} \, M$. Pure water is, by definition, a neutral solution, so this calculation shows that the definition of a neutral solution is not that $[H^+] = 1.0 \cdot 10^{-7}$ (which applies only at 25°C), but rather that $[H^+] = [OH^-]$, which applies at any temperature.

(c) ARRHENIUS, ON ACIDS AND BASES

Svante Arrhenius was the first scientist to recognize the presence of ions in aqueous solutions, back in 1889. Though this idea was slow in gaining acceptance, Arrhenius went right ahead to fit the properties of acids and bases into his scheme. He did the obvious thing, just as we did; he wrote down reactions (5-1), (5-2), (5-3), and (5-5), and then he associated acid properties (5-4) with the presence of $H^+(aq)$ and basic properties (5-6) with the presence of $OH^-(aq)$. Thus we would define an Arrhenius acid as a substance that can release H^+ in water and an Arrhenius base as a substance that can release OH^- in water:

Arrhenius acid: $\quad HX(aq) = H^+(aq) + X^-(aq)$ $\qquad$ (5-14)

Arrhenius base: $\quad MOH(aq) = M^+(aq) + OH^-(aq)$ $\qquad$ (5-15)

(d) BRØNSTED AND LOWRY, ON ACIDS AND BASES

Arrhenius' approach has two limitations. It is more or less restricted to aqueous solutions, though lots of related chemistry occurs in other solvents. Second, the theory only awkwardly accounts for the properties of some substances that do not contain H or OH, but give acidic or basic solutions anyway. For example, both sodium carbonate, Na_2CO_3, and ammonia, NH_3, dissolve in water to produce basic solutions. Yet neither contains an OH unit. However, each of these species, CO_3^{-2} and NH_3, can react with water to release OH^- from the solvent.

$$CO_3^{-2}(aq) + H_2O(\ell) = HCO_3^-(aq) + OH^-(aq) \qquad (5\text{-}16)$$

$$NH_3(aq) + H_2O(\ell) = NH_4^+(aq) + OH^-(aq) \qquad (5\text{-}17)$$

We can save the day for Arrhenius by saying that an acid is a substance that *forms* H^+ in water and a base is a substance that *forms* OH^- in water. Then carbonate ion and ammonia are bases through reaction with the solvent, even if they cannot simply dissociate into ions to release OH^-.

Two other scientists, J. N. Brønsted in Denmark and T. M. Lowry in England, pursued another aspect of reactions (5-16) and (5-17). They noted the fact that each reaction involves the transfer of a proton, H^+, from one substance to another. In 1923, they independently proposed a view of acid-base behavior that stressed this proton transfer. According to the Brønsted–Lowry definition, an acid can conveniently be defined as a substance that can donate a proton in a reaction like (5-16) or (5-17). Then a base will be a substance that can accept a proton.

> Brønsted–Lowry acid: a substance that can donate a proton   *(5-18)*

> Brønsted–Lowry base: a substance that can accept a proton   *(5-19)*

Now CO_3^{-2} ion and NH_3 are said to be bases because each can accept a proton from water, not because OH^- is formed. And H_2O has acted as an acid by donating a proton, even though $H^+(aq)$ does not appear. We are dealing with a theory in which an acid-base reaction could occur involving neither $H^+(aq)$ nor $OH^-(aq)$—only proton transfer. For example,

$$CO_3^{-2}(aq) + NH_4^+(aq) = HCO_3^-(aq) + NH_3(aq) \qquad (5\text{-}20)$$

Here again, CO_3^{-2} ion accepts a proton—it is a base. Now the acid is NH_4^+; ammonium ion donates the proton. Thus the acid-base idea is liberated to relate to a wider range of reactions.

The reaction (5-20) demonstrates another point. In (5-17), NH_4^+ ion was a product and we made no fuss about it. In (5-20), though, we see that NH_4^+ is a potential proton donor. In (5-17), the acid-base reaction between H_2O and NH_3 produced the potential acid, NH_4^+. And, we might add, it produced the prototype base, OH^-. So the acid-base reaction between NH_3 and H_2O produced another acid and another base.

With a moment's thought, this doesn't seem so marvelous, after all. Naturally, if a proton can be donated from one substance to another, it can be donated back. In fact, for either of the reactions (5-16) and (5-17) at equilibrium, the rate of proton transfer to the right must be exactly equal to the rate of proton transfer to the left. By the Brønsted–Lowry scheme, *every* acid-base reaction produces another acid and another base. Thus the base CO_3^{-2} accepts a proton to form its counterpart acid HCO_3^-, bicarbonate ion. These two substances are called a *conjugate acid-base pair*. The acid NH_4^+ and the base NH_3 are another conjugate acid-base pair. Every acid-base reaction is pictured as a proton transfer between two conjugate acid-base pairs.

This Brønsted–Lowry approach to acids and bases has some advantages and some liabilities. One of the liabilities is that the classical reaction of neutralization (5-7) doesn't seem to produce a conjugate acid-base pair, though we certainly can't legislate this reaction out of the acid-base category.

$$H^+(aq) + OH^-(aq) \longrightarrow H_2O(\ell)$$

That little problem is disposed of by inventing a new name for $H^+(aq)$. Instead of referring vaguely to the interaction between H^+ and the water around it, the Brønsted–Lowryites picture one molecule of water specially held.

$$H^+ + H_2O \longrightarrow H_3O^+ \qquad\qquad (5\text{-}21)$$

It is this species, H_3O^+, called hydronium ion, that tidies up the Brønsted–Lowry theory. When HCl gas dissolves in water, a proton transfer occurs from the acid HCl to the base H_2O to form Cl^-, the conjugate base of HCl, and H_3O^+, the conjugate acid of H_2O.

$$HCl(g) + H_2O(\ell) = H_3O^+(aq) + Cl^-(aq) \qquad\qquad (5\text{-}22)$$

When a base, like aqueous NaOH, is added to aqueous HCl, the reaction can be written

$$H_3O^+(aq) + OH^-(aq) = H_2O(\ell) + H_2O(\ell) \qquad\qquad (5\text{-}23)$$

Thus we see H_3O^+/H_2O as a conjugate acid-base pair and H_2O/OH^- as another conjugate acid-base pair. Again we conclude that the H_2O molecule acts both as an acid and as a base.

There is one more lingering problem about the $H_3O^+(aq)$ representation of $H^+(aq)$, (5-21). In dilute aqueous solutions of HCl or of HNO_3, we have difficulty finding any evidence that there is such a species as H_3O^+ present. This bothers some people more than others.

Exercise 5-3. In the reaction (5-16) CO_3^{-2} is a base and its conjugate acid is HCO_3^-. This ion, bicarbonate, HCO_3^-, can react with water to form carbonic acid, H_2CO_3. Write the balanced equation for the reaction and convince yourself that in this reaction HCO_3^- is a conjugate base. What are the two conjugate acid-base pairs in the reaction?

Exercise 5-4. In liquid ammonia, NH_3 self-ionizes to give a conducting solution due to ionic species $NH_4^+(am)$ and $NH_2^-(am)$. (This latter species is not detected in aqueous solutions, only in ammonia solutions.) Write the balanced equation for the reaction and identify the two conjugate acid-base pairs involved. Your results should show that, in this solvent medium, NH_3 can act either as an acid or a base, as water does in aqueous solutions.

(e) ENERGY AND $H^+(aq)$

We have already discussed the strong interactions that must exist between Na^+ ions and water and between Cl^- ions and water to account for the high solubility of NaCl solid. Figure 4-7 showed the energy aspect of this interaction: A mole of gaseous Na^+ and Cl^- ions releases 183 kcal of energy as it dissolves in water. Similar analyses for LiCl and KCl show that they release 207 and 168 kcal, respectively, upon aquation.

Like these salts, the gas HCl is extremely soluble in water—at 1 atm HCl pressure over water at room temperature, the gas will dissolve until the HCl concentration is 12 M! Again we can investigate the energy aspect of this solubility, with its formation of $H^+(aq)$ and $Cl^-(aq)$, by considering the process in steps. Just as we did for LiCl, NaCl, and KCl, we'll investigate the energy needed to form gaseous ions and then let them dissolve in water.

Our chemistry is as follows:

$$HCl(g) \xrightarrow[330 \text{ kcal}]{\substack{\text{heat} \\ \text{absorbed}}} H^+(g) + Cl^-(g) \xrightarrow[\text{on aquation}]{\substack{\text{heat} \\ \text{absorbed}}} H^+(aq) + Cl^-(aq) \qquad (5\text{-}24)$$

$$\text{over-all heat effect} = -17.9 \text{ kcal absorbed}$$
$$(+17.9 \text{ kcal released})$$
$$330 + \text{heat of aquation} = -17.9$$
$$\text{heat of aquation} = -348 \text{ kcal} \qquad (5\text{-}25)$$

The result, (5-25), tells us that a negative amount of heat is absorbed, that is, heat is evolved—a lot of it! The manner in which we deduced this is best shown by converting (5-24) to graphical form. Figure 5-1 shows the high hill that is climbed to get up to gaseous H^+ and Cl^- ions. But then, see what a joyride there is back down to the final products, 348 kcal lower!*

This energy of aquation of gaseous H^+ and Cl^- ions is even larger than that deduced for the salts—it is over twice as large as that for KCl! The difference must be attributed to the difference in aquation energy of K^+ and H^+, since the Cl^- aquation is common. Why does H^+ interact with water so much more strongly than K^+?

---

* As noted in Section 4-2(d), we are again using chemical thermodynamics when we draw the intuitively reasonable conclusion that the energy of the final state doesn't depend on how we got there, directly, or over the high hill.

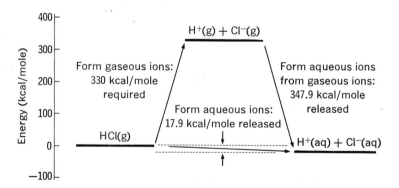

Figure 5-1 Energy diagram for the formation of gaseous and aqueous hydrogen and chloride ions.

A logical approach to this question would be to explore the simple electric dipole ion model that was shown in Figure 4-6. This is, after all, the picture most chemists have in mind when they think about the aquation of potassium or sodium or lithium ions—why not also hydrogen ions? There is, of course, a difference in size of the atoms that ought to matter. By inspection of ionization energies, it is clear that the proton likes its electron closer to it than is found for the most easily removed electron in potassium, sodium, or lithium. However, the simple electrostatic model used in Chapter Two (equation (2-11)) will give us a clue to the magnitude of that size effect.

If we picture the water dipoles orienting themselves around a proton just as they do around a potassium atom, then the relative energy of aquation will depend upon the tightness of packing. We will assume that the first ionization energy of the neutral atom gives a clue to the region of space that will be occupied by the water dipole electrons. The value of E_1 for hydrogen atoms is 314 kcal/mole and for potassium atoms, it is 100 kcal/mole. By the inverse relationship (2-11), we can conclude that the proton pulls the water dipoles in closer than does the potassium ion. We can even say how much:

$$\frac{r_H}{r_K} = \frac{E_K}{E_H} = \frac{100}{314} = 0.32 \qquad (5\text{-}26)$$

Now we can conclude that the aquation energies are themselves in inverse relation to the sizes:

$$\frac{E_{aq}(H^+)}{E_{aq}(K^+)} = \frac{r_K}{r_H} = \frac{1}{0.32} = 3.1 \qquad (5\text{-}27)$$

According to this simple approach, the energy of aquation of a proton should be about 3 times that of a potassium ion. The ratio of the aquation energies of HCl to KCl is not that large but, of course, each aquation energy includes a contribution from aquation of Cl^-. However, it should be a good approximation if we take the aquation energy of K^+ and Cl^- to be equal. These two ions have exactly the same number of electrons (the same number as argon) and each has a unit charge. With this reasonable approximation, we can estimate the individual aquation energies:

$$E_{aq}(K^+) = E_{aq}(Cl^-) = \tfrac{168}{2} = 84 \text{ kcal} \qquad (5\text{-}28)$$

$$E_{aq}(H^+) = 348 - E_{aq}(Cl^-) = 348 - 84 = 264 \text{ kcal} \qquad (5\text{-}29)$$

We conclude, with this very simple model, that the aquation energies of H^+

and K^+ are in the ratio 264 to 84

$$\frac{E_{aq}(H^+)}{E_{aq}(K^+)} = \frac{264}{84} = 3.1 \qquad (5\text{-}30)$$

This is exactly the ratio expected from size, as predicted in (5-27). For what it is worth, the argument tells us that the aquation of a proton is rather similar to the aquation of other positive ions, once size is taken into account. It is consistent with the suggestion that the proton is equivalently bound to several water molecules, perhaps four, to give a molecular species $H^+ \cdot 4\,H_2O$ or $H_9O_4^+$.

Unfortunately, this does not satisfy those chemists who find that the postulate of H_3O^+ helps them in their understanding of chemical phenomena (including acid-base reactions). Could we, then, build a comfortable explanation for the heat of aquation assuming that there is a preferential attachment of the proton to one H_2O molecule to form H_3O^+, followed by the aquation of this species? Let's try it.

Returning to the hypothetical, step-wise aquation process, (5-24) must be modified by adding another step. We must put in the formation of $H_3O^+(g)$ prior to the aquation step. This molecule is well known as a gaseous (though very reactive) ion and a fairly reliable value for its energy of formation in the gas phase is available; 151 kcal/mole of energy is released as $H^+(g)$ combines with $H_2O(g)$ to form $H_3O^+(g)$. Now we can write:

$$HCl(g) + H_2O(\ell) \xrightarrow[+330\ \text{kcal}]{\substack{\text{heat}\\ \text{absorbed}}} H^+(g) + Cl^-(g) + H_2O(\ell) \xrightarrow[+10.6\ \text{kcal}]{\substack{\text{heat}\\ \text{absorbed}}}$$

$$H^+(g) + Cl^-(g) + H_2O(g) \xrightarrow[-151\ \text{kcal}]{\substack{\text{heat}\\ \text{absorbed}}} H_3O^+(g) + Cl^-(g)$$

$$\xrightarrow[\text{on aquation}]{\substack{\text{heat}'\\ \text{absorbed}}} H_3O^+(aq) + Cl^-(aq) \qquad (5\text{-}31)$$

over-all heat effect $= -17.9$ kcal absorbed
$330 + 10.6 + (-151) +$ heat$'$ of aquation $= -17$
$$\text{heat}' \text{ of aquation} = -207 \text{ kcal} \qquad (5\text{-}32)$$

Figure 5-2 shows this information pictorially. Now the aquation of the presumed ions, H_3O^+ and Cl^-, releases only 207 kcal. Our task is to decide if this value is a reasonable enough figure that we feel comfortable assuming the presence of H_3O^+ in the water solution.

We can use (5-28) again to separate the aquation energies of $H_3O^+(g)$ and $Cl^-(g)$:

$$E_{aq}(H_3O^+) = 207 - E_{aq}(Cl^-) = 207 - 84 = 123 \text{ kcal} \qquad (5\text{-}33)$$

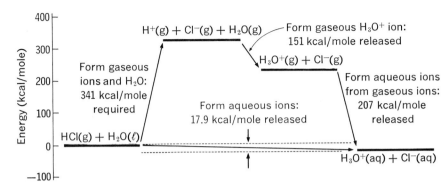

Figure 5-2 Energy diagram for formation of gaseous and aqueous H_3O^+ and Cl^- ions.

This aquation energy might be compared to that of $Na^+(g)$ and $K^+(g)$, derived from the aquation energy estimate for $E_{aq}(Cl^-) = 84$ kcal. They are, $E_{aq}(Na^+) = 99$ kcal and $E_{aq}(K^+) = 84$ kcal. Remembering that an oxygen atom bonded to two hydrogens "thinks it is neon" and that Na^+ has the neon electron population, we might guess that the aquation energy of Na^+ and H_3O^+ should be roughly the same. Our analysis indicates that the heat of aquation of H_3O^+ would have to be about 24% larger than that of Na^+. This is sufficiently close agreement that it does not constitute an embarrassment to the postulate that H_3O^+ is a dominant species in dilute aqueous solutions.* We are really back where we started. The evidence is not definitive, either for or against H_3O^+ as an important species in aqueous solutions. So chemists continue to choose the representation, $H^+(aq)$ or $H_3O^+(aq)$, that is more convenient for a given application.

(f) $H^+(aq)$, $H_3O^+(aq)$, AND EQUILIBRIUM EXPRESSIONS

This fickle attitude may seem strange in science, but it shows what a human endeavor we are conducting. Even so, it is legitimate to ask how equilibrium calculations are affected by the choice between $H^+(aq)$ and $H_3O^+(aq)$ as the representation for the aquated proton. An example from acid-base equilibria will illustrate.

Consider the effect of adding a sodium carbonate solution to a hydrochloric acid solution. The ion CO_3^{-2} is a base, so its addition will diminish the acidity of the acid solution. The chemistry can be written either in terms of $H^+(aq)$ or $H_3O^+(aq)$:

$$CO_3^{-2}(aq) + H^+(aq) = HCO_3^-(aq) \tag{5-34}$$

$$CO_3^{-2}(aq) + H_3O^+(aq) = HCO_3^-(aq) + H_2O(\ell) \tag{5-35}$$

* We shall see in Chapter Seventeen that the proposed species H_3O^+ would undoubtedly interact with the solvent water molecules through "hydrogen bonding" and the aquation energy would require three hydrogen bonds of energy 41 kcal each, stronger than any known hydrogen bond.

At equilibrium, (5-34) and (5-35) each give us an equilibrium expression:

$$K = \frac{[HCO_3^-]}{[CO_3^{-2}]\,[H^+]}$$

(5-36)

and

$$K' = \frac{[HCO_3^-]}{[CO_3^{-2}][H_3O^+]}$$

(5-37)

The equilibrium expression (5-37) does not include the H_2O concentration, of course. That is considered to be constant in dilute solutions, 55 M, and it is embodied in the constant K'. Now if we insert concentrations into the two expressions (5-36) and (5-37), we will obtain the same numerical magnitude for K and K'. This is because the H^+ concentration in (5-36) and the H_3O^+ concentration in (5-37) must be assigned as the number of moles of aquated protons we have per liter. The arithmetic is identical whether (5-36) or (5-37) is used—only the name of the aquated proton has been changed.

This is always the case. A reaction involving H^+(aq) can always be rewritten in terms of H_3O^+(aq) merely by adding H_2O to the reaction. Because the H_2O concentration is always incorporated into the equilibrium constant, there is no effect on equilibrium calculations for dilute aqueous solutions. Nevertheless, we see that there is a lot of interesting chemistry embodied in the innocent looking notation, "(aq)."[*]

5-2   Strong acid–strong base reactions

Acids, such as HCl and HNO_3, and bases, such as NaOH, KOH, and Ca(OH)$_2$, dissolve in water to give electrically conducting solutions: They are electrolytes. Furthermore, the substances mentioned seem to dissociate completely: There is no evidence in dilute solutions for undissociated molecules of HCl, HNO_3, NaOH, KOH, or Ca(OH)$_2$. To indicate this complete dissociation, these substances are called *strong electrolytes*. More specifically, the complete dissociation of an acid is indicated by calling it a *strong acid*. Hydrochloric acid and nitric acid are strong acids. The complete dissociation of a base is indicated by calling it a *strong base.* Sodium hydroxide, potassium hydroxide, and calcium hydroxide are strong bases. This is an important designation: Not all acids and bases can be said to be strong.

(a)   ACIDITY IN A STRONG ACID SOLUTION

Within the Arrhenius view of acids, we can ask, "What is the acidity in a 0.10 M HCl solution?" This question means, "What is the concentration

---

[*] We should note that there are other, still more abstract and more widely applicable theories of acid-base reactions than the two we've considered. The most used was proposed by G. N. Lewis in 1923 and it will be considered in Chapter Seventeen when our views on chemical bonding are more sophisticated.

of H$^+$(aq) in a 0.10 M HCl solution?" That is easy to answer. Since all of the HCl in water is completely dissociated into ions, we can immediately conclude that both [H$^+$] and [Cl$^-$] are equal to the total HCl concentration.

$$[H^+] = [Cl^-] = 0.10 \ M \qquad\qquad (5\text{-}38)$$

Now that we know [H$^+$], the self-ionization equilibrium of water permits calculation of [OH$^-$] by expression (5-13b).

$$[OH^-] = \frac{K_w}{[H^+]} = \frac{1.0 \cdot 10^{-14}}{0.10} = 1.0 \cdot 10^{-13} \ M \qquad\qquad (5\text{-}39)$$

We see that HCl solution contains both H$^+$(aq) and OH$^-$(aq), but it is also easy to see why the properties of H$^+$(aq) characterize the solution, not those of OH$^-$(aq). There is 10^{12} times more H$^+$(aq) than OH$^-$(aq).

(b) ACIDITY IN A STRONG BASE SOLUTION

"What is the acidity in a 0.10 M NaOH solution?" This seems, at first, a strange question—What is the acidity in a basic solution? However, if it means only, "What is [H$^+$]?" the answer is easy to find. Since NaOH is a strong base,

$$[OH^-] = [Na^+] = 0.10 \ M \qquad\qquad (5\text{-}40)$$

and

$$[H^+] = \frac{K_w}{[OH^-]} = \frac{1.0 \cdot 10^{-14}}{0.10} = 1.0 \cdot 10^{-13} \ M \qquad\qquad (5\text{-}41)$$

Again the solution contains both H$^+$(aq) and OH$^-$(aq), but now there is 10^{12} times more OH$^-$(aq) than H$^+$(aq). This is why the properties of OH$^-$(aq) characterize this solution.

(c) CHANGE OF ACIDITY WHEN BASE IS ADDED TO AN ACID

Now we can consider the change of acidity if a strong base is added to a strong acid solution. Suppose 0.20 mole of solid NaOH is slowly added to a liter of 0.10 M HCl. At first, the 0.10 M HCl has the H$^+$(aq) and OH$^-$(aq) concentrations calculated in (5-38) and (5-39): [H$^+$] = 0.10 M and [OH$^-$] = 1.0 · 10^{-13} M. Then, as NaOH is added, reaction (5-7) occurs, consuming H$^+$ and forming H$_2$O. The acidity drops. By the time all of the 0.20 mole of solid NaOH has been added, half of it has reacted to consume 0.10 mole of H^+ originally present in the liter of solution. We have, left over, the excess OH^- in amount, 0.10 mole, dissolved in one liter. So we end up with a 0.10 M OH^- solution. By (5-40) and (5-41), the $[H^+]$ is now $1.0 \cdot 10^{-13}\ M$.

Consider the enormous concentration changes associated with this simple experiment. As the NaOH is added to the HCl solution, the hydrogen ion concentration changes from 0.10 M to $1.0 \cdot 10^{-13}\ M$. That is a decrease by a factor of 10^{12}. At the same time, the hydroxide ion concentration increases by a factor of 10^{12}. The ability to control the concentration of $H^+(aq)$ over such a wide range is extremely important in many chemical reactions, including those that take place in living systems. We shall wish to study these changes systematically. But first, we need a convenient scale for such huge changes.

(d) pH SCALE

We shall often wish to examine graphically how acidity changes during chemical reactions and during chemical operations. We have just seen, however, that acidity can easily change by many powers of ten. In the simple process of adding 0.20 mole of NaOH to a liter of 0.10 M HCl, the H^+ concentration changed from $10^{-1}\ M$ to $10^{-13}\ M$. Halfway through, the H^+ concentration would be $10^{-7}\ M$. Imagine trying to decide on the scale for a graph that traced the course of this change.

Actually, there is an easy escape. As the concentration changes by a million million, from 10^{-1} to 10^{-13}, the exponential notation shows the exponent changing from -1 to -13. That exponent furnishes a more manageable measure of concentration changes when they extend over many orders of magnitude. The exponent is obtained mathematically by taking the logarithm.

$$\log 10^{-1} = -1$$
$$\log 10^{-13} = -13$$

This device is used by chemists to discuss and display acid and base concentrations. The resulting numbers, applicable to dilute solutions, would always be negative if the log itself were used. Consequently a scale has been defined, and is widely used, based upon the *negative* of the logarithm. This arbitrary, but useful, scale is called the pH scale.

$$\boxed{pH = -\log[H^+]} \qquad \qquad (5\text{-}42)$$

Figure 5-3 shows how this pH scale expands the H^+ concentration scale.

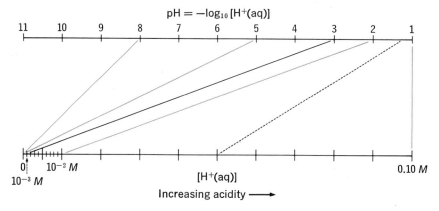

Figure 5-3 *Comparison of numeric and logarithmic (pH) concentration scales for hydrogen ions.*

The idea is useful for other ions, too, and the notation can be generalized. Thus, it can be applied to hydroxide concentration.

$$\text{pOH} = -\log[\text{OH}^-] \qquad\qquad (5\text{-}43)$$

It is even sometimes applied to equilibrium constants.

$$\text{p}K = -\log(K) \qquad\qquad (5\text{-}44)$$

In the 0.10 M HCl solution we considered earlier, we calculated [H$^+$] = 0.10 and [OH$^-$] = $1.0 \cdot 10^{-13}$. Now we can express these concentrations on the pH and pOH scales.

$$\text{pH} = -\log(0.10) = -\log(1.0 \cdot 10^{-1}) = -(-1) = +1$$
$$\text{pOH} = -\log(1.0 \cdot 10^{-13}) = -(-13) = +13$$

For another example, consider the pH and pOH in a 0.020 M HCl solution. We calculate [H$^+$] = $2.0 \cdot 10^{-2}$ and, by (5-13b), [OH$^-$] = $5.0 \cdot 10^{-13}$

$$
\begin{aligned}
\text{pH} &= -\log(2.0 \cdot 10^{-2}) = -(\log 2 + \log 10^{-2})\\
&= -(0.301 - 2) = -(-1.70)\\
&= +1.7 \qquad\qquad (5\text{-}45)
\end{aligned}
$$

$$
\begin{aligned}
\text{pOH} &= -\log(5.0 \cdot 10^{-13}) = -(\log 5 + \log 10^{-13})\\
&= -(0.70 - 13) = -(-12.3)\\
&= +12.3 \qquad\qquad (5\text{-}46)
\end{aligned}
$$

Exercise 5-5. Calculate pH and pOH in 50 ml of 0.20 M HNO$_3$ solution. Convince yourself that the pH is the same in 100 ml of 0.20 M HNO$_3$.

Exercise 5-6. Show that the pH changes by $+1$ if this 100 ml of 0.20 M HNO$_3$ is diluted with water to a volume of 1000 ml.

Exercise 5-7. By the definitions of pH, pOH, and pK_w, (5-42), (5-43), and (5-44), show that pH + pOH = pK = 14 (at 298°K).

Exercise 5-8. Use the pH you calculated in Exercise 5-5 and the result of Exercise 5-7 to calculate pOH. Compare to your earlier calculation of pOH in Exercise 5-5.

(e) A STRONG ACID–STRONG BASE TITRATION CURVE

With the benefit of this convenient, logarithmic scale, we might return to that experiment in which 0.20 mole of NaOH was added to one liter of 0.10 M HCl. We begin with $[H^+]$ = 0.10 M and end with $[H^+]$ = $1.0 \cdot 10^{-13}$ M. Halfway along, when the number of moles of base added exactly equals the number of moles of acid present, the $[H^+]$ becomes equal to that of water, $1.0 \cdot 10^{-7}$ M. Thus, the pH changes from 1.0 to 13.0, passing through the pH = 7 on the way.

If we had a way to keep track of pH during the addition of NaOH, this process would be quite useful. The point in the process at which pH = 7 marks the moment when the number of moles of added NaOH exactly equals the number of moles of HCl originally present. We could use such an experiment to learn the original HCl concentration, if it were unknown.

There are a variety of convenient ways to measure pH—we'll speak of them soon. Consequently chemists regularly use the addition of strong base to an acid solution to learn the acid concentration. This process is called a *titration*. The point at which (moles added base) = (moles acid originally present) is called the *equivalence point.*

In practice, it is more convenient to use a solution of NaOH with known concentration rather than solid NaOH. Let's explore the change of pH during the slow addition of 0.2000 M NaOH to 100.0 ml of 0.2000 M HCl solution. This titration isn't needed to reveal the HCl concentration—we already know it is 0.2000 M. But it will show us how pH changes during the titration, so we know what is going on in a real application of the method.

Initially, when no NaOH has been added, $[H^+]$ = 0.20 M and pH = 0.70. That is our starting point.

To get our feet wet, let's calculate the pH after 50.00 ml of 0.2000 M NaOH have been added. Since reaction (5-7) will occur, we need to know the number of moles of acid present and the number of moles of base added, so we can calculate which is in excess.

Let

$n_0(H^+)$ = number of millimoles of H^+ initially

$n_a(OH^-)$ = number of millimoles of OH^- added

$n(H^+) = n_0(H^+) - n_a(OH^-)$ = number of millimoles excess (5-47)
H^+ (before equivalence point is reached)

$n(OH^-) = n_a(OH^-) - n_0(H^+)$ = number of millimoles excess
OH^- (after equivalence point is passed)

V = total volume at any point in the titration

= initial volume HCl solution + added volume of
base solution

= $V_0(H^+) + V_a(OH^-)$

These definitions, (5-47), are expressed in millimoles and milliliters instead of moles and liters just as a matter of convenience. Concentrations are as easily calculated in one set of units as in the other:

$$\text{concentration} = \frac{\text{moles}}{\text{liters}} = \frac{\text{moles}/1000}{\text{liters}/1000} = \frac{\text{millimoles}}{\text{milliliters}} \qquad (5\text{-}48)$$

Now, back to our problem: to evaluate (5-47) after 50.00 ml of 0.2000 M NaOH have been added.

$n_0(H^+)$ = (initial volume)(initial H^+ concentration)

= $V_0(H^+)(0.2000) = (100.0)(0.2000) = 20.00$ millimoles H^+

$n_a(OH^-)$ = (added volume)(OH^- concentration)

= $V_a(OH^-)(0.2000) = (50.00)(0.2000) = 10.00$ millimoles OH^-

$n(H^+) = n_0(H^+) - n_a(OH^-) = 20.00 - 10.00 = 10.00$ millimoles
excess H^+

At this point in the titration, H^+ is still in excess, and the concentration of excess is

$$[H^+] = \frac{n(H^+)}{V} = \frac{n_0(H^+) - n_a(OH^-)}{V_0(H^+) + V_a(OH^-)} \qquad (5\text{-}49)$$

$$[H^+] = \frac{10.00}{100.0 + 50.0} = \frac{10.00}{150.0} = 0.066 \ M \qquad (5\text{-}50)$$

$$pH = -\log (6.6 \cdot 10^{-2}) = -[\log 6.6 + \log 10^{-2}]$$
$$= -[0.82 - 2.0] = 1.2 \qquad (5\text{-}51)$$

$$pOH = 14.0 - pH = 14.0 - 1.2 = 12.8$$

That is interesting! We have already added half of the amount of OH^- needed to reach the equivalence point, and the pH has changed only from 0.70 to 1.2.

Exercise 5-9. Show that after 90.00 ml of 0.2000 M NaOH have been added, the pH has only changed to 1.98.

We see that during the titration, the pH changes very slowly at first. The situation is different, however, when we near the equivalence point. Consider, for example, the conditions when we have added 99.0 ml of NaOH:

$$n_0(H^+) = (100.0)(0.2000) = 20.00 \text{ millimoles } H^+$$

$$n_a(OH^-) = (99.00)(0.2000) = 19.80 \text{ millimoles } OH^-$$

$$[H^+] = \frac{n_0(H^+) - n_a(OH^-)}{V_0(H^+) + V_a(OH^-)} = \frac{20.00 - 19.80}{100.0 + 99.0} = \frac{0.20}{199.0} \cong \frac{0.20}{200}$$

$$[H^+] = 1.0 \cdot 10^{-3}$$

$$pH = -\log[H^+] = -[\log 1.0 + \log 10^{-3}] = 3.0$$

Let's take stock of what is happening. Our first addition of 50 ml NaOH raised the pH from 0.70 to 1.2, a change of 0.5. Then the next 40 ml added gave a further rise from 1.2 to 1.98, a change of 0.8 pH units. Adding 9 ml more boosted the pH from 1.98 to 3.0, a change of 1.0 pH units. We see that the pH is changing more and more, per ml base added, as we near the equivalence point.

We might also look ahead to see where we are going. The addition of 1.0 ml more base will mean we have added exactly 100.0 ml base. Now, $n_0(H^+) = n_a(OH^-)$ and we have reached the equivalence point. We have, at this moment in the titration, a solution of NaCl in water. The H^+ concentration will be 10^{-7} M and pH = 7.0. Now there was a response! The last milliliter needed to reach the equivalence point sent the pH from 3.0 to 7.0, a change of 4.0! The pH changes very rapidly near the equivalence point.

Though it may seem anticlimactic, it is healthy to look beyond the equivalence point in the titration. Consider the situation when $V_a(OH^-) = 101.0$, 1.0 ml beyond the equivalence point.

$$n_0(H^+) = 20.00 \text{ millimoles } H^+$$

$$n_a(OH^-) = 20.20 \text{ millimoles } OH^-$$

$$n(OH^-) = n_a(OH^-) - n_0(H^+) = 20.20 - 20.00 = 0.20 \text{ millimole}$$
$$\text{excess } OH^-$$

$$[OH^-] = \frac{0.20}{100.0 + 101.0} = \frac{0.20}{201} \cong \frac{0.20}{200} = 1.0 \cdot 10^{-3}$$

$$pOH = 3.0$$

$$pH = 14.0 - pOH = 14.0 - 3.0 = 11.0$$

Again we have a very large change in pH, 4.0 pH units, with the addition of only one ml of NaOH beyond the equivalence point.

Exercise 5-10. Show that after 110 ml of 0.2000 M NaOH have been added, pOH = 2.0 and pH = 12.0.

We now have enough calculations to picture the course of the pH during the careful addition of the strong acid NaOH to the strong acid HCl. Such a plot is called a *titration curve* for 0.2000 M solutions, the curve is shown in Figure 5-4. The precipitous rise of pH as we come close to the equivalence point shows why titrations are such a useful way to learn acid concentrations. All we need is some signal system that rings a bell, lights a light, or otherwise announces that the pH has reached 7.0. With such a signal system, we can learn the equivalence point quite accurately in a strong acid–strong base titration.

(f) pH INDICATOR SYSTEMS

There is a signal system implicit in our original, experimental definitions of acidity and basicity. Look back to the array of properties (5-4) and (5-6) that tell us in an operational way if a solution is acidic or basic. One property listed is "causes the dye litmus to turn red (or blue)." That sounds like a useful signal! Why not just add some litmus (whatever that is) and use the red-to-blue color change to indicate the equivalence point. That is what is done in most acid-base titrations. A suitable dye is found that gives a

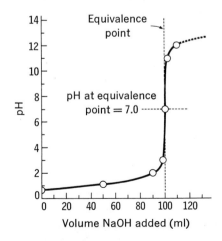

Figure 5-4 *Titration curve for titration of 100.0 ml 0.2000 M HCl with 0.2000 M NaOH.*

noticeable color change at the pH calculated for the equivalence point. Such a substance is called an *indicator*.

Of course, we would like to know how these dye indicators work. Why do they change color as pH changes? That is a matter we'll take up later on. For the moment, let's just enjoy the fact that litmus, or some other such dye, allows us to take advantage of the rapid change of pH near the equivalence point, so we can conduct an accurate titration.

5-3   Weak acids and bases

It would be a bit silly to identify HCl and HNO_3 as strong acids if the word "strong" didn't differentiate them from some other kind of acid, a "weak" acid. What is a weak acid, or a weak base? How do they behave in comparison to the strong acids and strong bases we considered in the previous section?

(a) WEAK ACID DISSOCIATION

Figure 5-5 shows our crude conductivity experiment again, this time carried out with 0.20 M sugar solution, 0.20 M acetic acid, CH_3COOH, and 0.20 M hydrochloric acid, HCl. The device shows that sugar dissolves without imparting electrical conductivity to its solution. We conclude that there are no charge carriers, ions, added to the very low concentrations of H^+ and OH^- already present in the water due to its self-ionization. The sugar must be present in solution as neutral molecules.

The middle bulb suggests that acetic acid can't make up its mind to be like sugar or like hydrochloric acid. The bulb glows dimly, so some charge carriers are present in a 0.20 M acetic acid solution, but by no means as many as in the 0.20 M HCl solution. We must conclude that acetic acid dissociates to ions only partly. Substances that do this are called *weak electrolytes*. In a solution of a weak electrolyte, the solute forms partly a

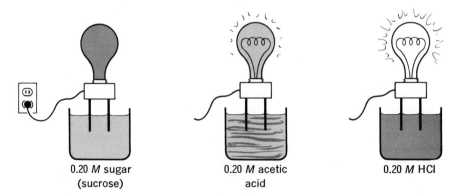

| 0.20 *M* sugar | 0.20 *M* acetic | 0.20 *M* HCl |
| (sucrose) | acid | |

Figure 5-5 Sugar forms a molecular solution that is nonconducting; 0.20 M acetic acid solution is a poor conductor; 0.20 M HCl is a good conductor.

molecular solution and partly an ionic solution. When one of the ions formed is H⁺(aq) or OH⁻(aq), the weak electrolyte is called a *weak acid* or a *weak base*, respectively.

(b) WATER: WEAK ACID AND WEAK BASE

The self-ionization of water has already been discussed. It furnishes our first example of a weak acid—and weak base as well. Most water molecules are present as water, but a few of them dissociate into H⁺ and OH⁻ ions. With the equilibrium expression (5-11), we calculated [H⁺] and [OH⁻] in pure water. Then, later, when we considered the strong acid–strong base titration, we kept in mind the fact that [H⁺] and [OH⁻] always maintain equilibrium concentrations that satisfy the relationship [H⁺][OH⁻] = $1.0 \cdot 10^{-14}$.

(c) ACETIC ACID, VINEGAR'S VITALITY

Acetic acid gives vinegar its tart, acidic taste. This compound has the formula CH_3COOH and the molecular structure shown in Figure 5-6. Three of the hydrogen atoms are attached quite firmly to a carbon atom. The fourth hydrogen atom is bonded to one of the oxygen atoms. This is the hydrogen that makes CH_3COOH an acid. In aqueous solution, some molecules release this atom as a proton, to give H⁺(aq), and to leave the rest of the molecule with a negative charge. This remaining ion, CH_3COO^-, is called *acetate* ion. The conductivity of acetic acid in water reveals the dynamic balance, the equilibrium, that exists between the

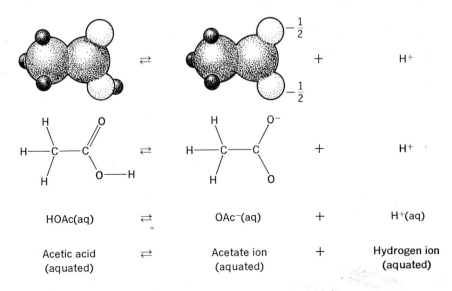

Figure 5-6 *The molecular structure of acetic acid and its ionization product, acetate ion.*

dissociation of dissolved neutral molecules to give ions and the recombination of these ions in the solution to restore the neutral molecules.

$$CH_3COOH(aq) = CH_3COO^-(aq) + H^+(aq) \qquad (5\text{-}52)$$

These chemical formulas are usually abbreviated to indicate only the key atoms, the hydrogen atom, the oxygen atom to which it is attached, and the rest of the molecule.

$$HOAc(aq) = OAc^-(aq) + H^+(aq) \qquad (5\text{-}53a)$$

or, in Brønsted–Lowry formulation,

$$HOAc(aq) + H_2O(\ell) = OAc^-(aq) + H_3O^+(aq) \qquad (5\text{-}53b)$$

The equilibrium expressions for (5-53a) and (5-53b) are, respectively,

$$K_a = \frac{[H^+][OAc^-]}{[HOAc]} \qquad (5\text{-}54a)$$

or

$$K_a = \frac{[H_3O^+][OAc^-]}{[HOAc]} \qquad (5\text{-}54b)$$

The constant for a reaction like (5-53) is called an acid dissociation constant and it is symbolized K_a. For the particular acid, acetic acid, the acid dissociation constant has the magnitude $1.78 \cdot 10^{-5}$, whether we choose to use (5-54a) or (5-54b). Acetic acid is only one among many substances, however, that act as weak acids in aqueous solution. Table 5-2 lists a flock of them. They are given the pompous name "monoprotic acids" to indicate that each of the molecules listed here can release only one proton, though it may contain several hydrogen atoms. Table 5-3 shows some bases; the constant K_b corresponds to the reaction (5-55a) or (5-55b).

$$MOH = M^+ + OH^- \qquad K_b \qquad (5\text{-}55a)$$

$$M + H_2O = MH^+ + OH^- \qquad K_b \qquad (5\text{-}55b)$$

(d) pH IN AN ACETIC ACID SOLUTION

Acetic acid is a fine example of a weak acid solution. Let's calculate the pH of an HOAc solution of analytical concentration $c = 0.20\ M$.* Reaction (5-53) produces acetate ion, OAc^-, and $H^+(aq)$ in equal concentration.

$$[OAc^-] = [H^+] = x \qquad (5\text{-}56a)$$

* The actual concentration of [HOAc] at equilibrium is slightly smaller than 0.20 M because of dissociation. However, if we were to analyze for total acid—perhaps by titrating with strong base—we would find 0.20 mole per liter. Hence the term "analytical concentration."

Table 5-2 Acid Dissociation Constants* of Monoprotic Acids in Water ($T = 298°K$)

Acid	Formula	K_a	Acid	Formula	K_a
perchloric acid	$HOClO_3$	strong acid	benzoic acid	C_6H_5COOH	$6.6 \cdot 10^{-5}$
hydrogen iodide	HI	strong acid	hydrazoic acid	HN_3	$1.91 \cdot 10^{-5}$
hydrogen bromide	HBr	strong acid	acetic acid	H_3CCOOH	$1.78 \cdot 10^{-5}$
hydrogen chloride	HCl	strong acid	propionic acid	CH_3CH_2COOH	$1.35 \cdot 10^{-5}$
nitric acid	$HONO_2$	strong acid	hypochlorous acid	$HOCl$	$2.95 \cdot 10^{-8}$
iodic acid	$HOIO_2$	$1.6 \cdot 10^{-1}$	hypobromous acid	$HOBr$	$2.51 \cdot 10^{-9}$
trichloroacetic acid	Cl_3CCOOH	$1.47 \cdot 10^{-2}$	ammonium ion	NH_4^+	$5.5 \cdot 10^{-10}$
monochloroacetic acid	ClH_2CCOOH	$1.35 \cdot 10^{-3}$	hydrocyanic acid	HCN	$4.8 \cdot 10^{-10}$
hydrofluoric acid	HF	$6.76 \cdot 10^{-4}$	hypoiodic acid	HOI	$4.8 \cdot 10^{-13}$
nitrous acid	$HONO$	$5.13 \cdot 10^{-4}$	water	H_2O	$1.8 \cdot 10^{-16}$
formic acid	$HCOOH$	$1.78 \cdot 10^{-4}$	methanol	H_3COH	$1.0 \cdot 10^{-17}$

* Extrapolated to zero ionic strength.

Table 5-3 Base Dissociation Constants* of Bases in Water ($T = 298°K$)

Base	Formula	K_b	Base	Formula	K_b
sodium hydroxide	NaOH	strong base	imidazole	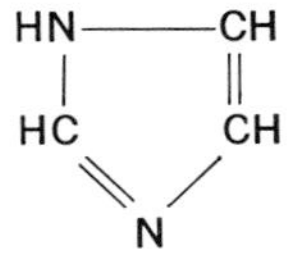	$8.1 \cdot 10^{-8}$
potassium hydroxide	KOH	strong base			
dimethylamine	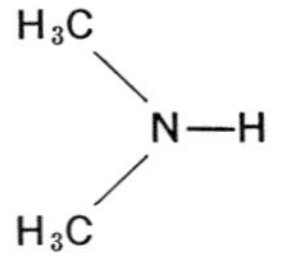 H_3C–N–H / H_3C	$1.18 \cdot 10^{-3}$		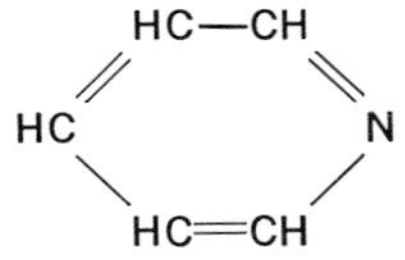	
methylamine	H_3CNH_2	$5.25 \cdot 10^{-4}$	pyridine		$1.5 \cdot 10^{-9}$
ethylamine	$CH_3CH_2NH_2$	$4.7 \cdot 10^{-4}$			
trimethylamine	$N(CH_3)_3$	$8.1 \cdot 10^{-5}$	aniline	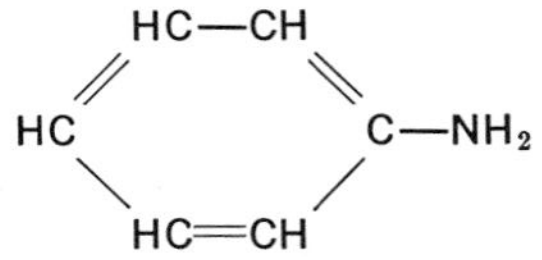	$4.2 \cdot 10^{-10}$
ammonia	NH_3	$1.78 \cdot 10^{-5}$			

* Extrapolated to zero ionic strength.

For simplicity, we will use x to represent these concentrations. Thus, every x moles/liter of HOAc that dissociates produces x moles/liter of aquated protons and x moles/liter of acetate ions. At the same time, the number of moles of undissociated HOAc is reduced by x moles/liter.

$$[HOAc] = (c - x) \text{ moles/liter} \tag{5-56b}$$

Substituting (5-56a) and (5-56b) into the equilibrium expression (5-54),

$$\boxed{K_a = \frac{x^2}{c - x}} \tag{5-57}$$

Expression (5-57) can be solved with the quadratic equation (4-58). Before doing so, we should consider simplifying approximations, just as we did in the last chapter. Perhaps x is negligible compared to c.

$$\text{Perhaps} \quad (c - x) \cong c \tag{5-58}$$

Expression (5-57) can now be easily solved.

$$K_a = \frac{x^2}{c}$$

$$x^2 = K_a \cdot c$$

$$x = \sqrt{K_a \cdot c} = \sqrt{(1.78 \cdot 10^{-5})(0.20)} = \sqrt{3.56 \cdot 10^{-6}}$$

$$x = [H^+] = [OAc^-] = 1.88 \cdot 10^{-3} \ M \tag{5-59}$$

Now we must check the simplifying approximation (5-58). The estimate (5-59) tells us how closely $(c - x)$ equals c.

$$c - x = 0.200 - 0.0019 = 0.1981$$

We see that neglecting x, 0.0019, compared to 0.200 gives a 1% error.

$$\frac{0.0019}{0.200} \cdot 100 = 1.0\%$$

For almost all purposes, a one percent error in $[H^+]$ can be neglected, so we can now calculate the pH.

$$pH = -\log [H^+] = -\log (1.88 \cdot 10^{-3}) = -\log 1.88 - \log 10^{-3}$$

$$pH = 3 - 0.27 = 2.73 \tag{5-60}$$

Exercise 5-11. Calculate $[H^+]$ in 0.10 M chloroacetic acid ($K_a = 1.35 \cdot 10^{-3}$) using the approximation $(c - x) \cong c$. Show that the calculated value of x is 12% of c.

Exercise 5-12. For many applications, a 12% error in $[H^+]$ is larger than desired. Use the value of x calculated in Exercise 5-11, 0.012 M, to make a second approximation $c - x \cong 0.100 - 0.012 = 0.088$ M. Calculate the value of x based upon this approximation and show that the second estimate involves only 1% error and that the pH $= 1.96$.

(e) pH IN AN ACETIC ACID–SODIUM ACETATE SOLUTION

The addition of the salt of a weak acid has an important effect on the acidity of a weak acid solution. Consider, for example, the effect of adding to a 0.200 M acetic acid solution enough solid sodium acetate to make the analytical concentration of NaOAc $= 0.200$ M as well. Again we shall represent the $H^+(aq)$ concentration as x moles/liter. Again the HOAc concentration will be decreased from 0.200 to 0.200 $- x$. The change in the problem relates to the sodium acetate concentration, which now comes from two sources, from dissociation and from the added salt.

$$HOAc(aq) = H^+(aq) + OAc^-(aq)$$

$$[H^+] = x$$

$$[HOAc] = 0.200 - x \tag{5-61}$$

$$[OAc^-] = 0.200 + x$$

The equilibrium expression (5-54) becomes

$$K_a = \frac{[H^+][OAc^-]}{[HOAc]} = \frac{(x)(0.200 + x)}{(0.200 - x)} \tag{5-62}$$

Again we'll try the approximation that x is negligible compared to 0.200

$$K_a = 1.78 \cdot 10^{-5} = \frac{(x)(0.200 + x)}{(0.200 - x)} \cong \frac{(x)(0.200)}{(0.200)} = x$$

$$[H^+] = 1.78 \cdot 10^{-5} \ M \tag{5-63}$$

The approximations $(0.200 - x) \cong 0.200$ and $(0.200 + x) \cong 0.200$ are obviously quite accurate: $1.78 \cdot 10^{-5}$ is negligible compared to 0.200.

$$pH = -\log[1.78 \cdot 10^{-5}] = 5 - 0.25 = 4.75$$

Exercise 5-13. Show that $[H^+] = 1.78 \cdot 10^{-5} \, M$ in a solution containing 0.0200 M HOAc and 0.0200 M NaOAc, the same as (5-63).

Exercise 5-14. Show that $[H^+] = 1.98 \cdot 10^{-6} \, M$ and pH = 5.70 in a solution containing 0.020 M HOAc and 0.180 M NaOAc.

(f) pH IN A SODIUM ACETATE SOLUTION

The acidity in a pure sodium acetate solution raises a new kind of equilibrium problem. A salt of a strong acid, like NaCl, has no effect upon pH. Sodium acetate, however, can react with water to form acetic acid and hydroxide ion.

$$OAc^-(aq) + H_2O(\ell) = HOAc(aq) + OH^-(aq) \tag{5-64}$$

In the Arrhenius scheme, reaction (5-64) is called hydrolysis; in the Brønsted–Lowry view, acetate ion acts as a proton acceptor, a base, to form its conjugate acid, HOAc. The equilibrium constant for (5-64) is the one we denoted K_b in (5-55b)

$$K_b = \frac{[HOAc][OH^-]}{[OAc^-]} \tag{5-65}$$

This constant K_b for acetate ion is simply related to the acid dissociation constant K_a of its conjugate acid acetic acid and to the water dissociation constant K_w. This can be seen by multiplying (5-65) in numerator and denominator by $[H^+]$:

$$K_b = \frac{[HOAc][OH^-][H^+]}{[OAc^-][H^+]} = \frac{[HOAc]}{[H^+][OAc^-]}[H^+][OH^-]$$

$$\boxed{K_b = \frac{K_w}{K_a}} \tag{5-66}$$

For acetic acid,

$$K_b = \frac{K_w}{K_a} = \frac{1.00 \cdot 10^{-14}}{1.78 \cdot 10^{-5}} = 5.6 \cdot 10^{-10} \tag{5-67}$$

Now we can calculate the hydroxide ion concentration. Reaction (5-64) produces HOAc and OH^- in equal quantities. We call these concentrations

x. Now it is the acetate ion concentration that is diminished from its initial concentration c to $(c - x)$.

$$[HOAc] = [OH^-] = x \frac{\text{moles}}{\text{liter}}$$

$$[OAc^-] = c - x$$

$$K_b = \frac{[HOAc][OH^-]}{[OAc^-]} = \frac{(x)(x)}{(c - x)}$$

$$K_b = \frac{x^2}{(c - x)} \tag{5-68}$$

Again, let's go through the arithmetic for a particular case, $c = 0.100\ M$. The approximation $(c - x) \cong c$ will be tried, as usual.

$$K_b = 5.6 \cdot 10^{-10} \cong \frac{x^2}{c} = \frac{x^2}{0.100}$$

$$x = \sqrt{5.6 \cdot 10^{-10}\ 0.100} = \sqrt{0.56 \cdot 10^{-10}} = 0.75 \cdot 10^{-5}\ M$$

This value of x is indeed negligible compared to $0.100\ M$. The calculation is valid.

$$[OH^-] = [HOAc] = 7.5 \cdot 10^{-6}\ M \tag{5-69}$$

$$[H^+] = \frac{K_w}{[OH^-]} = \frac{1.00 \cdot 10^{-14}}{7.5 \cdot 10^{-6}} = 1.33 \cdot 10^{-9}\ M \tag{5-70}$$

$$pH = -\log (1.33 \cdot 10^{-9}) = 9 - 0.12 = 8.88$$

Exercise 5-15. Show that $[OH^-]$ equals $8.6 \cdot 10^{-7}\ M$ in $0.100\ M$ sodium monochloracetate, the salt of chloroacetic acid, with $K_a = 1.35 \cdot 10^{-3}$.

(g) A WEAK ACID–STRONG BASE TITRATION CURVE

We have just about performed all of the work to portray the change of pH during a titration of 100.0 ml of $0.2000\ M$ acetic acid with $0.2000\ M$ sodium hydroxide. In Section 5-3(d), we found that the pH is 2.73 in a $0.200\ M$ acetic acid solution. Then, in Section 5-3(e), we calculated the pH to be 4.75 in a solution containing equal concentrations of acetic acid and sodium acetate. That is the situation when 50.0 ml of NaOH have been added to the 100.0 ml CH_3COOH. Exercise 5-14, with $0.020\ M$ HOAc and $0.180\ M$ NaOAc, has the same pH, 5.70, as our titration solution after 90.0 ml NaOH have been added. Then, Section 5.3(f) treats the equivalence point and it shows pH = 8.88.

To complete the titration curve, we really need only a couple of points beyond the equivalence point. Consider the addition of 101.0 ml 0.2000 M NaOH. We have added 1.0 ml of 0.2000 M NaOH over that needed to react completely with the acetic acid we started with. There is (1.0)(0.2000) = 0.2000 millimole excess NaOH. The volume will be 100 + 101 = 201 ml. Hence the OH⁻ concentration due to excess NaOH is

$$[OH^-] = \frac{0.2000 \text{ millimole}}{201 \text{ milliliters}} = 1.0 \cdot 10^{-3} \, M \qquad (5\text{-}71)$$

There is an additional source of OH⁻ from hydrolysis of the acetate ion, but that source of OH⁻ gave only $7.5 \cdot 10^{-6} \, M$ in sodium acetate. It will be even smaller in the presence of excess base, so it will be negligible.

This is an important result. Our calculation of the hydroxide ion concentration beyond the equivalence point was based entirely on the excess strong base. That means the titration curve becomes identical to that of the strong acid–strong base titration curve, once the equivalence point is passed. We can borrow from Figure 5-4 this part of the curve.

Figure 5-7 presents the acetic acid–sodium hydroxide titration curve. The dotted curve reproduces the HCl–NaOH titration curve for reference. The difference is typical of a weak acid–strong base titration. There are four distinctive features to note.

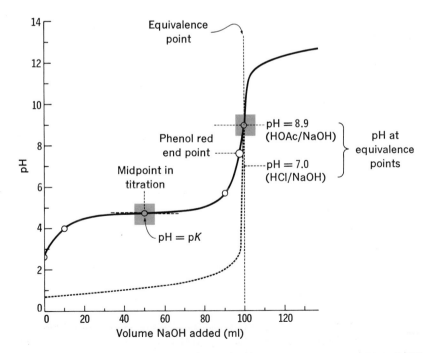

Figure 5-7 Titration curve for titration of 100.0 ml 0.2000 M acetic acid with 0.2000 M NaOH.

(i) The pH at the equivalence point is 8.9, higher than the neutral pH = 7.0 that characterizes the equivalence point in a strong acid–strong base titration. In general, the equivalence point of a weak acid–strong base titration depends upon the dissociation constant: the weaker the acid, the higher the equivalence point pH.

(ii) The pH beyond the equivalence point follows the titration curve of the strong acid–strong base titration curve.

(iii) The rise of pH at the equivalence point in a weak acid–strong base titration is less precipitous than in the strong acid–strong base titration. This means that the accurate determination of the equivalence point will be more difficult.

(iv) Halfway to the equivalence point, the pH changes rather slowly in the weak acid–strong base titration. This implies that a weak acid and its salt furnish an effective means of regulating or "buffering" the pH even at very low hydrogen ion concentrations.

(h) MORE ON INDICATORS

The acetic acid titration curve shows why it is necessary for us to work with a variety of indicators and to understand how they work. In the HCl/NaOH titration, within $\pm 1\%$ of the equivalence point, the pH changed from 3.0 to 11.0. In the HOAc/NaOH titration this change extends only from 6.75 to 11.0 and the equivalence point pH is 8.9. An indicator that accurately signals the equivalence point in the HCl case would not be suitable for the HOAc titration. We'd better look into their chemical nature.

To be a useful dye, a substance must be colored and a tiny amount of it must be sufficient to absorb light and impart color. These properties are important to us since we want a visible signal system when we add to our titration solution an amount of dye too small to interfere with the titration.

Some dyes are weak acids or weak bases. Any such dye could itself be titrated and it would have a titration curve qualitatively resembling the HOAc/NaOH curve in Figure 5-7. At the midpoint in the titration the pH = pK, and the acid form of the dye and the salt form are in equal concentration. At lower pH (higher acidity), the acid form will predominate. At higher pH (lower acidity), the base form will predominate. If the indicator acid and its base form are different in color, there will be a color change associated with the pH = pK.

For example, consider the dye phenolsulfonphthalein, which is also called, thank goodness, phenol red. This substance is a dye and a weak acid. Its acid form is yellow and its salt, or base form, is red. Figure 5-8 shows their structures. When base is slowly added to an acidic solution of phenol red, the color changes from yellow to orange to red. This color change is connected with the equilibrium between the cumbersome molecules shown in Figure 5-8. Despite the elephantine character of the ingredients, this equilibrium is as easy to understand as that of acetic acid. To

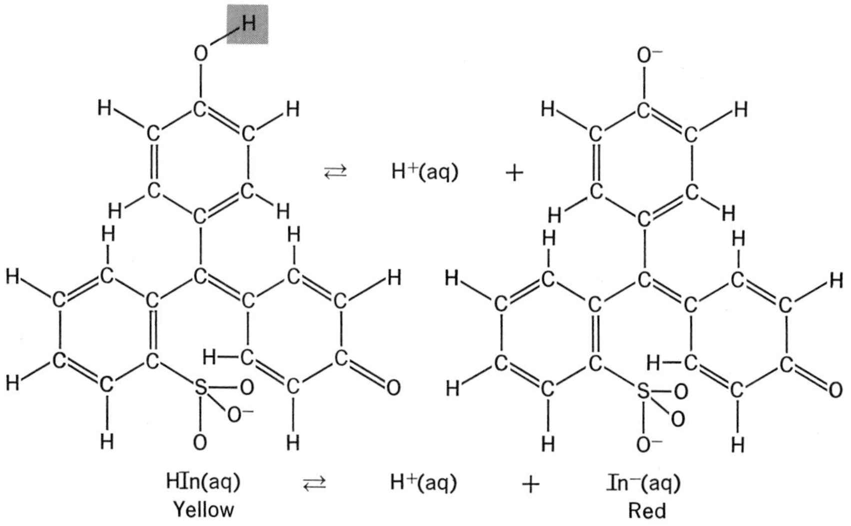

Figure 5-8 The molecular structure of phenol red in its acid and base forms.

prevent intimidation by the structures, we'll call the acid form HIn and the base form In⁻. Then we have a simple equilibrium,

$$HIn = H^+(aq) + In^-(aq)$$

$$K_{HIn} = \frac{[H^+][In^-]}{[HIn]} \tag{5-72}$$

We can solve this for the ratio of the concentrations of the two colored species:

$$\frac{[In^-]}{[HIn]} = \frac{K_{HIn}}{[H^+]} \tag{5-73}$$

This expression (5-73) gives us a pretty good basis for understanding the color change that was described. If the ratio $[In^-]/[HIn]$ is below 1/10, then the relative amount of HIn is high and its yellow color should predominate. If the ratio $[In^-]/[HIn]$ exceeds 10, the relative amount of In⁻ is high and its red color should predominate. Somewhere around the hydrogen ion concentration that makes these two concentrations equal, the color should be changing and the solution will be orange.

When $\dfrac{[In^-]}{[HIn]} = \dfrac{K_{HIn}}{[H^+]} = \dfrac{1}{10}$, the solution will be yellow.

$$\tag{5-74}$$

This will occur at $[H^+] = 10 \cdot K_{HIn}$ and above, or pH $\leq$ p$K_{HIn} - 1$.

When $\dfrac{[In^-]}{[HIn]} = \dfrac{K_{HIn}}{[H^+]} = 1,$ the solution will be orange.

(5-75)

This will occur at $[H^+] = K_{HIn}$ or pH $= pK_{HIn}$.

When $\dfrac{[In^-]}{[HIn]} = \dfrac{K_{HIn}}{[H^+]} = 10,$ the solution will be red.

(5-76)

This will occur at $[H^+] = K_{HIn}/10$ and below, or pH $\geq pK_{HIn} + 1$.

The particular indicator we selected, phenol red, has a value of $K_{HIn} = 2.5 \cdot 10^{-8}$ or a $pK = 7.6$. We can expect its color change to occur over the pH range 6.6 to 8.6 and to be centered near 7.6.

With this understanding, we can now see that an indicator's color change does not necessarily occur at the equivalence point. Consequently, the color change point in a titration is called the *end point*—that is where the dye signal system tells us to end the titration. The trick is to make the end point as close to the equivalence point as possible. Another lesson we have learned is that the change of indicator color consumes some of the hydroxide ion we are adding. To be able to neglect this complicating reaction, indicator concentration must be extremely small. Fortunately there are plenty of useful weak acid (or weak base) dyes that are so intensely colored that they give visible color at concentrations below $10^{-4}\,M$.

Just a glance at Figure 5-7 tells us that phenol red could be used as an indicator in the HCl/NaOH titration, but not in the HOAc/NaOH titration. In the first case, the pH changes so rapidly between pH $= 7$ (the equivalence point) and pH $= 7.6$ (the phenol red end point) that much less than one drop of NaOH beyond the equivalence point would cause the color change. That fraction of a drop would be the error in the signal system. For HOAc/NaOH, though, the color change would take place three or four drops before the equivalence point. That would be, for most purposes, an undesirable discrepancy between end point and equivalence point. Some other indicator dye would be better.

Table 5-4 lists some useful indicator dyes. To decide which to use, it is necessary to calculate the equivalence point pH. The indicator whose color change brackets that pH will give a satisfactory end point.

Exercise 5-16. Which indicator in Table 5-4 would be best for the acetic acid–sodium hydroxide titration?

(i) CONTROL OF pH: BUFFER SOLUTIONS

There are many chemical and biochemical processes that operate satisfactorily only if the pH is held within narrow limits. The human body abounds with interesting examples. The pH of the blood must be between

Table 5-4 *Some Indicators and the pH Ranges of Their Color Transitions*

Indicator, Common Name	pH Transition Range	Color Change, Acidic to Basic	
methyl orange	3.1 to 4.6	red	to orange-yellow
bromcresol green	3.8 to 5.4	yellow	to blue
methyl red	4.2 to 6.3	red	to yellow
bromthymol blue	6.0 to 7.6	yellow	to blue
phenol red	6.8 to 8.4	yellow	to red
phenolphthalein	8.0 to 9.8	colorless to red	
thymolphthalein	9.4 to 10.6	colorless to blue	

7.0 and 7.9—even slight deviation from these limits can be fatal. Plasma solutions are subject to even more stringent limits; they must have pH between 7.38 and 7.41! Saliva, on the other hand, is maintained at a pH of 6.8. The duodenum pH is between 6.0 and 6.5. In the stomach, the gastric juices are quite acidic—the pH must be between 1.6 and 1.8 to promote digestion of food materials. The body maintains these various pH ranges, as needed, by means of chemical constituents that resist pH change when small amounts of acid or base are added. Solutions with such pH regulatory power are called *buffer solutions.*

The titration curves shown in Figure 5-7 help us understand how to construct a buffer solution for a given pH range. For example, suppose a reaction proceeds to our satisfaction only if the pH is 5.0. Figure 5-7 shows that we could establish the pH at 5.0 by adding a bit of HCl—10^{-5} mole/liter, in fact. But the steepness of the HCl/NaOH titration curve at pH = 5 tells us that only a tiny amount of additional HCl or NaOH would cause a large change in pH. Contrast that situation to the HOAc/NaOH titration curve. The pH reached 5 in that titration when about 65% of the acetic acid had been converted to sodium acetate. Hence we could obtain the desired pH in our reaction solution merely by adding acetic acid and sodium acetate in this ratio, 1:2. Furthermore, the slope of the titration curve at pH = 5 is quite gentle. That means it takes quite a bit of added base to raise the pH, say, by one pH unit or quite a bit of added acid to lower the pH the same amount. Finally, if we hark back to Exercise 5-13, we are reminded that the pH does not change at all in such an HOAc/NaOAc solution when the solution is diluted even by a factor of ten or larger. Plainly such mixtures are buffer solutions. The useful pH range is also indicated by the titration curve. It is the relatively horizontal part of the curve. In general, a weak acid and its salt at equal concentration give a buffer pH = pK and by change of the relative amounts, they can furnish useful buffer action over the pH range of about p$K \pm 1$.

Exercise 5-17. Show that an equimolar mixture of formic acid and sodium formate would buffer the pH at 3.75.

5-4 Multiple equilibria

In this concluding section we will sharpen our newly acquired skills with dissociation equilibria by considering some problems of greater complexity. We will focus on solutions in which more than one equilibrium is important.

(a) VERY DILUTE SOLUTIONS

Every acid-base dissociation calculation in the preceding sections ignored the possible effect of the dissociation of the solvent itself.

$$H_2O(\ell) \rightleftharpoons H^+(aq) + OH^-(aq) \qquad (5\text{-}77)$$

We got away with this because the concentrations of protons or hydroxide ions from other equilibria were extremely large compared with those from water. This is not always so, however. Suppose, for example, that we have a solution of $1.30 \times 10^{-7}\ M$ HCl. This is a different kettle of fish, for proton concentrations of about the same magnitude, near $10^{-7}\ M$, result from the dissociation of pure water. We must consider the total concentration of hydrogen ions in solution, which equals the sum of contributions from all sources.

$$[H^+]_{total} = [H^+]_{H_2O} + [H^+]_{HCl} \qquad (5\text{-}78)$$

The protons from water dissociation are produced in quantity equal to the hydroxide ions from the same source.

$$[H^+]_{H_2O} = [OH^-]_{H_2O} \qquad (5\text{-}79)$$

The equilibrium constant K_w relates these concentrations at equilibrium.

$$K_w = [H^+]_{total} \cdot [OH^-]_{H_2O} = 1.00 \cdot 10^{-14} \qquad (5\text{-}80)$$

Inserting (5-79) into (5-80),

$$K_w = [H^+]_{total} \cdot [H^+]_{H_2O} \qquad (5\text{-}81)$$

Since we are interested in $[H^+]_{total}$, it is convenient to substitute (5-78) into (5-81) to eliminate $[H^+]_{H_2O}$:

$$K_w = [H^+]_{total} \cdot ([H^+]_{total} - [H^+]_{HCl}) \qquad (5\text{-}82)$$

Expression (5-82) can be expanded and rearranged into the quadratic form:

$$[H^+]^2_{total} - [H^+]_{total} \cdot [H^+]_{HCl} - K_w = 0 \qquad (5\text{-}83)$$

We can solve (5-83) with the quadratic formula (4-58) to give

$$[H^+]_{total} = 1.84 \cdot 10^{-7} \, M \qquad (5\text{-}84)$$

We find the hydrogen ion concentration, (5-84), to be 42% larger than the HCl would produce alone.

(b) POLYPROTIC ACIDS: SULFURIC ACID

Some substances that contain more than one hydrogen atom can release two or even three protons. Such a substance is called a *polyprotic acid*. Sulfuric acid, H_2SO_4, is an example—since it can release two protons, it is called, more specifically, a *diprotic acid*.

Sulfuric acid dissociates completely in dilute aqueous solution to form $H^+(aq)$ and hydrogen sulfate ions, $HSO_4^-(aq)$. In this first dissociation, sulfuric acid is a strong acid.

$$H_2SO_4(aq) \longrightarrow H^+(aq) + HSO_4^-(aq) \qquad (5\text{-}85)$$

The hydrogen sulfate ion is, however, a weak acid in its own right.

$$HSO_4^-(aq) \rightleftharpoons H^+(aq) + SO_4^{-2}(aq) \qquad (5\text{-}86)$$
$$K_a = 1.02 \times 10^{-2}$$

Equilibrium-wise, then, sulfuric acid acts like a weak monoprotic acid (HSO_4^-) in the presence of another strong acid (H_2SO_4). Calculations of the pH in a sulfuric solution need only be concerned with the second equilibrium, (5-86), taking $[H^+]$ as the sum of the proton concentration provided by (5-85) and that provided by (5-86).

(c) POLYPROTIC ACIDS: CARBONIC ACID

Most polyprotic acids are more complicated and more interesting than sulfuric acid because all of the acidic protons are released in weak acid equilibria. Carbonic acid is a good example.

Carbon dioxide dissolves in water to give a solution of carbonic acid, $H_2CO_3(aq)$. At one atmosphere CO_2 pressure, the total dissolved CO_2 is about 0.04 M.

$$CO_2(g) + H_2O(\ell) = H_2CO_3(aq) \qquad (5\text{-}87)$$

Carbonic acid is a weak diprotic acid that dissociates to form an even weaker acid, bicarbonate ion.

$$H_2CO_3(aq) = H^+(aq) + HCO_3^-(aq) \qquad K_1 = 4.2 \cdot 10^{-7} \qquad (5\text{-}88)$$

$$HCO_3^-(aq) = H^+(aq) + CO_3^{-2}(aq) \qquad K_2 = 4.8 \cdot 10^{-11} \qquad (5\text{-}89)$$

Any carbonic acid solution contains these two equilibria simultaneously and, of course, the water dissociation as well. Calculations required to cope with this busy situation are a bit complex. Nevertheless, they are worth a moment of our time because of the importance of this system of equilibria in biological systems. Not only does CO_2 determine the tartness of soda pop, it furnishes the buffer system that maintains the blood pH between 7.0 and 7.9.

There are some fairly easy conclusions that can be drawn about the principal species that must be present in a carbonate system at any given pH. Consider the equilibria appropriate to (5-88) and (5-89):

$$K_1 = \frac{[H^+][HCO_3^-]}{[H_2CO_3]} \qquad (5\text{-}90)$$

$$K_2 = \frac{[H^+][CO_3^{-2}]}{[HCO_3^-]} \qquad (5\text{-}91)$$

These are more usefully rearranged to show the concentration ratios of carbonate-containing species:

$$\frac{[HCO_3^-]}{[H_2CO_3]} = \frac{K_1}{[H^+]} = \frac{4.2 \cdot 10^{-7}}{[H^+]} \qquad (5\text{-}92)$$

$$\frac{[CO_3^{-2}]}{[HCO_3^-]} = \frac{K_2}{[H^+]} = \frac{4.8 \cdot 10^{-11}}{[H^+]} \qquad (5\text{-}93)$$

An interesting acidity is $[H^+] = 4.2 \cdot 10^{-7}$. This value of $[H^+]$ makes the HCO_3^- and H_2CO_3 concentrations equal. At the same time, (5-93) shows that $[CO_3^{-2}]$ is negligibly small.

At

$$[H^+] = 4.2 \cdot 10^{-7} \qquad \frac{[HCO_3^-]}{[H_2CO_3]} = \frac{K_1}{[H^+]} = \frac{4.2 \cdot 10^{-7}}{4.2 \cdot 10^{-7}} = 1$$
$$pH = 6.38$$

$$\frac{[CO_3^{-2}]}{[HCO_3^-]} = \frac{K_2}{[H^+]} = \frac{4.8 \cdot 10^{-11}}{4.2 \cdot 10^{-7}} = 1.1 \cdot 10^{-4}$$

Since the concentration of CO_3^{-2} is so small, the carbonate system at

$pH = 6.38$ and lower acts like a monoprotic acid with $K = K_1$. We can use this information to advantage to calculate, for example, the pH of a H_2CO_3 solution. The second dissociation can be ignored. The calculation becomes a familiar one. Consider a solution saturated with CO_2 at one atmosphere so that $[H_2CO_3] = 0.04$.

Let

$$[H^+] = [HCO_3^-] = x$$

$$[H_2CO_3] = 0.040$$

$$K_1 = 4.2 \cdot 10^{-7} = \frac{[H^+][HCO_3^-]}{[H_2CO_3]} = \frac{x^2}{0.040}$$

$$x = [H^+] = [HCO_3^-] = 1.3 \cdot 10^{-4}\,M \tag{5-94}$$

$$pH = 3.89$$

The result (5-94) enables us to verify the unimportance of $[CO_3^{-2}]$ by substitution into (5-91):

$$[CO_3^{-2}] = K_2\frac{[HCO_3^-]}{[H^+]} = (4.8 \cdot 10^{-11})\frac{1.3 \cdot 10^{-4}}{1.3 \cdot 10^{-4}} = 4.8 \cdot 10^{-11}$$

Now let's return to (5-93) to investigate another rather special value of $[H^+]$: $[H^+] = 4.8 \cdot 10^{-11}$, $pH = 10.32$. At this pH, $[CO_3^{-2}] = [HCO_3^-]$. At the same time, (5-92) shows that now $[H_2CO_3]$ is negligible.

$$\frac{[CO_3^{-2}]}{[HCO_3^-]} = \frac{K_2}{[H^+]} = \frac{4.8 \cdot 10^{-11}}{4.8 \cdot 10^{-11}} = 1$$

$$\frac{[HCO_3^-]}{[H_2CO_3]} = \frac{K_1}{[H^+]} = \frac{4.2 \cdot 10^{-7}}{4.8 \cdot 10^{-11}} = 8750 \tag{5-95}$$

$$[H_2CO_3] = \frac{[HCO_3^-]}{8750}$$

Again we can conclude that there is a pH range, $pH > 10$, within which the system can be considered to be monoprotic, due to HCO_3^-. This applies, for example, to the calculation of $[H^+]$ in a $0.20\,M$ Na_2CO_3 solution.

$$CO_3^{-2}(aq) + H_2O = HCO_3^-(aq) + OH^-(aq) \tag{5-96}$$

$$K_b = \frac{K_w}{K_2} = \frac{[HCO_3^-][OH^-]}{[CO_3^{-2}]}$$

Let

$$[HCO_3^-] = [OH^-] = x$$

$$[CO_3^{-2}] = 0.20 - x$$

$$K_b = 2.08 \cdot 10^{-4} = \frac{x^2}{0.20 - x} \cong \frac{x^2}{0.20}$$

$$x = [HCO_3^-] = [OH^-] = 6.4 \cdot 10^{-3} \, M \tag{5-97}$$

$$[H^+] = \frac{K_w}{[OH^-]} = \frac{1.0 \cdot 10^{-14}}{6.4 \cdot 10^{-3}} = 1.56 \cdot 10^{-12} \, M \tag{5-98}$$

$$pH = 11.81$$

We have now calculated the pH in a solution of H_2CO_3 of Na_2CO_3 in a H_2CO_3–HCO_3^- buffer and in a HCO_3^-–CO_3^{-2} buffer. There is one more special situation—that of a solution in which $[CO_3^{-2}] = [H_2CO_3]$. The $[H^+]$ at which this will occur is obtained by multiplying (5-90) by (5-91):

$$K_1 \cdot K_2 = \frac{[H^+][HCO_3^-]}{[H_2CO_3]} \cdot \frac{[H^+][CO_3^{-2}]}{[HCO_3^-]} = [H^+]^2 \frac{[CO_3^{-2}]}{[H_2CO_3]}$$

and if $[CO_3^{-2}] = [H_2CO_3]$,

$$[H^+]^2 = K_1K_2 = [4.2 \cdot 10^{-7}][4.8 \cdot 10^{-11}] = 20.2 \cdot 10^{-18}$$

$$[H^+] = 4.5 \cdot 10^{-9} \, M \tag{5-99}$$

$$pH = 8.35$$

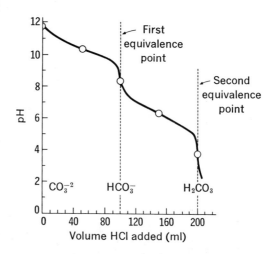

Figure 5-9 *Titration of 100.0 ml 0.2000 M Na_2CO_3 with 0.2000 M HCl.*

Table 5-5 Acid Dissociation Constants of Polyprotic Acids in Water ($T = 298°K$)

Acid	Formula	K	Acid	Formula	K
sulfuric acid	H_2SO_4	strong acid	carbonic acid	H_2CO_3	$4.2 \cdot 10^{-7}$
bisulfate ion*	HSO_4^-	$1.2 \cdot 10^{-2}$	bicarbonate ion*	HCO_3^-	$4.8 \cdot 10^{-11}$
oxalic acid	$HOOCCOOH$	$5.9 \cdot 10^{-2}$	hydrogen sulfide	H_2S	$1.1 \cdot 10^{-7}$
oxalate ion	$HOOCCOO^-$	$6.4 \cdot 10^{-2}$	bisulfide ion*	HS^-	$1.0 \cdot 10^{-15}$
sulfurous acid	H_2SO_3	$1.7 \cdot 10^{-2}$	phosphoric acid	H_3PO_4	$7.5 \cdot 10^{-3}$
bisulfite ion*	HSO_3^-	$6.2 \cdot 10^{-8}$		$H_2PO_4^-$	$6.2 \cdot 10^{-8}$
maleic acid	$H_2C_4H_2O_4$	$1.0 \cdot 10^{-2}$		HPO_4^{-2}	$1 \cdot 10^{-12}$
maleate ion	$HC_4H_2O_4^-$	$5.5 \cdot 10^{-7}$	citric acid	$H_3C_6H_5O_7$	$8.7 \cdot 10^{-4}$
phthalic acid	$C_6H_4(COOH)_2$	$1.3 \cdot 10^{-2}$		$H_2C_6H_5O_7^-$	$1.8 \cdot 10^{-5}$
phthalate ion	$HOOCC_6H_4COO^-$	$3.9 \cdot 10^{-6}$		$HC_6H_5O_7^{-2}$	$4.0 \cdot 10^{-6}$
tartaric acid	$H_2C_4H_4O_6$	$9.6 \cdot 10^{-4}$			
tartrate ion	$HC_4H_4O_6^-$	$2.9 \cdot 10^{-5}$			

* The names hydrogen sulfate, hydrogen sulfite, hydrogen carbonate, hydrogen sulfide, etc. are now preferred by many chemists to the more traditional names bisulfate, bisulfite, bicarbonate, bisulfide, etc.

This solution has special significance in a titration of Na_2CO_3 with HCl. When the first equivalence point has been reached, there will be just enough H^+ to convert all of the CO_3^{-2} to HCO_3^-. Of course, if some H_2CO_3 is formed, an equal amount of CO_3^{-2} must necessarily be left. We see that the condition $[H_2CO_3] = [CO_3^{-2}]$ corresponds to the pH of a pure $NaHCO_3$ solution and the midpoint in the CO_3^{-2}–HCl titration.

In Figure 5-9, all of this is translated into a titration curve. The curve applies to a 0.2000 M Na_2CO_3 solution titrated with 0.2000 M HCl. It begins at pH $= 11.81$, as we calculated for a pure Na_2CO_3 solution. At the 50.0 ml mark in the titration, $[CO_3^{-2}] = [HCO_3^-]$ and pH $= 10.32$. This is a buffer region. Then, at the first equivalence point, pH $= 8.35$. Next we reach the 150 ml mark, where $[HCO_3^-] = [H_2CO_3]$ and pH $= 6.38$. Again we have a buffer region (the one that stabilizes pH in the blood). Finally, the H_2CO_3 solution has a pH of 3.9 at the second equivalence point.

The complete curve of Figure 5-9 shows that Na_2CO_3 could be titrated either to an HCO_3^- equivalence point, by selecting an indicator with its color change at pH $= 8.35$, or to the H_2CO_3 equivalence point with an indicator that changes at pH $= 3.9$. Furthermore, the carbonate system has two distinct buffer regions, near pH $= 10.3$ and near pH $= 6.4$.

(d) POLYPROTIC ACIDS: OTHERS

Table 5-5 lists a number of polyprotic acids, most of which have some biological functions. One can generalize that a polyprotic acid can be treated as a sequence of independent and separately titratable monoprotic acids provided successive dissociation constants differ by factors as large as 10^3 to 10^4. This is true for several important acids, including H_2S and phosphoric acid; but it is notably not true for quite a few acids with very important biological roles—tartaric and citric acids included. Calculating the pH as a function of composition in these cases can be a bit of a chore, though nature seems to handle the arithmetic with no difficulty.

Problems

1. Calculate the H^+ and OH^- concentrations in a solution of $2 \cdot 10^{-4}$ mole of HCl dissolved in 100 ml of water.

2. Calculate the H^+ and OH^- concentrations in a solution of $2 \cdot 10^{-4}$ mole of solid $Ba(OH)_2$ dissolved in 100 ml of water.

3. Calculate $[H^+]$ and $[OH^-]$ in the solution obtained in Problem 1 if it is cooled to 0°C.

4. An aqueous solution has $[H^+] = 1.0 \cdot 10^{-7}$ M at 0°C. Is the solution neutral, acidic, or basic?

5. An aqueous solution has $[H^+] = 1.0 \cdot 10^{-7}$ M at 50°C. Is the solution neutral, acidic, or basic?

6. The dissociation constant for water has been measured at 230°C (503°K) under pressure, K_w (230°C) $= 9.0 \cdot 10^{-12}$. Calculate [H$^+$] and [OH$^-$] in pure water at this temperature.

8. Calculate [H$^+$] and [OH$^-$] in a solution made by mixing 100 ml of 0.30 M HCl and 200 ml of 0.45 M HNO$_3$.

8. Write the equation for the reaction between formic acid, HCOOH, and methyl amine, CH$_3$NH$_2$, to give formate ion, HCOO$^-$, and methyl ammonium ion, CH$_3$NH$_3^+$. Name the two acids involved in this equation and, for each, identify its conjugate base.

9. Calculate pH and pOH for Problem 1.

10. Calculate pH and pOH for Problem 2.

11. Calculate pH and pOH for pure water at 230°C under pressure (see Problem 6).

12. What is the pH of a 0.30 M HCl solution? $[H^+] = 3$

13. What is the pH of a 0.30 M acetic acid solution (CH$_3$COOH)?

14. Calculate the pH during the titration of 100.0 ml of 0.02000 M HCl with 0.02000 M NaOH after the following NaOH volumes have been added: 0, 50, 90, 99.0, 100.0, 101.0, 110 ml. Plot these along with the similar values calculated in Section 5-2(e) exactly as done in Figure 5-4.

15. Calculate [H$^+$], [OH$^-$], pH, and pOH in 0.0020 M acetic acid, CH$_3$COOH, to 10% accuracy.

16. The dissociation constant for acetic acid, CH$_3$COOH, has been measured in aqueous solution at 230°C (503°K) under pressure: K_a (230°C) $= 2.0 \cdot 10^{-6}$. Use this value and the data of Problem 6 to calculate [H$^+$], [OH$^-$], pH, and pOH in 0.0020 M CH$_3$COOH at 230°C.

17. What is the equilibrium constant for the reaction between sodium benzoate and water to give the conjugate acid, benzoic acid, and hydroxide ion:

$$C_6H_5COO^-(aq) + H_2O = C_6H_5COOH(aq) + OH^-(aq)$$

18. What is the [H$^+$] in 0.20 M sodium benzoate solution?

19. What are the pH and pOH in 0.20 M sodium acetate solution at 230°C under pressure (use the data of Problems 6 and 16).

20. Calculate the pH during the titration of 100.0 ml of 0.02000 M acetic acid with 0.02000 M NaOH after the following NaOH volumes have been added: 0, 50, 90, 100.0, 101.0, 110 ml. Plot these along with the similar values calculated in Section 5-3(g). Compare to the plot obtained in Problem 14.

21. Pyruvic acid, CH$_3$CCOOH, with a carbonyl (O double-bonded) group, is an intermediate species in the metabolism of carbohydrates, proteins, and possibly of fats. It is a monoprotic acid with dissociation constant $K_a = 6.6 \cdot 10^{-3}$. Calculate to 10% accuracy the H$^+$ concentration in a 0.10 M solution of pyruvic acid (HP) assuming the only equilibrium is the formation of the salt, P$^-$ (HP $=$ H$^+$ $+$ P$^-$). (See Exercise 5-12.)

22.   Measurements show that in a 0.10 M pyruvic acid solution, $[H^+] = 0.016\ M$ rather than the value calculated in Problem 21. This has been attributed to formation of pyruvic acid hydrate, $HP \cdot hy$:*

$$CH_3\overset{\overset{\displaystyle O}{\|}}{C}COOH + H_2O = CH_3\underset{\underset{\displaystyle OH}{|}}{\overset{\overset{\displaystyle OH}{|}}{C}}{-}COOH \qquad K_H = 1.5$$

or, in abbreviation,

$$HP + H_2O = HP \cdot hy \qquad K_H = 1.5$$

The hydrate is itself an acid, with dissociation constant a bit below that of pyruvic acid.

$$HP \cdot hy = H^+ + P \cdot hy^- \qquad K'_a = 2.5 \cdot 10^{-4}$$

(a) From the experimental value of $[H^+]$ calculate the concentration ratios

$$\frac{[P^-]}{[HP]} \quad \text{and} \quad \frac{[P \cdot hy^-]}{[HP \cdot hy]}$$

(b) Combine, as appropriate, K_a, K'_a, and K_H to calculate the equilibrium constant K'_H for the hydration of the salt, P^-

$$P^- + H_2O = P \cdot hy^- \qquad K'_H = ?$$

23.   Many nitrogen-containing organic acids act as bases in aqueous solution. Imidazole, whose structure (see Table 5-3) is found in some enzymes, is one of these. (a) Write a balanced equation for the reaction of imidazole with water to form the conjugate acid, imidazonium ion.

$$\begin{array}{ccc} HN & \!\!\!\!\!\!\!\!\text{---}\!\!\!\!\!\!\!\! & CH \\ | & & \| \\ HC & & CH \\ & \diagdown\ \diagup & \\ & N & \quad + \\ & | & \\ & H & \end{array}$$

(b) From the value of K_b for imidazole (Table 5-3) calculate K_a for its conjugate acid.
(c) Calculate $[OH^-]$ in 0.15 M imidazole solution.

24.   The ergot alkaloids are produced by a fungus known as ergot. One group of these compounds is derived from the indole derivative $C_{15}H_{15}N_2COOH$, lysergic acid. Lysergic acid was first prepared and its acid dissociation constant measured in 1938: $K_a = 6.3 \cdot 10^{-4}$. If 25.0 ml of lysergic acid, HLy, were titrated to a phenol red

end point, using 27.4 ml of 0.082 M NaOH, what was the original concentration of acid?

25. Which acid and conjugate base in Table 5-2 would provide the best buffer system at pH $= 3.2$? At pH $= 4.2$? At pH $= 8.7$?

26. Which acid and conjugate base in Table 5-2 would provide the best buffer system at pOH $= 11.2$? At pOH $= 9.8$? At pOH $= 4.7$?

27. What volume of 0.082 M NaOH should be added to the 25.0 ml of the solution titrated in Problem 24 to bring the pH to 3.2?

28. After eating too much pizza, an amateur doctor prescribes for himself a potion made by dissolving one-quarter teaspoon (0.70 gm) of baking soda ($NaHCO_3$) in a glass of water (250 ml).
(a) What is the pH in the potion?
(b) If this greedy pizza eater drinks 100 ml of the potion and it mixes in his tummy with 100 ml of 0.020 M HCl, what is the pH in there?

29. In Problem 22, the equilibrium concentrations [HP], [HP · hy], [P$^-$], and [P · hy$^-$] must add up to 0.10 M, the original pyruvic acid concentration. Combine this relationship with the equilibrium ratios calculated in Problem 22 to calculate the concentration of each species.

30. In a solution of *para*periodic acid, H_5IO_6, the following equilibria are important:

$$H_5IO_6 = H^+ + H_4IO_6^- \qquad K_1 = 5 \cdot 10^{-4}$$

$$H_4IO_6^- = H^+ + H_3IO_6^{-2} \qquad K_2 = 2 \cdot 10^{-7}$$

$$H_4IO_6^- = IO_4^- + 2 H_2O \qquad K = 40$$

(a) Calculate the concentration ratios $[H_5IO_6]/[H_4IO_6^-]$, $[H_4IO_6^-]/[H_3IO_6^{-2}]$ and $[IO_4^-]/[H_4IO_6^-]$ at pH $= 2.0$.
(b) If 0.25 mole of periodic acid is dissolved in one liter of water and solid NaOH is added until the pH $= 2.0$, how many moles of NaOH are needed?

six   spontaneous processes and electrochemistry

A river tumbling down a mountain valley expends an immense amount of potential energy, which is converted in a spendthrift way into the random motions of heat. As the water tumbles over rocks and splashes against canyon walls, the directed flow of the river becomes the disordered motion of heat: The water is warmed.

Because of the inspiration of a creative engineer, a series of dams is placed across the valley. Now, as the water makes its way down to the sea, it passes through a succession of turbines that convert much of its potential energy into electrical energy. This electrical energy can be converted into light, heat, chemical energy, or mechanical work, as man desires.

Chemical reactions often occur in ways as uncontrolled as the rush of an untamed river. Processes that take place without any outside assistance are called "spontaneous" changes. For example, a car will roll downhill spontaneously—it must be pushed back up. Similarly, carbon dioxide gas will expand spontaneously from a fire extinguisher—a compressor is needed to get the CO_2 back in. Iron will rust spontaneously in moist air—it takes a smelting plant to convert iron ore back to the metal. We must learn about spontaneous changes. They are responsible for what goes on around us and within us. We must find how to study them so we can find how to control them, as the engineer controlled the river torrent, to work to man's benefit.

In this chapter we will see how most spontaneous chemical processes can be studied profitably in electrochemical cells. We will find that the cell voltage is a measure of the reaction tendency, or degree of spontaneity. As well as giving us this important chemical information, electrochemical cells provide us with a bench-top equivalent of a hydroelectric dam. Spontaneous chemical reactions, like spontaneous mechanical processes, can be made to do work.

6-1   Spontaneous reactions in electrochemical cells

Figure 6-1 illustrates a whole bunch of simple reactions that we know from experience to be spontaneous. Ranging from the dispersion of a concen-

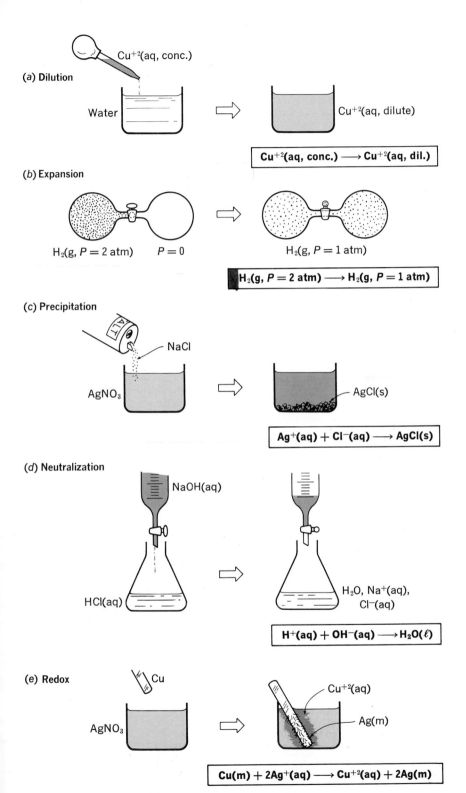

Figure 6-1   Five spontaneous reactions: no useful work done.

trated copper ion solution in water (a) to the plating of silver metal on a strip of copper (e), each of these processes takes place without assistance. Unfortunately, no useful work is done in any of the spontaneous processes as they are carried out in this figure. However, this wasteful situation can be remedied using electrochemical cells.

(a) CONCENTRATION CELLS: I. DILUTION

A typical electrochemical cell is shown in Figure 6-2. It has been set up to exploit the spontaneous tendency of a concentrated solution, given opportunity, to become diluted. We will first use this cell to illustrate the features essential to any electrochemical cell.

The cell has been set up in two distinct parts: the concentrated copper ion solution in the right-hand (RH) beaker, a dilute copper ion solution in the left-hand (LH) beaker. Electrodes of copper metal are placed in each beaker. Now we must connect the two beakers—or, as they are known, *half-cells*—electrically. The electrodes are connected by conducting wires, a switch, and a voltmeter to measure the voltage. That's the easy part— any electrons flowing between the electrodes will be carried by the wire.

In the liquid, however, electric current is not carried by electrons, but as Arrhenius showed, by charged ions. Since there is no way these charged molecules can travel along a wire, we must provide a more amenable path for them. One convenient device—a salt bridge—is shown in the figure. The salt solution in the tube allows ion current to flow. The porous plugs in the ends of the bridge prevent complete mixing of the three solutions.

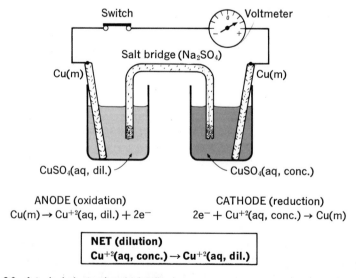

ANODE (oxidation)
$Cu(m) \rightarrow Cu^{+2}(aq, dil.) + 2e^-$

CATHODE (reduction)
$2e^- + Cu^{+2}(aq, conc.) \rightarrow Cu(m)$

NET (dilution)
$Cu^{+2}(aq, conc.) \rightarrow Cu^{+2}(aq, dil.)$

Figure 6-2 A typical electrochemical cell.

Before closing the switch, we should weigh the copper electrodes on a sensitive balance so that we will be able to detect any changes that may occur. Having done so, we hold our breath and go ahead. The switch is closed: The meter needle moves. Success! A current flows. Our electrochemical cell is acting like any well-behaved battery should. The meter tells us both the voltage and the direction of current flow: Electrons are being produced at the left-hand electrode (LHE) and traveling through the external circuit to the right-hand electrode (RHE).

But what is really happening? Is it really, as we have claimed, related to the spontaneous dilution in Figure 6-1(a)? If we reweigh the electrodes some time after closing the switch, we can learn the answer. This experiment shows that the LHE loses weight as current flows, while the RHE gains an *identical* amount. Consideration of the processes occurring at the individual electrodes will teach us a lot about electrochemical cells.

At the LHE, copper is dissolving and electrons are being produced according to the reaction

$$Cu(m) \longrightarrow Cu^{+2}(aq, \text{ dilute}) + 2\ e^- \tag{6-1}$$

At the RHE copper is plating out and electrons are being consumed.

$$2\ e^- + Cu^{+2}(aq, \text{ concentrated}) \longrightarrow Cu(m) \tag{6-2}$$

Neither of these reactions can possibly occur without the other. Reaction (6-1) alone would promptly accumulate a net negative charge on the electrode, which would discourage the production of more negatively charged electrons. It has to be a team effort. Electrons produced at one electrode must be used up at the other. These two reactions (6-1) and (6-2) are called *half-reactions*—they occur in the two half-cells that, together, make up our electrochemical cell.

To find out the net cell reaction, all we have to do is add together the two half-cell reactions—making sure that we do so in such a way that electrons are balanced.

| LHE | $Cu(m) \longrightarrow Cu^{+2}(aq, \text{ dilute}) + 2\ e^-$ |
| RHE | $2\ e^- + Cu^{+2}(aq, \text{ concentrated}) \longrightarrow Cu(m)$ |

$$\overline{\text{cell} \qquad Cu^{+2}(aq, \text{ concentrated}) \longrightarrow Cu^{+2}(aq, \text{ dilute})} \tag{6-3}$$

The net reaction (6-3) is very simple—it is just the spontaneous dilution process of Figure 6-1(a).

This is an interesting result. The simple, spontaneous, dilution reaction can be used to generate electricity—and electricity can be used to do work. Our first chemical "dam" is a great success. A cell that generates electricity as a result of a concentration difference is called, logically, a *concentration cell.* We will see many more of them.

Before galloping on, however, we should take time to learn the vocabulary of electrochemistry. If everyone uses a common language when talking about electrochemistry—or any other branch of science—communication is made much easier. Any half-cell reaction in which electrons are given up to the electrode is called an *oxidation reaction*, while any half-cell reaction in which electrons are consumed at the electrode is called a *reduction reaction*. The electrode at which oxidation occurs is called the *anode* and that at which reduction takes place is the *cathode*. Thus the LHE of Figure 6-2 is the anode and the RHE is the cathode. By these definitions, electrons are produced at the anode, so the flow of negative electricity in the circuit external to the cell is *always from the anode to the cathode.* Electrons produced at the anode by an oxidation half-reaction are consumed by a reduction process at the cathode.*

Now let's get quantitative about things for a moment. The amount of energy we can extract from a ton of water behind a hydroelectric dam depends on the difference in height between the top of the reservoir and the turbines. The greater the change in the potential energy of the water, the more useful work we can hope to extract. What about the driving force of our concentration cell? How does the operation of the cell change if we alter the concentration difference that causes it to work? Table 6-1 shows the answer for a variety of values of the concentration ratio, $[Cu^{+2}]_{anode}/[Cu^{+2}]_{cathode}$. We see that a concentration ratio of $1:1000$ leads to an observed voltage three times that for a concentration ratio of $1:10$. When the concentrations are identical, no dilution is possible and no electrical potential is developed in the cell. Then there is no tendency for any reaction to take place; the cell is in a state of equilibrium.

The data show that the operation of a concentration cell is dependent upon the concentration ratio. Furthermore, we are led to conclude that the voltage, or electric potential, measures the driving force for reaction in the same way that the height of the hydroelectric dam measures its potential for energy production. When the concentration ratio reaches unity, there

---

* This naming convention may be conveniently remembered by noting that Anode and Oxidation both begin with vowels, while Cathode and Reduction both start with consonants.

Table 6-1 *Measured Voltages: Copper Concentration Cell*

$\dfrac{[Cu^{+2}(aq)]_{anode}}{[Cu^{+2}(aq)]_{cathode}}$	Cell Voltage, $\Delta\varepsilon$ (millivolts)
10^{-3}	88.5
10^{-2}	59.0
10^{-1}	29.5
1	0.0

is no tendency for dilution, so the voltage reaches zero. No current will flow, and no change occurs. That describes equilibrium. At equilibrium, the cell voltage reaches zero.

(b) CONCENTRATION CELLS: II. GAS EXPANSION

A cell that makes use of the driving force of an expanding gas is a bit more complicated than the copper ion concentration cell. A suitable electrode employing hydrogen gas at a pressure of one atmosphere is shown in Figure 6-3. If the hydrogen ion concentration is 1.00 M, this half-cell is called a *standard hydrogen electrode*. As we shall see presently, this electrode plays an important role in electrochemistry. The electrode reaction is either

$$2 \, H^+(aq) + 2 \, e^- \longrightarrow H_2(g) \qquad\qquad (6\text{-}4)$$

or the reverse

$$H_2(g) \longrightarrow 2 \, H^+(aq) + 2 \, e^- \qquad\qquad (6\text{-}5)$$

depending on whether the electrode is acting as a cathode or anode, respectively. If the cell is open to atmospheric pressure, the H_2 pressure at the electrode is, of course, one atmosphere.

This simple design must be modified if hydrogen gas pressure greater than one atmosphere is wanted. A possible design for a complete cell is shown in Figure 6-4(a). The pressure of hydrogen at the electrode is controlled by the adjustable pressure relief valves, V_1 and V_2. Individual elec-

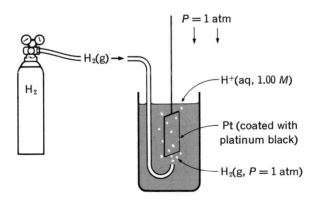

Figure 6-3 Standard hydrogen electrode. The platinum electrode is inert and does not take part in the electrode reaction. A coating of "platinum black" (finely divided Pt) acts as a catalyst to aid the reaction.

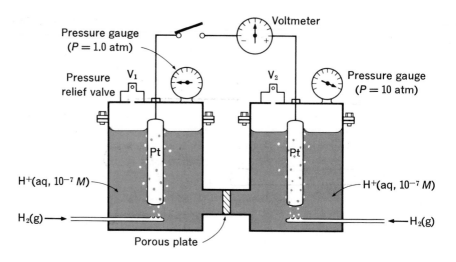

(a) Switch open

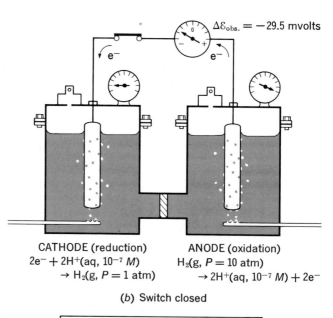

CATHODE (reduction)
$2e^- + 2H^+(aq, 10^{-7} M)$
$\rightarrow H_2(g, P = 1 \text{ atm})$

ANODE (oxidation)
$H_2(g, P = 10 \text{ atm})$
$\rightarrow 2H^+(aq, 10^{-7} M) + 2e^-$

(b) Switch closed

> **NET (expansion)**
> $H_2(g, P = 10 \text{ atm}) \rightarrow H_2(g, P = 1 \text{ atm})$

Figure 6-4 *Expansion cell.*

trode pressures are measured on the pressure gauges. Electrical continuity between electrode solutions is achieved through a porous plate.
This is a piece of ceramic material that allows ion current to flow while
preventing any large-scale mixing. A few drops of bromthymol blue will
allow us to monitor changes in hydrogen ion concentration in the two half-
cells. According to Table 5-4, this acid-base indicator is yellow at pH $= 6$
and blue at pH $= 7.6$.

With hydrogen ion concentrations of 10^{-7} in both half-cells (pure water)
and H_2 pressures of 1.0 atm (LHE) and 10 atm (RHE), a negative voltage of
about 30 millivolts is observed when the switch is closed, Figure 6-4(b). The
negative potential in our external circuit tells us that electrons are being
produced at the RHE and are traveling from right to left. The RHE electrode must be the anode (loss of electrons $=$ oxidation) and the LHE the
cathode (electron gain $=$ reduction). After current flows for a few minutes,
we observe color changes in the two cells. At the RHE the solution turns
yellow: [H$^+$] must be increasing there. At the LHE a blue color is observed:
[H$^+$] must be decreasing—the color shows that the solution is turning
basic. This is enough evidence to let us deduce the appropriate half-cell
reactions.

acid → RHE, anode $\qquad H_2(g, P = 10 \text{ atm}) \longrightarrow 2\,H^+(aq,\ 10^{-7}\,M) + 2\,e^-$

base → LHE, cathode $\quad 2\,H^+(aq,\ 10^{-7}\,M) + 2\,e^- \longrightarrow H_2(g, P = 1 \text{ atm})$

$\qquad$ CELL $\qquad\qquad\overline{\quad H_2(g, P = 10 \text{ atm}) \longrightarrow H_2(g, P = 1 \text{ atm})\quad}$

$$(6\text{-}6)$$

The similarity to the copper concentration cell is obvious. This, too, is a
concentration cell: As we found in Section 4-1(c), the solubility of a gas in a
liquid depends upon the pressure. So the pressure difference in the two
half-cells causes a concentration difference. In the over-all operation of the
cell, this concentration difference is maintained as long as gas pressure is
maintained. Hence, the net reaction is a simple one—the expansion of a
high-pressure gas.

(c) PRECIPITATION

The two concentration cells represented physical changes that might be
distinguished from chemical changes.* But chemical changes occur spontaneously—presumably they, too, can be put to work in an electrochemical
cell. For example, consider the formation and precipitation of a sparingly

---

* The distinction between "physical" and "chemical" change is one of degree, not of kind.
"Chemical" changes involve the making and breaking of chemical bonds—with energies
ranging from 30 to 250 kcal/mole. Forces between nonbonded atoms or molecules that must
be overcome to effect "physical" changes are normally less than 5 kcal/mole. As we will see
in Chapter Fifteen, the origin of *all* forces between atoms is the same, so the semantic
distinction on energy grounds is only one of convenience.

soluble salt, AgCl. The reaction of interest is, from Figure 6-1(c),

$$Ag^+(aq) + Cl^-(aq) \longrightarrow AgCl(s) \tag{6-7}$$

We might approach the experimental problem by putting 1.00 M solutions of the two ions, $Ag^+(aq)$ and $Cl^-(aq)$, in the two compartments of an electrochemical cell, as shown in Figure 6-5(a). Electrical continuity in the solutions may be made with a porous plate. Silver metal seems a good choice for the electrodes. Let's close the switch and see what happens.

Quite a lot of activity is immediately evident. With chloride ion at the LHE and silver ion at the RHE, a positive voltage is observed. A white precipitate of silver chloride forms at the LHE and whiskers of metallic silver appear on the RHE. These changes tell us the electrode reactions: From the observed direction of electron flow we know that the LHE is the anode and that silver chloride forms there. Silver ions are being formed and electrons are being produced.

LHE, anode $Ag(m) + Cl^-(aq) \longrightarrow AgCl(s) + e^-$ (6-8)

Meanwhile, back at the cathode, electrons are being consumed and silver ions are being reduced to metallic silver.

RHE, cathode $Ag^+(aq) + e^- \longrightarrow Ag(m)$ (6-9)

The cell reaction is the sum of these half-reactions

$$Ag(m) + Cl^-(aq) + Ag^+(aq) + e^- \longrightarrow AgCl(s) + Ag(m) + e^-$$

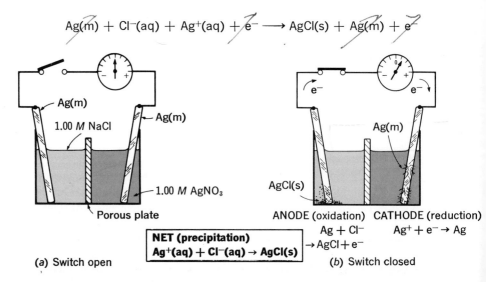

AgCl(s)

ANODE (oxidation)   CATHODE (reduction)
Ag + Cl⁻   Ag⁺ + e⁻ → Ag
→ AgCl + e⁻

NET (precipitation)
$Ag^+(aq) + Cl^-(aq) \rightarrow AgCl(s)$

(a) Switch open   (b) Switch closed

Figure 6-5 *Precipitation cell.*

which simplifies to

$$Ag^+(aq) + Cl^-(aq) \longrightarrow AgCl(s) \qquad (6\text{-}10)$$

So there we have it. Another simple spontaneous reaction has been put to work in an electrochemical cell. A little thought, however, shows that this cell is nothing more than a silver ion concentration cell. We could consider the anode reaction in two hypothetical stages. The first is the oxidation of silver metal.

$$Ag(m) \longrightarrow Ag^+(aq) + e^- \qquad (6\text{-}11a)$$

As silver ion is produced, it immediately reacts with chloride ion to precipitate silver chloride.

$$Ag^+(aq) + Cl^-(aq) \longrightarrow AgCl(s) \qquad (6\text{-}11b)$$

Taken together, (6-11a) and (6-11b) give us the net electrode reaction (6-8). There is, of course, a small concentration of silver ion in equilibrium with AgCl. Since

$$K_{sp} = [Ag^+][Cl^-] \qquad (6\text{-}12)$$

and since, initially

$$[Cl^-] = 1.00 \; M$$

the silver ion concentration just after the switch has been closed is

$$[Ag^+] = K_{sp}/[Cl^-] = K_{sp}/1.00$$
$$= K_{sp} \qquad (6\text{-}13)$$

At this point it is appropriate to introduce the standard shorthand notation for electrochemical cells. For our Ag/AgCl cell we would write

$$Ag(m)|Ag^+(aq, K_{sp} \; M), Cl^-(aq, 1.00 \; M)| \; |Ag^+(aq, 1.00 \; M)|Ag(m) \qquad (6\text{-}14)$$

Vertical lines indicate phase boundaries, a double line showing two boundaries at the salt bridge or porous plate. The anode is written at the extreme left and the cathode at the extreme right. A positive voltage indicates electron flow (in the external circuit) from left to right. A measured negative voltage tells us that the RHE is really the anode and that the cell has been written wrong-end-to.

In the notation, the cells of Figures 6-2 and 6-4 are written:

Cu(m)|Cu^{+2}(aq, 10^{-3} M)| |Cu^{+2}(aq, 1.00 M)|Cu(m)
anode   cathode
└──────────────electron flow──────────────↑

Pt(m)|H$_2$(g, P = 10 atm)|H$^+$(aq, 1.00 M)|
                                  |H$^+$(aq, 1.00 M)|H$_2$(g, P = 1 atm)|Pt(m)
anode   cathode
└──────────────────────electron flow────────────────↑

(d) **NEUTRALIZATION**

The neutralization of strong acid by strong base proceeds cheerfully and spontaneously. It is not too difficult to carry out this reaction in an electrochemical cell. The following cell will do the trick.

Pt|H$_2$(g, P = 1 atm)|OH$^-$(aq, 1.00 M)|
                                  |H$^+$(aq, 1.00 M)|H$_2$(g, P = 1 atm)|Pt

$$(6\text{-}15)$$

Two hydrogen electrodes are involved; a standard hydrogen electrode is the cathode and a hydrogen electrode with 1.00 M strong base the anode. Figure 6-6 illustrates the cell described in (6-15). The electrode reactions are:

anode	H$_2$(g, P = 1 atm) $\longrightarrow$ 2 H$^+$(aq) + 2 e$^-$	(6-16)
cathode	2 H$^+$(aq, 1.00 M) + 2 e$^-$ $\longrightarrow$ H$_2$(g, P = 1 atm)	(6-17)

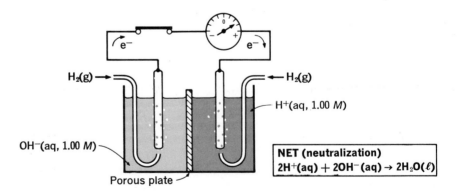

OH$^-$(aq, 1.00 M)

H$_2$(g) →   ← H$_2$(g)

H$^+$(aq, 1.00 M)

NET (neutralization)
2H$^+$(aq) + 2OH$^-$(aq) → 2H$_2$O(ℓ)

Porous plate

ANODE (oxidation)   CATHODE (reduction)
H$_2$(g) + 2OH$^-$(aq) → 2H$_2$O(ℓ) + 2e$^-$ 2e$^-$ + 2H$^+$(aq) → H$_2$(g)

Figure 6-6   Neutralization cell.

Of course, protons produced at the anode immediately react with hydroxide ions

$$2\ H^+(aq) + 2\ OH^-(aq,\ 1.00\ M) \longrightarrow 2\ H_2O(\ell) \tag{6-18}$$

The net anode reaction is (6-16) plus (6-18)

$$H_2(g,\ P = 1\ atm) + 2\ OH^-(aq,\ 1.00\ M) \longrightarrow 2\ H_2O(\ell) + 2\ e^- \tag{6-19}$$

and the net cell reaction is (6-17) plus (6-19)

$$2\ H^+(aq,\ 1.00\ M) + 2\ OH^-(aq,\ 1.00\ M) \longrightarrow 2\ H_2O(\ell) \tag{6-20}$$

which is exactly the reaction desired.

(e) REDOX REACTION

In some chemical reactions that can be carried out in an electrochemical cell, the net reaction involves the transfer of electrons, one or more, from one species to another. Such a reaction is colloquially called a *redox* reaction. The last example in Figure 6-1(e) is such a case; the plating of silver onto copper metal from a silver nitrate solution is representative of a classical redox process.

For such a reaction, life is simplified because we can tell by visual observation that silver ion is reduced and copper metal oxidized. Our cell will need a copper anode in contact with a copper sulfate solution and a silver cathode in contact with a silver nitrate solution. The simple cell illustrated in Figure 6-7 will do the trick. The electrode reactions are

$$\text{anode} \qquad Cu(m) \longrightarrow Cu^{+2}(aq) + 2\ e^- \tag{6-21}$$

$$\text{cathode} \quad Ag^+(aq) + e^- \longrightarrow Ag(m) \tag{6-22}$$

To find the cell reaction, we note that electrons produced at the anode *all* must be consumed at the cathode to prevent charge build-up. Twice (6-22) is needed.

$$\text{cell} \qquad Cu(m) + 2\ Ag^+(aq) \longrightarrow Cu^{+2}(aq) + 2\ Ag(m) \tag{6-23}$$

And so we end our foray into the electrical resources of spontaneous reactions—and a productive one it has been! We have found that quite a variety of simple spontaneous processes can be utilized as sources of electric power. We shall now investigate the significance of the measured voltage—it proves to be a way to study both chemical spontaneity and equilibrium.

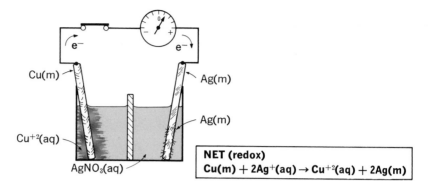

NET (redox)
$Cu(m) + 2Ag^+(aq) \rightarrow Cu^{+2}(aq) + 2Ag(m)$

ANODE (oxidation)   CATHODE (reduction)
$Cu(m) \rightarrow Cu^{+2}(aq) + 2e^-$ $Ag^+(aq) + e^- \rightarrow Ag(m)$

Figure 6-7   Redox cell.

6-2 $\Delta\mathcal{E}$ and concentration

Besides their obvious practical uses as batteries, electrochemical cells provide us with lots of useful chemical information. We have already noted that cell voltage, $\Delta\mathcal{E}$, seems to be a measure of chemical driving force or, as we shall call it, reaction tendency. The details of this concentration dependence, which we will explore in this section, will lead eventually to a relation between $\Delta\mathcal{E}$ and K, the equilibrium constant for the net reaction taking place in the cell. We will begin with a very simple cell, a silver ion concentration cell. First, however, we must give some attention to the measurement of $\Delta\mathcal{E}$.

(a) THE REVERSIBLE CELL

A cell is just a particular physical method for carrying out a chemical reaction. Whatever process is involved proceeds as long as the circuit is complete. Reactants are consumed and products appear. This means that the relative concentrations of the various species are continually changing. Therefore a $\Delta\mathcal{E}$ measured an instant after closing the switch will be different from one measured an hour or so later. Of course, the rate at which the cell reaction proceeds depends on the electrical resistance, R, in the external circuit. By Ohm's Law, the current flowing, i, is given by

$$\text{current} = i = \frac{\Delta\mathcal{E}}{R}$$

R : larger i : smaller
R : smaller i : larger   (6-24)

A high resistance leads to a low current. A low current implies a low rate of passage of electrons from anode to cathode, hence a slow rate of reaction.

This suggests that if we could measure $\Delta\mathcal{E}$ under conditions that permit *no* current flow, we would not have to worry about concentration changes at all, since no reaction would occur. This is easily done by *opposing* the cell voltage with a variable external voltage. When the external voltage $\Delta\mathcal{E}_{ext}$ is exactly equal to the cell voltage $\Delta\mathcal{E}_{cell}$, but opposite in sign, we have

$$-\Delta\mathcal{E}_{ext} = \Delta\mathcal{E}_{cell} \tag{6-25}$$

and

$$i = 0 \tag{6-26}$$

In practice, $\Delta\mathcal{E}_{ext}$ is adjusted to give a zero current reading on an ammeter. The cell voltage, by (6-25), is thus accurately determined at the same time.

A silver ion concentration cell operating under zero current conditions is illustrated in Figure 6-8(a). Such a cell is said to be operating *reversibly*. The reason for this name is straightforward. If the opposing voltage is decreased slightly (Fig. 6-8(b)), current flows and the cell discharges. A slight increase

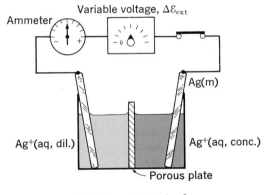

(a) $\Delta\mathcal{E}_{ext} = \Delta\mathcal{E}_{cell}$; $i = 0$

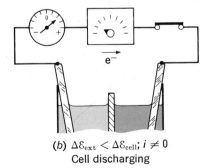

(b) $\Delta\mathcal{E}_{ext} < \Delta\mathcal{E}_{cell}$; $i \neq 0$
Cell discharging

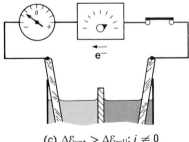

(c) $\Delta\mathcal{E}_{ext} > \Delta\mathcal{E}_{cell}$; $i \neq 0$
Cell charging

Figure 6-8   A reversible electrochemical cell.

in $\Delta\mathcal{E}_{ext}$ over the zero current requirement (Fig. 6-8(c)), *reverses* the direction of current flow and causes the cell to recharge. A cell is reversible, then, when a minute change in $\Delta\mathcal{E}_{ext}$ causes a reversal of the direction of current flow.

(b) SILVER ION CELL

The cell shown in Figure 6-8 can be represented in standard notation as

$$Ag(m)|Ag^+(aq, dilute)|\ |Ag^+(aq, conc.)|Ag(m) \qquad (6\text{-}27)$$

anode   cathode

e^-

The reversible $\Delta\mathcal{E}$'s observed for different values of the concentration ratio

$$\frac{[Ag^+(aq, dil.)]}{[Ag^+(aq, conc.)]} = \frac{[Ag^+(anode)]}{[Ag^+(cathode)]} \qquad (6\text{-}28)$$

are given in Table 6-2. Examination of the numbers shows a regular increase in $\Delta\mathcal{E}_{cell}$ every time the ratio decreases a factor of ten. This suggests a logarithmic relationship between $\Delta\mathcal{E}$ and the concentration ratio. When the logarithm of the ratio (column three, Table 6-2) is plotted against $\Delta\mathcal{E}_{cell}$ in Figure 6-9, a straight line is obtained.

Applying the general equation of a straight line

$$y = mx + b \qquad (6\text{-}29)$$

we can easily determine the slope m

$$m = \frac{\Delta y}{\Delta x} = -59 \text{ millivolts} \qquad (6\text{-}30)$$

The line intercepts both axes at the origin, so

$$b = 0 \qquad (6\text{-}31)$$

Table 6-2   Reversible $\Delta\mathcal{E}$'s: Silver Ion Concentration Cell

$\dfrac{[Ag^+(anode)]}{[Ag^+(cathode)]}$	$\Delta\mathcal{E}_{cell}$ (millivolts)	$\log_{10} \dfrac{[Ag^+(anode)]}{[Ag^+(cathode)]}$
10^{-3}	177	-3
10^{-2}	118	-2
10^{-1}	59	-1
1	0	0

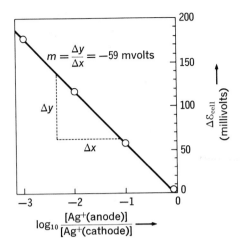

Figure 6-9 *Silver ion cell: concentration dependence.*

We have then, if Δℰ is expressed in volts

$$\Delta\mathcal{E}_{cell} = -0.059 \log_{10} \frac{[Ag^+(aq, dil.)]}{[Ag^+(aq, conc.)]}$$ (6-32)

The cell reaction is

$$Ag^+(aq, conc.) \longrightarrow Ag^+(aq, dilute)$$

We see that the logarithmic term in (6-32) has the form of an equilibrium expression for the cell reaction. Let's call it Q.

$$Q = \frac{[Ag^+(aq, dil.)]}{[Ag^+(aq, conc.)]}$$ (6-33)

so that (6-32) becomes

$$\Delta\mathcal{E}_{cell} = -0.059 \log_{10} Q$$ (6-34)

Of course Q will be numerically equal to K, the equilibrium constant, only when the cell is in chemical equilibrium. When will this happen? Why, when the silver ion concentrations in the two half-cells are equal.

At equilibrium

$$[Ag^+(aq, anode)] = [Ag^+(aq, cathode)]$$ (6-35)

Thus, for this special case,

$$Q = K = \frac{[Ag^+(aq, anode)]}{[Ag^+(aq, cathode)]} = 1$$

and finally, on substitution in (6-34)

$$\Delta\mathcal{E}_{cell} = -0.059 \log_{10} K = -0.059 \log_{10} 1$$
$$= 0$$

which is exactly what is observed.

(c) COPPER ION CELL

Let's check equation (6-34) against the results for the copper ion concentration cell in Table 6-1. Immediately we see that something is wrong. For each ratio measured, voltages are only half those predicted by (6-34). Do we have to discard our relationship between $\Delta\mathcal{E}$ and concentration? Some consideration of the details of the two cells suggests a modification we might make to equation (6-34). The reactions associated with the cells are summarized in Table 6-3. The two columns are almost exactly parallel— with one exception. Two electrons are transferred in the copper ion cell, but only one in the silver ion cell. Equation (6-34) would agree in *both* cases if we divided by n, the number of electrons transferred.

$$\Delta\mathcal{E} = -\frac{0.059}{n} \log_{10} Q \qquad (6\text{-}36)$$

This equation is found to hold for any concentration cell.

Exercise 6-1. Show that the concentration cell given below would show a voltage of 0.039 volt.

$$Al(m)|Al^{+3}(aq, 1.0 \cdot 10^{-3}\ M)|\ |Al^{+3}(aq, 0.10\ M)|Al(m)$$
anode   cathode
└─────────── electron flow ───────────↑

Table 6-3 *Comparison of Two Concentration Cells*

Ag(m)|Ag⁺(aq, dil.)| |Ag⁺(aq, conc.)|Ag(m)	Cu(m)|Cu⁺²(aq, dil.)| |Cu⁺²(aq, conc.)|Cu(m)
Anode (oxidation)	
$Ag(m) \longrightarrow Ag^+(aq, dil.) + e^-$	$Cu(m) \longrightarrow Cu^{+2}(aq, dil.) + 2\,e^-$
Cathode (reduction)	
$Ag^+(aq, conc.) + e^- \longrightarrow Ag(m)$	$Cu^{+2}(aq, conc.) + 2\,e^- \longrightarrow Cu(m)$
Cell	
$Ag^+(aq, conc.) \longrightarrow Ag^+(aq, dil.)$	$Cu^{+2}(aq, conc.) \longrightarrow Cu^{+2}(aq, dil.)$

(d) REDOX CELL

The copper–silver cell discussed in Section 6-1(e) has, as its net reaction

$$2 \, Ag^+(aq) + Cu(m) \longrightarrow Cu^{+2}(aq) + 2 \, Ag(m)$$

The equilibrium constant for this reaction is

$$K = \frac{[Cu^{+2}]}{[Ag^+]^2} \qquad (6\text{-}37)$$

We have found, for concentration cells, that $\Delta\varepsilon$ depends on a ratio of concentrations, Q, that has the form of an equilibrium expression. Let's examine this for the redox cell. Table 6-4 shows data for several concentrations. If the equilibrium expression for reaction (6-23) properly indicates the effect of concentration, we shall be interested in the quantity Q:

$$Q = \frac{[Cu^{+2}]}{[Ag^+]^2} \qquad (6\text{-}38)$$

Table 6-4   Concentration Dependence of Ag–Cu Cell

$[Cu^{+2}]$	$[Ag^+]$	$Q = \dfrac{[Cu^{+2}]}{[Ag^+]^2}$	$\Delta\varepsilon_{obs.}$ (volts)	$\log_{10} Q$
1.0	1.0	1.0	0.462	0
0.1	0.1	10	0.433	1
0.01	0.01	100	0.403	2

For a range of concentration levels extending from 1.0 to 0.01, the voltage changes from 0.462 to 0.403. This information is shown graphically in Figure 6-10, a plot of $\Delta\varepsilon$ versus $\log_{10} Q$. A straight line is observed—but this time

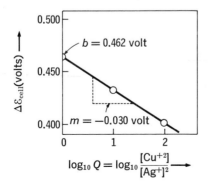

Figure 6-10   Copper–silver cell: concentration dependence.

the intercept b is not zero. In fact, when $Q = 1$ ($\log_{10} Q = 0$) we find

$$\Delta\mathcal{E}^0 = 0.462 \text{ volt} \tag{6-39}$$

The voltage, measured when all concentrations are 1.00 M, is called the standard cell voltage and is indicated by a superscript ^{0}.

The slope m of the straight line is $-0.030 = -0.059/2$ in accord with the fact that $n = 2$ since two electrons are transferred.

Now the equation of a straight line, $y = mx + b$, takes the following form for the concentration dependence of $\Delta\mathcal{E}$.

$$\Delta\mathcal{E} = \Delta\mathcal{E}^0 - \frac{0.059}{n} \log Q$$

$\Delta\mathcal{E}$ = voltage measured for concentrations represented in Q

$\Delta\mathcal{E}^0$ = voltage measured for all concentrations = 1 M

(6-40)

This important equation is usually called the Nernst Equation, after the scientist who first deduced it from experimental results. It is found to be applicable to all electrochemical cells.

(e) THE NERNST EQUATION AND CONCENTRATION CELLS

Let's reconsider concentration cells using the Nernst Equation in its complete form, (6-40). When the ionic concentrations in both half-cells of a concentration cell are identical, we have

$$Q = 1$$

thus

$$\log_{10} Q = 0$$

and because the cell is in an equilibrium state

$$\Delta\mathcal{E} = 0$$

This gives us the result that

$$\Delta\mathcal{E}^0 = 0$$

for any concentration cell. This is what led us to the simpler expression for the concentration dependence, (6-36).

6-3 Cell potential and reaction tendency

One of the cells discussed in Section 6-1 cannot be classified as a concentration cell. For this reaction we find that the constant term in the Nernst Equation, $\Delta\mathcal{E}^0$, is not equal to zero. We must examine $\Delta\mathcal{E}^0$ further to see what information it contains.

(a) TWO COPPER CELLS COMPARED

A piece of copper placed in a silver ion solution instantly begins to grow a hippie-like set of silver chin whiskers (Fig. 6-11(a)). Put another piece of copper in some zinc solution, however, and apathy reigns. No reaction is evident, even after a long wait (Fig. 6-11(b)). If we turn the tables by placing a zinc rod in some copper solution (Fig. 6-11(c)), copper metal plates out.

Let's put these spontaneous reactions to work for us in electrochemical cells. For convenience and simplicity we will choose standard conditions (all concentrations $= 1.00\ M$) and measure $\Delta\mathcal{E}^0$. We are already familiar with the copper–silver cell:

$$Cu(m)|Cu^{+2}(1.00\ M)|\ |Ag^+(1.00\ M)|Ag(m) \qquad (6\text{-}41)$$

anode   cathode

|_____electron flow_____↑

$$\Delta\mathcal{E}^0_{Cu|Ag} = 0.46\ V$$

Since copper is reduced in the copper–zinc case, the copper electrode

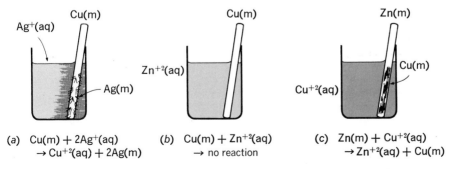

(a) $Cu(m) + 2Ag^+(aq)$ (b) $Cu(m) + Zn^{+2}(aq)$ (c) $Zn(m) + Cu^{+2}(aq)$
$\rightarrow Cu^{+2}(aq) + 2Ag(m)$ $\rightarrow$ no reaction $\rightarrow Zn^{+2}(aq) + Cu(m)$

Figure 6-11   (a) Copper reduces silver ion. (b) Copper doesn't reduce zinc ion. (c) Zinc reduces copper ion.

must be the cathode here:

$$Zn(m)|Zn^{+2}(1.00\ M)|\ |Cu^{+2}(1.00\ M)|Cu(m) \qquad (6\text{-}42)$$
anode   cathode

electron flow

The observed standard cell potential is

$$\Delta \mathcal{E}^0_{Zn|Cu} = +1.10\ V$$

Exercise 6-2. From the electron flow indicated in (6-42), deduce the two electrode reactions and verify that the cell actually represents the reaction taking place in the beaker of Figure 6-11(c).

By denoting the voltage $\Delta \mathcal{E}$ (or $\Delta \mathcal{E}^0$) we are implying that it is the *difference* between some electrical potentials $\mathcal{E}$ (or $\mathcal{E}^0$) at two electrodes. Since a voltage is generated only when two half-cells are hooked together, we cannot hope to measure the potential of an individual half-cell on an absolute scale. If we had a reference point, however, we could easily set up a voltage scale on which potential differences are represented. Figures 6-12(a) and (b) show, on the same scale, distances proportional to $\Delta \mathcal{E}^0_{Cu|Ag}$ and $\Delta \mathcal{E}^0_{Zn|Cu}$, respectively. In both examples the cathode half-cell is at the top, and the cell voltage $\Delta \mathcal{E}^0$ is represented by the difference between the cathode potential and the anode potential.*

$$\Delta \mathcal{E}^0 = \mathcal{E}^0_{cathode} - \mathcal{E}^0_{anode} \qquad (6\text{-}43)$$

Unfortunately, we don't yet have any way of assigning values to the individual $\mathcal{E}^0$'s.

Figure 6-12 becomes more informative if we combine (a) and (b) into a composite, Figure 6-12(c). From this diagram we can predict that a silver–zinc cell

$$Zn|Zn^{+2}(1.00\ M)|\ |Ag^+(1.00\ M)|Ag \qquad (6\text{-}44)$$
anode   cathode

---

* In a mechanical system, spontaneous reactions are accompanied by a *decrease* in potential energy. In an electrical system, a cell, electrons flow in the external circuit from the electrode of lower (more negative) potential to the electrode of higher (more positive) potential. The mechanical and electrical systems are consistent, since the electron's energy is lowest at the positive pole.

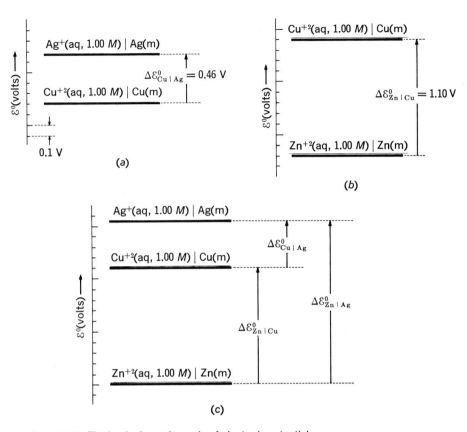

Figure 6-12 The beginnings of a scale of electrode potentials.

should have a potential given by

$$\Delta \mathcal{E}^0_{Zn|Ag} = \Delta \mathcal{E}^0_{Ag|Cu} + \Delta \mathcal{E}^0_{Cu|Ag} \tag{6-45}$$

$$= 1.10 + 0.46$$

$$= 1.56 \text{ V}$$

This is just the vertical difference between the zinc and silver lines. The net reaction will be

$$\text{Zn(m)} + 2\ \text{Ag}^+(1.00\ M) \longrightarrow \text{Zn}^{+2}(1.00\ M) + 2\ \text{Ag(m)} \tag{6-46}$$

We can test (6-45) qualitatively by sticking a zinc rod into a silver solution to see that reaction occurs. Then we can test quantitatively by preparing the appropriate cell. Both predictions are confirmed by the experiments,

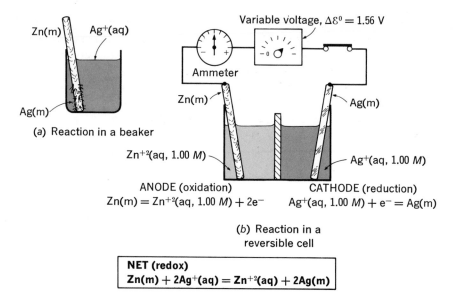

ANODE (oxidation) CATHODE (reduction)

$$Zn(m) = Zn^{+2}(aq, 1.00\ M) + 2e^- \qquad Ag^+(aq, 1.00\ M) + e^- = Ag(m)$$

(b) Reaction in a
reversible cell

NET (redox)
$Zn(m) + 2Ag^+(aq) = Zn^{+2}(aq) + 2Ag(m)$

Figure 6-13 *The silver–zinc reaction.*

as shown in Figure 6-13. The measured standard cell potential is

$$\Delta\mathcal{E}^0_{Zn|Ag}(obs.) = 1.56\ V$$

as first calculated by (6-45).

What we have said is quite simple. If a $1\frac{1}{2}$-volt battery is put end to end with a second $1\frac{1}{2}$-volt battery, three volts are available to run a flashlight. If a 1.10 V zinc–copper battery and a 0.46 V copper–silver battery are connected in series, a 1.56 V battery results, Figure 6-14. We might write down

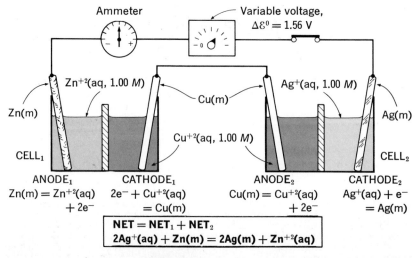

ANODE$_1$ CATHODE$_1$ ANODE$_2$ CATHODE$_2$
$Zn(m) = Zn^{+2}(aq)$ $2e^- + Cu^{+2}(aq)$ $Cu(m) = Cu^{+2}(aq)$ $Ag^+(aq) + e^-$
$\qquad + 2e^-$ $\quad = Cu(m)$ $\qquad + 2e^-$ $\quad = Ag(m)$

NET $= $ NET$_1 + $ NET$_2$
$2Ag^+(aq) + Zn(m) = 2Ag(m) + Zn^{+2}(aq)$

Figure 6-14 *Two reversible cells in series.*

all four electrode reactions to determine the cell reaction, but this is hardly necessary. Since the copper electrode is the cathode in the zinc–copper cell and the anode in the copper–silver cell, the copper half-reaction will cancel out, leaving us with the net reaction of the zinc–silver cell.

It is clear from (6-45) that knowledge of some $\Delta\mathcal{E}^0$'s allows us to calculate others. This would be most easily done if we had a scale of electrode, or half-cell, potentials. Such a scale is easily established if we pick some convenient half-cell as a reference point. The hydrogen electrode pictured in Figure 6-3 is the one used, and the scale of potentials so derived is called the standard half-cell potentials.

(b) STANDARD HALF-CELL POTENTIALS.

Half-cell reactions can be written in the form either of oxidation (release of electrons) or of reduction (consumption of electrons).

oxidation	$Ag(m) \longrightarrow Ag^+(aq) + e^-$	(6-47a)
reduction	$Ag^+(aq) + e^- \longrightarrow Ag(m)$	(6-47b)

We will use the latter convention.*

$$Cu^{+2}(aq) + 2\,e^- \longrightarrow Cu(m) \qquad\qquad (6\text{-}48)$$

$$Zn^{+2}(aq) + 2\,e^- \longrightarrow Zn(m) \qquad\qquad (6\text{-}49)$$

By (6-43), the cell potential $\Delta\mathcal{E}^0$ was shown to correspond to the difference between the (as yet unknown) cathode and anode potentials. Similarly, *when both electrode reactions are written as reductions,* the cell reaction is just the algebraic difference between the two half-cell reactions. For example, the zinc–silver cell, reaction (6-46), is expressed by the difference, (6-47b) minus (6-49), properly balanced to assure that all electrons produced at the anode are consumed at the cathode.

$$2[Ag^+(aq) + e^- \longrightarrow Ag(m)] \qquad\qquad \text{(cathode reaction)}$$
$$- [Zn^{+2}(aq) + 2\,e^- \longrightarrow Zn(m)] \qquad \text{(anode reaction)}$$
$$2\,Ag^+(aq) + 2\,e^- + Zn(m) \longrightarrow 2\,Ag(m) + Zn^{+2}(aq) + 2\,e^-$$
$$\text{(cell reaction)}$$

Now we can write these reduction half-reactions on our voltage scale,

---

* Many books and a huge volume of research literature use the opposite convention, presenting half-reactions in the oxidation sense, because of the dominant contribution to this field by Wendell M. Latimer and his students at the University of California. Latimer presented his authoritative work in the book *Oxidation Potentials,* which remains a prime reference source for values of cell potentials. The potentials found there need only be assigned the opposite algebraic sign to be applicable to the reduction potentials that we shall use to remain in vogue. We express our sincere apologies to Wendell M. Latimer, our beloved mentor.

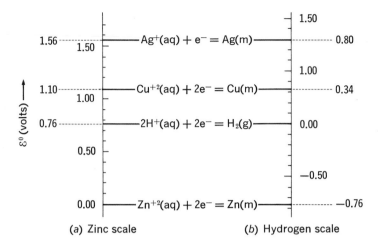

Figure 6-15 *Relative electrode potentials. (a) Zinc scale* $\mathcal{E}^0_{Zn} = 0$. *(b) Hydrogen scale* $\mathcal{E}^0_{H_2} = 0$.

Figure 6-12. Since we are interested only in differences, $\Delta\mathcal{E}^0$, we can arbitrarily assign a value, say 0.0 V, to one half-reaction. Designating a standard half-cell potential as $\mathcal{E}^0$, we might choose the zinc reduction, shown on the left-hand side of Figure 6-15, as our reference. The copper electrode then has an $\mathcal{E}^0$ of 1.10 V. The cell voltage of the zinc–copper cell is thus

$$\Delta\mathcal{E}^0_{Zn|Cu} = \mathcal{E}^0_{Cu} - \mathcal{E}^0_{Zn}$$

$$= (1.10 - 0.00)$$

$$= 1.10 \text{ V}$$

Zinc, however, is not the usual reference point. The standard hydrogen electrode is given this honor. Thus the reference potential $\mathcal{E}^0 = 0.00$ is applied to the half-reaction:

$$2 \text{ H}^+(aq, 1.00 \ M) + 2 \ e^- \longrightarrow H_2(g, P = 1 \text{ atm}) \tag{6-50}$$

Some standard electrode potentials on the hydrogen scale are shown on the right-hand side of Figure 6-15. A more complete list of $\mathcal{E}^0$'s is given in Tables 6-5 and 6-6. Since many half-reactions depend on pH, Table 6-5 compares some $\mathcal{E}^0$'s for acidic ($H^+(aq) = 1.00 \ M$) and basic ($OH^-(aq) = 1.00 \ M$) solutions.

Exercise 6-3. The standard reduction potential for the cadmium half-reaction $Cd^{+2}(aq) + 2 \ e^- \longrightarrow Cd(m)$ is -0.40 volt. Show that the voltage of a standard copper–cadmium cell would be 0.74 volt.

(c) DEPENDENCE OF ε ON CONCENTRATION

The Nernst Equation can be applied to half-cell reactions just as to cell reactions. It takes the form

$$\varepsilon = \varepsilon^0 - \frac{0.059}{n} \log Q \qquad (6\text{-}51)$$

For an example, let's calculate ε for the hydrogen half-cell in pure water, pH = 7. The half-reaction is

$$2\,H^+(aq,\ 10^{-7}\ M) + 2\,e^- = H_2(g,\ 1\ atm) \qquad (6\text{-}52)$$

By (6-51) we have

$$\varepsilon = 0.00 - \frac{0.059}{2} \log \frac{P_{H_2}}{[H^+(aq)]^2}$$

$$= 0.00 - \frac{0.059}{2} \log \frac{1}{(10^{-7})^2}$$

$$= -0.413\ V$$

Exercise 6-4. Show that $\varepsilon(pH = 7)$ is 0.81 volt for the half-reaction $\frac{1}{2}\,O_2(g,\ P = 1\ atm) + H_2O + 2\,e^- \longrightarrow 2\,OH^-(aq)$ $(\varepsilon^0 = 0.40)$.

We can apply (6-51) to a concentration cell. For example, let's take a standard copper half-cell

$$Cu^{+2}(aq,\ 1.00\ M) + 2\,e^- \longrightarrow Cu(m) \qquad \varepsilon^0 = +0.34\ V \qquad (6\text{-}53)$$

and a half-cell in which $[Cu^{+2}] = 10^{-3}\ M$

$$Cu^{+2}(aq,\ 10^{-3}\ M) + 2\,e^- \longrightarrow Cu(m) \qquad (6\text{-}54)$$

for which

$$\varepsilon = \varepsilon^0 - \frac{0.059}{2} \log \frac{1}{[Cu^{+2}]}$$

$$= 0.34 - \frac{0.059}{2} \log 10^3$$

$$= 0.251\ V$$

Table 6-5 Some Standard Reduction Potentials, $T = 298.16°K$

Acid Solution

Half-reaction	$\mathcal{E}^0$ (volts)
$F_2 + 2\,e^- \longrightarrow 2\,F^-$	$+2.65$
$H_2O_2 + 2\,H^+ + 2\,e^- \longrightarrow 2\,H_2O$	1.77
$MnO_4^- + 4\,H^+ + 3\,e^- \longrightarrow MnO_2 + 2\,H_2O$	1.695
$2\,HOCl + 2\,H^+ + 2\,e^- \longrightarrow Cl_2 + 2\,H_2O$	1.63
$BrO_3^- + 6\,H^+ + 5\,e^- \longrightarrow \tfrac{1}{2}\,Br_2 + 3\,H_2O$	1.52
$MnO_4^- + 8\,H^+ + 5\,e^- \longrightarrow Mn^{+2} + 4\,H_2O$	1.51
$Cl_2 + 2\,e^- \longrightarrow 2\,Cl^-$	1.3595
$Cr_2O_7^{-2} + 14\,H^+ + 6\,e^- \longrightarrow 2\,Cr^{+3} + 7\,H_2O$	1.33
$MnO_2 + 4\,H^+ + 2\,e^- \longrightarrow Mn^{+2} + 2\,H_2O$	1.23
$O_2 + 4\,H^+ + 4\,e^- \longrightarrow 2\,H_2O$	1.229
$IO_3^- + 6\,H^+ + 5\,e^- \longrightarrow \tfrac{1}{2}\,I_2 + 3\,H_2O$	1.195
$Br_2(\ell) + 2\,e^- \longrightarrow 2\,Br^-$	1.0652
$N_2O_4 + 4\,H^+ + 4\,e^- \longrightarrow 2\,NO + 2\,H_2O$	1.03
$HONO + H^+ + e^- \longrightarrow NO + H_2O$	1.00
$NO_3^- + 4\,H^+ + 3\,e^- \longrightarrow NO + 2\,H_2O$	0.96
$Cu^{+2} + I^- + e^- \longrightarrow CuI$	0.86
$Ag^+ + e^- \longrightarrow Ag$	0.7991
$Fe^{+3} + e^- \longrightarrow Fe^{+2}$	0.771
$O_2 + 2\,H^+ + 2\,e^- \longrightarrow H_2O_2$	0.682
$I_3^- + 2\,e^- \longrightarrow 3\,I^-$	0.536
$Cu^+ + e^- \longrightarrow Cu$	0.521
$Cu^{+2} + 2\,e^- \longrightarrow Cu$	0.34
$AgCl + e^- \longrightarrow Ag + Cl^-$	0.22
$Cu^{+2} + e^- \longrightarrow Cu^+$	0.153
$Sn^{+4} + 2\,e^- \longrightarrow Sn^{+2}$	0.15

Basic Solution

Half-reaction	$\mathcal{E}^0$ (volts)
$HO_2^- + H_2O + 2\,e^- \longrightarrow 3\,OH^-$	$+0.88$
$MnO_4^- + 2\,H_2O + 3\,e^- \longrightarrow MnO_2 + 4\,OH^-$	0.588
$ClO^- + H_2O + 2\,e^- \longrightarrow Cl^- + 2\,OH^-$	0.89
$BrO_3^- + 3\,H_2O + 6\,e^- \longrightarrow Br^- + 6\,OH^-$	0.61
$CrO_4^{-2} + 4\,H_2O + 3\,e^- \longrightarrow Cr(OH)_3 + 5\,OH^-$	0.13
$MnO_2 + 2\,H_2O + 2\,e^- \longrightarrow Mn(OH)_2 + 2\,OH^-$	-0.05
$O_2 + 2\,H_2O + 4\,e^- \longrightarrow 4\,OH^-$	$+0.41$
$Ag_2O + H_2O + 2\,e^- \longrightarrow 2\,Ag + 2\,OH^-$	0.345
$Fe(OH)_3 + e^- \longrightarrow Fe(OH)_2 + OH^-$	0.56
$O_2 + H_2O + 2\,e^- \longrightarrow HO_2^- + OH^-$	0.076
$Ag(NH_3)_2^+ + e^- \longrightarrow Ag + 2\,NH_3$	0.371
$2\,Cu(OH)_2 + 2\,e^- \longrightarrow Cu_2O + 2\,OH^- + H_2O$	-0.080
$Sn(OH)_6^{-2} + 2\,e^- \longrightarrow HSnO_2^- + H_2O + 3\,OH^-$	-0.90

Table 6-5 Some Standard Reduction Potentials, $T = 298.16°K$ (Continued)

Acid Solution		Basic Solution	
Half-reaction	ε^0 (volts)	Half-reaction	ε^0 (volts)
$S_4O_6^{-2} + 2\,e^- \longrightarrow 2\,S_2O_3^{-2}$	0.08		
$2\,H^+ + 2\,e^- \longrightarrow H_2$	0.00		
$Pb^{+2} + 2\,e^- \longrightarrow Pb$	−0.126		
$Sn^{+2} + 2\,e^- \longrightarrow Sn$	−0.136		
$PbSO_4 + 2\,e^- \longrightarrow Pb + SO_4^{-2}$	−0.356		
$Fe^{+2} + 2\,e^- \longrightarrow Fe$	−0.440		
$Zn^{+2} + 2\,e^- \longrightarrow Zn$	−0.763		
$Al^{+3} + 3\,e^- \longrightarrow Al$	−1.66		
$Mg^{+2} + 2\,e^- \longrightarrow Mg$	−2.37		
$Na^+ + e^- \longrightarrow Na$	−2.714		
		$2\,H_2O + 2\,e^- \longrightarrow H_2 + 2\,OH^-$	−0.828
		$HSnO_2^- + H_2O + 2\,e^- \longrightarrow Sn + 3\,OH^-$	−0.91
		$Fe(OH)_2 + 2\,e^- \longrightarrow Fe + 2\,OH^-$	−0.877
		$Zn(OH)_2 + 2\,e^- \longrightarrow Zn + 2\,OH^-$	−1.245
		$H_2AlO_3^- + H_2O + 3\,e^- \longrightarrow Al + 4\,OH^-$	−2.35
		$Mg(OH)_2 + 2\,e^- \longrightarrow Mg + 2\,OH^-$	−2.69

Table 6-6 Some Standard Reduction Potentials, Acid Solution, 298°K (Two Electrons Transferred Unless Indicated)

Group 7 Halogens

Reaction	E
$F_2 \longrightarrow 2 F^-$	2.65
$2 HOCl \longrightarrow Cl_2$	1.63
$H_5IO_6 \longrightarrow IO_3^-$	1.6
$2 HOBr \longrightarrow Br_2$	1.59
$2 HOI \longrightarrow I_2$	1.45
$Cl_2 \longrightarrow 2 Cl^-$	1.36
$2 IO_3^- \xrightarrow{10e} I_2$	1.20
$Br_2(\ell) \longrightarrow 2 Br^-$	1.07
$2 ICl_2^- \longrightarrow I_2$	1.06
$Br_3^- \longrightarrow 3 Br^-$	1.05
$I_3^- \longrightarrow 3 I^-$	0.54

Group 6 Oxygen Family

Reaction	E
$O_3 \longrightarrow O_2$	2.07
$S_2O_8^{-2} \longrightarrow 2 SO_4^{-2}$	2.01

Group 5 Nitrogen Family

Reaction	E
$NH_3OH^+ \longrightarrow NH_4^+$	1.35
$N_2H_5^+ \longrightarrow 2NH_4^+$	1.28
$\frac{1}{2} N_2O_4 \longrightarrow NO$	1.03
$HONO \xrightarrow{1e} NO$	1.00
$NO_3^- \xrightarrow{3e} NO$	0.96
$HN_3 \xrightarrow{8e} 3 NH_4^+$	0.69
$H_3AsO_4 \longrightarrow HAsO_2$	0.56
$HAsO_2 \xrightarrow{3e} As$	0.25
$\frac{1}{2} Sb_2O_3 \xrightarrow{3e} Sb$	0.15
$P \xrightarrow{3e} PH_3$	0.06
$\frac{1}{2} N_2 \longrightarrow \frac{1}{2} N_2H_5^+$	-0.23
$\frac{1}{2} N_2 \xrightarrow{3e} NH_4^+$	-0.27
$H_3PO_4 \longrightarrow H_3PO_3$	-0.28
$H_3PO_3 \longrightarrow H_3PO_2$	-0.50

Transition Metals d and f Orbital Elements

Reaction	E
$MnO_4^- \xrightarrow{3e} MnO_2$	1.70
$NiO_2 \longrightarrow Ni^{+2}$	1.68
$Ce^{+4} \xrightarrow{1e} Ce^{+3}$	1.61
$MnO_4^- \xrightarrow{5e} Mn^{+2}$	1.51
$Au^{+3} \xrightarrow{3e} Au$	1.50
$Cr_2O_7^{-2} \xrightarrow{6e} 2 Cr^{+3}$	1.33
$MnO_2 \longrightarrow Mn^{+2}$	1.23
$2 Hg^{+2} \longrightarrow Hg_2^{+2}$	0.92
$Cu^{+2} \xrightarrow{1e} CuI$	0.86
$Ag^+ \xrightarrow{1e} Ag$	0.80
$Hg_2^{+2} \longrightarrow 2 Hg$	0.79
$Fe^{+3} \xrightarrow{1e} Fe^{+2}$	0.77
$Ag_2SO_4 \longrightarrow 2 Ag$	0.65
$MnO_4^- \xrightarrow{1e} MnO_4^{-2}$	0.56

Group 3 Aluminum Family

Reaction	E
$Tl^{+3} \longrightarrow Tl^+$	1.25
$Tl^+ \xrightarrow{1e} Tl$	-0.34
$In^{+3} \xrightarrow{3e} In$	-0.34
$Ga^{+3} \xrightarrow{3e} Ga$	-0.53
$Al^{+3} \xrightarrow{3e} Al$	-1.66
$Sc^{+3} \xrightarrow{3e} Sc$	-2.08
$La^{+3} \xrightarrow{3e} La$	-2.52

Group 2 Alkaline Earth Elements

Reaction	E
$Be^{+2} \longrightarrow Be$	-1.85
$Mg^{+2} \longrightarrow Mg$	-2.37
$Ca^{+2} \longrightarrow Ca$	-2.87
$Sr^{+2} \longrightarrow Sr$	-2.89
$Ba^{+2} \longrightarrow Ba$	-2.90

$H_2O_2 \longrightarrow 2\,H_2O$ 1.77

$\tfrac{1}{2}\,O_2 \longrightarrow H_2O$ 1.23

$SeO_4^{-2} \longrightarrow H_2SeO_3$ 1.15

$H_2SeO_3 \xrightarrow{4e} Se$ 0.74

$O_2 \longrightarrow H_2O_2$ 0.68

$S_2O_6^{-2} \longrightarrow 2\,H_2SO_4$ 0.57

$4\,H_2SO_3 \xrightarrow{6e} S_4O_6^{-2}$ 0.51

$H_2SO_3 \xrightarrow{4e} S$ 0.45

$H_2SO_3 \longrightarrow \tfrac{1}{2}\,S_2O_3^{-2}$ 0.40

$HSO_4^- \longrightarrow H_2SO_3$ 0.17

$S \longrightarrow H_2S$ 0.14

$S_4O_6^{-2} \longrightarrow 2\,S_2O_3^{-2}$ 0.08

$Se \longrightarrow H_2Se$ -0.40

$Te \longrightarrow H_2Te$ -0.72

$H_3PO_2 \xrightarrow{1e} P$ -0.51

$Sb \xrightarrow{3e} SbH_3$ -0.51

$As \xrightarrow{3e} AsH_3$ -0.60

Group 4
Carbon Family

$PbO_2 \longrightarrow PbSO_4$ 1.69

$Sn^{+4} \longrightarrow Sn^{+2}$ 0.15

$C \xrightarrow{4e} CH_4$ 0.13

$Si \xrightarrow{4e} SiH_4$ 0.10

$Pb^{+2} \longrightarrow Pb$ -0.13

$Sn^{+2} \longrightarrow Sn$ -0.14

$GeO_2 \xrightarrow{4e} Ge$ -0.15

$CO_2 \xrightarrow{4e} C$ -0.20

$PbSO_4 \longrightarrow Pb$ -0.36

$SiO_2 \xrightarrow{4e} Si$ -0.86

$Cu^{+2} \xrightarrow{1e} CuCl$ 0.54

$Cu^+ \xrightarrow{1e} Cu$ 0.52

$Ag_2CrO_4 \longrightarrow 2\,Ag$ 0.45

$Fe(CN)_6^{-3} \xrightarrow{1e} Fe(CN)_6^{-4}$ 0.36

$Cu^{+2} \longrightarrow Cu$ 0.34

$Hg_2Cl_2 \longrightarrow 2\,Hg$ 0.27

$AgCl \xrightarrow{1e} Ag$ 0.22

$Cu^{+2} \xrightarrow{1e} Cu^+$ 0.15

$Ag(S_2O_3)_2^{-3} \xrightarrow{1e} Ag$ 0.01

$Ni^{+2} \longrightarrow Ni$ -0.25

$Cd^{+2} \longrightarrow Cd$ -0.40

$Cr^{+3} \xrightarrow{1e} Cr^{+2}$ -0.41

$Fe^{+2} \longrightarrow Fe$ -0.44

$Cr^{+3} \xrightarrow{3e} Cr$ -0.74

$Zn^{+2} \longrightarrow Zn$ -0.76

$Mn^{+2} \longrightarrow Mn$ -1.18

Group 1
Alkali Metals

$Na^+ \xrightarrow{1e} Na$ -2.71

$Cs^+ \xrightarrow{1e} Cs$ -2.92

$K^+ \xrightarrow{1e} K$ -2.93

$Rb^+ \xrightarrow{1e} Rb$ -2.93

$Li^+ \xrightarrow{1e} Li$ -3.05

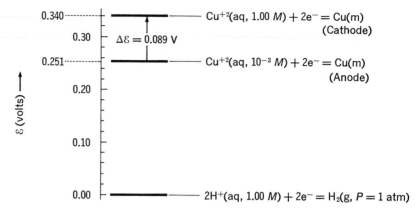

Figure 6-16 Potential diagram for copper ion concentration cell.

If we plot these half-cell potentials (Fig. 6-16) we see that (6-53) is the cathode and (6-54) the anode. The cell reaction is

$$[Cu^{+2}(aq, 1.00\ M) + 2\ e^- \longrightarrow Cu(m)]$$

$$- [Cu^{+2}(aq, 10^{-3}\ M) + 2\ e^- \longrightarrow Cu(m)]$$

$$\text{or}\quad Cu^{+2}(aq, 1.00\ M) \longrightarrow Cu^{+2}(aq, 10^{-3}\ M) \tag{6-55}$$

and, by (6-43), the cell voltage is

$$\Delta\mathcal{E} = \mathcal{E}_{cathode} - \mathcal{E}_{anode}$$

$$= 0.340 - 0.251$$

$$= 0.089\ V$$

These results are perfectly in accord with the quantitative findings shown in Table 6-1.

(d) ELECTRODE POTENTIALS AND REACTION TENDENCY

Now we have a quantitative way to assess potentially spontaneous processes. We should design an electrochemical cell employing the reaction in question. Then the half-cell potentials can be found: $\mathcal{E}^0$'s from tables and $\mathcal{E}$'s from the application of the Nernst equation to the appropriate $\mathcal{E}^0$. (Remember, by convention, $\mathcal{E}$'s refer to half-cell reactions written as reductions.) The more positive half-cell potential will be associated with the cathode; electrons flow *toward* the more positive potential. With the cathode reaction expressed as a reduction and the anode reaction as an oxidation the two half-cell reactions can be easily combined into a net equation representing a spontaneous reaction. If the calculated cell potential $\Delta\varepsilon$ is positive, the reaction has a tendency to proceed spontaneously.

But what about two reactions with different $\Delta\varepsilon$'s? Which one is the more spontaneous?

These questions are easily answered experimentally by setting up two cells in opposition, as shown in Figure 6-17 for the zinc–copper and copper–silver cells already shown in series (Fig. 6-14). As expected, the resultant voltage, 0.64 volt, is the difference between the two cell potentials,

$$\Delta\varepsilon = \Delta\varepsilon_{Zn|Cu} - \Delta\varepsilon_{Cu|Ag}$$

$$= (1.10) - (0.46) = 0.64 \text{ V}$$

Electron flow, as shown by the arrows, is toward the copper electrode in the copper–silver cell. Normally this electrode is the anode; in this arrangement it is not. In fact, if the cells are not operating reversibly—that is, if current is allowed to flow—the copper–silver cell will be driven in the direction opposite to its usual direction. Like a car battery being rejuvenated, it is being recharged.

The winner in the electron pushing game is clearly the zinc–copper cell ($\Delta\varepsilon^0 = 1.10$ V) not the copper–silver cell ($\Delta\varepsilon^0 = 0.46$ V). Cell voltage is shown to be a measure of the *relative* reaction tendency—a useful piece of information to have, indeed.

But wait a minute. Equilibrium constants are measures of reaction tendency also. How are they related to $\Delta\varepsilon$?

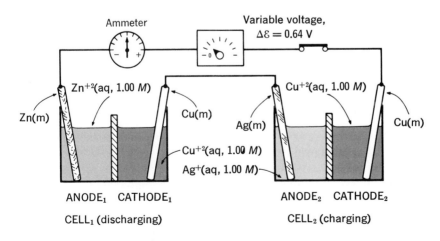

Figure 6-17 Two reversible cells in opposition.

(e) EQUILIBRIUM CONSTANTS AND CELL POTENTIALS

The Nernst Equation

$$\Delta \mathcal{E} = \Delta \mathcal{E}^0 - \frac{0.059}{n} \log Q$$

contains a term Q that has the form of an equilibrium constant, but may refer to any nonequilibrium set of concentrations. When all concentrations are unity, $Q = 1$ or $\log Q = 0$ and

at standard conditions $\Delta \mathcal{E} = \Delta \mathcal{E}^0$

If any cell is allowed to run down, eventually it will reach a state in which equilibrium has been achieved. At this point we can replace Q by K, the equilibrium constant. The Nernst Equation becomes

at equilibrium $\Delta \mathcal{E} = \Delta \mathcal{E}^0 - \dfrac{0.059}{n} \log K$ (6-56)

What is $\Delta \mathcal{E}$? Why that's easy. At equilibrium there is no tendency for any reaction to take place and so $\Delta \mathcal{E} = 0$. Equation (6-56) becomes

$$0 = \Delta \mathcal{E}^0 - \frac{0.059}{n} \log K$$

or

$$\Delta \mathcal{E}^0 = \frac{0.059}{n} \log K \tag{6-57}$$

or

$$\boxed{\log K = \frac{n \Delta \mathcal{E}^0}{0.059}} \tag{6-58}$$

This result is of enormous importance in chemistry. It allows us to measure a voltage, $\Delta \mathcal{E}^0$, in a cell made up with easily prepared solutions and calculate an equilibrium constant. The equilibrium concentrations to which this refers may be too small to measure analytically, yet with the aid of 1.00 M solutions, they can be precisely calculated.

For example, consider the zinc–silver cell. The net reaction is

$$2 \, Ag^+(aq) + Zn(m) = Zn^{+2}(aq) + 2 \, Ag(m)$$

for which

$$\Delta \mathcal{E}^0_{Zn|Ag} = 1.56 \text{ V}$$

By (6-58)

$$\log K = \frac{(2)(1.56)}{(0.059)} = 52.9$$

$$K = 10^{52.9} = 10^{0.9} \times 10^{52}$$

$$K = 7.94 \times 10^{52}$$

$$= \frac{[Zn^{+2}(aq)]}{[Ag^+(aq)]^2}$$

This enormous constant tells us that the reaction is essentially complete. The silver ion concentration at equilibrium is negligible—and impossible to measure. A simple voltage measurement in a standard cell with 1.0 molar concentrations, however, tells all.

Exercise 6-5. Show that the equilibrium constant for the reaction $Cu^{+2}(aq) + H_2(g) = Cu(m) + 2 H^+(aq)$ is $3 \cdot 10^{11}$.

6-4 Electrochemistry in action

It is probably impossible to find any single topic that has such wide practical importance as electrochemistry. Batteries and dry cells occupy a key place in twentieth century technology. Outcries against pollution are accelerating research for batteries or fuel cells to replace gasoline engines with electrically powered motors. Delicate instruments on space probes are powered by sophisticated fuel cells. Other techniques of electrochemistry help prevent corrosion. This section will deal with some of these important applications.

(a) ELECTROLYSIS

One electrochemical cell can charge another cell of lower voltage, as we found in Section 6-3(d). This process, the opposite of spontaneous discharge, is called *electrolysis*. Besides its everyday use in recharging automobile batteries, electrolysis is important in the industrial production of several elements (aluminum, magnesium, and elemental chlorine are prominent examples). The very simple electrolysis cell shown in Figure 6-18 is used commercially for the production of sodium metal. Fused sodium chloride (its melting point lowered to about 900°K by the addition of some sodium carbonate, Na_2CO_3) conducts electric current. Electrons flow from the battery toward the RHE, where liquid sodium metal can be observed rising to the surface. This electrode reaction must be the reduction

$$Na^+(NaCl\ melt) + e^- \longrightarrow Na(\ell) \tag{6-59}$$

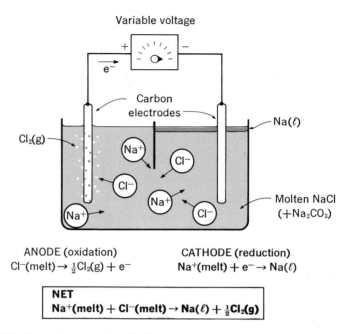

Figure 6-18 Electrolysis of molten NaCl.

and this electrode is the cathode. Bubbles of chlorine gas appearing at the LHE confirm the oxidation occurring at the anode to be

$$Cl^-(NaCl\ melt) \longrightarrow \tfrac{1}{2} Cl_2(g) + e^- \tag{6-60}$$

The net reaction produces liquid sodium and gaseous chlorine, and the required voltage is about 4 volts.

$$Na^+(melt) + Cl^-(melt) = Na(\ell) + \tfrac{1}{2} Cl_2(g) \qquad \Delta\mathcal{E}^0 \approx -4.1\ volts \tag{6-61}$$

Conduction within the melt results from the migration of charged ions; Na^+ toward the cathode and Cl^- toward the anode.

If a concentrated aqueous solution of NaCl (brine) is used in the electrolysis cell, a very different cathode reaction is observed, Figure 6-19. A gas is given off that proves to be hydrogen. If a few drops of acid-base indicator phenolphthalein are added to the solution, the pink color that indicates a basic solution (pH > 9) can be observed spreading from the cathode. The formation of hydrogen gas and hydroxide ions suggests that water is the reactant.

$$e^- + H_2O(\ell) \longrightarrow \tfrac{1}{2} H_2(g) + OH^-(aq) \qquad \mathcal{E}^0 = -0.83 \tag{6-62}$$

Since the anode reaction is the same for both cells, (6-60), the net electrolysis reaction for the brine solution is

$$Cl^-(aq) + H_2O(\ell) \longrightarrow \tfrac{1}{2} H_2(g) + \tfrac{1}{2} Cl_2(g) + OH^-(aq)$$
$$\Delta \mathcal{E}^0 = -2.19 \qquad (6\text{-}63)$$

The reason for the switch is easy to understand if we compare the minimum voltages required for (6-61) and (6-63). The reduction of Na^+ in aqueous solution has an $\mathcal{E}^0$ of -2.71 (see Table 6-5), much more negative than the $\mathcal{E}^0$ for (6-62). Consequently, reaction (6-63) can proceed at about half the voltage required for (6-61). In aqueous solution, with plenty of water molecules available, this is the process that takes place. It actually occurs at a voltage below that given, 2.19 volts, because the brine is more concentrated than $1\ M$.

What happens if we attempt to electrolyze pure water? Well, if it is really pure, not much. Conduction by pure water is too low to complete effectively the electrolysis circuit. A very small amount of sodium chloride dissolved in the water will solve the problem. The cell conducts and the formation of gases is observed at both electrodes (Fig. 6-20). When the gases are collected we immediately notice a difference from (6-63). Neither gas is colored, so chlorine, which is yellow, is not one of the products. Also, equation (6-63) predicts the production of equal quantities of $H_2(g)$ and

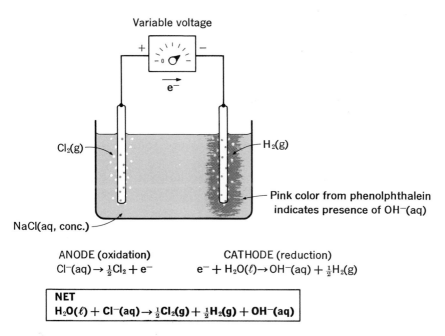

Variable voltage

$Cl_2(g)$

$H_2(g)$

Pink color from phenolphthalein indicates presence of $OH^-(aq)$

$NaCl(aq, conc.)$

ANODE (oxidation)
$Cl^-(aq) \rightarrow \tfrac{1}{2} Cl_2 + e^-$

CATHODE (reduction)
$e^- + H_2O(\ell) \rightarrow OH^-(aq) + \tfrac{1}{2} H_2(g)$

NET
$H_2O(\ell) + Cl^-(aq) \rightarrow \tfrac{1}{2} Cl_2(g) + \tfrac{1}{2} H_2(g) + OH^-(aq)$

Figure 6-19 Electrolysis of NaCl brine.

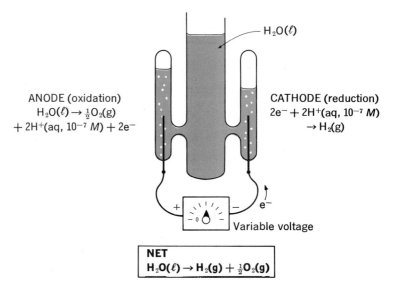

ANODE (oxidation)
$H_2O(\ell) \rightarrow \frac{1}{2}O_2(g)$
$+ 2H^+(aq, 10^{-7}\ M) + 2e^-$

CATHODE (reduction)
$2e^- + 2H^+(aq, 10^{-7}\ M)$
$\rightarrow H_2(g)$

NET
$H_2O(\ell) \rightarrow H_2(g) + \frac{1}{2}O_2(g)$

Figure 6-20 Electrolysis of water (small amount NaCl *added for conductivity).*

$Cl_2(g)$. Figure 6-20 shows that we have twice as much gas at the cathode as at the anode. By now everyone can guess what is happening. Water is being decomposed into hydrogen and oxygen.

anode $\quad H_2O(\ell) \longrightarrow O_2 + 2\,H^+ \qquad \mathcal{E}^0_{ox.} = -0.815$
$(10^{-7}\ M) + 2\ e^-$ $\qquad\qquad$ (6-64)

cathode $\quad 2\ e^- + 2\,H^+$
$\underline{\qquad\qquad (10^{-7}\ M) \longrightarrow H_2(g) \qquad\qquad \mathcal{E}^0 = -0.414 \quad \text{(6-65)}}$
cell $\qquad H_2O(\ell) \longrightarrow \frac{1}{2}\,O_2(g) + H_2(g) \quad \Delta\mathcal{E}^0 = -1.23\ V \quad \text{(6-66)}$

The results of these three electrolysis processes are summarized in Table 6-7. Three very different reactions take place, depending on the NaCl concentration.*

* In practice, electrolysis does not begin to produce O_2 until the voltage reaches about 1.7 volts. This phenomenon is called "overvoltage" and it is attributed to local concentration effects in the immediate vicinity of the electrode. Without overvoltage, electrolysis of the brine would also give O_2.

Table 6-7 Three Electrolysis Cells

Material	Produced at Anode	Produced at Cathode	Minimum Voltage Required
NaCl(melt)	$Cl_2(g)$	$Na(\ell)$	≈ 4.07
NaCl(brine)	$Cl_2(g)$	$H_2(g)$	< 2.19
NaCl(aq, dilute)	$O_2(g)$	$H_2(g)$	> 1.23

(b) EXPLOSIONS, ROCKETS, AND FUEL CELLS

Energy is required to decompose water into hydrogen gas and oxygen gas. The reverse process is obviously a source of energy. It is, however, immeasurably slow at room temperature. Walk into a room filled with this mixture at your peril, though. The slightest spark will set off a devastating explosion. The reaction

$$H_2(g) + \tfrac{1}{2} O_2(g) \longrightarrow H_2O(g) \tag{6-67}$$

releases nearly 60 kcal for every mole of water formed. The destructive power comes from the fact that once started, the reaction occurs almost instantaneously. The energy of the hydrogen–oxygen explosion is exploited in engines used in the first stage of the mighty Saturn V rockets. Liquid oxygen (b.p. 90°K) and liquid hydrogen (b.p. 20°K) combine to provide the enormous thrust needed.

Once in space, power is needed to run the spacecraft's electronic equipment. Now electrochemistry comes into its own—and again it is hydrogen and oxygen that lead the way.

A cell employing the reverse of reaction (6-66)

$$H_2(g) + \tfrac{1}{2} O_2(g) \longrightarrow H_2O(\ell) \tag{6-68}$$

would employy the two half-reactions

anode	$H_2(g) \longrightarrow 2\,H^+(aq) + 2\,e^-$
cathode	$\tfrac{1}{2} O_2(g) + 2\,H^+(aq) + 2\,e^- \longrightarrow H_2O(\ell)$

A schematic diagram of a commercial fuel cell based on this process is shown in Figure 6-21. Electrical continuity between electrodes is provided by a solid polymer electrolyte that also permits the transfer of protons from the anode to the cathode. Water, the product of the oxidation at the cathode, is a useful by-product, particularly in space applications. Fuel cells such as this normally produce about one volt (compare the standard voltage, $\Delta \mathcal{E}^0 = 1.23$ V). Many such cells are stacked together to form fuel cell battery systems.

For terrestrial applications, air is likely to be used as a source of oxygen gas. Much research is going into developing fuels less dangerous than hydrogen. Liquid hydrocarbons, methyl alcohol, ammonia, and hydrazine N_2H_4 are all possibilities. None of these share the great advantages of hydrogen as a fuel: high electrical efficiency and simplicity of reaction with oxygen. Anode reactions of hydrocarbons are complex and lead to many by-products.

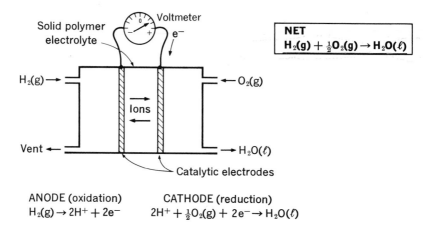

ANODE (oxidation)　　　CATHODE (reduction)
$H_2(g) \rightarrow 2H^+ + 2e^-$　　$2H^+ + \frac{1}{2}O_2(g) + 2e^- \rightarrow H_2O(\ell)$

Figure 6-21　Fuel cell (schematic). Several such cells assembled in series form a fuel battery. In zero gravity applications, water is removed by absorbent wicks.

(c) MAGNESIUM BURIED IN A SWAMPY PLACE

The corrosion of a metal, such as iron, is a redox process. When metallic iron is oxidized, either oxygen gas or hydrogen ion is reduced, forming ferrous ions and hydroxide ions or hydrogen gas.

$$Fe(m) \longrightarrow Fe^{+2} + 2\,e^- \tag{6-69}$$

$$\tfrac{1}{2}\,O_2(g) + H_2O + 2\,e^- \longrightarrow 2\,OH^- \tag{6-70}$$

$$2\,H^+ + 2\,e^- \longrightarrow H_2(g) \tag{6-71}$$

Whenever metals are wet or even damp, corrosion is possible. Estimates put the dollar loss from corrosion in the United States at $35 per year for every man, woman and child. Effective prevention and control of corrosion is obviously of great importance in any highly industrialized country. Anyone who has watched his car gradually disappearing during winter salting of the roads will surely agree wholeheartedly. (Salt makes water a better conductor, increasing the rate of electrolytic decomposition.)

A striking demonstration of the corrosion of an iron nail may be made as follows. Place a bent nail in a dish and cover it with a warm agar-agar solution (agar-agar is a gelatin-like substance extracted from seaweed) to which some potassium ferricyanide solution, $K_3Fe(CN)_6$, and some phenolphthalein have been added. When corrosion occurs, a blue color appears near the anodic regions due to the reaction of ferrous ions with $K_3Fe(CN)_6$ to form a deep-blue-colored compound, (6-72).

$$Fe \longrightarrow Fe^{+2} + 2\,e^-$$

$$3\,Fe^{+2} + 2\,Fe(CN)_6^{-3} \longrightarrow Fe_3[Fe(CN)_6]_2 \tag{6-72}$$

Meanwhile reduction of oxygen (6-70) occurs at cathodic regions on the nail. A pink phenolphthalein color signals the formation of hydroxide ions. Figure 6-22 shows a typical result: Corrosion occurs most readily at the end and at the bend. Corrosion is most likely to occur where the nail has been subjected to stress during manufacture or as a result of the bending process.

Protection against corrosion may be achieved by covering the surface with a coating of some sort (paint, for example). Better protection is possible if electrochemistry is brought into the picture. An iron surface can, for example, be forced to behave as a cathode instead of an anode as in (6-69) by creating an electrochemical cell with a metal of lower $\mathcal{E}^0$. Galvanized steel is a good example. The zinc coating effectively acts as the anode, making iron the cathode.

anode: oxidation

$$Zn(m) \longrightarrow Zn^{+2} + 2\,e^- \qquad \mathcal{E}^0 = +0.76 \qquad\qquad (6\text{-}73)$$

cathode: reduction

$$Fe^{+2} + 2\,e^- \longrightarrow Fe(m) \qquad \mathcal{E}^0 = -0.44 \qquad\qquad (6\text{-}74)$$

The standard cell potential is

$$\Delta\mathcal{E} = \mathcal{E}^0_{\text{cathode}} - \mathcal{E}^0_{\text{anode}}$$
$$= (-0.44) - (-0.76)$$
$$= +0.32 \text{ V}$$

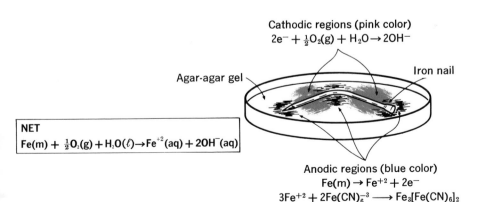

Cathodic regions (pink color)
$$2e^- + \tfrac{1}{2}O_2(g) + H_2O \rightarrow 2OH^-$$

Iron nail

Agar-agar gel

NET
$$Fe(m) + \tfrac{1}{2}O_2(g) + H_2O(\ell) \rightarrow Fe^{+2}(aq) + 2OH^-(aq)$$

Anodic regions (blue color)
$$Fe(m) \rightarrow Fe^{+2} + 2e^-$$
$$3Fe^{+2} + 2Fe(CN)_6^{-3} \longrightarrow Fe_3[Fe(CN)_6]_2$$

Figure 6-22 Demonstration of the corrosion of an iron nail.

Thus the spontaneous reaction is

$$Zn(m) + Fe^{+2}(aq) \longrightarrow Fe(m) + Zn^{+2}(aq) \qquad (6\text{-}75)$$

and oxidation of iron in the steel is prevented.

The common tin can is really steel that has been coated with tin. We can measure $\Delta\mathcal{E}^0$ for this pair of metals to see if tin offers anodic protection. The relevant half-reactions are

$$Sn^{+2} + 2\,e^- \longrightarrow Sn \qquad \mathcal{E}^0 = -0.14 \qquad (6\text{-}76)$$
$$Fe^{+2} + 2\,e^- \longrightarrow Fe \qquad \mathcal{E}^0 = -0.44$$

Since $\mathcal{E}^0$ (Fe^{+2}/Fe) is the more negative, tin will be the cathode. The spontaneous reaction is

$$Sn^{+2} + Fe \longrightarrow Fe^{+2} + Sn \qquad \Delta\mathcal{E}^0 = 0.30 \text{ V} \qquad (6\text{-}77)$$

But wait a minute! We want iron to form the cathode—but it refuses! This example illustrates the difference between a simple protective coat (tin on steel) and electrolytic anodic protection (zinc on steel). A scratched "tin" can will rust: A scratched piece of galvanized steel will not.

Underground or immersed steel structures and pipelines are particularly susceptible to corrosion. A piece of magnesium buried in a swampy place can be wired to an underground pipeline and, by acting as an anode, it will protect the steel while it itself oxidizes. The half-reactions are

$$Mg^{+2} + 2\,e^- \longrightarrow Mg \qquad \mathcal{E}^0 = -2.37 \qquad (6\text{-}78)$$
$$Fe^{+2} + 2\,e^- \longrightarrow Fe \qquad \mathcal{E}^0 = -0.44$$

They lead to a net reaction

$$Mg(m) + Fe^{+2}(aq) \longrightarrow Mg^{+2}(aq) + Fe(m) \qquad (6\text{-}79)$$
$$\Delta\mathcal{E}^0 = (-0.44) - (-2.37)$$
$$= +1.93 \text{ V}$$

In such an application magnesium is called a *sacrificial metal*. Blocks of magnesium attached to the hull of a seagoing ship perform the same function.

And so, we see that a combination of ingenuity and electrochemistry helps fight battles against corrosion.

6-5 Oxidation-reduction equilibria

Our aim up to now in this chapter has been to examine spontaneous reactions with the help of electrochemical cells. We found that many reactions can be performed in an electrochemical cell in such a way that electrons are transferred through an external circuit. Now it is time to focus our attention on oxidation-reduction reactions in general—those reactions that are classified as redox even when they don't take place in a cell.

(a) WHAT IS A REDOX REACTION, ANYWAY?

In Section 6-1(e) we considered a reaction in which the net effect is the transfer of electrons from one species to another. We called such a reaction a redox reaction. It is interesting to consider the possibility that this electron-transfer idea furnishes a useful, general definition of an oxidation-reduction reaction.

Perhaps we should define

> An **oxidation-reduction reaction** is a reaction in which electrons are transferred. $(6\text{-}80)$

This definition fits some reactions quite well. For example, consider

$$Co^{+3}(aq) + Cr^{+2}(aq) \longrightarrow Co^{+2}(aq) + Cr^{+3}(aq) \qquad (6\text{-}81)$$

Plainly, Cr^{+2} must lose an electron to form Cr^{+3}. Also, the Co^{+3} must acquire one electron to form Co^{+2}. The Cr^{+2} *donates* an electron to Co^{+3}, causing Co^{+3} to be reduced to Co^{+2}. Thus Cr^{+2} acts as a *reducing agent* as it donates electrons. In a reciprocal way, Co^{+3}, in accepting electrons from Cr^{+2}, causes chromium to be oxidized to Cr^{+3}. Thus Co^{+3} acts as an *oxidizing agent* as it accepts electrons. We see that the reaction involves—indeed, requires—both an electron donor (a reducing agent) and an electron acceptor (an oxidizing agent). The situation is reminiscent of the Brønsted–Lowry definition of an acid-base reaction as a proton transfer involving a proton donor (an acid) and a proton acceptor (a base). Indeed, the parallel is rather neat, as shown in Table 6-8. That makes the definition (6-80) quite palatable to many chemists and it can be found in quite a few textbooks.

Table 6-8 Acid-Base and Redox Compared

	Brønsted Acid-Base	Redox
species transferred	proton, H^+	electron, e^-
donor	acid	reducing agent
acceptor	base	oxidizing agent

Unfortunately, the definition, with its tidy parallel to proton-transfer reactions, has its difficulties. Consider these two reactions that involve elemental oxygen:

$$H_2(g) + \tfrac{1}{2} O_2(g) = H_2O(g) \tag{6-82}$$

$$\tfrac{1}{2} N_2(g) + \tfrac{1}{2} O_2(g) = NO(g) \tag{6-83}$$

Everyone familiar with the history of chemistry would agree that these are oxidation-reduction reactions. In fact, the term "oxidation" derives from reactions in which oxygen actually reacts with ("oxidizes") some other molecular species. But in reactions (6-82) and (6-83) it is surely not self-evident that electrons have been transferred and, if they were, what acted as the donor and what acted as the acceptor.

Consider the product H_2O in (6-82). It has a Lewis dot formula that displays electrons being *shared* by the hydrogen atoms and the oxygen atom, and it does not imply *transfer* of electrons from H to O or vice versa.

$$:\overset{\cdot\cdot}{\underset{\circ\,\circ}{O}}:H$$
$$H$$

Reaction (6-83) has the same problem, only more acute. Again if we are to describe this oxidation reaction as an electron-transfer process, then electrons must be transferred from the oxygen atom to the nitrogen atom, or vice versa. If a single electron charge were to be transferred, the molecule would have a molecular dipole moment given by the product of electron charge times the bond length:

$$\text{dipole moment (calc.)} = \mu(\text{calc.}) = (\text{charge})(\text{bond length})$$

$$= (4.8 \cdot 10^{-10}\ \text{esu})(1.15 \cdot 10^{-8}\ \text{cm})$$

$$= 5.5 \cdot 10^{-18}\ \text{esu-cm} \tag{6-84}$$

Unfortunately, the calculated result, (6-84), is 36 times larger than the actual observed value, $0.15 \cdot 10^{-18}$ esu-cm. The experimental number tells us that virtually no charge is transferred in this case, though again the reaction (6-83) must surely be classified as oxidation.

Now that this electron-transfer definition looks suspect, it isn't hard to find other weaknesses. Consider the half-reaction that describes the electrode reduction of nitrate ion in an electrochemical cell:

$$NO_3^-(aq) + 4\,H^+(aq) + 3\,e^- \longrightarrow NO(g) + 2\,H_2O(\ell) \tag{6-85}$$

The action of the cell and expression (6-85) show that three moles of electrons are consumed per mole of nitrate reduced. Yet, looking at the products, there is no way to tell where those electrons ended up, on nitrogen, oxygen, or hydrogen atoms. Even reactions like (6-81) prove, in some cases, to be more complicated than has been realized. The ions Co^{+3} and Cr^{+2} are each strongly bound to several water molecules, probably six. Hence the formula $Co^{+3}(aq)$ and $Cr^{+2}(aq)$ might better be written $Co(H_2O)_6^{+3}$ and $Cr(H_2O)_6^{+2}$. The relevance is that there is evidence that electron transfer in a reaction like (6-81) occurs through transfer of groups of atoms or ions. A particularly informative case is provided by the counterpart of reaction (6-81) involving the ammonia complex of cobalt, $[Co(NH_3)_5(H_2O)]^{+3}$. If the water molecule in this pentamine complex is isotopically labeled with ^{18}O, it is possible to see that the electron transfer is accompanied by quantitative transfer of the labeled water molecule.

$$[Co(NH_3)_5(H_2{}^{18}O)]^{+3} + [Cr(H_2O)_6]^{+2} \longrightarrow$$
$$[Co(NH_3)_5(H_2O)]^{+2} + [Cr(H_2{}^{18}O)(H_2O)_5]^{+3} \quad (6\text{-}86)$$

All of these deficiencies of the definition (6-80) require that we be introspective about it. Is there a better definition? Can we "save" the good aspects of the definition (6-80) without retaining the implications that are not realized in nature? Here is another opportunity to see the flow of science—its flexibility—and its frailty. In our effort to escape the problems of the electron-transfer definition, we introduce a formal bookkeeping device called "oxidation numbers."

(b) OXIDATION NUMBERS

Oxidation numbers are ficticious charges that are assigned to atoms in molecules (or ions) according to a set of rules. They have no physical significance—they do not represent real charges—but they are useful for categorizing redox processes.

In the early days of chemistry, the solid product of an oxidation process such as

$$2\ Fe(m) + \tfrac{3}{2}\ O_2(g) \longrightarrow Fe_2O_3(s) \tag{6-87}$$

was believed to consist of Fe^{+3} ions and O^{-2} ions in a regular array. Evidence now shows such an analysis to be an oversimplification of the true charge distribution. Nevertheless, the hypothetical minus two charge on oxygen is taken as the cornerstone of the oxidation number formalism. *Oxygen is generally assigned an oxidation number of* -2.

The second rule is that *the sum of the oxidation numbers assigned to the atoms in a molecule must add up to the electrical charge on that molecule.* Consider water, a neutral molecule. The sum of -2, the oxidation number of oxygen, added to twice the oxidation number of hydrogen (there are two H atoms in H_2O) must equal zero, the charge on a neutral molecule. We see

Table 6-9   Rules for Assignment of Oxidation Numbers

Rule	Examples	Exceptions
1. The oxidation number of any free element is zero.	H_2, O_2, Mg, Ne	
2. Oxygen has an oxidation number of -2.	H_2O, MgO	Peroxides (contain O–O link): H_2O_2 ox. no. $= -1$
3. Hydrogen has an oxidation number of $+1$.	H_2O, CH_4	Metallic hydrides: LiH ox. no. $= -1$
4. The sum of the oxidation numbers must be zero for a neutral molecule and equal to the net charge if an ion.	H_2O: $2(+1) + (-2)$ $= 0$ NO_3^-: $(+5) + 3(-2)$ $= -1$	

that the oxidation number of hydrogen must be $+1$. This is our second point of departure: *Hydrogen is generally assigned an oxidation number of $+1$.*

Why do we say "generally" with respect to the choice of oxidation numbers of oxygen and hydrogen? It is because most of the time we do, but sometimes we don't—in fact, sometimes we can't. Consider the elemental molecule O_2. Since it is neutral, the rule about the sum of the oxidation numbers prevents us from assigning to oxygen the value -2. We have the same sort of trouble with elemental hydrogen, H_2. Another rule, please! *All atoms in their elemental form are assigned oxidation numbers of zero.*

Now, for most cases, we have a complete set of rules, as summarized in Table 6-9.

Now some examples:

$Br_2(\ell)$	Br $=$ zero	Rule 1: element
$S_8(s)$	S $=$ zero	Rule 1: element
Al(metal)	Al $=$ zero	Rule 1: element
$Fe^{+3}(aq)$	Fe $= +3$	Rule 4: sum of ox. no.'s equals charge on ion
$Cl^-(aq)$	Cl $= -1$	Rule 4: sum of ox. no.'s equals charge on ion
NO	O $= -2$; N $= +2$	Rules 2, 4: $-2 + 2 = 0$
NO_2	O $= -2$; N $= +4$	Rules 2, 4: $2(-2) + 4 = 0$
MnO_4^-	O $= -2$; Mn $= +7$	Rules 2, 4: $4(-2) + 7 = -1$
$HONO_2$	O $= -2$; H $= +1$; N $= +5$	Rules 2, 3, 4: $3(-2) + (+1)$ $+ (+5) = 0$
$HCrO_4^-$	O $= -2$; H $= +1$; Cr $= +6$	Rules 2, 3, 4: $4(-2) + (+1)$ $+ (+6) = -1$

Exercise 6-6. Show that the oxidation number of nitrogen is $+3$ in both of the compounds N_2O_3 and nitrous acid, HONO.

Exercise 6-7. Show that in the compound hydrogen peroxide, either Rule 2 or Rule 3 must be ignored in order to satisfy Rule 4. What is the oxidation number of oxygen in H_2O_2 if Rule 3 is retained?

It is important to reiterate that oxidation numbers do *not* represent real charges actually centered on an atom. The positive charges on the aquated ferrous ion, $Fe^{+2}(aq)$, for example, are surely distributed over both the iron nucleus and all the water molecules bound to it. It would be more correct, but unconventional, to write

$$[Fe(H_2O)_x]^{+2} \quad \text{or} \quad [Fe(aq)]^{+2}$$

for this species.

For atoms in complex ions or molecules, the oxidation number or oxidation *state* is often shown by Roman numerals for the metal atoms. Thus we could write

$$Mn^{VII}O_4^-$$

to show quickly that the species contains manganese(VII). Some authors use Roman numerals for even simple aquated metal ions. Thus equation (6-81) might be written

$$Co^{III}(aq) + Cr^{II}(aq) \longrightarrow Co^{II}(aq) + Cr^{III}(aq)$$

In our study of the transition metals (Chapter Twenty-One) we will see that there is a periodic variation in possible oxidation states that helps us systematize the chemistry.

Now let's look at the examples of the last section to see what **changes** are observed in the oxidation numbers.

$$Co^{+3}(aq) + Cr^{+2}(aq) \longrightarrow Co^{+2}(aq) + Cr^{+3}(aq) \tag{6-88}$$

$$
\begin{aligned}
Co^{+3}&: +3 \longrightarrow +2 \quad &\text{change} &= -1 \\
Cr^{+2}&: +2 \longrightarrow +3 \quad &\text{change} &= +1 \\
& \text{net change} = -1 + 1 = 0
\end{aligned}
$$

$$H_2(g) + \tfrac{1}{2} O_2(g) \longrightarrow H_2O(g) \tag{6-89}$$

$$
\begin{aligned}
H&: 0 \longrightarrow +1 \quad &\text{change} &= 2(+1) = +2 \\
O&: 0 \longrightarrow -2 \quad &\text{change} &= -2 \\
& \text{net change} = +2 - 2 = 0
\end{aligned}
$$

$$2 \text{ Fe(m)} + \tfrac{3}{2} O_2(g) \longrightarrow Fe_2O_3(s) \qquad (6\text{-}90)$$

Fe: $0 \longrightarrow +3$ change $= 2(+3) = +6$
O: $0 \longrightarrow -2$ change $= 3(-2) = -6$
     net change $= +6 - 6 = 0$

$$NO_3^-(aq) + 4 H^+(aq) + 3 e^- \longrightarrow NO(g) + 2 H_2O(\ell) \qquad (6\text{-}91)$$

N: $+5 \longrightarrow +2$ change $= -3$
O: $-2 \longrightarrow -2$ no change
H: $+1 \longrightarrow +1$ no change
     net change $= -3$

$$Mn^{+2}(aq) + 2 H_2O \longrightarrow MnO_2(s) + 4 H^+ + 2 e^- \qquad (6\text{-}92)$$

Mn: $+2 \longrightarrow +4$ change $= +2$
O: $-2 \longrightarrow -2$ no change
H: $+1 \longrightarrow +1$ no change
     net change $= +2$

We can see a pattern appearing. In the balanced equations for reactions (6-88), (6-89), and (6-90) one element increases its oxidation state while another decreases and the net change is zero. In the reduction half-reaction (6-91) the oxidation state of nitrogen drops (is reduced) from $+5$ to $+2$. In the oxidation half-reaction (6-92) the oxidation state of manganese rises from $+2$ to $+4$. Oxidation implies an increase in oxidation state; reduction, a decrease. In a balanced equation any rise in oxidation number for one atom is counterbalanced by a reduction of oxidation number for another atom. The net change is zero.

Now we are ready to make a new try at a definition of a redox reaction, using oxidation numbers.

Oxidation numbers change during an oxidation-reduction process.

Exercise 6-8. Show by the oxidation number changes that ferrous ion, $Fe^{+2}(aq)$, is oxidized and iodate ion, $IO_3^-(aq)$, is reduced when they react to produce $Fe^{+3}(aq)$ and I_2.

(c) REDUCTION POTENTIAL DIAGRAMS

With the table of reduction potentials (Table 6-5), the cell potential $\Delta\mathcal{E}$ and hence the reaction tendency can quickly be calculated for a great many reactions. Often, however, we are concerned about the chemistry of only one or two elements. A useful way to display the reduction potential data for conversions between all the oxidation states of a single element is through what is called the *reduction potential diagram*. For example, we could consolidate all of the information about $\mathcal{E}^0$'s for the four half-reactions concerning iron in the form of a simple diagram, (6-93).

$$3\,e^- + Fe^{VI}O_4^{-2} + 8\,H^+ \longrightarrow Fe^{III} + 4\,H_2O \qquad \mathcal{E}^0 = 2.20$$

$$e^- + Fe^{III} \longrightarrow Fe^{II} \qquad \mathcal{E}^0 = 0.77$$

$$2\,e^- + Fe^{II} \longrightarrow Fe^0 \qquad \mathcal{E}^0 = -0.44$$

$$Fe^{III} + 3\,e^- \longrightarrow Fe^0 \qquad \mathcal{E}^0 = -0.04$$

$$Fe^{VI}O_4^{-2} \underset{\underset{\displaystyle -0.04}{\rule{5cm}{0.4pt}}}{\overset{2.20}{\rule{1.5cm}{0pt}}} Fe^{III} \overset{0.77}{\rule{1.5cm}{0pt}} Fe^{II} \overset{-0.44}{\rule{1.5cm}{0pt}} Fe^0 \qquad (6\text{-}93)$$

The terms are arranged in order of *decreasing* oxidation state and the number on the line connecting two states is $\mathcal{E}^0$ for the appropriate reduction half-reaction. The diagram for oxygen (in acid solution) appears as follows for the oxidation states 0, -1, and -2.

$$O_2 \underset{\underset{\displaystyle 1.23}{\rule{5cm}{0.4pt}}}{\overset{0.68}{\rule{1.5cm}{0pt}}} H_2O_2 \overset{1.77}{\rule{1.5cm}{0pt}} H_2O \qquad (6\text{-}94)$$

Consider now the two possible reactions of hydrogen peroxide: the reduction to water

$$2\,e^- + 2\,H^+ + H_2O_2 \longrightarrow 2\,H_2O \qquad \mathcal{E}^0 = 1.77 \qquad (6\text{-}95)$$

and the oxidation to oxygen

$$H_2O_2 \longrightarrow O_2 + 2\,H^+ + 2\,e^- \qquad \mathcal{E}^0_{ox.} = -0.68 \qquad (6\text{-}96)$$

The combination of (6-95) and (6-96) gives the net equation

$$2\,H_2O_2 \longrightarrow 2\,H_2O + O_2 \qquad \Delta\mathcal{E}^0 = 1.09 \qquad (6\text{-}97)$$

This reaction has a large, positive voltage—it is a spontaneous reaction. There we have the redox equivalent of the self-ionization of water. Hydrogen peroxide is behaving as both a reducing and an oxidizing agent. This *self-oxidation-reduction*, or *disproportionation* reaction, is possible whenever the $\mathcal{E}^0$ to the *right* of a species in a reduction potential diagram is *more positive* than the $\mathcal{E}^0$ immediately to the left. A glance at (6-93) tells us that both ferrous and ferric ions are stable with respect to disproportionation.

In practice, reaction (6-97) is very slow at room temperature, but we shall see that the reaction can be rapid under some circumstances.

Exercise 6-9. From the following reduction potential diagram for phosphorus in basic (1.0 M OH$^-$) solution, find any species that can disproportionate, write the appropriate reaction, and calculate $\Delta\mathscr{E}^0$. *2 species.*

$$\text{P}^\text{V}\text{O}_4^{-3}\xrightarrow{-1.12}\text{HP}^\text{III}\text{O}_3^{-2}\xrightarrow{-1.57}\text{H}_2\text{P}^\text{I}\text{O}_2^-\xrightarrow{-2.05}\text{P}_4^0\xrightarrow{-0.89}\text{PH}_3$$

with -1.18 spanning from $\text{H}_2\text{P}^\text{I}\text{O}_2^-$ to PH_3, and -1.31 spanning from $\text{HP}^\text{III}\text{O}_3^{-2}$ to P_4^0.

(d) BALANCING REDOX EQUATIONS: HALF-REACTIONS

Redox equations are often quite complicated—so complicated, in fact, that balancing by trial and error becomes almost impossible. Several formal methods of balancing redox equations have been devised. One of these is based on half-reactions, which we can demonstrate with the reaction of thiosulfate ion ($\text{S}_2\text{O}_3^{-2}$) with dichromate ($\text{Cr}_2\text{O}_7^{-2}$) in acid solution.

$$\text{S}_2\text{O}_3^{-2} + \text{Cr}_2\text{O}_7^{-2} \longrightarrow \text{SO}_4^{-2} + \text{Cr}^{+3} \tag{6-98}$$

First we have to find out what is oxidized and reduced. We can rewrite (6-98) labeling all the oxidation states:

$$\text{S}_2^\text{II}\text{O}_3^{-2} + \text{Cr}_2^\text{VI}\text{O}_7^{-2} \longrightarrow \text{S}^\text{VI}\text{O}_4^{-2} + \text{Cr}^\text{III} \tag{6-99}$$

Now it is clear: Chromium is reduced and sulfur is oxidized; dichromate is oxidizing thiosulfate. A good way to proceed is to write separate oxidation and reduction half-reactions, which can be balanced separately. Since the reaction is occurring in aqueous acid, we can use H$^+$(aq) and H$_2$O whenever necessary. First, oxidation:

$$\text{S}_2^\text{II}\text{O}_3^{-2} \longrightarrow \text{S}^\text{VI}\text{O}_4^{-2}$$

Balance sulfur atoms,

$$\text{S}_2^\text{II}\text{O}_3^{-2} \longrightarrow 2\,\text{S}^\text{VI}\text{O}_4^{-2}$$

then oxygen atoms, using H$_2$O

$$5\,\text{H}_2\text{O} + \text{S}_2^\text{II}\text{O}_3^{-2} \longrightarrow 2\,\text{S}^\text{VI}\text{O}_4^{-2}$$

finally hydrogen atoms, using H^+

$$5\,H_2O + S_2^{II}O_3^{-2} \longrightarrow 2\,S^{VI}O_4^{-2} + 10\,H^+$$

All that remains is to balance the charges—electrons can be used, as this is a half-reaction. At the moment there is a -2 charge on the left and a $+6$ charge on the right. Eight electrons will do the trick.

$$5\,H_2O + S_2^{II}O_3^{-2} \longrightarrow 2\,S^{VI}O_4^{-2} + 10\,H^+ + 8\,e^- \qquad (6\text{-}100)$$

Now the reduction half-reaction, easily balanced for chromium

$$Cr_2^{VI}O_7^{-2} \longrightarrow 2\,Cr^{III}$$

Balance oxygen, using H_2O

$$Cr_2^{VI}O_7^{-2} \longrightarrow 2\,Cr^{III} + 7\,H_2O$$

and hydrogen, using H^+

$$14\,H^+ + Cr_2^{VI}O_7^{-2} \longrightarrow 2\,Cr^{III} + 7\,H_2O$$

and finally, charge, using electrons

$$6\,e^- + 14\,H^+ + Cr_2^{VI}O_7^{-2} \longrightarrow 2\,Cr^{III} + 7\,H_2O \qquad (6\text{-}101)$$

We now combine (6-100) and (6-101) in the usual way—assuring that electrons balance. The lowest common multiple of 8 and 6 is 24, so we need 3 times (6-100) and 4 times (6-101).

$$15\,H_2O + 3\,S_2O_3^{-2} \longrightarrow 6\,SO_4^{-2} + 30\,H^+ + 24\,e^-$$
$$\underline{24\,e^- + 56\,H^+ + 4\,Cr_2O_7^{-2} \longrightarrow 8\,Cr^{+3} + 28\,H_2O}$$
$$15\,H_2O + 56\,H^+ + 3\,S_2O_3^{-2} + 4\,Cr_2O_7^{-2} \longrightarrow 8\,Cr^{+3} + 6\,SO_4^{-2}$$
$$+ 30\,H^+ + 28\,H_2O$$

Water and H^+ appear on both sides so some simplification is possible

$$26\,H^+ + 3\,S_2O_3^{-2} + 4\,Cr_2O_7^{-2} \longrightarrow 8\,Cr^{+3} + 6\,SO_4^{-2} + 13\,H_2O \quad (6\text{-}102)$$

Exercise 6-10. Balance the reaction considered in Exercise 6-8 to show that when one mole of iodate ion is reduced, six moles of H^+(aq) are consumed.

Reactions in basic media involve hydroxide ions rather than protons. There are several ways to modify the recipe for balancing equations to take account of this. One of the easiest of these is to proceed as though the solution is acidic. Then OH^- ions can be added to both sides of the balanced equation in sufficient number to convert all H^+ to H_2O. Let's see how it goes in the basic solution oxidation of Sn(II) to Sn(IV) by chromate, the form of Cr(IV) in basic solution. The reactants and products have the compositions:

$$CrO_4^{-2} + HSnO_2^- \longrightarrow Cr(OH)_3 + Sn(OH)_6^{-2}$$

or

$$Cr^{VI}O_4^{-2} + HSn^{II}O_2^- \longrightarrow Cr^{III}(OH)_3 + Sn^{IV}(OH)_6^{-2} \qquad (6\text{-}103)$$

First, reduction, a decrease in oxidation number,

$$Cr^{VI}O_4^{-2} \longrightarrow Cr^{III}(OH)_3$$

Balance oxygen by adding H_2O, hydrogen by adding H^+, and charge by adding electrons

$$Cr^{VI}O_4^{-2} + 5\,H^+ + 3\,e^- \longrightarrow Cr^{III}(OH)_3 + H_2O$$

Now we add 5 OH^- to each side

$$Cr^{VI}O_4^{-2} + 5\,H^+ + 5\,OH^- + 3\,e^- \longrightarrow Cr^{III}(OH)_3 + H_2O + 5\,OH^-$$

The 5 H^+ and 5 OH^- that appear as reactants can be considered to react to give 5 H_2O

$$Cr^{VI}O_4^{-2} + 5\,H_2O + 3\,e^- \longrightarrow Cr^{III}(OH)_3 + H_2O + 5\,OH^-$$

Now we have 5 molecules of H_2O consumed and one molecule of H_2O produced, a net of 4 molecules consumed. Our final, balanced half-reaction is:

$$Cr^{VI}O_4^{-2} + 4\,H_2O + 3\,e^- \longrightarrow Cr^{III}(OH)_3 + 5\,OH^- \qquad (6\text{-}104)$$

Exercise 6-11. (a) Show that balancing the equation for the half-reaction between $HSnO_2^-$ and $Sn(OH)_6^{-2}$ as though it occurs in acid solution gives the equation

$$HSnO_2^- + 4\,H_2O \longrightarrow Sn(OH)_6^{-2} + 3\,H^+ + 2\,e^-$$

(b) Show that adding OH^- and tidying up gives the balanced equation

$$HSnO_2^- + 3\,OH^- + H_2O \longrightarrow Sn(OH)_6^{-2} + 2\,e^- \qquad (6\text{-}105)$$

Given the balanced equations for the two half-reactions, (6-104) and (6-105), they are combined, as usual, to balance electron consumption and production:

$$2\ CrO_4^{-2} + 8\ H_2O + 6\ e^- \longrightarrow 2\ Cr(OH)_3 + 10\ OH^-$$

$$\frac{3\ HSnO_2^- + 9\ OH^- + 3\ H_2O \longrightarrow 3\ Sn(OH)_6^{-2} + 6\ e^-}{2\ CrO_4^{-2} + 3\ HSnO_2^- + 11\ H_2O + 9\ OH^- \longrightarrow 2\ Cr(OH)_3}$$
$$+ 3\ Sn(OH)_6^{-2} + 10\ OH^-$$

or

$$2\ CrO_4^{-2} + 3\ HSnO_2^- + 11\ H_2O \longrightarrow 2\ Cr(OH)_3$$
$$+ 3\ Sn(OH)_6^{-2} + OH^-$$
$$(6\text{-}106)$$

The recipes for balancing redox reactions by half-reactions are summarized in Table 6-10.

(e) BALANCING REDOX EQUATIONS: OXIDATION NUMBERS

Many chemists balance redox equations without ever writing out the half-reactions. Just the oxidation number changes are sufficient, since, in a balanced equation, the net change in oxidation numbers is zero. Let's try examples like those in the last section using this technique.

Consider the reaction between ferrous ion and dichromate to produce ferric ion and chromic ion:

$$Fe^{+2}(aq) + Cr_2O_7^{-2}(aq) \longrightarrow Fe^{+3}(aq) + Cr^{+3}(aq) \qquad (6\text{-}107)$$

Table 6-10 *Balancing Redox Equations*

1. Identify, with oxidation numbers, species being oxidized and reduced.
2. Write separate oxidation and reduction half-reactions.
3. Balance atoms, adding H^+, OH^-, and H_2O as required.
4. Balance charge with electrons.
5. Combine half-reactions, balancing over-all exchange of electrons.

Details of H,O Atom Balance

In acid solution: Balance oxygen with H_2O, then hydrogen with H^+.

In basic solution:

(a) Balance oxygen with H_2O, then hydrogen with H^+.
(b) Add OH^- on both sides in amount equal to H^+; convert $H^+ + OH^-$ to H_2O.
(c) Cancel H_2O if it appears on both sides of equation to show net effect.

The first step is to *assign oxidation numbers:*

$$Fe^{II} + Cr_2^{VI}O_7^{-2} \longrightarrow Fe^{III} + Cr^{III}$$

Now, we *chemically balance all atoms whose oxidation numbers change.* The iron atom changes from $+2$ to $+3$, but there is one atom on either side, so it is already balanced. The chromium atom changes from $+6$ to $+3$, and we find two atoms of Cr on the left, so we must have two on the right

$$Fe^{II} + Cr_2^{VI}O_7^{-2} \longrightarrow Fe^{III} + 2\ Cr^{III}$$

Now we *balance the change in oxidation numbers.* Each chromium atom drops from $+6$ to $+3$, so the total change by chromium is -6. Since each iron atom rises from $+2$ to $+3$, we need six of them to give a net change of zero.

$$6\ Fe^{II} + Cr_2^{VI}O_7^{-2} \longrightarrow 6\ Fe^{III} + 2\ Cr^{III}$$

Now we can proceed to *balance oxygen atoms, using H_2O, and then hydrogen atoms, using H^+.*

Balance oxygen atoms, using H_2O:

$$6\ Fe^{+2} + Cr_2O_7^{-2} \longrightarrow 6\ Fe^{+3} + 2\ Cr^{+3} + 7\ H_2O$$

Balance hydrogen atoms, using H^+:

$$6\ Fe^{+2} + Cr_2O_7^{-2} + 14\ H^+ \longrightarrow 6\ Fe^{+3} + 2\ Cr^{+3} + 7\ H_2O \qquad (6\text{-}108)$$

Now the equation should be balanced. The balance between rise and fall of oxidation numbers should take care of any electron-transfer aspect we care to read into the reaction (as is done in the half-reaction method). Hence there already should be a proper charge balance in (6-108). That furnishes a check that we didn't make any mistakes along the way:

Check, charge balance
$$6(+2) + 1(-2) + 14(+1) \overset{?}{=} 6(+3) + 2(+3)$$
$$12 \quad - \quad 2 \quad + \quad 14 \quad \overset{?}{=} \quad 18 \quad + \quad 6$$
$$24 \quad = \quad 24$$

If the charge balance doesn't check properly, a mistake was surely made.

Exercise 6-12. Using oxidation numbers only, balance the equation for the reaction (6-98) between thiosulfate, $S_2O_3^{-2}$, and dichromate, $Cr_2O_7^{-2}$, to produce sulfate, SO_4^{-2}, and chromic, Cr^{+3}. Compare the results obtained this way to that obtained using half-reactions, (6-102).

6-6 Understanding spontaneity

Electrochemical cells furnish a powerful technique for learning about reaction tendency. If a chemical change can be carried out in a cell, the voltage can be measured and translated into an equilibrium constant. With the equilibrium constant, we can predict whether a given mixture of reactants and products will proceed to form more products, consuming reactants, or just the opposite.

Unfortunately, electrochemical cells are not applicable to quite a number of interesting reactions. There is a more fundamental limitation, though. Even for the systems that are amenable to such study, the electrochemical cell gives no clue to *why* one chemical change proceeds spontaneously and another does not. If we are to control the myriad of possible reactions, we need an explanation that tells us what factors are influential in determining the tendency for change. The importance of such an understanding can hardly be overstated. As your eyes move over this page, as you think a thought, as you breathe a breath, as the sun rises and sets, spontaneous changes are at work, making the difference between an inert, drab universe and the glorious, living, evolving environment we find around us. What intricate clockwork motivates the atoms and molecules in their daily work, be it the mundane rusting of a nail or the delicate fabrication of a protein?

We shall begin our search for this explanation in the study of energy effects that accompany chemical changes. We won't restrict our interest to electrical energy—quite the opposite—we'll become preoccupied with heat. That is a fitting place to begin our search for understanding—fire was the first chemical reaction to command man's attention. Thus began and thus begins thermochemistry.

Problems

1. A standard cell involving $Ag/AgNO_3$ versus $Cd/CdCl_2$ shows a voltage of 1.20 volts, and cadmium metal dissolves when the circuit is closed. Draw a picture of the cell, using a salt bridge containing KNO_3 solution. Label the anode and cathode; indicate the direction of current flow; indicate the direction of charge migration of all ions in the solution (including those in the salt bridge) and indicate the balanced electrode reactions.

2. A standard cell involving $Ag/AgNO_3$ versus $Ni/NiCl_2$ shows a voltage of 1.05 volts, and nickel metal dissolves when the circuit is closed. Combine this information with that in Problem 1 to draw a picture of a $Cd/CdCl_2$ versus $Ni/NiCl_2$ cell. Label the anode and cathode; indicate the cell voltage, the direction of current flow, the direction of charge migration, and the balanced electrode reactions.

3. Current is allowed to flow in the cell pictured in Problem 1 until 0.0200 mole of electrons have moved through the external circuit. How many grams of cadmium have dissolved and how many grams of silver have been deposited?

4. Repeat Problem 3 for the cadmium–nickel cell of Problem 2.

5. A cell is prepared in which both electrodes are made of platinum (because it is inert) coated with MnO_2. In both beakers, the solution contains 0.030 M $KMnO_4$, but in the left half-cell $[H^+] = 0.10$ M and in the right half-cell the solution is buffered at pH = 5. The electrode reactions are

$$MnO_4^- + 4\ H^+ + 3\ e^- \rightleftharpoons MnO_2 + 2\ H_2O$$

Draw a picture of the cell; label the anode and the cathode; show the direction of current flow, the cell voltage and the direction of charge migration in the solution.

6. What would be the new voltage of the nickel–cadmium cell in Problem 2 if both $[Cd^{+2}]$ and $[Ni^{+2}]$ were changed to 0.10 M? What would be the voltage in the silver–cadmium cell in Problem 1 if both $[Cd^{+2}]$ and $[Ag^+]$ were reduced to 0.10 M?

7. Draw a picture of the Ag–Cd cell in Problem 1 placed beside the Ag–Ni cell mentioned in Problem 2, with the cadmium electrode electrically connected to the silver electrode of the Ag–Ni cell and the voltage measured between the nickel electrode and the silver electrode of the Ag–Cd cell. Label each electrode as an anode or a cathode, indicate the cell voltage and the direction of electron flow through each wire, and write each electrode reaction.

8. Repeat Problem 7, but turn the Ag–Cd cell to place its silver electrode next to the silver electrode of the Ag–Ni cell, connect electrically the two silver electrodes, and measure the voltage between the cadmium and nickel electrodes.

9. Acidic permanganate solutions, MnO_4^-, are used in quantitative analysis because the concentrations remain constant for several days. Yet there is a possible reaction with water to form MnO_2 and O_2. From the $\mathcal{E}^0$'s given in Table 6-5, calculate $\Delta\mathcal{E}^0$ for the reaction between MnO_4^- and H_2O. From $\Delta\mathcal{E}^0$, would one infer that MnO_4^- fails to react with H_2O because equilibrium favors the reverse reaction or because reaction is slow?

✱10. Some important aqueous oxidizing agents are: H_2O_2, I_3^-, Cl_2, MnO_4^-, $Cr_2O_7^{-2}$, Ce^{+4}, BrO_3^-, and IO_3^-. Some important aqueous reducing agents are I^-, Cr^{+2}, Sn^{+2}, Fe^{+2}, Hg_2^{+2}, and metallic zinc. On the basis of Tables 6-5 and 6-6, and for standard conditions, list
(a) every reagent that is unstable with respect to reduction by water and give the $\Delta\mathcal{E}^0$ values for each;
(b) every reagent that is unstable with respect to oxidation by water and give the $\Delta\mathcal{E}^0$ values for each;
(c) every reagent that is unstable with respect to oxidation by atmospheric O_2 and give the $\Delta\mathcal{E}^0$ values for each.

11. For the reagents listed in Problem 10, and under standard conditions,
(a) list every reagent that could be used to oxidize I^- to I_3^- and give the $\Delta\mathcal{E}^0$ values for each;
(b) list every reagent that could be used to reduce I_3^- to I^- and give the $\Delta\mathcal{E}^0$ values for each.

✗ 12. The separation of plutonium from uranium is one of the most important steps in the production of plutonium for nuclear power plants. Using the standard reduction potentials in Table 6-5, select two soluble reagents that could reduce plutonyl ion, PuO_2^{+2}, to Pu^{+4}, but not uranyl ion, UO_2^{+2} to U^{+4}.

$$PuO_2^{+2} + 4\ H^+ + 2\ e^- \longrightarrow Pu^{+4} + 2\ H_2O \qquad \mathcal{E}^0 = +1.04$$
$$UO_2^{+2} + 4\ H^+ + 2\ e^- \longrightarrow U^{+4} + 2\ H_2O \qquad \mathcal{E}^0 = +0.32$$

13. Calculate the equilibrium constant for the reaction discussed in Problem 9.

✗ 14. Dichromate ion, CrO_7^{-2}, can be used to oxidize various species in hydrochloric acid solutions as concentrated as 2 M. Reaction with Cl^- to produce Cr^{+3} and Cl_2 does not interfere.
(a) Calculate $\Delta\mathcal{E}^0$ for the reaction between $Cr_2O_7^{-2}$ and Cl^-, using Table 6-5.
(b) Calculate the equilibrium constant for the reaction.

15. Plutonium has three important oxidation states in acidic aqueous solution: Pu^{+3}, Pu^{+4}, and PuO_2^{+2}.

$$Pu^{+4} + e^- \longrightarrow Pu^{+3} \qquad\qquad \mathcal{E}_1^0 = +0.98$$
$$PuO_2^{+2} + 4\ H^+ + 3\ e^- \longrightarrow Pu^{+3} + 2\ H_2O \qquad \mathcal{E}_2^0 = +1.02$$
$$PuO_2^{+2} + 4\ H^+ + 2\ e^- \longrightarrow Pu^{+4} + 2\ H_2O \qquad \mathcal{E}_3^0 = +1.04$$

(a) What is $\Delta\mathcal{E}^0$ for the reaction between PuO_2^{+2} and H_2SO_3 to produce Pu^{+3} and HSO_4^-?
(b) By combining the first and third half-reactions so that the electrons cancel, balance the equation for the reaction between Pu^{+3} and PuO_2^{+2} to produce Pu^{+4}.
(c) What is $\Delta\mathcal{E}^0$ for the reaction in (b)?
(d) Calculate the equilibrium constant for the reaction in (b).

16. The chemistry of plutonium was first studied when only trace amounts were available—at concentrations below $10^{-10}\ M$—and there was some difficulty experienced in determining the existence of the Pu^{+4} state. Using the data of Problem 15, calculate the concentrations of Pu^{+3}, Pu^{+4}, and PuO_2^{+2} at equilibrium in a tracer solution of Pu^{+4} at a total plutonium concentration of $1.0 \cdot 10^{-12}\ M$. The acidity in this solution is adjusted by adding NaOH dropwise until bromcresol green indicator just turns from yellow to green (see Table 5-4). Repeat the calculation if HNO_3 is added until $[H^+] = 1.0 \cdot 10^{-1}\ M$.

✗ 17. What is the oxidation number of iodine in hypoiodous acid, HIO? In iodate ion, IO_3^-? In periodate ion, IO_4^-? In paraperiodic acid, H_5IO_6?

✗18. Decide the oxidation numbers of phosphorus in each of the following compounds and list them by name and by formula in order of decreasing oxidation number: phosphorus pentoxide, P_2O_5; phosphorus trioxide, P_2O_3; hypophosphoric acid, $H_4P_2O_6$; hydrogen diphosphide, P_2H_4; hypophosphorous acid, H_3PO_2; dihydrogen orthophosphate, $H_2PO_4^-$; phosphoric acid, H_3PO_4; phosphine, PH_3; phosphite, PO_3^{-3}; phosphorous acid, H_3PO_3; metaphosphoric acid, HPO_3; white phosphorus, P_4.

19. What is the oxidation number *change* when gallium(III), Ga^{+3}, is converted to $H_2GaO_3^-$? When americium(III), Am^{+3}, is converted to AmO_2^{+2}? When selenate, SeO_4^{-2}, is converted to selenous acid, H_2SeO_3? Which of these changes should be classified as oxidation? Which as reduction?

20. What is the oxidation number of iron in the iron ore magnetite, Fe_3O_4? Of iodine atom in I_3^-? Of sulfur in tetrathionate, $S_4O_6^{-2}$?

21. In HCl solution, the following reduction potential diagram has been deduced for ruthenium.

$$RuO_4 \xrightarrow{+0.9} RuO_4^- \xrightarrow{+1.6} RuO_4^{-2} \xrightarrow{+1.75} RuCl_5OH^{-2} \xrightarrow{+1.3} RuCl_5^{-2}$$

$$RuCl_5^{-2} \xrightarrow{+0.3} Ru^{+2} \xrightarrow{+0.45} Ru$$

(a) What is the oxidation number of ruthenium in each of the species given? (Assume the oxidation number of chlorine is -1.)
(b) Which oxidation states are unstable with respect to self–oxidation-reduction (disproportionation)?
(c) From Table 6-5, select a reagent that could oxidize $RuCl_5OH^{-2}$ to the gaseous tetraoxide, RuO_4.

22. Balance the acid solution half-reaction for the reduction of bromate, BrO_3^-, to hypobromous acid, $HOBr$; for the reduction of $HOBr$ to bromide, Br^-.

23. Balance the acid solution half-reaction for the reduction of the vanadium species pervanadyl, VO_2^+, to vanadyl, VO^{+2}; for the oxidation of the titanium species Ti^{+3} to TiO_2^{+2}.

24. Balance the basic solution half-reaction for the reduction of chromate, CrO_4^{-2}, to chromic hydroxide, $Cr(OH)_3$; for the reduction of chromate to chromite, CrO_2^- (which is obtained in excess NaOH).

25. Balance the equation for the reaction between Ti^{+3} and $RuCl_5^{-2}$ to give Ru and TiO_2^{+2}.

26. Balance the equations for the acid solution reactions:
(a) $H_3AsO_4 + Fe^{+2} \longrightarrow H_3AsO_3 + Fe^{+3}$
(b) $NO_3^- + Sn^{+2} \longrightarrow NO + Sn^{+4}$
(c) $H_2SO_3 + MnO_4^- \longrightarrow SO_4^{-2} + Mn^{+2}$

27. Balance the equations for the acid solution reactions:
(a) $MnO_4^- + H_2O \longrightarrow MnO_2 + O_2(g)$
(b) $Cl_2 + ClO_3^- \longrightarrow ClO^-$

28. Balance the equations for the reactions:
(a) $Ag + NO_3^- \xrightarrow{acid} Ag^+ + NO$
(b) $H_3AsO_3 + I_3^- \xrightarrow{acid} H_3AsO_4 + I^-$
(c) $MnO_4^- + H_2O_2 \xrightarrow{acid} Mn^{+2} + O_2$
(d) $N_2H_4 + 2Cu(OH)_2 \xrightarrow{base} N_2 + 2Cu + 4H_2O$
(e) $MnO_4^- + C_2O_4^{-2} \xrightarrow{base} Mn(OH)_3 + CO_3^{-2}$

29. Balance the equations for the reactions:
(a) $IO_3^- + I^- \xrightarrow{acid} I_3^-$
(b) $ClO_2 \xrightarrow{base} ClO_2^- + ClO_3^-$

30. Metallic copper is prepared from its chief mineral chalcopyrite, $CuFeS_2$, by four high-temperature reactions:
(a) "roasting" with O_2 to give Cu_2S, Fe_2O_3, and SO_2;

(b) addition of sand, SiO_2, to remove Fe_2O_3 as a low melting "slag" of iron silicate, $Fe_2(SiO_3)_3$;
(c) partial oxidation of Cu_2S to Cu_2O and SO_2 by O_2;
(d) reduction to Cu by the reaction between Cu_2S and Cu_2O.

The by-product SO_2 is further oxidized to SO_3 and hydrolyzed to give H_2SO_4, a useful chemical for making fertilizers. Balance the equations for reactions (b), (c), and (d).
(b) $Fe_2O_3 + SiO_2 \longrightarrow Fe_2(SiO_3)_3$
(c) $Cu_2S + O_2 \longrightarrow Cu_2O + SO_2$
(d) $Cu_2S + Cu_2O \longrightarrow Cu + SO_2$

31. Balance the equation for reaction (a) in Problem 30 using oxidation numbers,
(a) assuming that the oxidation number of iron is $+3$ and that of copper is $+1$ in both reactants and products;
(b) assuming that in $CuFeS_2$ the oxidation number of copper is $+2$ and that of iron is $+2$.

32. How many grams of copper can be obtained from 100 grams of chalcopyrite, $CuFeS_2$? (See Problem 30.) How many tons of copper can be obtained from 100 tons of chalcopyrite?

seven energy and chemical change

Heat has preoccupied man since his earliest conscious concern over survival. Countless men have shivered before their fires, entranced by the dancing flames, wondering how a dead tree limb could release this marvelous display with its life-sustaining warmth. Fire was given a prominent role in the alchemists' first rudimentary attempts to organize and apply experience with chemical change. They perpetuated the Greeks' view that matter was composed of the four elements—earth, air, fire, and water—up until the sixteenth century!

During this 2000-year period, all of science remained in the bud, awaiting an intellectual Spring before it could blossom. The end of this long Winter required the sunlight of experimentation—observational knowledge must guide ideas, and laboratory challenge must test them. This favorable climate finally developed about 250 years ago and only then did **thermodynamics,** *the study of heat and its transformations,* come into being as a systematic branch of science. Chemistry played its role in this development, and it has benefitted greatly therefrom.

Consider the central problem of chemistry, the control and understanding of chemical change. Some key questions that must be answered are as follows:

—When two substances are mixed, will they react?
—If reaction occurs, will the reaction be accompanied by energy release?
—If reaction begins, at what composition of reactants and products will reaction cease and equilibrium be established?
—If reaction can occur, how rapidly will it proceed?

Chemical thermodynamics is concerned with the first three questions. It neither explains nor predicts the rate of a chemical reaction, the substance of the fourth question. It does, however, provide us with an understanding of the factors that determine whether a given mixture of substances has a spontaneous tendency to react (rapidly or slowly) to form other substances. It tells us the conditions that prevail when chemical changes no longer occur and chemical equilibrium has been reached. It does these things through consideration of the heat effects in reactions. We will wish to study these energy changes, but first, let's focus our attention on events on the molecular scale* to increase our understanding of the origins of these energy effects.

7-1 Molecular energies

(a) KINETIC ENERGY

Energy of motion is called kinetic energy. In physics we learn that the kinetic energy of a body of mass m traveling with a velocity v is given by

$$KE = \tfrac{1}{2}mv^2 \tag{7-1}$$

Kinetic energy resulting from the over-all motion of a body is called energy of translation ($KE_{trans.}$). An orbiting satellite, a speeding automobile, a gas molecule, an atomic electron—all have translational kinetic energy.

What about motions other than translational? If we think carefully about our speeding automobile, we recognize that the car is alive with moving parts: Pistons are going up and down, wheels are going around, valves are opening and closing. Each of these makes its own contribution to the *total* kinetic energy of the moving automobile. We might write

$$(KE)_{total}^{auto} = KE_{trans.} + KE_{pistons} + KE_{wheels} + KE_{valves} + \cdots \tag{7-2}$$

Some of these contributions might be small compared with the over-all translational motion, but they still must be considered in a complete treatment. Some of them continue when the car is parked, if the engine is running.

Like automobiles, atoms and molecules have motions that are more complex than simple translations. A gaseous helium atom, for example, has two negatively charged electrons moving in the field of the positive nucleus. Each electron has kinetic energy of its own, so, for the total kinetic energy of a helium atom, we must write

$$(KE)_{He} = KE_{trans.} + KE_{electron\ 1} + KE_{electron\ 2} \tag{7-3}$$

The kinetic energy of the electrons is maintained even when the helium atom is "parked," that is, when it is at rest, so that $KE_{trans.} = 0$.

Molecules are even more complicated. In addition to the translational kinetic energies of the molecule and of the individual electrons, there are

---

* Throughout this book we will use the word "microscopic" to describe events involving individual atoms or molecules and "macroscopic" to describe events involving large numbers of atoms or molecules. Thus, a reaction involving a mole of gas would be called a macroscopic system.

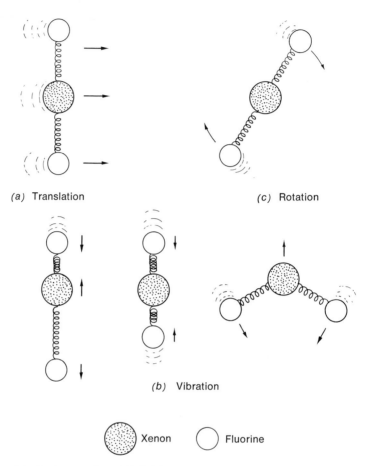

(a) Translation

(c) Rotation

(b) Vibration

Xenon   Fluorine

Figure 7-1   Nuclear motions of $XeF_2(g)$.

two further possibilities. In the gas phase, we must consider vibrations of the atoms relative to each other and rotation of the molecule as a whole. The three contributions from the atomic nuclei to the over-all kinetic energy are illustrated in Figure 7-1 for the XeF_2 molecule. For the total kinetic energy of any molecule, we can write

$$KE_{molecular} = KE_{trans.} + KE_{electrons} + KE_{vibration} + KE_{rotation} \qquad (7\text{-}4)$$

Of course, all these motions go on simultaneously, and even a simple molecule like XeF_2 would appear to be a busy little object were we able to see it.*

---

* The student may question the confidence with which we can attribute properties to an invisible object. In fact, the measurements of molecular spectroscopy show that the picture of molecular motions given above very accurately reproduces many measurable properties of substances.

In solids and liquids, the treatment of kinetic energy is not as simple as in the gaseous case discussed above. Tight packing and strong forces between molecules constrain the molecules to back-and-forth movement about fixed positions. The molecular vibrations are about the same, but the translations and rotations found in the gas phase are generally obstructed in condensed phases.

(b) POTENTIAL ENERGY

Picture a car parked on the side of a steep hill (Fig. 7-2). Should the brake suddenly fail, experience tells us that the car will spontaneously start to roll downhill, gathering momentum as it goes. With a radar trap at the bottom to measure its velocity, we could determine its kinetic energy of translation. Where did this energy come from? In order to satisfy the belief that energy is not created or destroyed in ordinary processes, but merely transformed, a new kind of energy was invented to deal with this question. In its original position, the car was "potentially" able to roll downhill and to gather kinetic energy as it went. Hence, we attribute to the car at the hilltop some stored energy that can be released when needed. The stored energy associated with its position on the hill is called *potential* energy. Detailed experiments would show that the final kinetic energy achieved by the car would be proportional to its original position atop the hill, or, in general,

$$PE = mgh \tag{7-5}$$

where h is the height above the bottom and g is the gravitational constant.

Atoms and molecules have potential energy also. Atoms in molecules are held together by forces related to the electrical attraction between positive nuclei and negative electrons. In a bonding situation, these attractive forces are exactly balanced by repulsive forces between like charges on pairs of electrons and pairs of nuclei. Thus, potential energy of position in molecules results from electrical interactions. Energy released or absorbed during chemical processes results from the making and breaking of chemical bonds and the resulting changes in this potential energy.

We can summarize the sources of potential energy in molecules as

$$PE_{molecule} = PE_{electron-nucleus} + PE_{electron-electron} + PE_{nucleus-nucleus} \tag{7-6}$$

$PE_{electron-nucleus}$ is associated with the attractive electron–nuclear interaction

$PE_{electron-electron}$ is associated with the repulsive electron–electron interaction

$PE_{nucleus-nucleus}$ is associated with the repulsive nuclear–nuclear interaction

We have not yet mentioned the enormous amounts of energy stored in atomic nuclei as a result of attractive forces between the nuclear particles,

Figure 7-2 (a) The car has potential energy only; (b) the car rolls spontaneously downhill; (c) all potential energy has been converted to kinetic energy.

neutrons and protons. (The nature of these forces is not fully understood even today.) The energies involved in ordinary chemical reactions are never sufficient to interfere with the nuclear bonding, so changes in nuclear potential energy do not normally take place. Systems in which changes in nuclear bonding do occur can release vast amounts of energy—such processes account for the energy coming from the sun, the work supplied by a nuclear power plant, and the fearsome destructiveness of a hydrogen bomb.

(c) TOTAL ENERGY

The total energy of a molecule is given by a summation of the potential and kinetic energy terms we have been discussing:

$$E_{total} = KE + PE_{molecule} + PE_{nuclei} \tag{7-7}$$

It should be clear that the measurement of E_{total} is a formidable task for one molecule, and a prodigious one in the case of a mole of molecules (one mole contains 6×10^{23} molecules). The branch of science that concerns itself with the calculation of absolute energies is called statistical mechanics or statistical thermodynamics.

The chemist, however, often does not need to know the *absolute* energy, E_{total}, of what he has in his beakers. Usually it suffices if he knows only the *change* in energy that takes place during the chemical processes that occur. He wants to know if the reaction absorbs heat so that the beaker has to be heated while the reaction is in progress, or, conversely, if heat is released, so that the reaction could serve as an energy source for some useful purpose. This more limited need simplifies things. All we have to do is find a way to measure the energy released or absorbed during the chemical process. We can escape from worrying about the exact manner in which the energy is distributed among the molecules of reactant and the molecules of product. The quantity of interest is called ΔE, where the symbol Δ, called "delta," means "change of."

energy change $= \Delta E$

where

$$\Delta E = [E_{total} \text{ (products)}] - [E_{total} \text{ (reactants)}] \tag{7-8}$$

$$= [(KE + PE)_{products}] - [(KE + PE)_{reactants}] \tag{7-9}$$

Since the nuclear energy does not change during the chemical reactions of interest to us, the term PE_{nuclei} of equation (7-7) is the same for both reactants and products and does not appear in (7-9).

(d) WHY DO WE CARE?

If chemical thermodynamics does not need to know the allotment of energy among the molecular claimants, why have we bothered to discuss molecular vibrations, rotations, electron kinetic energy, and the like? Why do chemists care? The answer is simple. Mere measurement of the quantities of energy exchanged during chemical processes does not bring with it any understanding of what is actually going on at the molecular level. Chemists, striving to explain and predict chemical reactions, find they need to know the nature of the changes involved, as well as their magnitudes. Yet, *a microscopic picture of chemical events is not required in chemical thermodynamics.*

7-2   Energy conservation

Energy is conserved. No single theory of physics is more widely accepted or more generally useful, yet the statement refers to an abstract concept about a quantity never measured directly. We measure velocity and mass to calculate energy of motion. We measure an altitude (from which valley?) to determine energy of position. We measure moles of a substance to infer its chemical energy. We measure the change in the density of mercury to infer transfer of heat. Frequently, the main evidence for the existence of a quantity or type of energy is that energy is apparently not conserved unless some unseen energy is assumed. A classical example was the postulate of an undetected nuclear particle, the neutrino, in some types of nuclear change to save the energy-conservation law. A more far-reaching instance was Einstein's deduction that $E = mc^2$, which requires that substance (mass) must also be recognized as energy.

To understand and work with this abstract and many-faceted idea, we will examine some simple processes. We will see how energy is observed and how it is recognized in different forms during these processes. First, however, we will need a simple vocabulary.

(a)   THE LANGUAGE OF THERMODYNAMICS

We almost always find it convenient to consider and study the properties of a limited part of our environment—one-half mole of ammonia gas in a 2-liter bulb, 200 milliliters of 1.5 molar potassium permanganate solution in a beaker, an eight-cubic-foot refrigerator and its contents. We call such a limited part of the things around us a *system.* If these limits are clearly stated and carefully controlled, another individual can study a like part of our environment, that is, another one-half mole of NH_3 in another 2-liter bulb, and so on. Then he can duplicate our measurements and observations to check our results and extend them to add to our knowledge. This limiting of our concern to a clearly stated part of the environment lends simplicity to our study and precision to our conclusions. If a system is not able to interact with any other system, we say it is *isolated.* All other systems with which our particular system might exchange energy are called the *surroundings.* (For example, the surroundings of our system might be a constant-temperature bath in which our 2-liter bulb of NH_3 is immersed; or a hot plate on which the beaker of permanganate is placed, plus the atmosphere around it; or, if we can't be more specific, the surroundings of a system can always be said to be the rest of the universe.)

(b) THE EXCHANGE OF ENERGY

Consider the simple mechanical system illustrated in Figure 7-3. When the billiard ball is struck by the cue, energy is exchanged. In the formal language of physics, work is done on the ball in propelling it (hopefully) on its intended path. Energy has been transferred to the ball. At the same time, the player's hand is slightly warmed by friction as the cue rubs his fingers. Some of the energy put into the cue by the player is returned to him, through friction, as heat. Energy is being exchanged, some as heat and the rest as work. These are the two forms of energy transfer that will concern us in our studies of chemical thermodynamics.

(c) HEAT AS ENERGY

We are quite used to the idea that energy dissipated through friction is lost as heat. However, only about 150 years ago, the view was that heat was a material substance called caloric, which was transferred from a hotter to a colder body. The science of thermodynamics, which is the study of heat, began only after Count Rumford observed in 1798 that an apparently inexhaustible amount of heat was produced when a cannon was bored. It seemed strange to him that a fixed amount of iron could contain apparently

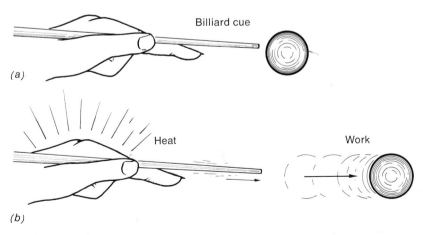

Figure 7-3 Exchange of energy. (a) Energy is stored in the muscles of player; (b) energy exchanged as heat (friction on hand) and work (on ball).

limitless quantities of caloric. These doubts led him to perform a controlled experiment in which the heat produced by a blunt borer was transmitted to a vessel of cold water. He observed:

". . . it was not long before the water which surrounded the cylinder began to be sensibly warm. . . . At two hours and twenty minutes it was at 200; and at 2 hours 30 minutes it ACTUALLY BOILED! . . . It is hardly necessary to add, that any thing which any *insulated* body, or system of bodies can continue to furnish *without limitation* cannot possibly be a *material substance:* and it appears to me to be extremely difficult, if not quite impossible, to form any distinct idea of any thing, capable of being excited and communicated in the manner the Heat was excited and communicated in these Experiments except it be MOTION."*

Now, over 150 years later, we find it difficult to appreciate the astonishment with which Rumford perceived the results of his observations. What is astonishing, however, is his intuitive conclusion that heat is *motion*. It is easy for us with our sophisticated equipment and detailed knowledge of atomic and molecular activity to understand that an increase in temperature is just the bulk manifestation of an increase in kinetic energy of the particles of a body. Rumford did not even dream of atoms and molecules!

(d) THE FIRST LAW OF THERMODYNAMICS

The First Law of Thermodynamics is merely the statement that energy is conserved during any change. Energy is neither created nor destroyed, but simply changed in form. Rumford's work shows that it is necessary to include heat within the boundaries of the energy-conservation law.

In the case of a chemical process, this tells us that whenever a certain reaction occurs, such as the combustion of propane gas,

$$C_3H_8(g) + 5\ O_2(g) \longrightarrow 3\ CO_2(g) + 4\ H_2O(g) \tag{7-10}$$

the difference between the potential energy held by the reactants (1 mole propane and 5 moles oxygen) and that held by the products (3 moles carbon dioxide and 4 moles water vapor) will result in energy being either released or absorbed. The amount of energy depends only on the chemical nature of the reactants and products. Some of the energy may be released as heat and some as work (perhaps work done in pushing back the atmosphere), but for a fixed amount of propane, the sum of the heat and the work will always be the same. This can be expressed quantitatively as follows:

$$\Delta E = E_{\text{final}} - E_{\text{initial}} \tag{7-11}$$

where ΔE is the change in energy in going from the initial to the final state.

* W. F. Magie, *A Source Book in Physics*, pp. 158–161. Harvard University Press: Cambridge, 1963.

First Law of Thermodynamics
$$\Delta E = q - w$$

(7-12)

w = work *done* by system

q = heat *absorbed* by system

The need for the minus sign is clear when we realize that work done by the system implies a loss of energy, while heat absorbed implies energy gain. Thus defined, the two quantities must be summed with opposite sign. The reason for this particular convention in signs is an historical one. Steam engineers, who pioneered the systematic study of thermodynamics, expressed the law in terms of the quantities they were accustomed to measuring. Naturally, they were interested in the heat that had to be put into a system (say, to heat a boiler) and in the work that would be delivered in return.

(e) MUST ENERGY BE CONSERVED?

The First Law is a statement encompassing the results of a vast accumulation of experimental evidence. We cannot say that energy *must* be conserved—however, we do know that, in the total experience of science to date, energy always *has been* conserved. To understand how our experience teaches us to believe in the First Law, we need only consider the effects of a violation of this law. Engineers have long hunted diligently for an engine that would perform more useful work than the energy supplied to make it run. Such an engine would, of course, violate the First Law, for it would "create" energy. The hunt was spurred on by the knowledge that such an engine would be more rewarding than a slot machine that clanged out two quarters every time one was dropped in. Our conviction of belief in the First Law is founded in the knowledge that such an engine has never been found —although not through want of trying. After all the past effort, we are convinced that it will be the same in the future. Energy having always been conserved provides a basis for expectation in experiments yet to be done. This is the route that scientific advance must take. We diligently accumulate a set of consistent facts, we build our theories to fit these facts, and then we base our expectations on these theories.

(f) CHEMICAL WORK

Energy can be transferred from one system to another either by means of heat flow or by means of work. It is important to bear in mind that the energy E is a characteristic property of a system (the summation of the potential and kinetic energy terms). Work and heat flow are *not* properties of a system, but they are the *means* by which energy is transferred when a change occurs.

There are many varieties of work, but in chemistry we are mainly concerned with electrical work and work done by expanding gases. Electrical work can be produced by an electrochemical cell. Expansion work results from a change in volume of the systems involved and is usually called pressure–volume work, or PV work. Expanding gases in the cylinder of an automobile engine do PV work on the piston; this work is harnessed to turn the wheels. Expanding gases from an open reaction vessel do PV work in pushing back the atmosphere; this work is lost to us. All kinds of work, whether useful or not, must be considered if our energy bookkeeping is to come out right.

Mechanical work is done when a force moves through a distance, and it is expressed numerically by

$$\text{work} = (\text{impressed force}) \times (\text{distance moved})$$
$$= f \cdot (\Delta r) \tag{7-13}$$

We can investigate pressure–volume work by considering what happens when gas contained in a cylinder is allowed to expand, pushing back a piston against some external force $f_{ex.}$. A simplified diagram of an apparatus to perform this experiment is shown in Figure 7-4.

At the beginning of the expansion, the force due to the internal pressure of the gas just equals the external force. Now, if we gently heat the cylinder, the internal pressure will increase and the piston will move until the force exerted on it by the gas again balances the external force. In Figure 7-4(b),

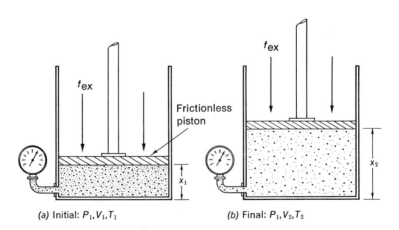

(a) Initial: P_1, V_1, T_1 (b) Final: P_1, V_2, T_2

Figure 7-4 *Pressure-volume work against constant external force (such as atmospheric pressure). The gas pressure as registered on the gauge remains constant throughout the process.*

the piston has moved from its original position x_1 to a new position x_2. The work done, as the piston pushed against the constant external force, is

$$\text{work} = f_{\text{ex.}} \cdot (\Delta x) \tag{7-14}$$

We may express this equation in terms of more easily measured quantities (always a primary aim in science) if we divide *and* multiply the right-hand side of (7-14) by A, the cross-sectional area of the piston:

$$\text{work} = f_{\text{ex.}} \cdot (\Delta x) \cdot \frac{A}{A}$$

$$= \frac{f_{\text{ex.}}}{A} \cdot [A \cdot (\Delta x)] \tag{7-15}$$

If we perform our experiment so that $f_{\text{ex.}}$ always exactly balances the internal pressure of the gas, the first term in (7-15), $f_{\text{ex.}}/A$ (the force per unit area), is just the internal pressure P of the gas. The second term $[A \cdot (\Delta x)]$ is the change in volume of the cylinder during the expansion, ΔV. Thus, (7-15) becomes

$$\text{work} = \frac{f_{\text{ex.}}}{A} \cdot [A \cdot (\Delta x)]$$

$$= P \cdot \Delta V$$

$$= P(V_2 - V_1) \tag{7-16}$$

The First Law of Thermodynamics can now be rewritten to include $P\Delta V$ (the explicit expression for PV work) and w' (all other forms of work).

At constant pressure,

$$\boxed{\Delta E = q - P\Delta V - w'} \tag{7-17}$$

or, if $w' = 0$

$$\boxed{\Delta E = q - P\Delta V} \tag{7-18}$$

This is the First Law in its most useful form to a chemist.

Exercise 7-1. In Chapter Four, we considered the heat effects caused by a salt dissolving in water. For example, when one mole of lithium chloride is dissolved in water, 8.9 kcal are transferred to the surroundings, whereas a mole of potassium chloride absorbs 4.0 kcal. There is a negligible volume change and no electrical work. Paying attention to algebraic sign, apply (7-18) to decide on the magnitude and sign of ΔE for each of these changes:

$$LiCl(s) \longrightarrow Li^+(aq) + Cl^-(aq) \qquad \Delta E = ?$$

$$KCl(s) \longrightarrow K^+(aq) + Cl^-(aq) \qquad \Delta E = ?$$

Use these values of ΔE to show that when 1.00 gram of LiCl dissolves in water, 210 calories of heat are released and that when 1.00 gram of KCl dissolves, 54 cal are absorbed.

Exercise 7-2.　When a mole of HCl(g) is dissolved in water, 18.0 kcal are released. In this case, however, there is a significant reduction in volume. At room temperature and one atmosphere pressure, one mole of HCl occupies 24.5 liters. As the HCl dissolves in water, the solution volume changes very little. Consequently $\Delta V = -24.5$ liters. Calculate ΔE with the aid of the energy conversion factor, 1 ml-atm $= 2.42 \cdot 10^{-5}$ kcal, for the constant pressure process and show that $\Delta E = -17.4$ kcal.

$$HCl(g) \longrightarrow H^+(aq) + Cl^-(aq) \qquad \Delta E = q - P\Delta V = ?$$

A graphical representation of PV work is shown in Figure 7-5(a). When the pressure is kept constant, as in the above examples, the PV

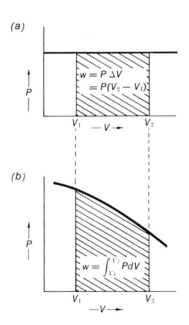

Figure 7-5　Pressure–volume work is given graphically by the area under P–V plot. (a) Constant pressure, $w = P\Delta V$; (b) pressure not constant, area given by the integral $\int_{V_1}^{V_2} PdV$ (see Appendix h).

work done is given numerically by the shaded area on the graph. Part (*b*) of this figure shows an example of an expansion during which the pressure was not constant. The work done (in going from V_1 to V_2) is still given by the magnitude of the shaded area. However, this value is not given by the simple expression (7-16), but must be worked out using the methods of integral calculus. A detailed knowledge of calculus is not necessary for us though. It is sufficient to recognize that it helps us calculate the shaded area in Figure 7-5(*a*) and (*b*). The process of calculating this area, the area under the curve, is called *integration*, and it is symbolized by the integral sign $\int$. The upper bound of the shaded area is fixed by the curve P, and the limits of the area are the initial and final volumes, V_1 and V_2. The shaded area, which is numerically equal to the work done, is then expressed as

$$w = \int_{V_1}^{V_2} P\,dV \qquad\qquad\qquad (7\text{-}19)$$

The term "dV" is merely another expression for ΔV in which the volume difference has become extremely small. Expression (7-19) is read aloud as "the work done equals the integral of $P\,dV$ between the limits $V = V_1$ and $V = V_2$." This process is more completely discussed in Appendix h for those who care.

(g) ENERGY, A FUNCTION OF STATE

Consider the situation shown in Figure 7-6. In part (*a*), one ball is sitting at the top of a hill, with a potential energy proportional to the height of the hill. If it is released, it rolls to the bottom, bashes into the other ball, and comes to a halt. All of its potential energy has been dissipated—some as heat, through frictional interaction with the hill, and some as work per-

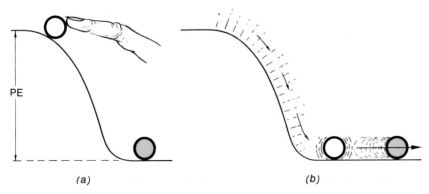

(a)   (b)

Figure 7-6 *Energy is a state function. Potential energy, PE, of the unshaded ball in (a) is completely transferred to hill (heat) and to the other ball (work) in (b).*

formed on the second ball. Our experience in these situations, as summed up in the First Law, tells us that the total amount of energy dissipated as work and heat will exactly equal the initial potential energy of the ball, measured relative to the bottom of the hill. If we were to polish the hill, thus reducing friction, more work would be done and less heat produced. *The total amount of energy exchanged would be the same.* In fact, any process that results in the ball moving from the top to the bottom of the hill will dissipate the same amount of energy, *no matter how this energy is distributed between heat and work.*

The potential energy of a molecule depends on the relative positions of the electrons and nuclei of the various atoms that make up the molecule. When two molecules react to form a third, the atoms are rearranged, their electrons are placed differently relative to the nuclei, and the potential energy of the products differs (usually) from that of the reactants. Thus, the energy change ΔE depends only on the nature of the products and the reactants. As in the above example of the ball rolling downhill, any energy released may be dissipated as work or heat in varying proportions, but ΔE is always the same for a given chemical change.

Quantities that depend only on the state of the system (and not on the path by which the system got there) are called *state functions.* Other state functions are pressure, temperature, and volume. The importance of this state function concept is clear. If we once measured the energy change ΔE for some process, we know that *any other* process effecting the same change in our system will have the same energy change. This is true even though the relative amounts of heat and work exchanged with the surroundings might be vastly different. Only one measurement of ΔE, then, need ever be made for a particular initial and final state.

It is simple to demonstrate that PV work is not a state function. Consider the PV plots in Figure 7-7. Suppose we have 0.01 mole of gas at an initial pressure $P_1 = 200$ torr and volume $V_1 = 20$ ml and expand it to a final pressure $P_2 = 800$ torr and a final volume $V_2 = 80$ ml. This change can be brought about as in Figure 7-7(a), first increasing the pressure to its final value, $P_2 = 800$ torr, at constant volume V_1, by heating the gas. No work is done in this step, because $\Delta V = 0$. Then at constant pressure $P_2 = 800$ torr, the gas is expanded to the final volume V_2 by heating the gas (Fig. 7-7(c)). The work done is given by the shaded area and is equal to

$$w_a = P_2 \Delta V$$
$$= P_2(V_2 - V_1)$$
$$= (800) \cdot (80 - 20)$$
$$= 48,000 \text{ ml torr}$$

In Figure 7-7(b) we approach the same final state by a different route.

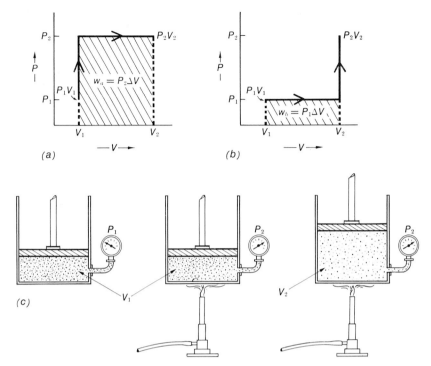

Figure 7-7 *Pressure–volume work is not a state function. Gas in a cylinder is taken from an initial state P_1V_1 to a final state P_2V_2 by two different routes (a) and (b). The work done in the two cases is different, as shown by the shaded areas. Thus work done depends on the path. Part (c) illustrates the process described graphically in (a). The gas in a volume V_1 registering a pressure P_1 is heated. If the piston is held fixed, the pressure rises to P_2. At this point the piston is allowed to move so that the pressure remains constant as the volume increases to V_2.*

The first step is a warming of the gas so it expands to the final volume V_2, at a constant pressure $P_1 = 200$ torr. After this step, the gas is warmed to increase the pressure at this constant volume until the final state P_2V_2 is reached. No work is done in the second step because the volume is constant. Hence the work done is given by the PV product in the first step, shown shaded on the graph:

$$w_b = P_1 \Delta V$$

$$= (200) \cdot (80 - 20)$$

$$= 12,000 \text{ ml torr}$$

In these two examples the over-all change was the same, yet the work done was different. In order to satisfy the First Law, it is necessary that q, the heat absorbed, also be different in these two cases. Only ΔE, the state function, is the same for both processes.

Exercise 7-3. By equation (7-18), the energy change in the process shown in Figure 7-7(a), $\Delta E_a = q_a - w_a$, is equal to the energy change in the process shown in Figure 7-7(b), $\Delta E_b = q_b - w_b$. By equating $\Delta E_a = \Delta E_b$, show that q_a exceeds q_b by 36,000 ml torr, which, by the energy conversion factor 1 ml-atm $= 2.42 \cdot 10^{-5}$ kcal, equals 1.15 cal.

7-3 Thermochemistry

The chemical engineer designing a chemical plant worries about the heat needed to keep a reaction going, or the cooling necessary to absorb heat evolved by a reaction. From a practical point of view, then, the part of the energy change dissipated as heat is an important part. In the next chapter we shall see how the heat of reaction also plays a crucial role in determining the position of equilibrium, hence, the extent of reaction. Generally, the value of q depends on how the reaction is carried out. However, q itself becomes a state function when a reaction is carried out either at constant volume or at constant pressure, two conditions that are easily maintained in practice. In this section we will investigate the nature of q under these special conditions.

(a) DEFINITIONS AND UNITS

To make measurements of heat changes, we need some consistent unit. Quantity of heat transferred is usually measured in terms of the *calorie*. One calorie is defined as the amount of heat needed to raise the temperature of one gram of water from 14.5°C to 15.5°C.* In chemistry, the energies involved when a single molecule reacts are immeasurably small, so it is necessary to deal with a convenient number of atoms or molecules, usually a mole. The heat evolved or absorbed during a chemical reaction is expressed in calories per mole of reactant, or, if the numbers are large, kilocalories per mole of reactant (1000 cal $=$ 1 kcal). A reaction that evolves heat is said to be *exothermic*, while one that absorbs heat is called *endothermic*.

(b) CONSTANT VOLUME PROCESSES

When a reaction is carried out in a closed vessel, the volume cannot change and the work done is zero; that is,

---

* For precision thermal measurements, a calorie is now defined with reference to electrical quantities: 1 calorie $=$ 4.1840 joules.

at constant volume,

$$\Delta V = 0 \tag{7-20}$$

and hence

$$w = P\Delta V = 0 \tag{7-21}$$

The First Law expression (7-12) simplifies to

$$\Delta E = q - w$$
$$= q - 0$$
$$\Delta E = q_V \tag{7-22}$$

The subscript V is used to indicate a constant volume process. In this case, the energy change ΔE, a function of state, is given by the heat released or absorbed during the process. It is important to realize that q is not generally a state function. It is equal to ΔE only when the process is carried out in one particular way—at constant volume. In Section 7-4, we will show how fixed-volume devices, known as "bomb calorimeters," are routinely used in the laboratory to measure q_V.

(c) CONSTANT PRESSURE PROCESSES

Much of chemistry is performed in beakers and other vessels that are open to the atmosphere. During any such process the volume of the system can change, but the pressure imposed by the atmosphere remains constant. Because of their great practical importance, much information on constant pressure processes is available in tabulated form.

Over one hundred years ago, J. Thomsen investigated the thermochemistry of the neutralization by sodium hydroxide of sodium bicarbonate and of carbon dioxide solutions. His experimental results were as follows:

I $CO_2(aq) + 2\,NaOH(aq) \rightleftharpoons Na_2CO_3(aq) + H_2O(\ell)$

$$q_P^I = -20.2 \text{ kcal/mole}$$

II $NaHCO_3(aq) + NaOH(aq) \rightleftharpoons Na_2CO_3(aq) + H_2O(\ell)$

$$q_P^{II} = -9.2 \text{ kcal/mole}$$

What about the direct conversion of CO_2 to $NaHCO_3$ as shown in equation III?

III $CO_2(aq) + NaOH(aq) \rightleftharpoons NaHCO_3(aq)$

$$q_P^{III} = ?$$

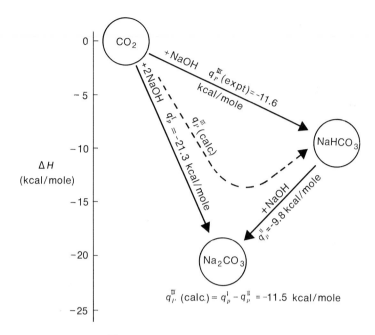

$$q_{P'}^{III} \text{ (calc.)} = q_p^I - q_p^{II} = -11.5 \text{ kcal/mole}$$

Figure 7-8   The value of q_P^{III} was first obtained from the experimental results for the two-step process in which Na_2CO_3 is an intermediate. This value agrees very well with the best experimental value for q_P^{III} presently available.

Thomsen estimated q_P^{III} in the following way. Consider the flow diagram Figure 7-8. The first step on the lower route to $NaHCO_3$ is reaction I. The second step is the reverse of reaction II. If reaction II is exothermic by 9.2 kcal/mole, the reverse reaction must *absorb* an identical amount of heat, or $-q_P^{II}$. The over-all heat of reaction at constant pressure for the two-step process is thus q_P^{III}, since the final result is the conversion of CO_2 to $NaHCO_3$. This is shown as a one-step process by the top arrow. Therefore

$$
\begin{aligned}
q_P^{III} &= q_P^{I} - q_P^{II} \\
&= -20.2 - (-9.2) \\
&= -11.0 \text{ kcal/mole}
\end{aligned}
$$

With this argument Thomsen predicted that reaction III should be exothermic by -11.0 kcal/mole. Shortly thereafter, M. Berthelot verified this result experimentally by measuring q_P^{III} directly.

As a result of this experiment, we can conclude that the heat of reaction at constant pressure for the conversion of CO_2(aq) into sodium bicarbonate is the same for the two different paths investigated. A vast amount of experimental data on countless reactions permit the generalization that q_P

is always independent of path; that is, q_P *is a state function.* This result allows an enormous reduction in the amount of experimental work that need be performed. As long as some step-wise route between products and reactions can be devised, in which q_P is already known for each step, q_P of the over-all process can be evaluated without further experiment. We shall devote most of the rest of this section to investigating the usefulness of this result.

(d) ENTHALPY

Let us now solve the First Law expression, (7-12), for q_P.

$$\Delta E = q - w$$

so, at constant pressure,

$$q_P = \Delta E + w$$
$$= \Delta E + P\Delta V \qquad (7\text{-}23)$$

or, when written out in full,

$$q_P = (E_2 - E_1) + (PV_2 - PV_1)$$

A simple rearrangement gives

$$q_P = (E_2 + PV_2) - (E_1 + PV_1) \qquad (7\text{-}24)$$

So we see that the heat associated with a process carried out at constant pressure can be given by a difference between two terms, each having the form

$$E + PV \qquad (7\text{-}25)$$

We have already noted the experimental observation that q_P is a state function. Hence we must be able to express it as the difference between some state property in the final and initial states. Expression (7-24) shows that q_P is given by the difference in the value of the function $(E + PV)$ evaluated at the final and initial states. Because of the importance of constant pressure processes, this function (7-25) is given the special symbol H, and is called the *heat content* or *enthalpy* of the system. For any constant pressure process, the heat evolved or absorbed is given by ΔH, the enthalpy change during the process.

$$\Delta H = q_P = H_2 - H_1 \qquad (7\text{-}26)$$

where

$$H = E + PV \tag{7-27}$$

Enthalpy changes are extremely important in the development of chemical thermodynamics. In fact, chemists tend to think in terms of ΔH rather than ΔE. Because of this central role, we will discuss a variety of enthalpy applications and measurements.

(e) HEAT OF FORMATION

Because the enthalpy H is a function of the state of the system only, we can calculate ΔH for any process, as long as we can find a route for which ΔH is known for each step. It is this result, based solidly on the sort of experimental data described in Section 7-3(c), that makes thermodynamics so useful. A relatively small number of experimental measurements permits the calculation of ΔH for vast numbers of different reactions. For convenience, the data are tabulated according to reaction type: heat of combustion, heat of vaporization, heat of fusion, etc. A particularly useful process to consider is the *formation* of the compound of interest from its constituent elements. The enthalpy change for this process is called the heat of formation and it is symbolized ΔH_f. For example, the heat of formation of water vapor is the heat evolved when the following reaction occurs:

$$H_2(g) + \tfrac{1}{2} O_2(g) \longrightarrow H_2O(g)$$

$$\text{heat of formation} = \Delta H_f(H_2O) \tag{7-28}$$

So that this number can be understood by every scientist, some standard set of experimental conditions must be specified. Heats of formation refer to a reaction carried out at 25°C (room temperature),* with all gases at a pressure of 1 atm and all solutions at a concentration of 1 mole/liter. All species are taken to be in their most stable state. For substances normally solid at room temperature, the crystal form must often be specified as well. A superscript ^{0} is used to indicate these standard conditions. Thus,

$$\Delta H_f^0(H_2O(g)) = -57.796 \text{ kcal/mole} \tag{7-29}$$

means that when 1 mole of water vapor is formed from 1 mole of hydrogen gas and $\tfrac{1}{2}$ mole of oxygen gas at room temperature (25°C = 298.16°K), 57.796 kcal of heat are given off.

Some representative values of ΔH_f^0 are collected in Table 7-1. (A much larger collection is given in Appendix d.) An element already in its standard

* 25°C = 298.16°K.

Table 7-1   Some Standard Heats of Formation, ΔH_f^0(kcal/mole)

Elements		Compounds	
$F_2(g)$	0.0	$CO(g)$	-26.4
$Cl_2(g)$	0.0	$CO_2(g)$	-94.1
$Br_2(\ell)$	0.0	$H_2O(\ell)$	-68.3
$Br_2(g)$	$+7.4$	$H_2O(g)$	-57.8
$I_2(s)$	0.0	$C_2H_6(g)$	-20.2
$I_2(g)$	$+14.9$	$C_2H_4(g)$	$+12.5$
$F(g)$	$+18.9$	$C_2H_2(g)$	$+54.2$
$C(s, graphite)$	0.0	$CaO(s)$	-151.9
$C(s, diamond)$	$+0.45$	$Ca(OH)_2(s)$	-235.8
$C(g)$	$+171.3$	$CaCO_3(s)$	-288.4
$O_2(g)$	0.0	$SO_2(g)$	-70.9
$S_8(s)$	0.0	$SO_3(g)$	-94.6

state naturally has a ΔH_f of zero. An element not in its standard state has $\Delta H_f \neq 0$.

Let's use some of these data to calculate ΔH^0 for the reaction

$$SO_2(g) + \tfrac{1}{2} O_2(g) \rightleftharpoons SO_3(g) \qquad (7\text{-}30)$$

The enthalpy change associated with this process can easily be calculated from $\Delta H_f^0(SO_2)$ and $\Delta H_f^0(SO_3)$, which refer to the following processes:

$$\tfrac{1}{8} S_8(s) + O_2(g) \rightleftharpoons SO_2(g) \qquad \Delta H_f^0(SO_2) = -70.9 \text{ kcal/mole} \qquad (7\text{-}31)$$

$$\tfrac{1}{8} S_8(s) + \tfrac{3}{2} O_2(g) \rightleftharpoons SO_3(g) \qquad \Delta H_f^0(SO_3) = -94.6 \text{ kcal/mole} \qquad (7\text{-}32)$$

Figure 7-9 illustrates the reasoning involved. We want to know ΔH_I. However, since H is a state function, ΔH_I must be equal to ΔH_{II}, the heat evolved when SO_2 is converted to SO_3 via the two-step process. The first step is the reverse of the heat of formation of SO_2 from the elements, and the second is ΔH_f^0 for SO_3. Thus, it follows that

$$\Delta H_I = \Delta H_{II} = \Delta H_f^0(SO_3) - \Delta H_f^0(SO_2) \qquad (7\text{-}33)$$

This may be verified from (7-31) and (7-32) by reversing (7-31) and algebraically adding them together:

$$SO_2 \rightleftharpoons \tfrac{1}{8} S_8 + O_2 \qquad \Delta H_1^0 = +70.9 = -\Delta H_f^0(SO_2) \qquad (7\text{-}34)$$
$$\tfrac{1}{8} S_8 + \tfrac{3}{2} O_2 \rightleftharpoons SO_3 \qquad \Delta H_2^0 = -94.6 = \Delta H_f^0(SO_3) \qquad (7\text{-}35)$$
$$\overline{SO_2 + \tfrac{1}{2} O_2 \rightleftharpoons SO_3 \qquad \Delta H_3^0 = -23.7 \text{ kcal/mole} \qquad (7\text{-}36)}$$

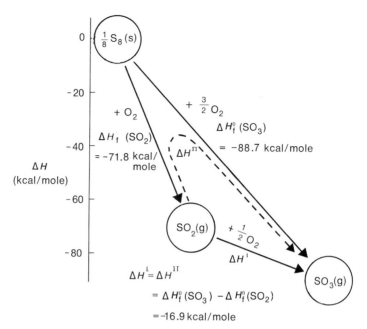

Figure 7-9 *Calculation of ΔH for the process* $SO_2(g) + \frac{1}{2} O_2(g) \rightleftharpoons SO_3(g)$ *from standard enthalpies of formation of SO_2 and SO_3.*

The enthalpy change ΔH_3^0, associated with reaction (7-36), is the algebraic sum of the enthalpy changes of the two component steps ΔH_1^0 and ΔH_2^0:

$$\Delta H_3^0 = \Delta H_1^0 + \Delta H_2^0 \tag{7-37}$$

We can substitute the two heats of formation for ΔH_1 and ΔH_2:

$$\Delta H_3^0 = \Delta H_f^0(SO_3) - \Delta H_f^0(SO_2) \tag{7-38}$$

Examination of (7-38) shows that the desired heat of reaction was obtained by taking the difference between the standard heat of formation of the products and the standard heat of formation of the reactants, remembering that the standard heat of formation of O_2, already in the standard state, is zero. This result is a perfectly general one that can be called the *Law of Additivity of Reaction Heats.*[*] The enthalpy change associated with any reaction is given, in general, by

$$\boxed{\Delta H^0 = \Sigma \Delta H_f^0(\text{products}) - \Sigma \Delta H_f^0(\text{reactants})} \tag{7-39}$$

[*] An old-fashioned and less self-explanatory name is the "Law of Constant Heat Summation."

where Σ means "sum of." By measuring and tabulating ΔH_f^0 for a few hundred compounds, it is possible to calculate ΔH for the thousands of reactions involving these compounds.

Exercise 7-4. Use the values of ΔH_f^0 in Table 7-1 to show that 67.7 kcal of heat is evolved when one mole of carbon monoxide is burned in oxygen to form carbon dioxide.

(f) HEAT OF COMBUSTION

It is not always possible to measure the heat of formation by the direct combination of elements—indirect methods must then be used. For example, it is experimentally impossible to measure directly the heat of the reaction

$$2\ C(graphite) + 2\ H_2(g) \rightleftharpoons C_2H_4(g) \tag{7-40}$$

Fortunately the hydrocarbon, ethylene (C_2H_4), can easily be burned in the presence of oxygen to form water and carbon dioxide:

$$C_2H_4(g) + 3\ O_2(g) \rightleftharpoons 2\ CO_2(g) + 2\ H_2O(\ell) \tag{7-41}$$

The heat associated with the combustion of ethylene is called the *heat of combustion* $\Delta H_{comb.}$ and is quite easily measured.* Knowledge of ΔH_f^0 for carbon dioxide and liquid water (Table 7-1) makes it possible to determine $\Delta H_f^0(C_2H_4)$ by an appropriate combination of the following set of equations:

I $2\ CO_2(g) + 2\ H_2O(\ell) \rightleftharpoons C_2H_4(g) + 3\ O_2(g)$ $-\Delta H_{comb.}(C_2H_4)$

II $2\ C(graphite) + 2\ O_2(g) \rightleftharpoons 2\ CO_2(g)$ $2\Delta H_f^0(CO_2)$

III $2\ H_2(g) + O_2(g) \rightleftharpoons 2\ H_2O(\ell)$ $2\Delta H_f^0(H_2O)$

Summing these three equations we get

$$2\ C(graphite) + 2\ H_2(g) \rightleftharpoons C_2H_4(g)$$

and

$$\boxed{\Delta H_f^0(C_2H_4) = 2\Delta H_f^0(CO_2) + 2\Delta H_f^0(H_2O) - \Delta H_{comb.}(C_2H_4)} \tag{7-42}$$

A flow chart showing our alternate route between graphite, hydrogen gas,

---

* In this section and in Table 7-6, heat given off during combustion will be given a negative sign in accord with usual conventions. Unfortunately exothermic heats of combustion are sometimes tabulated as positive quantities.

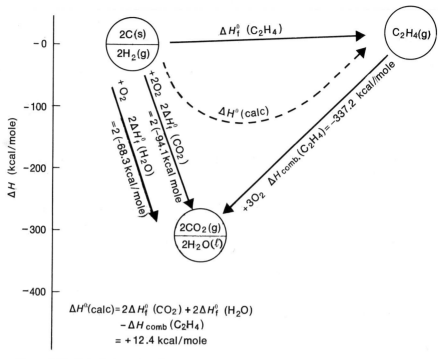

Figure 7-10 Calculation of ΔH_f^0 (ethylene) from the heats of combustion of ethylene and ΔH_f^0 of H_2O and CO_2.

and ethylene is given in Figure 7-10. Notice that reactions II and III are also combustion reactions. The heats of formation of carbon dioxide and water are identical with the heats of combustion of hydrogen gas and graphite.

Exercise 7-5. Balance the equation for the combustion of acetylene, C_2H_2, and show that the experimental heat of combustion, -310.62 kcal/mole C_2H_2, is in agreement with the value calculated by adding the standard heats of formation according to equation (7-39).

(g) BOND DISSOCIATION ENERGIES: DIATOMICS

When a chemical reaction occurs, reactant molecules are broken up and product molecules are formed. The energy changes that accompany chemical reactions result from changes in the relative positions of atoms. The potential energy associated with each chemical bond is unique to that bond. It is easy to see the effect of these changes in potential energy when we look in detail at a reaction involving diatomic molecules. Consider the formation of hydrogen fluoride gas from hydrogen gas and fluorine gas:

$$H_2(g) + F_2(g) \rightleftharpoons 2\ HF(g) \tag{7-43}$$

The enthalpy change associated with this reaction, as written, is twice the enthalpy of formation of HF (Appendix d):

$$\Delta H = 2\Delta H_f^0(HF) = -129.6\ \text{kcal/mole} \tag{7-44}$$

Remember that ΔH_f^0 is defined as the enthalpy change associated with the formation of *one* mole of product. Equation (7-43) involves the formation of *two* moles of HF.

We might think of the formation of HF in the following series of steps, which are illustrated in the flow chart of Figure 7-11. First, we must separate hydrogen and fluorine atoms from the molecules and let them recombine as HF.

I $H_2(g) \rightleftharpoons 2\ H(g)$ $\Delta H_I = 2\Delta H_f^0(H) = D_{H_2}$

II $F_2(g) \rightleftharpoons 2\ F(g)$ $\Delta H_{II} = 2\Delta H_f^0(F) = D_{F_2}$

III $2\ H(g) + 2\ F(g) \rightleftharpoons 2\ HF(g)$ $\Delta H_{III} = -2D_{HF}$

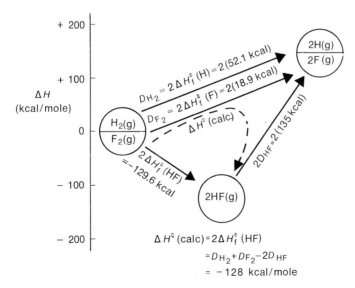

Figure 7-11   *Relationship between bond dissociation energies and enthalpies of formation for the process* $H_2(g) + F_2(g) = 2\ HF(g)$.

The enthalpy change associated with the net reaction (7-43) is given by the sum of the enthalpy changes of the three steps:

$$\Delta H = 2\Delta H_f^0(HF) = \Delta H_I + \Delta H_{II} + \Delta H_{III} \qquad (7\text{-}45)$$

Equation I represents the formation of hydrogen atoms from molecular hydrogen, the standard state. This involves breaking the H_2 chemical bond. The energy required to do this is equivalent to the potential energy stored in this bond and is generally called the *bond dissociation energy** and given the symbol D_{H_2}. The same is true for II, which represents the breaking of the F—F bond in F_2, and requires an amount of energy equal to the bond dissociation energy of F_2. In III, the combination of hydrogen atoms and fluorine atoms gives off an amount of energy related to the potential energy of the HF bond. In fact, the reverse of this process,

IIIa $2\,HF = 2\,H(g) + 2\,F(g)$ $\qquad (7\text{-}46)$

is an expression for the dissociation of two molecules of HF. The enthalpy change associated with III is then twice the negative (because we are considering the reverse reaction) of this bond dissociation energy of HF.

If we start with one molecule each of H_2 and F_2, we must first put in an amount of energy equivalent to the potential energy of the F—F and H—H bonds. Formation of two molecules of HF gives us back the potential energy of the HF bond for each molecule formed. We can thus rewrite (7-45) in terms of the bond dissociation energies:

$$\Delta H = 2\Delta H_f^0(HF) = D_{H_2} + D_{F_2} - 2D_{HF} \qquad (7\text{-}47)$$

(h) BOND DISSOCIATION ENERGIES: POLYATOMICS

The energy of a particular bond in a polyatomic molecule is not as easy to pin down as the bond dissociation energy of a diatomic molecule. For example, the energy (enthalpy) required to break a carbon–hydrogen bond in some organic compounds is tabulated in Table 7-2. The range is considerable, about 10 percent among the examples given.

It would be helpful, however, if some useful values of these bond energies were at hand. For example, there are over a million organic compounds that involve, at most, 25 different bonds. In order to calculate the enthalpy changes associated with the countless reactions these molecules undergo, it would be necessary to tabulate ΔH_f^0 for every one of these. This is a formidable task. However, if the heat of a given reaction could be related to

---

* It is customary to speak of bond dissociation *energies* even though the quantity we are actually measuring in this example is the bond dissociation *enthalpy*. A detailed consideration of the relation between ΔE and ΔH (taken up later in this chapter) shows that the two are nearly equal for processes such as these.

Table 7-2 Some Carbon–Hydrogen Bond Energies

Compound	$D_{\text{C}-\text{H}}$(kcal/mole)
CH_4	101
CH_3CH_3	96
$CH_3CH_2CH_3$	100(CH_3)
$CH_3CH_2CH_3$	94(CH_2)
C_6H_6	102
$CHCl_3$	90
CH_3Br	99

the energies involved in making and breaking bonds, a relatively small table of bond energies would suffice. For this purpose, collections of *average* bond energies have been prepared, which give representative bond energies based on a large selection of compounds in which that bond appears. A short list of these is given in Table 7-3. Calculations performed with these

Table 7-3 Some Average Bond Dissociation Energies

Bond	Average Energy (kcal/mole)
C—C	82.6
C=C	145.8
C≡C	199.6
H—H	104.2
O=O	118
N≡N	225.8
F—F	37
Cl—Cl	57.9
Br—Br	46.1
I—I	36.1
C—H	98.8
C—F	116
C—Cl	81
C—Br	68
C—I	51
C—O	85.5
C=O (aldehydes and ketones)	178
C=O (CO_2)	192.1
O—H	110.6
H—F	135
H—Cl	103.1
H—Br	87.4
H—I	71.4
N—H	93.4
N—F	65
N—Cl	46

values give only approximate ΔH's, but they provide useful indications when experimental values are not available.

For example, let's consider the reaction of hydrogen with carbon tetrachloride to give chloroform:

$$CCl_4(g) + H_2(g) \rightleftharpoons CHCl_3(g) + HCl(g) \tag{7-48}$$

A block diagram for this process is given in Figure 7-12. The first step is the breaking of a C—Cl bond and an H—H bond. This is followed by the formation of a C—H bond and an H—Cl bond. The enthalpy of this hypothetical process is given approximately by

$$\Delta H = \underset{\text{energy req.}}{(D_{C-Cl} + D_{H-H})} - \underset{\text{energy back}}{(D_{C-H} + D_{H-Cl})}$$
$$= (81 + 104) - (99 + 103)$$
$$= -17 \text{ kcal/mole} \tag{7-49}$$

The true ΔH obtained from measured heats of formation is -22.1 kcal/mole. So, while the answer obtained using the average bond energies is only approximate, it is fairly close to the correct answer and gives us a good indication of the magnitude of the energy changes involved.

Exercise 7-6. Show that ΔH is about $+30$ kcal for the reaction

$$H_3C-CH_3 \longrightarrow H_2C{=}CH_2 + H_2$$

by adding the energy absorbed to break all the bonds in ethane, C_2H_6, and the energy released to form the bonds in ethylene, C_2H_4, and hydrogen, H_2.

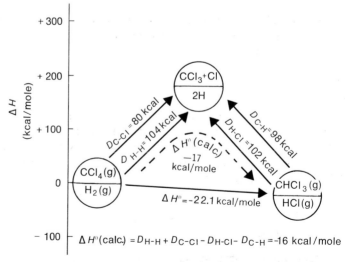

$$\Delta H''(\text{calc.}) = D_{H-H} + D_{C-Cl} - D_{H-Cl} - D_{C-H} = -16 \text{ kcal/mole}$$

Figure 7-12 *Bond energy route for estimation of* ΔH. *The value so derived* $(-17$ kcal/mole) *is a reasonable approximate value for* ΔH^0, *measured to be* -22.1 kcal/mole.

Compare this result to the accurate value from heats of formation (Table 7-1).

(i) PHASE CHANGES: LIQUID TO GAS

If a liquid is slowly heated, its temperature gradually rises until the boiling point is reached. The temperature then stays constant until all the liquid has been converted into vapor. The heat required to vaporize one mole of a liquid at its boiling point is called the heat, or enthalpy, of vaporization and is given the symbol $\Delta H_{\text{vap.}}$. This quantity is related to the energy required to pull a molecule out of the liquid and put it into the gas phase. Because there is a change in volume, some extra work has to be done by the escaping gas molecules to push back the atmosphere. Thus the measured quantity $\Delta H_{\text{vap.}}$ will be greater than the energy $\Delta E_{\text{vap.}}$ required to separate a molecule from the liquid.

It is quite easy to estimate the difference between $\Delta H_{\text{vap.}}$ and $\Delta E_{\text{vap.}}$. Let's do this for $H_2O(\ell) \rightleftharpoons H_2O(g)$. For a constant pressure process, we can write

$$\Delta H = \Delta E + P\Delta V \tag{7-50}$$

and solve for ΔE,

$$\Delta E_{\text{vap.}} = \Delta H_{\text{vap.}} - P\Delta V \tag{7-51}$$

The volume change is given by

$$\Delta V = V_{\text{vapor}} - V_{\text{liquid}} \tag{7-52}$$

and it is almost always a good approximation to assume that the gas volume of a given mass of a substance is much larger than the liquid volume of that same mass. If so, we can write

$$\Delta V \sim V_{\text{vapor}}$$

Let's check this assumption for water before going any further. The volumes occupied by one gram of liquid water and water vapor at 100°C are 1.04 ml and 1.68 liters, respectively. So we can see immediately that there is at least a one-thousand-fold increase in volume on vaporization. For one mole of water (18 g) we have

$$\Delta V = V_{\text{vapor}} - V_{\text{liquid}}$$

$$= 30.2 - 0.019$$

$$= 30.18 \text{ liters } (\approx 30.2 \text{ liters})$$

$$\cong V_{\text{vapor}} \tag{7-53}$$

If we assume that water vapor behaves as an ideal gas, we can write

$$PV_{\text{vapor}} = n_{\text{vapor}}RT \tag{7-54}$$

Since we are worrying about ΔH and ΔE on a one-mole basis,

$$n_{\text{vapor}} = 1 \text{ mole}$$

and the PV product is easily calculated:

$$
\begin{aligned}
PV_{\text{vapor}} &= RT \\
&= (1.987 \text{ cal/mole degK}) \cdot (373^\circ\text{K}) \\
&= 741 \text{ cal/mole}
\end{aligned}
$$

The observed enthalpy of vaporization of water at its boiling point is

$$\Delta H_{\text{vap.}} (\text{H}_2\text{O}) = 9820 \text{ cal/mole} \tag{7-55}$$

Thus the energy required for the vaporization process is

$$
\begin{aligned}
\Delta E_{\text{vap.}} &= 9820 - 741 \text{ cal/mole} \\
&= 9080 \text{ cal/mole} \tag{7-56}
\end{aligned}
$$

Thus only a relatively small fraction, less than 10 percent, of $\Delta H_{\text{vap.}}$ represents energy used to push back the atmosphere. We can feel quite relaxed, then, about using the easily determined quantity $\Delta H_{\text{vap.}}$ as an indication of the effort required to evaporate a liquid.

(j)   PHASE CHANGES: SOLID TO LIQUID

Energy is required to release molecules from the liquid into the gas phase. Energy is also required to convert a solid into liquid form. The enthalpy change associated with melting is called the enthalpy of fusion $\Delta H_{\text{fus.}}$. Once again it is ΔH, rather than ΔE, that is conveniently measured. In this case, however, the difference between them is indeed negligible. The expression for ΔE is

$$
\begin{aligned}
\Delta E_{\text{fus.}} &= \Delta H_{\text{fus.}} - P\Delta V \\
&= \Delta H_{\text{fus.}} - P(V_{\text{liquid}} - V_{\text{solid}}) \tag{7-57}
\end{aligned}
$$

The volume occupied by a mole of liquid is, in general, very close to the volume occupied by a mole of the same substance in the solid state. It is, therefore, a very good assumption that

$$\Delta V_{\text{solid} \rightarrow \text{liquid}} = 0 \qquad (7\text{-}58)$$

so that ΔH and ΔE are sensibly equal.

It is interesting to compare the processes of fusion and vaporization— Table 7-4 contrasts $\Delta H_{\text{fus.}}$ and $\Delta H_{\text{vap.}}$ for a number of substances. In every case, $\Delta H_{\text{vap.}}$ is much larger than $\Delta H_{\text{fus.}}$. The reasons are rather clear.

Table 7-4 Enthalpy of Fusion and Vaporization

Substance	$\Delta H_{\text{fus.}}$ (kcal/mole)	$\Delta H_{\text{vap.}}$ (kcal/mole)
HCl	0.48	3.9
HBr	0.58	4.2
H_2S	0.59	4.5
NH_3	1.35	5.6
CCl_4	0.60	7.3
C_6H_6	2.35	7.4
H_2O	1.44	9.8
H_2O_2	2.92	10.3

The change from solid to liquid requires that the tight and regular arrangement of molecules in the solid be loosened enough to permit the freedom of movement characteristic of fluids. The similarity of molar volumes tells us, however, that molecules in a liquid are still close together, hence held quite tightly. In the gas phase, on the other hand, molecules are so far apart that there are only very small forces between them (none in an ideal gas). Much more energy, then, has to be expended to remove a molecule from the liquid into the gas phase than just to loosen up the solid into a liquid.

7-4 Calorimetry

Until now we have blithely talked about thermochemistry without paying attention to the experimental details of the measurements involved. In this section, we will discuss the nature of the absorption of heat, and describe some of the experimental methods used.

(a) HEAT CAPACITY: MOLECULAR ORIGINS

Consider a bulb containing one mole of a monatomic gas at room temperature. Suppose the bulb is warmed in a constant-temperature bath. The absorption of heat by the gas and the bulb results in an increase in the average translational kinetic energy of the molecules. This increase in translational energy is manifested as a temperature rise. If the bulb is cooled, the molecules move more and more slowly until, at the absolute zero, there is no translational motion at all (the gas has by this time condensed to a solid).

Of course a polyatomic molecule has vibrational and rotational kinetic energy, as well as translational. When such a molecule is heated, some of the added energy goes toward increasing vibration and rotation, so that it will take more energy to achieve a given temperature rise in a polyatomic gas than in a monatomic gas. The *heat capacity* of a substance is defined as the amount of heat, in calories, necessary to effect a one-degree rise in temperature. In thermodynamics we will be interested primarily in processes carried out at constant pressure and will, as usual, refer our measurements to the standard quantity, the mole. Therefore, we will deal with the *molar heat capacity at constant pressure*, C_P. This is the heat in calories necessary to raise the temperature of one mole of substance one degree at constant pressure. Thus, the heat required for a temperature rise ΔT at constant pressure is given by

$$\Delta H = q_P = C_P \Delta T \tag{7-59}$$

$$\Delta T = (T_{final} - T_{initial})$$

where T is the absolute temperature in degrees K and C_P is the molar heat capacity at constant pressure.

It should be clear from what has been said that the molecular size and molecular geometry should have an important influence on the heat capacity. A molecule with lots of atoms has many more ways of absorbing added energy in vibration and rotation than does, say, carbon monoxide, a diatomic gas. Some representative values of C_P are given in Table 7-5.

The upper portion of this table shows the dependence of C_P on molecular complexity. The monatomic gases all have identical heat capacities— the natural result of the fact that all the absorbed energy must go into translational motion. (Only at very high temperatures do the effects of electronic kinetic energy become noticeable.) As the molecules become more and more complex, their heat capacities rise steadily. The extra energy goes into vibrational and rotational motion.

The lower half of the table illustrates another important point. For these similar pentatomic molecules, the heat capacity also increases as more and more heavy atoms are added. This is because the vibrations and rotations of molecules are *quantized* (the same effect that results in the line spectrum of the hydrogen atom). The heavier atoms in a molecule vibrate more slowly, so their vibrational energy levels are closer together. Hence, more vibrations can be excited at a given temperature in a "heavy-atom" molecule like CI_4 than in a "light-atom" molecule like CH_4. The accumulation of this effect can be seen most vividly in the sequence CH_4, CH_3Cl, CH_2Cl_2, $CHCl_3$, CCl_4.

Table 7-5 *Molar Heat Capacities at Constant Pressure* (cal/mole degK)

Monatomic Gases		Diatomic Gases		Triatomic Gases	
He	4.9	CO	7.0	H_2O	8.0
Ne	4.9	N_2	7.0	D_2O	8.2
Ar	4.9	F_2	7.5	CO_2	8.9
Kr	4.9	Cl_2	8.1	CS_2	10.9
Xe	4.9				

Tetratomic Gases		Polyatomic Gases	
NH_3	8.5	SF_6	29.0
H_2CO	8.5	UF_6	31.0
Cl_2CO	14.5	C_2H_6	12.6
		CH_3NHCH_3	16.9

Pentatomic Gases			
CH_4 (MW = 16)	8.4	CH_3Cl (MW = 51)	9.8
CF_4 (MW = 88)	14.6	CH_2Cl_2 (MW = 85)	12.3
CCl_4 (MW = 154)	19.9	$CHCl_3$ (MW = 120)	15.7
CI_4 (MW = 520)	22.9		

Exercise 7-7. Compare the heat needed to warm one mole of gaseous argon, Ar, from 25°C to 525°C (2.5 kcal) to the heat needed to warm one mole of gaseous water, H_2O, and that needed to warm one mole of gaseous chloroform, $CHCl_3$, over the same temperature interval.

(b) THE ICE CALORIMETER

The device illustrated in Figure 7-13 is known as an ice calorimeter and is commonly used for the measurement of the heat capacity of solids and liquids, or reaction heats. The operation of this calorimeter depends on the fact that the volume of a given mass of ice is *greater* than the volume of an equivalent mass of liquid water. In a container completely filled with liquid water and solid ice, any melting will increase the pressure in the container or, if there is a small opening, some liquid water will be forced out. The weight of the water expelled is determined by the volume change, hence the amount of ice melted. That gives us the heat.

Unfortunately, the practicalities are such that only a few drops of water will be obtained, so a small weight is implied. This difficulty is cleverly dispelled by arranging that the displaced liquid is mercury instead of water. With a density of 13.6 g/ml, mercury enhances the experimental accuracy so much that the ice calorimeter has wide and popular use.

At the beginning of the experiment, a mantle of ice is frozen around the sample tube by placing dry ice (solid CO_2) inside the tube. The mercury in the weighing beaker is accurately weighed. Then the hot sample is dropped

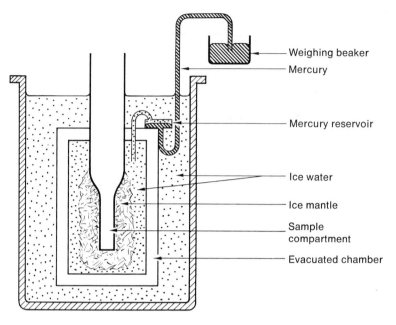

*Figure 7-13 Ice calorimeter. A hot sample placed in the sample compartment melts some of the ice mantle. The decrease in volume sucks mercury from the weighing beaker into the mercury reservoir. (This is a simplified drawing of an ice calorimeter described by D. C. Ginnings and R. C. Corruccini, Journal of Research of the National Bureau of Standards, **38**, p. 583, 1947.)*

into the container and allowed to cool to 0°C. As the sample cools, the ice melts and the total volume of ice and water decreases, thereby sucking mercury from the weighing beaker into the reservoir. When temperature equilibrium has been reached, the beaker is reweighed. The change in volume is proportional to the amount of ice melted and can be calculated from the weight of mercury pulled from the beaker. Heat losses to the surroundings are minimized by placing the entire apparatus in a large vessel containing an ice–water mixture.

The apparatus is calibrated by heating the sample tube with a heating coil to which an accurately known quantity of electrical power is supplied.

In a typical calibration experiment with an ice calorimeter, 61.4 grams of mercury were withdrawn from the weighing beaker. The heater was run for 7 minutes with an average current of 0.60 ampere and an average voltage of 66 volts. We want to find the calibration constant K_g of the calorimeter in calories per gram of mercury:

$$\text{electrical energy (joules)} = Vit \tag{7-60}$$

where V is volts, i is amperes, and t is time in seconds (1 cal = 4.2 joules).

Therefore

$$E_{\text{elect.}} = \frac{(66 \text{ V}) \cdot (0.60 \text{ A}) \cdot (420 \text{ sec})}{(4.2 \text{ joules/cal})}$$

$$= 3.96 \times 10^3 \text{ cal}$$

Now we can find the heat dissipated per gram of mercury withdrawn from the weighing beaker. We will call this quantity K_g, the calibration constant of the apparatus.

$$K_g = \frac{(3.96 \times 10^3 \text{ cal})}{(61.4 \text{ g Hg})} \qquad\qquad (7\text{-}61)$$

$$= 64.6 \text{ cal/g Hg}$$

Thus a sample must deliver 64.6 cal of heat to change the weight of the weighing beaker by one gram.

We can find C_P for carbon tetrachloride from the following experimental results: 103 ml of CCl_4 (3.0 moles) at 50°C (323°K) were placed in the ice calorimeter calibrated in the previous example. A total of 7.3 grams of mercury were sucked from the weighing beaker. The heat lost by the sample was

$$\Delta H = K_g \cdot \text{weight}$$

$$= (64.6 \text{ cal/g Hg}) \cdot (7.3 \text{ g Hg})$$

$$= 472 \text{ cal} \qquad\qquad (7\text{-}62)$$

The heat capacity for one mole is given by (7-59):

$$\Delta H = C_P \Delta T$$

When n moles are involved

$$\Delta H = n C_P \Delta T$$

Solving for C_P,

$$C_P = \frac{\Delta H}{n \Delta T}$$

$$= \frac{(472 \text{ cal})}{(3 \text{ moles}) \cdot (50°K)}$$

$$C_P = 31.5 \text{ cal/mole degK}$$

(c) THE BOMB CALORIMETER: HEATS OF COMBUSTION

The ice calorimeter is useful for measuring heat capacities and enthalpy changes of reactions occurring in solution. Another type of calorimeter in wide use is the bomb calorimeter, illustrated in Figure 7-14. This calorimeter is used for measuring heats of combustion. The sample is placed in a small cup inside the tightly sealed bomb, and a large pressure of O_2 is admitted. The bomb is placed in a vessel and covered with water. This part of the apparatus is then placed in a large insulated vessel to minimize heat losses.*
The sample is ignited by means of an electrical pulse through the ignition wires. The temperature rise of the water in the calorimeter is measured. Of course the apparatus has to be calibrated. This is usually done by igniting a sample with an accurately known heat of combustion. This allows the total heat capacity of the apparatus and water to be determined.

Any process carried out inside such a tightly closed bomb is obviously occurring at constant volume. We are measuring, then, the heat of combustion at constant volume, which is equivalent to the change in energy of

---

* The apparatus shown in Figure 7-14 is a simplified version of that normally used. A conventional bomb calorimeter keeps heat losses to a minimum by keeping the outer jacket of the apparatus at the same temperature as the calorimeter. This is done either by electric heating or by running warm water through the jacket.

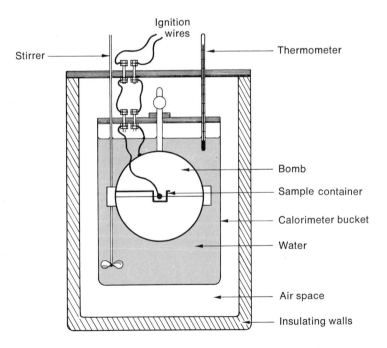

Figure 7-14 A simplified bomb calorimeter.

the system:

$$q_V = \Delta E \tag{7-63}$$

However, we want ΔH, the enthalpy change, not ΔE, if we are to keep all our thermochemical results consistent. Let's consider the difference between these two quantities. From the definition of enthalpy

$$H = E + PV \tag{7-64}$$

we find that the enthalpy *change* is given by

$$\Delta H = \Delta E + \Delta(PV) \tag{7-65}$$

If the products *and* reactants were all solids and liquids, the term $\Delta(PV)$ would be negligibly small and we could assume that

$$\Delta H = \Delta E \quad \text{(solids and liquids)} \tag{7-66}$$

If gaseous atoms are among the products or reactants, however, this is not the case. Suppose that any gaseous species involved behave as ideal gases and, thus, follow the ideal gas law

$$PV = nRT \tag{7-67}$$

We can then find $\Delta(PV)$ in terms of the other quantities as follows:

$$\Delta(PV) = \Delta n(RT) \tag{7-68}$$
$$= (n_2 - n_1)RT$$

where n_2 is the number of moles of gaseous products
$\quad n_1$ is the number of moles of gaseous reactants
$\quad R$ is the gas constant (1.987 cal/mole degK)
$\quad T$ is the average temperature (absolute) during measurement

The correction necessary to convert a measured ΔE into a ΔH depends, then, on the *difference* between the number of moles of gases appearing as products and reactants. As we shall see in the following example, the correction is usually not very large, even when several moles of gases are involved. Hence the use of an average temperature introduces a negligible error.

In a typical experiment* using a bomb calorimeter, a sample of biphenyl

* These results were obtained in a senior physical chemistry laboratory at the University of British Columbia.

$(C_6H_5)_2$ weighing 0.526 g was ignited in an excess of oxygen gas. The observed temperature rise was 1.91 degrees. In a separate calibration experiment, a sample of benzoic acid, C_6H_5COOH, weighing 0.825 g was ignited and produced a temperature rise of 1.94 degrees. The energy of combustion of benzoic acid (BA) is accurately known:

$$\Delta E_{comb.}(\text{benzoic acid}) = \Delta E_{comb.}(\text{BA}) = -771 \text{ kcal/mole} \qquad (7\text{-}69)$$

The water in the calorimeter was initially at room temperature 25°C (298°K). We can calculate $\Delta E_{comb.}$ and $\Delta H_{comb.}$ for biphenyl. The total heat capacity of the calorimeter and water must be calculated first from the known heat delivered by the combustion of benzoic acid.*

$$-\Delta E_{comb.}(\text{BA}) = C(\text{calorimeter}) \cdot \Delta T$$

$$C(\text{calorimeter}) = \frac{(771 \text{ kcal/mole}) \cdot (0.825 \text{ g})}{(1.94 \text{ degK}) \cdot (123 \text{ g/mole})}$$

$$= 2.66 \text{ kcal/degK}$$

Using this result, we can calculate the energy of combustion of biphenyl (BP):

$$-\Delta E_{comb.}(\text{BP}) = C(\text{calorimeter}) \cdot \Delta T$$

$$\Delta E_{comb.}(\text{BP}) = -\frac{(2.66 \text{ kcal/degK}) \cdot (1.91 \text{ degK}) \cdot (154 \text{ g/mole})}{0.526 \text{ g}}$$

$$= -1490 \text{ kcal/mole}$$

Now we must convert $\Delta E_{comb.}$ to $\Delta H_{comb.}$. The chemical equation is

$$(C_6H_5)_2(s) + \tfrac{29}{2} O_2(g) \rightleftharpoons 12 CO_2(g) + 5 H_2O(\ell) \qquad (7\text{-}70)$$

The change in moles of gaseous substances is

$$\Delta n \text{ (gases)} = (n_{\text{products}} - n_{\text{reactants}}) \qquad (7\text{-}71)$$
$$= 12 - \tfrac{29}{2}$$
$$= -\tfrac{5}{2}$$

---

* Heat given off by the combustion process is absorbed by the calorimeter. The heat absorbed is thus the negative of the heat of combustion, $-\Delta E_{comb.}$.

Substituting this into (7-68) and (7-65) gives us

$$\Delta H_{comb.} = \Delta E_{comb.} - \tfrac{5}{2}(1.987 \text{ cal/mole degK}) \cdot (298°K)$$
$$= (-1490 - 1.4) \text{ kcal/mole}$$
$$= -1491 \text{ kcal/mole}$$

Some experimentally measured heats of combustion are given in Table 7-6.

Table 7-6 Enthalpies of Combustion of Organic Compounds*

Name	Formula	$\Delta H_{comb.}$ at 298°K (kcal/mole)†
Methane	$CH_4(g)$	−212.8
Ethane	$C_2H_6(g)$	−372.8
Propane	$C_3H_8(g)$	−530.6
Ethylene	$C_2H_4(g)$	−337.2
Benzene	$C_6H_6(\ell)$	−781.0
Methanol	$CH_3OH(\ell)$	−173.6
Ethanol	$C_2H_5OH(\ell)$	−326.7
Dimethyl ether	$(CH_3)_2O(g)$	−347.8
Acetic acid	$CH_3COOH(\ell)$	−209.4
Glucose	$C_6H_{12}O_6(s)$	−673
Sucrose	$C_{12}H_{22}O_{11}(s)$	−1348.9

* The final products are $CO_2(g)$ and $H_2O(\ell)$; $\Delta H_f^0(CO_2(g)) = -94.1$ kcal; $\Delta H_f^0(H_2O(\ell)) = -68.3$ kcal/mole.
† Tables of combustion measurements are often reproduced with the opposite sign convention so that heats of combustion are listed as positive quantities.

(d) CALORIMETRIC DETERMINATION OF $\Delta H_{vap.}$

The energy (enthalpy) required to vaporize a mole of a pure liquid can be determined by supplying a measured amount of electric power. A suitable apparatus is illustrated in Figure 7-15. The liquid is placed in an insulated flask and an accurately known amount of electrical energy is supplied through the heating coil. The vapor is condensed in the water-cooled condenser and collected in the receiving flask.

A single measurement performed in this way would not be too satisfactory because of the heat losses to the apparatus. If measurements are made at several rates of evaporation, an accurate value of $\Delta H_{vap.}$ can be obtained.

The rate at which electrical energy is supplied in watts (joules per second) is given by the product:

power $= Vi$

where V is volts and i is amperes.

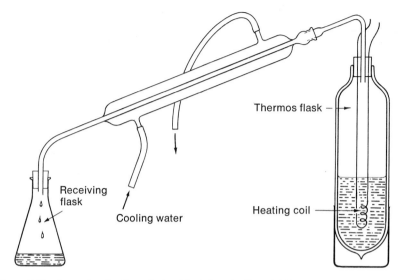

Figure 7-15 Apparatus for the calorimetric determination of $\Delta H_{\text{vap.}}$.

This is balanced by the evaporation process, which occurs at the rate

$$\text{rate of evaporation} = \frac{n\Delta H_{\text{vap.}}}{t}$$

where n is the number of moles
t is the time electrical power is supplied

and by the losses to the apparatus, which we shall call h. Thus we can combine these factors to give us

$$\text{power} = Vi = \frac{n\Delta H_{\text{vap.}}}{t} + h$$

If we perform experiments at several rates of evaporation (n/t), a plot of this rate against electrical power should give a linear plot of slope $\Delta H_{\text{vap.}}$ and intercept h.*

Using the apparatus described above,† the following results were

---

* This equation has the form of the general linear equation $y = ax + b$, in which a is the slope and b the intercept in the y axis.
† These data were obtained in a first-year chemistry lab at the University of British Columbia.

obtained for chloroform, $CHCl_3$. (1 cal = 4.18 joules)

	Power, Vi (joules/sec)	Rate of Distillation, n/t (mole/sec)
I	8.75	2.15×10^{-4}
II	6.84	1.47×10^{-4}

We can calculate $\Delta H_{vap.}$ for chloroform using only these two pieces of data. (A real determination would use several experimental rates and find $\Delta H_{vap.}$ from the slope of the graph described above.)

We can write

$$(Vi)^{\mathrm{I}} = \Delta H_{vap.}(n/t)_1 + h \qquad (7\text{-}72)$$

and

$$(Vi)^{\mathrm{II}} = \Delta H_{vap.}(n/t)_2 + h \qquad (7\text{-}73)$$

Subtracting equation (7-72) from equation (7-73), we have

$$\Delta(Vi) = \Delta H_{vap.}\Delta(n/t)$$

Thus,

$$\Delta H_{vap.} = \frac{\Delta(Vi)}{\Delta(n/t)}$$

$$= \frac{(8.75 - 6.84 \text{ joules/sec})}{[(2.15 - 1.47) \times 10^{-4} \text{ mole/sec}] \cdot (4.18 \text{ joules/cal})}$$

$$= 6.72 \text{ kcal/mole}$$

(e) THERMOCHEMISTRY AND MOLECULAR BEHAVIOR

To conclude this consideration of thermochemistry, let's return to the molecular level to probe the molecular origins of chemical heat effects. According to equation (7-9), it must all tie back to the changes in kinetic and potential energies that accompany the rearrangement of atoms into new molecules.

Consider again the hydrogen–fluorine reaction:

$$H_2(g) + F_2(g) \longrightarrow 2\,HF(g) \qquad \Delta H^0 = -129 \text{ kcal}$$

The quantity ΔH must equal the sum of $\Delta(KE)$ plus $\Delta(PE)$. The kinetic energy can be divided into its components, as expressed in (7-4):

$$\Delta(KE) = \Delta(KE)_{translation} + \Delta(KE)_{vibration} + \Delta(KE)_{rotation} + \Delta(KE)_{electrons}$$

The magnitudes of these quantities are all known. At constant temperature, the two moles of HF product have the same translational energy as the two moles of reactants ($H_2 + F_2$).

Hence

$$\Delta(KE)_{\text{translation}} = 0$$

The other components are known to have the following magnitudes for reaction (7-43):

$$\Delta(KE)_{\text{vibration}} = +0.5 \text{ kcal}$$

$$\Delta(KE)_{\text{rotation}} = -0.03 \text{ kcal}$$

$$\Delta(KE)_{\text{electrons}} = +128.5 \text{ kcal}$$

We see that the movement of the electrons accounts for almost all of the change in kinetic energy as HF is formed, but energy is absorbed as reaction proceeds.

The potential energy also can be divided into components, as expressed in (7-6),

$$\Delta(PE) = \Delta(PE)_{\text{electron−nucleus}} + \Delta(PE)_{\text{electron−electron}} + \Delta(PE)_{\text{nucleus−nucleus}}$$

This sum is known also: $\Delta(PE) = -258$ kcal. With this value, we can evaluate ΔH:

$$\Delta H = \Delta(KE) + \Delta(PE)$$

$$= (+0.5 - 0.03 + 128.5) + (-258)$$

$$\Delta H = -129 \text{ kcal}$$

We see that the exothermicity of the reaction is due to the drop in potential energy and that the kinetic energy rises as total energy falls. Furthermore, almost all of the kinetic energy change is provided by the electron movement. We will see that these conclusions apply to all exothermic reactions.

7-5 Energy and spontaneity—Why do things roll downhill?

In Figure 7-2, we saw a runaway car plummet down the hill. The brake having slipped, the car spontaneously began its merry trip. Figure 7-16 shows where it probably completed its thoughtless journey. At the beginning, atop the hill, the car held potential energy. As it neared the bottom, this energy changed into kinetic energy—to the dismay of casual pedestrians! Then everything ended in chaos.

Figure 7-16 End of a joyride.

(a) ROLLING DOWNHILL—A MODEL FOR CHEMISTRY?

Is there a model here concerning chemical events? Back in 1860, Berthelot thought there was, when he postulated that all chemical changes tend to liberate heat. Having carefully measured the heat released in dozens and dozens of exothermic reactions, he became so preoccupied with heat release that he decided it pointed in the direction of spontaneous change. He decided that chemical reactions are like the car in Figure 7-2: Reactions seem to "roll downhill" on a chemical energy landscape.

It was a strange conclusion for this brilliant scientist to reach. He had ample evidence at hand that contradicted this simple view. Many familiar changes *absorb* heat as they occur—salt dissolving in water is one example, and carbon dioxide expanding from a high-pressure cylinder is another. Both changes occur quite spontaneously, but obviously without releasing heat. Furthermore, chemists of the time knew of many reactions which, though exothermic, proceed to an equilibrium mixture in which some reactants remain unconsumed. Nevertheless, Berthelot's idea received a certain acceptance. After all, strongly exothermic reactions are often the most difficult to restrain, and the fact that balls do roll downhill seemed to offer some connection. To explore the idea, let's consider a laboratory version of the events of Figure 7-2.

Figure 7-17 shows a steel ball-bearing on a polished metal surface, position (a). Since it is on a slope, it begins to roll down the hill, picking up speed as it goes. By the time it is halfway down (position (b)), it has picked up considerable speed. Much of its potential energy (energy of position) has been converted into kinetic energy (energy of motion). But the ball is still on the hillside, so it accelerates still more. By the time it reaches the

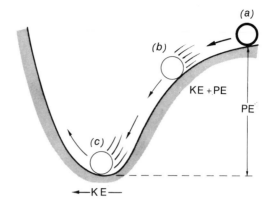

Figure 7-17   A ball-bearing placed on a polished metal hillside rolls downhill.

bottom of the hill (position (c)), all of the potential energy is expended and converted into kinetic energy. The ball has spontaneously rolled downhill.

Now wait a second—see Figure 7-18! Shortly after the ball reaches the bottom of the hill, it starts rolling *up* the hill on the other side. At (*d*) its velocity is reduced, but now the ball again possesses some potential energy. By the time it reaches (e), the energy of motion is gone and the ball again possesses only potential energy. So *the ball has now spontaneously rolled uphill.* Apparently it had just as strong a tendency to roll uphill, once it gained a good head of steam, as it had to roll downhill when it was parked on the hillside.

Figure 7-19 shows the track of the ball when it is left to roll back and forth on the hill. Each time it completes a tour, down on the right, up on the left, back down on the left and up on the right, it comes up a little shy of its starting point. As time goes on, the amplitude of its motion decreases more and more. Finally, it ends up at the bottom of the hill, at rest.

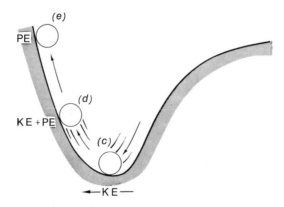

Figure 7-18   A ball-bearing can roll uphill, too.

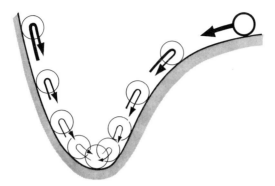

Figure 7-19 Gradually friction spoils the fun.

Where, now, is its energy? What happened to the tendency of the ball to roll back and forth, first down, then up? The answer is that friction took a hand. No matter how carefully we polish the hillside, a little friction remains, so the ball rubs against the surface and produces heat. Gradually all the energy gets converted into heat and, as it does, the ball reaches the bottom of the hill.

Recapitulating, we find that when the ball possesses potential energy, it tends to roll downhill. This merely generates kinetic energy, and then the ball tends to roll uphill. The energy of position is converted to directed energy of motion. What finally gets the ball to the bottom of the valley is the dissipation of this directed energy of motion into the random motion of heat.

(b) ATOMS REACTING—ROLLING DOWNHILL?

Figure 7-20 shows a plot of how the energy changes as two iodine atoms approach each other. The horizontal scale shows the internuclear distance, and the vertical scale shows the energy of the electrons as they sense the positive and negative charges on the other atom. The energy drops as they approach until it reaches a minimum at R_0, the equilibrium bond length for I_2.

However, the iodine atoms don't stop at R_0. They are accelerated as they approach each other, so they proceed right through the equilibrium distance and up the other side. On reaching point (e), the energy of the iodine atoms returns completely to potential energy. Then they turn around and come back down the hill, through the R_0 position and out again, flying apart. The iodine atoms fail to react!

What they need, according to Figure 7-19, is some friction. And this is true. The friction must be provided by collisions with other molecules, and it must be available in the short time before the iodine atoms separate again. If a third colliding molecule takes away a little of the energy, as in

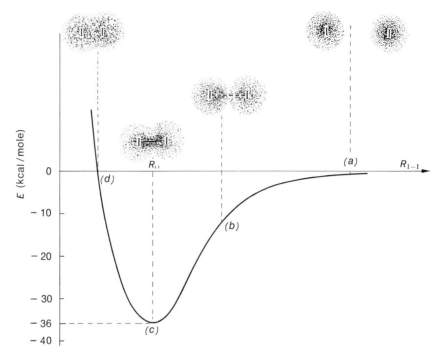

Figure 7-20 Energy of two iodine atoms as they collide.

Figure 7-21, the iodine atoms will not quite be able to separate. Later, per-haps another collision will drain off more energy, and the vibrational move-ment of the iodine atoms will be cooled still further. After a time and after many collisions, the iodine atoms will find themselves at the bottom of the valley, at the equilibrium distance. Their one time energy will have been transferred into the random energy (heat) of the collisional partners that made it all possible.

This is not just a fiction about friction. Experimentally, it is found that iodine atoms react very slowly, even though the formation of I_2 releases 36 kcal/mole as the reaction proceeds. They react slowly, that is, if the pres-sure is low. But as the number of collisions per second increases, the rate at which I_2 is formed increases. Molecular friction drains off the ordered energy and converts it into heat.

(c) SO WHAT ABOUT SPONTANEOUS CHANGES?

So why do our ball-bearing and our two iodine atoms end up at the bottom of the hill? Because friction tends to convert ordered motion into disordered motion. That seems to be involved in spontaneous change. Yet we still haven't grasped the essence of the matter, for some changes occur spon-taneously, absorbing energy as they take place. Can friction account for that? No, indeed.

So we conclude that the generation of disordered energy (heat) from ordered energy (either potential energy or kinetic energy) is somehow involved in spontaneous change, but not simply through the tendency of things to roll downhill. How do the molecules know what to do? Maybe we should choose some very simple spontaneous process and watch these intelligent molecules, so we can learn. The process we'll begin with is the expansion of a gas into a vacuum. This simple process gives insight into the mass psychology and combined intellect of a mole of molecules, as they inexorably head for equilibrium. We shall find that molecules *do* roll down-hill, but *not* on an energy landscape.

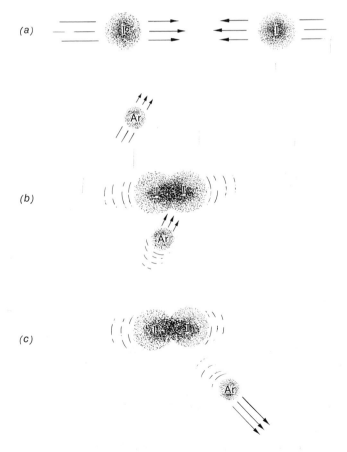

(a)

(b)

(c)

Figure 7-21 Collisional friction helps iodine atoms react: An argon atom will do.

Problems

1. Which kinds of kinetic energy, translational, rotational, and/or vibrational, does each of the following possess?
(a) a pitcher's curve ball en route to the plate
(b) a plucked guitar string
(c) a skydiver playing the bongo drums on the way down
(d) the earth in its orbital movement

2. A monatomic gas has only translational kinetic energy, and if it is a perfect gas, the average energy per mole depends only on temperature: $E_{av.} = \frac{3}{2}RT$ (see Section 1-2(d)). For 1.00 mole of such a gas,
(a) how much does the translational kinetic energy change when the gas is heated one degree?
(b) how much does the translational energy change if the gas is expanded by a factor of ten but at constant temperature?

3. In contrast to a monatomic gas, hydrogen, H_2, has translational, rotational, and vibrational kinetic energy. If 0.200 gram of H_2 is heated from 298 to 318°K in a closed bulb, 9.82 calories of heat is absorbed.
(a) Calculate the heat required to raise the temperature of one mole of H_2 one degree in a closed bulb. This is the heat capacity at constant volume, C_V.
(b) Assuming the translational energy rises by the same amount as a perfect monatomic gas (see Problem 2), calculate the rise in rotational plus vibrational kinetic energy when the H_2 is heated one degree.
(c) In (a), the energy absorbed for this one degree rise defines the heat capacity at constant volume, $\Delta E = C_V$. If the process occurred at constant pressure, then $\Delta H = C_P$. Use $\Delta H = \Delta E + \Delta(PV) = \Delta E + (P_2V_2 - P_1V_1)$ to show that $C_P = C_V + R$ for a perfect gas.

4. Answer questions (a) and (b) in Problem 3 for carbon tetrachloride, which has more rotational and vibrational motions than hydrogen. In the same experiment, again with 0.100 mole, the CCl_4 gas absorbs 35.8 calories when heated from 298 to 318°K.

5. Which of the following processes is endothermic, which is exothermic, and which is "thermoneutral" (i.e., involves no heat effects)?
(a) water being formed from H_2 and O_2 in the engine of a Saturn V rocket
(b) water droplets in the rocket's exhaust plume as they evaporate
(c) water droplets in the exhaust plume if they freeze
(d) water droplets in the exhaust plume if they freeze and then remelt, all at 0°C

6. Answer Question 5 for the following processes.
(a) A tree burns in a forest fire.
(b) A leaf grows through photosynthesis.
(c) A spoonful of jam falls to the floor.
(d) A perfect gas expands into a vacuum.
(e) A reversible glove is turned inside out, converting it from a right-hand glove to a left-hand glove.
(f) A rock-climber leaves camp, scales the vertical east wall of a mountain and returns to camp via the gentle, western slope.

7. (a) Trace the form of energy in the following processes, recognizing the law of conservation of energy.
   (i) Through thermal excitation, the atoms and ions in the sun emit light.
   (ii) Sunlight is absorbed by leaves; trees and plants grow.

(iii) A grazing cow eats grass and produces milk.
(iv) A child drinks milk and develops muscles.
(v) A man lifts a load of wood to carry to the fireplace.
(vi) Wood in the fireplace burns, emitting heat and light.
(b) Which of the processes in (a) depend, ultimately, on solar energy?

8. Limestone stalactites are formed in the reaction

$$Ca^{+2}(aq) + 2\ HCO_3^-(aq) \longrightarrow CaCO_3(s) + CO_2(g) + H_2O(\ell)$$

If one mole of $CaCO_3$ is deposited at 298°K under one atmosphere pressure, the reaction performs 590 calories of expansion work, pushing back the atmosphere as the gaseous CO_2 is formed. At the same time, 9310 calories of heat are absorbed from the environment. What is ΔE in kcal/mole $CaCO_3$?

9. When 0.00115 mole of zinc metal reacts in a beaker with 1 M silver nitrate to form silver metal and 1 M zinc nitrate, neither electrical nor expansion work is performed, but 0.100 kcal of heat is released.

$$Zn(s) + 2\ Ag^+(aq) \longrightarrow Zn^{+2}(aq) + 2\ Ag(s)$$

(a) What is ΔE for the reaction as it occurred?
(b) What is ΔE for one mole of reaction?

10. The reaction described in Problem 9 is carried out in an electrochemical cell and the voltage of the cell is 1.56 volts. As the 0.00115 mole of zinc dissolves, the electron current leaving the zinc electrode is passed through a motor so that 40 calories of electrical work is performed. Since ΔE is a state function, it must be the same as calculated in Problem 9. Calculate the heat absorbed by the cell to maintain constant temperature. What is the significance of its magnitude and sign relative to the value of q in Problem 9?

11. An expansion cell like that shown in Figure 6-4 has an over-all cell reaction that consumes H_2 gas at 10 atm pressure and produces H_2 gas at 1 atm pressure.

$$H_2(10\ atm) \xrightarrow{\ 298°K\ } H_2(1\ atm)$$

The voltage of the cell is 0.0295 volt.
(a) What is the expansion work at the cathode as one mole of H_2 is produced? Express your answer both in ℓ-atm and in calories (1 ℓ-atm = 24.2 cal). (Assume perfect gas behavior here and throughout this problem.)
(b) What is the expansion work at the anode?
(c) What is the net expansion work performed by the cell?
(d) What is ΔE for the process (see Problem 2)?
(e) What is the heat absorbed, q, if the circuit is closed with a short so that $w_{elect.} = 0$? (Assume cell resistance is negligible so that there is no electrical heating as the current flows.)
(f) The maximum possible electrical work that can be extracted equals the charge moved times the cell voltage. For this cell, it is 1.36 kcal/mole H_2. What will q become if the cell is discharged through a motor so that half of the possible work is extracted, i.e., if $w_{elect.} = +0.68$ kcal?

12. A liquid is placed in a cylinder at 298°K with a pressure on the piston equal to the equilibrium vapor pressure, 2.0 atm. Heat is slowly added to vaporize liquid at constant pressure and at constant temperature. When the gas volume has been increased by 0.245 liter, the amount of heat absorbed is 40.0 calories.
(a) How many moles of liquid were vaporized?

(b) How much work was performed? Express the work both in ℓ-atm and in calories (1 ℓ-atm = 24.2 cal). (Assume perfect gas behavior.)

(c) What is ΔE for the process, in kcal/mole?

(d) What is ΔH for the process, in kcal/mole? (This is called the heat of vaporization and symbolized $\Delta H_{vap.}$.)

13.   What is the value of ΔH for the stalactite reaction in Problem 8?

14.   The heats of reactions (c) and (d) in Problem 30, Chapter Six, (two successive reactions involved in producing metallic copper from its ore) are, respectively, -183.4 and $+27.7$ kcal.

$$2\ Cu_2S(s) + 3\ O_2(g) \longrightarrow 2\ Cu_2O(s) + 2\ SO_2(g) \qquad \Delta H = -183.4 \text{ kcal}$$

$$Cu_2S(s) + 2\ Cu_2O(s) \longrightarrow 6\ Cu(s) + SO_2(g) \qquad \Delta H = +27.7 \text{ kcal}$$

(a) What is the balanced equation for the over-all reaction?

(b) Using the Law of Additivity of Reaction Heats, calculate the value of ΔH for the over-all reaction.

15.   Use the heats of formation in Appendix d to calculate the ΔH for the over-all reaction in Problem 14 and compare to the answer obtained there.

16.   When sodium carbonate decahydrate, washing soda, is gently warmed, water vapor is formed, leaving the heptahydrate while absorbing 37.11 kcal of heat.

I $$Na_2CO_3 \cdot 10\ H_2O(s) \longrightarrow Na_2CO_3 \cdot 7\ H_2O(s) + 3\ H_2O(g) \qquad \Delta H = +37.11 \text{ kcal}$$

On more vigorous warming, the heptahydrate loses more water to give the mono-hydrate, while absorbing 76.51 kcal.

II $$Na_2CO_3 \cdot 7\ H_2O(s) \longrightarrow Na_2CO_3 \cdot H_2O(s) + 6\ H_2O(g) \qquad \Delta H = +76.51 \text{ kcal}$$

Continued heating finally gives the anhydrous salt, soda ash, while absorbing 13.70 kcal.

III $$Na_2CO_3 \cdot H_2O(s) \longrightarrow Na_2CO_3(s) + H_2O(g) \qquad \Delta H = +13.70 \text{ kcal}$$

(a) Use the Law of Additivity of Reaction Heats to calculate ΔH for the conversion of washing soda to soda ash.

IV $$Na_2CO_3 \cdot 10\ H_2O(s) \longrightarrow Na_2CO_3(s) + 10\ H_2O(g)$$

(b) What is the average heat absorbed per mole of water lost in reaction IV?

(c) What is the average heat absorbed per mole of water lost for the first three H_2O molecules lost? For the next six? For the last?

17.   Very strong heating of Na_2CO_3 causes it to lose gaseous CO_2, leaving solid sodium oxide, Na_2O.

$$Na_2CO_3(s) \longrightarrow Na_2O(s) + CO_2(g)$$

(a) From the heats of formation given in Appendix d, calculate ΔH^0 for the reaction hypothetically occurring at 298°K.

(b) Calculate the ratio of the enthalpy needed to remove CO_2 from Na_2CO_3 to the enthalpy needed to remove the last H_2O molecule from the monohydrate crystal, $Na_2CO_3 \cdot H_2O$ (see Problem 16).

18.   In the gas phase, HCl reacts with NO_2 to give an equilibrium mixture of H_2O, NO, and Cl_2.
(a) Balance the equation for the reaction.
(b) Calculate ΔH^0 for this reaction at 298°K (see Appendix d). Is the reaction exothermic or endothermic?

19.   The thermite reaction is so exothermic that it is used for welding massive units, such as propellers for large ships. Calculate the enthalpy change, ΔH^0, for the reaction if it were to occur at 298°K.

$$2 \; Al(s) + Fe_2O_3(s) \longrightarrow Al_2O_3(s) + 2 \; Fe(s)$$

20.   The heat of fusion of iron is 3.6 kcal/mole. How many grams of aluminum would have to react in the thermite reaction (see Problem 19) to melt 100 grams of iron?

21.   None of the following substances appears in Table 7-1. Which *need not* appear because their standard heats of formation are zero by definition?

$H_2(g)$, $H(g)$, $O_3(g)$, $H_2O(s)$, $Hg(\ell)$, $Hg(s)$, $Hg(g)$

22.   The heat of combustion to form $H_2O(\ell)$ and $CO_2(g)$ of cane sugar, $C_{12}H_{22}O_{11}$, is -1349.6 kcal/mole. Balance the reaction and calculate the heat of formation of sugar (sucrose) from the elements.

23.   Estimate the heat of hydrogenation of acetylene from average bond dissociation energies (see Table 7-3) and also from heats of formation (Appendix d). What is the percentage error in the bond energy estimate?

$$HC\equiv CH(g) + H_2(g) \longrightarrow H_2C=CH_2(g)$$

24.   A chemically pumped laser has been based upon the light-induced elimination of HCl or of HF from 1-fluoro-1-chloro-ethylene (see Berry and Pimentel, *Journal of Chemical Physics*, Vol. **51**, p. 2274, 1969). Unfortunately, the heat of formation of this compound is not known. Estimate ΔH for each of the two possible elimination reactions, using average bond dissociation energies (Table 7-3).

$$\longrightarrow Cl-C\equiv C-H(g) + HF(g)$$

25.   Calculate ΔE^0, q, and w for the constant temperature, constant pressure reaction whose ΔH^0 was calculated in Problem 17.

26.   Calculate ΔE^0, q, and w for the constant temperature, constant pressure reaction whose ΔH^0 was calculated in Problem 18.

27.   A volume 0.500 ml of 2.00 M NaOH was mixed with 0.500 ml of 2.00 M HCl or 2.00 M HNO_3 at 0°C in an ice calorimeter. The weight of ice melted was determined

several times:

$$
\begin{array}{lcccccc}
\text{acid:} & \text{HCl} & \text{HCl} & \text{HCl} & \text{HNO}_3 & \text{HNO}_3 & \text{HNO}_3 \\
\text{wt. ice} & & & & & & \\
\text{melted (g):} & 0.172 & 0.156 & 0.157 & 0.170 & 0.173 & 0.154
\end{array}
$$

(a) Calculate for each acid the average weight of ice melted and the average deviation from the average.

(b) Calculate, for each acid, the heat released and its uncertainty. Is the discrepancy between the two values significant compared to the uncertainty? ($\Delta H^0_{\text{fusion}}(\text{ice}) = 1.44$ kcal)

(c) Assuming both heats are due to the same reaction, $H^+(aq) + OH^-(aq) = H_2O(\ell)$, calculate ΔH^0 for this reaction in kcal/mole and the average deviation from the average.

eight randomness and chemical change

In Chapter Three we stressed the importance of the equilibrium state and, particularly, its composition. This is given by the equilibrium constant, which measures the extent of reaction. Chapter Seven dealt with the energy changes that accompany chemical changes, but did not come to grips with the question of when a reaction is spontaneous. Now we will probe into this problem and search for the factors that influence the position of the equilibrium state. We already know that a knowledge of the energy change ΔE or ΔH is not sufficient condition for predicting the direction of chemical change.

We will begin by looking at a very simple process: the expansion of a gas. The energy changes associated with such a process are extremely small (zero for a perfect gas), so we will gain some insight into the factors other than energy that affect chemical change.

8-1 Expansion of gas

Figure 8-1 shows two identical pieces of apparatus; each consists of a 1-liter bulb connected through a closed stopcock to a 3-liter bulb. In apparatus A, the 1-liter bulb A_1 contains Cl_2 at a pressure of 24 torr. Bulb A_2 is empty. In apparatus B, the 1-liter bulb B_1 has been filled with Cl_2 to a pressure of 3 torr, and the right-hand bulb B_2 has been filled to 7 torr.

Now, consider the question—what will happen in each of these systems if the stopcocks are opened? The situation is such a familiar one that none of us has the slightest difficulty in confidently predicting the result. In A, the Cl_2 gas will move from the smaller bulb A_1 into the larger bulb A_2 until the total pressure is uniformly 6 torr. In apparatus B, gas will move in the opposite direction, from the larger bulb to the smaller bulb, again until the pressure in the system is equalized at 6 torr. Everyone knows that a gas expands into a lower-pressure region until pressure equilibrium is reached. Despite our confidence, we try the experiment. The stopcocks are opened and the predicted changes occur, as shown in Figure 8-2. So our question was an easy one. Only a little thought told us what behavior to expect.

Now let us visualize this same "experiment" on a microscopic level.

Picture molecules of Cl_2 dashing about in the left-hand bulb of apparatus A and in the left-hand bulb of apparatus B. The stopcocks are opened. In a short interval of time, $\frac{3}{4}$ of the molecules in A_1 somehow decide to move into A_2, whereas, during the same period, molecules from B_2 move in the opposite direction into B_1. How did the molecules in A_1 *know* they should move into A_2? Were they, as we were, thinking back to past experience to decide? Not likely! Many scientists doubt that molecules think at all—yet molecules always do the right thing.

This poses a harder question. What governs the behavior of a mole of gas? How can we explain its reproducible performance in a given set of circumstances despite our general disrespect for the molecular mentality? *What are the factors that determine the direction of a spontaneous change?*

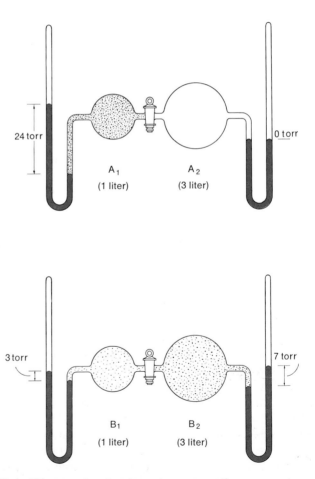

Figure 8-1   What will happen when the stopcocks are opened?

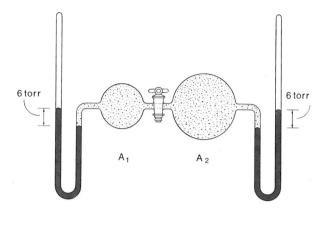

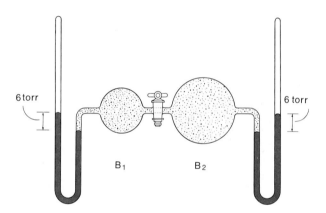

Figure 8-2 The gas pressure is uniform.

(a) AN EXAMPLE—RANDOMNESS AT WORK

Figure 8-3 shows a gambling game. The board has 25 depressions, numbered one to 25. A player bets one dollar and selects three numbers, say 3, 7, and 12. The operator then throws three ping-pong balls into the box. If the balls settle into the correct depressions, those numbered 3, 7, and 12, the house pays the player $2000. This enormous reward is, however, much less enticing than that in game 2 pictured in Figure 8-4. Here there are 100 depressions. This time the player who bets one dollar and picks the right three numbers wins $100,000. This is the way to get rich!

Not really! The man who runs the game is the only person who can count on getting rich. When the evening is over and thousands of bets have been made, the three ping-pong balls will have won the house 24 percent of all

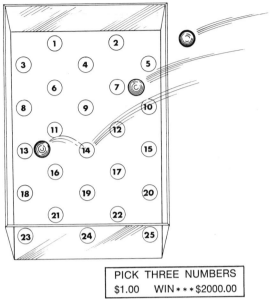

PICK THREE NUMBERS
$1.00   WIN ∗ ∗ ∗ $2000.00

Figure 8-3   Randomness game number one.

the money bet on game 1, shown in Figure 8-3. At the other table, with its enormous payoff, the three ping-pong balls will have won 40 percent of the players' money for the house!

Apparently, these ping-pong balls are intelligent, and if so, the operator is using the smarter balls in game 2. But you know and the operator knows that ping-pong balls cannot be trained; nor are they loaded or magnetized. The operator merely understands positional randomness. With this knowledge, he has carefully calculated the probability of any three particular numbers being obtained, to determine the payoff in each game and to guarantee his own success. Quite contrary to our implication of wile on the part of the ping-pong balls, the operator requires that they be stupid—they must be absolutely random in their choice of position, lest the players learn their preference. All the house wants is for the ping-pong balls to choose their positions at random and then the results of the game are perfectly predictable. Over time, the house always wins!

(b) BACK TO MOLECULES—ARE THEY STUPID TOO?

Could there be a connection between the experiments of Figure 8-1 and the performance of the ping-pong balls? Could molecules be as unenlightened as ping-pong balls, picking their positions at any given instant completely at random and without the slightest preference? As we ask this question, we are playing another game, the game of science. We have observed the behavior of a part of our environment, the expansion of a gas into a vacuum, and wondered about it. We then looked around for something else in our experience that is better understood, and came up with the gambler's lucrative game. With this possible analog, we have posed the scientist's question: "Does the ping-pong ball game furnish a *model* with which we can understand the expansion of a gas into a vacuum?"

What is the next step in this game of science? Having proposed a model, we must explore its implications—*we must test the model.* Without such tests, science would be only a meandering of the mind, as it was in the time of Aristotle—as it was until the time of Galileo, who not only observed and pondered, but also tested his ideas.

Suppose that at any instant of time, each molecule takes a position that is a completely random and impartial choice of all possible positions accessible to it. What would this model predict for the experiments in Figure 8-1? This question can be answered only in probability terms. If we had but one molecule, it would be found in the 1-liter bulb one quarter of the time and in the 3-liter bulb three quarters of the time. If we had four molecules to worry about, the probability that all four would be in the 1-liter bulb is $1/256$, while the probability that two would be in each bulb is $54/256$. An arrangement of special interest to us is that in which one molecule is found in A_1

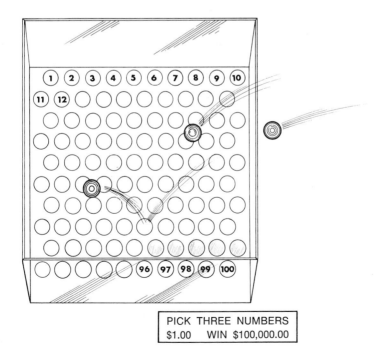

PICK THREE NUMBERS
$1.00   WIN $100,000.00

Figure 8-4 Randomness game number two.

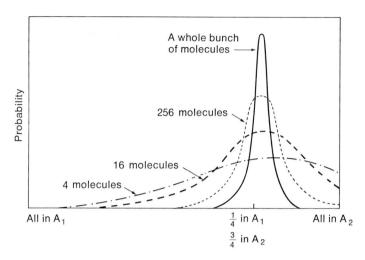

Figure 8-5   How the probability of molecular distribution depends on the number of molecules (random position model).

and three in A_2, which can be calculated (with our dumb-molecule model) to have a probability of 108/256. *This is the most probable arrangement.* If this type of calculation is made for 16 molecules, the most probable arrangement proves to have 4 molecules in A_1 and 12 in A_2 (again in the ratio 1:3). On a comparative basis, the arrangement of all 16 molecules being in A_1 is only one in 10^9 as likely. The arrangement of 2 molecules in A_1 and 14 in A_2 is only 54/91 as likely. If we take more and more molecules, the arrangement in which A_1 contains one third as many molecules as A_2 is *always* the most probable of *any* possible arrangement. Relative to this probability, all other arrangements become less and less probable, including the arrangements in which all the molecules are in A_1, or $\frac{1}{8}$ in B_1 and $\frac{7}{8}$ in B_2, the distributions shown in Figure 8-1. As the molecular population increases, the probability distribution becomes more and more narrow, always peaked at the distribution that has the molecules distributed in the ratio of the volumes of the two bulbs: 1:3, as shown in Figure 8-5.

Now we can see where this model will lead us. As we deal with larger and larger numbers of molecules (approaching Avogadro's number), *the probability* will close in on a situation in which *the molecules are distributed between the two bulbs in proportion to the bulb volumes.* This is the distribution that makes the number of molecules per milliliter the same in the two bulbs. This is exactly the condition needed to give equal pressures.

It is also profitable to examine the reverse of this process. Why don't the gas molecules of Figure 8-2 all rush back to the initial arrangement in Figure 8-1, at least occasionally? Once again we must answer in probability terms. For this event to occur, every single molecule must move in the same direction—through the stopcock into the left-hand bulb. It is clear that a situation in which several billion randomly moving molecules all pack up their bags and set off in the same direction is highly improbable. So, given a choice, molecules will find the state of greatest positional randomness and *stay there.*

So we can understand the changes that take place from the initial conditions in Figure 8-1 to the final ones in Figure 8-2 to be spontaneous changes toward the most probable positional arrangement. Our model, based upon positional randomness for each molecule, predicts the direction of spontaneous change for the expansion of a perfect gas.

8-2 An exothermic chemical reaction

Figure 8-6 shows a container that is separated into two halves by a thin membrane. The lower half contains hydrogen at a pressure of 10 torr, whereas above the membrane there is fluorine at a pressure of 10 torr. At the top there is a small weight held by an external magnet. When the magnet is removed, the weight drops, pierces the membrane, and allows the gases to mix. Will they spontaneously react to form hydrogen fluoride?

$$H_2(g) + F_2(g) \xrightarrow{\ ?\ } 2\,HF(g) \qquad\qquad (8\text{-}1)$$

Once again, many can answer this question without hesitation. Fluorine has a wide reputation as a fearsomely reactive chemical (iron ignites and

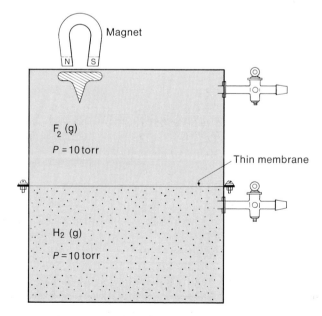

Figure 8-6   What happens when hydrogen and fluorine mix?

burns when in contact with liquid fluorine), and we found in Chapter Seven that its reaction with hydrogen liberates a large amount of heat (129 kcal per mole of fluorine reacted). Again, we have asked an easy question.

Once again, the more penetrating question is to query why the atoms spontaneously decide to rearrange to form the more stable molecules of hydrogen fluoride. Having decided that molecules are probably *non compos mentis*, it seems likely that atoms are no more intelligent. How can we understand their decisive behavior?

Let's take a closer look at this reaction mixture. As a mole of F_2 reacts with a mole of H_2, a large amount of energy is liberated as heat: The atoms take up more favorable positions, energy-wise, as they form HF molecules. We can think of the energy associated with a given molecular arrangement as potential energy (energy of position). The potential energy of two moles of HF molecules is lower than that of the same number of atoms arranged as a mole of H_2 molecules and a mole of F_2 molecules. As HF is formed, the atoms are "rolling downhill." The potential energy change results in the appearance of thermal energy (Fig. 8-7).

Yet again we encounter our question: "*Why* do molecules tend to roll downhill?" Why would the atoms end up in the energetically favorable positions? There is no energy "gain"—after all, energy is conserved. Someone merely has to figure out how to divide up the energy released among the random translational motions of the product molecules. Think of the confusion as this energy is somehow partitioned among the many molecular claimants.

In fact, here is the seed of an idea. Perhaps the model of random behavior that explained the spontaneous gas expansion will also furnish a basis for explaining why molecules "roll downhill."

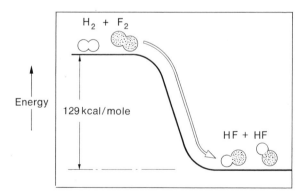

Figure 8-7 Potential energy change during the reaction

$H_2(g) + F_2(g) = 2\,HF(g)$

On an energy landscape the reactants roll downhill to form products.

(a)   A MODEL—ENERGY PACKETS RANDOMLY DISTRIBUTED

If one mole of F_2 and one mole of H_2 react according to (8-1), we find 129 kcal of energy released. Two moles of HF are formed, and these product molecules must divide up the 129 kcal into translational, rotational, and vibrational energy. One way this could be done would be to give every molecule an equal share—$129/2N_0$ (N_0 is Avogadro's number). In fact, the energy was almost divided in this way before the reaction, when it was in the form of potential energy. Every reactant molecule (H_2 or F_2) possessed the same potential energy as any other like reactant molecule. From a probability point of view, there would be little to choose between reactants and products if every product molecule possessed the same kinetic energy as every other.

There is another way in which energy of motion can be apportioned, however. Suppose the 129 kcal are divided into small increments, and that these energy packets are then distributed among the molecules at random, one at a time. This would be like collecting three coins each from 100 children and then redistributing these 300 coins by throwing them, one at a time, into the crowd. There is only a slight chance that each child would end up with three coins. It is far more likely that some children would catch several coins, and some unlucky ones would receive none, only one, or only two coins. From an economic point of view, this might be inequitable, but from a statistical slant, such a coin distribution is more probable. If the 129 kcal were similarly distributed in small increments to the molecules, paying no attention at all to a given molecule's previous energy "wealth" or "poverty," it is most probable that an "inequitable" distribution would occur. Some molecules would have lots of rotational, translational, and vibrational energy, many would have energies near the average, and some would have very little or none at all. According to this model, *motional energy is distributed in the most probable way.* Considering translational motion first, this model predicts a range of velocities for the molecules in a gas. This prediction can immediately be compared with the experimental evidence, which has been known for many decades. Molecular velocities of gas molecules *are* distributed over a range—they exhibit the famous "Boltzmann distribution" of velocities shown in Figure 8-8. At least in a qualitative way, we can see that a tendency towards motional randomness may be connected with spontaneous changes.

A more quantitative (though still approximate) check on this model can be developed. Let us consider the distribution of "packets" of translational energy among molecules.

(b)   A QUANTITATIVE EXAMPLE: VELOCITY DISTRIBUTION
   IN A GAS

The kinetic theory of gases accounts for the properties of a gas in terms of the directions and velocities of the molecular motions. According to this

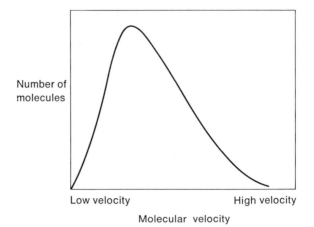

Figure 8-8 Experimentally determined velocity distribution of gas molecules.

theory, a mole of a perfect gas, at temperature $T(°K)$, has a translational energy of $\frac{3}{2}RT$.

Imagine that the gas is contained in a cubic box. The gas pressure is the same on every face, so that, on the average, the translational energy is $\frac{1}{2}RT$ in each of the three (x, y, z) directions. Since there are N_0 molecules in a mole, each molecule has, on the average, kinetic energy equal to

$$\frac{1}{N_0}(\tfrac{1}{2}RT) \text{ erg} \tag{8-2}$$

associated with motion in the x direction and the same amount, on the average, in each of the other two directions. Since we are often interested in the energy of a single molecule, it is convenient to define a new constant

$$k = \frac{R}{N_0} \tag{8-3}$$

which relates molecular energy directly to the temperature:

$$E_{\text{av.}} = \tfrac{3}{2}kT \text{ erg per molecule} \tag{8-4}$$

or

$$E_{\text{av.}}^x = E_{\text{av.}}^y = E_{\text{av.}}^z = \tfrac{1}{2}kT \text{ erg per molecule} \tag{8-5}$$

With this in mind, let's consider a simple system of four molecules that would have, at temperature T, a total translational energy of $4(\frac{3}{2}kT)$ erg. This amount averages out correctly to $\frac{3}{2}kT$ erg per molecule. Suppose, however, that we divide the $\frac{12}{2}kT$ erg into many packets, and then proceed to distribute these packets at random, "throwing them, one at a time, into the crowd." One can sense that the packets should be extremely small compared with the average energy. However, we'll make life simple by taking only a few packets, so we can keep track of our energy lottery. We'll take each packet to be $\frac{1}{2}kT$, so that only 12 packets must be accounted for. Even with these unrealistically large packets, there are many ways in which the four molecules might distribute the energy, but few enough for us to write down some of them in Table 8-1. We see that there are only four ways in which one molecule could get all 12 packets, but there are 12 ways in which one molecule could get 11 packets with some other molecule receiving the remaining one packet. The latter is a more probable distribution of the 12 packets. Still more probable is the distribution that gives one lucky molecule 10 packets, with two packets somehow doled out to the other three fellows—and so it continues. We can't say which distribution will occur; any one is as probable as any other, as far as our model goes.

Despite our lack of progress, let's persist and consider, now, Avogadro's number of molecules. This would really require too long a table to reproduce here, or, in fact, on all the paper in existence. However, we can approximate the situation by taking the N_0 molecules four at a time and,

Table 8-1   Some Ways in Which Twelve Energy Packets Can Be Distributed among Four Molecules

	Molecules		
A	B	C	D
12	0	0	0
0	12	0	0
0	0	12	0
0	0	0	12
	total = 4		
11	1	0	0
11	0	1	0
11	0	0	1
Plus 9 more in which B, C, or D has 11			
total = 12			
10	1	1	0
10	1	0	1
10	0	1	1
10	2	0	0
10	0	2	0
10	0	0	2
Plus 18 more in which B, C, or D has 10			
total = 24			

to keep track of energy, require that each group of four molecules divides up its "share" of $4(\frac{3}{2}kT)$ ergs of translational energy according to our model. If every set of four molecules averages $\frac{3}{2}kT$ erg per molecule, then surely the entire crowd of N_0 molecules will also average $\frac{3}{2}kT$ erg per molecule, or $\frac{3}{2}RT$ erg/mole.

Now statistics come to our rescue. If we have $\frac{1}{4} \cdot 6 \cdot 10^{23}$ quartets of molecules, we can be confident that every one of the arrangements shown in Table 8-1 will be represented many, many times. Since each single arrangement is as probable as any other arrangement, we can take the superposition of all of these arrangements to be an approximate prediction of the properties of the mole of gas. It will be approximate because we considered the molecules four at a time as a labor-saving device.

Now we can make a prediction: Our model predicts the distribution of molecular energies. If we reach into the gas and buttonhole a molecule at

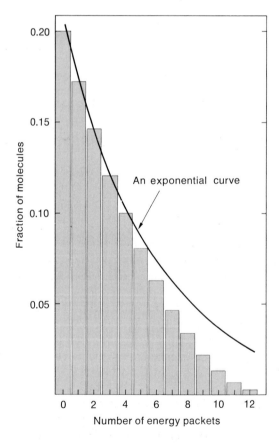

Figure 8-9 Molecular energy distribution corresponding to a random allotment of energy packets.

Table 8-2 Distribution of Molecular Energies: A Mole of Gas Considered
Four Molecules at a Time

Number of Energy "Packets" i	Number of Occurrences n_i	Fractional Number of Occurrences $n_i/\Sigma n_i$
0	364	0.200
1	312	0.171
2	264	0.145
3	220	0.121
4	180	0.099
5	144	0.079
6	112	0.062
7	84	0.046
8	60	0.032
9	40	0.022
10	24	0.013
11	12	0.006
12	4	0.002
	$\Sigma n_i = 1820$	

random, what is the probability that we will get one with no energy packets? Or with one energy packet? Or with any number from zero to twelve? The answer is obtained merely by counting the number of times each possible energy is found in a complete version of Table 8-1, since every arrangement will occur as often as any other. The result of this count is shown in Table 8-2. The most likely energy is predicted to be zero, and the least likely is 12—many arrangements give one or two molecules zero energy, whereas only the first four give one molecule as many as 12 packets of energy. We can perceive the results more easily in the graphical presentation in Figure 8-9. We see that the energy distribution resembles an exponential one.* In fact, if we consider six, eight, and ten molecules in the way we treated four in Table 8-1, the energy distributions approach more and more closely to an exponential curve.

Now we can return to the experimental data shown in Figure 8-8, but expressed in terms of molecular velocities. We see that there is no immediate resemblance between the predictions of our model and the experimental facts. The "theory" developed from our model does not agree with the facts of nature.

At this juncture in the development of a theory, a scientist has two options. He can cast about for an entirely different model ("theory"), or if he has sufficient confidence in the premises of the inadequate model, he can investigate minor changes in the model that still retain the desired

* Exponential curves are discussed more fully in Appendix h.

premises. In our problem, a useful theory was devised by the latter approach.

(c) A PATCH ON OUR MODEL OF RANDOM ENERGY DISTRIBUTION

The random distribution of small packets of energy leads to an exponential distribution. Could there be another factor at work that would convert our exponential curve of Figure 8-9 into the peaked and asymmetric experimental curve of Figure 8-8? We are looking for a factor that reduces the probability of low molecular energies and we would like it to be implied by the packet view of energy distribution.

The answer proves to be surprisingly simple—though conceptually quite significant. As a molecule gains "packets" of energy in our energy lottery, its velocity goes up in discrete jumps. However, the velocity of a molecule defines its momentum, since momentum equals mass times velocity. Hence, momentum must also come in "packets." To see how this affects our model, we must recall that for any given velocity, a molecule can move in a variety of directions. The variety of directions, which determines the probability of a given momentum, will be limited by the "packet" view we are assuming. This can be seen pictorially with a building-block model. In two dimensions, the accumulation of translational momentum might occur as in Figure 8-10. As we add the first momentum increment, we can "pack" it in any of several positions—eight are shown in part (b) of the figure. Adding another momentum increment increases the number of positions into which the second increment can be packed—there are more "directions" now—16 are shown in part (c). With three increments, the number of "directions" becomes 24. In two dimensions, the number of directions is proportional to the number of momentum increments.

Extending this same building-block model into three dimensions adds no new ideas. However, as shown in Figure 8-11, the number of directions goes up more rapidly as momentum increments are added. In fact, now *the number of velocity directions depends upon the square of the number of momentum increments.* Our exponential energy lottery model must be modified by a factor of v^2 to permit it to fit into the "momentum space" pictured in Figure 8-11. The probability $n(v)$ that a particle at velocity v will have a certain energy is a product, an exponential factor fixed by the random distribution of energy packets, but weighted by a v^2 multiplier that takes into account the available "momentum space." Hence,

$$n(v) = cv^2 \, e^{-[(1/2)mv^2/kT]} \tag{8-6}$$

This equation proves to be in exact agreement with the experimental velocity distribution shown in Figure 8-8. The proportionality constant c will not be

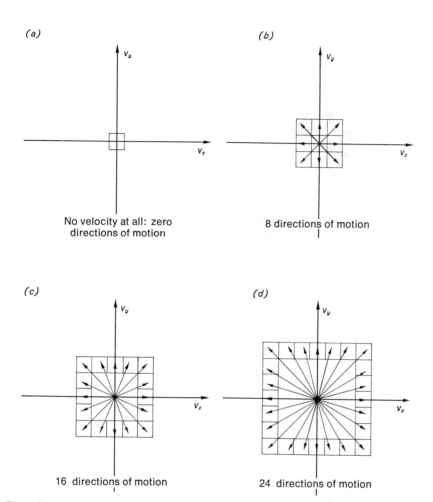

Figure 8-10 (a) A molecule with zero momentum; (b) a molecule with one increment of momentum; (c) a molecule with two increments of momentum; (d) a molecule with three increments of momentum.

discussed,* but it does not alter the conceptual content of our now-successful model.

(d) SUMMARY—RANDOMNESS IN A MONATOMIC GAS

Now our model fits the properties of a monatomic gas. This is a particularly simple system because there are only positional and translational move-

---

* Statistical mechanical arguments show that $c = 4\pi N (m/2\pi kT)^{3/2}$. This constant actually relates to the magnitude of the translational energy packets, and they prove to be small, indeed. Instead of $\frac{1}{2}kT$-sized packets, they must be 10^{-18} times smaller. Nevertheless, the model shows that translational energy must be quantized as required by quantum theory.

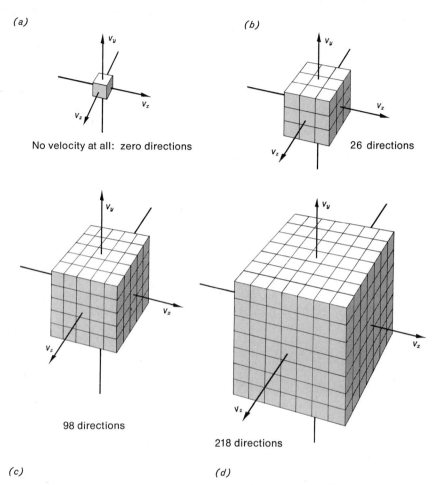

(a)

No velocity at all: zero directions

(b)

26 directions

98 directions

218 directions

(c)   *(d)*

Figure 8-11   (a) A molecule with zero momentum; (b) a molecule with one increment of momentum; (c) a molecule with two increments of momentum; (d) a molecule with three increments of momentum.

ments to worry about (no vibration or rotation). For such a simple system, energy distribution among the possible translational movements occurs in a random fashion, the ultimate distribution among a huge number of molecules being the most probable one. Again, as in the case of positional randomness, we find that a spontaneous change will take the direction that increases the randomness of energy distribution among the available repositories for energy—that is, among the available degrees of freedom.

Thus we have found an interesting parallel between positional and motional randomness. We could explain the spontaneous expansion of a gas if we assumed that all the positions in our three-dimensional positional space are equally probable and that the system tends to change to the most random (hence the most probable) positional arrangement. In the case of energy release, as in an exothermic chemical reaction, we found that the distribution of this energy could be explained in a similar way. Here we assumed that all "velocity positions" in a "three-dimensional velocity space" (as shown in Fig. 8-11) are equally probable and that the system tends to change to the most random (hence, the most probable) velocity arrangement. This most probable velocity arrangement also means the most probable energy arrangement.

This parallelism is usually expressed in terms of momentum rather than velocity. The momentum p of a particle is just its mass times its velocity:

$$p = mv \qquad\qquad (8\text{-}7)$$

Hence, our randomness picture for a monatomic gas can be expressed in a single expression if we consider a "six-dimensional space" for each atom whose six "Cartesian coordinates" are three positional coordinates x,y,z and three momentum coordinates p_x, p_y, p_z. In this space, called "phase space," there are six coordinates for each molecule. Now we can make a single statement about randomness: *A system tends to change spontaneously towards the most random (the most probable) arrangement in phase space.*

(e) RANDOMNESS IN A POLYATOMIC GAS

These same ideas of randomness apply equally well to polyatomic molecules, with their additional vibrational and rotational energy states. In fact, it is much easier to perceive, experimentally, the need for an energy packet approach to the distribution of energy into these types of motion. The reason is that the observed energy packets are large compared to kT at normal temperatures.

Consider vibration, first. It is discovered that the diatomic molecule I_2 can be vibrationally excited, but only in packets whose magnitude is about twice as large as $\frac{1}{2}kT$ when $T = 298°K$. That means that giving a molecule a vibrational packet is expensive, energy-wise. The situation is even worse for Cl_2. With its lighter atoms, Cl_2 vibrates more rapidly than I_2, and its energy packets are five times larger than $\frac{1}{2}kT$ at 298°K.

On reflection, though, we see that the development makes the coin distribution model even more directly applicable. A particle in a high energy state has a relatively small chance of being excited, for the same reason that finding all 12 energy packets cornered by one molecule was unlikely, in our energy lottery among four molecules. Concentration of the energy is not conducive to randomness. We can expect the exponential weighting factor again to determine the occupancy of these states. There is no analog to the directional v^2 factor that affected translation, so vibrational state occupancy merely follows the exponential curve on an intermittent

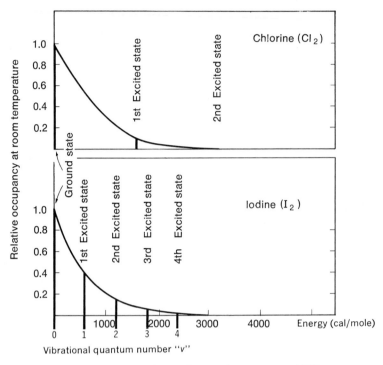

Figure 8-12 Relative occupancy of Cl_2 and I_2 vibrational states at 300°K.

basis. Figure 8-12 shows the relative occupancy at room temperature for the excited vibrational states of iodine, and of chlorine with a larger spacing. For iodine, the first excited state has about one third the occupancy of the ground (unexcited) state. In contrast, the vibrations of chlorine are barely excited at room temperature. This latter situation is more normal for simple molecules, but polyatomic molecules generally have a few vibrational movements with frequencies as low as those of iodine. Nevertheless, it is a reasonable approximation to assume that, at room temperature, the molecular vibrational states make little contribution to energy randomness.

The situation is quite different, and more interesting, for rotation. Experimentally, it is readily shown, again, that rotational energy is accepted by molecules in packets. The packets are small compared to kT at 298°K, but the size of a packet increases regularly as excitation of a given molecule increases. In other words, the more rotational energy a molecule possesses, the larger is the next energy packet it can possess. If we identify the number of energy packets already accepted by an integer number, J (the "quantum number"), the size of the next packet that molecule can acquire is proportional to $J(J + 1)$. Furthermore, there is a weighting factor analogous to the directional v^2 term (equation (8-6)) because each rotation axis can be oriented in $(2J + 1)$ directions. The product of the $(2J + 1)$ factor and the exponential gives an occupancy picture like that shown in Figure 8-13. This figure refers to the occupancy, at room temperature, of the HCl rotational levels. On a scale where the occupancy of the $J = 0$ state is called unity, the $J = 3$ state has an occupancy of four because the $2J + 1$ factor $(2J + 1 = 7)$ is more influential than the exponential part (the dashed curve in Fig. 8-13). Even states above $J = 6$ are more heavily populated than $J = 0$. Hence, the rotational states contribute very significantly to the energy randomness.*

The most obvious experimental evidence for the contribution of rotation and vibration to energy randomness is manifested in the heat capacity of the molecules. Table 8-3 gives the heat capacities of some gaseous substances and the breakdown into the various contributions. Plainly the rotational states play a heavy role in the energy lottery—as important as the translations—and for the heavier diatomic and for polyatomic molecules (which have low frequencies), the vibrational excitation also can be an important term.

---

* The vibrational and rotational results come directly from the application of quantum theory to the motions of atoms in molecules.

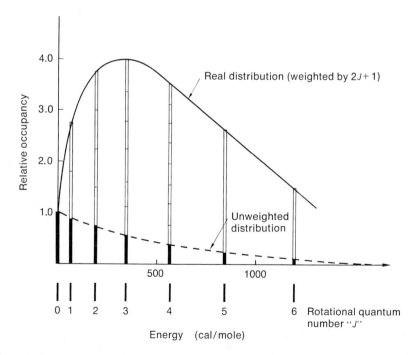

Figure 8-13 Relative occupancy of HCl rotational states at 300°K.

Table 8-3 *Heat Capacities at Constant Volume: Contributions by Rotation and Vibration* (cal/mole degK)

Molecule	$C_V(300°K)$	=	Transla-tional Part	+	Rotational Part	+	Vibrational Part
He	2.98	=	2.98	+	none	+	none
Ar	2.98	=	2.98	+	none	+	none
HCl	4.97	=	2.98	+	1.99	+	0.0004
Cl_2	6.12	=	2.98	+	1.99	+	1.15
I_2	6.82	=	2.98	+	1.99	+	1.85
CH_4	6.55	=	2.98	+	2.98	+	0.59
CCl_4	17.97	=	2.98	+	2.98	+	12.01
C_2H_6	10.60	=	2.98	+	2.98	+	4.64

In summary, when polyatomic molecules participate in a chemical change, any energy released must be distributed among the products, paying attention to all degrees of freedom available: translation, rotation, and vibration.* This distribution occurs with all possible energy states competing on a lottery basis, in accordance with the exponential energy–temperature dependence and with weighting factors appropriate to the degree of freedom. These considerations lead to the most random energy distribution—the one nature is searching for.

(f) ENERGY AND POSITION IN COMPETITION

Here are three exothermic chemical reactions:

$$NH_2 + NH_2 \longrightarrow N_2H_4 \qquad \text{heat released} = 56 \text{ kcal} \qquad (8\text{-}8)$$

$$NF_2 + NF_2 \longrightarrow N_2F_4 \qquad \text{heat released} = 20 \text{ kcal} \qquad (8\text{-}9)$$

$$NO_2 + NO_2 \longrightarrow N_2O_4 \qquad \text{heat released} = 14 \text{ kcal} \qquad (8\text{-}10)$$

In each reaction, the reactants have higher potential energy than the products. Each reaction is "rolling downhill" as far as energy is concerned (see Fig. 8-14), but now we know that randomness is what matters. Energy randomness is only one of the factors at work—positional randomness must also be considered.

We'll consider the energy randomness first. In each of these exothermic reactions, there is an increase in randomness connected with taking an amount of potential energy from each reactant and distributing it at random among translational, rotational, and vibrational degrees of freedom (i.e., releasing it as heat). Insofar as this effect is concerned, we are led to expect that all three reactions will occur.

---

* We have omitted reference to electronic excitation, but only a few molecules have such low energy excited electronic states that they, too, must be considered at other than very high temperatures.

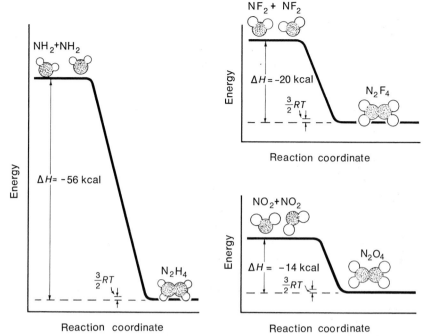

Figure 8-14   *Three similar exothermic reactions.*

Let's try the experiment. We place 0.40 mole of NH_2 into a 24.5-liter bulb (never mind how we do it) immersed in a constant temperature bath at 298°K. Instead of 0.40 atmosphere, the pressure is measured to be 0.20 atmosphere. This is exactly the pressure expected if all of the 0.40 mole of NH_2 molecules react according to (8-8) to form 0.20 mole of N_2H_4. Our expectation is realized. The very large increase in randomness associated with divvying up 56 kcal/mole among the translational, rotational, and vibrational degrees of freedom in the constant temperature bath causes the reaction to proceed with great, chaotic enthusiasm.

Now, let's do the same experiment with 0.40 mole of NF_2 molecules in a second 24.5-liter bulb and 0.40 mole of NO_2 in a third bulb of the same volume. In the second bulb, we measure a total pressure of 0.202 atmosphere, slightly above the 0.200 atmosphere pressure obtained with N_2H_4. Apparently the reaction is not quite complete! In the third bulb, the effect is even more significant. Here, the pressure is found to be 0.26 atmosphere, the pressure appropriate to the formation of only about 65 percent of the possible number of N_2O_4 molecules! Apparently, only about two thirds of the NO_2 molecules decided to "roll downhill" and the other one third refused to join them!

Our first thought might be to place the blame on the NO_2—perhaps we have the wrong reaction. Not so, however. Spectroscopic studies (the absorption of light by the gas mixture) clearly corroborate our interpretation of the pressure. The nitrogen dioxide *does* contain about equal numbers of the NO_2 and of N_2O_4 molecules; it reached two thirds completion and stopped rolling downhill! Hence, positional randomness must be at work.

To verify this interpretation, let's repeat all three experiments at 400°K. At this higher temperature, the NH_2 molecules again react almost completely to give hydrazine, N_2H_4, and the released energy is divided up into translational, rotational and vibrational energy packets. In the NF_2 bulb, however, we find that at 400°K only 60 percent of the N_2F_4 molecules have been formed! Now 40 percent of the NF_2 molecules refuse to roll downhill! In the NO_2 bulb, the pressure is very close to 0.40 atmosphere—almost no reaction occurs at all!

We see from these results that we cannot ignore positional randomness and that its importance increases as temperature goes up. Just as when a bachelor decides to take the big step, when two NO_2 molecules join together in the holy state of chemical bondedness, the loss of freedom of movement is considerable. Whereas before bonding, the two molecules could behave completely independently, both from a position and an energy point of view, as an N_2O_4 molecule they now must share the same energy bank account, live closely together in the same geographical location, go to the same molecular parties, and so on.

It is intrinsically more random to have many molecules rather than a few (for a given number of atoms). It is also intrinsically more random to have several kinds of molecules rather than one. These are the factors that caused the exothermic reactions (8-9) and (8-10) *not* to go to completion. Instead, the atoms lodge in the best (most probable) compromise, *taking all sorts of randomness into account*. This always leads to an equilibrium situation in which there is something of everything present, the relative amounts being determined by the specific energy and positional randomness contributions.

(g) WEIGHTING THE ENERGY RANDOMNESS

As we consider reactions (8-8), (8-9), and (8-10), our experimental results reveal the significant effect of temperature. At room temperature, reaction (8-8), with its 56 kcal/mole heat of reaction, caused almost complete formation of N_2H_4, whereas reaction (8-9) with its 20 kcal exothermicity, was incomplete. However, at 400°K, even reaction (8-9) was only 60 percent complete and reaction (8-10) barely took place at all. The relative importance of energy randomness compared with positional randomness decreases considerably as the temperature rises. The explanation can be based upon the "expense" of energy randomness. For example, if the average molecular translational energy is $\frac{3}{2}kT$, then the average translational energy per mole increases as T increases. In effect, our 56 kcal/mole become less and less effective as a means of increasing motional randomness as the average motional energy per mole becomes comparable to the heat of reaction. To be explicit, $\frac{3}{2}kT$ erg per molecule corresponds to $\frac{3}{2}RT$ kcal per mole, and we can make the following comparisons:

at $T = 300°K$ (room temperature) $\frac{3}{2}RT = 0.9$ kcal

at $T = 400°K$ $\frac{3}{2}RT = 1.2$ kcal

Table 8-4 shows the significance of these comparisons to reactions (8-8), (8-9), and (8-10). At both 300°K and 400°K, the 56 kcal of heat released in reaction (8-8) imply a very large number of translational energy packets if each is considered to have an average of $\frac{3}{2}RT$ kcal per mole. For the other two reactions, however, the magnitude of $\frac{3}{2}RT$ is not so very large compared with ΔH. In fact, the ratio $\Delta H/\frac{3}{2}RT$ for reaction (8-10) at 300°K has nearly the same value as for reaction (8-9) at the higher temperature 400°K. This correlates with the fact that NF_2 reacts at 400°K to about the same extent as NO_2 at 300°K. This emphasizes what was said earlier: As temperature rises, the energy randomness becomes less and less important compared with positional randomness. The ratio $\Delta H/RT$ measures the importance of energy randomness. If the temperature is raised to the point that ΔH is small compared with RT, energy randomness will be quite unimportant. Then, dividing a few kilocalories up among a group of energy-rich molecules affects their energy randomness about as much as dividing ten dollars among five millionaires affects their wealth.

We see that the randomness to be gained by converting the potential energy change (ΔH) into energy of motion is measured by the relative magnitude of ΔH and the average energy per degree of freedom, which is determined by the temperature. At low temperatures, the energy randomness will be quite important and equilibrium will favor the reaction in the exothermic direction (i.e., toward N_2H_4, N_2F_4, and N_2O_4 in our three examples). At high temperatures, energy randomness becomes less and less important, as $\Delta H/T$ becomes small. Then positional randomness will become dominant and the reaction may be favored in the endothermic direction (i.e., toward NH_2, NF_2, and NO_2). We shall expect to find $\Delta H/T$ an indicator of the temperature range at which this will begin to occur. In fact, $\Delta H/T$ is a measure of the energy randomness introduced by the heat of a reaction.

Table 8-4 *The Ratio of Heat of Reaction to Average Thermal Energy*

	Reaction	ΔH (kcal/mole)	$-\Delta H/\frac{3}{2}RT$ 300°K	400°K
(8-8)	$2\ NH_2(g) = N_2H_4(g)$	-56	62	50
(8-9)	$2\ NF_2(g) = N_2F_4(g)$	-20	22	17
(8-10)	$2\ NO_2(g) = N_2O_4(g)$	-14	16	12

8-3   Randomness, energy change, and equilibrium

We have before us now all of the ideas needed to understand the factors that motivate spontaneous change. On a macroscopic scale, events move in the direction of increasing randomness. To be sure, the concept of randomness is more sophisticated than the simple positional randomness of our ping-pong ball gambling game shown in Figure 8-3. Energy must be partitioned, as well. Energy lodged in chaotic translational, vibrational, and rotational movements constitutes heat, and in general, transfer of the potential energy of chemical bonds into heat increases the randomness. Thus, this energy randomization permits reactions to proceed in the exothermic direction. Positional randomness and molecular complexity also play their roles, however. Several molecular fragments can assume more positional arrangements than a single molecular cluster, even if the cluster has lower potential energy. The balance between all forms of randomness, including both reactants and products, determines the state of equilibrium.

We need to make all this quantitative. If randomness can be placed on a quantitative basis, we can predict reactions that might occur spontaneously —we can point the way toward equilibrium. We will do this in a rather laborious, but revealing, derivation. The treatment does not adhere to the historical development of the subject—that would take us afield from chemistry, toward steam engines, and it would require quite a bit of mathematics. Again, we will favor the logical rather than the chronological and the intuitively meaningful rather than the mathematically elegant. With apologies to Carnot and Clausius, we'll show in Chapter Nine how economy and exercise lead to entropy and how entropy leads to the Second Law of Thermodynamics.

Problems

1.   Suppose the gambling game shown in Figure 8-3 had only five depressions.
(a) Write down all the possible outcomes of throwing first a red ping-pong ball into the box and then a green ping-pong ball.
(b) Suppose that for a one-dollar bet you could choose in advance one of four possibilities, A, B, C, and D, and a correct choice wins the amount indicated.

A.	red, an odd number; green, an even number	win $2.75
B.	red, even; green, odd	win $3.00
C.	red, odd; green, odd	win $3.25 ✔
D.	red, even; green, even	win $8.00

After betting 80 times, which choice would most favor the player and which would most favor the house (assume statistical behavior)?

2.   Suppose that you discovered, by watching many throws, that the red ball does not behave in a random fashion because it does not fit in depression #3. Now which bet most favors the player and which most favors the house after $80.00 have been invested?

3. Calculate the number of blocks needed to expand Figure 8-11(d) to four increments of momentum. For example, we can add the number of blocks to build out
each of the six sides at its present dimension, then add a line of blocks along each
of the 12 edges, and, finally, fill in the eight corners.

4. For each value of n, the number of increments of momentum ($n = 1, 2, 3$),
plot the number of directions shown in Figure 8-11 versus n^2. Draw the best straight
line possible through the points and extrapolate to $n = 4$. Compare to the result in
Problem 3.

5. There are 15 ways to distribute four packets of energy ϵ among three molecules
A′, A″, and A‴. For example, A′ could have 4ϵ, then A″ and A‴ each would have
zero.
(a) Write down all 15 ways.
(b) If the three A molecules were "observed" many times, what would be the
percentage of times that A′ would have 4ϵ? The percentage of times A′ would have
3ϵ? 2ϵ? 1ϵ? 0ϵ?
(c) What is the average energy of any A molecule, based on your answer to (b)?

6. Four identical gaseous atoms A′, A″, A‴, and A‴′ must share two packets of
energy.
(a) Write down all of the ways the two packets could be distributed.
(b) If the molecules are "observed" on many separate occasions, so that all
arrangements are represented, what percentage of the observations will find
molecule A′ with both of the packets? What percentage with one packet? With no
packets?
(c) What is the average energy of any A molecule?

7. Suppose an A molecule in Problem 5 could turn into a molecule B of higher
energy in a reaction endothermic by 3ϵ.

$$A = B \qquad \Delta H = +3\epsilon$$

(a) Write down all the *new* ways that the 4ϵ of Problem 5 could be distributed.
(b) What is the probability that at any instant there would be one B molecule and
two A's, rather than three A's.
(c) What is the equilibrium constant for the endothermic reaction?

8. Suppose the molecule A‴′ in Problem 6 is removed and a B molecule is substituted, B being a molecule that can accept two packets or none, but it *cannot*
accept only one packet.
(a) Now how many ways can molecule A′ end up with both of the packets? With
one of the packets? With neither of the packets?
(b) What is the average energy of any A molecule in this mixture with the B
molecule?

9. Repeat Problem 7 for a lower temperature at which there are only three packets
of energy to share among the molecules. Which direction does Le Chatelier's
Principle (see Section 1-4(d)) predict that the equilibrium constant in Problem 7
would change as temperature is lowered?

10. Suppose that, in Problem 6, any two A atoms could react to form an A_2 molecule
in a reaction that is exothermic by two packets of energy.

$$A + A = A_2 \qquad \Delta H = -2\epsilon$$

Thus, if this system is isolated thermally, the four A atoms have 2 packets; if one A_2 molecule forms, the mixture has 2 additional packets, a total of 4; and if two A_2 molecules form, the mixture has a total of 6.

(a) Write down all of the ways two A_2 molecules, $A'A''$ and $A'''A''''$, can distribute the 6 packets they would share.

(b) Write down all of the distinct ways the four atoms can combine to form two A_2 molecules.

(c) Write down all of the ways the 4 packets could be shared if one A_2 molecule forms, leaving two A atoms uncombined.

(d) Write down all of the distinct ways four atoms can form one A_2 molecule and two free A atoms.

(e) What percentage of many observations of the mixture will find two A_2 molecules? What percentage will find one A_2 molecule? What will be the average number of A_2 molecules found in many observations? What will be the average number of A atoms found?

(f) What is the numerical magnitude of the equilibrium constant for the reaction, $K = (A_2)/(A)^2$?

11.   Repeat Problem 10 for a lower temperature at which the four atoms must share only one packet unless reaction occurs. Did the equilibrium shift in the direction dictated by Le Chatelier's Principle as the temperature was lowered?

12.   Vibrational excitation of iodine, I_2, occurs in packets as integer multiples of a "resonant vibrational frequency" with easily measured energy packets equivalent to 610 cal/mole. Thus $E_v = v \cdot 610$ cal/mole where $v = 0, 1, 2, 3 \cdots$. The integer v is called a "quantum number." The "random distribution of energy packets" model predicts energy level population proportional to $e^{-E_v/RT}$.

(a) Calculate the ratio of the population in level $v = 1$ to that of $v = 0$ at $T = 298°K$. (Compare to Fig. 8-12.)

(b) Repeat (a) for $v = 2$.

(c) Repeat (a) for both $v = 1$ and $v = 2$, but at $T = 596°K$.

13.   Repeat Problem 12 for Cl_2, whose "resonant vibrational energy" packets are equivalent to 1590 cal/mole and $E_v = v \cdot 1590$, $v = 0, 1, 2, 3 \cdots$.

14.   The rotational energy levels of a molecule depend upon an integer quantum number J, $E_J = J(J + 1)B$. The factor B is inversely proportional to the molecular moment of inertia. In addition, there is a directional randomness proportional to $(2J + 1)$. Therefore a given level has occupancy proportional to the product $(2J + 1) \cdot e^{-E_J/RT}$.

(a) Calculate the ratio at 298°K of the population of the $J = 4$ state for HCl to the population of the $J = 0$ state; B(HCl) = 30.3 cal/mole.

(b) Repeat (a) for deuterium chloride, DCl, which has a larger moment of inertia than HCl and, hence, a smaller value of B, B(DCl) = 15.6 cal/mole. Calculate the ratio for $J = 1, 2, 3, 4, 5$, and 6 and then plot the occupancies against rotational energy (in cal/mole). Compare to Figure 8-13.

nine   maximum work, entropy, and spontaneity

In Chapter Six, we saw that spontaneous processes can perform work. The amount of work depends upon external conditions and it can approach a maximum quantity that is characteristic of the process. The difference between this theoretical maximum work and the actual amount performed in a real change will prove to be important to us. This difference is a quantitative measure of the "driving force" or "motivation" for the spontaneous change. It will lead us logically to an understanding of how randomness determines the state of equilibrium. We will begin the chapter by exploring first the work performed in a simple gas expansion, then that done in the discharge of an electrochemical cell. We will conclude it by seeing how chemical thermodynamics is helping us solve the air pollution problem.

9-1   Expansion of a gas

Figure 9-1 shows the same changes of state that are pictured in Figure 8-1, but carried out in a new apparatus. Initially, N moles of chlorine gas are restrained in a 1-liter volume V_1 by the piston in the cylinder. The piston resists the gas pressure $P_1 = 24$ torr on its inner surface because of the weight M_1 on the suspension pan. This weight, M_1, is selected with just enough mass to hold the piston as shown in state I.

(a) EXPANSION INTO A VACUUM

In this apparatus we can easily bring about the expansion into a 3-liter volume (to give a final volume of $4V_1$). If the weight M_1 is removed from the suspension pan, the piston is no longer restrained, and the gas pressure on the left-hand side of the piston pushes it rapidly to the end of the cylinder. There it is stopped by the wall, with a final volume of $V_2 = 4V_1$ and a final pressure of $P_2 = 6$ torr. Since the chlorine gas is at a low pressure, it behaves like an ideal gas, so there is no temperature effect. The entire expansion occurs at a constant temperature T.

The advantage of this experimental setup is that it is easy to see how much work is done as the gas goes from state I (N, P_1, V_1, T) to state II

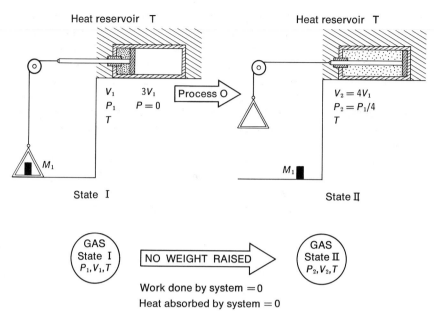

Figure 9-1 *Expansion of a gas into a vacuum: no work.*

(N, P_2, V_2, T). In the process just considered, the mass M_1 was removed from the suspension pan, and except for the negligible mass of the suspension pan, no mass was lifted. The change occurred like that of the spontaneous and unrestrained flow of a river to the sea—no work was done.

(b) LET'S GET SOME WORK

Figure 9-2 shows another process by which this change could be carried out. Instead of merely removing the mass M_1 from the pan, it is replaced by a mass $\frac{1}{4}M_1$. The restraining force on the piston is now less than the gas pressure on the piston, so it moves to the right. The weight $\frac{1}{4}M_1$ is lifted and the gas continues to expand until the gas pressure is reduced to $\frac{1}{4}P_1$. At this pressure, balance is restored. Again, the gas pushing to the right exerts a force on the piston just equal to the force pulling the piston to the left due to the smaller mass on the pan.

We see that, as far as the gas is concerned, the change shown in Figure 9-2 is the same as that in Figure 9-1. The gas expanded from the same initial state I (N, P_1, V_1, T) to the same final state II (N, P_2, V_2, T). However, work was done as the weight was lifted. The gas performed useful work. We can call this work w_1 for the moment. The mass $\frac{1}{4}M_1$ is at an altitude h, and its potential energy is increased by w_1, where

$$w_1 = (\tfrac{1}{4}M_1)gh \tag{9-1}$$

g = gravitational constant

There is, however, another way to express the work done. In the final state II, the balanced position of the piston shows that the weight $\frac{1}{4}M_1$ exerts just the same force on the piston as the pressure $P_2 = \frac{1}{4}P_1 = 6$ torr. As far as the piston is concerned, it might as well have the same pressure on both sides and no weight on the pan. Lifting the weight $\frac{1}{4}M_1$ is the same amount of work as if the piston had pushed, during its movement, against a constant pressure of $\frac{1}{4}P_1$. Hence, the work done in the process can be expressed in terms that relate directly to the actual change in state: the pressure and the volume.

If the work done is equal to the work of pushing back a constant pressure atmosphere, it is compression work. This we know about. In Chapter Seven, this work was shown to be merely $P\Delta V$ (equation (7-16)). It is work performed *by the system*, that is, by the gas pushing on the piston; we'll call it w_1.

$$w_1 = P\Delta V = \tfrac{1}{4}P_1(V_2 - V_1) = \tfrac{1}{4}P_1(4V_1 - V_1)$$

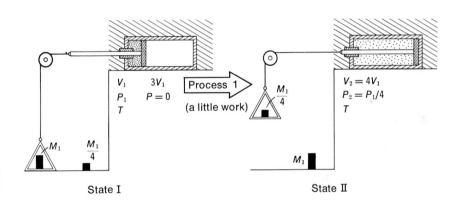

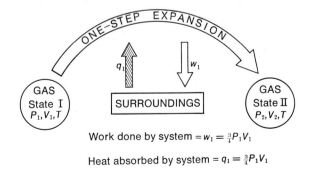

Work done by system $= w_1 = \tfrac{3}{4}P_1V_1$

Heat absorbed by system $= q_1 = \tfrac{3}{4}P_1V_1$

Figure 9-2 *Expansion of a gas: process 1, a little work.*

Hence,

$$w_1 = \tfrac{3}{4}(P_1V_1) \tag{9-2}$$

Compared with the process shown in Figure 9-1, we are much better off. In Figure 9-1 (process 0), no work was performed. In Figure 9-2 (process 1), a weight was lifted and work equivalent to $\tfrac{3}{4}(P_1V_1)$ was performed. The gas spontaneously made the same change of state, but in process 1 we got something out of it: The weight now has potential energy that can be released any time we wish to have a spot of energy to break a stone, cook an egg, or run an electric razor.

(c) STILL MORE WORK, PLEASE!

Even more work can be obtained in return for a little more attention. Consider process 2, shown in Figure 9-3. This time, we initially substitute for M_1 the mass $\tfrac{1}{2}M_1$. The gas pressure on the piston moves it to the right until balance is restored. This occurs when the pressure is halved and the volume doubled. An increment of work w_{2a} has been done, and we are in state Ia:

$$w_{2a} = P\Delta V = \tfrac{1}{2}P_1(V_{Ia} - V_I) = \tfrac{1}{2}P_1(2V_1 - V_1) = \tfrac{1}{2}P_1V_1 \tag{9-3}$$

At this juncture, the smaller weight $\tfrac{1}{4}M_1$ is substituted for $\tfrac{1}{2}M_1$, and the expansion proceeds the rest of the way to state II. Another increment of work w_{2b} has been done:

$$w_{2b} = P\Delta V = \tfrac{1}{4}P_1(V_2 - V_{Ia}) = \tfrac{1}{4}P_1(4V_1 - 2V_1) = \tfrac{1}{2}P_1V_1 \tag{9-4}$$

Now the total work done w_2 is simply the sum of w_{2a} and w_{2b}.

$$w_2 = w_{2a} + w_{2b} = \tfrac{1}{2}P_1V_1 + \tfrac{1}{2}P_1V_1$$
$$= P_1V_1 \tag{9-5}$$

In process 1 the system performed an amount of work equal to $\tfrac{3}{4}P_1V_1$. In process 2 the gas underwent the same expansion, but the work it performed went up to P_1V_1.

Exercise 9-1. Show that the work done by the gas is $1.08(P_1V_1)$ if the expansion is carried out in three steps, first lifting a weight $\tfrac{2}{3}M_1$, which is then replaced by $\tfrac{1}{2}M_1$, which is then replaced, finally, by $\tfrac{1}{4}M_1$.

(d) MAXIMUM WORK

Clearly this process could be modified further to gain still more work. Even more steps could be included as the expansion proceeds. The patience

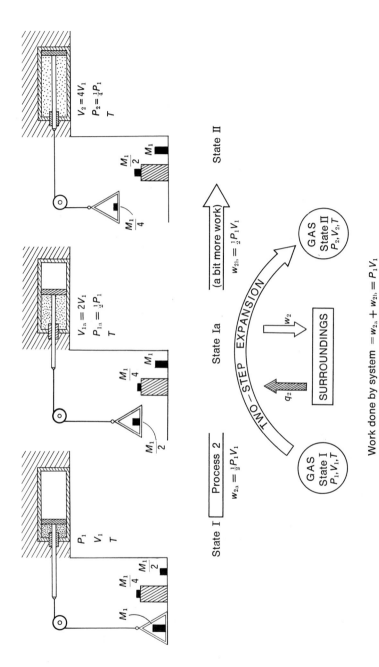

P_1
V_1
T

M_1

$\dfrac{M_1}{4}$ $\dfrac{M_1}{2}$

State I

$V_{1a} = 2V_1$
$P_{1a} = \frac{1}{2}P_1$
T

M_1

$\dfrac{M_1}{4}$

$\dfrac{M_1}{2}$

State Ia

$V_2 = 4V_1$
$P_2 = \frac{1}{4}P_1$
T

M_1

$\dfrac{M_1}{2}$

$\dfrac{M_1}{4}$

State II

| State I | Process 2 | State Ia | (a bit more work) | State II |
| $w_{2a} = \frac{1}{2}P_1V_1$ | | | $w_{2b} = \frac{1}{2}P_1V_1$ | |

TWO-STEP EXPANSION

GAS State I P_1, V_1, T

q_2

w_2

SURROUNDINGS

GAS State II P_2, V_2, T

Work done by system $= w_{2a} + w_{2b} = P_1V_1$

Heat absorbed by system $= q_2 = P_1V_1$

Figure 9-3 Expansion of a gas: process 2, more work.

required for this step-wise process is repaid in additional work, as can be seen in a graphical presentation of the work done. Each step in each process is a constant pressure step and contributes an increment of work $P\Delta V$. This is readily seen to be the area in a plot of P against V, as shown in Figure 9-4. The work done in each process is represented by the shaded area. The shaded area in process 4 is plainly larger than that in process 2; it includes all of the shaded area in process 2 and more. More work is done.

Obviously we obtain more and more work as weight is taken off the pan in smaller and smaller increments. If maximum work is our aim (and why shouldn't it be—remember the creative engineer and his hydroelectric-dam project), we should carry the incremental process right to the limit of extremely small mass increments, so that at every instant the expanding gas lifts a mass that is only very slightly under that needed for balance. Such small mass increments are called "differential masses" and are abbreviated dM. The implication is that every expansion occurs at a constant pressure only dP (differential of pressure) below the equilibrium or balance pressure. Of course, the volume then changes in each step by only a differential amount dV. As the differential gets smaller and smaller, the work increases, as shown in Figure 9-5, approaching, finally, the smooth curve on the right. In this limit, the summation of incremental contributions becomes a continuous process. Mathematicians have a special name, the integral, for such an infinite sum of infinitesimally small contributions. The maximum work is given by this limiting sum.

$$\text{work} = \int_{I}^{II} P dV \tag{9-6}$$

This esoteric equation is read: "Work equals the integral of the product PdV from state I to state II." The word "integral" means the limiting sum

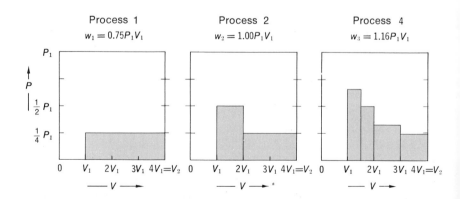

Figure 9-4 *Graphical presentation of work: $P\Delta V$.*

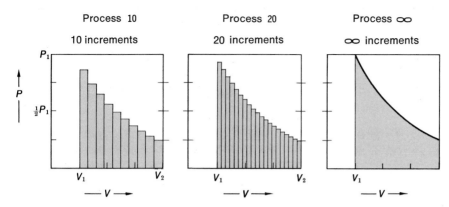

Figure 9-5 Expansion of a gas: approach to maximum work.

when the change from state I to state II is accomplished in smaller and smaller incremental steps.*

This limiting sum can be expressed analytically. Since the pressure is always kept extremely close to the pressure of the gas, we can apply the ideal gas law throughout the change.

$$P = \frac{nRT}{V} \tag{9-7}$$

where n is the number of moles of gas, R is the gas constant, and T is the absolute temperature.

Substituting (9-7) into (9-6) we obtain

$$\text{work} = \int_I^{II} \frac{nRT}{V} \, dV \tag{9-8}$$

The summation in (9-8) can be simplified since the conditions do not permit either the number of moles or the temperature to change. If these factors (and R) do not change, they do not affect the integral except as a constant factor. Hence, (9-8) can be rewritten as

$$\text{work} = nRT \int_I^{II} \frac{dV}{V} \tag{9-9}$$

In any elementary calculus class, this integral is one of the earliest considered. Its magnitude is relatively easy to calculate, since it is just a difference between two numbers fixed by the initial and final volumes V_1 and V_2. The

* Mathematical integration is discussed in Appendix h.

difference in question is the natural logarithm of the volume in the final state minus the natural logarithm of the volume in the initial state.* Thus,

$$\text{work} = nRT \int_{I}^{II} \frac{dV}{V} = nRT \int_{V_1}^{V_2} \frac{dV}{V}$$

$$= nRT(\log_e V_2 - \log_e V_1) \tag{9-10}$$

Since the difference between the logarithms of two numbers equals the logarithm of their quotient,

$$\text{work} = nRT \log_e \frac{V_2}{V_1} \tag{9-11}$$

Logarithms to the base e are less convenient than logarithms to the base 10, so for computational purposes, (9-11) is usually written as

$$\text{work} = 2.303nRT \log_{10} \frac{V_2}{V_1} \tag{9-12}$$

Substituting $V_2 = 4V_1$, (9-12) becomes

$$\text{work} = w_\infty = 2.303nRT \log_{10} \frac{4V_1}{V_1}$$

$$= 2.303nRT \log_{10} 4$$

$$= 2.303 \cdot 0.602nRT$$

$$= 1.40nRT \tag{9-13}$$

To compare this result with our earlier processes, let us use the perfect gas relation once more, $P_1V_1 = nRT$. Thus,

$$w_\infty = \text{maximum work} = 1.40P_1V_1 \tag{9-14}$$

Now we can sit back and assess the work done by the expanding gas in the various processes. Figure 9-6 shows a plot of the work as a function of the number of steps in the expansion. Both Figures 9-5 and 9-6 show that the process in which the pressure on the piston is almost balanced throughout the expansion gives the maximum work. That is, *maximum work is obtained when the system remains essentially at equilibrium throughout the process.*

---

* In Appendix h, the natural logarithm and the quantity e are defined, and the integral $\int dV/V$ is further discussed.

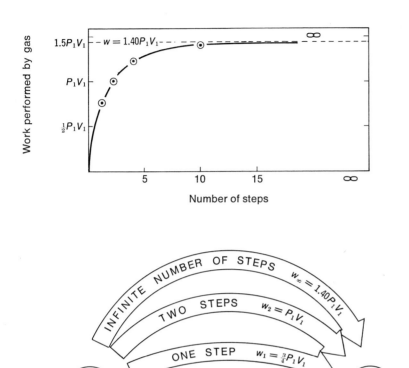

Figure 9-6 Work performed in a gas expansion.

It's too bad this last process is so tedious. Unfortunately, it takes an infinite time to expand a gas in an infinite number of steps.

(e) WHO PAID THE BILL?

In the several processes we have considered, all for the same change of state, only the work was discussed. It takes energy to do work—so who paid the energy bill? The answer is found in the First Law of Thermodynamics, the conservation law.

Our system is a quantity of gas in the state of a perfect gas; that is, a gas in which the molecules are so far apart that they do not interact noticeably.

Such a gas expands at constant temperature without any energy change. Hence, the energy change ΔE during the change of state is zero:

$$\Delta E = E_2 - E_1 = 0 \tag{9-15}$$

But also, the energy change must always equal the heat absorbed by the system, q, minus the work done by the system, w. That is,

$$\Delta E = q - w \tag{9-16}$$

and since $\Delta E = 0$,

$$q = w \tag{9-17}$$

So, for each of these processes, it was necessary to supply the gas with heat from a thermal reservoir.

Now we see who paid the bill. The surroundings (the thermal reservoir) had to give energy to the system so it could perform the appropriate amount of work. In each process, the amount of heat withdrawn equalled the amount of work performed. But we mustn't forget who was the entrepreneur. The motivation for the process was the tendency of a gas to expand, seeking to increase its positional randomness.

(f) NOW LET'S GET THINGS BACK

Having studied this expansion, it is now time to restore the gas to its initial conditions. Quite apart from tidiness, there is a lesson to be learned.

To restore our system to state I, the gas must be compressed back into volume V_1. This change does not occur spontaneously—muscle must be applied to the piston. The weight on the pan must exceed the gas pressure in the cylinder to move the piston. Once again there are many ways we could do this. To get it over quickly, we could lift a weight up onto the pan equivalent to the final pressure P_1. That weight would surely pull the piston back, thereby recompressing the gas. The volume would decrease until pressure balance is restored, and this pressure balance would be reached when the volume is again V_1.

There are two steps involved. First, *we do some work w' on the weight M_1* as we lift it. Then, as the pan lowers, compressing the gas, *the weight performs this same amount of work w' on the system*. We'll discuss the restoration of the gas in terms of w', the work that *we perform on* the system (we and the weight are part of the surroundings). However, we can always re-express the work transfer in terms of w, the work done *by* the system, since $w = -w'$. In other words, *any* communication of work between a system and its surroundings can be expressed equally well in terms of w, the work done by the system, or in terms of w', the work done by the surroundings. These must always be equal in magnitude and opposite in sign: $w = -w'$.

work we performed to $=$ work performed to lift $= w_1'$
recompress the gas M_1 up onto the pan

$$w_1' = P_1(4V_1 - V_1) = 3P_1V_1 \tag{9-18}$$

What's this? We have to perform an amount of work $3P_1V_1$ to go from state II to state I, when the gas would perform, at the very most, only $1.4P_1V_1$ of work as it went from state I to state II! Ouch, that was expensive.

(g) THERE MUST BE A CHEAPER WAY!

Indeed there is. Our technique in carrying out the expansion gives us the clue. Suppose the compression is also carried out in steps. Instead of placing the mass M_1 on the pan at the top of its rise, we could use $\frac{1}{2}M_1$. This mass would compress the gas only until its pressure was doubled, when the volume would be $2V_1$. Then the full weight M_1 could be used to finish the job. The work we have done is now the sum of two increments:

$$w_2' = \tfrac{1}{2}P_1(4V_1 - 2V_1) + P_1(2V_1 - V_1) = 2P_1V_1$$

These processes are shown in Figure 9-7. Plainly we are in the same sort of game as that pictured in Figures 9-1–9-3, except that now *we* are doing all the work. If we are to recompress the gas with still less effort, we shall again have to carry out the change in many small steps. Leaping to the obvious conclusion, the more steps employed, the less work we need do. The minimum work w_∞' will be expended when an infinite number of steps is employed, always keeping the mass on the pan no more than an infinitesimal amount above the gas pressure in the cylinder.

Exercise 9-2. Calculate the work done on the gas if the compression is carried out in four steps, first compressing with a weight $\frac{1}{3}M_1$, then $\frac{1}{2}M_1$, then $\frac{2}{3}M_1$, and finally, M_1.

If we return to a consideration of w_∞, work done by the system, it can be expressed again as an integral, except that now the volume is $4V_1$ in the initial state and V_1 in the final state.

$$w_\infty = \int_{V=4V_1}^{V=V_1} P\,dV$$

$$= nRT[\log_e V_1 - \log_e (4V_1)] \tag{9-19}$$

$$= -nRT[\log_e (4V_1) - \log_e V_1] \tag{9-20}$$

$$= -2.303nRT \log_{10} \frac{4V_1}{V_1}$$

$$= -1.40nRT = -1.40P_1V_1 \tag{9-21}$$

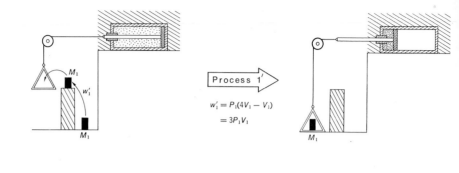

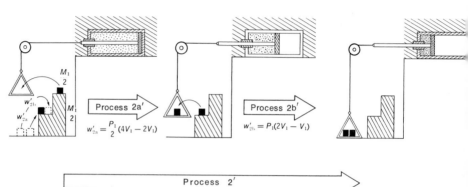

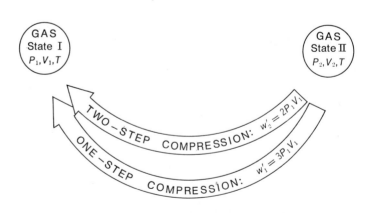

Figure 9-7 *Getting the system restored with a minimum amount of effort.*

Expression (9-21) is the same as (9-13) and (9-14), except that the algebraic sign is minus. This means negative work is done by the gas; that is, the *surroundings* must perform work on the gas. Therefore, our final result is as follows:

$$\text{minimum work done on the gas during recompression} = w'_\infty$$

$$= -w_\infty$$

$$= 1.40 P_1 V_1 \tag{9-22}$$

Unfortunately, history repeats itself. To perform a minimum amount of work during the compression, an infinite number of steps are needed and the process will take forever (literally!).

(h) NOW WHO PAID THE BILL?

Just as before, the energy of the perfect gas is constant because the compression occurs at constant temperature. By the First Law of Thermodynamics,

$$\Delta E = q - w = 0$$

so

$$q = w$$

As always, w is the work done by the system, which is given by (9-21)

$$q = w_\infty = -1.40 P_1 V_1 \tag{9-23}$$

Equation (9-23) says that negative heat is absorbed by the gas. In other words, the gas releases heat into the thermal reservoir. If one of the less efficient processes is used (as in Fig. 9-7), the heat transferred into the reservoir is even greater. Who pays the bill? Whoever does the work of compression! Hey, that's us!

Table 9-1 shows a box score of the energy-moving: the transfer of heat between the reservoir and the gas, as well as the athletics with the weights. In each case, the heat absorbed and the work done by the surroundings are just the negative of the corresponding values for the system. With this table, we can see at a glance the net effect of carrying out the expansion and then the compression by any combination of processes. For example, suppose the expansion is carried out in a single step (process 1), producing work $\frac{3}{4}PV$, and *also* the compression in a single step (process 1'), requiring work $3PV$. Table 9-1 shows that the net work done by the gas is $\frac{3}{4}PV - 3PV = -\frac{9}{4}PV$, and the heat absorbed by the gas is

$$\frac{3}{4}PV - 3PV = -\frac{9}{4}PV$$

Table 9-1   Summary View of Energy Transferred In and Out of the Gas

Number of Steps	w, Work Done by the Gas	q, Heat Absorbed by the Gas
Expansion		
0 (no weight)	0	0
1	$0.75PV$	$0.75PV$
2	PV	PV
4	$1.16PV$	$1.16PV$
∞	$1.40PV$	$1.40PV$
Compression	$(w = -w')$	$(q = -q')$
1'	$-3PV$	$-3PV$
2'	$-2PV$	$-2PV$
4'	$-1.67PV$	$-1.67PV$
∞'	$-1.40PV$	$-1.40PV$

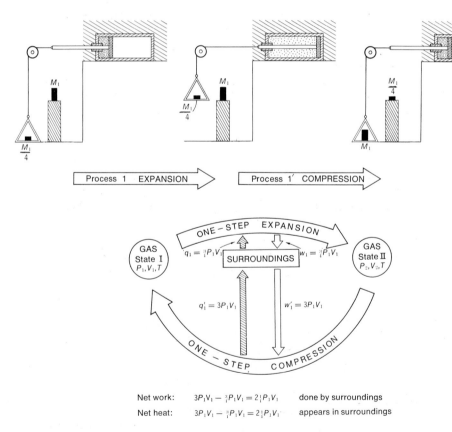

Net work: $3P_1V_1 - \frac{3}{4}P_1V_1 = 2\frac{1}{4}P_1V_1$ done by surroundings

Net heat: $3P_1V_1 - \frac{3}{4}P_1V_1 = 2\frac{1}{4}P_1V_1$ appears in surroundings

Figure 9-8   A possible expansion–compression cycle.

The negative signs mean that work was performed *on* the gas and heat was transferred *into* the surroundings. So the over-all result is as shown in Figure 9-8. The gas has undergone a cyclic process—it has gone through a change but it has been exactly restored to its initial state. The surroundings, however, have been permanently changed. As shown by the arrows, the surroundings performed an amount of work $\frac{9}{4}PV$, *all of which ended up in the surroundings as heat*. The net effect of the cyclic process is that work (ordered motion) was transformed into heat (disordered motion).

Exercise 9-3. Show that there is less net work done on the gas and less net heat produced in the cyclic process if the expansion is carried out in one step, as in Figure 9-8, but the recompression is carried out in two steps, by process 2′.

Now we are mainly interested in the expansion process. We can focus on it by considering the *hypothetical* cyclic process in which we carry out the expansion in a fixed number of steps (as in the top half of Table 9-1) and *then recompress in the cheapest possible way.* This last part makes the process hypothetical, because it would take an infinitely long time to complete the compression step. Nevertheless, it lets us look more closely at the nature of the expansion process.

For example, consider the expansion by process 2 and recompression by process ∞′. Figure 9-8 is changed to Figure 9-9. Again the gas is left in its initial state after the cyclic process. The route followed this time, however,

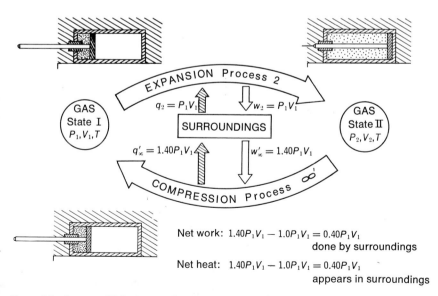

Net work: $1.40P_1V_1 - 1.0P_1V_1 = 0.40P_1V_1$
done by surroundings

Net heat: $1.40P_1V_1 - 1.0P_1V_1 = 0.40P_1V_1$
appears in surroundings

Figure 9-9 A more efficient expansion–compression cycle.

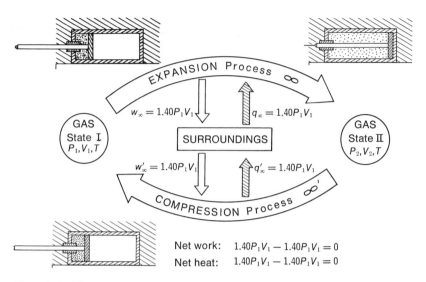

Net work: $1.40P_1V_1 - 1.40P_1V_1 = 0$
Net heat: $1.40P_1V_1 - 1.40P_1V_1 = 0$

Figure 9-10   The most efficient possible expansion–compression cycle.

requires that the surroundings perform a smaller net amount of work $0.4PV$, all of which is converted into heat.

Clearly, if we wish to complete the cyclic process with a minimum waste of ordered work degenerated into disordered heat, we should carry out the expansion in as many steps as time permits. In fact, the most efficient process is that shown in Figure 9-10.

Here is the lazy man's perfect process! In the first step, the surroundings receive an amount of work $1.40PV$, which just equals that expended in the second step. Not only is the gas carried through a cyclic process, but also the surroundings are unchanged at the end. During this tedious incremental process, we keep the system at pressure equilibrium throughout the expansion, and then maximum work is extracted from the gas as the change occurs. Then with an equally tedious compression, we can exactly reverse the changes *both* in the system and in the surroundings.

An equilibrium expansion is exactly reversible. Any nonequilibrium expansion process requires that work be converted into heat to restore the system to its initial state. We shall see that this important result applies to chemical reactions, as well as to gas expansions.

9-2   A chemical reaction in an electrochemical cell

When a piece of metallic zinc is immersed in an aqueous solution of silver nitrate, chemical reaction occurs. The zinc rod dissolves and bright silvery needles of metallic silver grow on the surface of the rod, as shown in Figure

9-11. Chemical analysis shows that the reaction is

$$Zn(m) + 2\ Ag^+(aq) \longrightarrow Zn^{+2}(aq) + 2\ Ag(m) \qquad (9\text{-}24)$$

Experiment shows that after 1.00 millimole of zinc has dissolved, 87 calories of heat have been released. However, this exothermic reaction has taken place in the beaker without the production of any useful work.

$$\Delta E = q - w$$

Since $w = 0$

$$\Delta E = q = -87\ \text{cal}$$

As this spontaneous change occurs, there is an energy decrease, but if no work is produced, all of this energy change is released as heat, merely heating up the solution and the surroundings.

We are reminded again of the creative engineer mentioned in Chapter Six who looked on a river flowing unimpeded to the sea and visualized a waterfall in which mechanical energy could be converted to electrical energy. This electrical energy can be used directly to run motors, light

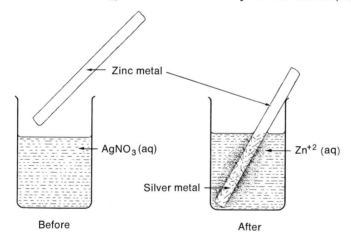

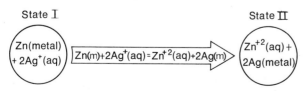

Figure 9-11 A spontaneous chemical change taking place in a beaker.

lamps, or heat ovens, but it can also be stored in an electrochemical cell—then, when needed, the electrical energy can be recovered in the form of useful work. Why don't we construct an electrochemical cell from zinc and silver nitrate solution? Then our reaction (9-24) could do work for us. We shall see how to do this, modelling our approach after that of extracting work from a gas expansion.

(a) THE OPERATION OF AN ELECTROCHEMICAL CELL

Figure 9-12 shows a zinc–silver electrochemical cell. On the left-hand side, the zinc rod is immersed in 1.0 M zinc nitrate solution while on the right-hand side, a silver rod is immersed in 1.0 M silver nitrate solution.*
Between the two beakers is a porous, ceramic wall to establish electrical continuity between the two solutions. Externally, the two metal rods are connected through a switch S and a current meter (ammeter) A.

When S is closed, a current flows. Electrons leave the zinc rod, pass

---

* The choice of concentration is, of course, arbitrary. The use of 1.0 M solutions gives a standard set of conditions commonly used as a reference point.

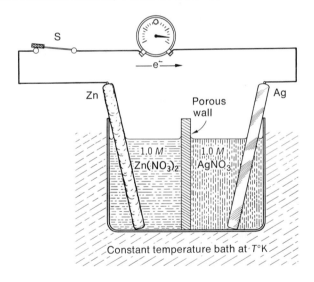

Figure 9-12 A chemical change in an electrochemical cell: no useful work.

through the ammeter, and enter the silver rod. We note that changes occur in the solutions. On the left-hand side, zinc metal dissolves to form Zn^{+2} ions. This is the source of the electrons that flow out of the zinc rod. On the left-hand side,

$$Zn(m) \longrightarrow Zn^{+2}(aq) + 2\,e^- \qquad (9\text{-}25)$$

On the right-hand side, silver metal is plated out, consuming both the Ag^+ ions and the electrons that flow into the silver rod from the external circuit:

$$Ag^+(aq) + e^- \longrightarrow Ag(m) \qquad (9\text{-}26)$$

Since the number of electrons leaving the zinc rod determines the amount of reaction (9-26) that can occur, we see that two moles of $Ag(m)$ are formed for every mole of $Zn(m)$ dissolved. The over-all chemistry of the cell is (9-24):

$$Zn(m) \longrightarrow Zn^{+2}(aq) + 2\,e^-$$
$$\underline{2\,Ag^+(aq) + 2e^- \longrightarrow 2\,Ag(m)}$$
$$Zn(m) + 2\,Ag^+(aq) \longrightarrow Zn^{+2}(aq) + 2\,Ag(m)$$

This is the same reaction that occurred in the beaker shown in Figure 9-11. Now we can study this reaction under controlled conditions.

For example, we can measure the current flow. As operated in Figure 9-12, the factors that determine the current are the voltage generated by the cell $\Delta\mathcal{E}$, 1.56 volts, and the internal resistance of the cell R, which we must measure. Suppose it proves to be 2.0 ohms. Then the current I is given by

$$I(\text{amperes}) = \frac{\Delta\mathcal{E}(\text{volts})}{R(\text{ohms})} \qquad (9\text{-}27)$$

$$= \frac{1.56}{2.0}$$

$$= 0.78\ \text{A}$$

One ampere is one coulomb of charge per second; so, if the switch were left closed for 250 seconds the charge that would leave the zinc rod and enter the silver rod is

$$\left(0.78\ \frac{\text{coulomb}}{\text{sec}}\right) \cdot (250\ \text{sec}) = 195\ \text{coulombs} \qquad (9\text{-}28)$$

As chemists, though, we'd like to express this in moles, so we divide 195 by the number of coulombs per mole of electrons, 96,500. Then, the moles of charge moved in 250 seconds is

$$\frac{195}{96,500} = 2.0 \cdot 10^{-3} \text{ mole of electrons} \tag{9-29}$$

We are now ready to calculate the energy dissipation. The power (energy per unit time) generated by a current I in a resistance R is I^2R. The energy released is the power multiplied by the time for which the switch is closed. Hence, the energy effect is

$$\text{energy} = (\text{power}) \cdot (\text{time})$$

$$= (I^2R) \cdot (t)$$

$$= \left(0.78 \frac{\text{coulomb}}{\text{sec}}\right)^2 \cdot (2.0 \text{ ohms}) \cdot (250 \text{ sec})$$

$$= 304 \text{ joules}$$

$$= (304 \text{ joules}) \cdot \left(0.239 \frac{\text{calorie}}{\text{joule}}\right) = 72.5 \text{ calories} \tag{9-30}$$

This is electrical work, though it is expended in the useless task of heating up the solution. We will call this work w_q to remind ourselves that it is destined for conversion into heat rather than for a more practical role, such as lifting weights or driving streetcars. The First Law becomes

$$\Delta E = q - w = q - w_q$$

$$-87 = q - 72$$

$$q = -15 \text{ calories}$$

Now we can assess the changes that occur during the 250 seconds that S is closed.

    (i)   2.0 millimoles of electrons moved from the zinc rod to the silver rod.

   (ii)   1.0 millimole of Zn metal dissolved.

  (iii) 2.0 millimoles of Ag metal deposited.

  (iv) ΔE, the energy change, is the same as in the beaker, -87 calories.

   (v) 72 calories of electrical work was performed as the current passed through the cell, but it all was immediately converted back into heat in the cell. No *useful* work was extracted.

Thus we have wasted the available energy. *No useful work was performed* even though the reaction (9-24) was carried out in an electrochemical cell. The situation can be likened to the expansion carried out in Figure 9-1, if the piston has a bit of friction. No weight would be lifted (so no useful work would be done), but the friction would generate non-useful heat, just as the internal cell resistance does. We have yet to extract work from the electrochemical cell.

(b) LET'S GET SOME WORK

Figure 9-13(a) shows again our zinc–silver cell, but this time with the current passing through the windings of a motor. When current flows, the armature of the motor turns and useful work is performed.

To calculate the work output, we will take a simplified view of the motor by representing it as an equivalent resistance R_M, as pictured in Figure 9-13(b). We will then equate the heat released in R_M, due to the flow of current, to the useful work performed by the motor M.

Suppose the magnitude of the resistance R_M is 18 ohms. Then the total resistance will be the sum of R_M plus the internal cell resistance, $18 + 2 = 20$ ohms. Consequently, the current will be lowered tenfold:

$$I = \frac{\Delta \mathcal{E}}{R} = \frac{1.56}{20} = 0.078 \text{ ampere}$$

To obtain the same charge movement, $2.0 \cdot 10^{-3}$ mole of electrons, at this lower current, we'll have to leave the switch closed ten times as long as before, 2500 seconds ($41\frac{2}{3}$ min). During this time, useful electrical work $w_{elect.}$ will be performed in the motor R_M, and electrical work w_q will be performed in the cell, but it will be degraded into heat.

$$w_{elect.} = (I^2 R_M)(\text{time})$$

$$= (0.078)^2(18)(2500)$$

$$= 274 \text{ joules}$$

$$w_{elect.} = 65 \text{ calories}$$

$$w_q = (I^2 R)(\text{time})$$

$$= (0.078)^2(2)(2500)$$

$$= 30 \text{ joules}$$

$$w_q = 7 \text{ calories}$$

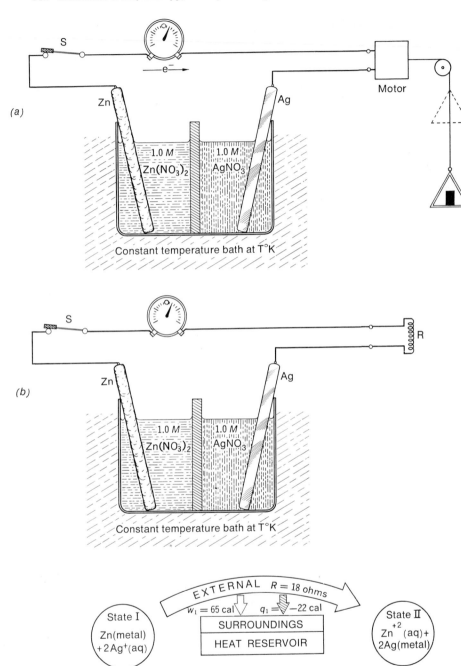

Figure 9-13 A chemical change in an electrochemical cell: process 1, a little work.

The First Law now gives us q:

$$\Delta E = q - w = q - w_q - w_{\text{elect.}}$$

$$-87 = q - 7 - 65$$

$$q = -15 \text{ calories}$$

The box-score this time is a bit more pleasing. The energy change is still -87 cal and we let some of this slip through our fingers to be used in heating the cell, $q - w_q = (-15) - (7) = -22$ cal. The remaining 65 calories we extracted in the motor as electrical work. And it didn't cost a bit more than previously when we got no useful work—well—except for one thing, it did take $41\frac{2}{3}$ minutes, ten times as long.

(c) STILL MORE WORK, PLEASE

Suppose a more powerful motor is used in Figure 9-13: let us say, a motor with an equivalent resistance of 78 ohms. Now, the total resistance, $78 + 2 = 80$ ohms, will lower the current by another factor of four, to $1.95 \cdot 10^{-2}$ ampere. Now it will take 10,000 seconds ($2\frac{3}{4}$ hours) for $2.0 \cdot 10^{-3}$ mole of electrons to flow through the motor. The useful work generated in the motor R_M will now be:

$$w_{\text{elect.}} = (I^2 R_\text{M})(\text{time})$$

$$= (1.95 \cdot 10^{-2})^2 (78)(10 \cdot 10^3)$$

$$= 296 \text{ joules}$$

$$= 70.2 \text{ calories}$$

The non-useful work converted into heat in the cell will be:

$$w_q = (I^2 R)(\text{time})$$

$$= (1.95 \cdot 10^{-2})^2 (2)(10 \cdot 10^3)$$

$$= 7.6 \text{ joules}$$

$$= 1.8 \text{ calories}$$

The First Law gives us q:

$$\Delta E = q - w = q - w_q - w_{\text{elect.}}$$

$$-87 = q - 1.8 - 70.2$$

$$q = -15 \text{ calories}$$

The amount of the $\Delta E = -87$ cal that was lost into heating is, this time, $q - w_q = (-15) - (1.8) = -16.8$ cal. The remaining 70.2 calories were extracted as useful work and, oh yes, it took more time—$2\frac{3}{4}$ hours, to be exact.

Exercise 9-4.   Show that if R_{M} is increased to 198 ohms, w_q will decrease to 0.7 calorie, so $w_{\mathrm{elect.}}$ becomes $72 - 0.7 = 71.3$ calories, and that this small additional improvement lengthens the time of the experiment to almost 7 hours.

(d) MAXIMUM WORK

Plainly we can get as high a fraction of the available work as we desire by merely increasing the magnitude of R_{M} relative to the intrinsic cell resistance. We also see that this available work is, at most, 72 calories. We can reduce w_q closer and closer to zero, but q always turns out to be -15 calories. Furthermore, to do this, to extract the maximum electrical work, R_{M} will have to be increased so much that it will take an infinite time. Thus it is possible to extract work from a spontaneous chemical change (via an electrochemical cell), but just as in the expansion of a gas, to obtain maximum work, the work must be removed at an infinitesimally slow rate.

We have considered the conversion of electrical work into heat as equivalent to its conversion into mechanical work via a motor. An alternate way to extract the available energy usefully is to transfer it to a second electrochemical cell, by charging that cell. Of course, the cell to be charged must itself generate a voltage below that of the discharging cell so that the current flows in the right direction. Whether driving a motor or charging another cell, maximum work is obtained from a chemical reaction if the reaction occurs in a cell working against an external voltage lower than its own by no more than an infinitesimally small amount.

We can obtain an analytical expression for the maximum useful work that can be extracted from a chemical reaction. Suppose we wish to extract maximum work from n moles of electrons (charge per electron $= e_0$) taken from a cell with constant voltage $\Delta\mathcal{E}$. The current I is (9-27),

$$I = \frac{\Delta\mathcal{E}}{R}$$

but also

$$I = \frac{\text{amount of charge moved}}{\text{time}} = \frac{nN_0e_0}{t} \tag{9-31}$$

where $N_0 =$ Avogadro's number. Equating (9-27) to (9-31) and solving for t, we obtain

$$t = nN_0 e_0 \frac{R}{\Delta\mathcal{E}}$$

$$w_{\text{elect.}} = [I^2 R](t)$$

$$= \left[\left(\frac{\Delta\mathcal{E}}{R}\right)^2 R\right]\left(nN_0 e_0 \frac{R}{\Delta\mathcal{E}}\right) = n\Delta\mathcal{E}N_0 e_0 = n\Delta\mathcal{E}\mathcal{F}$$

The quantity $\mathcal{F}$ is called the Faraday Constant. If we wish to express electrical work in joules, $\mathcal{F}$ has the magnitude 96,500; if we wish $w_{\text{elect.}}$ in kcal, $\mathcal{F} = 23.061$

$$\boxed{w_{\text{elect., max}} = n\Delta\mathcal{E}\mathcal{F}} \qquad\qquad (9\text{-}32)$$

$\mathcal{F} = 23.061 \text{ kcal/volt} \cdot \text{mole}$

$\mathcal{F} = 96,500 \text{ coulombs/mole}$

By (9-32), the maximum electrical work is fixed by the cell voltage $\Delta\mathcal{E}$ and by n, the number of moles of charge extracted from the cell. For our experiment, $n = 2.0 \cdot 10^{-3}$ mole, so

$$w_{\text{elect., max}} = (2.0 \cdot 10^{-3})(1.56)(23.061) = 72 \text{ cal}$$

Notice that although the energy of our system decreased by 87 calories, the maximum energy that can be withdrawn as electrical work is only 72 calories. Even by the infinitely slow discharge, we can free only 72 calories of energy to perform work. This portion of ΔE (or ΔH) that can be transformed into work—that is "freeable"—will be of particular interest to us. It is measured, according to (9-32), by the cell voltage under zero current.

(e)   NOW LET'S GET THINGS BACK

In each of the uses of the cell, we have considered the movement of the same amount of charge, $2.0 \cdot 10^{-3}$ mole of electrons. According to the over-all chemistry of the cell, we must have deposited this same number of moles, $2.0 \cdot 10^{-3}$, of metallic silver and dissolved half as many moles of metallic zinc.

$$Zn(\text{metal}) \longrightarrow Zn^{+2}(\text{aq}) + 2\,e^-$$
$$2\,Ag^+(\text{aq}) + 2\,e^- \longrightarrow 2\,Ag(\text{metal})$$

The net reaction is (9-24)

$$Zn(\text{metal}) + 2 \, Ag^+(aq) \longrightarrow Zn^{+2}(aq) + 2 \, Ag(\text{metal})$$

Therefore, to restore the cell, the current through the cell must be reversed, reversing the chemistry of (9-24). This must be continued until $2.0 \cdot 10^{-3}$ mole of electrons has passed through in this reverse direction, redepositing $1.0 \cdot 10^{-3}$ mole of zinc and redissolving $2.0 \cdot 10^{-3}$ mole of silver. This reversal can be accomplished with another electrochemical cell of higher voltage, as shown in Figure 9-14. If, for example, we use an

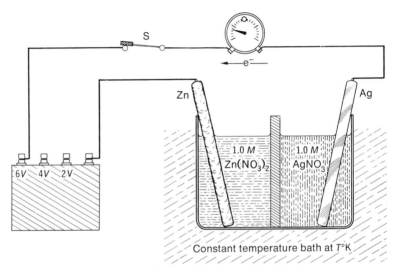

Net $\Delta \varepsilon = 6.0 - 1.56 = 4.44$ volts

Current $I = \dfrac{\Delta \varepsilon}{R} = \dfrac{4.44}{8} = 0.56$ ampere

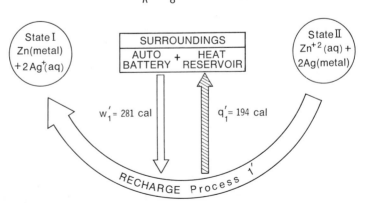

Figure 9-14   Recharging the cell: process 1'.

automobile battery with a voltage of 6.0 volts and an internal resistance of, say, 6 ohms, a large current will flow and our cell will be quickly restored. The net voltage available will be the difference between the two voltages, $6.0 - 1.56 = 4.44$ volts, since they are opposed to each other. The current that will flow (0.56 ampere) is determined by this net voltage and the sum of the two battery resistances, $6 + 2 = 8$ ohms. By our now-familiar calculations, we find that this current must flow for 348 seconds ($5\frac{2}{3}$ min), and during this time, non-useful heat will be generated in the two cells:

$$w'_q = (I^2R)(\text{time})$$

$$= (0.56)^2(8)(348)$$

$$= 873 \text{ joules}$$

$$w'_q = 209 \text{ calories}$$

Electrical work has been done *on* our cell, so $w_{\text{elect.}} = -72$ cal. By the First Law, $\Delta E = +87$ cal, so $q = \Delta E + w_{\text{elect.}} = +87 - 72 = +15$ cal. The positive sign shows that the cell absorbs heat from the surroundings. Meanwhile, the automobile battery, part of the surroundings, has performed the useful work $w' = -w_{\text{elect.}} = -(-72) = +72$ cal, but it has also provided all of the electrical work that ended up as heat, $w'_q = 209$ cal. The heat reservoir has received this 209 calories of electrical heating while it was giving up the 15 calories of heat absorbed by our cell during its recharge. Figure 9-14 shows the net result: The auto battery has performed $w'_1 = w_{\text{elect.}} + w'_q = 281$ calories and the reservoir has absorbed $q' = 209 - 15 = 194$ calories. The difference, $281 - 194 = 87$ calories, is exactly what is needed to restore the cell.

Ouch, again! That was expensive! Our cell is recharged—we raised its energy again by 87 calories and restored its capacity to deliver 72 calories of work, but we wasted 194 calories in useless heating to do so!

(f) THERE MUST BE A CHEAPER WAY!

Suppose we cut back the voltage used to recharge our cell. As shown in Figure 9-15, we might use only two of the cells of the automobile battery. This gives a voltage of 4.0 volts which, when set against the Zn–Ag cell, gives a net voltage of 2.44 volts. Now the current must flow for 476 seconds (almost 8 min), but less energy is wasted in heating the cells. The cell is recharged, but only 187 calories of work are required, 100 calories of which are dissipated into heat.

An even more economical recharge is obtained with only one of the automobile battery cells. Now the net charging voltage is quite small, $2.0 - 1.56 = 0.44$ volt. Of course, that means the current will be much smaller, 0.11 ampere, and it will take quite a while to move the desired

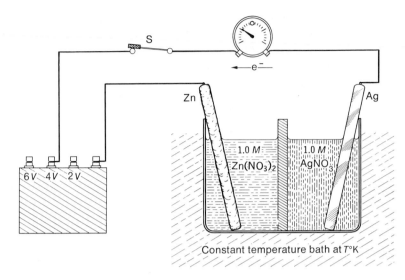

Net $\Delta\mathcal{E}$ = 4.0 − 1.56 = 2.44 volts

Current $I = \dfrac{\Delta\mathcal{E}}{R} = \dfrac{2.44}{6}$ = 0.41 ampere

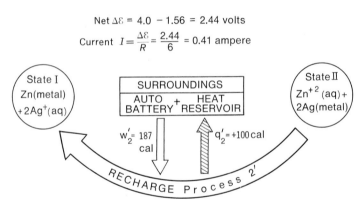

Figure 9-15 Recharging the cell: process 2′.

charge, $2.0 \cdot 10^{-3}$ mole of electrons. The recharge time becomes $29\frac{1}{2}$ minutes (1773 sec), but only 93 calories of work are required, only 6 calories of which are dissipated into heat.

Exercise 9-5. Show that if the driving battery voltage were reduced to 1.78 volts, so that the net voltage would be 0.22 volt, the electrical work converted into heat, w_q', would be only 10 cal, less than the 15 calories absorbed by the reaction as it proceeds. (Assume 4.0 ohms total resistance.)

Plainly, the way to recharge our cell with a minimum waste of energy is to recharge with a voltage source just barely above that of the Zn–Ag cell. This voltage source might be another cell, as pictured in Figures 9-14 and

9-15, or it might be a generator driven by a turbine in the engineer's hydraulic system. As the driving voltage is reduced, the electrical heating, w_q', becomes closer and closer to zero, until the only heat transfer left is the 15 calories given up by the heat reservoir to the reaction. As w_q' becomes negligible, the only electrical work that remains is the 72 calories performed on the Zn–Ag cell as it recharges. Under these limiting conditions, the Zn–Ag cell gains 15 calories of energy as heat and 72 calories as work, summing to 87 calories, exactly ΔE. Unfortunately, this most economical recharge again takes an infinite time.

(g) SUMMARY: A CYCLIC USE OF AN ELECTROCHEMICAL CELL

In the same way that we carried a gas through cyclic expansion and recompression, we have now considered the cyclic discharge and recharge of an electrochemical cell. Table 9-2 shows the energy balance in each of the several processes considered. Again we see that any combination of discharge and charge that takes a finite amount of time involves the waste of energy in undesired heating. For example, if the discharge requires 42 minutes, while 65 calories of work are obtained, 22 calories of energy are converted to unwanted heat. Then, if we recharge in 29 minutes to replace this 87 calories, 6 calories of useless heat are produced. The cyclic process

Table 9-2 Summary View of Energy Transferred During Discharge and Recharge of an Electrochemical Cell

Discharge

Process	External Resistance, R_M (ohm)	Time Required (min)	Heat Generated (cal)	Useful Work (cal)
0	0	$4\frac{1}{6}$	87	0
1	18	$41\frac{2}{3}$	22	65
2	78	167	17	70
3	198	417	15.7	71.3
∞	∞	∞	15	72

Recharge

Process	Excess Voltage (volts)	Time Required (min)	Heat Generated (cal)	Electrical Work Done (cal)
1'	4.44	$5\frac{2}{3}$	194	281
2'	2.44	8	100	187
3'	0.44	$29\frac{1}{2}$	6	93
4'	0.22	59	−5	82
∞'	0.00	∞	−15	72

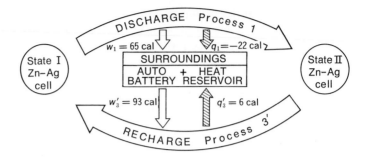

| Net work: | $93 - 65 = 28$ cal | done by surroundings |
| Net heat: | $22 + 6 = 28$ cal | appears in surroundings |

Figure 9-16   A possible discharge–recharge cycle.

takes 71 minutes and its net effect in the surroundings is as shown in Figure 9-16.

In this case, during the discharge process, both work (65 calories of it) and heat (22 calories) are transferred to the surroundings. This energy is all drawn from the exothermic chemical reaction between metallic zinc and aqueous silver ions in reaction (9-24). Then, in the recharge process, the surroundings must do work on our cell: 93 calories by process 3′, of which 6 calories are expended as useless heating and also end up in the surroundings. The net effects are the following.

(i)   Our cell is restored.

(ii)   The surroundings received 65 calories of work and delivered 93 calories, a net expenditure of 28 calories of work by the surroundings.

(iii)   Twenty-two calories of heat were generated during discharge and 6 more during recharge, a total of 28 calories, all of which appear in the surroundings.

We see that, in the 71-minute cyclic process, the over-all effect in the surroundings is that 28 calories of work are converted to heat.

Exercise 9-6.   Suppose that the cell is discharged by process 1 in 42 minutes, but to save time, it is recharged using the full 6 volts (by process 1′) in only $5\frac{2}{3}$ minutes. Show that in this 47-minute cyclic process the net energy effect converts 216 calories of work into useless heat in the surroundings.

Once again attention can be focussed on the discharge process by considering the hypothetical cyclic process in which the recharging process ∞′

is used, even though it would take an infinite time. Now the 42-minute discharge process performs 65 calories of useful work out of the possible 72. It wastefully heats the surroundings with the remaining 7 calories. By the time the reversible recharging is completed, the surroundings have transformed a net 7 calories of work into heat. If 167 minutes are allowed for discharge, then the cyclic process degrades only 2 calories of work into heat. If the discharge could take an infinite amount of time, then, and only then, could the cyclic discharge–recharge process exactly reverse the changes *both in the system and in the surroundings*. This would be accomplished by exactly balancing the cell voltage against an external voltage, so as to maintain electrical equilibrium at every moment in the cycle. *Such an equilibrium discharge of a cell is exactly reversible.* **Any nonequilibrium process,** *one that takes place spontaneously in a finite time,* **requires that work be converted into heat to restore the system to its initial state.**

(h) REVERSIBILITY

The result we have just obtained for the expansion of a perfect gas and for the discharge of an electrochemical cell proves to be true for any change of state. If a process can be found in which the change occurs entirely at equilibrium, the process is reversible. However, a reversible process can be closely approximated, but never actually carried out because it must take place infinitely slowly.

A change carried out reversibly can be restored in a reversible manner without any change in the surroundings. *In any real process* (one that takes place in a finite time), *the change can be restored only at the net expense of changing some work into heat*. Notice, however, that real processes are the ones that take place spontaneously. When the weight M_1 is removed from the pan, as in Figure 9-1 or Figure 9-2, the gas spontaneously pushes the piston to the right, lifting the pan. Then, no matter how carefully and slowly the gas is recompressed in the cyclic process, work will have been extracted from the surroundings and converted into heat in the surroundings. If a spontaneous chemical reaction is allowed to proceed in a beaker or in an electrochemical cell, as in Figures 9-11 or 9-12, it too can be reversed and the reactants restored. This can be done with another cell by reversing the current, as in Figure 9-14. However, as before, no matter how carefully and slowly the reaction is reversed in the cyclic process, work will have been extracted from the surroundings and converted into heat in the surroundings.

Thus, *the reversal of any spontaneous process always requires a change of work into heat*. The closer the process comes to an equilibrium process, the smaller will be the net change of work into heat when the process is reversed. At equilibrium, the energy degraded into heat will be a minimum. Later, we shall investigate the discrepancy between the actual heat q, in a spontaneous process, and the reversible heat q_{rev}, in a hypothetical, infinitely slow process. *We shall find this difference, $q_{rev} - q$, to be a measure of the tendency for a reaction to occur spontaneously.*

(i) THE SECOND LAW OF THERMODYNAMICS

Before exploring $(q_{rev} - q)$, we should note that the gas expansion and the electrochemical cell discharge have revealed an important principle of nature. The principle is so widely applicable that it is called the Second Law of Thermodynamics. (The First Law states that "Energy is conserved.") There are several equivalent statements of the Second Law; here are three of them.

The Second Law of Thermodynamics

   (i)   **All spontaneous processes are irreversible.**   (9-33)

   (ii) **After any spontaneous process, work must be converted to heat in order to restore the system to its initial state.**   (9-34)

Since work is a manifestation of ordered energy and heat is disordered energy, the essence of the Second Law is that a system changes spontaneously in the direction of increasing randomness. This gives us a third statement of this many-faceted law.

   (iii) **In a spontaneous process, randomness increases.**   (9-35)

9-3 Entropy and probability

In Chapter Eight it was concluded that a system tends to change spontaneously towards the most random arrangement. Then, in Section 8-2(g), we surmised that the quantity $\Delta H / T$, enthalpy change divided by temperature, might be a measure of randomness. Now we add that a spontaneous process involves a change of work into heat.

It is time to introduce a symbol to represent this change in randomness and to attempt a quantitative connection between randomness and heat effects. The name scientists have chosen is **entropy** and its change is symbolized ΔS. We shall now see how entropy is related to heat effects.

(a) ENTROPY AND ITS RELATION TO PROBABILITY

The concept of entropy is connected to the intrinsic probability of a system. Simple ideas of probability tell us the nature of S. What properties should entropy display? These can be written down intuitively.

(i) *Entropy should be a state function.* An entropy change ΔS must be the difference between the initial and final state entropies S_1 and S_2 and must not be dependent upon the process by which the change occurs.

$$\Delta S = S_2 - S_1 \tag{9-36}$$

(ii) *Entropy should be additive.* If a system is considered in parts, the total entropy S should be the sum of the entropies of the parts, say S' and S''.

$$S_{\text{total}} = S' + S'' \tag{9-37}$$

In probability terms, the second condition is easily assured. Suppose a system can be considered to be made up of two parts. If the probability of finding part one in a given situation is designated W' and the probability of finding part two in a given situation is W'', then the over-all probability W of finding *both* parts one and two in the given situations is the product of W' and W'',

$$W = W' \cdot W'' \tag{9-38}$$

The combination of (9-37) and (9-38) tells us that if entropy is a function of probability (that is, if $S = S(W)$), the mathematical form of $S(W)$ must be such that

$$S(W) = S(W') + S(W'')$$

or

$$S(W' \cdot W'') = S(W') + S(W'') \tag{9-39}$$

There is, in fact, only one function that satisfies (9-39), and a mathematician recognizes it at once. The logarithm has the property we desire.

$$\log (a \cdot b) = \log a + \log b$$

So if we are to have the mathematical property of (9-39), entropy must have a logarithmic dependence upon probability.

$$\boxed{S = \log W} \tag{9-40}$$

So, if $W = W'W''$,

$$S = \log (W'W'') = \log W' + \log W'' = S' + S''$$

(b) ENTROPY AND ITS RELATION TO HEAT IN A PERFECT GAS EXPANSION

The perfect gas expansion was first considered in Section 8-1(b) in terms of randomness. It was considered again in Section 9-1 with attention to the work and heat effects. Let us summarize the results.

Randomness: We found in Section 8-1(b) that "*probability leads to a situation that distributes molecules between two bulbs in proportion to their volumes.*" We can say this mathematically in terms of a proportionality:

W is proportional to V

or

$$W = aV \tag{9-41}$$

where a is some proportionality constant we don't yet know. If we wish to consider an expansion from state I with volume V_1 to state II with volume V_2, (9-41) tells us that

$$W_1 = aV_1$$
$$W_2 = aV_2$$

Each of these probabilities can be substituted into (9-40):

$$S_1 = \log W_1 = \log (aV_1) \tag{9-42}$$
$$S_2 = \log W_2 = \log (aV_2) \tag{9-43}$$

Now it is time to return to property (9-36). The *change* in entropy ΔS in the expansion must simply be

$$\begin{aligned} \Delta S &= S_2 - S_1 \\ &= (\log aV_2) - (\log aV_1) \\ &= \log \frac{aV_2}{aV_1} = \log \frac{V_2}{V_1} \end{aligned} \tag{9-44}$$

Oh, wonder of wonders! Expression (9-44) is immediately reminiscent of the maximum work that can be obtained in a reversible expansion, as given by (9-11):

$$\begin{matrix}\text{"reversible"} \\ \text{work done}\end{matrix} = \begin{matrix}\text{"reversible"} \\ \text{heat absorbed}\end{matrix} = nRT \log_e \frac{V_2}{V_1}$$

By using $\log_e$ in (9-44), we can substitute (9-44) into (9-11) and obtain

$$\frac{\text{reversible heat}}{\text{absorbed}} = nRT \log_e \frac{V_2}{V_1} = nRT\Delta S$$

or, rearranging,

$$\Delta S = \frac{\text{reversible heat}}{nRT} \qquad\qquad (9\text{-}45)$$

Expression (9-45) can be simplified by considering the entropy on a "per mole" basis; that is, by choosing $n = 1$.

$$\Delta S = \frac{(\text{reversible heat})}{RT} = \frac{q_{\text{rev}}}{RT} \qquad\qquad (9\text{-}46)$$

Expression (9-46) is usually written without the constant term $1/R$ (R = gas constant), but this is only a matter of traditional definition. If we had decided to define $S = R \log_e W$, then (9-46) would appear in the form usually seen:

$$\boxed{\Delta S = \frac{q_{\text{rev}}}{T}} \qquad\qquad (9\text{-}47)$$

This can be taken as the definition of entropy. We have derived it for a simple process, the expansion of a perfect gas. The result (9-47) was obtained long ago in the nineteenth century, at a time when heat was treated as though it were a substance (called "caloric") and before any connection with probability had been postulated. The usefulness of the concept of entropy has been verified in innumerable applications of chemistry and physics. Hence, the probability significance of entropy and the validity of the entropy concept throughout all chemical phenomena are now firmly based in experience and are accepted by all practicing scientists.

(c) ENTROPY AND THE SECOND LAW

We see that the probability of a given state is measured by the quantity called entropy. Since increasing randomness implies increasing probability, the Second Law (as expressed in form (iii, 9-35)) can be restated in terms of entropy. Spontaneous changes occur in the direction that increases randomness in the universe, so entropy must also increase in the universe during any spontaneous change. Now the First and Second Laws of Thermodynamics can be stated in parallel and concise forms.

> **First Law**
> **The energy of the universe is constant.** (9-48)
> **Second Law**
> **The entropy of the universe increases.** (9-49)

9-4   Free energy and spontaneous change

We are on the verge of something great. The entropy, which has a simple microscopic significance, probability, can be measured through heat effects. Also, entropy tells us the direction of spontaneous change! If someone asks whether a given system will spontaneously undergo a certain change, the answer is to be found in the total effect on the entropy of the universe. If we can see that the combined entropies of the system and of the surroundings will increase, the answer is "Yes." If the combined entropies decrease, the change does not occur spontaneously. If the combined entropies remain constant, the system is at equilibrium. Obviously, to a chemist entropy is as important as energy.

It would be more convenient, however, if this spontaneity criterion could be re-expressed in terms of properties of the system alone. The surroundings, after all, are the rest of the universe, which can present quite a bookkeeping problem. Fortunately this bookkeeping is readily done for changes that occur at constant temperature and constant pressure, the common laboratory situation of interest to us.

(a)   REVERSIBLE HEAT AND SPONTANEOUS CHANGE

In Sections 9-1 and 9-2 we looked closely at two constant temperature processes. In each case we found that the difference between the maximum possible heat absorbed q_{rev} and the actual heat absorbed q is specially informative. For these constant temperature processes, the difference $(q_{rev} - q)$ measures the net change of work into heat when we try to restore the system reversibly. This net change of work into heat represents the "extent of irreversibility."

$$\text{extent of irreversibility} = q_{rev} - q \qquad (9\text{-}50)$$

We shall see that this quantity, expression (9-50), expresses the Second Law criterion for spontaneous change.

Therefore, let's return to Figure 9-6, which shows several of the ways in which the change in question can be brought about. First, consider process ∞, during which the maximum possible work $1.40PV$ was extracted from the expanding gas. This process is reversible, so the heat effect can be labeled q_{rev}. This heat effect immediately allows the specification of the entropy change for the *system*:

$$\text{heat absorbed} = 1.40PV = q_{\text{rev}} \qquad (9\text{-}51)$$

$$\text{entropy change} = \frac{q_{\text{rev}}}{T} \qquad (9\text{-}52)$$

As the system absorbed q_{rev}, the surroundings had to give up that same amount of heat. We can say, then, that the surroundings *absorbed* $(-q_{\text{rev}})$ calories of heat. Now the entropy change in the *surroundings* can be given:

$$\text{heat absorbed} = (-q_{\text{rev}}) \qquad (9\text{-}53)$$

$$\text{entropy change} = -\frac{q_{\text{rev}}}{T} \qquad (9\text{-}54)$$

The entropy change of the entire universe is the sum of (9-52) and (9-54). Adding these, we obtain

$$\Delta S_{\text{universe}} = \Delta S_{\text{system}} + \Delta S_{\text{surroundings}}$$

$$= \left(\frac{q_{\text{rev}}}{T}\right) + \left(-\frac{q_{\text{rev}}}{T}\right)$$

$$= 0 \qquad (9\text{-}55)$$

The reversible process causes no change in the entropy of the universe. The criterion for a spontaneous change is not met—entropy does not increase. Instead, it is exactly zero. This identifies an equilibrium process.

Now consider one of the irreversible paths in Figure 9-6. In any one of these processes, the work extracted will be less than the maximum possible. The heat effect q will be less than q_{rev}. Let's look again at the entropy changes in the system, the surroundings, and then their sum, that is, the entropy change of the universe.

First consider the system. We already know its entropy change—entropy is a state function, $\Delta S = S_2 - S_1$, so it depends only upon the initial and final states. The entropy change of the system must be $q_{\text{rev}}/T = 1.40PV/T$, as given in (9-52), no matter how the expansion is carried out.

The entropy of the surroundings, however, is something else. As far as the heat reservoir is concerned, a certain amount of heat q was extracted at constant temperature. The surroundings (the reservoir) do not care whether the q calories went to a system changing reversibly or irreversibly.

As long as the heat is extracted at constant temperature, the heat reservoir has changed in a reversible manner. Hence $-q/T$ *is* the entropy increase in the surroundings during the system's irreversible step.

We see that in an irreversible process, q/T is *not* the entropy change of the system, but it *is* the entropy change of the surroundings. This is not a contradiction, though at first glance it may seem so. Within the universe, a part is changing irreversibly (the system), and the rest is changing reversibly (the surroundings). That this is a correct analysis can be most clearly seen for the limiting irreversible process: the expansion of the gas into a vacuum. For this change, "process 0," no work at all is performed and $q = 0$. Yet the initial and final states of the system are the same as for any of the processes pictured in Figure 9-6. For *all* these expansions,

$$\Delta S = S_2 - S_1$$

$$= \frac{q_{\text{rev}}}{T}$$

$$= \frac{1.40 P V}{T}$$

However, consider the surroundings for this case. For process 0, no heat at all is extracted from the heat reservoir and no work is delivered to the surroundings. There is no doubt at all that the entropy change of the surroundings is zero: *Since process 0 causes no change whatsoever in the surroundings, the entropy (randomness) of the surroundings cannot have changed.* We see that ΔS_{surr} must be correctly given by $\Delta S_{\text{surr}} = q/T = 0$.

With this reassurance that $\Delta S_{\text{surr}} = q/T$, let us recapitulate our entropy ledger for any one of the irreversible expansions.

For the system:

heat absorbed $= q$

entropy change $= \dfrac{q_{\text{rev}}}{T}$ (9-56)

For the surroundings:

heat absorbed $= -q$

entropy change $= -\dfrac{q}{T}$ (9-57)

For the universe:

$$\Delta S_{\text{universe}} = \Delta S_{\text{system}} + \Delta S_{\text{surroundings}}$$

$$= \left(\frac{q_{\text{rev}}}{T}\right) + \left(-\frac{q}{T}\right)$$

$$= \frac{q_{\text{rev}} - q}{T} \qquad\qquad (9\text{-}58)$$

This expression is reminiscent of what we called the "extent of irreversibility," expression (9-50). We see that $(q_{\text{rev}} - q)$ is a measure of tendency for spontaneous change in a constant temperature process because, when divided by T, it becomes the Second Law criterion for spontaneous change (9-49). If $(q_{\text{rev}} - q)$ is positive, the entropy of the universe increases and the change can take place spontaneously.

> For a spontaneous change at constant T $\qquad q_{\text{rev}} - q > 0$ $\qquad$ (9-59)

Exercise 9-7.   Figure 9-6 shows that the one-step expansion performs work equal to $\frac{3}{4}PV$ and the gas absorbs an equal amount of heat from the surroundings. Show that this spontaneous process has a positive value of $(q_{\text{rev}} - q)$ and that it causes an entropy change of the universe equal to $+0.65PV/T$.

(b) CONSTANT PRESSURE–CONSTANT TEMPERATURE PROCESSES

The result (9-59) is particularly valuable when applied to a constant pressure–constant temperature process. This type of change is one of the most important to a chemist because so many chemical changes take place in open beakers. The pressure then is automatically constant at one atmosphere, and to keep the temperature constant is very easy. For these interesting processes, the expansion work is extremely easy to calculate. At constant pressure,

$$\text{expansion work} = P(V_2 - V_1) = P\Delta V \qquad\qquad (9\text{-}60)$$

Further, if the reaction proceeds merely in a beaker (not in an electrochemical cell), expansion work is the only kind of work performed. As we have shown in Chapter Seven (see Section 7-3(d)),

$$\Delta H = q_P$$

Substituting q_P into our measure of reaction tendency $(q_{rev} - q) > 0$, we obtain:

$$q_{rev} - q > 0 \qquad \text{for a spontaneous change at constant } T$$

$$q_{rev} - q_P > 0 \qquad \text{for a spontaneous change at constant } T \text{ and } P$$

$$(9\text{-}61)$$

Also, by substituting (7-26) and (9-47) into (9-61), we obtain

$$T\Delta S - \Delta H > 0 \qquad \text{for a spontaneous change at constant } T \text{ and } P$$

or

$$T(S_2 - S_1) - (H_2 - H_1) > 0$$

or

$$(TS_2 - H_2) - (TS_1 - H_1) > 0 \qquad (9\text{-}62)$$

Since both S and H are "functions of state," $(TS - H)$ must be also. This quantity, or rather its negative, is called the *free energy* and is designated these days by the symbol G.

$$\boxed{\text{free energy} = G = H - TS} \qquad (9\text{-}63)$$

Substituting (9-63) into (9-62), we obtain

$$(-G_2) - (-G_1) > 0$$

or

$$G_2 - G_1 < 0.$$

$$\boxed{\begin{array}{l} \textbf{For a spontaneous change} \\ \textbf{at constant } T \textbf{ and } P \qquad \Delta G < 0 \end{array}} \qquad (9\text{-}64)$$

We see that a spontaneous change (constant T,P) is accompanied by a decrease in G. The connection to $(q_{rev} - q)$ shows that when ΔG is negative, the entropy of the universe increases. *The free energy is a quantitative measure of reaction tendency.* Its advantage is that $\Delta G = \Delta H - T\Delta S$ gives us a criterion for spontaneous change in which we need look only at properties of our system. This is sufficient to take account of the rest of the universe.

Free energy change has one more important property. Remember that we define ΔG in such a way that, for a change at constant pressure and temperature (with expansion work only), we would have

$$\Delta G = q_P - q_{\text{rev}} \tag{9-65}$$

If this relation is applied to a change that occurs *at equilibrium,* while maintaining constant pressure and temperature, then q_P becomes equal to q_{rev}.

For an equilibrium process $\Delta G = q_{\text{rev}} - q_{\text{rev}} = 0$ (9-66)

This would be the case, for example, when a liquid vaporizes at a constant pressure equal to its vapor pressure at the temperature of the thermostat. It would be applicable to a chemical reaction occurring hypothetically in an equilibrium mixture. Most important of all, it is a crossover point between conditions in which a reaction will take place spontaneously as written and conditions in which the reaction will take place spontaneously in the reverse direction.

If ΔG is negative, $\Delta G < 0$ change will occur spontaneously   (9-67)

If $\Delta G = 0$ equilibrium exists   (9-68)

If ΔG is positive, $\Delta G > 0$ change cannot occur spontaneously as written—reverse change will occur spontaneously   (9-69)

(c) THE MEANING OF ΔG

We have achieved an important advance. The Second Law of Thermodynamics tells us unequivocally that *all* processes proceed spontaneously in the direction that raises the entropy (randomness) of the universe. However, the universe is a big place, and it's hard to keep track of its parts. Now, with the aid of free energy, we can look at only properties of the system. For changes that occur at constant temperature and pressure, we can be sure that the entropy of the universe goes up if the free energy of the system goes down.

The free energy criterion is the one that chemists most commonly use, and it is often posed as a measure of competition between an energy effect and a randomness effect. In our new expression for reaction tendency,

$$\Delta G = \Delta H - T \Delta S \qquad (9\text{-}70)$$

$$\Delta G < 0 \quad \text{indicates a spontaneous reaction} \qquad (9\text{-}71)$$

Notice that ΔG is made more negative (spontaneous direction) by a *negative* value of ΔH (exothermic reaction) and it is made more negative by a *positive* value of ΔS (more randomness in the system). Hence, the free energy criterion seems to raise a conflict of desires between a reaction's tendency to proceed in the exothermic direction (that was Berthelot's proposal) and the reaction's tendency to proceed to the most random state. This is a workable basis upon which to predict chemistry, and it will be done in this fashion for some time to come. Nevertheless, it is desirable to remember that the Second Law makes no reference whatsoever to energy effects—*spontaneity is governed solely by tendency to randomness.*

Let us be reminded again of the way in which $\Delta G = \Delta H - T \Delta S$ implements the randomness principle.

First, let us divide each term in (9-70) by $-T$:

$$\left(-\frac{\Delta G}{T} \right) = -\frac{\Delta H}{T} + \Delta S \qquad (9\text{-}72)$$

Now we see a sum of quantities, all with the dimensions of entropy (calories/mole degK). On the left-hand side we have the negative of $\Delta G/T$. Remembering that ΔG must be negative for a spontaneous change, we see that $(-\Delta G/T)$ must be positive. On the right-hand side we see two contributions; the second being clearly the entropy of the system. Its contribution is obvious—if the entropy of the system goes up (counting both the system's positional and motional randomness), the entropy of the universe tends to go up. A positive ΔS tends to urge the reaction to proceed spontaneously. All we need look into now, is the entropy change of the rest of the universe. That must be embodied in the $-\Delta H/T$ term. How does the exothermic heat of a reaction reflect the entropy of the rest of the universe? It does so for the special case of a constant temperature–constant pressure process, because, if temperature is to remain constant, the exothermic enthalpy must be taken up as heat in the surroundings. Converting an amount of potential energy ΔH into heat creates motional randomness in the surroundings in amount $(-\Delta H/T)$. Thus $(-\Delta G/T)$ equals the entropy change of the universe (for constant T,P processes), and it is a direct implementation of the Second Law.

$$\left(-\frac{\Delta G}{T}\right) \;=\; \left(-\frac{\Delta H}{T}\right) \;+\; \Delta S \tag{9-73}$$

entropy rise for the universe	=	entropy rise due to motional ran- domness in the surroundings	+	entropy rise due to all kinds of randomness in the system

If $(-\Delta G/T)$ is positive, the entropy of the universe increases, the universe becomes more random, and the change will occur spontaneously. In order for $(-\Delta G/T)$ to be positive, ΔG must be a negative quantity. Hence, $\Delta G < 0$ for the system is a criterion for spontaneous change.

9-5 Some examples: Enthalpy, entropy, and free energy

Having concluded that free energy ΔG is a measure of reaction tendency, it behooves us to look at a few examples. Free energy is a composite of enthalpy and entropy, and for reactions at constant temperature and pressure, these are both simply related to thermal effects. The enthalpy change can be directly measured in a calorimeter, as described in Section 7-4, and it is the heat at constant pressure, q_P, that is desired,

$$\Delta H = q_P \tag{7-26}$$

The entropy change is not so simply measured in a calorimeter, because it is the *reversible* heat that is needed, and reactions that proceed spontaneously in a calorimeter do not proceed reversibly. Nevertheless, there are several good ways to measure ΔS and some of them will be mentioned in the next chapter. For our purposes here, we'll merely observe that there are lengthy tables of measured entropies available for our use. If we can find a tabulation of measured enthalpies and entropies, then we should be able to make quick predictions concerning reaction tendency. Here are some examples.

(a) THE MELTING OF ICE: $H_2O(s) \longrightarrow H_2O(\ell)$

At 0°C (273°K), the heat of melting (fusion) of ice has been measured innumerable times and is known with high accuracy. This constant pressure heat equals the enthalpy of melting:

$$\Delta H_{273} = q_{\text{melting}} = +1436 \text{ calories/mole} \tag{9-74}$$

Equally well known is the entropy of melting:

$$\Delta S_{273} = +5.257 \text{ calories/mole degK}$$

According to equation (9-70), the free energy change associated with melting can now be calculated:

$$\Delta G_{273} = \Delta H - T\Delta S$$

$$= (+1436) - (273.16)(5.257)$$

$$= +1436 - 1436$$

$$= 0$$

Since the free energy change is zero, we see that ice at 273°K and water at 273°K are at equilibrium.

Now let us consider the same change, but at 25°C (298°K). Here again, experimental measurements show that

$$\Delta H_{298} = +1669 \text{ calories/mole}$$

$$\Delta S_{298} = +6.076 \text{ calories/mole degK}$$

We can calculate the free energy change:

$$\Delta G_{298} = \Delta H_{298} - T\Delta S_{298}$$

$$= +1669 - (298.16)(6.076)$$

$$= -143 \text{ calories/mole}$$

The free energy change is negative—the process will occur spontaneously. The free energy change tells us what we already knew—ice melts at 25°C.

It is instructive to examine the signs and magnitudes of the contributions to ΔG. The value of ΔH is positive—heat is absorbed, so the ice, as it melts, is "rolling uphill" as far as potential energy is concerned. The surroundings must surrender heat, so the entropy of the surroundings opposes reaction. Nevertheless, the entropy of the system more than overcomes this effect at 25°C, and impels the melting process. The positive sign of ΔS implies that the melting process tends to a state of higher probability. Our knowledge of the regular nature of a solid and the random disorder of a liquid makes this intuitively reasonable. This situation *always* exists for a melting process; the regular crystal lattice has lower potential energy, so ΔH is positive, but the positional randomness of the liquid, expressed in a positive ΔS, favors melting.

Exercise 9-8.   Solid chloroform, $CHCl_3$, has a melting point of $-63.5°C$ and a heat of fusion equal to 2.2 kcal/mole. Show that the increase in randomness (ΔS_{fusion}) associated with melting one mole of chloroform is about double that associated with melting one mole of ice.

Now we can look at the melting point in a new light. As the temperature rises, both ΔH and ΔS change very little. There is, however, the additional rapid dependence upon temperature in the entropy term $T\Delta S$. Because of this multiplication by T the $T\Delta S$ term becomes unimportant at low temperatures. The ΔH term then dominates ΔG, and the solid is the stable form. As T rises, there will be a temperature at which $T\Delta S$ exactly equals the potential energy contribution to ΔG; then $\Delta G = 0$. At this temperature, both solid and liquid are stable and they can coexist. This is the melting point. At still higher temperatures, the $T\Delta S$ term dominates, ΔG is negative, and the liquid becomes the thermodynamically stable form.

(b) VAPORIZATION OF LIQUID WATER

Table 9-3 shows the values of ΔH and ΔS for the vaporization of water, at both 100°C (373°K) and 25°C (298°K). In each case, we must specify the pressure of the water vapor, since the entropy of a gas depends significantly upon its pressure. As the pressure of a gas rises, its intrinsic randomness decreases. Putting this in more intuitive terms, as the pressure rises, a gas becomes more and more like a liquid, and the difference in randomness between a gas and its liquid form tends to decrease.

Table 9-3 Thermodynamics of the Vaporization of Water $H_2O(\text{liquid}) \longrightarrow H_2O(\text{gas})$
T, P

Temperature	100°C(373°K)	25°C(298°K)	25°C(298°K)
H_2O vapor pressure (torr)	760	760	23.8
ΔH (cal/mole)	+9721	+10,489	+10,489
ΔS (cal/mole degK)	+26.05	+28.26	+35.18
$\Delta G = \Delta H - T\Delta S$ (cal/mole)	0	+2063	0

Consulting the last line of Table 9-3, we see that at 100°C the vaporization of liquid water to gas at a pressure of 760 torr is an equilibrium process: $\Delta G = 0$. Liquid water and water vapor at one atmosphere can coexist. However, the same change carried out at 25°C, also at one atmosphere pressure, has a large, *positive* ΔG. A spontaneous change will occur when ΔG is negative; so to obtain a spontaneous change, we must consider the the reverse change!

If

$$H_2O(\ell) \xrightarrow{298°K} H_2O(g) \qquad \Delta G = +2063$$
$$760 \text{ torr}$$

then

$$H_2O(g) \xrightarrow{298°K} H_2O(\ell) \qquad \Delta G = -2063$$
760 torr

We see that the *free energy predicts the spontaneous condensation of water vapor to the liquid.* The last column of Table 9-3 indicates how long this condensation process will continue to occur spontaneously. As condensation proceeds, the vapor pressure drops, and the entropy difference between liquid and gas increases. Finally, at 23.8 torr, the entropy change rises to the point at which $T\Delta S$ becomes as large as ΔH. Now $\Delta G = 0$, hence equilibrium is restored. The pressure 23.8 torr is the equilibrium vapor pressure of water at 25°C.

Exercise 9-9.   Liquid chloroform, $CHCl_3$, boils at $+61.2°C$ and its heat of vaporization is 7.02 kcal/mole. Show that the increase in randomness $(\Delta S_{vap.})$ associated with vaporizing one mole of chloroform at its normal boiling point is about 25% lower than the corresponding randomness increase for one mole of water (for which $\Delta S_{vap.} = +26$ cal/mole degK at 100°C). Compare this result to that obtained in Exercise 9-8.

Let us consider again the magnitudes of ΔH and ΔS. Notice that the heat effect in vaporization is several time larger than the heat effect in the melting process. This is because the melting process leaves the molecules close together, at distances fixed by attractive forces—it costs only the extra potential energy of the long-range crystal regularity. Vaporization, however, requires that the molecules be pulled away from each other, a more expensive process, energy-wise.

The randomness effects in vaporization are also larger. The gas phase, characteristically, gives the molecules more positional randomness than the liquid, because of the molecular freedom of movement in a gas. This factor is, however, sensitively dependent upon the pressure, and at any temperature, there is a particular pressure at which $\Delta H = T\Delta S$, so that $\Delta G = 0$. This criterion fixes the equilibrium vapor pressure.

(c) SOME CHEMICAL CHANGES INVOLVING SOLIDS

Here are two familiar chemical reactions involving both gases and solids.

$$CaCO_3(solid) \xrightarrow{298°K} CaO(solid) + CO_2(gas) \qquad (9\text{-}75)$$
$$\phantom{CaCO_3(solid) \xrightarrow{298°K} CaO(solid) + } 1 \text{ atm}$$

$$C(graphite) + O_2(gas) \xrightarrow{298°K} CO_2(gas) \qquad (9\text{-}76)$$
$$ 1 \text{ atm} \phantom{(gas) \xrightarrow{298°K}} 1 \text{ atm}$$

Equation (9-75) is concerned with the stability of limestone (calcium carbonate), a stable solid found in nature all about. Equation (9-76) relates to
the oxidation of carbon at 25°C, obviously a reaction connected to the flam-
mability of carbon and the contrasting stability of the graphite writing core
of an ordinary pencil. Table 9-4 shows the thermodynamic factors in each
reaction.

The first reaction has a positive ΔG, and the significance of this is in
accord with experience. A change with positive ΔG proceeds spontaneously
in the *reverse* direction. In other words, calcium oxide reacts with carbon
dioxide at one atmosphere pressure to give calcium carbonate. This occurs
in spite of the large positive ΔS that accompanies the formation of a gaseous
product, with its high positional randomness. Of course, the effect of this
entropy factor will increase markedly if the temperature is raised. There is
some temperature above which limestone becomes unstable with respect
to reaction (9-75). Experiments, including thermodynamic measurements,
show that this temperature is near 850°C. Calculations of this kind are
helpful in designing a furnace to convert limestone ($CaCO_3$) into lime (CaO)
for use in plaster. In fact, lime is manufactured in this way, by heating
limestone to temperatures in excess of 800°C (to the tune of millions of
tons per year in the U.S. alone).

The second reaction has a startling result. Since ΔG is negative, our
thermodynamic reasoning indicates a tendency for reaction to occur.
Graphite has lots of tendency to react (to "burn") with the oxygen in the
atmosphere. For example, if the graphite in your pencil understood its
thermodynamics, it would react promptly and release a large amount of
heat.

Here we see one of the most important distinctions that a chemist must
keep in mind. There are two facets of chemical change that interest him,
that must be understood, and that can be manipulated to advantage. First,
there is the direction of spontaneous change, and secondly, the rate of the
change. Thermodynamic arguments are essentially equilibrium arguments;
they give us a signpost toward equilibrium. They tell nothing, however, of
the mountains that will be encountered on the way. These interesting
mountains (free energy mountains) determine the rate. Obviously it is to
our advantage to know that graphite is "ready" to react with air to release
energy generously, if we also know how to speed up the reaction when the
energy is needed.

Table 9-4 *Thermodynamic Properties of Two Chemical Reactions Involving Solids*
($T = 298°$K)

$CaCO_3(s) \longrightarrow CaO(s) + CO_2(1$ atm, g)		$C(s) + O_2(1$ atm, g) $\longrightarrow CO_2(1$ atm, g)	
ΔH(kcal/mole)	+42.55	−94.05	
ΔS(cal/mole degK)	+38.4	+ 0.73	
ΔG(kcal/mole)	+31.10	−94.27	

There is another valuable aspect to the graphite oxidation. Observe, in Table 9-4, the very small entropy contribution to the reaction tendency, particularly compared with that of the limestone decomposition. This small value of ΔS reflects the fact that one mole of gas is consumed as one mole is produced. The high positional randomness associated with a gaseous product is balanced almost exactly by the high positional randomness of an equal number of moles of a gaseous reactant.

(d) MORE CHEMICAL CHANGES INVOLVING GASES ONLY

Table 9-5 lists four gas-phase reactions that provide some interesting contrasts. First, note that the four reactions have been selected for their similarity. In each, two moles of a triatomic molecule react to produce three moles of diatomic molecules. The production of three gaseous molecules of two types from only two gaseous molecules of a single type should increase the randomness considerably. The values of ΔS should be all large and positive—as they are.

The ΔH column shows dramatic variation, however, and these differences are influential in determining the sign and magnitude of ΔG, hence the reaction tendency. The large positive ΔH for decomposition of carbon dioxide causes the free energy change to be positive, so the reaction proceeds spontaneously to the left if all constituents are at 1 atmosphere pressure. The opposite is true for N_2O. Here the negative ΔH causes ΔG to be negative, and we conclude that N_2O is unstable with respect to decomposition. Its availability as a common laboratory chemical depends upon kinetic effects. Under suitable conditions or in the presence of a catalyst, the spontaneous tendency for N_2O to decompose can be used to advantage.

The important conclusion is that the reaction heat (the enthalpy change) is of crucial importance to us. The macroscopic effect, the difference between reactions (9-77) and (9-78), must be understood on the microscopic level in order to understand and predict chemical reactions. It is because of differences such as in the ΔH values for CO_2 and N_2O in Table 9-5 that chemists are so concerned about the chemical bonding in molecules.

The last pair of reactions in Table 9-5, (9-79) and (9-80), also contains an

Table 9-5 *Thermodynamic Properties of Some Simple Reactions Involving Gases**

Reaction (298°K)	ΔH (kcal/mole)	ΔS (cal/mole degK)	ΔG (kcal/mole)	
$2\ CO_2(g) \longrightarrow 2\ CO(g) + O_2(g)$	+135.3	+41.5	+122.9	(9-77)
$2\ N_2O(g) \longrightarrow 2\ N_2(g) + O_2(g)$	−39.0	+35.4	−49.5	(9-78)
$2\ NO_2(g) \longrightarrow 2\ NO(g) + O_2(g)$	+27.0	+34.7	+16.7	(9-79)
$2\ NO_2(g) \longrightarrow 2\ O_2(g) + N_2(g)$	−16.2	+28.8	−24.8	(9-80)

* All pressures one atmosphere.

interesting lesson. Here we see a molecule that might decompose in two ways: either to NO plus O_2 or to N_2 plus O_2. Thermodynamic measurements show that NO_2 will not spontaneously form NO plus O_2 at one atmosphere pressure for each constituent—no, the opposite reaction might occur. On the other hand, NO_2 *is* unstable with respect to formation of N_2 and O_2; the free energy change for reaction (9-80) is negative. Again the explanations of these differences must be sought in the bonding of the molecules. For the moment we can observe the potential value of thermodynamic arguments in terms of a practical and terribly pressing human problem. One of our growing concerns is the air pollution in cities. Gasoline-powered automobiles daily spew tons of nitric oxide into the atmosphere we breathe. The reactions of Table 9-5 might contain a solution. According to Table 9-5, nitric oxide can react spontaneously with oxygen to form NO_2. In turn, NO_2 is unstable with respect to decomposition into N_2 and O_2, both of which are quite acceptable atmospheric constitutents. Thus, our attention is focussed on finding conditions that would accelerate reaction (9-80). With disconcerting confidence, we can predict that smog, like other threats to human existence, will ultimately yield in the face of our ever-increasing knowledge of ourselves and our environment, in this case, through understanding of thermodynamics, reaction kinetics, and chemical bonding.

Problems

1. A balloon used for upper atmospheric research is filled from a bank of high-pressure cylinders at essentially constant pressure of 100 atm ($\sim$1500 pounds per square inch) through a pressure reduction valve. The balloon bag has a volume of $2 \cdot 10^5$ liters when fully inflated. At sea level, this balloon is slowly inflated with helium to one half its capacity at constant T (298°K) and P (1 atm). Calculate ΔE, q, w, and ΔH done by the helium as it inflates the bag. Express your answer both in ℓ-atm and in kcal. (1 ℓ-atm $= 24.2$ cal) Let's assume helium is a perfect gas, with average energy per mole $\bar{E} = \frac{3}{2}RT$.

2. Evaluate the work performed in each of the expansion steps in process 4 shown in Figure 9-5 and verify that their sum is $1.16P_1V_1$. The four expansion steps are carried out at constant pressures of, successively, $\frac{2}{3}P_1$, $\frac{1}{2}P_1$, $\frac{1}{3}P_1$, and $\frac{1}{4}P_1$.

3. The half-inflated balloon in Problem 1 (vol $= 1 \cdot 10^5 \ell$) slowly rises to an altitude at which the pressure is 0.50 atm and, as it does so, the balloon bag slowly expands to full volume, $2 \cdot 10^5$ liters. Calculate ΔE, q, w, and ΔH done by the helium during this ascent, assuming atmospheric temperature is constant.   *max work.*

4. Figure 9-9 shows the net work done by and heat produced in the surroundings for the cyclic process in which the gas is expanded by process 2 (Table 9-1) and compressed by process ∞'. Repeat this calculation for processes 0, 1, 3 (Exercise 9-1), and 4, always coupled with process ∞'. For these cyclic processes, make a graph (like Fig. 9-6) showing the net work converted into heat versus number of expansion steps.

5. Suppose the helium in the half-inflated balloon in Problem 1 is reversibly recompressed to the initial pressure of 100 atm.

(a) What are q and w for the system and what are q' and w' for the surroundings in this compression? (Assume perfect gas behavior for helium even though it is not a very good model at 100 atm.)

(b) What is the net amount of work converted into heat in the surroundings in the cyclic process of the inflation followed by the recompression?

6. What is the net amount of work converted into heat in the surroundings in the cyclic process of the balloon expansion during its slow ascent (as in Problem 3) and then its recompression by an equally slow descent back to firm footing and $P = 1$ atm?

7. For the electrochemical cell considered in Section 9-2, calculate the discharge time and the values of $w_{elect.}$, w_q, and $(q - w_q)$ if $R_M = 8$ ohms and the current flows until $1.0 \cdot 10^{-3}$ mole of zinc has dissolved.

8. For the electrochemical cell considered in Section 9-2, plot on a single graph the discharge time on the horizontal axis and on the vertical axis all of the following quantities: the useful electrical work, $w_{elect.}$; the electrical work dissipated as heat, w_q; the total electrical work, $(w_{elect.} + w_q)$; and the total heat transferred to the surroundings, $-(q - w_q)$.

9. In Section 9-2, ΔE for the cell in each of the discharge processes is -87 cal and, in each of the recharge processes, it is $+87$ cal. Use the data in Table 9-2 to construct a new table applicable to the surroundings in which you tabulate for each discharge and recharge process the values of q', w', and $\Delta E' = q' - w'$ where $q' = -(q - w_q)$, $w' = -w_{elect.}$, and $\Delta E'$ is the energy change in the surroundings.

10. What is the entropy change associated with the inflation of the balloon in Problem 3 during its ascent? (Express your answer in cal/degree.)

11. Calculate the entropy change of the gas considered in Section 9-1 as it is expanded by the reversible process ∞. What is ΔS for the gas as it is expanded by the two-step process 2 in which less than maximum work is obtained? (Remember, entropy is a state function!)

12. What is the entropy change (in cal/degree) associated with the first inflation of the balloon in Problem 1 to its launching volume of $1 \cdot 10^5$ liters? (See also Problem 5.) Calculate the ratio of the entropy change in the balloon bag during the first half of its inflation to the entropy change during the second half of its inflation.

13. Calculate the entropy change (in cal/deg) of the electrochemical cell considered in Section 9-2 as it is discharged reversibly. What is ΔS for the cell as it is discharged through the motor with equivalent resistance $R_M = 78$ ohms?

14. At the high altitude reached in Problem 3, the balloon bag rips. In the next, exciting ten seconds, indicate for each of the quantities listed below whether it will increase, decrease, or not change, and explain why. Assume that just before the fatal moment the helium P and T equals the atmospheric P and T.
(a) kinetic energy of the balloon and its instrument payload
(b) potential energy of the balloon and payload
(c) total energy of the balloon and payload
(d) enthalpy of the helium
(e) entropy of the helium
(f) entropy of the universe

15. What happens to the entropy of the system and of the universe in each of

these processes? (Notice that all of these processes actually occur, so they must be spontaneous.)

System	Immediate Surroundings	Process
(a) gin, vermouth	chemist, jigger, shaker	quantitative preparation of a potion
(b) crude oil	chemist, refinery	distillation to extract gasoline fraction
(c) automobile battery	absent-minded professor plus garage	prof. leaves battery on all night, battery discharges, becoming warm
(d) automobile battery	absent-minded prof. plus service-station facilities	prof. has battery recharged

✕ 16. What happens to the entropy of the system and of the universe in each of these processes?

System	Immediate Surroundings	Process
(a) plant	sun, earth, rain, atmosphere	plant grows
(b) plant	sun, earth, lack of rain, atmosphere	plant dies, leaves fall, decay
(c) student's brain	pages of this book, student's body, food, respiration	student reads book, becomes remarkably intelligent
(d) student's brain	hallucinatory drug	part of memory is damaged during a bad trip
(e) lettered blocks	mother and child	mother arranges blocks to spell CAT
(f) lettered blocks arranged to spell CAT	mother and child	child arranges blocks to spell A ⊣Ɔ

17. From the results of Problem 13 and of Problem 9, Chapter 7, calculate the values of ΔH, ΔS, and ΔG for the cell reaction $Zn(m) + 2\ Ag^+(aq) \longrightarrow Zn^{+2}(aq) + 2\ Ag(m)$ both for $1.0 \cdot 10^{-3}$ mole and for one mole of Zn dissolved. Specify units.

18. For a constant volume process, $\Delta E = q_V$. Use the condition that $(q_{rev} - q)$ must be positive for a spontaneous reaction at constant temperature, together with the definitions $\Delta H = \Delta E + \Delta(PV)$ and $\Delta G = \Delta H - T\Delta S$ to show that a constant T process will occur spontaneously in a closed bulb provided $(\Delta G - V\Delta P)$ is negative.

19. The reaction that occurs in an automobile battery during discharge is $Pb(m) + PbO_2(s) + 4\ H^+(aq) + 2\ SO_4^{-2}(aq) \longrightarrow 2\ PbSO_4(s) + 2\ H_2O(\ell)$. The value of ΔH for this reaction has been measured to be -76.3 kcal/mole Pb. Since $\Delta(PV) = 0$, $\Delta H = \Delta E$. The voltage generated by one of the six cells of the battery is 2.07 volts.

Calculate for the battery the heat absorbed from the surroundings, the useful electrical work performed, the time required, ΔS, and ΔG as 20.7 grams of Pb are consumed *in each cell* while the battery is discharging at constant T and P in each of the following ways. (Assume the internal battery resistance is 4 ohms.)
(a) reversibly
(b) through the headlights, with external resistance 8 ohms
(c) through the headlights, the radio, and the air conditioner in parallel, with combined external resistance of 2.0 ohms.

20.   When a solid melts at its normal freezing point, $\Delta G = 0$. Here are some heats of fusion and freezing points. Calculate the entropies of fusion (which measure the randomness increase on melting) and compare to the values calculated in Exercise 9-8.

	Ethane	Chlorine	Sulfur Dioxide	Benzene	Sodium	Magnesium
substance	C_2H_6	Cl_2	SO_2	C_6H_6	Na	Mg
m. p. (°K)	89.9	172	198	278.7	371	923
ΔH_{fusion} (kcal/mole)	0.68	1.53	1.77	2.35	0.63	2.2

21.   When a liquid boils under its equilibrium vapor pressure, $\Delta G = 0$. Here are some heats of vaporization and normal boiling points. Calculate the entropies of vaporization (which measure the randomness increase on vaporization) and compare to the values calculated in Exercise 9-9.

	Argon	Ethane	Chlorine	Sulfur Dioxide	Benzene	n-Penta-decane	Sodium	Mag-nesium
substance	Ar	C_2H_6	Cl_2	SO_2	C_6H_6	$C_{15}H_{32}$	Na	Mg
b. p. (°K)	87.3	184.5	238.9	263	353	543	1162	1393
$\Delta H_{vap.}$ (kcal/mole)	1.59	3.52	4.88	6.00	7.35	11.8	24.1	31.5

22.   Calculate ΔS and ΔG for the process described in Problem 12, Chapter 7.

23.   Which reaction will have the most positive entropy change and which the most negative?
(a) $S_8(s) + 8\,O_2(g) \longrightarrow 8\,SO_2(g)$
(b) $XeF_4(s) \longrightarrow Xe(g) + 2\,F_2(g)$
(c) $2\,H_2O_2(g) \longrightarrow 2\,H_2O(g) + O_2(g)$
(d) $NH_3(g) + HCl(g) \longrightarrow NH_4Cl(s)$

24.   When one gram of salt (NaCl) dissolves in one liter of water isolated in a thermos bottle, the solution is slightly cooled.
(a) Is ΔS for the process positive or negative?
(b) Is ΔH for the process positive or negative?
(c) If the process were carried out at constant temperature, would ΔG be positive or negative and would ΔH or ΔS dominate in fixing that sign?

25.   If the NaCl in Problem 24 is replaced by NaI, the solution is slightly warmed. Answer questions (a), (b), and (c) for sodium iodide.

ten free energy and equilibrium

At constant pressure and temperature, spontaneous processes seek the lowest possible free energy. In such cases, lowering the free energy of a system corresponds to raising the entropy of the universe. Hence, we can say that on a free energy landscape, chemical changes tend to "roll downhill." If a postulated change has positive ΔG, the reaction is heading uphill, and we can be sure the reaction will in fact go in the reverse direction (reaction rate being willing). When $\Delta G = 0$, there is no tendency to go in either direction. At the bottom of the free energy valley, equilibrium exists.

10-1 Free energy—the reaction direction signpost

Knowing that reactions proceed downhill on the free energy landscape, it is clear that free energy will help us make predictions about chemistry. We can answer such a question as "Is ammonia a potential fuel?" If we can learn the free energy change associated with the combustion reaction, its sign will tell us whether the path is downhill. If ΔG is large and negative, the reaction tendency favors products and ammonia may be a useful fuel. The reaction must also be rapid, of course, at an attainable temperature, so thermodynamics doesn't tell us everything we need to know. It also cannot tell us the economics of the process. Nevertheless, thermodynamics is a logical place to start, for *thermodynamics defines the possible.*

(a) STANDARD FREE ENERGIES OF FORMATION

The results of equilibrium and calorimetric studies are extensively tabulated. One such tabulation, used internationally, is the "Selected Values of Chemical Thermodynamic Properties" published by the U.S. National Bureau of Standards.* In this valuable collection are found quantities called the "standard free energy of formation" and "standard enthalpy of forma-

---

* "Selected Values of Chemical Thermodynamic Properties, Part I: Tables," National Bureau of Standards Circular 500—Part I, reprinted July 20, 1961. This reference has recently been superseded, in part, by National Bureau of Standards Technical Note 270-3, January 1968.

Table 10-1 Standard Enthalpies and Free Energies of Formation ($T = 298.16°K$ (25°C))

	ΔH_f^0 (kcal/mole)	ΔG_f^0 (kcal/mole)
N_2(g, 1 atm)	(0.000)	(0.000)
O_2(g, 1 atm)	(0.000)	(0.000)
H_2(g, 1 atm)	(0.000)	(0.000)
H_2O(g, 1 atm)	−57.7979	−54.6357
NH_3(g, 1 atm)	−11.04	−3.976
NO(g, 1 atm)	21.600	20.719
NO_2(g, 1 atm)	8.091	12.390
N_2O_4(g, 1 atm)	2.309	23.491
N_2O(g, 1 atm)	19.49	24.76

tion." Table 10-1 lists some entries from these tables that are relevant to our question about the fuel potentialities of ammonia.

Obviously some advance knowledge was needed to select just nine compounds from the thousands of entries in the 822-page document. Merely to open the reference book meaningfully, it was necessary to write down some possible reactions. Drawing upon previous knowledge of combustion processes, probably learned from a textbook, but ultimately by experiment, we decided to consider the following possibilities:

$$NH_3(g) + \tfrac{7}{4} O_2(g) \longrightarrow NO_2(g) + \tfrac{3}{2} H_2O(g) \qquad (10\text{-}1)$$

$$NH_3(g) + \tfrac{5}{4} O_2(g) \longrightarrow NO(g) + \tfrac{3}{2} H_2O(g) \qquad (10\text{-}2)$$

$$NH_3(g) + O_2(g) \longrightarrow \tfrac{1}{2} N_2O(g) + \tfrac{3}{2} H_2O(g) \qquad (10\text{-}3)$$

Each of these three reactions is a candidate for our consideration. In each, ammonia reacts with oxygen. Which one might be important? Perhaps the data in Table 10-1 will tell.

Just a glance at Table 10-1 reveals the surprising fact that the first three molecules, N_2, O_2, and H_2, have identical entries in both columns. How does it happen that both the enthalpy and the free energy of nitrogen are exactly zero at 298°K, and that oxygen and hydrogen are the same? Our answer ought to be found in a microscopic look at a mole of nitrogen molecules.

Consider the enthalpy of this mole of nitrogen molecules. By our examination of molecular activities in Chapter Seven, the enthalpy must be a composite of many energy contributions. First, the molecules have a variety of translational energies that average out to $\tfrac{3}{2}RT$ ergs per mole. The molecules also rotate and vibrate; so there are two more energy contributions. In addition to these "energy of motion" quantities, each nitrogen molecule possesses "energy of position;" that is, each molecule consists of two atoms spaced at the equilibrium distance at which their energy is at a minimum. Probing further, the electrons in the atoms have energy of motion and energy of position. And finally, to compound the problem to the

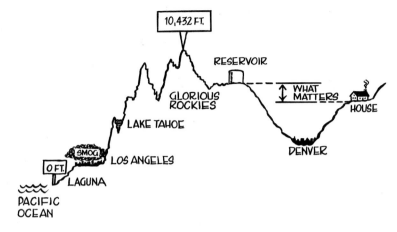

Figure 10-1   It's all relative anyway.

utmost, the nuclei have enormous, though little understood, binding energies holding each one together.

So just to express the enthalpy of a mole of nitrogen relative to its primordial constituents at $0°K$, we set off on quite an odyssey. The situation can be likened to a building contractor in Denver, Colorado, who wonders whether his tract site is above or below the water reservoir on the other side of the valley (see Fig. 10-1). He could try to determine the altitude of the reservoir relative to sea level out in California. On the other hand, he might decide he hasn't time to triangulate his way over the Rockies and all the way to Laguna Beach. It will suffice to pick some convenient local reference point and forget about sea level. In fact, the reservoir itself might as well be taken as the reference altitude.

We will follow the home-builder's lead. For the present purpose, let's calculate all enthalpies relative to the elements as they are normally found —for N_2, O_2, and H_2, that means as gases at some convenient pressure and temperature. Convenience alone dictated the choice that chemists have agreed upon: the elements as they exist at 25°C and at pressures of one atmosphere. It is a matter of necessity, however, to have some reference point. **Only** *changes* **in enthalpy and free energy are known** (or needed, for a chemist's purposes). Since the elements are selected for reference, they are *assigned* enthalpies and free energies of zero.

(b) THE USE OF ΔH_f^0 AND ΔG_f^0

The use of these tabulated values depends upon the fact that enthalpy and free energy are both functions of state. This means that the change of enthalpy (or free energy) in a change of state is independent of the path (see Section 7-2(g) if you wish to brush up on this point). Because of this property, we can consider a chemical change in a step-wise way if we wish. Thus, the change (10-1) can be considered in two steps as shown:

$$
\begin{bmatrix}
\text{1 mole NH}_3(g) \\
\text{1 atm} \\
\tfrac{7}{4} \text{ mole O}_2(g) \\
\text{1 atm} \\
25°C
\end{bmatrix}
\xrightarrow[\Delta G_{\mathrm{I}}]{\Delta H_{\mathrm{I}}}
\begin{bmatrix}
\text{1 mole NO}_2(g) \\
\text{1 atm} \\
\tfrac{3}{2} \text{ mole H}_2O(g) \\
\text{1 atm} \\
25°C
\end{bmatrix}
$$

$$
\Delta H_{\mathrm{II}} \quad \Delta G_{\mathrm{II}}
\qquad
\begin{bmatrix}
\tfrac{1}{2} \text{ mole N}_2(g) \\
\tfrac{3}{2} \text{ mole H}_2(g) \\
\tfrac{7}{4} \text{ mole O}_2(g) \\
\text{all 1 atm} \\
25°C
\end{bmatrix}
\qquad \Delta H_{\mathrm{III}} \quad \Delta G_{\mathrm{III}}
\tag{10-4}
$$

If ΔH_{I} depends only on the initial and final state, it can be found by summing ΔH_{II} and ΔH_{III}, the enthalpy changes connected with forming the elements from the reactants followed by combining the elements into the products. Thus we can write

$$
\Delta H_{\mathrm{I}} = \Delta H_{\mathrm{II}} + \Delta H_{\mathrm{III}}
\tag{10-5}
$$

We note that ΔH_{III} is merely the sum of the enthalpies of formation of one mole of NO_2 and $\tfrac{3}{2}$ moles of H_2O from the elements. The quantity ΔH_{II} is the *negative* of the heat of formation of one mole of NH_3. Hence we obtain

$$
\Delta H_{\mathrm{I}} = [-\Delta H_f^0(NH_3)] + [\Delta H_f^0(NO_2) + \tfrac{3}{2}\Delta H_f^0(H_2O)]
\tag{10-6}
$$

From Table 10-1,

$$
\Delta H_{\mathrm{I}} = [-(-11.04)] + [8.091 + \tfrac{3}{2}(-57.7979)]
$$
$$
= -67.56 \text{ kcal}
$$

We conclude that reaction (10-1) is exothermic by 67.56 kcal per mole of NH_3 consumed. Surely this suggests that reaction (10-1) is a possible energy (heat) source. We determined this merely by consulting a reference table of enthalpies of formation and by using the "function of state" property. Thus *any* reaction enthalpy change can be calculated in a few minutes if the tabulated enthalpies of formation are available.

Exercise 10-1. Use the tabular values of ΔH_f^0 in Table 10-1 and a step-wise process like (10-4) to show that reaction (10-2) is exothermic by 54 kcal.

Thermodynamics can also tell us the reaction tendency, the information contained in ΔG. If ammonia is to be used as a fuel, it must not only produce heat, but also the reaction must occur spontaneously. We learned in Chapter Nine that reactions that occur at constant pressure and constant temperature will occur spontaneously if ΔG is negative.

A tabulation of free energies of formation makes the ΔG for any reaction easily accessible. This is because *free energy is also a function of state.* It is such because each of its two parts, ΔH and $T\Delta S$, are themselves functions of state. (Recall that H was shown to be a function of state by the law of energy conservation. Entropy S was defined so as to be a function of state, determined solely by the intrinsic probabilities of initial and final states.) The sum (or difference) of two functions of state (as in $G = H - TS$) must also be a function of state.

Referring to the diagram (10-4) and to Table 10-1, we obtain

$$\Delta G_I = \Delta G_{II} + \Delta G_{III}$$
$$= [-\Delta G_f^0(NH_3)] + [\Delta G_f^0(NO_2) + \tfrac{3}{2}\Delta G_f^0(H_2O)]$$
$$= [-(-3.976)] + [+12.390 + \tfrac{3}{2}(-54.6357)]$$
$$= -65.59 \text{ kcal/mole of } NH_3 \qquad (10\text{-}7)$$

We see that ΔG^0 for reaction (10-1) is negative. Therefore, the reaction between ammonia and oxygen to form nitrogen dioxide and water vapor can proceed spontaneously. If it does, it will release 67.56 kcal of heat per mole of ammonia burned. We have said "if it does" because thermodynamics does not tell us the rate at which spontaneous changes occur. In fact, we all know that ammonia does not burst into flame when it is used as a household cleaner, although it is exposed to oxygen in the air. The reaction is too slow. You may not know, however, that ammonia can be passed into an ordinary bunsen burner instead of methane and, once lit, the ammonia burns quite like a normal burner fuel.

Exercise 10-2. Use the tabular values of ΔG_f^0 in Table 10-1 and a step-wise process like (10-4) to show that reaction (10-2) can proceed spontaneously, because $\Delta G = -57$ kcal.

In this manner, we can calculate ΔH^0 and ΔG^0 for each of the three possible combustion reactions (10-1), (10-2), and (10-3). The results are compiled in Table 10-2. All three reactions are exothermic, and each one

Table 10-2 Enthalpy and Free Energy Changes for Possible Ammonia Combustion Reactions (All Constituents at 298.16°K and 1 atm Pressure)

	ΔH^0 (kcal)	ΔG^0 (kcal)
$NH_3(g) + \tfrac{7}{4}O_2(g) \longrightarrow NO_2(g) + \tfrac{3}{2}H_2O(g)$	-67.56	-65.59
$NH_3(g) + \tfrac{5}{4}O_2(g) \longrightarrow NO(g) + \tfrac{3}{2}H_2O(g)$	-54.06	-57.26
$NH_3(g) + \quad O_2(g) \longrightarrow \tfrac{1}{2}N_2O(g) + \tfrac{3}{2}H_2O(g)$	-65.90	-65.60

has a tendency to proceed spontaneously, since all three ΔG^0's are negative! Yet we note again that ammonia does not burst into flame on exposure to air at room temperature. There must be impeding energy barriers that cause all three reactions to be extremely slow at room temperature (Fig. 10-2). It is found by experiment that when ammonia does burn, at a high temperature at which reaction rates are high, NO_2, NO, and N_2, but *not* N_2O, are formed in a mixture, the relative amounts of each being dependent upon the amount of oxygen available. Thermodynamics tells us (see Table 10-2) that any one of the three reactions (10-1), (10-2), or (10-3) *could* occur and, if it did, heat would be released. However, it cannot tell us the speed of reaction for any of the three.

Figure 10-2 Standard free energy signposts: Downhill is the way to go.

10-2 Free energy and equilibrium

At first glance, these considerations seem to contradict what we know about equilibrium. At equilibrium, both reactants and products are present, in dynamic balance. Yet in Section 10-1, we tried to answer our question about ammonia merely by asking whether ΔG is positive or negative. The negative ΔG implies that NH_3 and O_2 can react—it doesn't seem to say anything about the reverse reaction. Yet we shall see that there is no incompatibility between the ΔG criterion for spontaneous reaction and our dynamic view of equilibrium—as long as we know what the ΔG criterion means.

(a) REACTION UNDER STANDARD CONDITIONS

Reaction (10-4) specifies plainly that the change considered is that of one mole of ammonia and $\frac{7}{4}$ moles of oxygen reacting at fixed pressures, each at one atmosphere. Furthermore, the products are formed at fixed pressures, each at one atmosphere. To carry out such a reaction in the laboratory, all the reactants and products would have to be mixed in a reaction vessel, each at one atmosphere pressure. For this "standard" reaction mixture, the free energy change is called the "standard free energy" and it is designated ΔG^0. The algebraic sign of the "standard" free energy tells us the "direction" chemical change will take from these starting conditions —whether reactants, each at one atmosphere, will be consumed at the specified temperature to form products at one atmosphere, or vice versa.

Equilibrium, on the other hand, indicates that there must be a particular mixture of reactants and products for which free energy is a minimum. For example, an equimolar mixture of H_2 and I_2, each at 1 torr pressure, will react at 700°K to form the product HI. Beginning with H_2 and I_2 each at 1 torr, downhill on the free energy landscape heads us toward HI. However, experiment shows that reaction stops when the mole fraction of product, HI, is 0.81. But, if we begin with pure HI at 2 torr and 700°K, decomposition to H_2 and I_2 begins—now downhill is the other direction, as far as ΔG is concerned. Again the reaction proceeds until the mole fraction of HI is 0.81. At this composition, equilibrium exists—there remains no net tendency for change.

Our tabular values of ΔG^0 merely "point" toward equilibrium, the free energy minimum. We want to know more than this—we want to know how far away it is. Surprisingly, it turns out that ΔG^0 tells us that, as well! Here is how the story goes.

(b) REACTION UNDER NONSTANDARD PRESSURE CONDITIONS

Given ΔG^0, the free energy change under standard conditions (1 atm and a specified temperature), it would be helpful to be able to calculate ΔG

for some other set of conditions that might interest us. For the moment, let's consider changing only the pressures. For example, consider the oxidation of nitric oxide NO to nitrogen dioxide NO_2.

$$NO(g) + \tfrac{1}{2} O_2(g) \xrightarrow[\Delta G^0]{\Delta H^0} NO_2(g) \qquad\qquad (10\text{-}8)$$

$$\begin{array}{ccc} 1\ \text{atm} & 1\ \text{atm} & 1\ \text{atm} \\ 25°C & 25°C & 25°C \end{array}$$

Table 10-1 contains the necessary information for us to calculate the free energy and enthalpy changes under these standard conditions: $\Delta H^0 = -13.51$ kcal and $\Delta G^0 = -8.33$ kcal. The reaction is exothermic and has a tendency to react as shown.

Now suppose our interest is in the reaction as written, except for one change—we would like to consider the conditions under which NO_2 is formed at a pressure of *one-half* atmosphere.

$$NO(g) + \tfrac{1}{2} O_2(g) \xrightarrow[\Delta G]{\Delta H} NO_2(g) \qquad\qquad (10\text{-}9)$$

$$\begin{array}{ccc} 1\ \text{atm} & 1\ \text{atm} & \tfrac{1}{2}\ \text{atm} \\ 25°C & 25°C & 25°C \end{array}$$

In determining ΔH and ΔG for such nonstandard conditions, the traditional strategy is to try to devise a step-wise way of accomplishing the change. If a path can be found for which ΔH and ΔG for each step are easily learned, the over-all ΔH and ΔG can be evaluated, since they are independent of path. In our case, we would like a step-wise process that accomplishes (10-9), but includes as one of its steps the change under standard conditions, since we know ΔH^0 and ΔG^0. Here is such a two-step process.

$$\begin{bmatrix} NO(g) + \tfrac{1}{2} O_2(g) \\ 1\ \text{atm} \qquad\quad 1\ \text{atm} \\ 25°C \qquad\quad 25°C \end{bmatrix} \xrightarrow[\Delta G]{\Delta H} \begin{bmatrix} NO_2(g) \\ \tfrac{1}{2}\ \text{atm} \\ 25°C \end{bmatrix}$$

$$\begin{array}{cc} \Delta H^0 & \Delta H' \\ \Delta G^0 & \Delta G' \end{array} \qquad\qquad (10\text{-}10)$$

$$\begin{bmatrix} NO_2(g) \\ 1\ \text{atm} \\ 25°C \end{bmatrix}$$

Plainly,

$$\Delta H = \Delta H^0 + \Delta H' \qquad\qquad (10\text{-}11)$$

and

$$\Delta G = \Delta G^0 + \Delta G' \qquad\qquad (10\text{-}12)$$

We have already looked up ΔH^0 and ΔG^0 in the tables, so all we need to do is calculate $\Delta H'$ and $\Delta G'$.

The first quantity $\Delta H'$ is the easier of the two. Remember that

$$H = E + PV$$

so

$$\Delta H' = \Delta E' + \Delta (PV)'$$

But for a perfect gas, E is a function of temperature only. This means that the energy content of a perfect gas is constant at a fixed temperature, so $\Delta E' = 0$. Also for a perfect gas $PV = nRT$ and, if T is constant, the PV product is constant. So $\Delta (PV)' = 0$, and hence,

$$\Delta H' = 0 \qquad\qquad (10\text{-}13)$$

$$\boxed{\Delta H = \Delta H^0} \qquad\qquad (10\text{-}14)$$

The second quantity $\Delta G'$ is more interesting. Since $G = H - TS$,

$$\Delta G' = \Delta H' - \Delta (TS)' \qquad\qquad (10\text{-}15)$$

Substituting (10-13) into (10-15) and remembering that T is constant, (10-15) becomes

$$\Delta G' = 0 - \Delta (TS)' = -T\Delta S' \qquad\qquad (10\text{-}16)$$

The quantity $\Delta S'$ is just the entropy change accompanying a constant temperature expansion. Increasing the volume occupied by a mole of gas increases positional randomness, so we cannot expect $\Delta S'$ to be zero. However, it is readily evaluated.

Entropy is a function of state. Hence the entropy change $\Delta S'$ that accompanies expansion of the mole of NO_2 from 1 atm to $\frac{1}{2}$ atm does not depend upon how the expansion is carried out. To evaluate $\Delta S'$, then, we can consider the process most convenient for that purpose, a reversible one. Such an expansion gives us the maximum possible work and the reversible heat, as was shown in Section 9-1.

For a reversible expansion,

$$q = q_{rev}$$

and

$$w = w_{max} = nRT \log_e \frac{V_2}{V_1}$$

therefore,

$$q_{\text{rev}} = nRT \log_e \frac{V_2}{V_1} \qquad (10\text{-}17)$$

For the change as specified in (10-10), it is more convenient to express (10-17) in terms of pressure, using the perfect gas relation $P_1 V_1 = P_2 V_2$.

$$q_{\text{rev}} = nRT \log_e \frac{P_1}{P_2} \qquad (10\text{-}18)$$

Expression (10-18) is just what we need to calculate $\Delta S'$ since, for a constant temperature process,

$$\Delta S = \frac{q_{\text{rev}}}{T} \qquad (10\text{-}19)$$

so

$$\Delta S' = \frac{nRT \log_e (P_1/P_2)}{T}$$

$$= nR \log_e \frac{P_1}{P_2} \qquad (10\text{-}20)$$

$$\Delta G' = -T\Delta S' = -T\left(-nR \log_e \frac{P_2}{P_1}\right)$$

$$= nRT \log_e \frac{P_2}{P_1} \qquad (10\text{-}21)$$

and

$$\Delta G = \Delta G^0 + \Delta G'$$

so

$$\Delta G = \Delta G^0 + nRT \log_e \frac{P_2}{P_1} \qquad (10\text{-}22)$$

Expression (10-22) can be simplified even further because P_1 is the standard pressure, one atmosphere, and $\log_e 1 = 0$, so

$$\Delta G = \Delta G^0 + nRT \log_e \frac{P_2}{1} = \Delta G^0 + nRT(\log_e P_2 - \log_e 1)$$

$$\boxed{\Delta G = \Delta G^0 + nRT \log_e P_2} \qquad (10\text{-}23)$$

In (10-23), n refers to the number of moles of NO_2, the gas that expanded to P_2.

Exercise 10-3. Show that ΔG for reaction (10-9) differs from ΔG^0 by -0.41 kcal. (Note: $R = 1.99$ cal/deg mole; $\log_e P_2 = 2.303 \log_{10} P_2$.)

(c) REACTION UNDER EQUILIBRIUM CONDITIONS

It is clear that we can compute ΔG for *any* set of pressure conditions exactly as we did (10-23). For example, consider reaction (10-8) occurring at generalized pressures p_{NO}, p_{O_2}, and p_{NO_2}. Our multistep process would be the following:

$$
\begin{bmatrix} NO(g) \\ p_{NO} \\ 25°C \end{bmatrix} + \begin{bmatrix} \frac{1}{2} O_2(g) \\ p_{O_2} \\ 25°C \end{bmatrix} \xrightarrow[\Delta G]{\Delta H} \begin{bmatrix} NO_2(g) \\ p_{NO_2} \\ 25°C \end{bmatrix}
$$

$$
\Big\downarrow {}^{\Delta H_1}_{\Delta G_1} \qquad\qquad \Big\downarrow {}^{\Delta H_2}_{\Delta G_2} \qquad\qquad \Big\uparrow {}^{\Delta H_3}_{\Delta G_3}
$$

$$
\begin{bmatrix} NO(g) \\ 1\,atm \\ 25°C \end{bmatrix} + \begin{bmatrix} \frac{1}{2} O_2(g) \\ 1\,atm \\ 25°C \end{bmatrix} \xrightarrow[\Delta G^0]{\Delta H^0} \begin{bmatrix} NO_2(g) \\ 1\,atm \\ 25°C \end{bmatrix} \qquad (10\text{-}24)
$$

Thus

$$\Delta H = \Delta H_1 + \Delta H_2 + \Delta H^0 + \Delta H_3 \qquad (10\text{-}25)$$

and

$$\Delta G = \Delta G_1 + \Delta G_2 + \Delta G^0 + \Delta G_3 \qquad (10\text{-}26)$$

Just as in (10-13), at constant temperature, $\Delta H_1 = \Delta H_2 = \Delta H_3 = 0$, so

$$\boxed{\Delta H = \Delta H^0} \qquad (10\text{-}27)$$

The quantities ΔG_1, ΔG_2, and ΔG_3 are all evaluated using (10-21).

$$\Delta G_1 = n_{NO}\, RT \log_e \frac{1\,atm}{p_{NO}}$$

$$\Delta G_2 = n_{O_2}\, RT \log_e \frac{1\,atm}{p_{O_2}}$$

and

$$\Delta G_3 = n_{NO_2}\, RT \log_e \frac{p_{NO_2}}{1\,atm}$$

hence,

$$\Delta G = \Delta G^0 + n_{NO}RT(-\log_e p_{NO}) + n_{O_2}RT(-\log_e p_{O_2}) + n_{NO_2}RT(\log_e p_{NO_2})$$

or, since $n\log p = \log p^n$,

$$\Delta G = \Delta G^0 - RT\log_e(p_{NO})^{n_{NO}} - RT\log_e(p_{O_2})^{n_{O_2}} + RT\log_e(p_{NO_2})^{n_{NO_2}}$$

that is,

$$\Delta G = \Delta G^0 + RT\log_e \frac{(p_{NO_2})^{n_{NO_2}}}{(p_{NO})^{n_{NO}}(p_{O_2})^{n_{O_2}}}$$

Inserting $n_{NO_2} = 1$, $n_{NO} = 1$, and $n_{O_2} = \frac{1}{2}$ from the balanced equation (10-9),

$$\boxed{\Delta G = \Delta G^0 + RT\log_e \frac{p_{NO_2}}{p_{NO}p_{O_2}^{1/2}}} \qquad (10\text{-}28)$$

Equation (10-28) can be applied to calculate ΔG for any set of NO_2, NO, and O_2 pressures. For example, it can be applied at any composition on a free energy plot, such as the one shown in Figure 10-3. In fact it can be applied even at the minimum in the free energy curve, the point at which equilibrium exists. Under these equilibrium conditions, a minute change either toward reactants or toward products causes no change in free energy. At equilibrium, $\Delta G = 0$. This special relationship really makes equation (10-28) take on its greatest meaning. When the gas pressures reach equilibrium values,

$$\Delta G = 0 = \Delta G^0 + RT\log_e \frac{(p_{NO_2})_{eq}}{(p_{NO})_{eq}(p_{O_2})_{eq}^{1/2}}$$

so

$$\Delta G^0 = -RT\log_e \frac{(p_{NO_2})_{eq}}{(p_{NO})_{eq}(p_{O_2})_{eq}^{1/2}} \qquad (10\text{-}29)$$

Expression (10-29) can be rearranged as follows:

$$\log_e \frac{(p_{NO_2})_{eq}}{(p_{NO})_{eq}(p_{O_2})_{eq}^{1/2}} = -\frac{\Delta G^0}{RT} \qquad (10\text{-}30)$$

or

$$\frac{(p_{NO_2})_{eq}}{(p_{NO})_{eq}(p_{O_2})_{eq}^{1/2}} = e^{-\Delta G^0/RT} \qquad (10\text{-}31)$$

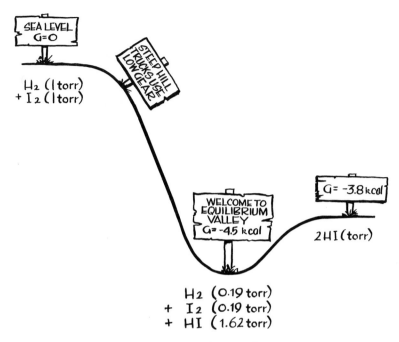

Figure 10-3 Down the free energy hill to equilibrium valley.

On the left-hand side in (10-31), we see a relation among concentrations (pressures) of reactants and products—the familiar equilibrium expression! On the right, we find an exponential, the exponent of which is fixed by ΔG^0 and T. Since ΔG^0 is just a number characteristic of the reaction, $\Delta G^0 / RT$ is a constant at any given temperature. Hence the right-hand side of (10-31) must be a constant, and this constant must be the equilibrium constant! Hence (10-31) can be rewritten as

$$\frac{(p_{NO_2})_{eq}}{(p_{NO})_{eq}(p_{O_2})_{eq}^{1/2}} = K = e^{-\Delta G^0/RT} \qquad (10\text{-}32)$$

and (10-30) becomes

$$\Delta G^0 = -RT \log_e K = -2.303 \, RT \log_{10} K \qquad (10\text{-}33)$$

$$K = e^{-\Delta G^0/RT} = 10^{-\Delta G^0/2.303RT} \qquad (10\text{-}33a)$$

at 298°K
$$\Delta G^0 = -1364 \log_{10} K \qquad (10\text{-}33b)$$

Expression (10-32) is the familiar equilibrium law, the one we obtained in Chapter Three through reaction rate arguments. Now we have seen its origin in thermodynamics. Expression (10-33) tells us how to calculate K from standard free energies. Obviously these are extremely valuable relationships.

(d) THE ΔG REACTION SIGNPOST AT WORK

We might exercise relations (10-28) and (10-33) by applying them to the H_2–I_2 reaction mentioned earlier. First, ΔG^0 is needed and, once again, we can obtain this from the standard free energies of formation of the reactants and products, as given in the National Bureau of Standards tables.

$$H_2(g) + I_2(g) = 2\,HI(g) \tag{10-34}$$

$$\Delta G^0 = 2\Delta G_f^0(HI) - \Delta G_f^0(H_2) - \Delta G_f^0(I_2)$$

$$= 2(0.31) - (0.00) - (4.63)^*$$

$$= -4.01 \text{ kcal} \tag{10-35}$$

Now we can apply (10-33) to calculate the equilibrium constant at 298.16°K for reaction (10-34):

$$K = e^{-\Delta G^0/RT} = e^{-(-4010)/(1.99 \cdot 298)} = e^{+6.76}$$

$$= 10^{+6.76/2.303} = 10^{+2.93}$$

$$= 8.6 \cdot 10^{+2} \tag{10-36}$$

With this equilibrium constant, we can calculate the equilibrium composition that would result if H_2 gas and I_2 gas, each at 1 torr pressure, were to react at room temperature (perhaps in the presence of a catalyst). In terms of our quantity x, the equilibrium pressures would be

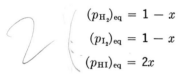

$$(p_{H_2})_{eq} = 1 - x$$

$$(p_{I_2})_{eq} = 1 - x$$

$$(p_{HI})_{eq} = 2x$$

These can now be substituted in the equilibrium expression

$$K = \frac{(p_{HI})_{eq}^2}{(p_{H_2})_{eq}(p_{I_2})_{eq}}$$

$$= \frac{(2x)^2}{(1-x)^2} \tag{10-37}$$

---

* Note that in the tables, the standard free energy of formation of gaseous I_2 is not zero because the standard state is taken to be *solid* iodine at 298.16°K. The $\Delta G_f^0(I_2)$ given refers to the hypothetical iodine gas at one atmosphere pressure formed from solid iodine, the reference state.

The solution to this expression is $x = 0.94$. At room temperature, 94 percent of the H_2 will react to reach the equilibrium situation. Remember that at the higher temperature, 700°K, the equilibrium mixture involved an 81 percent conversion. This is the appropriate direction of change, as suggested by Le Chatelier's Principle, which indicates that, since the reaction releases heat, a rise in temperature will shift equilibrium in the endothermic direction, toward reactants. Thermodynamics tells us why and exactly how much.

Exercise 10-4. Calculate the equilibrium constant for reaction (10-1), using ΔG^0 from Table 10-2. Notice that the very high value, $K = 10^{48}$, implies that at equilibrium there would be practically no reactants remaining, yet ammonia and oxygen react very slowly at room temperature.

Now we might apply relation (10-28) at other than equilibrium conditions, for example, on either side of equilibrium. Let's begin with $x = 0.50$, so that $p_{H_2} = p_{I_2} = 0.50$ torr and $p_{HI} = 1.00$ torr. Now

$$\Delta G_{x=0.50} = \Delta G^0 + RT \log_e \frac{(1.00)^2}{(0.5)(0.5)}$$

$$= -4010 + (2.303)(1.99)(298.16) \log_{10} 4$$

$$= -4010 + 820$$

$$= -3190 \text{ cal}$$

We see that at $x = 0.50$, ΔG is still negative, so equilibrium lies ahead in the direction of more reaction. If we now consider $x = 0.95$, so that $p_{H_2} = p_{I_2} = 0.05$ torr and $p_{HI} = 1.90$ torr, the calculation changes to

$$\Delta G_{x=0.95} = \Delta G^0 + RT \log_e \frac{(1.90)^2}{(0.05)(0.05)}$$

$$= -4010 + (2.303)(1.99)(298.16) \log_{10} 1444$$

$$= -4010 + 4310$$

$$= +300 \text{ cal}$$

Now ΔG is positive, so our direction signpost says that equilibrium lies behind us. At $x = 0.95$, the downhill direction is back toward reactants.

We see that the free energy change ΔG^0 tells us the direction of equilibrium beginning with the very special (and usually uninteresting) mixture of all reactants and products at the same pressure of one atmosphere. But ΔG, calculated via expression (10-28) tells us the direction of equilibrium from *any* starting mixture. If ΔG turns out to be negative, reaction can proceed toward products. If ΔG turns out to be positive, the mixture is already beyond equilibrium, and it can only proceed in the reverse direction, consuming the products and producing the reactants expected from the reaction as written.

10-3 Temperature dependence of ΔH, ΔS, and ΔG

The temperature dependencies of thermodynamic properties are obviously important. Most important is the fact that equilibrium conditions are observed to change rapidly with temperature, implying that temperature provides a powerful measure of control over chemical reactions. We shall explore these dependencies.

(a) ENTHALPY AND ENTROPY—TEMPERATURE INDEPENDENT (ALMOST)

Experimentally, we find that for most reactions, neither ΔH nor ΔS changes dramatically with temperature. Table 10-3 shows this for three typical gas-phase reactions and a couple of typical reactions involving solids. The first three columns show the enthalpy change at the particular temperature 298°K, the change at twice that temperature, 596°K, and the percentage difference in ΔH. The last three columns do the same for entropy change.

In the typical examples given, the value of ΔH changes by, at most, a few percent as the absolute temperature doubles. *To a very good and useful approximation, we can say that ΔH is temperature independent.*

The value of ΔS changes rather more—up to 10 percent for the ethylene–hydrogen reaction (C_2H_4–H_2). Nevertheless, the most striking property of ΔS is that it changes characteristically only a few percent per hundred degree change in temperature, making it unimportant for most considerations.

It is not too difficult to perceive why these two thermodynamic quantities should be relatively insensitive. The enthalpy change and the entropy change express the energy and randomness properties of the reaction products *relative* to the reactants. When the temperature is raised, the enthalpy of the products increases, but so also does the enthalpy of the reactants. Only if they go up by different amounts does a change in ΔH occur. The same is true for randomness. While the intrinsic randomness of the reactants increases as the temperature rises (there is more motional randomness), it also increases for the products. The net change in ΔS is the difference between these two values, and as for ΔH, there is a tendency for these changes to be of similar magnitude and therefore to cancel.

In conclusion, *it is a useful approximation to assume that ΔH and ΔS are temperature independent.* For temperature changes of the order of 100°, ΔH remains unchanged to less than one percent, and ΔS usually changes no more than two or three percent.

Table 10-3 Temperature Dependence of ΔH and ΔS

Reactions Involving Gases	ΔH^0(298°K) (kcal)	ΔH^0(596°K) (kcal)	Percentage Change	ΔS^0(298°K) (cal/degK)	ΔS^0(596°K) (cal/degK)	Percentage Change
$NO(g) + \frac{1}{2}O_2(g) = NO_2(g)$	−13.5	−14.0	3.5	−17.4	−18.5	6.3
$CO(g) + \frac{1}{2}O_2(g) = CO_2(g)$	−67.6	−68.1	0.7	−20.7	−21.8	5.3
$C_2H_4(g) + H_2(g) = C_2H_6(g)$	−32.7	−34.0	3.8	−28.8	−31.6	9.7
Reactions Involving Solids						
$CaCO_3(s) = CaO(s) + CO_2(g)$	+42.6	+42.1	1.1	+38.4	+37.4	2.7
$C(s) + \frac{1}{2}O_2(g) = CO(g)$	−26.4	−26.0	1.6	+21.4	+22.4	4.5

Table 10-4   Temperature Dependence of ΔG

	ΔG^0(298°K) (kcal)	ΔG^0(596°K) (kcal)	Percentage Change
Reactions Involving Gases			
$NO(g) + \frac{1}{2} O_2(g) = NO_2(g)$	− 8.33	− 2.97	+64
$CO(g) + \frac{1}{2} O_2(g) = CO_2(g)$	−61.4	−55.1	+10.3
$C_2H_4(g) + H_2(g) = C_2H_6(g)$	−24.1	−15.2	+37
Reactions Involving Solids			
$CaCO_3(s) = CaO(s) + CO_2(g)$	+31.2	+19.8	−36.6
$C(graphite) + \frac{1}{2} O_2(g) = CO(g)$	−32.8	−39.4	−20.0

(b)   FREE ENERGY CHANGE WITH TEMPERATURE—ANOTHER STORY

Table 10-4 shows ΔG for the same reactions listed in Table 10-3. Here, the *smallest* change is 10 percent, and ΔG for nitric oxide changes by a factor of 2.8. For some reactions, a temperature change can actually change the sign of ΔG from positive to negative. The free energy is by no means a temperature-independent quantity. This is graphically evident in Figure 10-4; over the temperature range shown, ΔG changes by a factor of two and plainly will change sign if the temperature is raised much more.

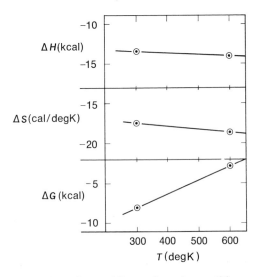

Figure 10-4   Temperature dependence of thermodynamic quantities

$$NO(g) + \frac{1}{2} O_2(g) = NO_2(g)$$

Again it is readily seen why ΔG changes so much with temperature, although ΔH and ΔS do not. Remembering the definition of ΔG for a reaction occurring at a fixed temperature T_1, the free energy change will be

$$\Delta G_1 = \Delta H - T_1 \Delta S \qquad (10\text{-}38)$$

At a different temperature T_2, ΔG will become

$$\Delta G_2 = \Delta H - T_2 \Delta S \qquad (10\text{-}39)$$

Expressions (10-38) and (10-39) show that the total randomness contribution to ΔG changes in proportion to temperature because of the T factor in $T\Delta S$. In fact, to a good approximation, a plot of ΔG against temperature gives a straight line. Figure 10-5 shows typical data for the reaction

$$M^+(FHF)^-(s) \rightleftharpoons M^+F^-(s) + HF(g) \qquad (10\text{-}40)$$

Obviously a straight line fits the experimental points quite well.

Exercise 10-5. For reaction (10-34), $\Delta H^0 = -2.4$ kcal and $\Delta S^0 = +5.1$ cal/deg mole. Calculate ΔG and K at 700°K, assuming ΔH and ΔS do not change with temperature.

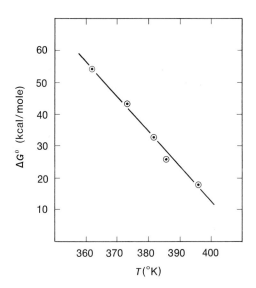

Figure 10-5 Variation of ΔG_2 with temperature for the dissociation reaction

$$M^+(FHF)^-(s) \rightleftharpoons M^+F^-(s) + HF(g)$$

(The abbreviation M^+ is used for the tetramethyl ammonium ion $(CH_3)_4N^+$.)

(c) ΔG^0 AGAINST T: AN EASY MEASUREMENT OF ΔH^0

The standard free energy is related simply to the equilibrium constant (as shown in Section 10-3):

$$\Delta G^0 = -RT \log_e K = -2.303RT \log_{10} K \qquad (10\text{-}41)$$

Consequently, expressions (10-38) and (10-39) can be rewritten in terms of the equilibrium constants K_1 and K_2 (at temperatures T_1 and T_2), provided the ΔH and ΔS values refer to standard conditions (1 atmosphere pressure for every constituent).

$$-RT_1 \log_e K_1 = \Delta H^0 - T_1 \Delta S^0 \qquad (10\text{-}42)$$

and

$$-RT_2 \log_e K_2 = \Delta H^0 - T_2 \Delta S^0 \qquad (10\text{-}43)$$

A more convenient form is obtained by dividing by $-RT$,

$$\log_e K_1 = -\frac{\Delta H^0}{RT_1} + \frac{\Delta S^0}{R} \qquad (10\text{-}44)$$

and

$$\log_e K_2 = -\frac{\Delta H^0}{RT_2} + \frac{\Delta S^0}{R} \qquad (10\text{-}45)$$

These two expressions can be combined to relate the temperature change in K to ΔH. Subtracting (10-45) from (10-44), the ΔS^0 term cancels out completely:

$$\log_e K_1 - \log_e K_2 = -\frac{\Delta H^0}{RT_1} + \frac{\Delta H^0}{RT_2}$$

or, more tidily,

$$\boxed{\log_e \frac{K_1}{K_2} = -\frac{\Delta H^0}{R}\left(\frac{1}{T_1} - \frac{1}{T_2}\right)} \qquad (10\text{-}46)$$

The relationship (10-46) is very important to chemists because it generally provides the easiest possible measurement of ΔH^0. The two equilibrium constants can be measured merely through constituent analysis at equilibrium. While this is not child's play and often requires quite sophisticated experiments, it is generally much more reliable than a direct calorimetric measurement of ΔH. It is, of course, obvious that measurement of K gives ΔG^0 directly through (10-33), and once ΔH^0 is known, ΔS^0 can also be calculated from (10-42). Hence, *a measurement of the temperature dependence of the equilibrium constant suffices to give all the thermodynamic functions, ΔG^0, ΔH^0, and ΔS^0, over that temperature range.*

10-4 Free energy and electrochemical cells

In Chapter Nine we saw how a spontaneous chemical change could be employed to produce useful work in an electrochemical cell. The amount of work that can be extracted is maximized when the cell is operated reversibly, that is, when the current is minimized. Now that we have the free energy concept at our disposal, it is timely to return to the electrochemical cell and investigate this maximum work. We shall see that cells provide an easy way to measure free energy.

(a) FREE ENERGY AND MAXIMUM WORK

Electrochemical cells are usually operated in open beakers, hence, at constant pressure and thermostatted at constant temperature. Under such common conditions, the free energy change that accompanies the cell reaction is simply related to the electrical work that can be extracted. This is readily seen by the use of the definitions of ΔG, ΔS, and ΔH.

At constant temperature,

$$\Delta G = \Delta H - T\Delta S \tag{10-47}$$

and, at constant pressure,

$$\Delta H = \Delta E + P\Delta V \tag{10-48}$$

Inserting (10-48) into (10-47),

$$\Delta G = \Delta E + P\Delta V - T\Delta S \tag{10-49}$$

Now we recall the relationships that ΔS and ΔE have to reaction heats:

$$\Delta E = q - w \tag{10-50}$$

and

$$\Delta S = \frac{q_{rev}}{T} \tag{10-51}$$

Substituting (10-50) and (10-51) into (10-49),

$$\Delta G = q - w + P\Delta V - q_{rev}$$

or

$$\Delta G = (q - q_{rev}) - w + P\Delta V \tag{10-52}$$

Now, in an electrochemical cell there are two kinds of work. The first is the $P\Delta V$ expansion work associated with pushing back the atmosphere at constant pressure. The second we will call $w_{elect.}$, the electrical work done by the cell. Putting these explicitly into (10-52), we obtain

$$\Delta G = (q - q_{rev}) - (P\Delta V + w_{elect.}) + P\Delta V$$

or

$$\Delta G = (q - q_{rev}) - w_{elect.} \tag{10-53}$$

In Section 9-2, the succession of ways in which the electrochemical cell was discharged showed that there is a maximum amount of electrical work $w_{elect.,max}$ that can be obtained. However, this maximum work is obtained only if the cell is discharged infinitely slowly. Under these reversible conditions, q becomes equal to q_{rev}:

$$\Delta G = (q_{rev} - q_{rev}) - w_{elect.,max}$$

or

$$\boxed{\Delta G = - w_{elect.,max}} \tag{10-54}$$

This result is the reason ΔG is called the *free* energy. At constant pressure and temperature, ΔG measures the maximum work other than the pressure–volume expansion work. In an open beaker, this $P\Delta V$ work is wasted; it merely pushes back the atmosphere. The rest of the possible work, measured by ΔG, is "available" or "free" for our use.

(b) FREE ENERGY AND CELL VOLTAGE

In Section 9-2(d), the relationship between work performed and cell voltage was developed. We found that the maximum electrical work that could be extracted is proportional to the cell voltage, $\Delta \mathcal{E}$:

$$\boxed{w_{elect.,max} = n\Delta\mathcal{E}\mathcal{F}} \tag{9-32}$$

where n is the number of moles of electrons exchanged in reaction, $\Delta\mathcal{E}$ is the voltage generated by the cell at zero current, $\mathfrak{F}$ is equal to 96,500 if the work is expressed in joules, equal to 23,061 if the work is expressed in calories, and equal to unity if the work is expressed in electron volts.

This result, (9-32), can be combined with (10-54) to lead to a powerful conclusion. If

$$\Delta G = -w_{\text{elect.,max}} \tag{10-54}$$

and

$$w_{\text{elect.,max}} = n\Delta\mathcal{E}\mathfrak{F} \tag{9-32}$$

we have

$$\boxed{\Delta G = -n\Delta\mathcal{E}\mathfrak{F}} \tag{10-55}$$

We seem to have encountered a new way to measure free energy changes! Merely by preparing an electrochemical cell and measuring its voltage under reversible conditions, we can determine ΔG. The expression "under reversible conditions" means merely the "zero-current conditions" that we mentioned in Chapter Six. Let's review how it is done.

Figure 10-6 shows a Zn–Ag cell set in opposition to a fraction of the voltage $\Delta\mathcal{E}'_0$ of a calibrated cell. The fraction is fixed by the ratio of R/R_0, which depends upon the position of the slide contact. For small values of R (Fig. 10-6(a)), the Zn–Ag voltage will exceed the opposing voltage $(R_a/R_0) \cdot \Delta\mathcal{E}'_0$, and current will flow in the direction that discharges our cell. For large values of R (Fig. 10-6(b)), the Zn–Ag voltage will be less than the opposing voltage $(R_b/R_0) \cdot \Delta\mathcal{E}'_0$, and the electron current will be reversed, flowing so as to recharge our cell. At the slide-contact position at which no current flows at all (Fig. 10-6(c)), the measured voltage $(R_c/R_0) \cdot \Delta\mathcal{E}'_0$ exactly equals the Zn–Ag voltage. Since the opposition of equal voltages stops current flow through the cell, its voltage is measured under reversible conditions. The ease with which a reaction can be studied under reversible conditions in an electrochemical cell is the reason for the particular charm these devices have for a chemist. *Cell-voltage measurements provide one of the best means of determining free energy changes for reactions that occur in aqueous solutions.* Of course, the absence of electrical conductivity limits the usefulness of such measurements in other solvents and for entirely gas-phase reactions.

(c) FREE ENERGY AND THE NERNST EQUATION

If we can determine ΔG through a cell-voltage measurement, we can apply the measurement to the special case of a cell operating with all concentra-

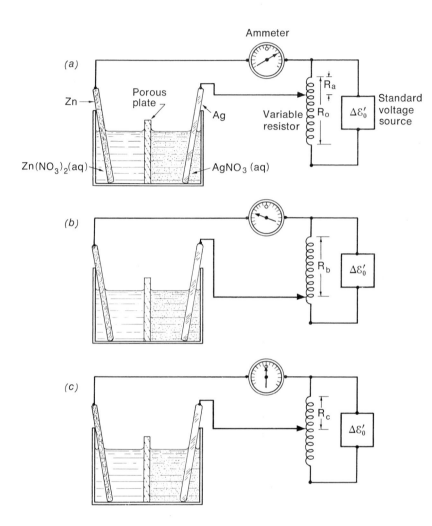

Figure 10-6   Measurement of ε of an electrochemical cell.

tions set at the standard states. This means that all the gases involved should be at one atmosphere pressure, soluble substances should be at one mole per liter concentration,* and solids of low solubility should be

---

* A more rigorous statement would specify "unit activity" instead of "one mole/liter concentration." The "activity" differs from the concentration by a multiplicative factor that corrects for restrictions of mobility due to intermolecular interactions, as discussed in Section 4-4(b).

present in excess. Also, the temperature should be controlled at a specified, constant temperature. Usually, this will be 25°C, 298.16°K, the standard "room temperature." Under these standard state conditions, the cell voltage measures ΔG^0, the standard free energy change. Correspondingly, this cell voltage is designated $\Delta \mathcal{E}^0$, the standard voltage, or as it is usually expressed, the "standard e.m.f." (e.m.f. $\equiv$ emf $\equiv$ "electromotive force"). Yet another term used for voltage (or emf) is "potential." In this usage, "potential" is a noun, but its meaning ties closely to the adjectival use in "potential energy."

So, at standard cell conditions,

$$\Delta G^0 = -n\Delta \mathcal{E}^0 \mathcal{F} \qquad (10\text{-}56)$$

and since standard free energy determines the equilibrium constant, (relation (10-33))

$$\Delta G^0 = -RT \log_e K = -n\Delta \mathcal{E}^0 \mathcal{F}$$

Thus we derive, by thermodynamic arguments, the relationship between $\Delta \mathcal{E}^0$ and the equilibrium constant:

$$\boxed{\Delta \mathcal{E}^0 = \frac{RT}{n\mathcal{F}} \log_e K} \qquad (10\text{-}57)$$

Of course, R and $\mathcal{F}$ must be selected in the same units: If R is in calories, $\mathcal{F}$ should be the number 23,061. It is convenient to remember the value of $RT/\mathcal{F}$ for normal room temperature, 298.16°K, and with the factor 2.303 inserted to convert to logarithms to the base 10:

$$\boxed{\Delta \mathcal{E}^0 = \frac{0.059}{n} \log_{10} K} \qquad (10\text{-}58)$$

That sounds familiar! 'Way back in Chapter Six, we deduced expression (6-58) in an empirical way, relating $\Delta \mathcal{E}^0$ to K. But expression (10-58) is just (6-58) rearranged! So now we have a thermodynamic understanding of why the voltage measured in a reversible electrochemical cell can be used to measure the equilibrium constant (see Section 6-3(e) and Exercise 6-5). But wait, the best is yet to come! Earlier in this chapter, we found how to calculate ΔG from ΔG^0, with expression (10-28). Now we can apply that expression to a general reaction and connect it to electrical measurements through (10-55) and (10-56). For a general reaction,

$$xA + yB + \cdot\,\cdot\,\cdot\, = wC + zD + \cdot\,\cdot\,\cdot \qquad \text{(10-59)}$$

therefore,

$$\Delta G = \Delta G^0 + RT \log_e \frac{(C)^w (D)^z \cdot\,\cdot\,\cdot}{(A)^x (B)^y \cdot\,\cdot\,\cdot} \qquad \text{(10-60)}$$

The concentration quotient on the right-hand side has the same form as the equilibrium expression, but it involves whichever concentrations we wish to consider. This is the quantity we called the concentration coefficient, Q,

$$\Delta G = \Delta G^0 + RT \log_e Q \qquad \text{(10-61)}$$

Now, substituting (10-55) and (10-56) into (10-61), we obtain

$$-n\Delta\mathcal{E}\mathcal{F} = -n\Delta\mathcal{E}^0\mathcal{F} + RT \log_e Q$$

or

$$\boxed{\Delta\mathcal{E} = \Delta\mathcal{E}^0 - \frac{RT}{n\mathcal{F}} \log_e Q} \qquad \text{(10-62)}$$

With $T = 298.16°K$ and inserting the factor 2.303,

$$\boxed{\Delta\mathcal{E} = \Delta\mathcal{E}^0 - \frac{0.059}{n} \log_{10} Q} \qquad \text{(10-63)}$$

This is the famous expression for the dependence of cell voltage on concentration that we called the Nernst Equation (see Section 6-2(d)).

We have already seen how useful the Nernst Equation can be to calculate $\Delta\mathcal{E}^0$ from an experimental value of $\Delta\mathcal{E}$. An electrochemical cell is assembled with carefully measured concentrations. Convenience often dictates that these concentrations be other than standard conditions. The experimental value of $\Delta\mathcal{E}$ is then measured as shown in Figure 10-6. The value of Q can be calculated, and the value of $\Delta\mathcal{E}$ has already been measured, so $\Delta\mathcal{E}^0$ is the only unknown quantity in the Nernst Equation (10-62). Thus $\Delta\mathcal{E}^0$ is determined, and hence also the equilibrium constant via (10-57), and ΔG^0 via (10-56). Of course the temperature dependence of $\mathcal{E}^0$ can also be measured and, hence, the temperature dependence of ΔG. This, in turn, can be used to establish ΔH^0 and ΔS^0, as described in Section 10-3. Clearly, electrochemical cells have great importance in chemistry as a means of studying equilibrium and determining thermodynamic quantities.

10-5 Solution thermodynamics

Theoretical discussions are often limited to considerations of species in the gas phase. This is because the gas phase is the simplest of all states. Individual gas molecules are far enough separated to eliminate any significant forces between them. Practical chemistry, on the other hand, is definitely not limited to gas-phase reactions. In fact, most of the chemistry we see around us occurs in solution and most of that in the especially abundant solvent, water. That is why the study of equilibrium in ionic aqueous solutions forms a major part of any introductory chemistry course. In this section we will discuss the thermodynamics of some simple ionic, aqueous solutions, paying special attention to the extremely important role played by the solvent water. Table 10-5 lists some thermodynamic properties of ions in aqueous solution and, for comparison, some gas-phase values as well. The solution values are referred to the properties of H^+(aq) as a reference state. Accordingly, *the thermodynamic properties of* H^+(aq) *are defined to be zero.* Remember that we are always concerned with differences, so that our choice of a reference point does not matter (recall the building contractor in Denver).

Table 10-5 *Thermodynamic Properties of Selected Gaseous and Aqueous Ions and Their Compounds*

	ΔH_f^0 (kcal/mole)	ΔG_f^0 (kcal/mole)	S^0 (cal/mole degK)
H^+(aq)	0	0	0
(g)	$+367.2$		
Na^+(aq)	-57.4	-62.6	14.1
(g)	$+146.0$		
Ca^{+2}(aq)	-129.8	-132.2	-13.2*
Ag^+(aq)	25.2	18.4	17.4
Cl^-(aq)	-39.9	-31.4	13.5
(g)	-58.8		
OH^-(aq)	-54.9		
(g)	-33.7		
F^-(aq)	-79.5	-66.6	-3.3
HF(g)	-64.2	-64.7	$+41.5$
HCl(g)	-22.1	-22.8	$+44.6$
$NaOH$(s)	-101.7		
CaF_2(s)	-290.3	-277.7	16.46
$AgCl$(s)	-30.4	-26.2	23.0
$NaCl$(s)	-98.2	-91.8	17.3

* Negative entropies occur because of the choice of $S^0[H^+(aq)] = 0$.

(a) THERMODYNAMICS OF SOLUBILITY

Chemical substances known as "salts" are strongly bonded, ionic solids. Most salts that dissolve in water are totally dissociated into ions. The variation of solubility encountered in different salts is high, however. What factors affect the solubility? Well, from the previous discussion we can expect the solvent to play a very large role. Consider the two salts, sodium chloride and calcium fluoride. Considerable energy is needed for the dissociation into ions in the gas phase, yet both dissolve in water with the absorption of only a small amount of heat.

$$NaCl(s) \longrightarrow Na^+(aq) + Cl^-(aq) \tag{10-64}$$
$$\Delta H^0 = +1.3 \text{ kcal/mole}$$

$$CaF_2(s) \longrightarrow Ca^{+2}(aq) + 2 F^-(aq) \tag{10-65}$$
$$\Delta H^0 = +1.5 \text{ kcal/mole}$$

The experimental solubilities of these two salts are very different: Sodium chloride dissolves to a concentration above five moles per liter, while calcium fluoride has an equilibrium solubility below 0.001 mole per liter. Entropy changes associated with the two processes must account for the difference, since the enthalpy changes are nearly identical. We would probably predict that ΔS^0 would be greater in the CaF_2 dissociation, since three ions are being produced rather than two, as in the case of NaCl. This would cause the CaF_2 solubility to be the higher of the two. In fact, just the opposite is true. Sodium chloride is quite soluble, whereas calcium fluoride is barely soluble at all.

Turning to the thermodynamic properties summarized in Table 10-6, we see that the entropy of CaF_2 does not meet our expectations. Not only is the entropy change less for CaF_2 than for NaCl, it is actually negative! The aqueous, ionized state is less random than the reactants, pure solid and pure water. This points out dramatically the important effect of the solvent on solution processes. The small fluoride ions and highly charged Ca^{+2} ions are tightly surrounded by several water molecules, in arrangements that are more ordered than their environments in the crystal! Apparently the larger size of the chloride ions and the lower charge of the sodium ions makes these effects much less pronounced for NaCl.

Table 10-6 *Thermodynamics of Salt Solubilities in Water* (298°K)

	ΔH^0 (kcal/mole)	ΔS^0 (cal/mole degK)	ΔG^0 (kcal/mole)
$NaCl(s) = Na^+(aq) + Cl^-(aq)$	+1.3	+10.3	−1.8
$CaF_2(s) = Ca^{+2}(aq) + 2 F^-(aq)$	+1.5	−36.3	+12.3

Similar results are found for gas solubilities. Hydrogen chloride gas is roughly one thousand times more soluble in water than is hydrogen fluoride gas under comparable HX gas pressures. The thermodynamics of these processes is given in Table 10-7. We see that not only is HF less soluble in water than HCl, but it is also not significantly dissociated into ions! The reason is apparent in the thermodynamics. After dissolving in water, the HF would decrease the solution entropy by another 23.5 cal/degK if it were to dissociate into ions. Again, the small size of the F$^-$ ion causes it to arrange solvent molecules strongly around itself to produce a highly ordered solution. So entropy effects cause the behavior of aqueous HF and aqueous HCl to be entirely dissimilar. Hydrofluoric acid is a weak acid with only moderate solubility in water, whereas hydrochloric acid is a very strong acid that dissolves to concentrations above 10 moles per liter.

Table 10-7 *Thermodynamics of Gaseous Solubility (298°K)*

	ΔH^0 $\left(\dfrac{\text{kcal}}{\text{mole}}\right)$	ΔS^0 $\left(\dfrac{\text{cal}}{\text{mole degK}}\right)$	ΔG^0 $\left(\dfrac{\text{kcal}}{\text{mole}}\right)$	K
HF(g) = HF(aq)	−11.7	−20.3	−4.6	$2.3 \cdot 10^3$
HF(aq) = H$^+$(aq) + F$^-$(aq)	−4.0	−23.5	+3.2	$4.5 \cdot 10^{-3}$
HF(g) = H$^+$(aq) + F$^-$(aq)	−15.7	−43.8	−1.4	10.4
HCl(g) = H$^+$(aq) + Cl$^-$(aq)	−17.8	−31.1	−8.6	$2.0 \cdot 10^6$

(b) REACTIONS IN SOLUTION

Solvent effects have an equally important influence on the outcome of chemical reactions occurring in solution. One of the crucial decisions in bringing about a reaction is the choice of a suitable solvent. Water, our most abundant chemical, is frequently our choice, particularly when the reactions involve ions.

Two pairs of reactions are contrasted in Table 10-8. Reaction (i), an endothermic reaction, results in the formation of one product species from two reactant ones. By itself, this should make the system less random (two going to one). In fact, the expected decrease in entropy does not occur. Quite the opposite, the entropy *increases* enough, in fact, to overcome the endothermicity and produce a negative free energy change. The reason for this effect is, again, the highly ordered hydration around the small fluoride ion. This order is lost (the system becomes more random) as the larger HF$_2^-$ ion is formed. The resulting increase in randomness more than overcomes the fact that two particles combine to form one.

Reaction (ii) shows the opposite effect. In this reaction, two reactant molecules give three product species—randomness should increase.

Table 10-8 *Thermodynamics of Ionic Reactions (298°K)*

	ΔH^0 $\left(\dfrac{\text{kcal}}{\text{mole}}\right)$	ΔS^0 $\left(\dfrac{\text{cal}}{\text{mole degK}}\right)$	ΔG^0 $\left(\dfrac{\text{kcal}}{\text{mole}}\right)$	K
(i) $HF(aq) + F^-(aq) = HF_2^-(aq)$	$+0.7$	$+4.2$	-0.7	3.3
(ii) $HOCl(aq) + OCl^-(aq)$				
$= H^+(aq) + Cl^-(aq) + ClO_2^-(aq)$	-1.3	-6.3	$+9.6$	9.1×10^{-8}
(iii) $NH_3(aq) + HF(aq)$				
$= NH_4^+(aq) + F^-(aq)$	-14.8	-24.0	-8.35	1.4×10^6
(iv) $OH^-(aq) + H_3O^+(aq)$				
$= 2\,H_2O$	-13.3	$+19.3$	-19.1	1.0×10^{14}

Instead, ΔS is negative! Again the effect is due to the production of three ions, each of which creates an ordered hydration environment around itself. This means an entropy decrease despite the production of three molecules from two. Now negative entropy overpowers an exothermic ΔH to make ΔG positive and the equilibrium constant very small.

These effects are strikingly underscored by reactions (iii) and (iv) in Table 10-8. These processes involve identical numbers of product and reactant species and are exothermic by nearly the same amount. However, reaction (iii), which involves the formation of ions from neutral species, has a negative ΔS, while reaction (iv), in which ions are converted into neutral water molecules, has a positive ΔS. The difference, entirely due to positional randomness, causes a factor of 10^8 difference in the equilibrium constants!

Thus we see that aqueous solution chemistry cannot be predicted on the basis of enthalpy considerations alone. The hydration entropy of ions is often important, and sometimes dominant, in determining the reactions that occur.

10-6 Some things about entropy

The entropy of a mole of molecules contributes importantly to the reactions of those molecules. What factors determine the entropy of those particular molecules? According to Chapter Eight, it is connected with their positional and motional randomness at the temperature in question. The positional randomness depends upon the state: gas, liquid, solution, or solid. The motional randomness depends upon the number of degrees of freedom: translational, rotational, and vibrational—and their energy spacing. We can learn much about the entropy of a substance by merely considering how all of these degrees of freedom are affected as the substance is cooled, first as a gas, then, after condensation, as a liquid, and finally, as a solid.

(a) TEMPERATURE AND RANDOMNESS

The average translational energy of a gaseous molecule is $(\frac{3}{2})kT$. Hence, as the temperature drops, the energy to be divided among the translational states is less and less. Thinking of the energy as being in packets, we have fewer and fewer energy packets to distribute among the molecules. Their motional randomness, hence their entropy, drops. The same qualitative effect of lowered temperature is felt among the rotational and vibrational degrees of freedom. Fewer and fewer states have substantial occupancy, so randomness drops. *The entropy of any gas drops as it is cooled.*

At some temperature the gas will liquefy. Then the translational motion of the molecules is very much restricted, and positional randomness is drastically curtailed. The molecules still move about, randomly vibrating and tumbling against each other, but without anything like the positional freedom of the gas phase. Randomness drops precipitously as the gas changes to the liquid state. *The entropy of any gas drops sharply when it liquefies.*

Cooling the liquid below the condensation temperature lowers still further the amount of molecular motion. Again fewer and fewer energy states are occupied—motional randomness is decreased as temperature continues to fall. *The entropy of any liquid drops as it is cooled.*

When the freezing point is reached, molecules settle into the regular positions of the crystal lattice, sacrificing the orientational disorder of the liquid state and more of the motional freedom. To be sure, the crystal has its own vibrational modes, but these reflect the rigidity and regularity of the solid state. *The entropy of any liquid drops sharply when it solidifies.*

Finally, what happens as we continue to cool the solid? As the temperature drops still further, there is less and less vibrational movement of the molecules about their equilibrium positions. Positional randomness is now virtually nonexistent, and motional randomness is disappearing. In fact, as absolute zero is approached, there is less and less energy available to excite any kind of molecular movement. It is intuitively reasonable that all random motion should stop at absolute zero. Then every molecule is fixed in an equilibrium lattice site, and there are no random vibrations excited. This describes a system without any randomness at all! That means the entropy, which measures disorder, has reached zero! Thinking of the relationship of entropy to probability, $S = R\log_e W$, at absolute zero, there is only one arrangement, that of the perfect crystal, so $W = 1$ and $S = R\log_e 1 = 0$. *At the absolute zero of temperature, the entropy of all substances becomes zero.* This generalization can be taken as an experimental fact as well as an intuitively reasonable proposal. The experimental evidence is so complete and so convincing that this statement is known as the Third Law of Thermodynamics.

> **The Third Law of Thermodynamics:**
> **The entropy of any substance is zero**
> **at absolute zero temperature.**

(10-66)

(b) THE THIRD LAW AND ABSOLUTE ENTROPIES

With this universal reference point, we can specify entropies on an absolute scale. Each molecular type has a natural reference state, so we need no arbitrary choice. Hence tabulated entropies are specified as "absolute entropies." The entropy is simply measured by heating the solid from the lowest accessible temperature up to the temperature and state of interest, always keeping track of the quantity q_{rev}/T. Figure 10-7 shows a typical experimental entropy plot, measured from 4°K up to room temperature. Two of the sharp rises are caused by melting and vaporization. These changes are characteristic, as mentioned in Section (a), and as shown for a number of substances in Table 10-9.

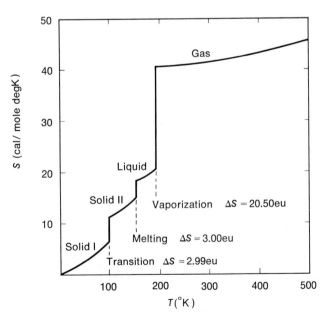

Figure 10-7   Experimental measurement of the entropy of SO_3 as a function of temperature.

Table 10-9 Dependence of Entropy on State (cal/mole degK)

Substance	$S^0_{298}(s)$	$S^0_{298}(\ell)$	$S^0_{298}(g)$
SO_3	12.5	22.9	61.3
S	7.6		40.1
I_2	27.8		62.3
Br_2		36.4	58.6
BrF_3		42.6	69.9
CH_3OH		30.3	57.3

10-7   Conclusion

In this chapter we have seen how the Equilibrium Law can be derived from thermodynamic arguments. The free energy change accompanying a reaction tends toward a minimum, and at this minimum, equilibrium prevails. The equilibrium constant that expresses the equilibrium conditions depends upon ΔG^0, the free energy change under standard conditions. Furthermore, the relation between free energy and electrical work provides a thermodynamic basis for the Nernst Equation. Finally, the temperature dependence of the equilibrium constant provides a convenient measure of ΔH^0 and ΔS^0.

We can now see the great importance of thermodynamic considerations in chemistry. The motivation for spontaneous change is the tendency to maximum randomness. For the familiar "open beaker" experiments conducted at constant pressure and constant temperature, this tendency to maximum randomness in the universe is connected with a tendency to minimum free energy in the system. Hence the direction and distance to equilibrium are measured by ΔG. These powerful principles are applicable throughout chemistry and they guide chemists in their thinking about chemical processes.

Problems

1. In addition to (10-1), (10-2), and (10-3), there is a fourth possible reaction between NH_3 and O_2:

$$NH_3(g) + \tfrac{3}{4}O_2(g) \longrightarrow \tfrac{1}{2}N_2(g) + \tfrac{3}{2}H_2O(g)$$

Experimentally, it is found that this reaction is as important as (10-1) and (10-2) at flame temperatures. Use the standard heats of formation in Table 10-1 to calculate ΔH^0 for this reaction.

2. By the First and Second Laws of Thermodynamics, for any reaction that proceeds spontaneously at constant T and P, which of the following *must* be positive? Which *must* be negative? Which *must* be conserved?

for the system: $\Delta E,\ \Delta H,\ \Delta S,\ \Delta G$

for the universe: ΔE ΔS

3. Use the standard free energies of formation in Table 10-1 to calculate ΔG^0 for the reaction in Problem 1.

✦ 4. If positional randomness was the main factor that determined the entropy change in reactions (10-1), (10-2), and (10-3), ΔS would depend primarily on the change in number of moles going from reactants to products. On this assumption, predict for each reaction whether it will have a positive entropy change, a negative entropy change, or ΔS near zero.

⊢ 5. Use the values of ΔH^0 and ΔG^0 compiled in Table 10-2 to calculate ΔS^0 for each reaction, (10-1), (10-2), and (10-3). Compare these experimental values to your predictions in Problem 4.

6. Heats of combustion can be measured very accurately, so they are an important source of thermodynamic information. For example, the heat of combustion of carbon in the form of graphite is -94.0518 kcal/mole and that of carbon in the form of diamond is -94.5051 kcal/mole.

$$C(gr) + O_2(g) \longrightarrow CO_2(g) \quad T = 298.16°K \quad \Delta H^0 = -94.0518 \text{ kcal}$$

$$C(d) + O_2(g) \longrightarrow CO_2(g) \quad T = 298.16°K \quad \Delta H^0 = -94.5051 \text{ kcal}$$

(a) What is the heat of transition ΔH^0 from graphite to diamond? Which is energetically more stable?

$$C(gr) \rightleftharpoons C(d)$$

(b) The absolute entropies of graphite and diamond are, respectively, 1.372 and 0.568 cal/deg. What is ΔS^0 for the transition?
(c) Calculate ΔG^0 for the transition, predict whether a diamond ring is stable or unstable at 298°K, and decide how ΔH^0 and ΔS^0 contribute to this stability (or instability).

➤ 7. The dimerization of gaseous NO_2 is accompanied by an enthalpy change of -13.6 kcal and an entropy change of -41.9 cal/deg at standard conditions:

$$\begin{array}{c} & 298°K & \\ 2\ NO_2(g) & \xrightarrow{\hspace{1cm}} & N_2O_4(g) & \Delta H^0 = -13.6 \text{ kcal} \\ 1 \text{ atm} & & 1 \text{ atm} & \Delta S^0 = -41.9 \text{ cal/deg} \end{array}$$

What are ΔH and ΔS for this reaction (per mole of N_2O_4) if the pressure is lowered so that both pressures are 0.010 atm? (Assume perfect gas behavior.)

$$\begin{array}{c} & 298°K & \\ 2\ NO_2(g) & \xrightarrow{\hspace{1cm}} & N_2O_4(g) \\ 0.010 \text{ atm} & & 0.010 \text{ atm} \end{array}$$

✗ 8. Gaseous nitrous acid decomposes to nitric oxide, nitrogen dioxide, and water vapor, and the reaction is accompanied by an enthalpy change of $+10.19$ kcal and an entropy change of $+33.84$ cal/deg if it can be carried out under standard conditions. (Notice that $H_2O(g)$ would condense at 298°K and 1 atm.)

$$\begin{array}{c} & 298°K & \\ 2\ HONO(g) & \xrightarrow{\hspace{1cm}} & H_2O(g) + NO(g) + NO_2(g) \\ 1 \text{ atm} & & 1 \text{ atm} \quad 1 \text{ atm} \quad 1 \text{ atm} \end{array}$$

$$\Delta H^0 = +10.19 \text{ kcal}$$
$$\Delta S^0 = +33.84 \text{ cal/deg}$$

What are ΔH and ΔS for this reaction (per two moles of HONO) if the pressure is lowered so that all pressures are 0.010 atm? (Assume perfect gas behavior.)

9. Gaseous nitrous acid exists in two structures, each with all four atoms in the same plane. The bent form is called *cis*-nitrous acid and the extended form is called *trans*-nitrous acid.

cis-HONO *trans*-HONO

$\Delta H^0 = -0.51$ kcal

$\Delta S^0 = +0.11$ cal/deg

The conversion of the *cis*- to the *trans*- form is accompanied by an enthalpy change of -0.51 kcal and an entropy change of $+0.11$ cal/deg. What are ΔH and ΔS for this reaction if the pressure is lowered so that both pressures are 0.010 atm? (Assume perfect gas behavior.) no change because no molecule is added.

10. Calculate ΔG for the dimerization of NO_2 (see Problem 7) both under standard conditions and with all pressures equal to 0.010 atm. From these values, predict the direction in which the reaction will tend to occur under each of the starting conditions.

11. Answer Problem 10 for the decomposition of gaseous nitrous acid, HONO (see Problem 8).

12. What is the equilibrium constant for the dimerization of NO_2 (see Problems 7 and 10)? What N_2O_4 pressure would be in equilibrium with an NO_2 partial pressure of 0.010 atm? Is this calculation consistent with the reaction direction predicted in Problem 10, with N_2O_4 pressure equal to 0.010 atm?

13. What is the equilibrium constant for the decomposition of gaseous nitrous acid (see Problems 8 and 11)? What HONO pressure would be in equilibrium with the products if each of their partial pressures were 0.010 atm? Is this calculation consistent with the reaction direction predicted in Problem 11, with HONO pressure equal to 0.010 atm?

14. Normal octane, a constituent of petroleum, has the formula n-C_8H_{18} and an extended structure, $CH_3CH_2CH_2CH_2CH_2CH_2CH_2CH_3$. It explodes in an automobile engine rather than burning, hence causes "engine knock" and contributes to smog. A hydrocarbon that burns more smoothly is iso-octane, i-C_8H_{18}, with the branched structure

$$CH_3CH_2CH_2CH_2CH_2CHCH_3$$
$$|$$
$$CH_3$$

From the thermodynamic properties of these two compounds, calculate ΔH^0, ΔG^0, ΔS^0, and the equilibrium constant for the conversion of the undesirable n-C_8H_{18} into the useful fuel, i-C_8H_{18}.

	ΔH_f^0	ΔG_f^0	S^0
n-C_8H_{18}(g)	-49.82	3.95	111.55
i-C_8H_{18}(g)	-51.50	3.06	108.81

15. 1-Octene, $1\text{-}C_8H_{16}$, is a hydrocarbon that does not cause "engine knock," hence it is a desirable fuel. From its thermodynamic properties, together with those of normal octane (see Problem 14) and of H_2 (see Appendix d), calculate ΔH^0, ΔG^0, ΔS^0, and the equilibrium constant for the conversion of the undesirable $n\text{-}C_8H_{18}$ into the useful fuel, $1\text{-}C_8H_{16}$: $n\text{-}C_8H_{18}(g) \longrightarrow 1\text{-}C_8H_{16}(g) + H_2(g)$.

$1\text{-}C_8H_{16}$: $CH_3CH_2CH_2CH_2CH_2CH_2CH{=}CH_2$

$\Delta H_f^0 = -19.82$

$\Delta G_f^0 = 24.96$

$S^0 = 110.55$

16. Unfortunately, the conversion of n-octane into i-octane is very slow at room temperature (see Problem 14). Reaction rate is rapid at 1000°K, however. Calculate the equilibrium constant at 1000°K, assuming ΔH^0 and ΔS^0 are temperature independent.

17. The conversion of n-octane into 1-octene is very slow at room temperature (see Problem 15). Reaction rate is rapid at 1000°K, however. Calculate the equilibrium constant at 1000°K, assuming ΔH^0 and ΔS^0 are temperature independent.

18. Answer the first two questions of Problem 12, except at $T = 250°K$. Assume ΔH^0 and ΔS^0 are temperature independent.

19. Answer the first two questions of Problem 13, except at $T = 250°K$. Assume ΔH^0 and ΔS^0 are temperature independent.

20. The decomposition equilibrium of HONO (see Problem 8) shifts toward products when pressure is lowered (Problem 11) and toward reactants as temperature is lowered (Problem 19), both in accord with Le Chatelier's Principle (see Section 1-4(d)). From Problem 11, decide which thermodynamic quantity, ΔH or ΔS, accounts for the success of Le Chatelier's Principle, for each variable, P and T.

21. What is the percentage of nitrous acid in the *cis*- form due to the rapid $cis \rightleftharpoons trans$ equilibrium at 298°K? At 250°K? At 600°K? (See Problem 9.)

22. (a) Assuming ΔH^0 and ΔS^0 in Problem 6 are temperature independent, calculate ΔG for the reaction at 1000°K, at which temperature the transition from one form to the other is rapid.
(b) Explain why, after a house burns down, the ruins are full of charcoal, not diamonds.
(c) The density of graphite is 2.25 g/cc and that of diamond is 3.51 g/cc. By Le Chatelier's Principle, (see Section 1-4(d)), does ΔG become more positive or more negative as the pressure is raised? Explain why diamonds are found in the cores of extinct volcanos in which temperature and pressure were once very high.

23. The U.S. and Soviet exploration of the planet Venus has led to a possible atmospheric model having a temperature and pressure at the surface of 700°K and 75 atm with a composition of almost pure CO_2. On earth, we have lots of lime-stone, $CaCO_3$, which decomposes on heating to lime, CaO, and carbon dioxide. This raises the question of whether the equilibrium vapor pressure of limestone on Venus could account for this enormous atmospheric pressure.

$CaCO_3(s) \longrightarrow CaO(s) + CO_2(g)$
limestone   lime

(a) From Appendix d, calculate ΔH^0, ΔG^0, and ΔS^0 for this reaction under standard conditions.

(b) From the heat capacity, C_p^0, calculate the values of ΔH and ΔS associated with raising the temperature of $CaCO_3$ to 700°K; of CaO to 700°K; of CO_2(g) at 1 atm pressure to 700°K. (Note that $\Delta S = \int dq/T = \int C_p dT/T = C_p \int dT/T = C_p \ln (T_2/T_1)$.)

(c) Combine (a) and (b) to calculate ΔH and ΔS for the reaction at 700°K to give CO_2 at 1 atm pressure. From these, calculate ΔG.

(d) From ΔG (700°K), calculate the equilibrium constant at 700°K. Could the limestone equilibrium vapor pressure account for the high atmospheric pressure on Venus at temperatures near 700°K?

24. Here is another tack on the problem of the origin of the high atmospheric pressure on Venus (see Problem 23). A plentiful carbonate mineral on earth is siderite, $FeCO_3$. It decomposes, on heating, to wustite, FeO, and CO_2.

$$FeCO_3(s) \longrightarrow FeO(s) + CO_2(g)$$
$$\text{siderite} \qquad \text{wustite}$$

(a) From Appendix d, calculate ΔH^0, ΔG^0, and ΔS^0 for this reaction under standard conditions.

(b) Calculate the values of ΔH, ΔS, and ΔG associated with compression of CO_2 to 75 atm pressure at 298°K. (Assume a perfect gas.)

(c) Assuming $\Delta H = \Delta G = \Delta S = 0$ for raising the pressure of the solids to 75 atm, combine (a) and (b) to calculate ΔH, ΔS, and ΔG for the reaction

$$FeCO_3(s) \xrightarrow{298°K} FeO(s) + CO_2(g)$$
$$\text{75 atm} \qquad \text{75 atm} \quad \text{75 atm}$$

(d) By setting ΔG (75 atm) $= 0$, calculate the temperature at which the equilibrium vapor pressure of siderite is 75 atm. Could siderite account for the high atmospheric pressure on Venus at temperatures near 700°K?

25. The equilibrium vapor pressure over solid ammonium perchlorate has been measured to be 0.026 torr at 520°K and 2.32 torr at 620°K. Adding NH_3 gas represses the vaporization, suggesting that the equilibrium under study is

$$NH_4ClO_4(s) \rightleftharpoons NH_3(g) + HClO_4(g)$$

(a) With this interpretation, calculate the equilibrium constant at each temperature. (Express pressures in atmospheres.)

(b) For each value of K, calculate ΔG^0.

(c) From the temperature dependence of K, calculate ΔH^0.

(d) Calculate ΔS^0.

26. Repeat Problem 25 assuming that the ammonia repression of vaporization had not been discovered, so the equilibrium was incorrectly assumed to be

$$NH_4ClO_4(s) \rightleftharpoons NH_4ClO_4(g)$$

27. From the tabulated data, calculate ΔG^0 and the voltage that would be measured if the reaction between ferric and hydrazinium ion could be obtained in a reversible cell under standard conditions.

$$2\,Fe^{+3}(aq) + 2\,N_2H_5^+(aq) \longrightarrow 2\,Fe^{+2}(aq) + N_2(g) + 2\,NH_4^+(aq) + 2\,H^+(aq)$$

	Fe^{+3}	$N_2H_5^+$	Fe^{+2}	N_2	NH_4^+	H^+	
ΔH_f^0	−11.4	−1.7	−21.0	(0.00)*	−31.74	(0.00)*	kcal
S^0	−70.1	31.0	−27.1	45.8	27.0	(0.00)*	cal/deg

* By definition.

28. From the tabulated data, calculate ΔH^0, ΔS^0, and ΔG^0 for the cell reaction studied at length in Section 9-2:

$$Zn(s) + 2\,Ag^+(aq) \longrightarrow Zn^{+2}(aq) + 2\,Ag(s)$$

	Zn	Ag$^+$	Zn^{+2}	Ag
ΔH_f^0	(0.00)*	25.31	-36.43	(0.00)*
S^0	10.0	17.67	-25.45	10.2

* By definition.

Explain why only a fraction of the exothermic energy release could be "freed" for electrical work in this cell. From the relation of ΔG, ΔH, and ΔS, what condition would permit the maximum electrical work that could be extracted from a reaction to *exceed* the energy release in an exothermic reaction?

29. (a) For each of these three reactions, calculate $\Delta \varepsilon^0$ and ΔG^0 from the reduction potentials given in Table 6-5.

$$Cu^+(aq) + Fe^{+3}(aq) \longrightarrow Cu^{+2}(aq) + Fe^{+2}(aq)$$

$$2\,Fe^{+2}(aq) + Tl^{+3}(aq) \longrightarrow 2\,Fe^{+3}(aq) + Tl^+(aq)$$

$$2\,Cu^+(aq) + Tl^{+3}(aq) \longrightarrow 2\,Cu^{+2}(aq) + Tl^+(aq)$$

(Note: $Tl^{+3} + 2\,e^- \longrightarrow Tl^+$, $\varepsilon^0 = +1.25$.)

(b) From the following heats of formation (relative to $\Delta H_f^0(H^+) = 0.00$), calculate ΔH^0 for each reaction.
(c) For each reaction, calculate the fraction of the reaction tendency (ΔG^0) associated with reaction exothermicity.
(d) Combine (a) and (b) to calculate ΔS^0 for each reaction.
(e) Use the appropriate ionization energies (Appendix b) to calculate ΔH^0 for each of these reactions if they occur in the gas phase.
(f) Approximately what entropy change is expected for each of these reactions in the gas phase? Notice that the absence of correlation between the results in aqueous solution and in the gas phase shows that *both* energy and orientational aspects of the interaction between solvent and ions are important in aqueous chemistry.

ΔH_f^0	Cu$^+$(aq)	Cu^{+2}(aq)	Fe^{+2}(aq)	Fe^{+3}(aq)	Tl$^+$(aq)	Tl^{+3}(aq)
(kcal/mole)	12.4	15.4	-21.0	-11.4	1.4	27.7

30. Figure 4-6 shows a simple model of the interaction between an ion and water molecules to explain the stability of ions in water. Does this orientation tend to raise or to lower the entropy of the solution? Since the arrangement is highly ordered what causes the orientation to occur spontaneously? What will be the effect of raising the temperature upon the extent of this orientation?

31. (a) What are the probable signs (positive or negative) for ΔH and ΔS when paraffin (a solid hydrocarbon) dissolves in kerosene (a liquid hydrocarbon)?
(b) Answer (a) when butane (a gaseous hydrocarbon) dissolves in kerosene.

32. In Problem 21, Chapter Nine, it was discovered that many (in fact, most) liquids have an entropy of vaporization at the normal boiling point equal to 21 ± 2 cal/deg.

$$\text{liquid} \longrightarrow \text{gas (1 atm)} \qquad \Delta S = 21 \pm 2 \text{ cal/deg}$$

However, Table 9-3 shows that the entropy of vaporization of water at its normal boiling point is 26.05 cal/deg. Discuss the probable significance of this result in terms of the randomness in pure liquid water compared to that in other pure liquids.

33. The fuel used for the descent and ascent rocket engines of the Apollo 11 and 12 landing modules is the combination of dimethylhydrazine $(CH_3)_2NNH_2$ and dinitrogen tetroxide. These two substances are hypergollic (they ignite on contact), burning to give mainly gaseous N_2, H_2O, and CO_2 as products.
(a) Balance the equation for the reaction.
(b) How many kilograms of N_2O_4 are needed to react with one kilogram of dimethylhydrazine? (1 kgm = 2.20 pounds)
(c) The heats of formation of $(CH_3)_2NNH_2$ and of liquid N_2O_4 are, respectively, +11.8 and −4.66 kcal/mole. Calculate the heat of the reaction per mole of dimethylhydrazine consumed. Calculate the heat per kilogram of fuel mixture (at the stoichiometric ratio).
(d) The absolute entropy of liquid N_2O_4 is about 52 cal/deg, but that of dimethylhydrazine is not well known. Suppose the latter is about the same as liquid isobutene, $(CH_3)_2CCH_2$, which is 51.7 cal/deg. Estimate ΔS^0 per mole of $(CH_3)_2NNH_2$ consumed.
(e) From (c) and (d) calculate ΔG^0 and the equilibrium constant at 298°K for the reaction.

34. The thermodynamic properties of liquid hydrazine, N_2H_4, are known: $\Delta H_f^0 = 12.10$, $\Delta G_f^0 = 35.67$, and $S^0 = 28.97$.
Answer all of the questions in Problem 33 for N_2H_4 in place of $(CH_3)_2NNH_2$.
(a) What is the ratio of ΔH of reaction for $(CH_3)_2NNH_2$ and N_2H_4 on a per mole basis?
(b) What is the ratio on the basis of one kilogram of fuel mixture?

35. The entropy change calculated in Problem 33 was based on a guess of the entropy of dimethylhydrazine, but in Problem 34 it was based on tabulated quantities. Since liquid reactants give gaseous products in each case, positional randomness of the products dominates ΔS. Compare the ratio of the ΔS values in Problems 33 and 34 to the ratio of the number of moles of gaseous products to see if ΔS in Problem 33 is reasonable.

the rates of chemical reactions

The business of a chemist is to understand and control chemical reactions. Thermodynamics gives us signposts, but it indicates nothing about the road conditions on the way to the destination. One reaction with a favorable free energy change of -10 kcal may reach equilibrium in a millisecond, while another reaction with the same ΔG might take a thousand years. The -45.6 kcal free energy change associated with the reaction of hydrogen and chlorine indicates that the product, hydrogen chloride, is thermodynamically favored. The ΔG contains no clue, however, to the experimental observation that a hydrogen–chlorine mixture will not react noticeably in a week in a dark room, but it will explode violently if momentarily exposed to a bright light. The study of such reaction rate behavior is called *reaction kinetics*.

11-1 General characteristics of reaction rates

A gaseous mixture of hydrogen bromide and oxygen reacts to form bromine and water vapor:

$$4\,\text{HBr(g)} + \text{O}_2\text{(g)} \longrightarrow 2\,\text{H}_2\text{O(g)} + 2\,\text{Br}_2\text{(g)} \tag{11-1}$$

At a temperature of 700°K and reactant pressures of a few torr, the reaction might take a few hours to proceed to the extent of 1%. Furthermore, it is easy to measure the progress of the reaction. The total pressure changes as reaction proceeds because four molecules are produced as five are consumed, so pressure change per unit time gives a measure of reaction speed. Bromine is colored, so simple color photometry provides a second avenue. Hydrogen bromide and water vapor have characteristic infrared spectra, so either of these substances can be measured quantitatively as a function of time. These features make reaction (11-1) a good example for our consideration; we will use it to explore the properties of chemical reaction rates, the properties we shall wish to explain.

(a) WHAT IS A REACTION RATE?

After mixing HBr and O_2, we can follow the progress of the reaction by measuring periodically the Br_2 color intensity and, simultaneously, the time. Figure 11-1(a) shows a plot of typical experimental measurements. At any time during the reaction, we can tell how fast the reaction is proceeding by measuring the slope at that time. Figure 11-1(a) shows that the slope decreases with time as reactants are consumed and products accumulate. If we plot the slope as a function of time, as in Figure 11-1(b), the intercept at zero time represents the slope at the temperature and reactant pressures of the initial mixture. This slope at zero time is the initial reaction rate.

The rate can be expressed in a variety of units. For example, the reaction of HBr and O_2 at 750°K and pressures of 8 torr HBr and 2 torr O_2 produces 0.0016 torr Br_2 per minute. This rate might be expressed equally well in moles HBr/cc consumed per second or molecules HBr/cc consumed per second according to convenience. To convert to these units, we need to calculate the number of moles/cc per torr pressure at 750°K. Rearranging the perfect gas expression,

$$\frac{n}{V} = \frac{P}{RT} = \frac{(1/760 \text{ atm})}{[82.05(\text{cc-atm}/\text{degK-mole})](750°K)} = 2.14 \cdot 10^{-8} \frac{\text{mole/cc}}{\text{torr}}$$

$$(11\text{-}2)$$

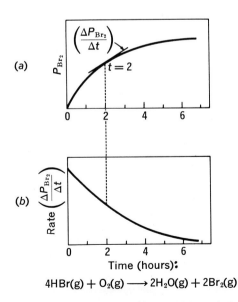

(a)

(b)

Time (hours):

$$4HBr(g) + O_2(g) \longrightarrow 2H_2O(g) + 2Br_2(g)$$

Figure 11-1 Time behavior of the oxidation of HBr at 750°K. (a) Partial pressure of Br_2 versus time. (b) Rate of increase of Br_2 versus time.

Hence, if

$$\frac{\Delta P(\text{Br}_2)}{\Delta t} = 0.0016 \text{ torr/min} = 2.7 \cdot 10^{-5} \text{ torr/sec}$$

then

$$\frac{\Delta n(\text{Br}_2)}{\Delta t} = \left(2.7 \cdot 10^{-5} \frac{\text{torr}}{\text{sec}}\right)\left(2.14 \cdot 10^{-8} \frac{\text{mole}}{\text{cc-torr}}\right) = 5.7 \cdot 10^{-13} \frac{\text{mole/cc}}{\text{sec}}$$

and

$$\frac{\Delta N(\text{Br}_2)}{\Delta t} = \left(5.7 \cdot 10^{-13} \frac{\text{mole}}{\text{sec-cc}}\right)(6.023 \cdot 10^{23}) = 3.4 \cdot 10^{11} \frac{\text{molecules/cc}}{\text{sec}}$$

Here are three exactly equivalent statements about the rate of formation of bromine in a mixture of 8 torr HBr and 2 torr O_2 at 750°K:

$$\frac{\Delta P(\text{Br}_2)}{\Delta t} = 2.7 \cdot 10^{-5} \frac{\text{torr}}{\text{sec}}$$ (The partial pressure of Br_2 rises $2.7 \cdot 10^{-5}$ torr during one second of reaction.) (11-3)

$$\frac{\Delta n(\text{Br}_2)}{\Delta t} = 5.7 \cdot 10^{-13} \frac{\text{mole/cc}}{\text{sec}}$$ (There is $5.7 \cdot 10^{-13}$ mole of Br_2 produced in each cc during one second of reaction.) (11-4)

$$\frac{\Delta N(\text{Br}_2)}{\Delta t} = 3.4 \cdot 10^{11} \frac{\text{molecules/cc}}{\text{sec}}$$ (There are $3.4 \cdot 10^{11}$ molecules of Br_2 produced in each cc during one second of reaction.) (11-5)

Exercise 11-1. Show that the concentration of molecules in one cc of gas at 298°K and one torr pressure is $5.38 \cdot 10^{-8}$ mole/cc. How many molecules are there in one cc under these conditions?

Exercise 11-2. Iodine atoms at 1.00 torr pressure in the presence of 500 torr argon react very rapidly at room temperature to form gaseous I_2. If 0.020 torr I_2 is formed in 1.0 microsecond (10^{-6} sec), show that the rate of formation of I_2 is $\Delta[I_2]/\Delta t = 1.1 \cdot 10^{-3}$ mole/cc-sec.

(b) HOW IS THE INITIAL RATE MEASURED?

The progress of the reaction can be followed by measuring either the growth of products or the loss of reactants. Of course, the loss of HBr will be four times faster than the loss of O_2, because reaction (11-1) tells us that four molecules of HBr are consumed for every molecule of oxygen used up. On the other hand, the loss of HBr will be double the growth of Br_2, again because reaction (11-1) insists.

To determine an initial rate we need to measure the change of pressure of any reactant or product, for example, Δp_{Br_2}, over some time interval Δt sufficiently short so that a negligible fraction of the reactants are consumed or products formed. When only a few percent of the reaction occurs, we can take the ratio of Δp_{Br_2} to Δt to be equal to the initial rate.

$$\frac{\Delta p_{Br_2}}{\Delta t} = \text{initial rate} \qquad\qquad (11\text{-}6)$$

(c) HOW DOES THE INITIAL RATE DEPEND UPON REACTANT PRESSURES?

This question must be answered experimentally. A succession of reaction mixtures with various compositions are prepared, and the initial rates are measured. If the compositions are varied, one component at a time, the dependence of the rate upon concentration of each constituent can be discovered. In general, the dependence is *not* that suggested by the balanced chemical reaction.

To use our sample reaction (11-1), if the rate is measured at 750°K for a mixture of 4 torr HBr and 1 torr O_2, the initial rate of formation of Br_2 is found to be 0.0004 torr/min. If the initial rate is measured for a mixture of 8 torr HBr and the same amount of oxygen, 1 torr, the reaction rate is doubled, 0.0008 torr/min. The rate depends linearly upon HBr pressure. Now if we try our original mixture of 8 torr HBr and 2 torr O_2, the rate is again doubled, 0.0016 torr/min. It depends linearly upon O_2 pressure as well. Thus, for reaction (11-1), the rate depends linearly upon both reactant pressures, even though four times as many HBr molecules are consumed as O_2 molecules. This dependence cannot be predicted by looking at the equation for the over-all reaction; it must be determined experimentally.

We can express this information in terms of proportionalities.

$$\text{rate} = \frac{\Delta p_{Br_2}}{\Delta t} \text{ is proportional to } p_{HBr}$$

and

$$\text{rate} = \frac{\Delta p_{Br_2}}{\Delta t} \text{ is proportional to } p_{O_2}$$

or, defining a proportionality constant k,

$$\frac{\Delta p_{Br_2}}{\Delta t} = k(p_{HBr})(p_{O_2}) \qquad\qquad (11\text{-}7)$$

Expression (11-7) is called the *rate law* for reaction (11-1); it is experimentally determined. The proportionality constant k is called the *rate constant*. Explaining these rate laws will be one of our major interests in this chapter.

Exercise 11-3. Show that, if the mixture of 4 torr HBr and 1 torr O_2 is compressed to one fifth the volume at 750°K, the rate of formation of Br_2 increases from 0.0004 torr/min to 0.01 torr/min.

(d) HOW DOES THE INITIAL RATE DEPEND UPON TEMPERATURE?

Obviously this question can be answered by a succession of studies like that just described. It is discovered that commonly the rate law remains the same, but the proportionality constant k, the rate constant, can change enormously.

For example, the rate constant for reaction (11-1) increases by a factor of 10 when the temperature is raised only 75°. It takes 25 minutes for 1% reaction in an 8 torr HBr–2 torr O_2 mixture at 750°K, but when the temperature is raised to 825°K, the 1% reaction occurs in 152 seconds.

Our everyday experience suffices to conclude that this extreme sensitivity of reaction rates to temperature is common. Almost every combustion process speeds up remarkably as temperature increases, as any Boy Scout can testify. Consider the fact that some of the oldest trees are over 2000 years old, despite the favorable free energy for the reaction between wood and the ever-present oxygen. Yet one of these trees can be burned to carbon dioxide and water in a few minutes when exposed to the intense heat of a forest fire. Explaining this sensitivity to temperature will be our second major goal.

11-2 Reaction rates and molecular collisions

The most important idea in chemical kinetics is that molecules can react with each other only when they are close to each other. We say that "molecules must collide." In the gas phase, we visualize molecules bumping into each other as they move about independently. In liquids, the molecules come close together through diffusional movements through the solvent. Plainly this view is consistent with, indeed suggested by, our picture of the molecule with a characteristic size. This size is indicated by the closeness of packing in condensed phases, particularly in the solid phase. Let's explore the consequences of this approach for reaction (11-1).

(a) THE SIZES OF HBr AND O_2

Figure 11-2 shows cutaway sections of crystals of solid HBr and solid oxygen as learned from X-ray studies. The dashed circles represent the approximate volume occupied by each molecule in its crystal phase. Each HBr molecule

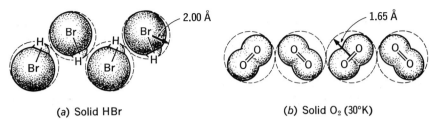

(a) Solid HBr   (b) Solid O_2 (30°K)

Figure 11-2   Crystal structures of solid hydrogen bromide and oxygen.

occupies a sphere approximately 2.0 Å in radius. The O_2 molecule needs a sphere about 1.65 Å in radius. The relative incompressibility of each of these solids shows that it is very difficult to force the molecules closer together than the radii dictate. We shall use these radii as a measure of molecular size.

(b) THE COLLISION RATE BETWEEN HBr AND O_2

Visualize a gaseous oxygen molecule rocketing around in a bottle full of hydrogen bromide gas. How many times a second does it get close enough to an HBr molecule to enjoy a collision? There are three factors that are obviously important: the molecular speed as the O_2 goes looking for trouble, the sizes of the colliding molecules, and the number of HBr molecules per cubic centimeter with which it may collide. Let's take up each in turn.

The average translational energy of a molecule at temperature T was explored in Chapter One. By examining a billiard ball model of gas pressure, we obtained expression (1-11):

$$PV = n \cdot \tfrac{2}{3}(KE) \qquad\qquad (1\text{-}11)$$

Then, by defining temperature in terms of perfect gas behavior, we converted (1-11) to the perfect gas law,

$$PV = nRT \qquad\qquad (1\text{-}13)$$

These two expressions constitute our definition of absolute temperature:

$$\frac{\text{av. KE}}{\text{mole}} = \tfrac{3}{2}RT$$

or, in terms of the kinetic energy per molecule,

$$\frac{\text{av. KE}}{\text{molecule}} = \frac{3}{2} \cdot \frac{R}{N_0}T = \tfrac{3}{2}\mathbf{k}T \qquad\qquad (11\text{-}8)$$

The constant **k**, called the *Boltzmann Constant,* is just the gas law constant expressed on a per molecule basis.

$$\mathbf{k} = \frac{R}{N_0} = 1.38 \cdot 10^{-16} \frac{\text{erg}}{\text{degree-molecule}} \tag{11-9}$$

Corresponding to the average kinetic energy per molecule is some average square velocity,* v^2.

$$\text{av. KE} = \tfrac{1}{2}mv^2 = \tfrac{3}{2}\mathbf{k}T \tag{11-10}$$

Solving (11-10) for v, we can decide how fast an oxygen molecule is moving at 750°K.

$$v = \sqrt{\frac{3\mathbf{k}T}{m}} \tag{11-11}$$

$$= \sqrt{\frac{(3)(1.38 \cdot 10^{-16})(750)}{32.0/6.023 \cdot 10^{23}}} = 7.6 \cdot 10^4 \text{ cm/sec}$$

This velocity is about 1700 miles per hour—it means that during each second, the oxygen molecule travels $7.6 \cdot 10^4$ cm, about half a mile. How many HBr's is it liable to encounter during that second? This is easily calculated—it is the number that are found in the "collision volume" swept out by this boisterous molecule.† The effective volume for a collision is the volume of a cylinder whose length is $7.6 \cdot 10^4$ centimeters and whose radius is the sum of the collision radii of HBr and O_2, $2.0 + 1.6$ Å $= 3.6$ Å. Any HBr molecule whose center is in this cylinder is close enough to the passing O_2 to deflect it. The volume of a cylinder is the cross-sectional area times its length.

Volume = (area)(length)

$$= [\pi(r_1 + r_2)^2] \cdot \sqrt{3\mathbf{k}T/m} \text{ cm}^3 \tag{11-12}$$

$$= \pi(3.6 \cdot 10^{-8})^2 \cdot (7.6 \cdot 10^4)$$

$$V = 3.09 \cdot 10^{-10} \frac{\text{cc}}{\text{sec}}$$

---

* There is a difference between the average square velocity, $(\overline{v^2})$ and the square of the average velocity $(\bar{v})^2$, which, for our purposes, can be ignored.

† The calculation presented is correct if the HBr molecules are not moving. The motion of both molecules participating in a collision is readily taken into account merely by replacing the mass term m in (11-11) by the "reduced mass" μ, where $1/\mu = 1/m_1 + 1/m_2$. When m_2 is very much larger than m_1, $\mu = m_1$. When $m_1 = m_2$, $\mu = m_1/2$. For O_2–HBr collisions, $\mu = 22.93$ gms and $m(O_2) = 32.00$ gms. This changes the velocity by the factor $\sqrt{32.0/22.9} = 1.18$, or 18%.

The number of HBr molecules in this volume depends, of course, upon the partial pressure of HBr. Since we know the reaction rate for a mixture of 8 torr HBr and 2 torr O_2, let's use these conditions. In a volume of 1 cc at 750°K and 8 torr pressure, the number of moles is eight times the number at 1 torr pressure, calculated in expression (11-2): $8(2.14 \cdot 10^{-8}) = 17.1 \cdot 10^{-8}$ mole/cc. In $3.09 \cdot 10^{-10}$ cc, the number of moles is $(17.1 \cdot 10^{-8}$ mole/cc) $(3.09 \cdot 10^{-10}$ cc$) = 5.28 \cdot 10^{-17}$ mole.

$$N = n \cdot (\text{Avogadro's number})$$

$$= (5.28 \cdot 10^{-17} \text{ mole})(6.02 \cdot 10^{23} \text{ molecules/mole})$$

$$= 3.18 \cdot 10^{7} \text{ molecules}$$

This is the number of collisions an O_2 molecule must suffer in one second, unless it reacts in one of them. For this molecule, the collision rate is

$$\begin{array}{l} \text{collisions/sec per} \\ \quad \text{cc for each} \\ \text{oxygen molecule} \end{array} = (\text{volume swept/sec})(\text{molecules HBr/cc})$$

$$= 3.18 \cdot 10^{7} \frac{\text{collisions}}{\text{sec}} \qquad (11\text{-}13)$$

If this is the number of collisions per second for one O_2 molecule, the total number of O_2–HBr collisions per second in one cc is obtained by multiplying by the number of O_2 molecules in one cc. The O_2 pressure is 2 torr, so the number of moles per cc is just two times the 1 torr figure,

$$\text{moles } O_2/\text{cc} = 2(2.14 \cdot 10^{-8}) = 4.28 \cdot 10^{-8}$$

$$\text{molecules } O_2/\text{cc} = (4.28 \cdot 10^{-8})(6.02 \cdot 10^{23}) = 2.58 \cdot 10^{16}$$

The total collision rate per cc is then

$$\frac{\text{total collisions/sec}}{\text{cc}} = \left(\begin{array}{c}\text{volume} \\ \text{swept} \\ \text{sec}\end{array}\right)\left(\begin{array}{c}\text{molecules} \\ \text{HBr} \\ \text{cc}\end{array}\right)\left(\begin{array}{c}\text{molecules} \\ O_2 \\ \text{cc}\end{array}\right)$$

$$= [\pi(r_1 + r_2)^2] \sqrt{3kT/m} \, [\text{HBr}][O_2] \qquad (11\text{-}14)$$

$$\frac{\text{collisions/sec}}{\text{cc}} = (3.18 \cdot 10^{7})(2.58 \cdot 10^{16}) = 8.2 \cdot 10^{23} \qquad (11\text{-}15)$$

Now we can compare the measured reaction rate, which we expressed in several types of units in (11-3), (11-4), and (11-5):

$$\frac{\text{molecules reacting}}{\text{sec}} = \frac{\Delta N}{\Delta t} = 3.4 \cdot 10^{11} \frac{\text{molecules/sec}}{\text{cc}} \qquad (11\text{-}5)$$

The number of O_2–HBr *collisions* in one second is $8.2 \cdot 10^{23}$, while the number of O_2–HBr *reactions* in one second is only $3.4 \cdot 10^{11}$. We find an enormous discrepancy, a factor larger than 10^{12}! It demands a reassessment of our model!

We could, of course, throw our hands in the air and discard the collisional model. Having nothing better, however, we look for some aspect of the model in which we can reasonably modify an assumption. Two possibilities come to mind. Perhaps we have overestimated the size of the molecules, or perhaps only a minute fraction of the collisions are actually effective in bringing about a reaction. We can hardly blame our size estimate, however, since the discrepancy can be explained only if the molecules are each a million times smaller than concluded from the crystal structure. *Hence our collisional model can fit the experimental facts only if very few of the collisions cause reaction.*

Exercise 11-4.   Calculate the average distance traveled in one second by an oxygen molecule at room temperature, 300°K. Show that a hydrogen molecule travels four times as far, $1.9 \cdot 10^5$ cm/sec.

Exercise 11-5.   Although a mixture of 10 torr H_2 and 10 torr Cl_2 will explode if ignited, there is negligible reaction at room temperature. Calculate the collisions per second in one cc if the H_2 mass determines collision rate and if the radius of an H_2 molecule is 1.0 Å and that of Cl_2 is 2.8 Å.

(c) THE VELOCITY EFFECT OF TEMPERATURE ON REACTION RATE

Despite this bad news about the collisional model, we might persist long enough to see what it predicts about the temperature effect on reaction rates. Temperature gets into the act in expression (11-11), where the velocity is shown to depend on the square root of the temperature. That means a given O_2 molecule travels further in a second if temperature is raised, so it will have more collisions per second. The reaction rate should rise as T goes up.* That agrees, at least qualitatively, with the facts.

Let's pursue the temperature effect quantitatively. If temperature is raised from 750°K to 825°K, the average molecular velocity should change by $\sqrt{825/750} = 1.049$. Thus we predict a 4.9% increase. But we observed experimentally that the rate increases by a factor of 10! Again our collisional model has failed to come even close to the observed facts!

(d) THRESHOLD ENERGY AND REACTION RATE

Now there are two discordant notes in our collisional model of molecular reactions. First, only a small fraction of the collisions seems to be effective,

---

* This statement is based on the premise that the *concentration* of molecules is held constant as T is raised. If *pressure* were held constant as T increased, the molecules per cc would decrease, so the conclusion would be reversed!

and second, the temperature effect on average molecular velocity does not explain at all the sensitivity of rates to temperature rise. Perhaps both of these difficulties can be resolved by adding to our model the reasonable assumption that the *average* molecular velocity is not sufficient to cause reaction. Perhaps only the molecules that by chance acquire much larger than average energy can actually react. Perhaps there is a threshold energy ϵ_a required to "activate" the molecules to permit reaction.

To explore this possible change in our theory, we must measure or calculate the distribution of molecular velocities in a gas at 750°K and, again, at 825°K. This subject came up earlier in our consideration of gas behavior in Chapter Eight. There we found that molecular velocities have the Boltzmann distribution, pictured in Figure 8-8. If most molecules do not possess enough energy to react, it must be because the energy threshold, or activation energy ϵ_a, is far out on the tail of the Boltzmann curve. We can use this curve to estimate the probability of reaction, since the area under the curve beyond ϵ_a is proportional to the number of molecules N_a with E greater than ϵ_a. The total area under the curve is proportional to the total number of molecules in the system, N. Hence, their ratio N_a/N is the fraction of the molecules that have enough energy to react if they properly collide with a reaction partner.

The shape of the Boltzmann curve changes with temperature, as seen in Figure 11-3. The area under the tail of the curve is particularly sensitive, and it can be expressed analytically if ϵ_a is much larger than the average molecular energy, which is determined by $\mathbf{k}T$. In fact, it can be shown that, if ϵ_a is much larger than $\mathbf{k}T$ (i.e., $x_a = \epsilon_a/\mathbf{k}T \gg 1$), then

$$\frac{N_a}{N} \approx 2\sqrt{\frac{x_a}{\pi}} \cdot e^{-x_a} \qquad\qquad (11\text{-}16)$$

We must appeal to experiment to search for this exponential dependence of N_a/N upon $\epsilon_a/\mathbf{k}T$.

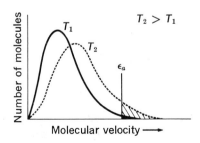

Figure 11-3   Effect of temperature upon molecular velocities. The numbers of molecules with velocities greater than that necessary for the activation energy ϵ_a are given by the shaded areas.

(e) RATE CONSTANT AND TEMPERATURE:
     THE EXPERIMENTAL FACTS

We hypothesize that reaction between two molecules depends not only upon the occurrence of a collision, but also upon the molecules possessing energy in excess of some ϵ_a. If so, then the expected reaction rate will be lower than the rate of collisions by the factor N_a/N. The easiest way to see whether this model really corrects the discrepancy between calculated collision rate and observed reaction rate is to examine the model's implied temperature dependence.

We propose

$$\left(\frac{\text{reactions}}{\text{sec}}\right) = \left(\frac{\text{collisions}}{\text{sec}}\right) \cdot \frac{N_a}{N} \qquad (11\text{-}17)$$

Substituting (11-16) for N_a/N and the complete expression (11-14) for the collision rate, (11-17) becomes

$$\left(\frac{\text{reactions}}{\text{sec}}\right) = \pi(r_1 + r_2)^2 \sqrt{\frac{3kT}{m}} \cdot [\text{HBr}][\text{O}_2] \cdot 2\sqrt{\frac{x_a}{\pi}}\, e^{-x_a} \qquad (11\text{-}18)$$

$$x_a = \epsilon_a/kT$$

Before examining (11-18), however, let's separate the terms that depend only on concentrations from the terms that are characteristic of the reaction and that include the temperature. This returns us to the form (11-7), in which we now can evaluate the proportionality constant k that was named the *rate constant*.

$$\left(\frac{\text{reactions}}{\text{sec}}\right) = k \cdot [\text{HBr}][\text{O}_2] \qquad (11\text{-}19)$$

where

$$k = \pi(r_1 + r_2)^2 \sqrt{\frac{3kT}{m}} \cdot 2\sqrt{\frac{\epsilon_a}{\pi kT}}\, e^{-\epsilon_a/kT} \qquad (11\text{-}20)$$

In (11-20) we immediately see that the velocity term introduces a temperature-dependent factor $\sqrt{T}$, but the $\sqrt{\epsilon_a/\pi kT}$ term introduces a $1/\sqrt{T}$ factor that cancels the velocity effect. The rest of the terms preceding the exponential we will group together and call the *collision number Z*.

$$k = Z\, e^{-\epsilon_a/kT} \qquad (11\text{-}21)$$

$$Z = [\pi(r_1 + r_2)^2] \sqrt{\frac{12\epsilon}{\pi m}} \; \frac{\text{cc}}{\text{molecule-sec}} \qquad (11\text{-}22)$$

Notice that (11-22) differs from (11-12) in the replacement of $\sqrt{3kT/m}$, the *average* molecular velocity, by the term $\sqrt{12\epsilon/\pi m}$, which is proportional to the velocity of a molecule with translational energy, $v_\epsilon = \sqrt{2\epsilon/m}$. We are no longer concerned with the collision rate of an average molecule; we are now interested in the collision rate of those molecules with sufficient energy to exceed the proposed threshold energy.

With this simpler notation, we can return to the burning question: Does the threshold energy idea, the ϵ_a embodied in (11-21), fit the facts any better than our unadorned collisional model? To check this point, we need only look at the experimental evidence to see if the observed temperature dependence matches that proposed in (11-21). To make the comparison, it is convenient to take the logarithm of (11-21).

$$\log_e k = \log_e Z - \epsilon_a/kT \tag{11-23}$$

Also, it is convenient to deal with energy per mole instead of energy per molecule, so we define

$$E_a(\text{kcal/mole}) = N_0\epsilon_a \tag{11-24}$$

We can substitute (11-24) into (11-23) and use (11-9), $k = R/N_0$, to obtain

$$\log_e k = \log_e Z - \frac{E_a/N_0}{kT} = \log_e Z - \frac{E_a}{(kN_0)T}$$

$$\log_e k = \left(-\frac{E_a}{R}\right) \cdot \frac{1}{T} + \log_e Z \tag{11-25}$$

Expression (11-25) shows the kind of measurement and test we need. At a known set of reactant concentrations, we must measure the reaction rate at several temperatures. Then, with (11-19), we can calculate k, the rate constant, at each of these temperatures. These data can be compared to the predicted dependence (11-25). The logarithm of the rate constant should vary in a linear fashion if plotted against the reciprocal of the absolute temperature. The algebraic relation is

$$y = mx + b$$

$$y = \log_e k \qquad x = 1/T \tag{11-26}$$

$$m = -\frac{E_a}{R} \qquad b = \log_e Z$$

Figures 11-4 and 11-5 show real experimental plots of this sort as published in research journals to verify the linear relationship. *Clearly there is an exponential relationship between reaction rate and temperature.* The activation energy or threshold energy model is supported, at least for these

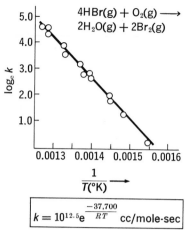

$$k = 10^{12.5}e^{\frac{-37,700}{RT}} \text{ cc/mole-sec}$$

Figure 11-4 Temperature dependence of reaction rate, HBr + O₂ reaction. (Journal of Physical Chemistry, 63, p. 1753, 1959.)

three reactions. Such behavior is extremely common and the activation energy idea is now an essential ingredient in our collisional explanation of reaction rate behavior.

More specific information can be extracted from the three plots in Figures 11-4 and 11-5. The slope of each curve, the quantity m in expression (11-26), is determined by E_a/R. Careful measurements of these slopes give the following results:

$$4\ HBr + O_2 \longrightarrow 2\ H_2O + 2\ Br_2 \qquad E_a = 37,700\ \text{cal/mole} \qquad (11\text{-}27)$$

$$2\ HCl + NO_2 \longrightarrow H_2O + NO + Cl_2 \qquad E_a = 23,400\ \text{cal/mole} \qquad (11\text{-}28)$$

$$2\ HBr + NO_2 \longrightarrow H_2O + NO + Br_2 \qquad E_a = 13,000\ \text{cal/mole} \qquad (11\text{-}29)$$

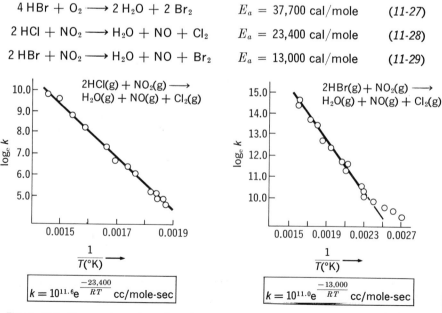

Figure 11-5 Temperature dependence of HX + NO₂ reactions. (Journal of Physical Chemistry, 64, p. 602, 1960.)

Table 11-1 The Activation Energy Factor in Reaction Rates

| Reaction | E_a (kcal/mole) | $e^{E_a/RT}$ | | Rate (750°K) |
		$T = 300°K$	$T = 750°K$	Rate (300°K)
4 HBr + $O_2 \longrightarrow$	37,700	$5.0 \cdot 10^{-28}$	$1.2 \cdot 10^{-11}$	$2.5 \cdot 10^{16}$
2 HCl + $NO_2 \longrightarrow$	23,400	$1.3 \cdot 10^{-17}$	$1.7 \cdot 10^{-7}$	$1.3 \cdot 10^{10}$
2 HBr + $NO_2 \longrightarrow$	13,000	$4.0 \cdot 10^{-10}$	$1.7 \cdot 10^{-4}$	$4.2 \cdot 10^{5}$

The first use we might make of these threshold energies is to calculate the fraction of the collisions that are effective in causing reaction, insofar as this exponential term determines it. Thus we can get a feel for the effect of the activation energy requirement. Let's do it for room temperature, 300°K, and for the elevated temperature at which our experiment was conducted, 750°K. Table 11-1 shows the results. We see how this activation energy concept fits the experimental facts. First, at 300°K, even for the lowest value of E_a, only one collision in 10^{10} (10 billion) will have enough energy to react. For reaction (11-27), the energy factor is $5.0 \cdot 10^{-28}$, which clearly prevents any reaction in a finite time. Taken literally, this factor indicates that if the collision rate in a cc volume is $8.2 \cdot 10^{23}$ collisions/sec (see (11-15)), it takes about 10^4 seconds (almost 3 hours) for a *single* molecule to react. There would be no measurable change in a million years! Furthermore, when the temperature is raised to 750°K (this is an easily attained laboratory temperature, 477°C), the activation energy factor is down to about 10^{-11}. Now we have 10^{13} molecular reactions per second in a cc volume, and the rate of reaction is readily measurable during a single day. Thus both the inefficiency of collisions and the extreme temperature sensitivity of reaction rates are explained.

Notice too that the E_a's in reactions (11-27), (11-28), and (11-29) differ considerably even though they apply to rather similar reactions. As a direct result, the reaction between HCl and NO_2 is quite slow at room temperature, while the similar reaction between HBr and NO_2 is practically instantaneous. The temperature sensitivity is correspondingly different. Changing the temperature to 750°K accelerates reaction (11-27) by the factor 10^{16}, while (11-28) speeds up by 10^{10} and (11-29) by only 420,000.

Exercise 11-6. Show that the reaction rate of HCl and NO_2 (see (11-28)) is increased by a factor exceeding 10,000 if the temperature is raised from 300°K to 400°K.

(f) IS IT ALL REASONABLE?

For one last assessment, we can ask whether the picture makes sense. Why is this threshold activation energy needed? Is the magnitude understandable?

A good analogy is the collision of one automobile with another. If they are both moving slowly, they'll bounce off each other and there will be no more damage than, possibly, heated emotions. If one of the cars is moving quite rapidly, however, there may be a significant rearrangement of the parts of each car. A bumper may be knocked off, the door from car A may leave the scene embedded in the hood of car B, and so on. It takes some minimum amount of energy though to break off a bumper and some different energy to wrench a car door loose from its hinges. These would be the threshold energies for those particular automotive changes.

In a similar way, a reactant molecule is held together with chemical bonds. These must be broken, at least partially, before the atoms can be rearranged into the product molecules. Therefore we can expect activation energies to be related to chemical bond energies. For example, in reaction (11-28), the HCl bond must be broken as new bonds are formed. The bond energy of HCl is 103 kcal/mole, so the value of E_a indicates that only about one fifth of the bond energy is needed before the formation of new bonds makes the rest of the reaction downhill. To contrast the similar reaction with HBr, we note that the HBr bond energy is only 87 kcal, 15 kcal lower than that of HCl. This surely must be a part of the explanation of the fact that the reaction of HBr with NO_2 has an activation energy 10.4 kcal lower than that of the analogous reaction of HCl with NO_2.

(g) ONE MORE FACTOR: COLLISIONAL GEOMETRY

With some confidence that the threshold energy idea explains the huge temperature effects in reaction rates, we can reassess the collisional model. Our initial view, that collisions are necessary for reaction, was modified by adding an exponential temperature factor that expresses the need for a minimum energy E_a for reaction. With this new factor, the calculated and experimental rates again can be compared. Table 11-2 shows, for reactions (11-27), (11-28), and (11-29), this comparison after the exponential factor has been taken into account. Each reaction is considered at a temperature at which the rate can be conveniently measured. The last column in Table 11-2 shows that there is still a discrepancy to be accounted. The three experimental rates are still slower than calculated by factors of the order of 100 to 10,000.

Table 11-2   Reaction Probability and Collisional Geometry

Reaction	Temperature	Experimental Rate Constant*	Calculated Rate Constant*	Experimental Calculated
HBr + O_2	750°K	38	12,650	$0.0030 = \frac{1}{330}$
HCl + NO_2	500°K	24	72,600	$0.00033 = \frac{1}{3000}$
HBr + NO_2	300°K	36	289,000	$0.00012 = \frac{1}{8000}$

* Dimensions for k are cc/mole-sec.

Once again the discrepancy could be attributed to an erroneous estimate of the size of molecules, insofar as reactions are concerned. To account for a factor of $1/100$, however, we would have to picture the molecules smaller by a factor of ten. This unsavory prospect is avoided, however, by a much more reasonable and intuitive alternative. Chemists speak of the need for a favorable *collisional geometry*. It is quite sensible to picture some collisional orientations as much less desirable for reaction than others. Figure 11-6 shows some possible molecular encounters and, without needing to know which are the best, it is obvious that some are much better than others. A head-on collision might make the most effective use of whatever energy is available to exceed E_a, but even then it might be that only one of the three arrangements shown is actually effective in bringing about reaction. Perhaps the hydrogen atom must contact the oxygen atoms, and of all the collisions, only those like the first shown are likely to undergo reaction.

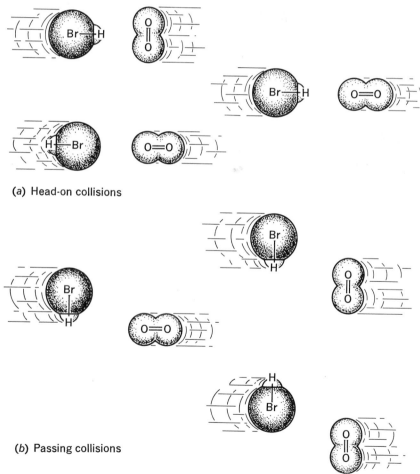

(a) Head-on collisions

(b) Passing collisions

Figure 11-6 *Collisional geometry and reaction probability.*

To incorporate this idea into the collisional view of reaction rates, we will introduce a probability term p into expression (11-21). The term p incorporates all the remaining discrepancy, and it is attributed to the probability that the collision geometry is favorable. At last we have a useful view of what factors fix the reaction rate constant k.

$$k = p \cdot Z \cdot e^{-E_a/RT} \qquad (11\text{-}30)$$

p = probability of a favorable collision geometry

Z = rate of collisions assuming thermal velocities and normal molecular size (see 11-22)

$e^{-E_a/RT}$ = exponential factor connected with the fraction of collisions possessing the minimum threshold energy

11-3 Reaction rates and thermodynamics

The expression (11-30) offers an intuitively acceptable view of the factors that determine reaction rates. It provides a framework within which reactions can be contrasted and classified, according to the magnitude of E_a, Z, and p. That is not to say there are no difficulties. Chemists still feel rather impotent in their ability to estimate in advance either p or E_a. Furthermore, at this stage of development, the reaction rate theory makes no contact with equilibrium thermodynamics. Fortunately such contact can be made, and it brings us another step closer to fulfilling our desire to understand and predict reaction rates.

(a) EQUILIBRIUM—A DYNAMIC BALANCE BETWEEN OPPOSING RATES

Let us consider the simple two-molecule (bimolecular) reaction between ozone and nitric oxide—simple because it probably occurs through an oxygen atom transfer.

$$O_3(g) + NO(g) \xrightarrow{k_f} O_2(g) + NO_2(g) \qquad (11\text{-}31f)$$

The subscript f indicates we are considering the reaction to the right, the *forward* direction. The rate of loss of O_3 through reaction (11-31f) has the following time dependence:

$$\left(\frac{\Delta p_{O_3}}{\Delta t}\right)_f = -k_f(p_{O_3})(p_{NO}) \qquad (11\text{-}32)$$

and

$$k_f = p_f Z_f \cdot e^{-E_{a,f}/RT} = \text{rate constant for the } forward \text{ reaction} \qquad (11\text{-}33)$$

The negative sign in (11-32) merely indicates that the ozone concentration is decreasing, because it is a reactant. At the same time, the products O_2 and NO accumulate, so the reverse reaction will increase in speed.

$$O_3(g) + NO(g) \xleftarrow{k_r} O_2(g) + NO_2(g) \qquad (11\text{-}31r)$$

We'll designate this reaction as (11-31r) with rate constant k_r to signify that it is the reverse of reaction (11-31f). This reaction forms O_3 through (11-31r) with the following time dependence:

$$\left(\frac{\Delta p_{O_3}}{\Delta t}\right)_r = +k_r(p_{O_2})(p_{NO_2}) \qquad (11\text{-}34)$$

Now the rate has a positive sign because ozone is a product in the reverse reaction. Again,

$$k_r = p_r Z_r \, e^{-E_{a,r}/RT} \qquad (11\text{-}35)$$

Of course, the actual change of ozone pressure with time will be the sum of (11-32) and (11-34):

$$\frac{\Delta p_{O_3}}{\Delta t} = \left(\frac{\Delta p_{O_3}}{\Delta t}\right)_f + \left(\frac{\Delta p_{O_3}}{\Delta t}\right)_r \qquad (11\text{-}36)$$

$$\frac{\Delta p_{O_3}}{\Delta t} = -k_f(p_{O_3})(p_{NO}) + k_r(p_{O_2})(p_{NO_2}) \qquad (11\text{-}37)$$

The equilibrium state, then, can be represented as the dynamic balance between forward and reverse reactions, (11-31f) and (11-31r), that is obtained when $\Delta p_{O_3}/\Delta t$ becomes equal to zero. At this point, the ozone is being formed exactly as fast as it is being lost. Equation (11-37) becomes:

$$\frac{\Delta p_{O_3}}{\Delta t} = 0 = -k_f(p_{O_3})(p_{NO}) + k_r(p_{O_2})(p_{NO_2})$$

or

$$k_f(p_{O_3})(p_{NO}) = k_r(p_{O_2})(p_{NO_2}) \qquad (11\text{-}38)$$

Expression (11-38) can be rearranged to place all concentration-dependent terms on the right and the rate constants on the left:

$$\text{at equilibrium} \qquad \frac{k_f}{k_r} = \frac{(p_{O_2})(p_{NO_2})}{(p_{O_3})(p_{NO})} \qquad (11\text{-}39)$$

But at equilibrium, we already know that the right-hand side of (11-39) gives the equilibrium constant:

$$\frac{(p_{O_2})(p_{NO_2})}{(p_{O_3})(p_{NO})} = K = \frac{k_f}{k_r} \tag{11-40}$$

Exercise 11-7. The rate constant k_f for the reaction between NO and O_3 (reaction 11-31f) is $1.6 \cdot 10^{10}$ cc/mole-sec at 300°K. The equilibrium constant for reaction (11-31) can be calculated from the standard free energies, which give $\Delta G^0 = -47.4$ kcal/mole. Show that k_r for reaction (11-31r) between O_2 and NO_2 is $4.8 \cdot 10^{-25}$ cc/mole-sec.

Now we can connect rate behavior with thermodynamic arguments by substituting (11-33) and (11-35) on the left and the free energy dependence of K on the right, $\Delta G^0 = -RT \ln K$ or,

$$K = e^{-\Delta G^0/RT} = \frac{p_f \cdot Z_f \cdot e^{-E_{a,f}/RT}}{p_r \cdot Z_r \cdot e^{-E_{a,r}/RT}} \tag{11-41}$$

Expression (11-41) can be simplified a bit by combining the exponentials on the right

$$e^{-\Delta G^0/RT} = \left(\frac{p_f Z_f}{p_r Z_r}\right) \cdot e^{-\frac{E_{a,f} - E_{a,r}}{RT}} \tag{11-42}$$

and, taking the logarithm of both sides,

$$-\frac{\Delta G^0}{RT} = \ln \frac{p_f Z_f}{p_r Z_r} - \frac{(E_{a,f} - E_{a,r})}{RT} \tag{11-43}$$

Now, by the definition of ΔG^0 for a constant temperature reaction, $\Delta G^0 = \Delta H^0 - T\Delta S^0$, (11-43) becomes

$$\left(-\frac{\Delta H^0}{R}\right)\frac{1}{T} + \frac{\Delta S^0}{R} = \left(-\frac{(E_{a,f} - E_{a,r})}{R}\right)\frac{1}{T} + \ln \frac{p_f Z_f}{p_r Z_r} \tag{11-44}$$

Both left and right sides of (11-44) contain a term sensitively dependent upon temperature, but each side also contains a term that depends very little, if at all, upon temperature. The first and third terms of (11-44) *must* correspond to each other, as must the second and last:

$$\left(-\frac{\Delta H^0}{R}\right)\frac{1}{T} = \left(-\frac{E_{a,f} - E_{a,r}}{R}\right)\frac{1}{T} \tag{11-45}$$

and

$$\frac{\Delta S^0}{R} = \ln \frac{p_f Z_f}{p_r Z_r} \tag{11-46}$$

(b) THE ENERGY PART—WHAT DOES IT MEAN?

Equation (11-45) needn't involve RT; it can be written

$$\Delta H^0 = E_{a,f} - E_{a,r} \qquad (11\text{-}47)$$

It clearly has sensible meaning. Consider a graphical representation of the enthalpy as the reaction proceeds, shown in Figure 11-7. In the forward reaction, the reactants begin at an intrinsic enthalpy H_1 and must climb an activation enthalpy hill $\Delta H_f^{\ddagger} = E_{a,f}$ to proceed toward products. In the reverse reaction, the products begin at an intrinsic enthalpy H_2 and must climb an activation enthalpy hill $\Delta H_r^{\ddagger} = E_{a,r}$ to go on to reform reactants. The activation energy hilltop is the same, going left or right, so we conclude

$$H_1 + \Delta H_f^{\ddagger} = H_2 + \Delta H_r^{\ddagger}$$

or

$$H_2 - H_1 = \Delta H_f^{\ddagger} - \Delta H_r^{\ddagger} = E_{a,f} - E_{a,r} \qquad (11\text{-}48)$$

And, since $H_2 - H_1 = \Delta H^0$, we have obtained expression (11-47) from the reasonable diagram of energy change during the course of reaction, that is, along the "reaction coordinate."

Exercise 11-8.   The activation energy for the reaction between O_3 and NO (reaction 11-31f) is 2.55 kcal. Use the standard heats of formation of O_3, NO, O_2, and NO_2 to calculate ΔH^0 and show that $\Delta H^{\ddagger}$ for the reverse reaction between O_2 and NO_2 is 50.4 kcal.

(c) THE ENTROPY PART—WHAT HERE?

An exactly parallel, though less obvious, argument can be made for expression (11-46). First, observe that $p_f Z_f$ is a probability term; it measures the

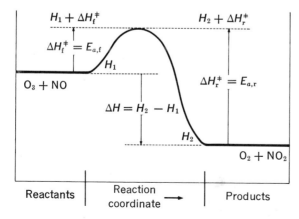

Figure 11-7   Enthalpy change as reaction proceeds.

probability that a collision will occur with a particular geometry. We can write, if we wish,

$$W_f = p_f Z_f \tag{11-49f}$$

and

$$W_r = p_r Z_r \tag{11-49r}$$

Now (11-46) can be written

$$\Delta S^0 = R \,\ell n \, \frac{W_f}{W_r} = R \,\ell n \, W_f - R \,\ell n \, W_r \tag{11-50}$$

This result carries us all the way back to Chapter Nine, in which entropy change was seen to be related to the intrinsic probabilities of initial and final states. There we found that the change in entropy ΔS is equal to the difference $S_2 - S_1$ in which the initial and final entropies are logarithmically fixed by the initial and final probabilities W_2 and W_1:

$$\Delta S^0 = S_2 - S_1 = R \,\ell n \, W_2 - R \,\ell n \, W_1 \tag{11-51}$$

With expression (11-51), it is logical to consider an activation *entropy* drawing like the energy drawing, as pictured in Figure 11-8. Here we see that we have the same hilltop to cross going forward or reverse, so

$$S_1 + \Delta S_f^{\ddagger} = S_2 + \Delta S_r^{\ddagger}$$

$$S_2 - S_1 = \Delta S_f^{\ddagger} - \Delta S_r^{\ddagger}$$

$$\Delta S^0 = \Delta S_f^{\ddagger} - \Delta S_r^{\ddagger} = R \,\ell n \, p_f Z_f - R \,\ell n \, p_r Z_r \tag{11-52}$$

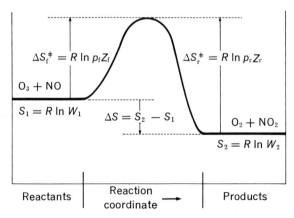

Figure 11-8 *Entropy change as reaction proceeds.*

Thus we are led to consider the activation process in thermodynamic terms. It is not just an energy hill that obstructs and thus controls the progress of the reaction. There is also an entropy-like factor involved.

(d) THE FREE ENERGY HILL—THE ONE THAT MATTERS

With this new idea in mind, we can return to the free energy relation (11-43), rewritten for convenience in the form

$$\Delta G^0 = (E_{a,f} - E_{a,r}) - RT \ln \frac{p_f Z_f}{p_r Z_r} \tag{11-43}$$

Substituting (11-47) and (11-52) into (11-43), we obtain

$$\Delta G^0 = (\Delta H_f^{\ddagger} - \Delta H_r^{\ddagger}) - R(\Delta S_f^{\ddagger} - \Delta S_r^{\ddagger})$$

or

$$\Delta G^0 = (\Delta H_f^{\ddagger} - T\Delta S_f^{\ddagger}) - (\Delta H_r^{\ddagger} - T\Delta S_r^{\ddagger}) \tag{11-53}$$

Plainly, the right-hand side has the form of the difference between two free energies, in this case, free energies of activation,

$$\Delta G^0 = \Delta G_f^{\ddagger} - \Delta G_r^{\ddagger} \tag{11-54}$$

We might even go the next step and draw a free energy diagram like Figures 11-7 and 11-8. Figure 11-9 is such a drawing, illustrating how free

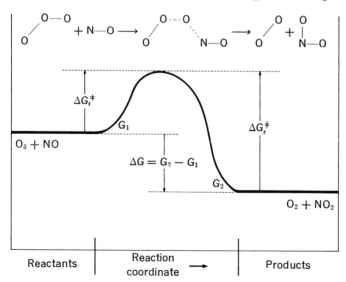

Figure 11-9 Free energy change as reaction proceeds: the activated complex and free energy of activation.

energy might change during the course of a reaction. At some point in the collision, there is a point at which the free energy hill reaches its crest. This collisional arrangement is called the "activated complex." This transient configuration is a molecular species of particular importance.

We now see that the rate at which reaction occurs is determined by the necessity to climb a hill, but it is a hill on the free energy landscape. Thus free energy not only acts as a reaction-direction signpost, pointing the way toward equilibrium where free energy is a minimum, it also reveals the critical topography along the route. Equilibrium is determined by an interplay between the energy change of the system during reaction, ΔH^0, and the randomness change during reaction, ΔS^0. The minimum value of $\Delta H^0 - T\Delta S^0$ for the system corresponds to maximum randomness in the universe. Similar statements can be made now for reaction rate phenomena. In order to proceed toward equilibrium, a system may have to cross a free energy hill $\Delta G^{\ddagger}$. If so, this free energy hill is in part due to an energy hill $\Delta H^{\ddagger}$ and in part due to a randomness hill $\Delta S^{\ddagger}$. In a bimolecular collision, two molecules held close together are *less* random than the separated molecules. Hence $\Delta S^{\ddagger}$ can be expected to be negative. Since $\Delta G^{\ddagger} = \Delta H^{\ddagger} - T\Delta S^{\ddagger}$, the negative value of $\Delta S^{\ddagger}$ causes it to raise the free energy hill above what $\Delta H^{\ddagger}$ alone would make it. Of course, there is no reason to be confident that activation enthalpy and activation entropy will peak at the same value of the reaction coordinate. Somewhere, however, activation free energy $G^{\ddagger} = H^{\ddagger} - TS^{\ddagger}$ reaches a crest and this determines the rate at which the reaction proceeds.

(e) THE FREE ENERGY HILL AND RANDOMNESS
    IN THE UNIVERSE

One final note gives meaning to it all. A minimum free energy for a system implies equilibrium because it corresponds to maximum randomness for the universe. How, then, does the universe look at our need to climb a free energy hill in order to bring about a reaction? If free energy of the system is rising, then the randomness of the universe must be decreasing!

The answer again lies in probability. The universe is dynamic and it is constantly rearranging itself, just as the gas molecules in a bulb are constantly changing positions and exchanging energy. The condition of maximum randomness is maintained only statistically—momentarily there can (and will) be fluctuations from this most probable state. To climb a free energy hill, we need a momentary decrease in the randomness of the universe, counter to the general flow of things. How long we will have to wait for such an unlikely situation depends upon how much of a decrease we need. How fitting! But we can be sure that if we wait long enough, sooner or later random events will conspire unwittingly to give us the configuration and energy that will permit reaction. It is all a matter of probability.

So now we find a fundamental sameness in the motivation toward equilibrium and the time it takes to get there. Probability tells the universe the direction its history must take: Over time, randomness increases. But probability also governs the pace. When change is blocked on the site of a local randomness hill, progress depends upon momentary deviations from maximum randomness to leave that hill so that even higher ground nearby can be reached. Such deviations ultimately will happen: They are not in violation of the laws of statistics—they are merely improbable, but in strict accordance with these laws.

11-4 Reaction mechanisms

The first reaction mentioned in this chapter, the oxidation of hydrogen bromide by oxygen, has a dependence upon reactant concentrations that is not obviously connected to the balanced equation. The same is true for most reactions, as the following examples show.

reaction $\quad$ $4\,HBr(g) + O_2(g) \longrightarrow 2\,H_2O(g) + 2\,Br_2(g)$ $\qquad$ (11-27)

rate law $\quad$ $\dfrac{\Delta O_2}{\Delta t} = -k_{27}[HBr][O_2]$ $\qquad$ (11-55)

reaction $\quad$ $2\,HBr(g) + NO_2(g) \longrightarrow H_2O(g) + NO(g) + Br_2(g)$ $\quad$ (11-29)

rate law $\quad$ $\dfrac{\Delta NO_2}{\Delta t} = -k_{29}[HBr][NO_2]$ $\qquad$ (11-56)

reaction $\quad$ $2\,HCl(g) + NO_2(g) \longrightarrow H_2O(g) + NO(g) + Cl_2(g)$ $\quad$ (11-28)

rate law $\quad$ $\dfrac{\Delta NO_2}{\Delta t} = -k_{28}[HCl][NO_2]$ $\qquad$ (11-57)

Reaction (11-27) requires the consumption of four HBr molecules per O_2 molecule, but the rate depends upon the O_2 pressure just as much as on the HBr pressure. Reaction (11-29) requires two HBr molecules per NO_2 molecule, but again the rate law shows first-power dependences upon reactant pressures. There is one conclusion that we can reach without doubt. Reaction (11-27) cannot occur through a simultaneous collision of four HBr molecules and one O_2 molecule. If it did, the rate law would depend much more heavily upon the HBr pressure, to the fourth power, in fact. This is not such a surprising result. After all, a simultaneous collision among five gaseous molecules must be an exceedingly rare event. The reaction must take place in a sequence of simpler, hence individually more probable, collisional steps.

This type of argument leads chemists to the formulation of likely reaction mechanisms—sequences of reactions, each involving only two or, at most, three molecules in collision. These sequences must agree, of course, with the observed rate law. Sometimes this criterion leads to a clear and unambiguous mechanism, but often two or three possibilities exist. Reaction (11-27) is a case in point.

(a) THE REACTION MECHANISM FOR THE HBr-O₂ REACTION

The fact that the rate law (11-55) depends equally upon HBr and O_2 pressure suggests that the first step in reaction (11-27) is a simple collision between HBr and O_2. In fact, we glibly pictured it that way in Figure 11-6. Such a collision might, for example, form HOOBr, a molecule that would be similar in structure to hydrogen peroxide HOOH. This substance, HOOBr, is unknown, so it must be extremely reactive. Perhaps HOOBr, once formed, initiates a rapid sequence such as:

$$HBr + O_2 \xrightarrow{k} HOOBr \qquad\qquad slow \qquad\qquad (11\text{-}27a)$$

$$HOOBr + HBr \longrightarrow HOBr + HOBr \qquad fast \qquad\qquad (11\text{-}27b)$$

$$HOBr + HBr \longrightarrow H_2O + Br_2$$

and $\qquad\qquad\qquad\qquad\qquad\qquad\qquad\qquad fast \qquad\qquad (11\text{-}27c)$

$$HOBr + HBr \longrightarrow H_2O + Br_2$$

The sum of these four reactions is just the over-all reaction (11-27), consuming four molecules of HBr and one of O_2. If the first reaction (11-27a) is much slower than the subsequent reactions, then it will act as a "bottleneck" and the entire sequence will be rate-controlled by that single reaction. That reaction is called the *rate determining step*. Since it involves one molecule each of HBr and O_2, it will have the observed rate law.

$$\frac{\Delta p_{O_2}}{\Delta t} = -k[HBr][O_2] \qquad\qquad\qquad (11\text{-}55)$$

Now the typical chemist will try to picture the possible course of the reaction in detail, as shown in Figure 11-10. The third structure shown might be the activated complex, the structure with maximum free energy in which the partial rupture of the HBr bond raises the energy by 37,700 cal. Further movement of the atoms to the last structure would then be downhill on the free energy surface. It would become an energetically favorable path because the new O–H and O–Br bonds are sufficiently formed to dominate the situation.

But wait a minute! Before becoming too enamored with this intimate description of reaction (11-27a), we should wonder if there isn't some entirely different mechanism that needs to be considered. Strangely, the completely reasonable reaction sequence (11-27a), (11-27b), and (11-27c) is not the one given in the research literature by the scientists who studied

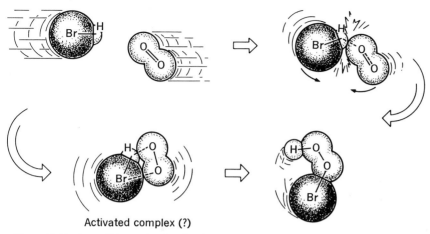

Activated complex (?)

Figure 11-10   A possible mechanism for the rate controlling step in the HBr + O$_2$ reaction.

the rate of reaction (11-27).* They propose, instead,

$$HBr + O_2 \xrightarrow{k'} HOO + Br \qquad \text{slow} \qquad (11\text{-}27a')$$

$$HOO + HBr \longrightarrow HOOH + Br \qquad \text{fast} \qquad (11\text{-}27b')$$

$$HOOH \longrightarrow HO + OH \qquad \text{fast} \qquad (11\text{-}27c')$$

$$OH + HBr \longrightarrow H_2O + Br \qquad \text{fast} \qquad (11\text{-}27d')$$

$$Br + Br \longrightarrow Br_2 \qquad \text{fast} \qquad (11\text{-}27e')$$

Adding (11-27a'), (11-27b'), (11-27c'), two times (11-27d'), and two times (11-27e') gives the over-all reaction (11-27). This sequence would also have the simple rate law (11-55), because again the first reaction can act as a bottleneck. Rate considerations are not sufficient to choose between the sequence beginning with (11-27a) and that beginning with (11-27a'). Sometimes other considerations give clues. For example, bond energy arguments indicate that reaction (11-27a') is endothermic by 42 kcal (we must break the HBr bond and half of the O$_2$ double bond, and in return, we form only one OH bond). Reaction (11-27b') is known experimentally to be exothermic by only 3 kcal. Then reaction (11-27c') is known experimentally to be endothermic by 51 kcal. It is difficult to rationalize these facts with an experimental activation energy of 37.7 kcal and the simple rate law governed by reaction (11-27a'). Hence the sequence (11-27a) to (11-27c) seems a much more reasonable choice. Notice, though, that the choice between these two mechanisms could not be made from the rate law alone. Notice, also, that chemists have not yet agreed which mechanism is to be preferred.

* W. A. Rosser and H. Wise, *Journal of Physical Chemistry*, Vol. **63**, p. 1753, 1959.

Exercise 11-9.   Write a sequence of reactions involving only two-molecule collisions that accomplishes reaction (11-29) and in which nitrous acid HONO is formed.

Exercise 11-10.   Repeat Exercise 11-9 except involving HOBr and not HONO.

(b) FIRST ORDER REACTIONS

Some reactions have rate laws that depend upon one constitutent concentration only and that to the first power. Such a reaction is called a *first order reaction*. An example often quoted is the decomposition of N_2O_5:

$$2\,N_2O_5(g) \longrightarrow 4\,NO_2(g) + O_2(g) \tag{11-58}$$

with experimental rate law

$$-\frac{\Delta[N_2O_5]}{\Delta t} = k[N_2O_5] \tag{11-59}$$

This rate law says that the number of molecules of N_2O_5 disappearing per second depends only upon the number of N_2O_5 molecules present. This is the same rate law that governs radioactive decay; the number of nuclei that decompose per second is simply proportional to the number of radioactive nuclei present. In a first order reaction, the N_2O_5 disappears at any instant as given by (11-59), and after some time interval t, the amount left has an exponential relation to the amount present at $t = 0$.

$$[N_2O_5]_t = [N_2O_5]_0\,e^{-kt} \tag{11-60}$$

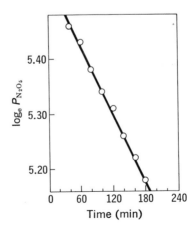

Figure 11-11   *Decomposition of N_2O_5: Evidence for first order dependence of rate on $p_{N_2O_5}$ (298°K). (Journal of the American Chemical Society,* **43**, *p. 53, 1921.)*

A possible mechanism for (11-58) is as follows:

$$N_2O_5 \longrightarrow NO_2 + NO_3 \qquad \text{slow} \qquad (11\text{-}58a)$$

$$NO_3 \longrightarrow NO + O_2 \qquad \text{fast} \qquad (11\text{-}58b)$$

$$NO + N_2O_5 \longrightarrow NO_2 + N_2O_4 \qquad \text{fast} \qquad (11\text{-}58c)$$

$$N_2O_4 \longrightarrow 2\,NO_2 \qquad \text{fast} \qquad (11\text{-}58d)$$

Evidence that a reaction has a first order rate law usually is obtained by plotting the measured concentration as a function of time in a logarithmic plot. If N_2O_5 pressure depends upon e^{-kt}, then the log of $p(N_2O_5)$ gives a straight line if plotted against time as shown in Figure 11-11. Figure 11-12(a) shows a similar plot given in the research literature for the decomposition of chloroformic acid. This molecule has an extremely transient existence: Notice that the time scale in Figure 11-12(a) does not extend as high as 1 millisecond. From such measurements, the first order rate law was deduced and the energy of activation was found to be 14 kcal. The magnitude of $\Delta H^{\ddagger}$ suggests that the mechanism involves internal rotation around the OH bond. Then the proximity of the hydrogen and chlorine atoms allows the elimination of HCl, in part because of the size of chlorine and in part because of the stable HCl bond formed (see Fig. 11-12(b)).

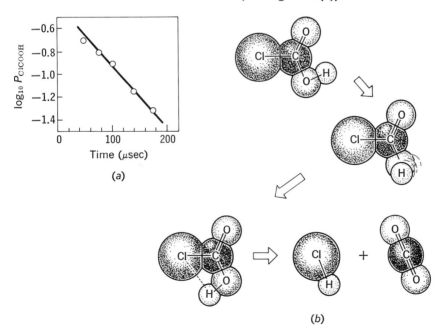

Figure 11-12 Decomposition of ClCOOH. (a) First order rate dependence. (b) Detailed mechanism.

(c) SECOND ORDER REACTIONS

We have already learned experimentally that reaction (11-27) has the rate law

$$\frac{\Delta[O_2]}{\Delta t} = -k[HBr][O_2]$$

This reaction is said to be a *second order reaction* because its rate law depends upon the pressures (or concentrations) of *two* constituents, each to the first power.

Another second order reaction is:

$$2\,HONO \longrightarrow H_2O + NO + NO_2 \tag{11-61}$$

$$\frac{\Delta[HONO]}{\Delta t} = -k[HONO]^2 \tag{11-62}$$

This time the reaction is second order because the rate law depends upon the pressure (or concentration) of a single constituent, but to the second power.

Second order reactions such as (11-27), (11-28), and (11-29) are fairly common. Such a rate law is observed, even for a complicated reaction mechanism, if the first step involves a simple two-molecule collision that is the slowest step in the sequence.

(d) THIRD ORDER REACTIONS

An important smog reaction is the oxidation of nitric oxide by oxygen:

$$2\,NO(g) + O_2(g) \longrightarrow 2\,NO_2(g) \tag{11-63}$$

Its rate law is well known to depend upon the square of the NO pressure and the first power of the O_2 pressure:

$$\frac{\Delta[NO]_2}{\Delta t} = k[NO]^2[O_2] \tag{11-64}$$

Such a reaction is called a third order reaction: Its rate law depends upon molecular concentrations to the third power.

Some third order reactions require a simultaneous collision of three reactant molecules, but this is not necessarily so. For example, the mechanism of reaction (11-64) could be a sequence of two-molecule collisions.

$$NO + O_2 \rightleftharpoons NO_3 \qquad \text{fast} \tag{11-64a}$$

$$NO_3 + NO \longrightarrow 2\,NO_2 \qquad \text{slow} \tag{11-64b}$$

If reaction (11-64a) is very rapid (compared to the loss of NO_3 in (11-64b)), it quickly reaches equilibrium.

$$K_a = \frac{[NO_3]}{[NO][O_2]} \tag{11-65}$$

If reaction (11-64b), on the other hand, proceeds slowly, its rate will be dictated by collisions between NO_3 and NO, so

$$\frac{\Delta[NO_2]}{\Delta t} = k_b[NO_3][NO] \tag{11-66}$$

Since the NO_3 concentration is extremely small, it cannot be readily measured. That makes it difficult to use (11-66). The situation can be remedied, however, by substituting (11-65) into (11-66) so as to eliminate the NO_3 concentration:

$$\frac{\Delta[NO_2]}{\Delta t} = k_b \cdot K_a[NO][O_2] \cdot [NO]$$

or

$$\frac{\Delta[NO_2]}{\Delta t} = k[NO]^2[O_2] \tag{11-67}$$

in which

$$k = k_b \cdot K \tag{11-68}$$

The expression (11-67) is exactly what is observed experimentally, so we see that the mechanism (11-64a)–(11-64b) is compatible. Furthermore, the experimentally observed rate constant k is seen to be a composite of the equilibrium constant for reaction (11-64a) and the rate constant k_b for the rate determining step, reaction (11-64).

Exercise 11-11. Show that if reaction (11-27b) were slow and all other reactions fast in the reaction sequence (11-27a) to (11-27c), the reaction would be third order with rate law

$$\frac{\Delta[O]_2}{\Delta t} = -k[HBr]^2[O_2]$$

(e) EXPLOSIONS

At the beginning of this chapter, we noted that a gaseous H_2–Cl_2 mixture will not react noticeably in a week in a dark room, but it will explode violently if momentarily exposed to a bright light. This system provides one of the simplest examples of a reaction mechanism that leads to an explosion.

Thermodynamically, the reaction has everything going for it. Using the standard thermodynamic tables, ΔH^0, ΔS^0, and ΔG^0 are readily calculated:

$$H_2(g) + Cl_2(g) \rightleftharpoons 2\, HCl(g) \qquad\qquad (11\text{-}69)$$

$$\Delta H^0 = -44.2 \text{ kcal/mole}$$

$$\Delta S^0 = +4.7 \text{ cal/deg mole}$$

$$\Delta G^0 = -45.6 \text{ kcal/mole}$$

The reaction is quite exothermic, the entropy increases, and the negative value of ΔG^0 corresponds to an equilibrium constant of $4 \cdot 10^{+34}$. There is an activation energy barrier, however, that exceeds 50 kcal/mole, so the reaction is exceedingly slow if it must depend upon energetic collisions between H_2 and Cl_2 molecules.

Exercise 11-12. Use the results of Exercise 11-5 to calculate the number of molecular reactions per cc per second if the factor p is guessed to be 0.001 and if $E_a = 50,000$ cal.

Things really change though if a very bright light is flashed into the gas mixture. Chlorine absorbs light and the energetic chlorine molecules dissociate into atoms.

$$Cl_2(g) + \text{light} \longrightarrow 2\, Cl \qquad\qquad (11\text{-}70)$$

The presence of these chlorine atoms starts an entirely new chain of chemical reactions. Since a chlorine atom does not have the inert gas electron population (see Section 2-4), it is extremely reactive. These species react with hydrogen with only a small activation energy, 5.5 kcal/mole. The products are HCl and H atoms, and the reaction is almost thermoneutral.

$$Cl + H_2 \longrightarrow HCl + H \qquad\qquad (11\text{-}71)$$

$$\Delta H^0 = +1.0 \text{ kcal/mole}$$

$$E_a = 5.5 \text{ kcal}$$

Insofar as reaction (11-71) takes place, another reactive species is added to the mixture, H atoms. Now their chemistry must be considered. Hydrogen atoms react with chlorine molecules with even lower activation energy, about 2.5 kcal. This reaction is extremely exothermic, since it breaks the relatively weak bond of Cl_2 and forms the strong bond of HCl.

$$H + Cl_2 \longrightarrow HCl + Cl \qquad\qquad (11\text{-}72)$$

$$\Delta H^0 = -45.2 \text{ kcal/mole}$$

$$E_a = 2.5 \text{ kcal}$$

Now we have the interesting result that reaction (11-72) was initiated by (11-71), but it, in turn, reciprocates by initiating (11-71) again. Together, these reactions furnish a repetitious, cyclic reaction. Such a cyclic process is called a *chain reaction*.

$$
\begin{array}{llllll}
 & Cl + H_2 \longrightarrow HCl + H & \Delta H = +1.0 & E_a = 5.5 & (11\text{-}71) \\
\text{net} & H + Cl_2 \longrightarrow HCl + Cl & \Delta H = -45.2 & E_a = 2.5 & (11\text{-}72) \\
\text{reaction} & H_2 + Cl_2 \longrightarrow 2\,HCl & \Delta H = -44.2 & E_a > 50 & (11\text{-}69)
\end{array}
$$

We see that the net reaction of the chain, the sum of (11-71) and (11-72) is just (11-69). However, the chain reaction achieves the over-all reaction in a mechanism whose highest activation energy is 5.5 kcal in competition with the direct reaction with activation energy above 50 kcal. This explains why adding a few chlorine atoms here and there in the reaction vessel makes such a tremendous change in the rate of formation of product HCl.

There remains the question of why the mixture explodes. After all, reaction (11-69) consumes two molecules and it produces two—the pressure won't rise on that account. Of course, the temperature will rise as the exothermic reaction (11-72) releases its heat. We can calculate easily the final temperature of the gas if one mole of H_2 reacts with one mole of Cl_2, releasing 44.2 kcal of heat. The heat capacity of HCl(g) at constant volume is 5.0 cal/°K, so for two moles of product, the temperature will rise accordingly till 44,200 calories have been consumed.

$$\Delta T(^\circ K) \cdot C_p \left(\frac{\text{cal}}{\text{deg}}\right) = 44,200 \text{ cal}$$

$$(T - 300)(2 \cdot 5.0) = 44,200$$

$$T = 300 + \frac{44,200}{10} = 4720^\circ K \qquad\qquad (11\text{-}73)$$

This huge temperature rise *will* raise the pressure—by a factor of 4720/300 = 15.7.

The most important effect of temperature rise is yet to be mentioned, though. Whatever the reaction rate of the chain reaction (11-71) and (11-72), the effect of temperature is bound to speed it up. Thus the chain is not only *self-propagating* (once started, it keeps going), it is also *self-accelerating*. As reaction proceeds, heat is released, the reaction speeds up, more heat is released, the reaction speeds up, more heat is . . . BOOM!

Such a chain reaction has its own natural enemies. Any process other than reactions (11-71) and (11-72) that removes either a Cl or an H atom will kill off a cyclic chain. If there are too many such chain terminations, the whole process can be stopped before it gets going. Here are some chain terminating reactions:

$$Cl + Cl + M \longrightarrow Cl_2 + M \tag{11-74}$$

$$H + H + M \longrightarrow H_2 + M \tag{11-75}$$

$$H + Cl + M \longrightarrow HCl + M \tag{11-76}$$

In each reaction, there is the enigmatic symbol M. These equations imply that three-molecule collisions are needed. That is what is observed experimentally. Chlorine and hydrogen atoms won't combine to form Cl_2, H_2, or HCl unless some third species M—any molecule will do—drains off some of the energy. That's just what we concluded 'way back in Section 7-5(b) (see Figs. 7-20 and 7-21). If two chlorine atoms collide, they may momentarily form Cl_2, but the molecule forms with energy equal to the bond energy, just enough to cause it to fly apart again. Molecular friction—a collision with some other molecule M—is needed before these chain terminating steps can interfere with our jolly explosion. Since three-molecule collisions are rare, chain termination is not very effective.

Exercise 11-13.   Show that reaction (11-71) is speeded up by a factor of 100 as the temperature rises from 300°K to 600°K.

11-5   Catalysis

A concentrated hydrogen peroxide solution can be stored for many weeks. However, if a few drops of ferrous nitrate solution are added, bubbles of oxygen begin to form in a few seconds. In a few minutes, most of the hydrogen peroxide is gone, decomposed to water and oxygen.

$$2\,H_2O_2(aq) \longrightarrow 2\,H_2O(\ell) + O_2(g) \tag{11-77}$$

Although the decomposition reaction (11-77) does not involve ferrous ion, somehow the $Fe^{+2}(aq)$ presence speeds it up. *A substance that accelerates a reaction without being consumed is called a* **catalyst.**

(a) HOMOGENEOUS CATALYSIS

The case just discussed is called *homogeneous catalysis* because both the reactant H_2O_2 and the catalyst $Fe^{+2}(aq)$ are present in the same phase, in solution. We will contrast this later with reactions that are accelerated locally at the interface between two phases (liquid–solid or gas–solid); this is called *heterogeneous catalysis*.

Why does hydrogen peroxide decompose in the presence of a small amount of $Fe^{+2}(aq)$? Before trying to answer this question, we should consider why hydrogen peroxide decomposes at all. That is a question about equilibrium, one for thermodynamics. The free energies of formation give the free energy change of reaction (11-77) and, hence, the equilibrium constant:

$$\Delta G^0 = 2\Delta G_f^0(H_2O,\ell) + \Delta G_f^0(O_2) - 2\Delta G_f^0(H_2O_2,aq)$$

$$= 2(-56.7) + 0 - 2(-31.6)$$

$$\Delta G^0 = -50.2 \text{ kcal} \tag{11-78}$$

$$K = e^{-\Delta G^0/RT} = 6.2 \cdot 10^{36} \tag{11-79}$$

The negative value of ΔG^0 and the high equilibrium constant show that H_2O_2 is quite unstable with respect to O_2 and H_2O. It doesn't decompose because of the rate of reaction. Apparently the activation free energy for decomposition is very high.

The function of the ferrous ion must be to provide a new path with a lower free energy of activation. Thinking of our free energy topography, $Fe^{+2}(aq)$ must furnish an alternate route toward the products. The mechanism of the catalysis is, in this case, quite straightforward. Consider the decomposition as an oxidation–reduction reaction, using half-cell reactions.

Figure 11-13   A catalyst furnishes a low pass through the free energy mountains.

The H_2O_2 half-cell reduction potentials can be contrasted with that of the Fe^{+2}–Fe^{+3} half-cell reactions.

$$H_2O_2 + 2\,H^+ + 2\,e^- \longrightarrow 2\,H_2O \quad \mathcal{E}^0 = +1.77 \qquad (11\text{-}80)$$

$$Fe^{+3} + e^- \longrightarrow Fe^{+2} \quad \mathcal{E}^0 = +0.77 \qquad (11\text{-}81)$$

$$O_2 + 2\,H^+ + 2\,e^- \longrightarrow H_2O_2 \quad \mathcal{E}^0 = +0.68 \qquad (11\text{-}82)$$

These potentials tell us that, equilibrium-wise, H_2O_2 can react with *both* $Fe^{+2}(aq)$ and $Fe^{+3}(aq)$. Combining half-reactions (11-80) and (11-81),

$$
\begin{array}{ll}
H_2O_2 + 2\,H^+ + 2\,e^- \longrightarrow 2\,H_2O & \mathcal{E}^0 = +1.77 \\
2\,Fe^{+2} \longrightarrow 2\,Fe^{+3} + 2\,e^- & \mathcal{E}^0 = -0.77 \\
\hline
H_2O_2 + 2\,Fe^{+2} + 2\,H^+ \longrightarrow 2\,Fe^{+3} + 2\,H_2O & \Delta\mathcal{E}^0 = 1.77 - 0.77 \quad (11\text{-}83) \\
& \Delta\mathcal{E}^0 = +1.00
\end{array}
$$

and by relation (10-58)

$$K = 10^{n\Delta\mathcal{E}^0/0.059} = 10^{2 \cdot 1.00/0.059} = 5.7 \cdot 10^{+33} \qquad (11\text{-}84)$$

And, by combining half-reactions (11-81) and (11-82)

$$
\begin{array}{ll}
2\,Fe^{+3} + 2\,e^- \longrightarrow 2\,Fe^{+2} & \mathcal{E}^0 = +0.77 \\
H_2O_2 \longrightarrow O_2 + 2\,H^+ + 2\,e^- & \mathcal{E}^0 = -0.68 \\
\hline
H_2O_2 + 2\,Fe^{+3} \longrightarrow 2\,Fe^{+2} + O_2 + 2\,H^+ & \Delta\mathcal{E}^0 = 0.77 - 0.68 \quad (11\text{-}85) \\
& = +0.09
\end{array}
$$

$$K = 10^{2 \cdot 0.09/0.059} = 1.1 \cdot 10^{+3} \qquad (11\text{-}86)$$

These constants, (11-84) and (11-86), show that equilibrium strongly favors products in both reactions (11-83) and (11-85). Together, these reactions make up a "chain" process that has as its net reaction the decomposition of H_2O_2 reaction (11-77):

$$
\begin{array}{ll}
H_2O_2 + 2\,Fe^{+2} + 2\,H^+ \longrightarrow 2\,Fe^{+3} + 2\,H_2O & (11\text{-}83) \\
H_2O_2 + 2\,Fe^{+3} \quad\quad\quad\quad \longrightarrow 2\,Fe^{+2} + O_2 + 2\,H^+ & (11\text{-}85) \\
\hline
\text{net reaction} \quad\quad\quad\quad 2\,H_2O_2 \longrightarrow 2\,H_2O + O_2 & (11\text{-}77)
\end{array}
$$

The situation is reminiscent of the explosion chain, (11-71) and (11-72). If reactions (11-83) and (11-85) have low activation energies, decomposition will occur rapidly via the intermediaries $Fe^{+2}(aq)$ and $Fe^{+3}(aq)$. Experiment shows that this is the case.

Exercise 11-14. Sodium bromide acts as a catalyst in the decomposition of H_2O_2. From the half-cell potential $Br_2 + 2\,e^- = 2\,Br^-$, $\mathcal{E}^0 = 1.09$, deduce the reaction sequence that facilitates the decomposition, and calculate the equilibrium constant for each reaction in the sequence.

(b) HETEROGENEOUS CATALYSIS

There are many analogous cases of catalysis at solid–liquid or at solid–gas interfaces. For example, if a lump of solid manganese dioxide is added to a solution of hydrogen peroxide, again decomposition becomes rapid. After a time, the evolution of oxygen gas subsides, but only because the hydrogen peroxide is consumed. The manganese dioxide lump remains, its weight undiminished. It has accelerated reaction (11-77) without being consumed; solid MnO_2 is a catalyst.

An example of catalysis at a solid–gas interface is provided by the reaction of hydrogen and oxygen in the presence of a platinum surface. The H_2–O_2 reaction is like the H_2–Cl_2 reaction. A gaseous mixture shows no appreciable reaction unless it is ignited by a spark. Then an explosion occurs based upon chain processes analogous to the sequence (11-71) and (11-72). The net reaction is:

$$2\,H_2(g) + O_2(g) \longrightarrow 2\,H_2O(g) \qquad \Delta H = -115.6 \text{ kcal} \qquad (11\text{-}87)$$

This reaction can be controlled, however, so that it proceeds rapidly, but not explosively, by streaming the mixture through a mesh of platinum gauze. The mixture ignites when it impinges on the gauze and burns enthusiastically, but in a mannerly way.

One mechanism by which platinum could catalyze (11-87) involves reaction between oxygen and platinum at the metallic surface. Figure 11-14

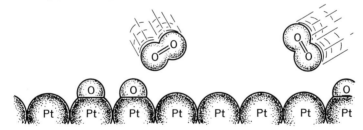

(a) Oxygen molecules react with the platinum surface forming atoms

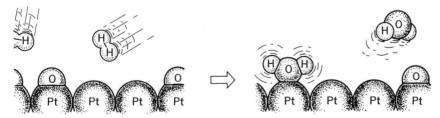

(b) Hydrogen molecules react with oxygen atoms forming water

Figure 11-14   A possible mechanism for the platinum catalyzed H_2–O_2 reaction.

shows what is thought to occur. An oxygen molecule attaches itself to the surface and reacts, breaking the oxygen–oxygen bond in order to form two platinum–oxygen bonds. Now if a hydrogen molecule happens to encounter one of these oxygen atoms, it finds the dirty work already done. The oxygen–oxygen bond is already broken, and the platinum surface furnishes a congenial spot for the reaction between H_2 and O to form two O–H bonds, while breaking the platinum–oxygen bond to release H_2O.

(c) CATALYSTS AT WORK

It would be difficult to exaggerate the importance of catalysts, either in their industrial applications or in biological processes. Most of the petroleum refining now done involves heterogeneous catalysts to restructure the hydrocarbons found in crude oils into configurations that have better combustion properties. The syntheses of many polymers depend upon catalytic assistance. There are some catalysts that not only facilitate making and breaking of particular bonds, but that also generate products in special geometric configurations (i.e., *stereospecific catalysts*).

In the catalyst department, however, Nature really puts us in the shade. Almost all biological processes depend upon specific, natural catalysts that are delicately tuned to their function. These natural catalysts, called *enzymes*, are, in most instances, much more specific and much more effective than the best that man has turned up on his own.

11-6　Conclusion

Many aspects of reaction rates have not been mentioned here—rates of heterogeneous reactions, rates in solution, diffusion-controlled processes, etc. Some of these will be taken up in later chapters, some not at all. Enough has been said, however, to show that we have a substantial understanding of the factors that determine how rapidly a chemical change will take place. This understanding, placed alongside the introduction to chemical thermodynamics, indicates that we have come a long way toward the control of the macroscopic world around us. We can develop reliable expectations about changes that may occur, where equilibrium will be found, and now, how rapidly we will approach it.

That is half the world of chemistry—the real half. The other half is no less important, though it exists only in the mind of man. It is the microscopic world of atoms and molecules—a model world that we have developed to explain the macroscopic observations that we can see and measure. This microscopic model of the universe complements our chemical thermodynamics and reaction rate theory—together they carried chemistry out of alchemy into a modern science. Let's turn to it now.

Problems

1. The rate of reaction between nitric oxide and chlorine to form nitrosyl chloride

$$2 \, NO(g) + Cl_2(g) \longrightarrow 2 \, ClNO(g)$$

was studied by measuring the infrared absorptions of ClNO as a function of time (*Journal of Physical Chemistry*, Vol. **73**, p. 2980, 1969). Gaseous NO at 5.0 torr pressure was quickly mixed with 400 torr Cl_2 at 298°K. After 5.0 minutes, the absorption of ClNO showed that its pressure had reached 0.18 torr.
(a) Over this 300-second period, what is the reaction rate, $\Delta(ClNO)/\Delta t$, expressed in torr/sec?
(b) Re-express the answer to (a) in moles/cc-sec.
(c) Re-express the rate of the reaction in terms of the loss of Cl_2, $\Delta(Cl_2)/\Delta t$, expressed in moles/cc-sec.

2. For the reaction studied in Problem 1, doubling the Cl_2 pressure caused the reaction to proceed twice as fast, but doubling the NO pressure increased the reaction rate by a factor of four. What is the expression for the concentration dependence of the reaction rate?

3. From the answers to Problems 1 and 2, deduce the magnitude of the rate constant, expressed in $cc^2/mole^2$-sec.

4. The gas phase reaction $NO_2 + 2 \, HCl \longrightarrow H_2O + NO + Cl_2$ has an experimental rate law $\Delta[NO_2]/\Delta t = -k[HCl][NO_2]$. If the initial pressures of NO_2 and HCl are each 10 torr, it is found that at 600°K the reaction consumes 1 torr NO_2 in 4.5 minutes.
(a) What is the reaction rate, $\Delta[NO_2]/\Delta t$, expressed in torr/sec?
(b) What is the reaction rate, expressed in moles/cc-sec?
(c) How many minutes would be required to consume 1 torr NO_2 if the gas mixture were compressed, to raise NO_2 and HCl pressures each to 20 torr?
(d) What are the magnitude and the units of the rate constant k if the reaction rate is expressed in torr/sec and [HCl] and [NO_2] are expressed in torr?
(e) What are the magnitude and the units of k if the rate is expressed in moles/cc-sec and [HCl] and [NO_2] are expressed in moles/cc?

5. The experimental rate law for the hydrogenation of ethylene, $C_2H_4(g) + H_2(g) \longrightarrow C_2H_6(g)$, is $\Delta[C_2H_6]/\Delta t = k[C_2H_4][H_2]$. At fixed temperature, the reacting gas is compressed to triple the total pressure. As the gas is compressed, by what factor does the reaction rate change?

6. The aqueous oxidation of chromic ion, Cr^{+3}, to chromate, CrO_4^{-2}, can be accomplished in buffered H^+ solution with ceric ion, Ce^{+4}, as oxidizing agent. The rate of the reaction is found to depend upon concentrations as follows:

$$\frac{\Delta[Cr^{+3}]}{\Delta t} = -k \, \frac{[Ce^{+4}]^2 \, [Cr^{+3}]}{[Ce^{+3}]}$$

(a) Balance the reaction between Cr^{+3} and Ce^{+4} to produce Ce^{+3} and CrO_4^{-2}.
(b) How is the rate of the reaction, expressed in terms of the production of $Ce^{+3}(aq)$, $\Delta[Ce^{+3}]/\Delta t$, related to the rate of consumption of $Cr^{+3}[aq]$?
(c) If all concentrations were diluted by a factor of ten, by what factor would the rate change?

7. The rate constant k for the reaction in Problem 4 drops to $2.0 \cdot 10^{-13} \, torr^{-1} \, sec^{-1}$ at 300°K. Now how many minutes will be needed for the consumption of one

torr of NO_2 for the mixture of 10 torr NO_2 and 10 torr HCl? Express your answer in years, as well.

8. Repeat Exercise 11-5 for D_2, assuming it has the same size as H_2 (though double the mass, of course).

9. In the upper atmosphere at about 25 miles altitude, ultraviolet photolysis of O_2 gives oxygen atoms, which react with O_2 to give ozone, O_3. This ozone absorbs ultraviolet light, thereby protecting living organisms on earth from this lethal radiation. Calculate the number of oxygen molecule collisions per second experienced by an average oxygen atom if the temperature is $500°K$, if the O_2 pressure is 10^{-6} torr, and if the O atom and O_2 molecule can be represented as spheres of radii 1.4 Å and 2.0 Å, respectively.

10. The ethylene hydrogenation considered in Problem 5 has an activation energy of 43.2 kcal. By what factor does the rate change if the temperature is doubled, from $300°K$ to $600°K$?

11. The rate constant for the reaction studied in Problem 1 was also measured at $0°C$ and $50°C$, giving, respectively, $k(273°K) = 0.97 \cdot 10^7$ and $k(323°K) = 3.5 \cdot 10^7$ $cc^2/mole^2sec$. Assuming that k has the form $k = A\ e^{-E_a/RT}$, use the ratio of $k(323°K)/k(273°K)$ to deduce the activation energy, E_a. ($A = pZ$)

12. Free radicals react very rapidly because of their unsatisfied bonding capability. For example, the rate constant for the reaction between two CF_3 radicals has been measured on the microsecond time scale with infrared spectroscopy (see *Journal of Physical Chemistry*, Vol. **74**, p. 2090, 1970).

$$2\ CF_3(g) \xrightarrow{k} C_2F_6(g)$$

The rate constant was measured at $25°C$ and at $60°C$, with results $k(298°K) = 5.9 \cdot 10^{12}$ and $k(333°K) = 7.1 \cdot 10^{12}$ cc/mole-sec.
(a) Assuming that k has the form $k = A\ e^{-E_a/RT}$ (where $A = pZ$), use the ratio $k(333)/k(298)$ to deduce the activation energy, E_a, for this reaction between two free radicals.
(b) Making measurements on the microsecond time scale is a tough business. How much would the time for 1% consumption of CF_3 be lengthened if the temperature were halved, i.e., lowered to $T = 149°K$? How much would the time be lengthened if CF_3 detection could be improved so that the initial pressure of CF_3 could be reduced by a factor of five?

13. The ethylene hydrogenation considered in Problems 5 and 10 occurs with a collision factor, $A = pZ = 4 \cdot 10^{13}$ cc/mole-sec. Assuming H_2 and C_2H_4 can be represented by spheres with radii 1.0 Å and 2.2 Å, respectively, calculate Z (using expression (11-22)) and deduce p for this reaction. (Notice that if $E_a = 43.2$ kcal/mole, $\epsilon = 3.00 \cdot 10^{-12}$ erg/molecule).

14. When hydrogen iodide is irradiated with ultraviolet light, the molecules are dissociated into atoms.

$$HI(g) + light \longrightarrow H(g) + I(g)$$

If a mixture of $HI(g)$ and $I_2(g)$ is irradiated, the resulting hydrogen atoms can react either with HI or with I_2, with rate constants k_1 and k_2.

$$H + HI \xrightarrow{k_1} H_2 + I$$

$$H + I_2 \xrightarrow{k_2} HI + I$$

(a) Use the table of bond dissociation energies (Table 7-3) to calculate ΔH for each of these hydrogen atom reactions.

(b) Calculate the number of collisions per second (Z_1) experienced by an average H atom at 300°K with HI molecules at 0.10 torr. Assume the radius of an H atom to be 0.79 Å and that of HI is about equal to the radius of an iodine atom, 2.2 Å.

(c) Repeat (b), but for I_2 molecules (Z_2) at 0.10 torr. Assume the radius of an I_2 molecule equals the radius of an iodine atom plus one half the I–I bond length, 3.5 Å. What is the ratio Z_2/Z_1?

(d) Suppose that an H atom can react if it collides at either end of an I_2 molecule, but only at the hydrogen end of an HI molecule. If other factors are equal, what will be the ratio p_2/p_1?

(e) Studies of the rate of H_2 formation as a function of temperature show that $k_2/k_1 = 4.9\ e^{+640/kT}$. By what factor does your estimate of p_2Z_2/p_1Z_1 (from (c) and (d)) differ from the experimental value? Also, what does the experimental ratio k_2/k_1 tell us about $E_{a,2}$ and $E_{a,1}$?

15. Use the value of E_a deduced in Problem 11, combined with either value of k, to calculate $A = pZ$.

16. At 1000°K, $H_2(g)$ reacts with $O_2(g)$ to produce $H_2O(g)$ according to the rate equation

$$\frac{\Delta[O_2]}{\Delta t} = -k_1[H_2][O_2]^{4/3}$$

The rate law for the reverse reaction, the decomposition of H_2O vapor, has not been studied. Suppose we give it a hypothetical rate law

$$\frac{\Delta[O_2]}{\Delta t} = k_1[H_2O]^x[H_2]^y[O_2]^z$$

At equilibrium, the net value of $\Delta[O_2]/\Delta t = 0$ and k_1/k_1 equals the equilibrium constant K. Use these facts to determine x, y, and z.

17. The rate constant for the reaction $H_2(g) + I_2(g) \longrightarrow 2\ HI(g)$ has been measured as a function of temperature and its activation energy is 40 kcal/mole. Use the data in Appendix d to calculate ΔH^0 and then the activation energy for the reverse reaction.

18. The gas-phase reaction $2\ HBr + NO_2 \longrightarrow H_2O + NO + Br_2$ has a $\Delta H^0 = -19.6$ kcal and a rate constant $k = 10^{11} \cdot e^{-13000/RT}$ cc/mole-sec. Suppose the mechanism below is correct:

$$HBr + NO_2 \xrightarrow{k_1} HOBr + NO \qquad \Delta H = +1\ \text{kcal} \qquad \text{slow}$$

$$HBr + HOBr \xrightarrow{k_2} H_2O + Br_2 \qquad \Delta H = -20.6\ \text{kcal} \qquad \text{fast}$$

(a) Draw an energy–reaction coordinate diagram, showing plateaus for reactants (2 HBr + NO_2), for intermediates (HOBr + NO + HBr), and for products (NO + H_2O + Br_2). Take the energy of reactants as zero and show all known reaction heats and activation energies.

(b) What can be said about the activation energy for reaction 2?

(c) What is the activation energy for the reverse reaction? Show it on your diagram.

19. The activation energy for the unimolecular decomposition of borine carbonyl, H_3BCO, was measured to be 24 kcal (*Journal of Chemical Physics*, Vol. **73**, p. 873, 1969).

$$H_3BCO \underset{E_{a,r}}{\overset{E_{a,f}\,=\,24}{\rightleftharpoons}} BH_3 + CO \qquad \Delta H_I$$

At the time, there was reported in the literature a reliable measurement of ΔH_{II}, but there were three indirect and conflicting estimates of ΔH_{III}, given below:

$$B_2H_6 + 2\,CO = 2\,H_3BCO \qquad \Delta H_{II} = -9.1 \text{ kcal}$$

$$B_2H_6 = 2\,BH_3 \qquad \Delta H_{III} = 28,\ 35,\ \text{or}\ 59 \text{ kcal}$$

(a) The compound BH_3 is thought to react with activation energies like those of free radicals, i.e., between 0 and 2 kcal (see Problem 12). If this is correct, what is ΔH_I?

(b) With this value of ΔH_I, the new rate data positively eliminate one of the estimates of ΔH_{III}. By calculating the implied value of $E_{a,r}$ from ΔH_I, ΔH_{II}, ΔH_{III}, and $E_{a,f}$, decide which estimate of ΔH_{III} is unreasonable and which is to be preferred.

20. The collision factor $A_f = p_f Z_f$ in the rate constant for the reaction studied in Problem 17 is measured to be $1.0 \cdot 10^8$ cc/mole-sec. Use the data in Appendix d to calculate ΔS^0 and the collision factor $A_r = p_r Z_r$ for the reverse reaction, $2\,HI(g) \longrightarrow H_2(g) + I_2(g)$.

21. The gas-phase decomposition of ozone has the stoichiometry

$$2\,O_3(g) \longrightarrow 3\,O_2(g)$$

Experimental measurements of $\Delta[O_3]/\Delta t$ at a suitable temperature were made at various pressures of O_3 and O_2 with the following results:

$p[O_3]$(torr)	0.20	0.20	0.40
$p[O_2]$(torr)	0.50	1.0	1.0
$\Delta[O_3]/\Delta t$	6.0	3.0	12

What is the rate law?

22. Chloroformic acid, ClCOOH, can be formed by the flash photolysis of Cl_2 in the presence of formic acid, HCOOH, through the following reactions:

$$Cl_2(g) + \text{light} \longrightarrow 2\,Cl(g)$$

$$Cl(g) + HCOOH(g) \longrightarrow HCl(g) + COOH(g)$$

$$COOH(g) + Cl_2(g) \longrightarrow ClCOOH(g) + Cl(g)$$

Then, more slowly, the chloroformic acid decomposes in a unimolecular decomposition, probably in two steps.

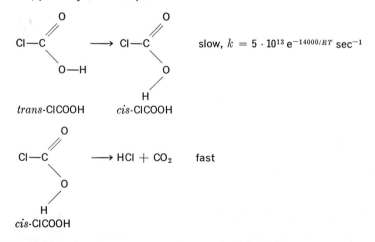

trans-ClCOOH → cis-ClCOOH, slow, $k = 5 \cdot 10^{13}\,e^{-14000/RT}$ sec^{-1}

cis-ClCOOH → HCl + CO$_2$, fast

If a 20-microsecond flash produces 0.10 torr pressure of ClCOOH, in how many microseconds will this pressure be reduced to 0.050 torr at 298°K? (Use expression (11-60).)

23. In a unimolecular reaction, there is a simple relationship between the rate constant k and the time needed for half of the reactant concentration to be consumed (the "half-time," $\tau_{\frac{1}{2}}$). Substitute into expression (11-60) the concentration $[N_2O_5]_t = \frac{1}{2}[N_2O_5]_0$ at $t = \tau_{\frac{1}{2}}$ to derive this relationship. Use the relationship to calculate the time needed to reduce 10 torr pressure of ClCOOH to 5 torr at 298°K. Compare your answer to that in Problem 22.

24. Which of the following reaction mechanisms is consistent with the rate law deduced for the third order reaction studied in Problems 1 and 2?

$$2 \text{ NO} + \text{Cl}_2 \xrightarrow{k} 2 \text{ ClNO} \qquad \text{Mechanism A}$$
three body
collision

	Mech. B	Mech. C
$2 \text{ NO} \xrightarrow{k_1} \text{N}_2\text{O}_2$	slow	fast
$\text{N}_2\text{O}_2 + \text{Cl}_2 \xrightarrow{k_2} 2 \text{ ClNO}$	fast	slow

	Mech. D	Mech. E
$\text{NO} + \text{Cl}_2 \xrightarrow{k_1} \text{ClNO} + \text{Cl}$	slow	fast
$\text{Cl} + \text{NO} \xrightarrow{k_2} \text{ClNO}$	fast	slow

25. Write the rate law for each of the following possible mechanisms for the gas-phase reaction considered in Problem 21. Which could account for the experimental facts?

Mechanism A: $\text{O}_3 + \text{O}_3 \xrightarrow{k_1} 3 \text{ O}_2$

Mechanism B: $\text{O}_3 \xrightarrow{k_1} \text{O}_2 + \text{O}$ slow

  $\text{O} + \text{O}_3 \xrightarrow{k_2} 2 \text{ O}_2$ fast

Mechanism C: $\text{O}_3 \xrightarrow{k_1} \text{O}_2 + \text{O}$ fast

  $\text{O} + \text{O}_3 \xrightarrow{k_2} 2 \text{ O}_2$ slow

Mechanism D: $\text{O}_3 + \text{O}_3 \xrightarrow{k_1} \text{O}_2 + \text{O}_4$ slow

  $\text{O}_4 \xrightarrow{k_2} 2 \text{ O}_2$ fast

Mechanism E: $\text{O}_3 + \text{O}_3 \xrightarrow{k_1} \text{O}_2 + \text{O}_4$ fast

  $\text{O}_4 \xrightarrow{k_2} 2 \text{ O}_2$ slow

26. The mechanism deduced for reaction (11-29) in Exercise 11-9 is

$\text{HBr} + \text{NO}_2 \xrightarrow{k_1} \text{HONO} + \text{Br}$ slow

  $2 \text{ Br} \xrightarrow{k_2} \text{Br}_2$ fast

  $2 \text{ HONO} \xrightarrow{k_3} \text{H}_2\text{O} + \text{NO} + \text{NO}_2$ fast

Suppose the third reaction were much the slowest of the three reactions so that the first two reactions remained near equilibrium throughout the reaction.

(a) What would the rate law be now?

(b) Express the over-all rate constant for this mechanism in terms of k_3 and the equilibrium constants K_1 and K_2.

27. What would be the rate law for the reverse reaction in Problem 18? Assume that the fast reaction remains near equilibrium, throttled by the slow reaction. Call the rate constants for the reverse reactions k_{-1} and k_{-2}.

28. Gaseous chlorine and carbon monoxide react to form the very toxic compound phosgene, Cl_2CO.

$$Cl_2(g) + CO(g) \longrightarrow Cl_2CO(g)$$

The rate law for this reaction is $\Delta[CO]/\Delta t = -k[Cl_2]^{3/2}[CO]$. Which of the following mechanisms are consistent with the observed rate law?

	Mech. A	Mech. B	Mech. C
$Cl_2 + CO \xrightarrow{k_1} ClCO + Cl$	slow	fast	fast
$Cl + CO \xrightarrow{k_2} ClCO$	fast	slow	fast
$ClCO + Cl_2 \xrightarrow{k_3} Cl_2CO + Cl$	fast	fast	slow

	Mech. D	Mech. E	Mech. F
$Cl_2 \xrightarrow{k_1} 2\ Cl$	slow	fast	fast
$Cl + CO \xrightarrow{k_2} ClCO$	fast	slow	fast
$ClCO + Cl_2 \xrightarrow{k_3} Cl_2CO + Cl$	fast	fast	slow

	Mech. G	Mech. H	Mech. I
$Cl_2 \xrightarrow{k_1} 2\ Cl$	slow	fast	fast
$Cl + Cl_2 \xrightarrow{k_2} Cl_3$	fast	slow	fast
$Cl_3 + CO \xrightarrow{k_3} Cl_2CO + Cl$	fast	fast	slow

29. Which of the following reaction mechanisms are consistent with the observed rate law for the reaction discussed in Problem 6?

	Mech. A	Mech. B
$Ce^{+4} + Cr^{+3} \xrightarrow{k_1} Ce^{+2} + Cr^{+5}$	slow	fast
$Ce^{+4} + Ce^{+2} \xrightarrow{k_2} 2\ Ce^{+3}$	fast	fast
$Ce^{+4} + Cr^{+5} + 4\ H_2O \xrightarrow{k_3} Ce^{+3} + CrO_4^{-2} + 8\ H^+$	fast	slow

	Mech. C	Mech. D	Mech. E
$Ce^{+4} + Cr^{+3} \xrightarrow{k_1} Ce^{+3} + Cr^{+4}$	slow	fast	fast
$Ce^{+4} + Cr^{+4} \xrightarrow{k_2} Ce^{+3} + Cr^{+5}$	fast	slow	fast
$Ce^{+4} + Cr^{+5} + 4\ H_2O \xrightarrow{k_3} Ce^{+3} + CrO_4^{-2} + 8\ H^+$	fast	fast	slow

30. For Problem 29, what are the rate laws for the mechanisms that do *not* fit the observed rate law?

31.   Chloroformic acid is produced in high yield through the photochemically induced reactions given in Problem 22 because a chain reaction is involved. Which reactions make up the chain? Write the equations for three chain terminating reactions.

32.   Because chain reactions are involved, none of the mechanisms A, D, and G in Problem 28 will have the simple behavior of a bimolecular reaction. Instead, the reactions will accelerate. Explain.

twelve quantum mechanics and the hydrogen atom

Chemistry is primarily concerned with the making and breaking of chemical bonds. Consequently rules of combination, systematics of reactivity, and, ultimately, theories of chemical bonding have occupied central positions in the activity of chemists and in their scientific literature.

12-1 The beginnings

Rudimentary bonding concepts actually predated the atomic theory by many centuries. The postulate of four "elements," earth, air, fire, and water, contains the germ of the compositional idea. Furthermore, the theme of chemical bonding can be seen in early writings about the "modes of combination" expressed in the three principles called sulfur, mercury, and salt. Consider these words by Paracelsus, written in about 1525.*

"As to the manner in which God created the world, take the following account. He originally reduced it to one body, while the elements were developing. This body He made up of three ingredients, Mercury, Sulphur, and Salt, so that these three should constitute one body. Of these three are composed all the things which are, or are produced, in the four elements [earth, air, fire, and water]. These three have in themselves the force and the power of all perishable things. In them lie hidden the mineral, day, night, heat, cold, the stone, the fruit, and everything else, even while not yet formed."

That this early scientist was groping for principles of constitution is evident in his writings about how these three "principles" act together to make up the "element" air.

"Mercury, Sulphur, and Salt are so prepared as the element of air that they constitute the air, and make up that element. Originally the sky is nothing but while Sulphur coagulated with the spirit of Salt and clarified by Mercury, and the hardness of this element is in this pellicle and shell thus formed from it. Then, secondly, from

* H. M. Leicester and H. S. Klukstein, *Source Book in Chemistry*, p. 19, McGraw-Hill, N.Y., 1952.

456

the three primal parts it is changed into two—one part being air and the other chaos—in the following way. The Sulphur resolves itself by the spirit of Salt and in the liquor of Mercury, which of itself is a liquid distributed from heaven to earth, and is the albumen of the heaven, and the mid space. It is clear, a chaos, subtle, and diaphanous. All density, dryness, and all its subtle nature are resolved, nor is it any longer the same as it was before. Such is the air."

Such efforts to clarify chemical constitution can be seen, in retrospect, to have been abortive in absence of the atomic theory. Until the existence of atoms was recognized, chemistry was destined to remain a collection of recipes and empirical prescriptions, only encumbered by the embryonic and metaphysical theories of the time.

(a) THE ATOMIC HYPOTHESIS: CORNERSTONE OF CHEMISTRY

Nevertheless, chemical analysis became such a well-developed and powerful technique that it furnished the first compelling evidence that matter is particulate. Thus, in the first half of the nineteenth century, the accumulating knowledge of chemical composition led Dalton to propose the existence of atoms. This proposal liberated and stimulated views of chemical bonding whose usefulness can still be seen today.

Bonding rules—called "valence rules"—began to evolve during the second half of the nineteenth century. These empirical rules were given systematic foundation when the importance of the Periodic Table as a regularizing guide became evident. A rich harvest of progress followed— the preparation, characterization and practical use of new compounds not found in nature came to be a day-to-day activity. This success, in turn, confronted the valence rules with an enormous and growing volume of new descriptive chemical facts. To cope with them, the valence rules became more and more ornate. Without a basic unifying theory, chemists resorted to classification according to "bond type" and, as the first half of the twentieth century drew to its close, chemists manipulated with dexterity a complex catalog of covalent bonds, ionic bonds, metallic bonds, coordinate bonds, charge-transfer bonds, dative bonds, chelate bonds, bridge bonds, one-electron bonds, and hydrogen bonds. Chemical bonding was attributed to electron sharing, to "exchange" forces, to magnetic spin, "overlap," available space, and lowered kinetic energy. All of it was kept in some sort of workable order with the aid of the Periodic Table. This powerful ordering device came to be engraved in the mind of every working chemist, and it gave him at least an intuitive basis for predicting what chemical composition and which type of chemical bond to anticipate as new preparations were attempted. The lack of a "first principles" approach drew the scorn of theoretical physicists, but no apology was either forthcoming or needed. Chemists had prepared almost a million compounds by the year 1950 and the number of new compounds was growing at the rate of two or three hundred per day.

(b)   THE DAWN OF QUANTUM MECHANICS

We now realize that the evolution of a single, unifying theory of chemical bonding had to await the development of quantum mechanics. Yet it was a giant step forward when, at the turn of the century, Rutherford determined that the atom consisted of negatively charged electrons moving around a small, massive, positively charged nucleus. This atomic structure suggested that chemical bonds might be associated with electron sharing. Electron counting became a popular game, for stable compounds seemed to be connected to the specially stable electron populations of the inert gases. G. N. Lewis observed that a large part of chemical bonding could be explained in terms of atoms striving for these inert gas populations through the sharing of pairs of electrons. A variety of useful schemes developed— the octet rule and the electron-dot diagram are two—these and other schemes accelerated the juggernaut of chemical progress.

Then the quantum mechanical picture of the atom began to emerge. Physicists found they now possessed a description of the atom that could accurately predict atomic properties whenever the mathematics could be solved. By 1930 the scene was set for a true understanding of chemical bonding. It was a while coming, however, because of mathematical difficulties. Calculations for the simplest molecules H_2 and H_2^+ proved that quantum mechanics did contain the explanation of chemical bonding. Urged on by a few bold individuals, like Linus Pauling and Robert Mulliken, chemists began incorporating quantum mechanical ideas into their bonding theories, even though quantitative calculations were quite out of reach. Then, as ever is so, each human problem is finally overtaken by a solution. The electronic computers appeared and they multiplied many, many fold our computational ability. Now even polyatomic molecules with many electrons can be handled; our quantum mechanical understanding of bonding is firm.

So now we will examine what quantum mechanics tells us about chemical bonds. We will review the chemist's valence rules in the light of these "first principles." But first we will examine the quantum mechanical view of the atom. Here can be seen most clearly the qualitative differences and likenesses between physical behavior on the scale of atomic dimensions and that of the macroscopic world we sense directly around us. These differences and likenesses furnish the setting needed to understand what bonding is all about.

12-2   An intuitive basis for quantum theory

Toward the close of the nineteenth century, the two brightest areas in physics were separable and not in conflict. The pre-1900 classical laws of mechanics accurately described the motions of particulate bodies from pebbles to planets. In a similar way, electromagnetic theory was entirely successful in explaining electrical and magnetic phenomena, including the behavior of light. Both theories were, of course, based solidly in experimental knowledge.

Within the next three decades, the foundations of both areas were shaken by crucial failures. In retrospect, one experiment can be seen to be the experimental key that pointed inexorably toward quantum mechanics, though it is not usually presented in that way. This experiment was the Michelson–Morley search for an "electromagnetic ether" to explain the propagation of light.

(a) THE ELECTROMAGNETIC ETHER THAT WASN'T THERE

All knowledge of how acoustic waves are propagated through gas, liquid, or solid media, led quite naturally to the assumption that light, a wave phenomenon, must have some sort of propagating medium. Nineteenth century physicists called this hypothetical medium an "ether," a term that causes doubt about their chemical familiarity. A. Michelson, an American physicist, set out in 1881 to verify this confident model with an experiment that would measure sensitively any motion of the earth through the expected "ether." His negative result challenged our most fundamental descriptions of the spatial and temporal relationships between objects and events.* Then two decades passed before a significant attempt to adjust to Michelson's surprising result was published by a Dutch scientist named H. A. Lorentz. He showed how the description of positions and times of events had to be changed if light is propagated in a vacuum at a fixed velocity. Thus, the stage was set for a young German scientist named Albert Einstein to make a revolutionary advance.

(b) $E = mc^2$

In 1905, a year after Lorentz's work, Einstein reopened another subject with a previous history—relativity. He considered the implications of these new views of space and time on the description of dynamic events (events that involve changes in positions, times, velocities, momenta, and energies) in a moving laboratory. The new model led to the then startling result that energy and mass are merely different manifestations of a single entity. Every energy transfer is accompanied by a mass transfer and every mass, whether moving or not, represents a store of energy. Mass is usually measured with certain devices and is expressed in grams. Energy is measured with other probes and is expressed in ergs. The proportionality constant

* Six years later, in 1887, Michelson and a colleague from chemistry, E. W. Morely, verified the experimental result with improved accuracy. Nevertheless, a quarter of a century later, some reputable scientists were still repeating the experiment with the conviction that the result might not actually be negative. In fact, the experiment has been considered worthy of repetition and confirmation even within the last few years using lasers and the most precise measuring devices available.

between these units is c^2, the square of the velocity of light:

$$E = mc^2 \qquad\qquad (12\text{-}1)$$

The appearance of c in this relationship is a logical consequence of the Lorentz modification of our views of space and time, as required by the Michelson—Morley experiment. Good Dr. Einstein showed this to be a natural result of our use of light as a communication link when measuring energy relations between moving masses.

Exercise 12-1. When one gallon of gasoline burns, 30,000 kcal of heat is released. The reaction can be typified by combustion of octane:

$$C_8H_{18}(\ell) + \tfrac{25}{2}\, O_2(g) \longrightarrow 8\, CO_2(g) + 9\, H_2O(\ell)$$

$$\Delta H = -30,000 \text{ kcal/gallon}$$

Calculate the mass associated with this energy from Einstein's relation $E = mc^2$ and the conversion factor 1 kcal $= 4.2 \cdot 10^{10}$ ergs. Assuming that the gallon contains 23 moles of octane, show that the products of the reaction weigh less than the reactants by 0.06 microgram per mole of octane.

Exercise 12-2. If one mole of deuterium nuclei (^{2}H) could be "fused" with one mole of tritium nuclei (^{3}H), the products, a mole of helium nuclei (^{4}He) and a mole of neutrons (^{1}n) weigh less than the reactants by 0.019 gram. How many kcal are released per mole?

(c) ENERGY HAS MASS: MASS IS ENERGY

There were some immediately obvious implications that didn't sit well with scientists of the time. The physicists saw their precious Law of Conservation of Energy thus married to the chemists' Law of Conservation of Mass. Neither law was correct as it had been applied earlier. The two laws became one and the same statement, since energy and mass are the same thing. Mass is conserved only if its energy forms are taken into account, and vice versa.

Quite as significant and more germane to quantum theory are the implications of $E = mc^2$ on the relationship between light and mass. Light, then recognized as a wave phenomenon, was already known to be a form of energy. Absorption of light can be measured with traditional energy-measuring devices, such as calorimeters, thermometers, thermocouples, and the like. Now light must also have mass and, presumably, all the properties of mass. Mass must also do its part and possess the properties of energy. Neither classical mechanics nor electromagnetic theory could provide these links.

(d) $E = mc^2$ AND THE IMPLICATION OF THE ATOMIC THEORY

Light is an energy form. Energy must have the properties of mass. What can we learn about light from what we know about mass? Let's see.

Suppose a gram of table salt is ground so finely that it blows in the wind. Now examine it under the very best microscope you can find. Even under the highest magnification, you'll see only smaller and smaller crystals of the same stuff, sodium chloride. Matter is continuous on a macroscopic scale. All our evidence that matter is made up of tiny, discrete particles called atoms is indirect. Nevertheless, we are firmly convinced of the atomic theory because this model is so widely useful in explaining the properties of matter. *Matter looks continuous, but its properties are explained with a particulate model.*

Light emitted by a hot radiating source can be passed through a prism to produce a spectrum. The color spectrum we see is continuous—apparently every color is represented. There is a smooth variation of intensity that peaks at a wavelength fixed by the source temperature. If the source is moved away from the prism, the intensities of all colors decrease together uniformly until the whole spectrum gradually becomes too weak to see. Light energy, like mass, appears continuous.

Now we can ask about the implications of the identity of energy and mass. If many properties of matter are best described with a particulate model, should not the same be true for light energy? This is exactly what Max Planck decided (without the benefit of Fig. 12-1 or $E = mc^2$) as he struggled with the problem of explaining the intensity–wavelength relationship described above. This was the first property of light that displayed real inadequacy of the wave description; Planck found it necessary to postulate the existence of little light "packets" or "quanta" to explain the radiation from a hot source. This was the beginning of the *quantum theory of light* and, in retrospect, we see this as the first fulfillment of the parallelism implied by $E = mc^2$.*

(e) HOW BIG IS A QUANTUM?

This particulate model of light quantitatively connects the energy of a light packet to its frequency or wavelength. Light waves are usually described in terms of either the frequency (the number of electromagnetic waves per second, designated by ν, the Greek letter "nu"), or the wavelength (designated λ, the Greek letter "lambda"). To convert to the equivalent energy

---

* Planck was not aided at the time by $E = mc^2$; in fact, he proposed his view about six years before Einstein published his results on relativity. Planck was wholly concerned with the failure of electromagnetic theory to explain the experimental observations of the intensity–color relationship.

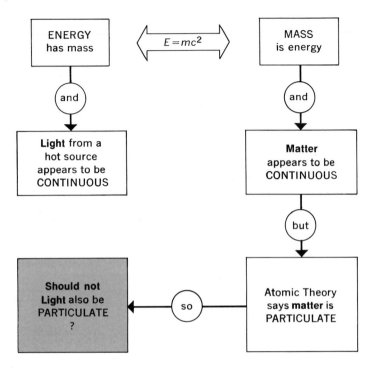

Figure 12-1 Energy has mass: Mass is energy. The beginnings.

in ergs, Planck found that he needed a conversion factor of $6.6 \cdot 10^{-27}$ erg-second. He called this factor h:

$$E = h\nu = h(c/\lambda)$$
$$h = 6.6 \cdot 10^{-27} \text{ erg-sec}$$

(12-2)

Thus Planck's constant h tells the "size" of these "particles" or quanta of light, just as Avogadro's number N tells us the size atoms must have. Everyone senses how small an atom is—after all, a big gulp of water contains about a mole, $6 \cdot 10^{+23}$ molecules! Suppose we tried to heat this 18 gram gulp of water with an infrared lamp to make a cup of tea. How many quanta of light would we need? Here's how the calculation goes:

(i) How many calories are needed to heat 18 grams of water from 0°C to 100°C?

heat = (mass)(heat capacity)(temperature change)

$$= (18 \text{ grams}) \left(\frac{1.0 \text{ calorie}}{\text{gram deg}} \right) (100°) = 1800 \text{ calories}$$

(ii) How many ergs is 1800 calories of heat?

$$\text{ergs} = (\text{calories})(\text{conversion factor, ergs to calories})$$
$$= (1800 \text{ calories})(4.2 \cdot 10^7 \text{ ergs/calorie}) = 7.6 \cdot 10^{10} \text{ ergs}$$

(iii) How many quanta of infrared light of wavelength 10 microns (frequency $3 \cdot 10^{13}$ cycles/sec) are needed to give $7.6 \cdot 10^{10}$ ergs?

$$\text{number of quanta} = (\text{ergs needed})/(\text{ergs per quantum})$$
$$= (7.6 \cdot 10^{10} \text{ ergs})/h\nu$$
$$= \frac{(7.6 \cdot 10^{10} \text{ ergs})}{(6.6 \cdot 10^{-27} \text{ erg sec})(3.0 \cdot 10^{13} \text{ cycles/sec})}$$

$$\text{number of quanta} = 3.8 \cdot 10^{23} \text{ quanta}$$
$$\text{number of moles of quanta} = \frac{3.8 \cdot 10^{23}}{6.0 \cdot 10^{23}} = 0.63 \text{ mole}$$

We see that a mole of water can be warmed from 0°C to 100°C with about $\frac{2}{3}$ of a mole of infrared light quanta. We conclude that quanta or packets of light energy are indeed small compared with our macroscopic measuring devices—they are appropriate in magnitude to the molecular scale of things. How appropriate!

Exercise 12-3. Show that only 0.032 mole of quanta would be needed to heat 18 grams of tea from 0°C to 100°C if we used green light (wavelength = 5000 Å = 0.5 μ) instead of infrared light with wavelength 10 μ.

(f) MASS AND MOMENTUM OF A LIGHT QUANTUM

It is an easy arithmetical task to combine (12-1) and (12-2) to calculate the mass associated with a single quantum of light energy with frequency ν (nu). In general

$$E = mc^2$$

and, for light,

$$E = h\nu$$

hence

$$h\nu = m_\nu c^2$$

or

$$m_\nu = \frac{h\nu}{c^2} \tag{12-3}$$

Exercise 12-4. Show that the mass of a mole (the molecular weight) of ultraviolet quanta ($\lambda = 3000$ Å, $\nu = 10^{15}$ cycles/sec) is $4.4 \cdot 10^{-9}$ g/mole.

Momentum p is just mass times velocity. Expression (12-3) gives the mass of a quantum. Since light is propagated with velocity c, we can calculate the momentum p_ν of a quantum.

$$p_\nu = m_\nu c = \frac{h\nu}{c^2} c$$

$$p_\nu = \frac{h\nu}{c} \qquad (12\text{-}4)$$

and since $\lambda = c/\nu$,

$$\boxed{p_\nu = \frac{h}{\lambda}} \qquad (12\text{-}5)$$

Experimental verification of this momentum is provided by the scattering of light. In 1922, A. H. Compton found that X rays, when scattered by electrons, suffer a momentum change, just as would two colliding billiard balls. He measured the resulting frequency shift as a function of the scattering angle. The shift was accurately explained by the laws of conservation of energy and momentum using the particulate model of the light quantum.

(g) $E = mc^2$ AND THE IMPLICATION OF THE WAVE NATURE OF LIGHT

This parallelism of energy and mass is, of course, a two-way street. If light must (and does) have the particulate properties of mass, is it not necessary to expect mass to possess properties that would normally be attributed to light? In particular, we are used to describing light as an electromagnetic wave. It has a frequency of ν cycles per second, and a corresponding wavelength λ. What "wavelength" should be ascribed to a particle to make the parallelism of Figure 12-1 complete?

Equations (12-4) and (12-5) provide the link we need. We know how to express the momentum of a mass particle, $p_m = mv$, and expression (12-5) shows how a light wave connects momentum and wavelength. Why not assume the same relationship for mass? Since $p_\nu = h/\lambda$, perhaps

$$p_m = \frac{h}{\lambda_m}$$

or

$$\lambda_m = \frac{h}{p_m} = \frac{h}{mv} \qquad (12\text{-}6)$$

(a)

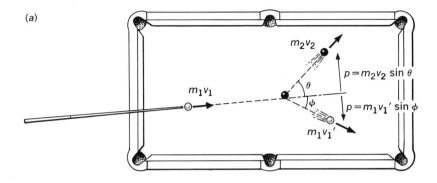

(b)

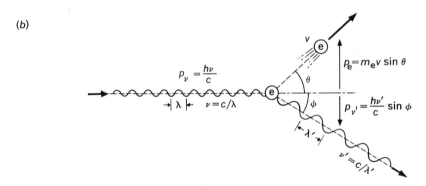

Figure 12-2 Light ball in the corner pocket. (a) The momenta p of the two balls after collision are correctly predicted by the Law of Conservation of Momentum. (b) Optical billiards. The frequency of the light wave, hence the momentum of the photon, is changed after collision with an electron. The same Law of Conservation of Momentum applies to photons as to billiard balls, reinforcing our belief in the particle properties of light.

If this connection is a valid one, as suggested by $E = mc^2$, then we should find characteristic wave phenomena associated with a moving particle. Diffraction of light as it strikes a grating-like surface is one such example. To display diffraction, the grating spacing must be comparable to the wavelength. What grating spacing is needed to test the wave properties of a particle? Expression (12-6) gives the answer.

Consider an electron (mass = $9.1 \cdot 10^{-28}$ gram) moving with a velocity of 0.01 times the velocity of light ($0.01 \cdot c = 3 \cdot 10^8$ cm/sec is the velocity of an electron accelerated through 26 volts).

$$p_m = mv = (9.1 \cdot 10^{-28})(3 \cdot 10^8)$$

$$= 27.3 \cdot 10^{-20} \text{ gm cm/sec}$$

$$\lambda = \frac{h}{p_m}$$

$$= \frac{6.6 \cdot 10^{-27}\ \text{erg sec}}{27.3 \cdot 10^{-20}\ \text{gm cm/sec}}$$

$$= 0.24 \cdot 10^{-7}\ \frac{\text{gm cm}^2}{\text{sec}}\ \frac{\text{sec}}{\text{gm cm}}$$

So,

$$\lambda = 2.4 \cdot 10^{-8}\ \text{cm} = 2.4\ \text{Å}$$

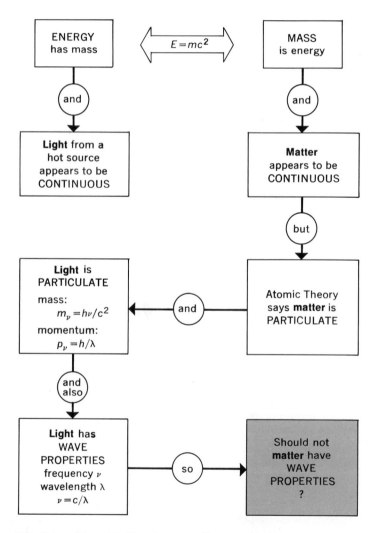

Figure 12-3 Energy has mass: Mass is energy. Part-way there.

To demonstrate diffraction, then, all we need is a surface with regular grooves cut into it at intervals of about 2.4 Å. Such a cutting job would be beyond our very best milling machines, but fortunately, nature abounds with crystals that have atoms regularly aligned with spacings of this order. So electrons that are bounced off the face of a crystal should show the same sort of diffraction patterns as light reflected from a grating! So they do, as Davisson and Germer showed at the Bell Laboratories in 1927 (see Fig. 12-4)! This result confirmed the reality of the wave properties of a moving particle, as expressed in (12-6) and as proposed in 1925 by a French scientist, Louis de Broglie.

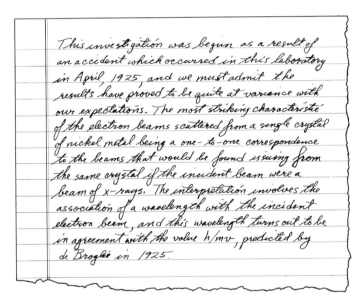

This investigation was begun as a result of an accident which occurred in this laboratory in April, 1925, and we must admit the results have proved to be quite at variance with our expectations. The most striking characteristic of the electron beams scattered from a single crystal of nickel metal being a one-to-one correspondence to the beams that would be found issuing from the same crystal if the incident beam were a beam of x-rays. The interpretation involves the association of a wavelength with the incident electron beam, and this wavelength turns out to be in agreement with the value h/mv, predicted by de Broglie in 1925.

Figure 12-4   Electrons behave like waves. This quotation, taken from the 1927 paper of Davisson and Germer, illustrates a frequent route to scientific advance. Some unexpected results when followed up lead to important discoveries. In this case, a laboratory mishap led to the conversion of some polycrystalline nickel into a crystalline form. Diffraction patterns similar to those observed with X-radiation were observed. De Broglie's predictions were in this way given their first experimental verification.

Exercise 12-5. Show that the wavelength of a proton with velocity $0.01\ c = 3 \cdot 10^8$ cm/sec is 0.0013 Å. (A proton weighs $1.7 \cdot 10^{-24}$ g, 1836 times the mass of an electron.)

This demonstration of the diffraction of "particle waves" completes our parallelism between energy and matter. The relationships qualitatively portrayed in Figure 12-5 finally took form in the first three decades of this astonishing twentieth century. The stage was set for quantum mechanics.

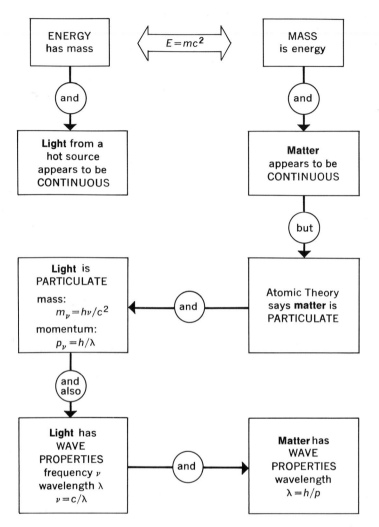

Figure 12-5 Energy has mass: Mass is energy. The story is complete.

12-3 Quantum mechanics

To understand atomic structure, we begin with the simplest atom, the hydrogen atom. We would like to describe the dynamics of motion of an electron moving in the vicinity of a proton. The "dynamics of motion" means the electron energy, momentum, trajectory, and the changes in these quantities as the hydrogen atom makes its way through life. Ultimately we might hope that this description would tell us why two H atoms bond to form a molecule H_2, but three H atoms do not form H_3. These were questions without answer in the pre-$E = mc^2$ classical physics. The chemists had to "go it alone" with no mathematical model of chemical bonding to support the burgeoning knowledge of chemical changes. Energy relationships were hung upon the sturdy frame of thermodynamics; chemical bonding was an empirical and mysterious art hunting for its guiding rationale. Again one particular experimental result provided the key to this mystery, the spectrum of the hydrogen atom.

(a) THE LINE SPECTRUM OF A HYDROGEN ATOM

Hydrogen atoms are produced when an electric discharge is struck through gaseous hydrogen. Their presence is revealed in the light emitted, a fact recognized long before there was any explanation for the color spectrum of this light. Figure 12-6 shows the hydrogen atom spectrum. Unlike the radiation emitted from hot sources, the electric discharge glow emits only special colors in a "line" spectrum. Interpreted in terms of expression (12-2), $E = h\nu$, *only selected energies can be emitted by the hydrogen atom in the form of light quanta.* As research study of this phenomenon progressed, it developed that the photon energy of $15.35 \cdot 10^{-11}$ erg was observed (which is 235.2 kcal per mole of quanta) and also photon energy $19.36 \cdot 10^{-11}$ erg (278.8 kcal/mole), but no quanta of energy in between. To heighten the interest, photons of energy exactly equal to the difference $(19.36 - 15.35) \cdot 10^{-11} = 4.01 \cdot 10^{-11}$ erg (43.6 kcal/mole) were also observed.

These observations were quite unexplainable within classical physics. Consider the classical, planetary model of the atom, the view popular just

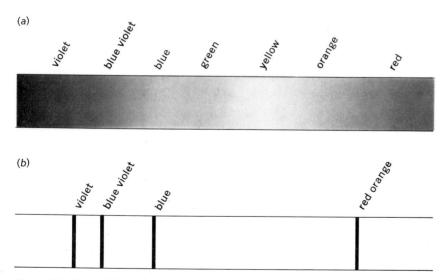

Figure 12-6   The visible region of the spectrum (a) and the line spectrum of the hydrogen atom in the visible region (b).

after the turn of the century. The electron was pictured to be orbiting around the nucleus with the centrifugal force just balanced by the coulombic attraction. Such an atom should be able to absorb or release any energy whatsoever, the change merely altering the orbital radius. Scientists of the day were forced to consider either an unpalatable modification of this model or its complete abandonment. Rarely is a theory abandoned as long as no alternative is available. In fact, this planetary model, despite its obvious inadequacy, was not abandoned until, at last, the quantum mechanical model was developed.

It seemed unlikely to propose that the atom could contain any energy (as a planetary model dictates) and yet absorb or release energy only in special, metered amounts. The alternative was to postulate that the atom could *hold* only special energies. This would automatically imply that only special energies $(E_1 - E_2)$ could be released as the atom changed from one of its "allowed" energies E_1 to another of its "allowed" energies E_2. What was needed was a rationale upon which to hang this then unlikely proposal.

In 1913, Niels Bohr furnished a possible rationale. He discovered that the hydrogen atom spectrum could be explained if, in Nature's hydrogen atom, the electron moved in those special planetary orbits that limit the angular momentum to integer multiples of $h/2\pi$. This rather arbitrary proposal was made acceptable by its tie-in with other phenomena (the radiation from a hot surface, through h), plus the opportunity to save, in part, the planetary model of the atom. It is an interesting commentary on the nature of scientific activity that this proposal by Bohr opened the way to quantum mechanics despite the fact that it proved to be incorrect in almost every detail. Accepting, as everyone now does, the quantum mechanical view of the atom, the following failures of the Bohr planetary atom can be cataloged:

—The electron does *not* have a planetary trajectory.
—The integer-related momenta proved to be incorrect, as shown in Table 12-1.
—The momentum criterion used for the one-electron hydrogen atom failed to explain the observed energy levels of *any* atom with two or more electrons.
—The model proved to give no clue whatsoever to the origin of chemical bonding.
—The model provided no basis for understanding *why* quantization occurs or *why* an orbiting electron would fail to radiate its energy as electrodynamics dictates; both properties were imposed without justification.

Despite this dismal record, Bohr's courage in recognizing the need for a departure from classical physics won for him a place in history. Still today,

Table 12-1 *Comparison of the Momenta of Hydrogen Atom States: Bohr's Planetary Model and the Quantum Mechanical Atom*

Hydrogen Atom State	Planetary Atom $p = n(h/2\pi)$ $n =$ integer	Q. M. Atom $p = \sqrt{\ell(\ell+1)}(h/2\pi)$ $\ell =$ integer
1s	$1(h/2\pi)$	0
2s	$2(h/2\pi)$	0
2p	$2(h/2\pi)$	$1.41(h/2\pi)$
3s	$3(h/2\pi)$	0
3p	$3(h/2\pi)$	$1.41(h/2\pi)$
3d	$3(h/2\pi)$	$2.45(h/2\pi)$
4s	$4(h/2\pi)$	0
4p	$4(h/2\pi)$	$1.41(h/2\pi)$
4d	$4(h/2\pi)$	$2.45(h/2\pi)$
4f	$4(h/2\pi)$	$3.46(h/2\pi)$

the "allowed" energy states of an atom are called "stationary states," as named by Bohr. These stationary states were and still are characterized by "quantum numbers," integers that account for the marvelous pattern in the hydrogen atom spectrum (see Fig. 12-7). It took a few decades for the limitations of the Bohr atom to be admitted, but Bohr's contribution was the crucial break from the physics of the macroscopic world.

In 1926, things finally fell into place when a German scientist, Erwin Schroedinger, recognized the connection between Bohr's stationary states and de Broglie's wave properties of the electron. This step in the development of quantum mechanics must be considered as one of the most important ever made in science—to be compared with the contributions of Galileo, Newton, and Maxwell. To understand this connection, we will consider the "wave properties" of two other physical systems, a vibrating string and a vibrating drum. Their tones are in harmony with the music of the atom!

(b) THE LINE SPECTRUM OF A GUITAR STRING

There are innumerable systems around us that can accept or release energy in any amount: An automobile can be accelerated to any velocity; a baseball can be bunted, smashed out of the park, or anything in between; a mass can be lifted any height. Are there macroscopic systems around us that behave like atoms, absorbing or releasing energy in special amounts? Yes, there are—the guitar string is one of them.

A plucked guitar string produces a musical sound of definite pitch and quality. If the string is held down against one of the frets, thereby changing

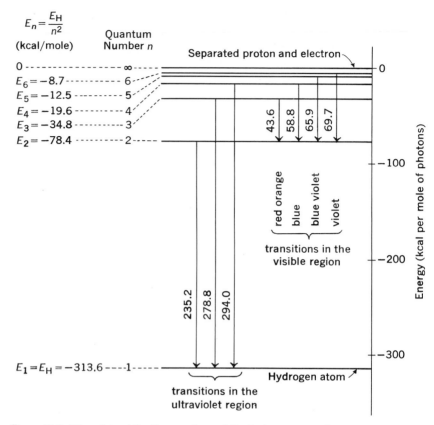

Figure 12-7 *The origin of the line spectrum of the hydrogen atom. The allowed electronic energies of the stationary states are shown by horizontal lines. The energies of these states are related by a simple formula involving the quantum number n. Some of the many possible transitions are shown by vertical arrows. The energy of the emitted light is just the difference between the energies of the two stationary states involved. The visible lines of Figure 12-6 result from transitions to the n = 2 level.*

the string's effective length, the pitch is changed. In this familiar and aesthetic example, we find the elements of quantum mechanics and the essence of the atomic stationary states.

Obviously the pitch of a plucked string is determined by its vibrational frequencies, which set up acoustic disturbances in the air. As the string vibrates back and forth, its movement at any point along its length and its frequency of oscillation characterize the vibration. If we examine this movement, we find two spots at which the string cannot move—at the ends. The string is fixed at the bridge, and again along the neck by one of the frets (according to the fingering). These restraints on the guitar string's movement are called *boundary conditions*. The *boundaries*, in this case, are the ends of the string, and the *conditions* are that the string displacement at the ends must always be zero. We shall see that *because of the boundary conditions, only special vibrations can occur: The vibrations are quantized.*

Figure 12-8 shows some of the ways in which the string can vibrate. The simplest vibration (i) is called the fundamental and it provides the principal tone, or pitch. The next simplest vibration (ii) distorts the string like a sine wave. As the string vibrates back and forth, every point on the string passes periodically through a zero displacement. At the center, however, there never is any displacement. This spot is called a "node." The spots where the string has maximum displacement are called "antinodes." We see that vibration (i) has two nodes, those at the ends forced by the boundary conditions, and one antinode. Vibration (ii) has three nodes, the two ends and one in the middle, and two antinodes. This vibration divides the length L into one wavelength. Vibration (iii) has four nodes and three antinodes, and includes one and a half wavelengths. The vibrations of type (ii), (iii), and higher are called "harmonics" in musical theory, and they provide the "quality" or "timbre" of the guitar's notes. For our purposes, however, we see that the guitar string vibrates only in special frequencies and, hence, emits only special sounds. Each of these special frequencies can be characterized by the fundamental frequency and an integer quantum number n. The quantum number n gives the number of half-wavelengths in the vibration and $(n + 1)$ is the number of nodes (including the nodes at the ends).

One more aspect of these vibrations interests us. Figure 12-8 shows that in vibration (ii) the two antinodes displace the string at the same *time* but in opposite sense. When the string is up at the left antinode, the string is down on the right. Later in the vibration the opposite is true. The two antinodes are said to be "out of phase." There is a change in phase every time we pass through a node.

Figure 12-8(*b*) shows some ways in which the string *cannot* vibrate. These displacements would require the string to move up and down at the base of the guitar or at the fret, but at both these places the string is held. *Thus, the boundary conditions are the origin of the quantization.* With boundary conditions, the wave motion of the vibrating string has a line spectrum whose frequencies are characterized by integer quantum numbers, a nodal pattern, and phase relationships. All of these characteristics appear in the quantum mechanical description of the atom.

(c) THE LINE SPECTRUM OF A DRUM

The same concepts and language of the guitar string carry over to the vibrations of a drum. Most of us think of a drum as a musical instrument with little personality, but it isn't so. By analyzing the motions of a drumhead, you will see that it too has both pitch and quality.

Around the circumference of the drum the leather of the drumhead is tightly held so it cannot move. This is the boundary condition of a drum.

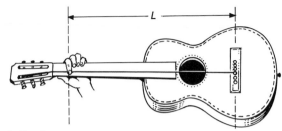

(a) allowed vibrations

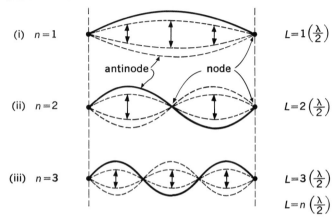

(i) $n=1$ $L=1\left(\frac{\lambda}{2}\right)$

antinode   node

(ii) $n=2$ $L=2\left(\frac{\lambda}{2}\right)$

(iii) $n=3$ $L=3\left(\frac{\lambda}{2}\right)$

$L=n\left(\frac{\lambda}{2}\right)$

(b) forbidden vibrations

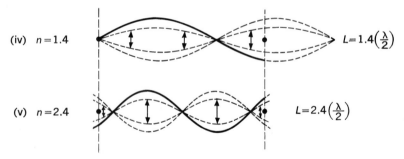

(iv) $n=1.4$ $L=1.4\left(\frac{\lambda}{2}\right)$

(v) $n=2.4$ $L=2.4\left(\frac{\lambda}{2}\right)$

(c) the line spectrum

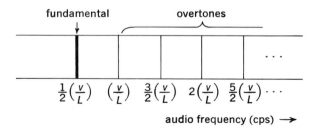

fundamental   overtones

. . .

$\frac{1}{2}\left(\frac{v}{L}\right)$ $\left(\frac{v}{L}\right)$ $\frac{3}{2}\left(\frac{v}{L}\right)$ $2\left(\frac{v}{L}\right)$ $\frac{5}{2}\left(\frac{v}{L}\right)$ $\cdots$

audio frequency (cps) $\longrightarrow$

Figure 12-8   The allowed (a) and unallowed (b) vibrations of a guitar string and its spectrum (c). The wavelength of the string is related to its length by an integer number n: $\lambda = (2/n)L$. The audio frequency is given by: ν (audio) $= v/\lambda$, in which v is the velocity of sound.

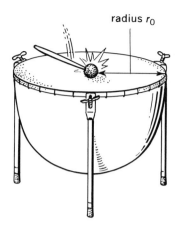

radius r_0

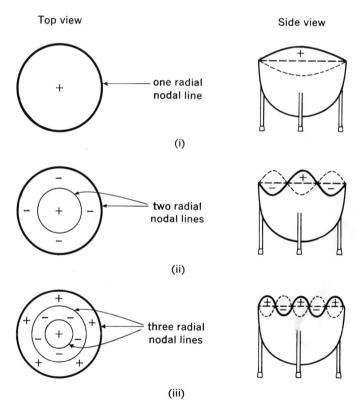

Top view

Side view

one radial
nodal line

(i)

two radial
nodal lines

(ii)

three radial
nodal lines

(iii)

Figure 12-9   Radial vibrations of a drum: radial nodes.

Thus, the two-dimensional vibrator has a nodal *line* just as the one-dimensional vibrator (the string) has a nodal *point*. This boundary condition limits the ways in which the drumhead can vibrate and gives the drum a specific set of vibrational frequencies.

Suppose the drum is struck just in the center, as shown in Figure 12-9. There are then special ways in which the leather can move. In vibration (i), the fundamental, there is a single antinode and a single nodal line. In (ii) there are two nodal lines, the outer one being the circle defined by the drum edge and the second being a circular nodal line at an intermediate radius. There are also two antinodal lines, and the entire movement maintains circular symmetry. Across each nodal line there is a change of phase: If the leather is displaced upward on one side of the line (designated in Fig. 12-9 by a plus sign), it is displaced downward on the other side (shown by a negative sign).

The drum vibrates differently, however, if it is struck off center. This displaces the leather skin, but without the circular symmetry. Again it turns out, though, that the vibrations are quantized and that they have nodal lines, now extending across the drum, as shown in Figure 12-10. In vibration (iv) there is one nodal line across the drum dividing it into two halves. In (v) the vibration divides the head into quadrants, two going up while the other two are displaced downward. In such motions, there is a disturbance of the circular symmetry by the nodal lines across the drum. If we imagined a 0 to 360° scale around the drum's edge, the displacement would vary with angle (except at the edge where the boundary conditions still hold). Hence, the nodes shown in Figure 12-10 are called *angular nodes*.

(d) MUSIC AND THE ELECTRON

What we have learned is that musical instruments have line spectra like those of an atom. The wave-like vibrations of a guitar string or a drumhead are quantized, apparently like the dynamic motion of an electron in an atom. These musical motions can be described in mathematical detail by a characteristic mathematical equation, the "wave equation."

Along comes Erwin Schroedinger, a mathematical physicist who could write down and solve the wave equation for a vibrator on a napkin during lunch. He heard tell that a hydrogen atom has a line spectrum (like a vibrating string) and that an electron can be diffracted like a wave (the de Broglie prediction). Two and two equals four, said Dr. Schroedinger, and the line spectrum of the hydrogen atom shows that its equation of motion must be a wave-type equation with boundary conditions that fix the possible energy values. This bold decision was the real birth of quantum mechanics.

The classical equation of motion did not fit this description. It can be simply stated in terms of the sum of kinetic and potential energy, as follows:

kinetic energy + potential energy = total energy

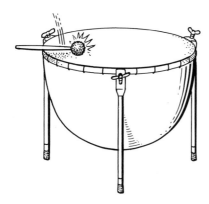

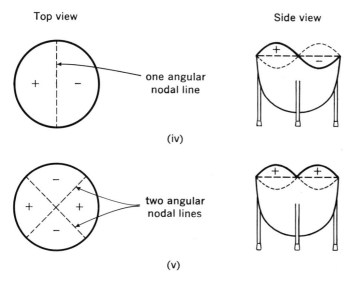

Figure 12-10 *Angular vibrations of a drum: angular nodes.*

or, in conventional symbols,

$$T + V = E \tag{12-7}$$

The kinetic energy T is just $\frac{1}{2}mv^2$. The potential energy is the electrostatic energy of an electron attracted to a proton at some distance r, $V = -e^2/r$. Hence (12-7) can be rewritten as

$$\tfrac{1}{2}mv^2 - \frac{e^2}{r} = E \tag{12-8}$$

To get this equation into momentum form, as Niels Bohr preferred it, we substitute $p = mv$.

$$\frac{1}{2}\frac{p^2}{m} - \frac{e^2}{r} = E \qquad\qquad (12\text{-}9)$$

That's no wave equation, thought Erwin. He saw how to make it into one, though. Schroedinger replaced the classical momentum p by a mathematical "operator" (which is merely an instruction to carry out a mathematical operation upon some mathematical quantity). He selected, of course, an operator that converted (12-9) into a wave equation. This required the invention of a new function, called the *wave function* and symbolized ψ (psi). This wave function was inserted to give the operator something to operate on. Now the "Schroedinger Equation" looks like this:

$$\left(\frac{1}{2}\frac{p^2}{m} - \frac{e^2}{r}\right)\psi = E\psi \qquad\qquad (12\text{-}10)$$

and we must remember that in this new "wave mechanics," p is no longer simply m times v, it is an instruction to "operate" upon the function ψ.* It developed that ψ contains our knowledge of the electron in the atom.

Now we need some boundary conditions—they give us the line spectrum. Physical reality must be the guide here. If the function ψ contains our knowledge of the whereabouts of the electron in the hydrogen atom, the boundary conditions must be built on limiting conditions that make sense for the problem. One of these boundary conditions can be likened to the edge of the drumhead, where the displacement must always be zero. The edge of the atomic drum is at infinity—if the electron reached an infinite separation from the proton, it would no longer be attached. So we place the mathematical restriction on ψ that it always must be zero at infinity. This and two other physically reasonable boundary conditions† give equation (12-10) a line spectrum, and it turns out to be exactly the hydrogen atom spectrum.

This achievement was only the beginning of a stunning series of successes for the new quantum mechanics. Without modification, it explains the energy levels of many-electron atoms. More important, it has proved to be in quantitative accord with the properties of molecules: Calculations of bond energies, bond lengths, molecular vibration frequencies, and energy levels agree with experiment as accurately as calculational approxi-

---

* Some typical mathematical operators that might be applied to a function ψ are "find the slope of ψ" and "find the curvature of ψ." The last one is the one that Schroedinger substituted for p^2 to obtain a wave-type equation.

† These other two conditions are a bit more mathematical. In essence, they guarantee that ψ remains sensibly finite and does not hop about, changing in a discontinuous fashion. These, too, are based in the physical reality that prevents electrons from doing things at infinite velocity.

mations and experimental uncertainty permit test. Only a few years after Schroedinger proposed his equation, it was found to be applicable to the simplest molecules, H_2 and H_2^+, but mathematical difficulties stood in the way of similar tests for more complicated molecules. Approximations were needed, but gradually it became clear that *all of chemistry is explained in quantum mechanics*. As computers have come onto the scene, calculations can now be valuable for molecules with as many electrons as carbon monoxide, methane, water, and ammonia. In principle, all chemical changes could be predicted if the mathematical obstacles were not so insurmountable. Nevertheless, with the advent of quantum mechanics, chemical bonding stepped out of empiricism. Now we have a firm theoretical framework to guide us in understanding and predicting chemical phenomena.

(e) WHAT DOES THE WAVE FUNCTION TELL US?

The solution of the wave equation gives us immediately two quantities of interest: E, the "allowed" energy levels of the atom or molecule, and, for each allowed energy, a function ψ of the electron's position coordinates. What information is contained in ψ?

We have two models with which to relate. The energy conservation equation (12-7), $T + V = E$, relates to the classical description of the motion of the particles in the system. The Schroedinger equation (12-10) is a wave-type equation and it relates to the vibratory motion of a string or a drumhead. In classical mechanics, particles move in trajectories, whereas wave motion is described by a displacement function that gives the nodal pattern, the amplitudes of displacement, and the phases. Examination of ψ reveals greater similarity to the latter, the wave model. The function ψ has nodal surfaces and phase changes across these nodal surfaces. When two wave functions interact, they interact either constructively or destructively, as waves do, depending on whether they interact in phase or out of phase. On the other hand, there is no information whatsoever in ψ about the trajectory of an electron. In fact, on the microscopic scale, the *concept* of electron trajectory in an atom or molecule has lost significance. Instead, we find that the function ψ, or rather, its square ψ^2, gives only the probability of finding the electron within a given volume. Where ψ^2 is high, the probability of finding the electron is high. Where ψ^2 is low, the electron is rarely found. On a nodal surface the value of ψ is zero, so the electron is never found there.

The information contained in ψ^2 can be likened to the information contained in the holes in a dart board. The pattern of holes shows the result of many individual dart throws—it also would guide a gambler wagering on the next throw. There is a high probability that the dart will land within a circle containing many dart holes and a low probability for a circle of equal area but containing few dart holes. The density of dart holes gives probability information. If we were to express mathematically the density of dart holes as a function of position on a dart board, we would have a function exactly analogous to ψ^2.

To take full advantage of the easily interpreted dart board analogy, we might make a similar graph for the wave function of a hydrogen atom. For any set of coordinates x, y, z, the function ψ has a particular numerical value. We could make a graph in which dots are shown with density (dots per unit volume) proportional to the square of this numerical value. Figure 12-11 shows such a plot for a 1s wave function, this being the ψ that corresponds to the lowest allowed energy state of hydrogen. Of course, the electron in the hydrogen atom occupies space in three dimensions around the proton, so Figure 12-11 shows only a cross-sectional cut through the atom. Nevertheless, the picture gives an immediate qualitative picture of where the electron is most likely to be found—somewhere reasonably near the nucleus.

This is, of course, only one way to represent the probability picture of how the electron moves in the vicinity of the proton. Figure 12-12 shows three additional ways. Figure 12-12(b) shows a graph of the dot density (ψ^2) along some line radiating from the proton. Along the horizontal axis we have plotted the value of r, the distance from the nucleus, and along the vertical axis we plot ψ^2.

Figure 12-12(c) shows density contours on a cross-sectional cut through the probability dot diagram of Figure 12-12(a). Figure 12-12(d) shows the significance of the density contours: They present a contour map of where the electron spends its time. Such a map has the advantage that it displays qualitatively the shape of the distribution (in two dimensions) and it displays quantitatively how rapidly the probability drops off far from the nucleus. This representation is much used because it is easy to calculate.

Figure 12-12(e) is probably the most informative representation. The innermost contour line is that line that *encloses* 10 percent of the probability

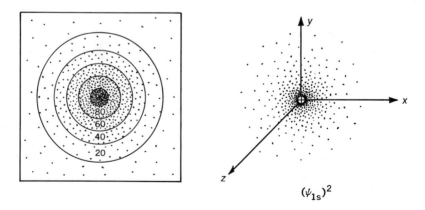

$(\psi_{1s})^2$

Figure 12-11 The dart board and a hydrogen 1s function: probability density plots.

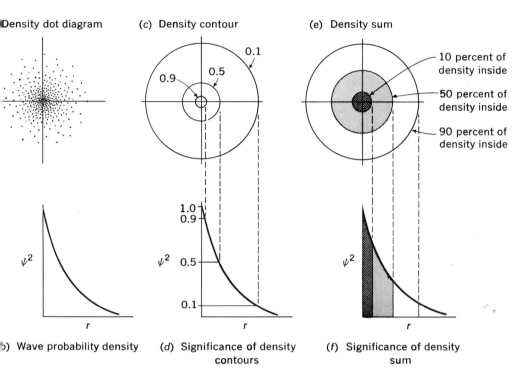

Figure 12-12 Representations in two dimensions of the probability distribution for a hydrogen 1s wave function.

in the smallest possible volume. The next contour shows the line that encloses 50 percent in the smallest possible volume. Finally, the third contour shows the 90 percent enclosure line. These contours are most useful to the gambler, whether betting on where the dart will fall or where the electron will be. They not only show where the electron is most likely to be found, but also tell us something about how "large" the atom is. Plainly, the size is not well defined—*there is no boundary to the atom.* One could arbitrarily take its size to be the 0.5 contour in Figure 12-12(c) or one could take the 0.5 density-sum contour in 12-12(e). Some workers might prefer the 0.9 density-sum contour in 12-12(e) on the argument that the electron spends only 10 percent of its time outside this 0.9 line, a small enough fraction to be unimportant. At this moment we needn't choose between these alternatives. Our main interest is in the picture of the atom given by quantum mechanics. It differs greatly from the classical view, but since this quantum mechanical picture agrees with all the experimental facts about atoms and molecules, it surely must be preferred. In the quantum mechanical description of the atom,

—the electron trajectory is completely unspecified;

—the electron position is known only through a probability pattern;

—though the position of the electron at any instant is not known, it is at some *point;* the electron should *not* be thought of as being atomized into a "cloud" with charge distributed according to the probability function;*

—there is a variety of ways to designate the probability pattern that shows where the electron is likely to be found;

—this probability pattern shows that there are no boundaries to the atom—it extends to infinity.

(f) THE UNCERTAINTY PRINCIPLE

Before proceeding to examine other properties of the quantum mechanical atom, we should look briefly at the Uncertainty Principle—because some people are uncertain about the meaning of this principle. About the same time that Schroedinger was framing the wave picture of the atom, a German scientist named Heisenberg discovered a limitation on our possible knowledge of the atom. He realized that a consequence of the particulate nature of light is that light disturbs the object viewed. If a photon can be represented as a bullet with momentum appropriate to its frequency, then as this photon bullet ricochets off an object, there will be some recoil momentum transferred to the object. This is the same momentum transfer that causes a change in the photon frequency, as pictured in Figure 12-2.

Because of the low electron mass, this recoil effect is not small. Furthermore, the higher the photon energy, the larger will be its momentum (see expression (12-4)) and, hence, the greater the recoil effect. Unfortunately, the high-energy photons give the sharpest probes for fixing the electron's position.

These facts present a dilemma if we try to measure simultaneously *both* the position and the momentum of an electron. To pinpoint the electron's position accurately, we need a high-energy photon, but the high-energy photon disturbs the electron's momentum. If we use a low-energy photon to permit an accurate momentum measurement, the photon is too "soft" to locate the electron precisely. We can either optimize conditions to measure position or to measure momentum, but we cannot measure *both* simultaneously to any accuracy we wish.

This qualitative statement of the Uncertainty Principle is sufficient for our needs. It says that we cannot precisely locate the electron and at the same time measure its momentum. However, it places no limit on how closely we can locate the electron. It *is* somewhere and we can find its

---

* Such a hypothetical electron cloud has a potential energy associated with the repulsion of its parts that does not appear in the potential function used in the Schroedinger Equation. Nevertheless, this "electron-cloud model" has significant use merely because it has some computational simplicities.

position to any degree of certainty we wish. The Uncertainty Principle tells us only that the closer we determine position, at a given instant, the less we will be able to learn about its momentum at that same instant.

12-4 The hydrogen atom

With the assurance that quantum mechanics explains the properties of atoms and molecules, it behooves us to adapt our thinking to this model. The hydrogen atom is our most informative example because an exact solution of its Schroedinger Equation is possible. No such exact solution is possible for any atomic (or molecular) system with two or more electrons, although approximations can be made that allow us to come very close to the correct solutions. These many-electron atoms will be considered in Chapter Thirteen when we have the benefit of a clear understanding of the hydrogen atom.

(a) QUANTUM NUMBERS

We have learned that only particular energies can be held by a hydrogen atom. These particular energies, called energy levels, and the electron probability pattern are given by the Schroedinger Equation. Both the energy and the electron probability distribution depend upon integer numbers that are closely related to the integer numbers that identify the nodal properties of a vibrating string or drum. For a string, there is displacement in only one dimension (the y direction), and we need only one number to describe its permitted motions. The two-dimensional vibrations of the drum, on the other hand, need two numbers, one of which describes the radial nodal properties and the other of which describes the angular nodal properties.

The Schroedinger wave equation for the hydrogen atom describes the electron wave in *three dimensions*. It is entirely appropriate, then, that *three* integer numbers are necessary to describe fully each energy state of the hydrogen atom. These numbers are called quantum numbers. Each set of quantum numbers fixes a possible energy of the atom and also the probability pattern that describes what we know of the electron's location. Figures 12-11 and 12-12 show the atom in its lowest energy state. More energetic energy levels correspond to more complex spatial distributions. These spatial distributions are counterparts to the orbital trajectories that describe the classical motions of planets in a solar system. If we picture a solar system that could be shrunk as much as desired, by the time the sun had shrunk to the mass of a proton, the orbital trajectory would have become the quantum mechanical probability distribution expressed by ψ^2. Because of this correspondence, scientists refer to a probability pattern as an "orbital." We must remember, though, that the word orbital now refers to a picture like that in Figure 12-11 and the trajectory meaning is lost.

The three quantum numbers that define a particular probability distribution (a particular orbital) are symbolized n, ℓ and m. The first of these, n, is the most important, so it is called the *principal* quantum number. This number n can have any integer value, 1, 2, 3, 4, ... That integer value determines the energy of the hydrogen atom according to the formula

$$E_n = -R \cdot \frac{Z^2}{n^2}$$ (12-11)

where R = a constant = $313.6 \dfrac{\text{kcal}}{\text{mole of atoms}}$

Z = charge on the nucleus ($+1$ for a hydrogen atom)

n = principal quantum number = 1, 2, 3, 4, . . .

The negative sign in (12-11) indicates that the energy is lowered relative to the reference state, a proton and an electron separated from each other at an inifinite distance.

Exercise 12-6. Calculate the energy in kcal/mole of hydrogen atoms ($Z = 1$) with $n = 2$. Calculate the energy difference between the $n = 2$ and $n = 1$ states and compare to the line spectrum of the H atom, as pictured in Figure 12-7.

The principal quantum number also fixes the nodal characteristics of an orbital. While a vibrating string has nodal *points* (see Fig. 12-8) and a vibrating drum has nodal *lines* (see Fig. 12-9), a hydrogen atom has nodal *surfaces*. At such a surface the wave function ψ changes phase, like the up-down phase change at a node in the guitar string. At this nodal surface, ψ^2 is zero; the electron is not found at this location. *The number of nodal surfaces is equal to* n, one of which is the boundary condition nodal surface at $r =$ infinity.

The spatial extent of the orbital is determined by two of the quantum numbers, n and ℓ. For example, quantum mechanics tells us the average distance $\bar{r}$ of the electron from the nucleus:

$$\bar{r} = a \frac{n^2}{Z} \left[\frac{3}{2} - \frac{\ell(\ell + 1)}{2n^2} \right]$$ (12-12)

where a = a constant = 0.529 Å

Z = charge on the nucleus ($+1$ for a hydrogen atom)

n = principal quantum number

ℓ = angular quantum number

The second quantum number ℓ can have integer values beginning at zero but not exceeding $(n - 1)$. Thus, if n is 3, ℓ can have any of the values 0, 1, or 2, but no others. If n is 1, ℓ can have no value other than 0.

Exercise 12-7. Calculate the average distance of the electron from the nucleus in the three states shown in Figure 12-13: the 1s state $(n = 1, \ell = 0)$, the 2s state $(n = 2, \ell = 0)$ and the 2p state $(n = 2, \ell = 1)$.

The quantity ℓ also relates to the nodal pattern of the orbital. There are ℓ nodal surfaces with angular dependence. Since there is a total of n nodal surfaces, there must be $(n - \ell)$ nodal surfaces without angular dependence, that is, spherically symmetric. For example, if $n = 2$, then ℓ can be either 0 or 1. If $\ell = 0$, there are no nodal surfaces with angular dependence; the two radial nodal surfaces are both spherical in shape. One of these is the surface at $r = $ infinity. If $\ell = 1$, there is one angularly dependent nodal surface, a plane. The second nodal surface $(n - \ell = 2 - 1 = 1)$ is a sphere, again the nodal surface at infinity. Figure 12-13 shows cross-sectional views of the probability distributions corresponding to $n = 2$. The change of phase at the nodal surface is designated by plus and minus signs. These signs have significance related to the up-down displacements in wave phenomena (as portrayed in Figs. 12-9 and 12-10) and do not signify electric charge.

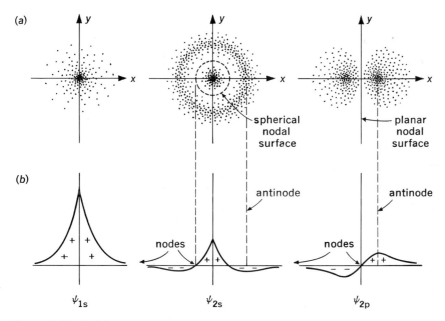

Figure 12-13 Nodal surfaces for the hydrogen atom. (a) Electron probability density. (b) Electron wave functions (plus and minus signs designate phase). Notice that the 2p ($n = 2$, $\ell = 1$) function changes phase at the nucleus. The two lobes of the p functions are out of phase, like a guitar string vibrating in the first overtone.

The $n = 2$, $\ell = 1$ orbital is shown in Figure 12-13 to be directed along the x axis, and the yz plane is its nodal surface. This brings to mind the possibility that this orbital could be otherwise oriented. There are three dimensions, why should the x axis be preferred? Indeed, it is not. There are three orbitals with $n = 2$, $\ell = 1$, and these can be considered to be directed along the three axes, as shown in Figure 12-14. These three orbitals are identified by the third quantum number m. This number is called the *magnetic quantum number* and it can take integer values either positive or negative in sign, but not exceeding ℓ:

$$m = \ell, \ell - 1, \ell - 2 \cdot \cdot \cdot 1, 0, -1, -2 \cdot \cdot \cdot -\ell \qquad (12\text{-}13)$$

Again, using the example of $n = 2$, we now find the following set of orbitals, all with the same energy,

$$E = -R\frac{Z^2}{n^2} = -R\frac{1}{2^2} = -\tfrac{1}{4}R$$

			E	$\bar{r}$	$(12\text{-}14)$
$n = 2$	$\ell = 0$	$m = 0$	$-\tfrac{1}{4}R$	$6a$	
$n = 2$	$\ell = 1$	$m = +1$	$-\tfrac{1}{4}R$	$5a$	
$n = 2$	$\ell = 1$	$m = 0$	$-\tfrac{1}{4}R$	$5a$	
$n = 2$	$\ell = 1$	$m = -1$	$-\tfrac{1}{4}R$	$5a$	

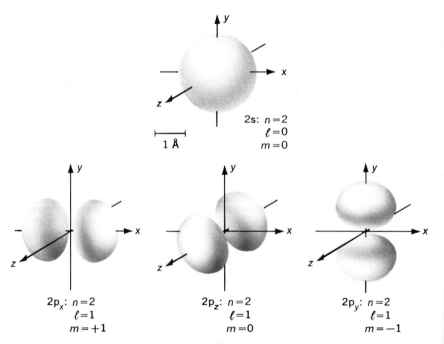

Figure 12-14 The $n = 2$ orbitals for the hydrogen atom: orientations and nodal surfaces.

These four orbitals have already been pictured in Figure 12-14. We note that with $n = 1$ there is only one orbital ($n = 1$, $\ell = 0$, $m = 0$). With $n = 2$ there are four orbitals (12-14) and four is equal to n^2. This is so for any choice of n; there are n^2 orbitals with that value of the principal quantum number.

(b) THE HYDROGEN ATOM ENERGY LEVEL DIAGRAM

In summary, the three quantum numbers, n, ℓ and m, fix the possible energy states of the hydrogen atom and the corresponding probability distributions or orbitals. The energy and the number of orbitals with a particular energy are fixed by the value of n. The nodal patterns depend as well upon ℓ and m.

Table 12-2 shows how the number of states rises as n rises. It also shows the popular notation used to designate each orbital. Instead of indicating three numbers n, ℓ and m, an orbital is designated first by a number, the value of n, and then by a letter, s, p, d, or f, to indicate the value of ℓ. Thus the orbital $n = 2$ and $\ell = 0$ is called the 2s orbital. We shall see that all orbitals with $\ell = 0$ have spherical symmetry, so the symbol s can be considered to stand for "spherical."* An orbital with $n = 2$ and $\ell = 1$ is called

* Actually the letters s, p, d, and f were chosen before the hydrogen atom problem was solved and the letters were descriptive of characteristic spectral behaviors. Hence the original abbreviations no longer have value.

Table 12-2 The Hydrogen Atom Quantum Numbers

n	$E = -\dfrac{313.6}{n^2}$ kcal/mole	$\ell = 0, 1, 2, 3,$ $\ldots, n-1$ s, p, d, f, $\ldots$	$m = \ell, \ell - 1, \ldots,$ $1, 0, -1, \ldots -\ell$	Number of Orbitals n^2
1	$-\dfrac{313.6}{1^2} = -313.6$	0(1s)	0	1
2	$-\dfrac{313.6}{2^2} = -78.4$	0(2s) 1(2p)	0 +1, 0, −1	1 + = 4 3
3	$-\dfrac{313.6}{3^2} = -34.8$	0(3s) 1(3p) 2(3d)	0 +1, 0, −1 +2, +1, 0, −1, −2	1 + 3 = 9 + 5
4	$-\dfrac{313.6}{4^2} = -19.6$	0(4s) 1(4p) 2(4d) 3(4f)	0 +1, 0, −1 +2, +1, 0, −1, −2 +3, +2, +1, 0, −1, −2, −3	1 + 3 + = 16 5 + 7

a 2p orbital. There are three of these, corresponding to the three values of m ($+1$, 0, -1). Since the three 2p orbitals can be pictured to be oriented at right angles to each other, the symbol p can be related to the descriptive word "perpendicular" and the three p orbitals can be designated p_x, p_y and p_z.

The information in Table 12-2 can be represented pictorially through an energy level diagram, as in Figure 12-15. Horizontally, we represent an orbital as a pigeonhole that might be occupied by an electron. The orbital is placed on a vertical energy scale to indicate how much the energy is lowered as an electron is captured into that orbital by a proton.

Such a diagram simplifies discussion of the spectra of atoms. Picture an electron in a 2p orbital of hydrogen. This atom has an energy lower than

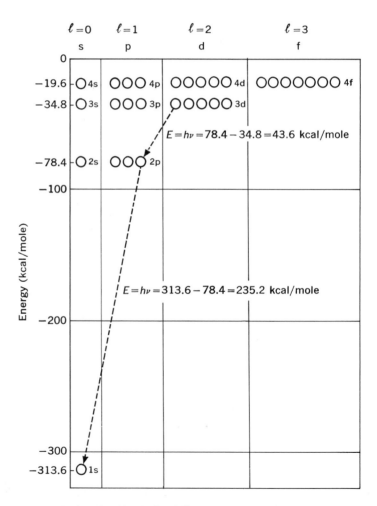

Figure 12-15 Hydrogen atom energy level diagram.

that of a separated electron and proton by 78.4 kcal/mole. If, now, this atom were to release energy and drop to a lower state, quantum mechanics and experiment tell us there is only one such state, the 1s orbital. This energy level, the lowest, is 313.6 kcal/mole below our zero reference energy. Hence the atom can make the transition 2p → 1s only if it can dispose of the energy difference

$$(-78.4) - (-313.6) = 235.2 \text{ kcal/mole}$$

This is, of course, one of the energy changes actually observed in the hydrogen atom line spectrum. *All of the lines in the hydrogen atom are accounted for* by energy differences like the two shown in Figure 12-15.

(c) OTHER ONE-ELECTRON ATOMS

The ion He$^+$ consists of a nucleus with charge $+2$ and a single electron. Its Schroedinger Equation is, mathematically speaking, identical to that of a hydrogen atom with only one change, the nuclear charge Z is doubled. Equations (12-11) and (12-12) apply, provided we make $Z = 2$ instead of $Z = 1$. The energy level scheme looks identical to that in Figure 12-15 except that every energy is multiplied by Z^2, or 4. To remove the electron from a hydrogen atom requires 313.6 kcal. To remove the electron from He$^+$ requires $(313.6)(4) = 1254.4$ kcal. Thus we see that the higher positive charge on the nucleus holds the electron more tightly—an entirely reasonable result.

It is also reasonable to expect that the $+2$ nuclear charge will pull the electron in closer. Equation (12-12) shows this is also true. Whereas the 1s state of the H atom has an average radius of $\frac{3}{2} \cdot 0.529 = 0.79$ Å, the He$^+$ ion has an average radius of 0.39 Å, half as large.

On the other hand, the orbital distributions are the same for He$^+$ as for H (see Fig. 12-16), except smaller in all dimensions by the $1/Z = \frac{1}{2}$ factor.

In a similar way, the energy levels and orbital characteristics of the doubly charged lithium ion, Li^{+2}, are like those of the hydrogen atom except for the effect of nuclear charge. A Li^{+2} ion consists of a nucleus with $Z = 3$ and a single electron. Its ionization energy is $(313.6)(3^2) = 2822.4$ kcal and the spatial extent of its orbitals is one-third that of H.

Exercise 12-8. Use expressions (12-11) and (12-12) to calculate the energies and average radii of the $n = 2$ states (2s and 2p) for the one-electron atoms, He$^+$ and Li^{+2}. Compare to the values calculated for the H atom (Exercises 12-6 and 12-7).

Figure 12-17 contrasts the energy level diagrams predicted for these three one-electron atoms, H, He$^+$ and Li^{+2}. These energy level diagrams

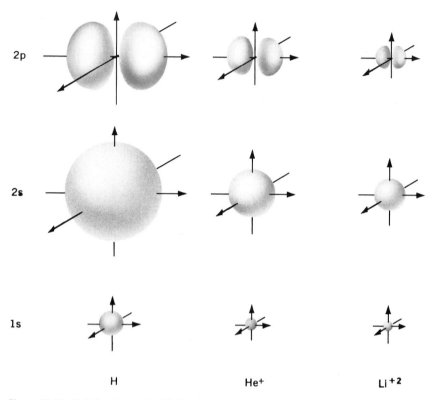

2p

2s

1s

H

He⁺

Li⁺²

Figure 12-16 Relative sizes of orbitals of one-electron atoms and ions, H, He⁺ and Li⁺².
Boundary surfaces represent the 90 percent density sum.

are in perfect accord with the experimental spectra emitted by these species in glow discharge experiments. Quantum mechanics scores again!

Qualitatively we see two important results. First, as nuclear charge increases, electrons are held more tightly. Second, as nuclear charge increases, the electrons are pulled closer—the atom gets smaller.

(d) CONTRAST TO THE PLANETARY ATOM

Figure 12-18 pictures the planetary and the quantum mechanical views of the atom. The planetary model is simple. The electron endlessly circles about its proton sun, always at a fixed radius (in the 1s state). It doesn't radiate, despite instructions to do so by classical electromagnetic theory, because Niels Bohr said it had better not. Otherwise, the electron would spiral into the nucleus and all matter would collapse. It has a particular energy because one of the integer-fixed momenta corresponds to that energy. Momenta that would give other energies, lower and higher, were again arbitrarily forbidden by Bohr's authoritarian model. The planetary concept gave no clue as to why these other energies are not possible; quantization was imposed on the model to force it to agree with the facts.

The quantum mechanical view is less readily assimilated. The picture doesn't even show the electron; it shows a probability pattern instead. This pattern only gives us a basis for betting on where the electron would be if we did an experiment to look for it. The probability per unit volume is largest near the nucleus and it dwindles off as we look farther away. There is no electron trajectory shown at all. That was lost in the mathematics when the kinetic energy term was modified to convert the energy conservation law into a wave-type equation. The electron seems to move about—but the model includes no trajectory! This is a new way to have to think!

Indeed it is; we must be reminded again and again of the guitar string and put the solar system out of our minds. But why? Why use the compli-

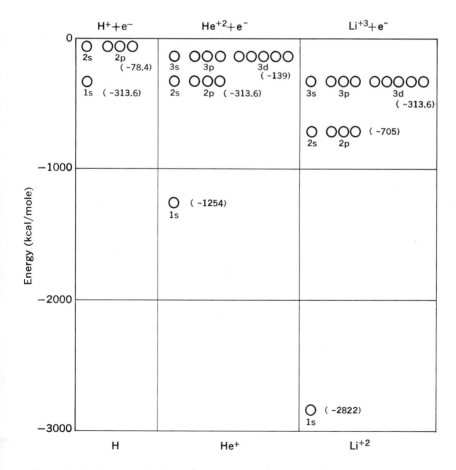

Figure 12-17 Energy level diagrams of one-electron atoms and ions.

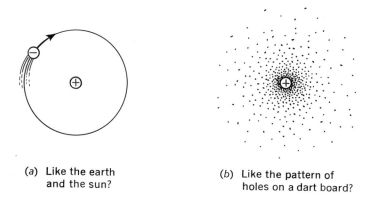

(a) Like the earth and the sun?

(b) Like the pattern of holes on a dart board?

Figure 12-18 The planetary (a) and quantum mechanical (b) hydrogen atom.

cated, abstract model instead of the simple, planetary one? The answer lies in science's implacable test of usefulness. The probability view fits many different kinds of observations. The planetary model fits only one kind: the energy levels of the one-electron atoms. That is not enough when a better model is at hand.

A striking difference between the two pictures in Figure 12-18 is found in the electron–proton distance. The planetary electron never comes closer to the nucleus than dictated by its fixed orbital radius. In contrast, in Figure 12-18(b) the probability density near the nucleus is greater than anywhere else. Are there any observations that test these two conflicting conclusions? Yes, there are several, three of which we will mention here.

(i) *Electron capture by the nucleus.* A few nuclei are unstable with respect to a change in which an electron is captured into the nucleus, decreasing its positive charge by one unit. This process requires the electron to be extremely close to the nucleus part of the time—in accord with the quantum mechanical view.

(ii) *Nuclear magnetic splittings.* Most nuclei, including protons, are tiny magnets. In a strong magnetic field, these nuclear magnets can be lined up in a quantized way. The magnets can then be reoriented by the absorption of radiation (light) whose frequency just matches the energy levels of the magnets in the field. However, the electron is itself a tiny magnet, so it, too, communicates with the nuclear magnets. Certain aspects of the nuclear-magnet energy level diagrams require that the electron magnet come very close to a proton magnet part of the time. In fact, the quantum mechanical atom correctly tells how much.

(iii) *Spectral splittings in the H atom spectrum.* When the hydrogen atom spectral lines are examined with the most powerful optical equipment, each line is found to have two components. The minute frequency difference is explained with the aid of the quantum mechanical view of the H atom. The splitting turns out to be due to the part-time proximity of the electron and the nucleus, again in accord with Figure 12-18(b).

With this reassurance, let's return to the troubles of the planetary atom. First, Bohr had to impose fixed energy levels (through his momentum restriction) on an unreceptive model. But Schroedinger's guitar-string model is perfectly tuned to the quantization of energy. A wave phenomenon with boundary conditions *naturally* leads to special energy levels. Instead of appearing as a strange constraint, special energies are an intrinsic part of a wave description.

The second headache of the planetary model is that radiation and loss of energy are implied by the movement of an electron charge in a curved trajectory. Quantum mechanics simply does not have this difficulty since the trajectory concept is gone. The wave description of the electron in an atom does not describe a trajectory, so it cannot be accused of implying energy loss through radiation. Furthermore, it cannot be said that there is a trajectory, but quantum mechanics doesn't give it. The change of the classical equation of motion to the quantum mechanical one specifically altered the kinetic energy term. We can paraphrase this change into an instruction—"Stop thinking about trajectories—we can't fit Nature with a trajectory-type equation of motion." And why should we complain? Without a trajectory, the electron need not radiate its energy and we know that it does not.

But the most compelling nails are yet to be driven into the planetary model's coffin. This simple picture never explained *any* many-electron atom, beginning with helium, nor did it provide even a tiny step forward in explaining chemical bonding. Quantum mechanics deals successfully with both. The many-electron atoms are considered in Chapter Thirteen, and

Figure 12-19 The Planetary Atom: R.I.P.

the next five chapters explore chemical bonding, all illuminated with the aid of our quantum mechanical light.

Problems

1. Which of the following true statements show, directly or indirectly, that light is a form of energy?
(a) It's cold at night.
(b) Smart bedouins wear white robes.
(c) Green plants don't grow in the cellar.
(d) The clothes hanging on the part of the clothesline in the shade dry more slowly.
(e) With a magnifying glass, a child can focus the sun's rays and burn a hole in his mother's favorite rug.

2. By what percentage has the mass of an electron increased when it has been accelerated to one tenth the velocity of light, $0.10\ c$?

3. If you could hold an electron in your hand and move it back and forth, rhythmically, an oscillating electric disturbance (light) would be propagated at the speed of light ($c = 3 \cdot 10^{10}$ cm/sec). With what frequency ν would you have to move the electron to give a wavelength λ equal to the diameter of the earth, 8000 miles (one mile = 1.6 km)?

4. What is the minimum wavelength of light (in Å, $1\ \text{Å} = 10^{-8}$ cm) needed to provide enough energy per quantum to rupture the chemical bond in hydrogen iodide? The bond dissociation energy of HI is given in Table 7-3 as 71.4 kcal/mole.

5. The earth absorbs about 10^{21} kcal of energy per year in the form of light from the sun. If it lost none of this, by how many tons would the mass of the earth increase per year? (1 kcal = $4.18 \cdot 10^{10}$ ergs; 1 pound = 454 grams)

6. If the earth absorbs 10^{21} kcal of energy per year in the form of light from the sun, think how much solar energy goes everywhere else! Calculate this *total* energy radiation per year by the sun if our tiny disc of cross-sectional area $5 \cdot 10^7$ square miles (that's the earth) placed at a distance of 93 million miles (that's the approximate radius of the earth's orbit) receives that much solar energy. (Area of a sphere $= 4\pi r^2$.) Convert the result into the number of grams of mass lost per year by the sun. If the sun has been frittering itself away at this rate since the earth was born (about 10^9 years ago), how many earth-masses have been lost by the sun during this period? (The earth weighs $6.00 \cdot 10^{27}$ gm.)

7. The Soviet Union and the U.S. together (though apart) probably explode about ten hydrogen bombs underground per year for the pure joy of testing them. If each H bomb explosion converts about 10 grams of matter into an equivalent amount of energy, how many kcal of energy are released per bomb (1 kcal = $4.18 \cdot 10^{10}$ ergs)? If the energy of these ten bombs could be used for propulsion to substitute for gasoline combustion, how many gallons of gasoline would not have to be converted into smog? (One gallon of gasoline releases about $3 \cdot 10^4$ kcal during combustion.)

8. The intense red light from a powerful ruby laser (wavelength = 6940 Å) will not dissociate Cl_2 molecules into atoms (the bond dissociation energy of Cl_2 is 57 kcal/mole). When this laser light passes through a perfect crystal of potassium dihydrogen phosphate, KH_2PO_4, some of it emerges as blue light of exactly double the frequency (wavelength = 3470 Å). This blue light is much weaker in intensity, but it *does* dissociate Cl_2 into atoms. Explain with the aid of $E = h\nu$.

9. Potassium metal can be used as the active surface in a photodiode because electrons are relatively easily removed from a potassium surface: The energy needed is 51.7 kcal per mole of electrons removed. What is the longest wavelength light (lowest frequency) with quanta of sufficient energy to eject electrons from a potassium photodiode surface?

10. The pressure on a surface equals the momentum transferred to the surface per second. If one square centimeter of a spacecraft passing the planet Mercury receives and absorbs about 10^{18} quanta per second from the sun and if the average wavelength of these quanta is 2500 Å (ultraviolet light), calculate the "light pressure" in torr on that square centimeter (one atm pressure $= 1.01 \cdot 10^6$ dynes/cm^2 $= 1.01 \cdot 10^6$ gm cm/sec^2).

11. There is a theorem in classical physics, called the Virial Theorem, that simply relates the kinetic energy $\bar{T}$ of an electron in an atom to its potential energy $\bar{V}$ and to its total energy $\bar{E}$: $\bar{E} = -\bar{T} = +\frac{1}{2}\bar{V}$. According to this theorem, if it takes 313.6 kcal/mole to remove an electron from a hydrogen atom, then the classical kinetic energy of the electron in the atom is also 313.6 kcal/mole.
(a) Calculate the velocity such an electron would possess (1 kcal/mole $= 6.95 \cdot 10^{-14}$ erg/molecule).
(b) Calculate the wavelength of such an electron.
(c) Calculate the ratio of the wavelength to the apparent size of a hydrogen atom, 0.79 Å. (This is one way to show that a wave description of the atom is needed.)

12. A 160-pound astronaut on his way to Mars is buzzing along at about 10,000 miles per hour.
(a) Express his weight in grams (one pound $=$ 454 grams) and his velocity in cm/sec (one mile/hr $=$ 44.7 cm/sec) so that you can calculate the astronaut's wavelength. (This is one way to show that a wave description of an astronaut is not needed.)
(b) Suppose the astronaut detects outside his spacecraft a proton ($m = 1.67 \cdot 10^{-24}$ g) from the "solar wind" moving at exactly the spacecraft velocity. What is the proton's wavelength in Å? (1 Å $= 10^{-8}$ cm)

13. How fast is an electron moving if it has a wavelength equal to the distance it travels per second?

14. The light from the sun includes extremely sharp "dark lines" at certain wavelengths superimposed on a bright continuum at all other wavelengths. One of these "dark lines," discovered by J. von Fraunhofer around 1825, occurs in the orange range and another in the blue. Fraunhofer measured their wavelengths to be 6562 Å and 4860 Å, respectively. With the aid of Figure 12-7, show that these are spectral lines from the hydrogen atom spectrum (they are called the Hα and Hβ Fraunhofer lines). Such lines provided us with the first clues to the chemical composition of the sun.

15. How much energy is required to remove an electron from a hydrogen atom in the 4s state (in kcal/mole)? In the 4p state? In the 4d state? In the 4f state?

16. What is the average radius of an electron in a hydrogen atom in the 4s state? In the 4p state? In the 4d state? In the 4f state?

17. Which answers are correct conclusions to the following sentence? "Energy must be added to a hydrogen atom in the 2s state in order to
(a) increase the average radius."
(b) increase the quantum number n."
(c) increase the quantum number ℓ."
(d) increase the quantum number m."

(e) increase the number of angular nodal surfaces in its probability distribution.''
(f) increase the total number of nodal surfaces in its probability distribution.''
(g) remove the electron from the atom.''

18. Calculate the energy required to remove an electron from He$^+$ ion in the 2s state. Compare to the energy required to remove an electron from H in the 1s state.

19. Calculate the average radius of an electron in He$^+$ ion in the 2s state. Compare to the average radius of an electron in H atom in the 1s state.

thirteen many-electron atoms

The hydrogen atom is simple because it contains only one electron—there is but one electrostatic attraction contributing to the potential energy. This simplicity is lost even in the next element, the helium atom. With two electrons, this atom has a potential energy made up of three contributions: Each electron is attracted to the nucleus and the electrons are repelled by each other. Immediately the mathematical difficulty increases to the point at which laborious approximation methods are required. The situation is magnified manyfold if we delve further into the Periodic Table. Just consider the oxygen atom which, with its eight electrons, has eight electron–nucleus attractive terms and *twenty-eight* separate electron–electron repulsive terms. No wonder it takes a digital computer to keep track of such calculations!

Yet, in Chapter Two we learned much about the structure of an atom just by looking at ionization energies. The quantities E_m/m seemed to divide electrons into types, characterized by integers. We called these integers "quantum numbers," the same term used in Chapter Twelve to describe the energy levels of the hydrogen atom. The usage suggests that our naive examination of E_m/m revealed some quantum mechanical aspects of the many-electron atom. Perhaps we can understand these now in terms of corresponding aspects of the much simpler hydrogen atom.

Nature is kind to us here. Astonishingly, the energy levels of the many-electron atoms are obviously related to those of hydrogen. The connection is so clear that chemists use the hydrogen atom quantum number designations, 1s, 2s, 2p, . . . , for atoms with many electrons. We have already "discovered" the principal quantum number in Figure 2-6 by plotting E_m/m. With such a head start on the energy level diagram, disturbances due to electron–electron repulsions can be identified and understood. Then we can elaborate the structure of many-electron atoms under the guidance of our quantum mechanical description of the one-electron atom.

13-1 The energy levels of many-electron atoms

Atoms in an electrical discharge emit light, and the colors seen reveal that atom's energy level scheme. Like the hydrogen atom, many-electron atoms also emit line spectra—particular energies only are observed. A quantum mechanical model is needed. However, no atom with two or more electrons has energy level spacings as simply related as the integer-connected levels of the H atom. The rather complicated frequency patterns are understandable, though, with the aid of the H atom quantum numbers when electron–electron repulsions are considered. How this works out can be seen by looking at the energy level diagrams of the next two elements, helium and lithium.

(a) HELIUM ATOM

A naive way to approach the two-electron helium atom would be to assume that somehow we could turn off the electron–electron repulsion. Then each electron would move around the nucleus oblivious to the presence of the other. Each electron would occupy the 1s orbital of the He$^+$ ion, the lowest energy level in Figure 12-17. Each electron would be bound by 1254 kcal, as is the one electron of He$^+$. Each electron would occupy an orbital with an average radius of 0.39 Å.

With this start, let's turn up the knob on the electron–electron repulsion. How will the movement of the electrons and their energy levels change? First, their repulsion will tend to keep them apart. If one electron is on the north side of the nucleus, the other will tend to be on the south side. *Their motions will tend to correlate.* Second, the energy levels will change because the electrons repel each other. How much? Well, we can make a very rough guess in a simple way. Suppose the two electrons were each to occupy a 1s orbital (average radius 0.39 Å) and to stay, by mutual agreement, on opposite sides of the nucleus. They would then be, on the average, two times 0.39 Å apart. The repulsion energy of two electrons 0.78 Å apart is just 420 kcal. This crude estimate tells us that if two electrons occupy the He 1s orbital, it will be easier to remove one of them because of the electron–electron repulsion. How much easier? Well, something like 420 kcal easier. The ionization energy might be near 1254 − 420 ≅ 800 kcal.

In fact, the ionization energy of a neutral helium atom is not 1254 kcal (as it would be if there were no electron–electron repulsion), but 567 kcal. As our simple model predicted, electron repulsions are quite substantial—in fact, our naive model underestimated their effect by 60 percent.

Suppose we apply the "turned-off electron" model to a helium atom that has one electron in a 1s orbital and the other in a 2s orbital, He(1s,2s). Now our naive model predicts average radii of 0.39 Å and 1.59 Å for these two electrons (see Exercise 12-8). The 2s electron will be easier to remove than the 1s electron; the 2s electron energy, without electron repulsions, would be $(-313.6) \cdot Z^2/n^2 = -313.6 \cdot 4/4 = -313.6$ kcal.

As before, if we assume perfect correlation of electron positions, we can estimate the effect of electron repulsions. Perfectly correlated, the electrons would remain on opposite sides of the nucleus, so they would be separated by 0.39 + 1.59 = 1.98 Å. The repulsion energy would be 168 kcal at this separation. With this value, we can improve our estimate of the 2s electron energy. Instead of -313.6 kcal, it will be $-313.6 + 168 = -146$ kcal.

Contrast this result for He(1s,2s) to the same calculation for He(1s,2p). In the "turned-off electron" model, the electrons occupy orbitals of average radii 0.39 Å and 1.32 Å; their energies are -1254 and -313.6 kcal. But when we try to estimate the electron repulsion effect, assuming perfect correlation, the electrons have an average separation of $0.39 + 1.32 = 1.71$ Å. This is a smaller separation than obtained for He(1s,2s) and the repulsion energy is correspondingly larger, 194 kcal. Our improved estimate of the 2p electron energy is $-313.6 + 194 = -120$ kcal. This means that because of electron repulsions, the He(1s,2s) and He(1s,2p) ionization energies are not the same! It requires about 146 kcal to remove the 2s electron from He(1s,2s), but only 120 kcal to remove the 2p electron from He(1s,2p). The energies in this two-electron atom depend upon ℓ as well as upon n!

Figure 13-1 shows the experimental facts. With two electrons in the helium 1s orbital, He(1s,1s) has an ionization energy of 567 kcal. If one of

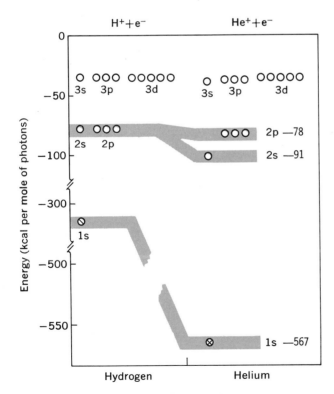

Figure 13-1   *Energy level diagram for hydrogen and helium atoms. (Notice breaks in the vertical scale.)*

these electrons is excited to the 2s orbital, He(1s,2s) has an ionization energy of 91 kcal. If an electron is excited to the 2p orbital, He(1s,2p) has an ionization energy of 78 kcal. Thus, we experimentally corroborate the conclusion that the 2p electron is easier to remove than the 2s electron, 13 kcal easier. Our "perfect correlation" model, which approximated this difference, $146 - 120 = 26$ kcal, tells us why. *Electron repulsions are smaller in He(1s,2s) than in He(1s,2p)*.

The arithmetic of our model shows how this comes about. The average radius of a 2s orbital is larger than that of a 2p orbital of the same energy. A 2s orbital can have a larger *average* radius than a 2p orbital of the same energy because *some of the time* the 2s electron penetrates very close to the nucleus. A 2p orbital has a nodal surface through the nucleus, so it is seldom found in that neighborhood. Its *average* radius must be smaller to make up for the exclusion of this energetically favorable region. The smaller average radius of the 2p orbital causes the 1s–2p electron–electron repulsion to be larger than the 1s–2s electron–electron repulsion.

The most important conclusion of this discussion is that the energy levels of the two-electron helium atom depend upon ℓ as well as n. This splitting of the energy levels with the same principal quantum number is present in all many-electron atoms, not only between s and p orbitals, but even more so between p and d orbitals. It arises because of electron repulsions, which are ℓ-dependent because angular nodal surfaces keep the electron away from the region very near the nucleus. The resulting ℓ-dependent splitting of energy levels has enormous significance: It shapes the Periodic Table.

(b) LITHIUM ATOM

If we returned to our naive model, the "turned-off electron" repulsion model, three electrons might try to crowd into the lithium 1s orbital. Each would be bound with an energy of 2822 kcal (see Fig. 12-17) and would have an average radius of 0.26 Å. There would, of course, be very large electron–electron repulsions with three electrons in the same neighborhood. This situation need not be pursued, however, for all of our experimental evidence indicates that *no orbital ever accommodates more than two electrons*. This is an empirical result whose explanation has *not* been found to lie in electron–electron electrostatic repulsion. It is called the *Pauli Principle* and it is the key to the structure of the Periodic Table. We will see its impact immediately when we apply this rule to the neutral lithium atom.

According to the Pauli Principle, if two electrons already occupy the 1s orbital, a third cannot be accommodated, so it must enter some higher energy state. The lowest one available is the 2s state. We will designate this state as we did for He, listing the orbital occupancies of the electrons: Li(1s²2s). The notation indicates that two electrons are in the 1s orbital (1s²) and one electron is in the 2s orbital (1s²2s). How tightly is the third

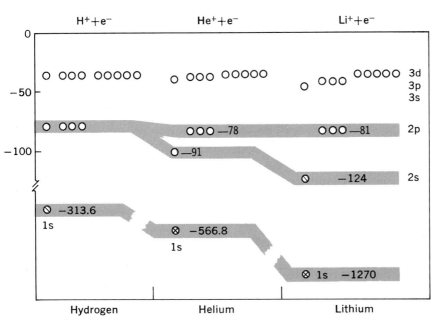

Figure 13-2   *Energy level diagram for hydrogen, helium and lithium atoms. (Notice break in the vertical scale.)*

electron bound in this case? Figure 13-2 shows that the (1s²2s) state has an ionization energy of 124 kcal/mole.

$$\text{Li}(1s^2 2s) \longrightarrow \text{Li}^+(1s^2) + e^- \qquad E_{\text{ion'n}} = 124 \text{ kcal/mole} \qquad (13\text{-}1)$$

In contrast, the next higher state of the neutral lithium atom, Li(1s²2p) has an ionization energy of only 81 kcal/mole.

$$\text{Li}(1s^2 2p) \longrightarrow \text{Li}^+(1s^2) + e^- \qquad E_{\text{ion'n}} = 81 \text{ kcal/mole} \qquad (13\text{-}2)$$

Just as in the helium atom, the 2s and 2p states no longer have the same energies. *Electron repulsion splits energy levels of the same principal quantum number.*

It is informative to investigate in an approximate way the significance of these ionization energies. Suppose we were to construct a one-electron atom with ionization energy of a 2s electron equal to 124 kcal/mole. Equation (12-11) tells us the magnitude of the nuclear charge Z^* we would need.

$$E = -313.6 \frac{Z^{*2}}{n^2} \qquad (13\text{-}3)$$

so, $E = -124 \text{ kcal} = -313.6 \cdot \dfrac{Z^{*2}}{2^2}$

$Z^* = 1.26$

Furthermore, with this effective nuclear charge, the average orbital radius, according to equation (12-12), would be

$$\bar{r} = 0.529 \cdot \frac{n^2}{Z^*} \left\{ \frac{3}{2} - \frac{\ell(\ell + 1)}{2n^2} \right\} \tag{13-4}$$

$$\bar{r} = 0.529 \, \frac{2^2}{1.26} \left\{ \frac{3}{2} - 0 \right\} = 2.52 \, \text{Å}$$

This gives us a simple model of the atom that incorporates the effects of electron repulsions. We see that the outermost electron acts as though the nuclear charge is much smaller than its actual magnitude. For lithium, the 2s electron "feels" the $+3$ nucleus through the "shield" provided by the two tightly held 1s electrons. The 2s electron feels an effective nuclear charge of 1.26.

We can make the same argument for the lithium state Li(1s^{2}2p). The 2p electron is more easily removed—its ionization energy is only 81 kcal. Calculating effective nuclear charge, this time we obtain $Z^* = 1.02$. The average radius of such an electron is calculated as follows:

$$\bar{r} = 0.529 \cdot \frac{2^2}{1.02} \left\{ \frac{3}{2} - \frac{1 \cdot (2)}{2 \cdot 2^2} \right\}$$

$$\bar{r} = 2.59 \, \text{Å}$$

The 2p electron feels a lower effective nuclear charge than the 2s electron, despite the fact that the average radius seems to be about the same. The explanation of this difference is evident in the orbital pictures in Figure 12-13. The 2s orbital penetrates right down to the nucleus, whereas the 2p orbital has a nodal plane at the nucleus. Thus, the 2s electron has the opportunity to "penetrate" the electron shield provided by the two 1s electrons in the lithium atom. Evidently the 2p electron does this much less effectively. Because of its nodal pattern, the probability of finding the electron at the nucleus is zero; hence we find the 2p electron barely penetrates the 1s electron shield—it feels effectively only the difference between the $+3$ charge of the nucleus and the -2 charge of the two 1s electrons. Again we see that electron repulsions are larger for the 2p than for the 2s state.

This penetration effect explains the 2s–2p splitting effect seen in the energy levels of both helium and lithium. As remarked earlier, this same effect persists and is even larger in the higher states, such as 3s, 3p and 3d. In fact, we shall see that the combination of the Pauli Principle (two electrons per orbital) and the splitting of s, p, and d orbitals enables us to understand the regularities of the entire Periodic Table.

Exercise 13-1. Extend Figure 13-2 to show the orbital occupancy of beryllium and boron. Compare the implied grouping of electrons in beryllium to the grouping inferred in Section 2-3(b) from E_m/m. Make the same comparison for boron using E_m/m as calculated from the measured ionization energies $E_1 = 191$, $E_2 = 580$, $E_3 = 875$, $E_4 = 5980$, $E_5 = 7843$ kcal/mole.

Exercise 13-2. The lithium orbital occupancy pictured in Figure 13-2 is symbolized Li($1s^2 2s$). In the lowest state, the beryllium orbital occupancy pictured in Exercise 13-1 should be symbolized Be($1s^2 2s^2$). How would that of boron be symbolized?

Exercise 13-3. In the simple electrostatic model used in Section 2-3(b), ionization energy was presumed to vary inversely with average electron radius. Use expressions (13-3) and (13-4) to calculate Z^* and $\bar{r}$ for the first electron removed from Be ($E_1 = 215$ kcal) and also for the first electron removed from B ($E_1 = 191$ kcal). What part of expression (13-4) causes $\bar{r}$ for boron to be smaller than $\bar{r}$ for beryllium despite the fact that the ionization energy of boron is less than that for beryllium?

(c) ON TO NEON

We now have the guidelines needed to look at the trends across an entire row of the Periodic Table from lithium to neon. Figure 13-3 shows the important orbitals, the 2s and 2p orbitals, for these eight atoms. For each atom the 1s orbital is shown to be fully occupied (two electrons only, according to the Pauli Principle!), but its energy is shown on a compressed scale. Also, the higher orbitals 3s, 3p, 3d, 4s, . . . are not shown; they are vacant and need not concern us now.

The orbitals that *are* shown suffice to hold all the electrons in excess of the two in the tightly bound 1s orbital. The orbital occupancy, as we proceed across the Periodic Table, is simple to understand, given the Pauli Principle. Moving from lithium to beryllium, the second electron can be accommodated by the 2s orbital, so the 2p orbitals need not be used. The ionization energy is, of course, higher than that of lithium, because beryllium has a higher nuclear charge. Proceeding to boron, however, the fifth electron is denied occupancy in either the 1s orbital or the 2s orbital. Consequently the lowest energy state for boron is B($1s^2 2s^2 2p^1$), with one electron in the higher-energy 2p orbital. Hence, despite the higher nuclear charge of boron, its ionization energy is lower than that of beryllium. The reason is obvious from Figure 13-3—the nuclear charge effect is more than offset by the fact that the most easily removed electron is in a 2p orbital instead of a 2s orbital.

Proceeding now to carbon, the six electrons will be distributed as follows. Two electrons occupy the 1s orbital, two occupy the 2s orbital, and the remaining two electrons make themselves comfortable among the three 2p orbitals as they wish. Two of them *could* occupy the $2p_z$ orbital, as far as Pauli is concerned. However, that would place them in the same region of

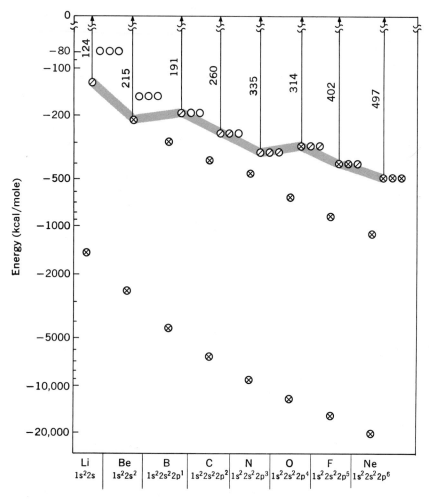

Figure 13-3   Energy levels, orbital occupancy and ionization energies for first-row elements. (Notice scale compression with rising energy.)

space, concentrated along the x axis, resulting in high electron–electron repulsion. A better situation results if one electron goes into the $2p_x$ orbital and the second enters either the $2p_y$ or $2p_z$ orbital. That keeps the two electrons in different neighborhoods and reduces electron repulsion. So the lowest energy state of carbon atom is $C(1s^2 2s^2 2p_x^1 2p_y^1)$. Its ionization energy exceeds that of boron because the nuclear charge of carbon is higher than that of boron, and for both atoms a 2p electron must be removed.

Nitrogen is more of the same. By the Pauli Principle, three electrons are left for the 2p orbitals and, to keep apart, they occupy all three, one in each of $2p_x$, $2p_y$ and $2p_z$. The ionization energy increases over carbon because of nuclear charge.

With oxygen, however, the last electron no longer finds a vacant 2p orbital in which to lodge. It must enter a 2p orbital already containing an electron. Hence electron repulsion shows up and, in fact, dominates the increase in the nuclear charge. The ionization energy of oxygen is less than that of nitrogen.

Fluorine and neon introduce no new ideas. Their added electrons must, in turn, enter half-occupied orbitals, but the increasing nuclear charge effect continues to increase the ionization energy.

We have added, now, to our understanding of the trends in ionization energies across the first two rows of the Periodic Table, as pictured in Figure 2-3 and discussed in Section 2-2(a). Going from H to He, the ionization energy increases very markedly because of the doubling of the nuclear charge. Lithium then drops precipitously because the Pauli Principle forces the third electron up into a high-energy, 2s orbital. Across the first row, the next seven electrons all enter 2s–2p orbitals. There is a general rise in ionization energy due to nuclear charge, but with a couple of minor jogs due, first to the 2s–2p energy difference, and then to the first double occupancy of the 2p orbitals with the higher electron repulsion.

Figure 2-3 goes on to show that going from neon to sodium results in another precipitous drop, just like that of lithium. The reason is obvious. Once again, the filling of the 2s and 2p orbitals forces the next electron up into a higher-energy orbital, the 3s orbital.

(d) EFFECTIVE NUCLEAR CHARGE ACROSS THE FIRST ROW

It is informative to use our one-electron approximation to examine the effective nuclear charge felt by the most easily removed electron and the effective orbital radius for that electron. For each element we can calculate Z^*, the nuclear charge that would give rise to the observed ionization energy for a 2s or 2p electron $(n = 2)$, using our modified form of (12-11).

$$E = -E_{\text{ion'n}} = -313.6 \frac{Z^{*2}}{n^2} \tag{13-5}$$

Then with this Z^* we can calculate the average radius that one-electron orbital would have, using the modified form of (12-12), with $\ell = 0$ for the 2s orbital and $\ell = 1$ for the 2p.

$$\bar{r} = \frac{0.529 \cdot n^2}{Z^*} \left\{ \frac{3}{2} - \frac{\ell(\ell + 1)}{2n^2} \right\} \tag{13-6}$$

The results are summarized in Table 13-1. We see that the extent to which the most easily removed electron feels the nuclear charge increases across

Table 13-1 Effective Nuclear Charge and Orbital Radius in the One-Electron Approximation

	Z^*	$\bar{r}(\text{Å})$
H	1.00	0.79
He	1.34	0.59
Li	1.26	2.52
Be	1.66	1.92
B	1.56	1.70
C	1.82	1.45
N	2.07	1.28
O	2.00	1.32
F	2.26	1.17
Ne	2.52	1.05
Na	1.84	3.88
Mg	2.25	3.18

a row of the Periodic Table. This increasing attraction causes the orbital to shrink—the average radius decreases, just as pictured in Figure 2-7. But now we can draw a more detailed picture of the electron distribution in the first-row atoms. including nodal properties as well as size (see Fig. 13-4). Much of chemistry is explainable in terms of the two trends displayed in Table 13-1: the rise in ionization energy and decrease in orbital size across a row of the Periodic Table.

(e) ELECTRON SPIN AND THE PAULI PRINCIPLE

Careful measurements show that the electron has magnetic as well as electrostatic properties! If a beam of one-electron atoms is passed through an uneven magnetic field, it is split into two streams. The amount of the deflection indicates that each atomic electron acts like a small magnet. One of the atomic streams bends in a fashion appropriate to being attracted by the imposed field and the other appropriate to an equal and opposite repul-

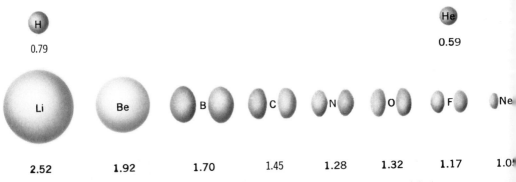

Figure 13-4 Average radii of atoms: one-electron, effective-charge model.

sion. The evidence is that each tiny magnet is obliged (by Nature) to make a yes-or-no choice—it must be aligned either completely *with* the imposed field or *against* it. *The magnetic interaction is quantized and there are only two possible states.*

From where does this electron magnetic field come? On a macroscopic scale, a magnetic field is generated when electric charge moves in a circular path. An instinctive interpretation, then, is that there must be some way in which the charge within an electron moves in a circular path. *Perhaps the electron is spinning!*

This explanation is the one now accepted. This simple model permits a calculation of the magnetic field that would result from a given spin angular momentum. However, such a calculation gives a result that is double the observed value. To interpret this, we must recall that, as the electron spins, the distribution of the mass will determine the angular momentum and the distribution of charge will determine the magnetic field. Now the discrepancy can be given an intuitive meaning. If the mass and charge are not distributed identically, the classical calculation will predict the wrong relationship between angular momentum and magnetic field—as it does.

We needn't dwell on this dilemma. It suffices to say that the electron spin (or magnetic field) is said to be able to take on only two values, $+\frac{1}{2}$ and $-\frac{1}{2}$. The $+$ and $-$ signs denote the empirical observation that is depicted in Figure 13-5; the electron must be *for* or *against* the magnetic field, nothing in between. The $\frac{1}{2}$ factor is the momentum/field discrepancy factor between the simple calculation and the observed result. The importance of all this to chemistry is that magnetic measurements show a relation between these spins and the Pauli Principle. Such magnetic experiments coupled with spectroscopic studies of orbital occupancy show that the Pauli Principle can now be stated in a more sophisticated form: Only two electrons can occupy a single orbital and *those two electrons must have opposite spins.*

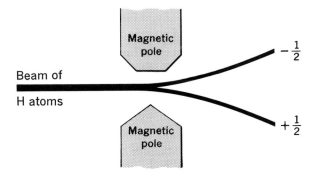

Figure 13-5 Electron spin revealed in an uneven magnetic field produced by pole faces of different shapes.

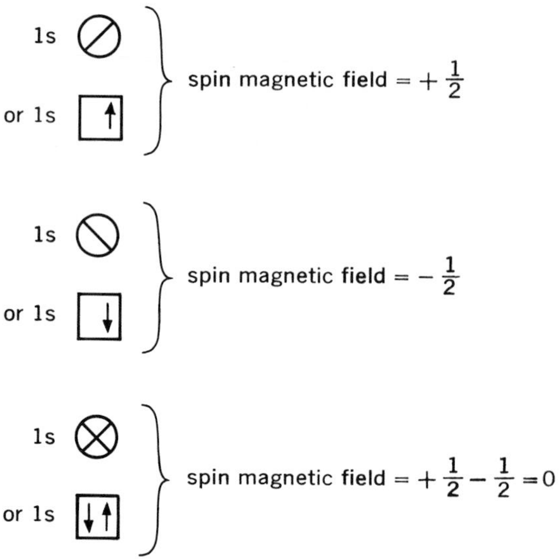

1s $\oslash$

or 1s $\boxed{\uparrow}$ $\Bigg\}$ spin magnetic field $= +\frac{1}{2}$

1s $\obslash$

or 1s $\boxed{\downarrow}$ $\Bigg\}$ spin magnetic field $= -\frac{1}{2}$

1s $\otimes$

or 1s $\boxed{\downarrow\uparrow}$ $\Bigg\}$ spin magnetic field $= +\frac{1}{2} - \frac{1}{2} = 0$

Figure 13-6 *Representations of electron spin in filled and partially filled orbitals.*

This idea is simply represented by the slant of the slash across a pigeon-hole representation of an orbital. When one electron is placed in the orbital, shown by a slash from left to right, $\oslash$, a second can be added, shown by a slash from right to left, and the resulting symbol $\otimes$ designates a filled orbital. Alternate representations use arrows, pointed up and down. It is immaterial which pictorial representation is used (see Fig. 13-6).

Unfortunately, the origin of this rule of Nature is essentially as empirical as the Pauli Principle itself. We observe that the orbital occupancy and the magnetic properties of atoms (and molecules) are all consistent with this rule: Two electrons, at most, are allowed per orbital and they must have opposite spin. The explanation is apparently *not* found in the magnetic attraction that occurs between two adjacent magnets oppositely oriented and the repulsion that results from parallel orientations. The explanations now offered delve into relativistic effects, and these sophisticated arguments shed no light on chemistry that isn't already brought there by the Pauli Principle itself.

13-2 Ionization energy and valence electrons

There is a strong connection between the ease of removal of an electron from an atom and the chemistry of that atom. The electrons that are most easily removed are most likely to become involved in chemical bonding. Let's look at the successive ionization energies of the atoms lithium through neon—the first row of the Periodic Table.

(a) LITHIUM AND BERYLLIUM

It is entirely reasonable that it should be more difficult to remove a second electron from an atom than the first. The first is pulled away from a neutral atom, whereas the second (and successive ones) must be removed from a positively charged ion. Even with this expectation, there are some crucially important variations from atom to atom. Compare, for example, the ionization of gaseous lithium and beryllium atoms.

$$Li \longrightarrow Li^+ + e^- \qquad E_1 = 124 \ \ kcal/mole$$
$$Li^+ \longrightarrow Li^{+2} + e^- \qquad E_2 = 1744 \qquad ``$$
$$Li^{+2} \longrightarrow Li^{+3} + e^- \qquad E_3 = 2824 \qquad ``$$

$$Be \longrightarrow Be^+ + e^- \qquad E_1 = 215 \qquad ``$$
$$Be^+ \longrightarrow Be^{+2} + e^- \qquad E_2 = 420 \qquad ``$$
$$Be^{+2} \longrightarrow Be^{+3} + e^- \qquad E_3 = 3548 \qquad ``$$
$$Be^{+3} \longrightarrow Be^{+4} + e^- \qquad E_4 = 5020 \qquad ``$$

It is more difficult to remove one electron from beryllium than from lithium—as noted in Figure 13-3. The nuclear charge on the beryllium nucleus is larger than that of lithium, so this should be so. The surprise is that it is *easier* to remove a second electron from beryllium ($E_2 = 420$ kcal) than from lithium ($E_2 = 1744$ kcal). The situation then reverses again for the third electron!

There is, however, nothing mysterious about these ionization energies when the orbital occupancies are taken into account. Figure 13-7 shows the ionization processes in terms of energy level diagrams. The huge energy jump for lithium between E_1 and E_2 is simply determined by the need to pull the second electron from the tightly held 1s orbital. In beryllium, the second electron must occupy the 2s orbital, so E_2 for beryllium exceeds E_1 only because of the extra electron repulsion in the neutral beryllium atom with its two 2s electrons. The actual nuclear charge is the same and so is the orbital. The first electron leaves from a neutral atom in which four electrons repel each other. The second electron leaves from a positively charged atom in which only three electrons repel each other.

However, E_3 is enormous for beryllium, since now an electron must be removed from the 1s orbital. The magnitude of E_3 is enhanced, and E_4 still more, by the increasing positive charge of the ion (decreased electron repulsion with constant nuclear charge), but the reason for the large jump between E_2 and E_3 is due to the need to reach down to a 1s electron for E_3.

Exercise 13-4. Use the ionization energies, E_m, for boron (given in Exercise 13-1) to picture its ionization energy and orbital occupancy as done in

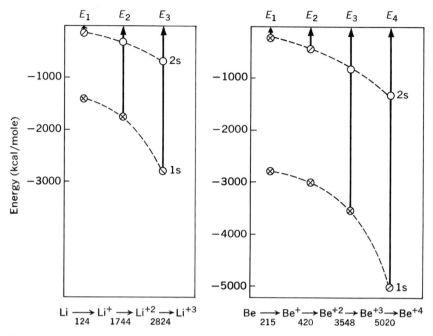

Figure 13-7 Orbital occupancy and ionization energy for Li and Be.

Figure 13-7 for Li and Be. Use E_m/m to establish the energy level spacing for the neutral atom and draw a smooth curve for each orbital.

It is informative to apply the one-electron, effective-charge model to the second and third most easily removed electrons of Li and Be. From the values of E_2, the one-electron values of Z^* and $\bar{r}$ are contrasted in Table 13-2 with the corresponding values for the first electron of each atom. There are two interesting points made in this table. First, note again the very much smaller average radius associated with the 1s electrons, both in Li^+ and Be^{+2}. This accounts for their effective shielding of the nuclear charge insofar as the electrons in the 2s orbital are concerned. The other contrast is the value of Z^* for the first 2s and the second 2s electron ionizations of beryllium. Here we see that the second 2s electron "feels" an effective nuclear charge $2.31 - 1.66 = 0.65$ higher than the first. Then the third electron, an electron in the 1s orbital, feels an effective nuclear charge $3.36 - 2.31 = 1.05$ higher than the second. In turn, the fourth electron, again a 1s electron, feels an effective nuclear charge $4.00 - 3.36 = 0.64$ higher than the third. We see that as successive s electrons are removed, the effective nuclear charge rises by about two thirds of a proton charge if the electrons have the same principal quantum number. If the principal quantum number changes, then Z^* rises by a full proton charge. *Inner electrons shield the nucleus more fully than do electrons in the same orbital.*

Table 13-2 Effective Nuclear Charge and Orbital Radii Contrasted for First and Second Ionizations of Li and Be

	Electron Removed	Z^*		$\bar{r}(\text{Å})$
Li	2s	1.26		2.52
			1.10	
Li$^+$	1s	2.36		0.34
			0.64	
Li^{+2}	1s	3.00		0.26
Be	2s	1.66		1.92
			0.65	
Be$^+$	2s	2.31		1.37
			1.05	
Be^{+2}	1s	3.36		0.24
			0.64	
Be^{+3}	1s	4.00		0.20

(b) THE FIRST ROW OF THE PERIODIC TABLE

The factors at work in Li and Be are visible in the successive ionization energies of all the atoms in the first row of the Periodic Table. Table 13-3 and Figure 13-8 show these atoms and their ionization energies related according to orbital occupancy.

It will suffice to examine the successive ionizations of fluorine to see the significance of the table. Of course, the neutral fluorine atom orbital occupancy is $F(1s^2 2s^2 2p_x^2 2p_y^2 2p_z^1)$. Hence the first two electrons are removed from doubly occupied 2p orbitals, requiring 402 and 807 kcal/mole, respectively. Then the third electron, removed from a half-occupied orbital,

Table 13-3 Ionization Energies of the First-Row Elements (kcal/mole)

	Li	Be	B	C	N	O	F	Ne
E_1	124	215	191	260	335	314	402	497
E_2	1744*	402	579	562	683	809	807	945
E_3	2824	3548	875	1102	1098	1271	1445	1463
E_4		5020	5980	1487	1785	1782	2010	2235
E_5			7845	9040	2257	2622	2634	2916
E_6				11,200	12,700	3180	3620	3620
E_7					15,300	17,000	4270	4470
E_8						20,000	21,900	5490
E_9							25,400	27,400
E_{10}								31,100

*A sharp jump in ionization energies occurs on removal of the first 1s electron (shaded portion).

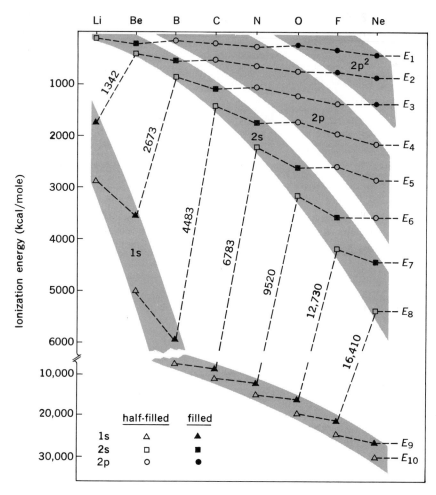

Figure 13-8   Ionization energies and valence electrons for the first-row atoms. The four shaded bands refer to removal of an electron from the 1s orbital ($\triangle$), the valence 2s orbital ($\square$), and a valence 2p orbital ($\bigcirc$). Open symbols refer to singly occupied orbitals and solid symbols to doubly occupied orbitals. (Notice the large energy gap between the valence orbitals and the 1s orbitals.)

requires 1445 kcal/mole, 638 kcal more than the second electron. The next big jump comes between E_5 and E_6, since $E_5 = 2634$ kcal removes the last 2p electron and $E_6 = 3620$ kcal removes one of the 2s electrons at the cost of an extra 1000 kcal.

The biggest point made by Figure 13-8 and Table 13-3 is the huge gap between the energy needed to remove the second 2s electron and the first 1s electron. For fluorine, these energies are 4270 and 21,900 kcal/mole, respectively. Plainly the first seven electrons, requiring from 402 to 4270 kcal for removal, are qualitatively different from the last two electrons, requiring 21,900 and 25,400 kcal. We see a gap like this for every atom. These energy differences account for the chemistry of each atom and the trend across a row of the Periodic Table. For the second row, the 2s and 2p orbitals are comparable in energy—their electron occupancy determines the chemistry of the elements lithium to neon. Hence these orbitals are given a special name.

—*All the uppermost occupied or partially occupied orbitals of comparable energy are called the* **valence orbitals.**

—*Electrons that occupy the valence orbitals are called* **valence electrons.**

We see that lithium has one valence electron and has valence oribtal capacity for an additional seven electrons. Carbon has four valence electrons and valence orbital capacity for four additional electrons. Fluorine is a sort of mirror image of lithium—F has seven valence electrons and valence orbital capacity for only one more. These are the factors we'll use to explain chemical bonding—the availability of valence electrons and valence orbital vacancies determine the number and strength of the chemical bonds that an atom can form.

(c) ELECTRON AFFINITY

It is found experimentally that some gaseous, negative ions are energetically stable, though extremely reactive. The halogen atoms are the most thoroughly studied; for example, the energy required to remove an electron from a negative fluorine ion is 79.5 kcal/mole.

$$F^-(g) \longrightarrow F(g) + e^-(g) \qquad \Delta H = +79.5 \text{ kcal/mole}$$

Plainly this energy is analogous to an ionization energy, and it might logically be denoted E_0. However, such quantities are usually written in terms of the reverse process, the capture of an electron by a neutral atom. The energy released is called the electron affinity. It is usually given as a positive number, an arbitrary, but widely used exception to normal thermodynamic practice (scientists are fickle, too!).

$$F(g) + e^- \longrightarrow F^-(g) \qquad \Delta H = -79.5 \text{ kcal/mole}$$
$$\text{electron affinity } E = +79.5 \text{ kcal/mole}$$

The reason a fluorine atom can capture an electron is suggested by our one-electron interpretation of the ionization energy. Figure 13-9 shows a plot of the effective nuclear charge felt by the successive valence electrons of fluorine as they are removed one by one. The plot is extrapolated back to "E_0." This extrapolation shows that the most easily removed electron

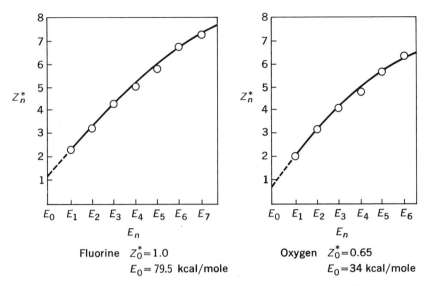

Figure 13-9   Effective nuclear charge for removal of the nth electron (Z_n^*): fluorine and oxygen. When these curves are extrapolated back (dashed lines) a value for the ionization energy of the negative ion (E_0) is obtained.

from the negative ion feels an effective nuclear charge of about 1.0. This shows that even for the negative ion, the nuclear attraction exceeds the electron repulsion. Figure 13-9 also shows a Z^* plot for oxygen. This time, the extrapolated value of Z^* for O^- is only 0.65 and the electron affinity is 34 kcal/mole. We see that electron affinity, like ionization energy, drops off as we move to the left in the Periodic Table.

Electron affinities are extremely difficult to measure; yet they are as important as ionization energies if we try to understand chemical bonding. Figure 13-10 shows, via the Periodic Table, the few that are known.

Exercise 13-5.   Calculate Z^* for the first three electrons removed from neon ($E_1 = 497$, $E_2 = 947$, and $E_3 = 1500$ kcal), and determine Z_0^* and the electron affinity E_0 that neon would have if the Pauli Principle did not prevent neon from accepting another 2p electron. (This plot can be interpreted as evidence that the Pauli Principle does not arise from electron–electron repulsions.)

An atom's electron affinity is important to its bonding because a positive E_0 implies that the energy of the neutral atom is lowered as another electron approaches. If this extra electron happens to be attached to a second atom, the two atoms will be held together—bound—by the energy lowering that accompanies this electron sharing. This is a chemical bond.

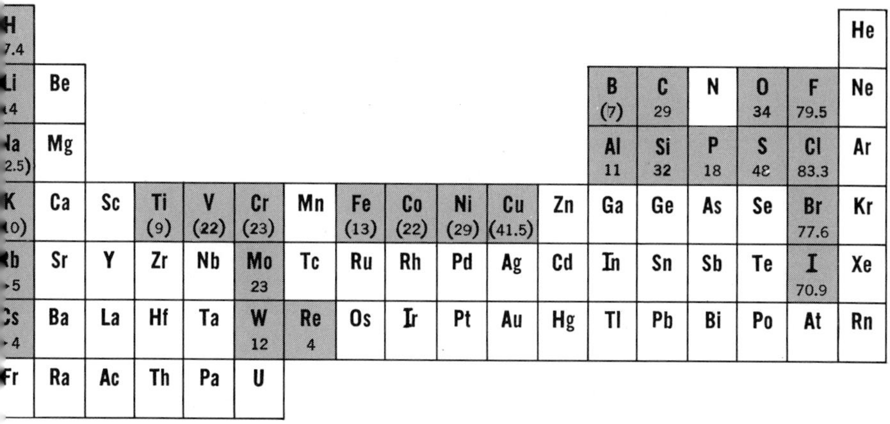

H 7.4																	He
Li _4	Be											B (7)	C 29	N	O 34	F 79.5	Ne
Na 2.5)	Mg											Al 11	Si 32	P 18	S 4ε	Cl 83.3	Ar
K 0)	Ca	Sc	Ti (9)	V (22)	Cr (23)	Mn	Fe (13)	Co (22)	Ni (29)	Cu (41.5)	Zn	Ga	Ge	As	Se	Br 77.6	Kr
Rb 5	Sr	Y	Zr	Nb	Mo 23	Tc	Ru	Rh	Pd	Ag	Cd	In	Sn	Sb	Te	I 70.9	Xe
Cs 4	Ba	La	Hf	Ta	W 12	Re 4	Os	Ir	Pt	Au	Hg	Tl	Pb	Bi	Po	At	Rn
Fr	Ra	Ac	Th	Pa	U												

Figure 13-10   Electron affinities (kcal/mole). (Parenthetical values are calculated, others are experimental.)

(d) THE SECOND AND THIRD ROWS OF THE PERIODIC TABLE

Figure 13-11 shows the empirical grouping of the elements in the Periodic Table. Each vertical column relates elements whose chemistry is similar enough to define a familial relationship. We have been talking about the first row, the eight elements lithium to neon, whose chemistry is fixed by the 2s and 2p orbitals. Now let's see how the second- and third-row chemistries are fixed by ionization energies and orbital occupancies. For a starter, we note that the second row, like the first, encompasses eight elements, sodium through argon, whereas the third row includes eighteen elements, potassium through krypton.

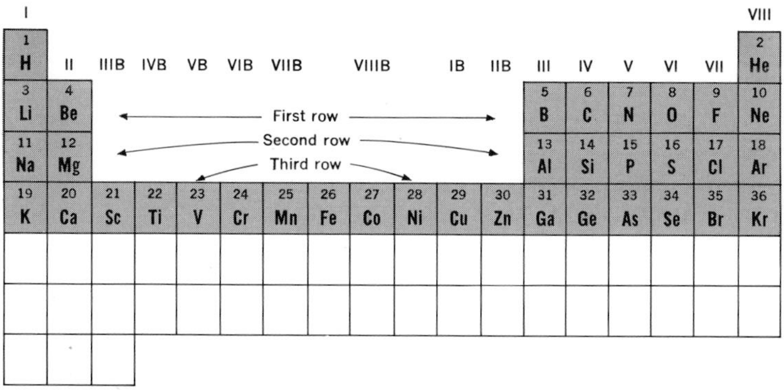

I																	VIII
1 H	II	IIIB	IVB	VB	VIB	VIIB		VIIIB		IB	IIB	III	IV	V	VI	VII	2 He
3 Li	4 Be			←———— First row ————→								5 B	6 C	7 N	8 O	9 F	10 Ne
11 Na	12 Mg			←—— Second row ——→								13 Al	14 Si	15 P	16 S	17 Cl	18 Ar
19 K	20 Ca	21 Sc	22 Ti	23 V	24 Cr	25 Mn	26 Fe	27 Co	28 Ni	29 Cu	30 Zn	31 Ga	32 Ge	33 As	34 Se	35 Br	36 Kr

Figure 13-11   The first three rows of the Periodic Table.

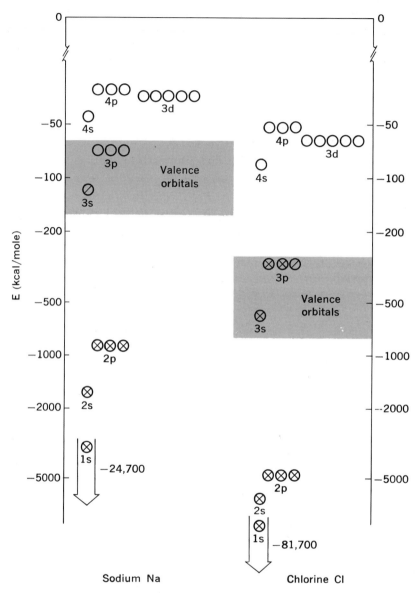

Figure 13-12   Energy levels and orbital occupancy for sodium and chlorine atoms. (Notice vertical scale compression as energy becomes more negative.)

Referring back to the hydrogen atom energy level diagram, Figure 12-15, we would conclude that the second row of the Periodic Table should have eighteen elements. The three sets of orbitals 3s, 3p and 3d have an electron capacity of eighteen. However, the eight-element second row shows that only the 3s and 3p orbitals act as valence orbitals. The reason for this behavior is evident in Figure 13-12. The electron shielding by the inner electrons (the 1s, 2s and 2p electrons) separates the 3s, 3p and 3d orbital energies. The d orbitals are least penetrating, hence they see a much smaller effective nuclear charge than either the 3s or 3p orbital electrons. The effect is so large that the 3d orbitals are raised in energy up to the 4s and 4p orbitals, leaving an energy gap between the 3s–3p orbitals and the 4s–4p–3d cluster.

Exercise 13-6. Assuming that the qualitative spacing of levels shown in Figure 13-12 does not change, draw the orbital occupancies for potassium, calcium, and scandium. Compare these to sodium (Fig. 13-12), magnesium, and aluminum.

The first ionization energies of the third-row elements confirm these energy relationships. Figure 13-13 extends Figure 2-3 to include the third row. In this figure, the orbital from which the first ionization occurs is indicated (see legend). The 4s orbitals fill first (elements potassium and calcium), as is to be expected from Figure 13-12. Next, the d orbitals fill and ionization energies remain reasonably constant all the way from scandium to zinc. The crosses indicate that the 3d orbitals remain higher in energy than the 4s orbitals from scandium to manganese, at which element they are just half filled. Thereafter, succeeding electrons enter 3d orbitals until they are completely filled (elements iron to zinc), but the 3d orbitals drop below the 4s orbitals.

Only after the 4s and 3d orbitals are completely filled at zinc, do valence electrons begin to occupy the 4p orbitals. Thereafter, from gallium to

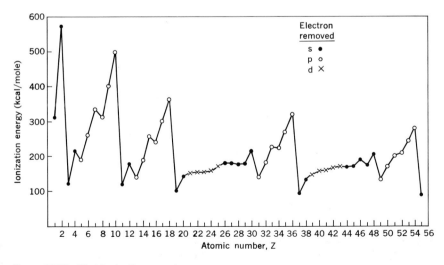

Figure 13-13 First ionization energies.

krypton, the ionization energies rise in the pattern evident in the first and second rows and characteristic of the p orbitals. The ionization energy rises rapidly for the three elements gallium, germanium and arsenic. There is a jog at element selenium because of the first double occupancy of a 4p orbital. Then the ionization energy rises rapidly again for bromine and krypton, completing the row.

Thus we see that the chemistry of the elements, as systematically represented in the Periodic Table, is determined by the energy clustering of orbitals. These orbitals can be directly related to those of the hydrogen atom provided electron repulsions are considered. These electron effects are themselves understandable in terms of the spatial distribution and nodal patterns of the hydrogen atom orbitals. The s orbitals penetrate right to the nucleus, so the s electrons more effectively sense the nuclear charge, despite electron repulsions. The p orbitals and, more so, the d orbitals, have nodal surfaces that pass through the nucleus, so a p or d electron never gets very close to the nucleus. Hence, the electron repulsions shield the nuclear charge more effectively, and p orbital energies are above the s orbitals. In turn, the d orbital energies are above the p orbitals, so much so that the 3d orbitals cluster with the 4s and 4p orbitals, and the 4d orbitals cluster with 5s and 5p, and so on. The explanation for the Periodic Table is taking form before our eyes. Its cornerstone is the quantum mechanical view of the hydrogen atom.

Problems

1.   A "partially turned-off" model for beryllium can be based on the assumption that the two 1s electrons shield perfectly so that the 2s electrons feel a nuclear charge of $Z^* \cong 2.00$, but they do not repel each other.
(a) Calculate the resulting value of E and $\bar{r}$ for the "turned-off" 2s electrons.
(b) Calculate the electron–electron repulsion, $V = +332/r_{12}$ assuming that the electrons perfectly correlate in position, so that $r_{12} = 2\bar{r}$.
(c) Combine (a) and (b) to get a better approximation to E_1. The result obtained in (a) gives directly an approximation to E_2 since there is no 2s–2s repulsion in Be^+. Compare the better estimates of E_1 and E_2 to the experimental values. (Notice that E_1 contains two errors that tend to cancel. The E_2 estimate shows that $Z^* \cong 2.00$ overestimates the effectiveness of the 1s shielding and the helium example shows that perfect correlation underestimates the electron–electron repulsion.)

2.   Calculate Z^* for the neutral argon atom using the one-electron approximation (see Appendix b for ionization energies).

3.   Calculate Z^* for the neutral chlorine atom in the one-electron approximation.

4.   Estimate $\bar{r}$ for the neutral argon atom using Z^* calculated in Problem 2.

5.   Estimate $\bar{r}$ for the neutral chlorine atom using Z^* calculated in Problem 3.

6.   The orbital occupancy (or electron configuration) of a neutral chlorine atom can be written $1s^2 2s^2 2p^6 3s^2 3p^5$. What is the electron configuration of a neutral argon atom?

7. What is the electron configuration of Cl^-? From this point of view, is it more like neutral Cl or neutral Ar?

8. What is the electron configuration of Ar^+? From this point of view, is it more like neutral Cl or neutral Ar?

9. Make a graph of E_1, like that shown in Figure 2-3, for the first 12 elements, but on a compressed energy scale extending from 0 to 1800 kcal. Add to this graph the E_2 ionization energy values for the same elements (see Appendix b).

10. Here are some answers looking for questions. Find questions below to which these answers fit in a causal way (not a *casual* way!), using your plot in Problem 9 to help. List as many correct answers to each question as you can find.

Answers

A. A 1s electron must be removed.
B. Z is higher.
C. A 2p electron is easier to remove than a 2s electron.
D. A 1s orbital is much closer to the nucleus.
E. Z^* is higher.
F. None of the other answers apply.

Questions

(a) The reason there is a general rise in E_1 from Li to Ne (ignoring minor jogs) is because, for Ne, _____.
(b) E_1 for boron is below E_1 for beryllium because, for B, _____.
(c) E_1 for helium is very much larger than E_1 for lithium because, for He, _____.
(d) E_2 for boron is above E_2 for beryllium because, for B, _____.
(e) E_2 for helium is smaller than E_2 for lithium because, for Li, _____.
(f) E_2 for lithium is very much larger than E_2 for beryllium because, for Li, _____.
(g) E_2 for fluorine is below E_2 for oxygen because, for F, _____.

11. Here are some second-row ionization energies:

	Na	Mg	Al	Si	P	S	Cl	Ar	(K)
E_1	119	176	138	188	242	239	300	363	100
E_2	1091	347	434	377	453	540	549	637	734

Explain why:
(a) $E_2(Na)$ exceeds $E_1(Na)$.
(b) $E_2(Mg)$ exceeds $E_1(Mg)$.
(c) $E_2(Al)$ exceeds $E_2(Si)$, but $E_1(Si)$ exceeds $E_1(Al)$.
(d) $E_2(S)$ exceeds $E_2(P)$, but $E_1(P)$ and $E_1(S)$ are almost identical.
(e) $E_2(S)$ and $E_2(Cl)$ are almost identical, but $E_1(Cl)$ exceeds $E_1(S)$.
(f) $E_2(K)$ exceeds $E_2(Ar)$, but $E_1(Ar)$ exceeds $E_1(K)$.

12. Gaseous sodium atoms absorb light quanta with energies, per mole of quanta, of 48.5, 73.6, 83.4, and 86.6 kcal. The electronic configurations of the states so obtained are as follows.

Ground State,	0 kcal	$1s^2$	$2s^2 2p^6$	$3s^1$
	$+48.5$ kcal	$1s^2$	$2s^2 2p^6$	$3p^1$
	$+73.6$ kcal	$1s^2$	$2s^2 2p^6$	$4s^1$
	$+83.4$ kcal	$1s^2$	$2s^2 2p^6$	$3d^1$
	$+86.6$ kcal	$1s^2$	$2s^2 2p^6$	$4p^1$

(a) Calculate the ionization energy for each of these states from $E_1 = 118.5$ kcal for the ground state.

(b) Draw to scale an energy level diagram for these uppermost orbitals, displacing the d orbitals to the right of the p orbitals, and these to the right of the s.
(c) Calculate Z^* for each state.
(d) Calculate $\bar{r}$ for each state.
(e) What is the s–p splitting for $n = 4$? Why is it so much smaller than the s–p splitting for $n = 3$ (48.5 kcal/mole)?

13. Gaseous beryllium atoms absorb quanta with energies, per mole of quanta, of 77.5, 152.3, 170.0, and 180.8 kcal.
(a) Make a table like that in Problem 12 in which the orbital occupancies are specified. (Note: for Be, the 3d state is lower in energy than the 4s state.)
(b) Calculate the ionization energy for each of these states from $E_1 = 214.9$ kcal for the ground state.
(c) Draw to scale an energy level diagram for these uppermost orbitals, displacing the d orbitals to the right of the p orbitals and these to the right of the s.
(d) Calculate Z^* for each state.
(e) Calculate $\bar{r}$ for each state.

14. Calculate Z^* for each of the ionization energies of fluorine given in Appendix b. Now examine the differences between successive values of Z^* and show that shielding by an electron occupying a filled 2p orbital is more effective than by an electron occupying a half-filled 2p orbital.

15. A method for estimating electron affinities alternate to that shown in Figure 13-9 is to extrapolate Z^* values for atoms and ions that contain the same number of electrons as the negative ion of interest. This tends to give smoother plots because orbital occupancy effects are constant.
(a) Calculate Z^* for Ne, Na+, Mg+2, and Al+3 from the appropriate value of E_m (see Appendix b).
(b) Plot Z^* for these elements (as in Fig. 13-9) and extrapolate to Z^* for F−.
(c) Calculate from Z^* (F−) the electron affinity and compare to the experimental value.

16. Repeat Problem 15 for O−, using isoelectronic ions F, Ne+, Na+2, and Mg+3. Plot on the same graph as Problem 15 if possible.

17. At present, there does not seem to be a reliable experimental value for the electron affinity of nitrogen atom. Repeat Problem 15, using isoelectronic ions O, F+, Ne+2, and Na+3. Plot on the same graph as Problems 15 and 16 if possible.

18. Indicate for each statement all of the answers that correctly finish the statement, in reference to the most easily removed electron.

Answers

A. n is higher.	E. n is lower.
B. Z^* is higher.	F. Z^* is lower.
C. Z is higher.	G. Z is lower.
D. ℓ is higher.	H. ℓ is lower.

Statements
(a) Sodium atom is larger than lithium atom because, for Na, _____.
(b) Chlorine atom is smaller than sodium atom because, for Cl, _____.
(c) Sodium atom is larger than Na+ ion because, for Na, _____.
(d) E_1 for Be is higher than E_1 for boron because, for B, _____.
(e) The average radius of an H atom in the 3d state is smaller than that of an H atom in the 4s state because, for the 3d state, _____.

19. Calculate the average radius $\bar{r}$ of a *vacant* argon atom 4s orbital assuming $Z^*(4s)$ will be equal to Z^* calculated for the neutral Ar atom from E_1 (see Problem 2). Draw a picture, to scale, of two argon atoms separated 3.8 Å, the internuclear separation in crystalline argon. For each atom, draw a circle with radius $\bar{r}$.

20. The Ar^+ ion is isoelectronic with Cl and has some chemistry in common (for example, it forms a monohydride, HAr^+). Use E_2 for argon to calculate Z^* and $\bar{r}$ for Ar^+. Compare to the corresponding values for Cl (see Problems 3 and 5).

21. Calculate $\bar{r}$ for F^- from its experimental electron affinity, 79.5 kcal. Compare to the average radii $\bar{r}$ for neutral atoms of the next two elements, Ne and Na.

22. X-ray diffraction studies of crystals are often based upon the radiation emitted when a copper atom is excited so that one of its 2p electrons can drop to a half-occupied 1s orbital. The wavelength emitted is 1.54 Å. Calculate the 2p $\longrightarrow$ 1s energy difference in a Cu atom. What is the nuclear charge of Cu? What would be the 2p–1s energy difference in the one-electron atom, Cu^{+28}?

fourteen quantum mechanics and chemical bonds

Quantum mechanics successfully explains the observable properties of single atoms and ions. Fortunately, it also applies to the stable aggregates of atoms, called molecules. Thus, quantum mechanics gives us a unifying theory by which we can understand the existence, stability, and reactions of the many hundreds of thousands of compounds known to chemists.

We've already complained about the mathematical obstacles in the quantum mechanical treatment of many-electron atoms. Compared with atoms, even the simplest molecules are much more complicated. Exact calculations can be performed only for a handful of them, even with the aid of the largest computers. For these simple molecules, however, the results are in almost perfect agreement with experimental measurements of bond energies, bond lengths, vibrational frequencies, and the other observables that fix molecular properties. On this foundation, we confidently assert the general usefulness of quantum mechanics for explaining and predicting chemistry.

Even though the mathematics remains a serious block to the precise use of quantum mechanics, the theory permits us to refine and use more sensibly our existing schemes of chemical bonding. These schemes, though approximate and often empirical, have proved to be amazingly successful in leading chemists to new discoveries. Now they can be unified, inter-related, and made more quantitative with the benefit of a single guiding theoretical foundation.

14-1 What is a molecule?

Before proceeding, we'd best agree on the meaning of the word "molecule."

A molecule is an aggregate of atoms that possesses distinctive and distinguishing properties.

This definition is quite general—but deliberately so. There are some very important omissions:

—*It does not specify what properties shall be measured.* Any properties that suffice to identify and characterize a particular aggregate of atoms will also make it convenient and useful to recognize that aggregate as an entity—a molecule.

—*It does not limit the number of atoms in the aggregate.* There is no restriction that prevents a molecule from including 30,000 atoms, as long as the 30,000-atom aggregate has distinctive properties (as does the DNA molecule). It is a bit ambiguous when we consider a crystal of salt, a nugget of gold, or a strand of nylon. Any of these may be considered to be molecules if convenience dictates—our discussions of chemical bonding will not be disturbed. A single strand of nylon is usually called a molecule, but a single crystal of salt is not. Yet they have the same ambiguity: Each involves particular atoms in definite, simple ratios, and in special geometrical arrangements, but with widely varying total numbers of atoms. A strand of nylon can have a carbon–nitrogen skeletal chain 100 or 10,000 atoms long. A salt crystal can come through a salt shaker or weigh 500 grams. No important distinction is made if we insist either that the nylon and salt must be called molecules or that a semantic law be passed against such a usage.

—*It does not say that the molecule must exist under normal conditions*—at room temperature and one atmosphere pressure. We'll talk about garden-variety molecules like hydrogen H_2 and carbon dioxide CO_2 and, in the same breath, about molecules that have never been detected at room temperature, like gaseous LiF or the inert gas compound KrF_2. The diatomic LiF molecule is detectable only in ovens at temperatures around 1000°K, and the triatomic KrF_2 molecule decomposes spontaneously if it is as warm as melting ice.

—*It does not say that the molecule must be neutral.* The principles by which we explain the bonding in polyatomic ions ("molecule-ions") are the same as those deduced for neutral molecules. A molecule-ion is merely a special case of a molecule—one that possesses a net negative or positive charge.

—*It requires stability, but not lack of reactivity.* Thus, the two molecules nitric oxide, NO, and methyl, CH_3, are both stable and each possesses many distinctive and identifying characteristics. In air, however, nitric oxide is extremely reactive, and an average NO molecule might join with oxygen to form nitrogen dioxide NO_2, another molecule, in a few thousandths of a second. Methyl, on the other hand, would react even in the absence of air and even more rapidly to form the more complex ethane molecule, C_2H_6.

Thus we distinguish "stable" and "unreactive." A molecule is considered to be stable if the molecular group of atoms does not spontaneously rearrange or fall to pieces; a molecule is said to be unreactive under a given set of conditions if it does not undergo rearrangements (chemical changes) involving itself and other molecules under these conditions. At room temperature, a molecular aggregate of eight hydrogen atoms flies apart into four stable molecules of H_2. In contrast, an aggregate of eight sulfur atoms remains together in a stable ring molecule S_8, with fixed molecular structure and distinctive properties. On the other hand, S_8 becomes unstable at an elevated temperature; the S_8 molecule will separate into four molecules of S_2 at a temperature of 1200°K.

With this definition of a molecule—a definition of convenience—we can proceed to discuss chemical bonding. We would like to understand why a particular molecule hangs together and why that atomic cluster displays its own peculiar properties. A complete theory of bonding must tell us which molecules can exist and which cannot. It must also tell us the properties each molecule will display, including the most important property, reactivity with other molecules. Such a theory would encompass chemistry: substances and their changes. We will begin our study of bonding with the simplest cases—diatomic molecules in the gas phase.

14-2   Why do chemical bonds form?

If gaseous fluorine is exposed to an intense burst of light, a large concentration of fluorine atoms can be produced. The fluorine atoms persist, however, for only a small fraction of a second. As collisions occur, fluorine atoms recombine to form stable molecules of F_2. A chemical bond is formed.

$$F(g) + F(g) = F_2(g) \tag{14-1}$$

(a) POSITIONAL OR MOTION RANDOMNESS?

Thermodynamics tells us that chemical changes occur spontaneously only if the randomness, the entropy, of the universe increases. Entropy can rise either because of increased positional randomness or because of increased motional randomness. In a reaction like (14-1), the positional randomness of the system is always *decreased*, since two particles are being locked in the blissful state of chemical bondedness. Certainly it is more random to let each atom have its own carefree and independent existence. So chemical bonds do *not* form because of positional randomness.

The alternative, increased motional randomness, requires that the reaction be exothermic. Energy is then released and distributed among the myriad, chaotic motions we identify as heat, or thermal energy. A reaction is exothermic if the energy of the products is below the energy of the reactants. Therefore, *chemical bonds must form because the energy of the system decreases*, this energy appearing as heat in the surroundings.

(b) KINETIC OR POTENTIAL?

With that confident conclusion, we can proceed to a more detailed question. What *causes* the energy to be lowered? Is it connected with kinetic energy or potential energy or can we be that specific? It turns out that we can.

Quantum mechanics allows us to calculate the average kinetic energy and the average potential energy of the electrons and nuclei in a molecule. We will designate them as follows:

$\bar{T}$ = average kinetic energy

$\bar{V}$ = average potential energy

These two averages must, of course, sum to the average of the total energy $\bar{E}$.

$$\bar{E} = \bar{T} + \bar{V} \tag{14-2}$$

Is it possible that one of these terms, $\bar{T}$ or $\bar{V}$, is more important than the other in explaining the formation of chemical bonds? If so, we could focus our consideration on that energy term.

To investigate the relative importance of electronic, kinetic and potential energies, we can use a quantum mechanical theorem known as the *Virial Theorem*. The Virial Theorem says that there is a remarkably simple relationship between the average potential energy $\bar{V}$ and the average kinetic energy $\bar{T}$. When all the forces acting are simple electrostatic attractions and repulsions, this relation is

$$\bar{T} = -\tfrac{1}{2}\bar{V} \tag{14-3}$$

Since the forces in atoms and molecules are dominated by the electric charges, equation (14-3) is applicable in chemistry.* It says that the average kinetic energy within the molecule is opposite in sign and only half as large as the average potential energy.

Exercise 14-1. Substitute (14-2) into (14-3) to show that the Virial Theorem requires that $\bar{E} = -\bar{T}$ and $\bar{E} = +\tfrac{1}{2}\bar{V}$.

Exercise 14-2. The total energy $\bar{E}$ of a hydrogen atom in the 1s state is -313.6 kcal/mole. Use the Virial Theorem to calculate the average potential energy $\bar{V}$ and the average kinetic energy $\bar{T}$ of the electron. (We used this result in Problem 12-11.)

Exercise 14-3. Repeat this calculation for the electron in a helium ion, He^+, in the 1s state and also in the 2p state (remember, for one-electron atoms, $\bar{E} = -313.6 \cdot Z^2/n^2$ kcal/mole).

Let us apply this relation to the formation of a chemical bond between our two fluorine atoms. The initial situation, which we will call state 1,

---

*There are magnetic forces in atoms and molecules due to nuclear and electron spin, but their energy effects are extremely small. The same is true, but more so, for the gravitational forces between the particles in the atom.

involves two widely separated fluorine atoms, each with an average kinetic energy $\bar{T}_F$ and potential $\bar{V}_F$. Since there are two of them, the state 1 kinetic and potential energies will be $\bar{T}_1 = 2\bar{T}_F$ and $\bar{V}_1 = 2\bar{V}_F$. The final situation in reaction (14-1), state 2, is the stable fluorine molecule F_2. This molecule determines the kinetic and potential energies after reaction, so $\bar{T}_2 = \bar{T}_{F_2}$ and $\bar{V}_2 = \bar{V}_{F_2}$.

For both the initial and final states, we can apply the Virial Theorem:

$$\bar{T}_1 = -\tfrac{1}{2}\bar{V}_1 \tag{14-4a}$$

and

$$\bar{T}_2 = -\tfrac{1}{2}\bar{V}_2 \tag{14-4b}$$

Now we can investigate the energy changes that occur as the system goes from state 1 to state 2, that is, as the fluorine atoms form a chemical bond in F_2.

$$\Delta\bar{T} = \bar{T}_2 - \bar{T}_1 \tag{14-5a}$$
$$\Delta\bar{V} = \bar{V}_2 - \bar{V}_1 \tag{14-5b}$$

Substituting (14-4a) and (14-4b) into (14-5a), we obtain

$$\Delta\bar{T} = \bar{T}_2 - \bar{T}_1 = (-\tfrac{1}{2}\bar{V}_2) - (-\tfrac{1}{2}\bar{V}_1) = -\tfrac{1}{2}(\bar{V}_2 - \bar{V}_1) = -\tfrac{1}{2}\Delta\bar{V}$$

$$\boxed{\Delta\bar{T} = -\tfrac{1}{2}\Delta\bar{V}} \tag{14-6}$$

This Virial Theorem result can be inserted into the total-energy expression to relate $\Delta\bar{E}$ to either $\Delta\bar{T}$ or to $\Delta\bar{V}$:

$$\Delta\bar{E} = \Delta\bar{T} + \Delta\bar{V}$$
$$\Delta\bar{E} = (-\tfrac{1}{2}\Delta\bar{V}) + \Delta\bar{V} = +\tfrac{1}{2}\Delta\bar{V}$$

or

$$\Delta\bar{E} = \Delta\bar{T} + (-2\Delta\bar{T}) = -\Delta\bar{T}$$

or summarizing,

$$\boxed{\Delta\bar{E} = +\tfrac{1}{2}\Delta\bar{V} = -\Delta\bar{T}} \tag{14-7}$$

The change in total energy must always carry the same sign as the change in potential energy. Therefore, $\Delta\bar{E}$ can decrease (a bond can form) only if $\Delta\bar{V}$ decreases. The kinetic energy will change at the same time, but it will always change in the opposite direction and by half as much. We can get to the heart of chemical bonding by investigating which changes can lower the potential energy as two atoms come together to form a bond. *Bonds can form only if the potential energy of the electrons and nuclei decreases as the atoms come together.*

14-3 The simplest molecule, H_2^+

There is no simpler molecule than two protons bonded together by one electron. This molecule (or molecule-ion) H_2^+ exists in a high-voltage glow discharge through hydrogen gas. Since it is an ion, it is extremely reactive, but its spectral properties reveal its bond energy, the energy needed to pull the atoms apart, and its equilibrium bond length. The molecule H_2^+ is stable.

(a) POTENTIAL ENERGY OF H_2^+

Let's apply the Virial Theorem to this molecule to see why the potential energy drops and the bond forms. All the potential energy terms are electrostatic: Each term depends on the charges on two of the particles, q_1 and q_2, and the distance between them, r_{12}.

$$V_{12} = \frac{q_1 q_2}{r_{12}} \qquad (14\text{-}8)$$

Figure 14-1 shows graphically the reaction between a neutral hydrogen atom made up of a proton A and an electron, and a second proton B, to form H_2^+. All the charges are the same in magnitude, a positive or a negative q_e. Table 14-1 tabulates the averaged contributions to potential energy before and after bond formation. It is clear how $\bar{V}$ goes down as the bond

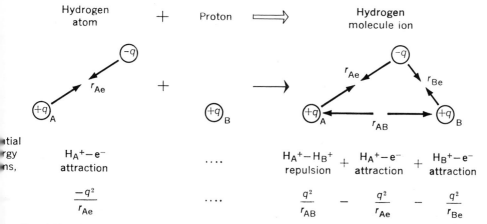

Figure 14-1   The formation of the simplest molecule—H_2^+.

Table 14-1   Potential Energy Change* in the Formation of H_2^+

	Before Reaction	**After Reaction**
Terms that lower V	$V_{Ae} = -\dfrac{q^2}{r_{Ae}}$	$V_{Ae} = -\dfrac{q^2}{r_{Ae}}$
		$V_{Be} = -\dfrac{q^2}{r_{Be}}$
Terms that raise V	none	$V_{AB} = +\dfrac{q^2}{r_{AB}}$

	Before Reaction	**After Reaction**	**Change**
$\bar{V}$	$\bar{V}_{Ae} = -627$ kcal	$\bar{V}_{Ae} = -534.5$ $\bar{V}_{Be} = -534.5$ $\bar{V}_{AB} = +313$	$\Delta V = \bar{V}_2 - \bar{V}_1 = -129$
$\bar{T}$	$\bar{V}_1 = -627$ $\bar{T}_1 = +313.5$ $\bar{E}_1 = -313.5$	$\bar{V}_2 = -756$ $\bar{T}_2 = +378$ $\bar{E}_2 = -378$	$\Delta\bar{T} = \bar{T}_2 - \bar{T}_1 = +64.5$ $\Delta\bar{E} = \bar{E}_2 - \bar{E}_1$ $= -64.5$ kcal/mole

* Note: Calculations refer to the experimental value of r_{AB} in H_2^+, 1.06 Å. The numerical magnitudes are averaged over all electron positions as dictated by the wave function, which gives the electron probability distribution. The final bond energy, $\Delta\bar{E}$, differs from the experimental value, 61.1 kcal, exactly by the minimum vibrational energy (the "zero point energy"), 3.4 kcal, since vibration is not taken into account here.

forms: There are two new contributions to $\bar{V}$, one that raises the energy (the nuclear–nuclear repulsion) and one that lowers the energy (the new electron–nucleus attraction). The only possible cause for the decrease in energy is the second term; *the electron is now near two nuclei at the same time.*

Exercise 14-4.   If the two protons in H_2^+ could be pushed all the way together, the electron would find itself in a He^+ ion. Thus, if nuclear–nuclear repulsions are removed, leaving only the electron's contribution to energy, H_2^+ is "between" H and He^+. Fill in the following table, extracting the desired numbers from Table 14-1 and Exercises 14-2 and 14-3.

	H atom 1s	**H_2^+ molecule-ion**	**He^+ ion 1s**
$\bar{E}-\bar{V}_{AB}$			
$\bar{V}-\bar{V}_{AB}$			
$\bar{T}$			

(b) THE FORCES IN H_2^+

It is also interesting to consider the forces that exist in the hydrogen molecule-ion. Of course the electron position cannot be specified with certainty

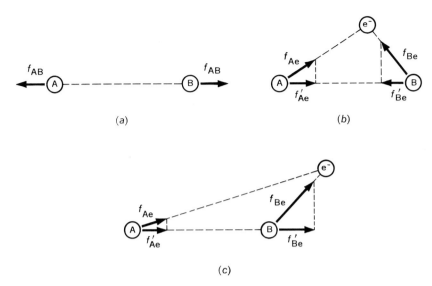

Figure 14-2 The forces in H_2^+. (a) The ever present nuclear–nuclear repulsion works against binding the nuclei together. (b) A possible electron position that contributes to binding the nuclei together. (c) A possible electron position that works against binding.

in a molecule any more than it can in an atom. All that quantum mechanics gives us is a probability picture. At any instant, the electron might be in a position such as that shown in Figure 14-2(b); at another instant it will be in a new position, perhaps the one shown in Figure 14-2(c). Each of these configurations contributes to the average forces felt by the nuclei.

Figure 14-2(a) shows, first, the nuclear–nuclear force f_{AB}. It is directed along the AB axis and it tends to force the nuclei apart. This force must be offset somehow by the electron–nuclear forces if there is to be a stable molecule in which the average net forces are zero.

Consider the configuration (b). The electron is attracted to both nuclei with forces f_{Ae} and f_{Be} that become larger as the distance gets smaller. Each of these two forces has a component directed along the AB axis, f'_{Ae} and f'_{Be}. The force f'_{Ae} pulls A toward B and the force f'_{Be} pulls B toward A. This is the type of thing we need to counteract the nuclear–nuclear repulsion; configuration (b) contributes to the binding in the molecule.

Things aren't as good in configuration (c). When the electron is out on the periphery of the molecule, it exerts attractive forces that pull both nuclei to the right. Of course it pulls nucleus B more strongly than A because it is much closer to B. The forces along the AB axis, f'_{Be} and f'_{Ae}, give a net force that tends to pull B away from A. This hurts as far as neutralizing the nuclear–nuclear repulsion is concerned. This configuration works *against* binding.

Obviously these ideas can be made quantitative. A straightforward

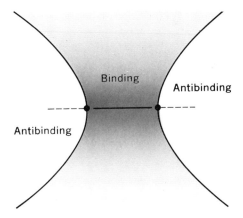

Figure 14-3 Binding and antibinding regions in any homonuclear diatomic molecule.

calculation indicates, for every possible electron position, whether that position causes the electron to pull the nuclei together or away from each other. Figure 14-3 shows the result for H_2^+. Anywhere in the shaded region, the forces bind the nuclei together—this is called the *binding* region. Outside the shaded region, the opposite is true: The net force pulls the nuclei apart. Since this works against binding, this region is called the *antibinding* region.

Examining Figure 14-3, we see that consideration of forces in a molecule leads to the same conclusion arrived at by the energy arguments. If the nuclei are to be held together despite the repulsive forces between them, electrons must preferentially occupy the binding region, the volume between the nuclei. The closer the electrons come to the line between the nuclei, the better. This is, of course, exactly the region in which the electron is simultaneously near both nuclei. *Bonds form because electrons are simultaneously near two or more nuclei.*

(c) THE CORRELATION DIAGRAM FOR H_2^+

It is extremely informative to examine the evolution of each of the energy terms shown in Figure 14-1 as the two protons approach each other. In fact, in Exercise 14-4, we considered pressing the two nuclei right up against each other. This process requires a lot of muscle because of the proton–proton repulsion, but it is very educational, hence worth the effort. When the two protons are very close together, the electron sees a helium nuclear charge, so the energy levels must approach those of a helium ion, He^+. That is convenient. Ignoring the rising proton–proton energy, the energy levels must smoothly connect the levels of two separated hydrogen atoms to the levels of a He^+ atom. Both extremes are one-electron atoms, so their energy levels are exactly known, as shown in Figure 14-4.

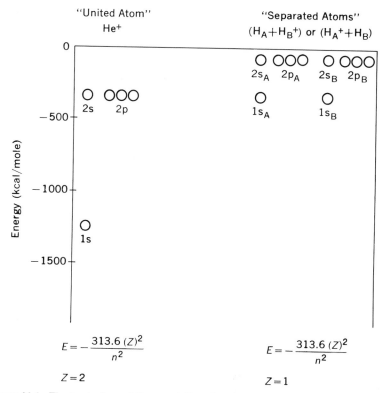

Figure 14-4   *The beginnings of the correlation diagram for H_2^+.*

On the right we have the energy levels of the separated atoms. We'll consider only the lowest energy levels, $1s_A$ and $1s_B$. These two orbitals correspond to the two possible electron occupancies when the atoms are separated by an infinite distance. The electron either can be near proton A, in orbital $1s_A$ with energy -313.6 kcal/mole, or it can be near proton B, in orbital $1s_B$, again with energy -313.6 kcal/mole.

As these two protons approach each other, the single electron has no basis for preference between occupying orbital $1s_A$ or $1s_B$. If we looked at many such pairs of protons, in some pairs we'd find the electron in $1s_A$, but in an equal number of others, we'd find the electron in $1s_B$. In any of the pairs, however, as the protons near each other, we would find the electron moving in a fashion that takes equal account of the presence of both nuclei. Now, instead of occupying one of the atomic orbitals, $1s_A$ or $1s_B$, the electron occupies a *molecular orbital* whose parentage lies equally in $1s_A$ and $1s_B$. Then, as the protons are pressed very close together, this molecular orbital must evolve again into an atomic orbital belonging to the helium ion.

To decide how this parentage reflects into the resultant molecular orbital, we must "think waves." Waves are characterized by nodal patterns and these nodal patterns furnish a guide to the manner in which the separated atom orbitals, $1s_A$ and $1s_B$, connect, through molecular orbitals, to particular atomic orbitals of the "united atom" He^+. Consider, for example, the nodal properties of the lowest He^+ orbital, $1s$. It has no nodal surfaces other than the inevitable one at infinity. As the protons separate, first just a little, then more and more, this nodal pattern is retained. At intermediate separations, we must think of a molecular orbital that has no nodal surfaces except the one at infinity. The upper part of Figure 14-5 pictures this evolution as it must occur. Figure 14-5(a) shows the probability representation and Figure 14-5(b) shows the magnitude of ψ along the line of the nuclei. The latter representation is the molecular orbital counterpart to the representations of atomic ψ's in Figure 12-13(b). We see that there is no nodal surface between the two protons at any separation. As the separation approaches infinity, the wave function begins to resemble two $1s$ functions with the same phase.

Now we note that the He^+ $1s$ orbital evolved through a molecular orbital into an in-phase combination of *two* separated atom orbitals. Reversing the direction of our thinking, two proton orbitals, $1s_A$ and $1s_B$, are parentally related to the single He^+ $1s$ orbital. It is intuitively reasonable to question how two orbitals could suddenly become only one. Even more reasonable, however, is that we should expect two separated atom orbitals to give rise to *two* molecular orbitals and connect to *two* united atom orbitals. This proves to be a general quantum mechanical principle—as atoms approach each other, orbitals do not suddenly disappear or appear.

So our in-phase combination of $1s_A$ and $1s_B$ (that connects to He^+ $1s$) is only one of two molecular orbitals with the $1s_A$, $1s_B$ parentage. This phase relationship gives us a clue to the nature of the second. If one molecular orbital is like the in-phase combination of $1s_A$ and $1s_B$, perhaps the other is like the out-of-phase combination. This possibliity is represented in Figure 14-5(c) and 14-5(d). The opposite phase relationship implies a nodal surface halfway between the protons. If this nodal pattern persists as the protons approach each other, it implies that the united atom will have a nodal plane through the nucleus. Of course, there is always a phase change at a nodal surface. Hence the united atom distribution is just like a $2p_x$ orbital (see Fig. 12-13) of He^+.

These are the two molecular orbital offspring of $1s_A$ and $1s_B$; they have the nodal properties of and they connect to the $1s$ and $2p_x$ orbitals of the united atom He^+. These nodal properties immediately tell us how the energy will change as the molecular orbital forms. The in-phase combination concentrates the electron probability between the nuclei (see Fig. 14-5(b)) in the binding region. This will lower the energy, tending to form a bond between the protons. This is a *bonding molecular* orbital. In contrast, when the orbitals interact out of phase, they possess a nodal surface

nding orbital

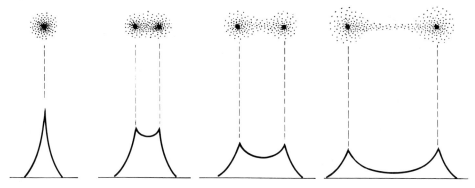

tibonding orbital

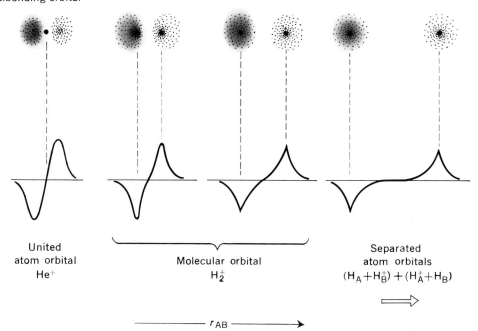

United
atom orbital
He$^+$

Molecular orbital
H$_2^+$

Separated
atom orbitals
$(H_A + H_B^+) + (H_A^+ + H_B)$

r_{AB}

Figure 14-5 The evolution of orbitals as He$^+$ is separated into two protons plus an electron: (a) and (c) are probability patterns; (b) and (d) show magnitudes and phases. (Phase in the probability patterns is shown by shading.)

between the nuclei and the probability that the electron will be found there is zero. This moves the electrons out of the binding region to the periphery of the molecule and into the antibinding region. The energy is raised, working against forming a bond between the protons. This is an *antibonding orbital*.

These two molecular orbitals, though having different nodal properties (one has no new nodal surface, the other has one between the nuclei) are both directed along the line AB. Such orbitals are designated σ (sigma) orbitals, the Greek letter for s. The antibonding orbital is identified by an asterisk as σ^* with axial symmetry. Such nodal properties tell us how to connect energy levels at the extremes shown in Figure 14-4. The in-phase orbital σ has no nodal surfaces (except at infinity). The He^+ ion has only one such orbital, its 1s orbital. Hence σ must connect to the He^+ 1s orbital. The out-of-phase orbital has a nodal surface in the yz plane, exactly like the He^+ $2p_x$ orbital. It must connect to that tie-point. The result is shown in Figure 14-6, first omitting the nuclear–nuclear repulsion (dashed curves) and then adding it on (solid curves).

Plainly, σ has the energy properties needed for chemical bonding. The energy drops as the nuclei near each other because the electron is near two nuclei simultaneously. At too close range, however, the energy rises because of nuclear–nuclear repulsion. There is a value of r_{AB} at which the energy is a minimum. This is the equilibrium bond length.

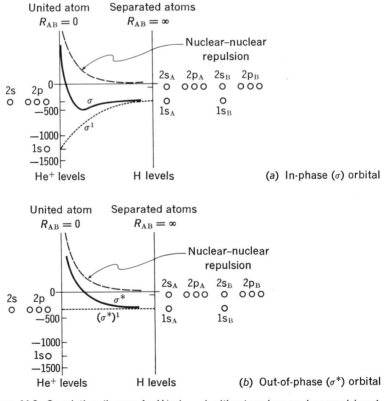

Figure 14-6 Correlation diagram for H_2^+: (------) without nuclear–nuclear repulsion; (———) with nuclear–nuclear repulsion.

In contrast, σ^* has no tendency for the energy to drop as the nuclei approach and, instead, the nuclear–nuclear repulsion causes it to rise steadily. At any value of r_{AB}, the energy would be lowered if the nuclei were to separate again.

Exercise 14-5. The nuclear–nuclear repulsion at 1.06 Å separation is $+313$ kcal/mole. What would it be if the internuclear distance were halved?

(d) THE ENERGY LEVEL DIAGRAM FOR H_2^+

Now we can draw a pigeonhole diagram for H_2^+ like those for atoms. For zero energy, we can take the separated atoms. On such a scale, the σ orbital lowers the energy and σ^* raises it, as shown in Figure 14-7. If the electron occupies σ, a bond should be able to form. Experiment shows that H_2^+ is stable; 64.5 kcal/mole is required to break the bond and pull the nuclei apart to form a proton and a neutral hydrogen atom. The experimental bond length is found to be 1.06 Å. Both these numbers can be calculated exactly by the methods of quantum mechanics.

There are, of course, higher energy pigeonholes associated with H_2^+; states with their parentage in the higher orbitals of the component hydrogen atoms (2s, 2p, 3s, etc.). The most important pigeonholes, however, are the lowest in energy and their occupancy will determine the important properties of the molecule formed: the bond energy, the bond length, and, as we shall see, the chemistry.

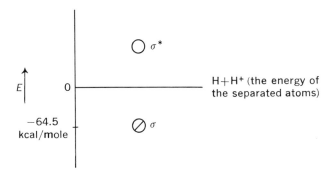

Figure 14-7 The lowest orbitals of H_2^+.

14-4   Molecules and the Pauli Principle

The molecular parallel to the quantum mechanics of atoms has been so close, it is natural to expect it to continue. For example, consider the **Pauli Principle**—*only two electrons can occupy a given atomic orbital*. We can expect this principle to apply to molecules as well. Let's see what it implies.

(a) THE HYDROGEN MOLECULE

Figure 14-7 shows the pigeonhole diagram for H_2^+ and, for the orbital occupancy shown (one electron in the bonding σ orbital), the energy is lowered by 64.5 kcal. If the Pauli Principle applies, this orbital should be able to accommodate a second electron. The molecule would then be converted to a neutral hydrogen molecule H_2, and the orbital diagram suggests that the bond between the two protons should be strengthened. If one electron in the σ orbital lowers the energy by 64.5 kcal, two electrons might lower the energy by double that amount, $2 \cdot 64.5 = 129$ kcal. Of course we recognize this again as our "turned-off" electron repulsion approximation. Actually, the energy won't be lowered as much as 129 kcal because of a new term in the energy box score due to the electron–electron repulsion. Experimentally, it is found that the bond in hydrogen, H_2, has an energy of 104 kcal/mole, not quite double the 64.5 kcal but close to it. The difference is due to the new potential energy term—a term that works against bonding. Nevertheless, to a first approximation, the second electron in a bonding molecular orbital strengthens the bond by a factor of nearly two.

(b) MORE ELECTRONS AND THE PAULI PRINCIPLE

We see that the H_2^+ orbital diagram, with the aid of the Pauli Principle, explains the bonding in H_2. Perhaps a more severe test would be to consider the potential bonding in the helium analogs to H_2^+ and H_2, that is, to the possible molecules $(He)_2^+$ and $(He)_2$.

A naive application of the Pauli Principle (two electrons, at most, per orbital) to the molecular orbital pigeonhole diagram is pictured in Figure 14-8. For $(He)_2^+$, the third valence electron must occupy the σ^* antibonding orbital. Instead of strengthening the bond, this electron should weaken it, effectively negating the beneficial effect of one of the bonding electrons. Sure enough, calculations show that the bond energy of the molecular ion He_2^+ is 55 kcal/mole, quite close to that of H_2^+. Continuing on to the hypothetical molecule $(He)_2$, the fourth electron also must occupy the σ^* antibonding orbital. Now, with two bonding electrons and two antibonding electrons, there should be no bond at all. Again experiment is consistent with this view. Two helium atoms attract each other very weakly—effectively there is no chemical bond. We see that the Pauli Principle is valid for molecules as well as for atoms. It helps us understand and predict chemical bonds.

(c) BONDING ELECTRONS AND BOND ORDER

The molecule H_2, with two electrons in a bonding orbital, proves to be a prototype of the bonding in many familiar, garden-variety substances. Molecules like H_2O (water), NH_3 (ammonia), and Cl_2 (chlorine) have bonds

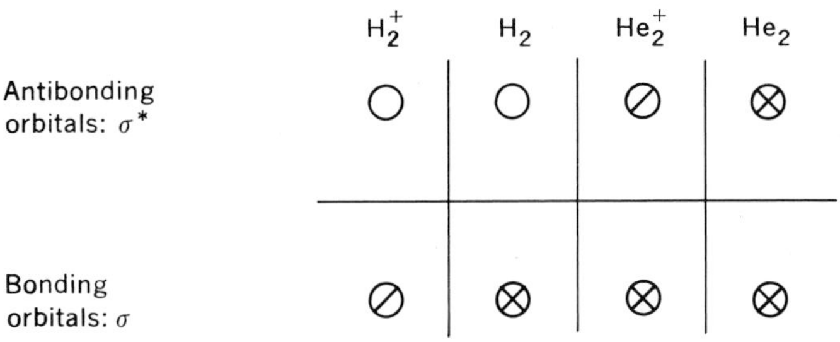

	H_2^+	H_2	He_2^+	He_2
Antibonding orbitals: σ^*	○	○	⊘	⊗
Bonding orbitals: σ	⊘	⊗	⊗	⊗
Number of bonding electrons: N_b	1	2	2	2
Number of antibonding electrons: N_a	0	0	1	2
Bond order $\frac{1}{2}(N_b - N_a)$	$\frac{1}{2}$	1	$\frac{1}{2}$	0
Experimental bond energy (kcal/mole)	64.5	104	55	0.02

Figure 14-8 Electron occupancy of the lowest H_2^+ molecular orbitals for some simple molecules.

each associated with a pair of electrons. About 40 years ago, G. N. Lewis postulated that a normal chemical bond was caused by the sharing of two electrons between two atoms. Such a bond became known as a "single bond" and it was assigned a "bond order" of one. Lewis' brilliant hypothesis can now be evaluated in the light of quantum mechanics. A "single bond" formed by the sharing of two electrons corresponds to the full use of a bonding molecular orbital in accordance with the Pauli Principle.

Extending this idea, then, a bonding orbital containing only one electron should be called a "half-order bond." Now a new idea appears; one that amplifies the Lewis concept of the bond—an electron in an antibonding orbital contributes a minus-one-half bond order. Hence, the combination of two electrons in a bonding orbital and one electron in the associated antibonding orbital again constitutes a "half-order bond." Figure 14-8 includes appropriate indications of the bond order for each of the molecules considered.

(d) EXPERIMENTAL MEASURE OF BOND ORDER

We have already been using bond energy as a measure of bond order. The H_2^+ molecule, with a one-half bond order, releases 64.5 kcal/mole when the bond is formed. With two bonding electrons and a full single bond, H_2 releases 104 kcal/mole when the bond is formed. In each case, as the reac-

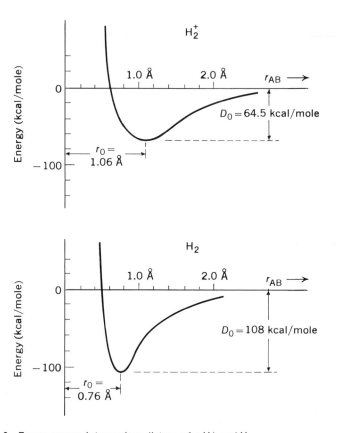

Figure 14-9   Energy versus internuclear distance for H_2^+ and H_2.

tants approach each other, the energy drops, but more in the case of H_2 than H_2^+. This is shown pictorially in Figure 14-9. To pull each of these molecules apart again requires that we pay back the price to climb the energy hill—we must put in 64.5 kcal/mole to break the bond in H_2^+ and 104 kcal/mole to break the stronger bond in H_2. *These bond dissociation energies are a primary measure of bond order and bond strength.*

Figure 14-9 reveals another characteristic of chemical bonds. The stronger bond in H_2 pulls the two protons closer together than in H_2^+. Whereas in the half-order bond of H_2^+, the nuclei find their energy minimum (equilibrium bond length) at 1.06 Å, the single bond of H_2 moves the energy minimum to 0.76 Å. This is not too surprising, since two negative electron charges placed in the binding region between the protons provide more electrostatic "glue" to offset the nuclear–nuclear repulsion.

This relationship is a general and most useful one. For similar atoms, *as bond order increases, bond length decreases.* Bond length measures bond order, too.

Figure 14-9 has one more bit of gold for us to mine. There is one other obvious difference between the energy curves of H_2^+ and H_2: the curvature at the bottom of the energy valley. For H_2^+, the energy hills near the equilibrium distance rise gently. In contrast, the energy minimum for H_2 lies in a rather steep-walled valley. This curvature difference has an important experimental consequence that provides a third measure of bond order.

The significance of the energy curvature at $r = r_0$ can be seen with a simple analog. It takes energy to change the proton–proton distance. If, for example, we try to stretch the H_2^+ bond by 0.1 Å, we must roll up its energy curve by 1.0 kcal/mole. If we stretch H_2 by the same amount, 0.1 Å, almost four times as much energy is required because of the steeper valley walls—this time 3.9 kcal/mole. Having stretched each of these two bonds, if we now let go of the molecule, the protons will spring back together. Of course, the H_2 will spring back faster because it is rolling down a steeper hill. Each molecule will reach the valley floor with some kinetic energy, and it will climb up the inside wall. Then the compressed bond will expand again, so that the atoms roll back and forth in the valley.

This is analogous to the behavior of two weights hooked together by a spring. There is an equilibrium distance at which the spring is neither stretched nor compressed. If the spring is stretched and then released, the weights vibrate back and forth at a frequency characterized by the masses of the weights and the strength of the spring. Molecules are just the same. The atomic masses determine the "masses of the weights" and the curvature of the bond energy plot determines the "strength of the spring." Figure 14-10 shows the connection chemists make between the energy–distance relation and the ball-and-spring model.

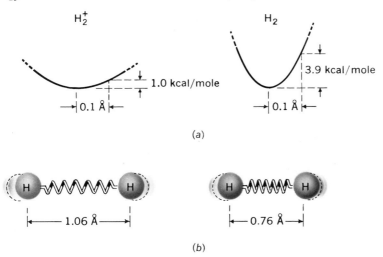

(a)

(b)

Figure 14-10 *Molecular vibrations—a measure of bond order. (a) Greatly expanded view of the bottom of the energy curves of Figure 14-9 showing the energy required for a 0.1 Å change in bond length. (b) Ball-and-spring model shows a weak spring in H_2^+ and a stiff spring in H_2.*

It should be no surprise that the vibrational energy levels are quantized, just as are the energy levels in which electrons are excited to high-energy orbitals. The vibrational energies are quite low, however, compared with bond energies. The frequency at which H_2^+ absorbs corresponds to only 6.2 kcal/mole. Quanta with energies this low have frequencies (colors) well below the deepest red color discernable to the human eye. For this reason, the light absorbed by molecules as their vibrations are excited is called "infrared" (beyond-the-red) light. With a suitable "eye" (an infrared-sensitive detector) and appropriate optical equipment (a spectrometer), the infrared "colors" absorbed by a molecule can be measured. These colors tell us about the bonds in molecules. The information we desire is in the spring strength. With the known atomic masses, it is an easy matter to calculate the spring strength, or *force constant*, from the observed vibrational frequency. Force constants have the dimensions of force per unit length and can be expressed in dynes per centimeter. The dimensions are unimportant and we'll choose millidynes per Angstrom (mdyne/Å), because the magnitudes are then typically between 1 and 20.

Table 14-2 gives our three measures of bond order for H_2^+ and H_2. These will be useful as we proceed to the study of larger and more complicated molecules.

Table 14-2 Bond Order Measures in H_2^+ and H_2

	H_2^+	H_2
D_0, energy needed to break bond	64.5	104 kcal/mole
r_0, equilibrium bond length	1.06	0.76 Å
k_0, force constant for vibration	1.4	5.1 mdyne/Å
bond order	$\frac{1}{2}$	1

14-5 Molecules with more electrons

The correlation diagram is applicable to molecules with many electrons, though it is more complex. The principles, however, guide us as we frame molecular orbitals and use the Pauli Principle to decide on their occupancy. This can be illustrated with some simple, first-row diatomic molecules, beginning with Li_2.

(a) DILITHIUM, Li_2

In the gas phase at moderate temperatures (say, 500–600°K), the vapor of lithium contains a substantial fraction of Li_2 molecules. Spectroscopic and thermodynamic studies show that Li_2 has the bond properties shown in Table 14-3. Let's see how the molecular orbitals and orbital occupancies in this molecule help us understand these properties.

Table 14-3 The Properties of Dilithium, Li_2

bond energy	$D_0 = 25$ kcal/mole	
bond length	$r_0 = 2.68$ Å	
force constant	$k_0 = 0.25$ mdyne/Å	

Figure 14-11 shows the energy level diagrams of two lithium atoms, and how molecular orbitals form as the two atoms approach the equilibrium separation 2.68 Å. At the bottom we find that the 1s orbitals form a bonding molecular orbital 1σ, and an antibonding molecular orbital $1\sigma^*$. There is very little energy difference between 1σ and $1\sigma^*$ and the reason for this is apparent in Figure 14-12. This figure reproduces the size of the 1s orbitals as given in Table 13-2, $\bar{r} = 0.34$ Å. At the equilibrium distance, these two orbitals are so small they barely interact at all. Of course these two M.O.'s (M.O. = molecular orbital) are the lowest in energy so, according to the Pauli Principle, they will be occupied first. As shown, four electrons can be accommodated, but these electrons will not contribute to the bonding. Since 1σ and $1\sigma^*$ are both fully occupied, the antibonding orbital $1\sigma^*$

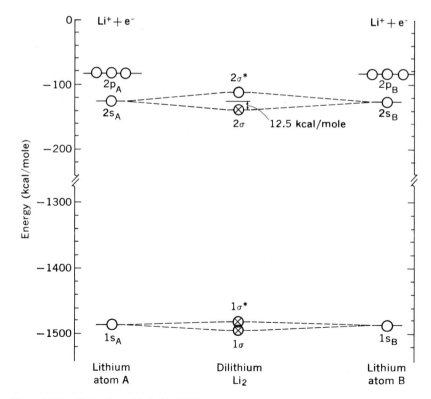

Figure 14-11 Molecular orbitals in dilithium, Li_2.

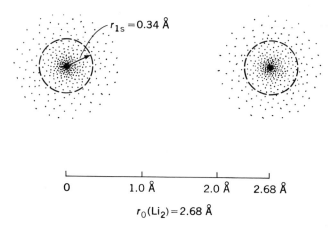

$r_{1s} = 0.34$ Å

$r_0(Li_2) = 2.68$ Å

0   1.0 Å   2.0 Å   2.68 Å

Figure 14-12   The size of the lithium 1s orbitals in Li$_2$. The atoms are placed at the inter-nuclear distance of Li$_2$ and the average radius of the 1s orbitals is shown to scale by the dashed circles.

neutralizes any bonding tendency furnished by 1σ. Quite apart from this, however, the interaction is very slight because of the small orbital size relative to the bond length. This illustrates our first guiding rule in predicting bonding in many-electron atoms. *Orbitals inside the valence orbitals must be occupied* (two electrons each), *but they do not affect the bonding.*

Proceeding upward in Figure 14-11, we next encounter the M.O.'s formed predominantly from the in-phase and out-of-phase interactions of the 2s orbitals. The 2p orbitals are about 43 kcal higher and they, too, interact, but to a first approximation, we can consider only the 2s orbitals as we discuss 2σ and 2σ*. The M.O. 2σ accommodates the remaining two electrons possessed by Li$_2$ and it, then, must account for the bonding. Figure 14-11 shows this doubly occupied orbital to be 25 kcal below the energy of the separated atoms, as indicated by the bond energy (Table 14-3). Counting all the electrons, we find, net, one pair of electrons in a bonding orbital. The bond in Li$_2$ is a single bond.

The validity of this description has been confirmed by detailed computer calculations. Figure 14-13(a) shows probability-density contours calculated for the 1σ, 1σ* and 2σ orbitals, and these three superimposed, giving the total electron distribution for Li$_2$. Each contour line shows a line along which electron probability density is constant. These plots show, for individual M.O.'s, the spatial extent of density contours differing successively by factors of four. The diagrams for 1σ and 1σ* show that over 90 percent of the probability density is contained within the 0.053 e$^-$/Å^3 density contour, which has a radius of 0.7 Å. For the 2σ orbital, this same density contour occurs *between* the nuclei. The computer's plot shows that the 2σ orbital concentrates electrons between the nuclei and that, going outward, the

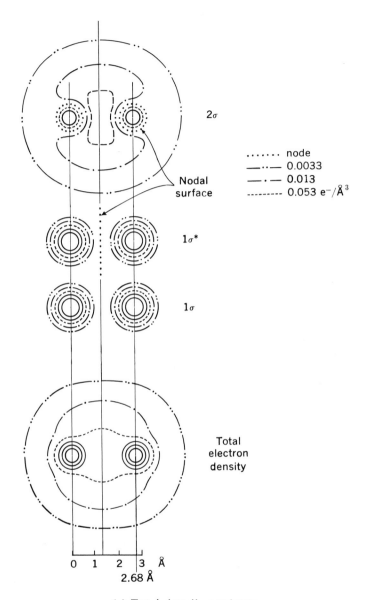

node
0.0033
0.013
0.053 e$^-$/Å^3

2σ

Nodal
surface

$1\sigma^*$

1σ

Total
electron
density

0 1 2 3 Å
2.68 Å

(a) Total density contours

Figure 14-13 (a) Probability density contours for Li$_2$. Identical contours are drawn in each diagram: Successive contours differ by a factor of four. (Units are electrons per cubic angstrom.)

density distribution drops off at a much lower rate. This orbital distributes electron probability both between and around the lithium nuclei, placed at the equilibrium distance 2.68 Å. Hence the 2σ orbital, occupied by two electrons (the valence electrons) determines the bonding.

Figure 14-13(*b*) shows the difference between the calculated total electron density shown in (*a*) and the density of two lithium atoms placed 2.68 Å apart. That is, it shows the *shift* in electron density on bond formation. The dotted lines mark the boundaries between the binding and antibinding regions. The diagram shows that, when the bond forms, electron density decreases in the antibinding regions (dashed contours). At the same time it *increases* in the binding region (solid contours). This is exactly what is needed for bond formation. The energy is lowered because the electrons are shared by both nuclei and the bond is stable because the electrons spend more of their time in between the nuclei—in the binding region.

With this conclusion, we can contrast the Li_2 single bond with the single bond in H_2. Contrasting the data in Tables 14-3 and 14-2, we see that the single bond in Li_2 is much weaker than the single bond in H_2—it has a much lower bond energy, a longer bond length, and a lower force constant. Why is this? The answer is found first in Figure 13-2 where the ionization energies are compared. We see that the H atom attracts its 1s valence electron quite strongly. It takes 313.6 kcal/mole to remove that 1s electron. Lithium, however, will let go of its 2s valence electron for only 124 kcal/mole. Now, bonding arises from the sharing of electrons as they occupy valence orbital space of two atoms at once. Clearly there is less to be gained when two lithium atoms share electrons than when two hydrogen atoms share electrons.

This explanation is given more meaning in Table 13-2. The one-electron approximation indicates that the lithium 2s electron feels an effective nuclear charge of 1.26, higher than the $Z = 1.00$ felt by the hydrogen 1s electron. However, this lithium valence electron must occupy a 2s orbital and, with principal quantum number $n = 2$, its average radius is 2.52 Å. This is to be compared with the average radius of the H atom 1s orbital, 0.79 Å. Obviously the lithium electron will be more weakly bound at such a large radius, even though the effective nuclear charge is a bit larger than that in the H atom. This is generally true: *As atomic size increases, chemical bonds tend to weaken.*

As a matter of fact, it is revealing to compare the Li_2 bond length to the one-electron size estimate for lithium. The average radius of the lithium atom is 2.52 Å. Where would we expect a second lithium atom to lodge as it forms a bond? The bond forms because lithium atom A places its valence electron near lithium nucleus B, and conversely. In other words, each lithium nucleus wants to be immersed in the valence electron probability density of its partner. This suggests that the bond length should be reasonably close to the average radius of the valence electron. This proves to be true, within 10 to 20 percent, for all single bonds in the first two rows of the Periodic Table. The results for Li_2 and H_2 are shown in Table 14-4.

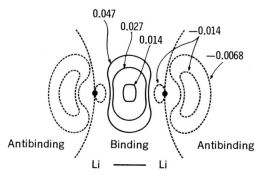

Antibinding   Binding   Antibinding

Li ———— Li

(b) Difference density contours

Figure 14-13 (b) *Difference density contours: total density minus density of two lithium atoms at Li_2 internuclear distance. Notice shift of electron density from antibinding region into binding region.*

Exercise 14-6. Though bond energies tend to weaken as atomic size increases, all energy terms become larger. Hence the exact calculation of bond energy, a small difference among very large numbers, becomes more and more difficult. To sense this difficulty, show that the nuclear–nuclear repulsion in Li_2 is 1114 kcal. Use H_2^+ as a reference ($\bar{V}_{AB} = 313$ kcal at $r_{AB} = 1.06$ Å), calculating the effects of increasing r_{AB} to 2.68 and Z to 3.

(b) LITHIUM HYDRIDE

So far, all the molecules considered are of the type called "homonuclear." Each contains two identical nuclei. Life becomes more interesting when we turn to the gaseous molecule, lithium hydride. It is the simplest example of a "heteronuclear" compound—one with two different kinds of atoms. This particular molecule is difficult to study—though stable, it is extremely reactive. Yet its properties are very well known. Because of the importance of this prototype heteronuclear molecule, much experimental and theoretical effort has been focussed upon it. The properties listed in Table 14-5 are all well established.

Table 14-4 Bond Length and Average Valence Orbital Radius (One-Electron Approximation)

Molecule	Average Radius of Valence Orbital, $\bar{r}$	Bond Length, r_0	$\dfrac{r_0}{\bar{r}}$
H_2	0.79 Å	0.76 Å	0.96
Li_2	2.52 Å	2.68 Å	1.06

Table 14-5 The Properties of Lithium Hydride, LiH

bond energy	$D_0 = 58$ kcal
bond length	$r_0 = 1.61$ Å
force constant	$k_0 = 0.96$ mdyne/Å

Consider first the orbital situation as shown in Figure 14-14. Obviously the lithium 1s orbital is in a class by itself. As is always the case, this 1s orbital plays no direct role in the bonding of lithium. These two electrons are tucked away and, except for their shielding of the nucleus, they are forgotten. The next orbitals, going upward in the diagram, are the hydrogen 1s and the lithium 2s pigeonholes. As usual, bonding and antibonding M.O.'s are formed. Orbital occupancy, given four electrons (three electrons from lithium and one electron from hydrogen) places two electrons in the Li 1s orbital, leaving two electrons for the σ bonding orbital. A single bond is formed.

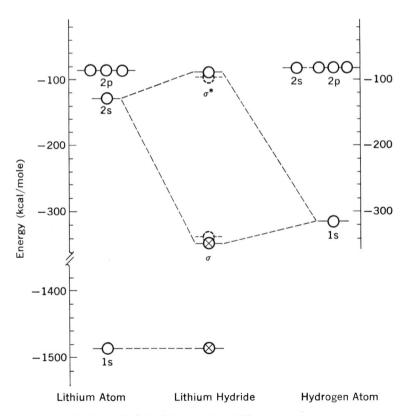

Figure 14-14 Molecular orbitals in lithium hydride, LiH.

The experimental bond energy, 58 kcal/mole, is much stronger than that of Li_2, but not as strong as that of H_2. Well, that seems reasonable. The bond energy should be between that of Li_2 and that of H_2—the two bonding electrons, if perfectly shared, would look in one direction and think they are in Li_2, and looking the other way, the poor dumb things would think they are in H_2. As a first guess, we might estimate that the bond energy of LiH should be some average of the Li_2 and H_2 bond energies.* The average we want is called the geometric mean, defined as the square root of the product of the two bond energies. Let's call this average $\bar{D}(LiH)$, the "perfect sharing" bond energy.

$$\bar{D}(LiH) = \sqrt{D_0(Li_2) \cdot D_0(H_2)} \qquad (14\text{-}9)$$

$$= \sqrt{(25)(104)}$$

$$= 51 \text{ kcal}$$

Actually $\bar{D}(LiH)$ underestimates the observed dissociation energy. The observed bond is stronger by 7 kcal.

$$D_0(LiH) - \bar{D}(LiH) = 58 - 51 \text{ kcal} = 7 \text{ kcal} \qquad (14\text{-}10)$$

Figure 14-14 shows the discrepancy (14-10) pictorially. The dashed pigeon-hole shows where the bonding M.O. would be in the perfect sharing situation; the one that would give a bond energy $\bar{D}(LiH)$. It is actually down closer to where the hydrogen atom energy level wants it. What does this mean in terms of electron probability distribution and the shape of the σ M.O.?

Figure 14-15 shows the answer to our question in terms of plots of the amplitudes of the wave functions as a function of position along the Li–H line of centers. If the separated atom probability distributions (Fig. 14-15(a)) are superimposed at the observed internuclear distance (1.61 Å), the "perfect sharing" molecular orbital is just their sum. This gives a first approximation to the bonding M.O., which always pulls the electrons toward the center of the bond, into the binding region (as shown by the shading in Fig. 14-15(b)). However, since the hydrogen atom attracts the bonding electrons more strongly than the lithium, there may

---

* Here are two ways to average two numbers X and Y:

Arithmetic mean $= \dfrac{X + Y}{2}$

Geometric mean $= \sqrt{(XY)}$

These two averages differ very little if X and Y have about the same magnitude. For example, if $X = 100$ and $Y = 104$, the two averages are, respectively, 102 and 101, only 1 percent different. They behave very differently, however, if X is large and Y is small. If $X = 100$ and $Y = 4$, the arithmetic mean is 52 and the geometric mean is only 20, a factor of $2\frac{1}{2}$ smaller. In fact, if Y approaches zero, the arithmetic mean approaches $X/2$, whereas the geometric mean approaches zero. Experience shows us that we want the latter.

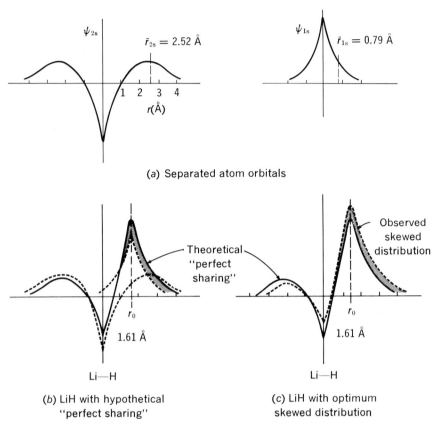

(a) Separated atom orbitals

(b) LiH with hypothetical
"perfect sharing"

(c) LiH with optimum
skewed distribution

Figure 14-15 Skewing of the molecular wave function in LiH: *(a) separated atom wave functions; (b) "perfect sharing" wave function (shading shows concentration of electron probability near the bond center); (c) skewed distribution (shading indicates skewing toward the hydrogen atom).*

be an additional lowering in energy to be gained by skewing the probability distribution toward the hydrogen atom. Notice in Figure 14-15(c) that this does not require the valence electron to be removed from lithium. Its average radius was 2.5 Å before hydrogen arrived on the scene. Concentrating the electron probability near the H atom at the equilibrium distance tends to concentrate the bonding electrons even closer to the lithium than in the separated atoms. The extra 7 kcal of bonding (14-10) exist because this skewed distribution is energetically more favorable than the "perfect sharing" distribution. It tends to concentrate the electrons near the atom that attracts them more strongly.

Exercise 14-7. Calculate $\bar{D}$ for HF from the bond energies of H_2 and F_2, respectively, 104 and 37 kcal, and compare $\bar{D}$ to the measured bond energy of HF, 135 kcal.

(c) THE CHARGE DISTRIBUTION IN LITHIUM HYDRIDE

The lithium hydride molecule is sufficiently simple that modern computers can adequately cope with its quantum mechanical complexity. Consequently, we know quite a lot about the probability distributions partially represented in Figure 14-15. Again, to give a more complete picture, a contour diagram is needed in which each contour line shows a line along which probability density is constant. Just as in reading a map, we can then see the regions in which electron probability is high and where it is low. Figure 14-16(a) shows such a contour map for LiH, taken from recent

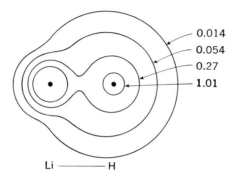

0.014
0.054
0.27
1.01

Li ——————— H

(a) Total electron distribution

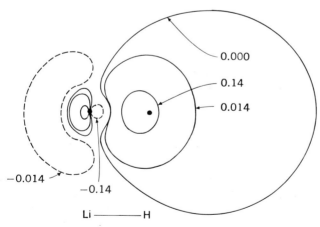

0.000
0.14
0.014
−0.014
−0.14

Li ——————— H

(b) Difference electron distribution (total distribution minus "perfect sharing" distribution)

Figure 14-16 The electron distribution in LiH (electrons per cubic angstrom). (a) Total distribution: The two 1s electrons cause the density to rise rapidly in the vicinity of the lithium nucleus. (b) Difference electron distribution: The dashed contours indicate an electron density less than that predicted by the "perfect sharing" model. Electron density has moved from the regions indicated by these contours into regions marked by positive contours. Thus the distribution is skewed toward the hydrogen nucleus.

research literature. Even more informative, though, is Figure 14-16(b) in which something approximating the "perfect sharing" probability has been subtracted. This contour map shows where the LiH molecule has higher electron probability (solid contours) and lower probability (broken contours) than that dictated by superimposing the H atom and Li atom distributions. Figure 14-16(b) clearly shows that the electrons have moved toward the H atom and away from the more symmetrical "perfect sharing" distribution.

One would think that this migration of electrons toward the H atom would have electrical consequences. The positive charge, all located at the nuclei, hasn't moved, but the negative charge has. This skewing of electron distribution causes the center of negative charge to no longer be at the same spot as the center of positive charge. Such a charge distribution is called an *electric dipole*. Figure 14-17 shows how an electric field exerts a torque on this charge distribution. Whether the field is imposed by condenser plates with an external battery, or by another molecule, a molecular dipole tends to orient into the position most favorable from an energy point of view. This orienting response to an applied field lets us measure these electric dipoles. The corresponding response to a nearby molecule implies that these dipoles affect interactions between molecules. We already know that the skewed electron distributions also affect the bond energy. So molecular dipoles are extremely important in chemistry.

The magnitude of the torque felt by a molecular dipole depends on two factors: how much charge has been displaced and how far it has been moved. The product of these two, (amount of charge) times (distance moved), is called the dipole moment. These quantities are so important

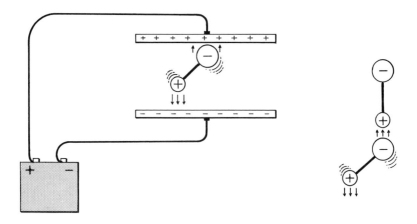

(a) A field imposed by
an experimenter

(b) A field imposed by another
molecular dipole

Figure 14-17 A molecular dipole responds to an electric field. (Molecules are represented by a highly simplified ball-and-stick model.)

that scientists have developed a variety of ways to measure them. That of LiH is $5.9 \cdot 10^{-18}$ esu-cm or, in more conventional units, 5.9 D (1 D = one "Debye" = 10^{-18} esu-cm).

Now let's return to our immediate interest, the bond in lithium hydride. We should recall that the charge movement that caused the dipole moment also caused the bond to be stronger than first predicted. Hence the charge movement changes the character of the bond. Chemists say that a bond with a dipole moment has "ionic character." The amount of ionic character is connected to the deviation from the symmetrical sharing of the bonding electrons. We have two measures of it—the magnitude of the dipole moment and the energy excess $D_0 - \bar{D}$, given by (14-10). The estimating of extent of ionic character has been the subject of volumes and volumes of controversy—a measure of the importance of a concept.

(d) A CONTRAST OF THREE MOLECULES WITH IONIC CHARACTER

With the three kinds of atoms, hydrogen, lithium and fluorine, we can form three heteropolar diatomic molecules, LiH, HF and LiF. The two lithium compounds are extremely reactive and can be obtained in the gas phase only at high temperature and under carefully controlled conditions. The third, hydrogen fluoride, is a common laboratory chemical. It is reactive compared with other compounds on the laboratory shelf, but gaseous HF is easily obtained. All three of these molecules have been well studied, again both experimentally and theoretically. We might compare them.

As a starter, consider the ionization energies for the first electron in each atom: Li, 124 kcal/mole; H, 313.6 kcal/mole; F, 402 kcal/mole. Hydrogen holds its valence electron more tightly than lithium, but fluorine holds its most easily removed valence electron more tightly still. If hydrogen can pull electrons toward itself in LiH, then fluorine should pull the electrons its way, and away from hydrogen, in HF. Even larger charge movement, or ionic bond character, is to be expected in LiF.

Table 14-6 shows the relevant quantities: $D_0 - \bar{D}$ and the dipole moments, which are symbolized μ (mu). As expected, both $D_0 - \bar{D}$ and μ for lithium fluoride are higher than the corresponding values for both HF and LiH. Figure 14-18 completes the description with contour maps for HF and LiF, showing both the total electron probability distributions and the

Table 14-6   *Evidence of Skewed Charge Distributions (Ionic Bonds) in LiH, HF and LiF*

	D_0 (kcal/mole)	$D_0 - \bar{D}$ (kcal/mole)	μ (Debye)	$\delta = \dfrac{\mu}{r_0 e}$
LiH	58	7	5.9	0.76
HF	135	73	1.9	0.43
LiF	137	107	6.3	0.84

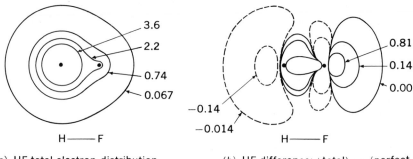

(a) HF total electron distribution

(b) HF difference: (total) — (perfect sharing) distribution

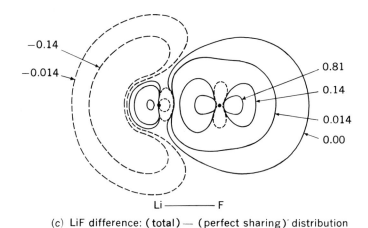

(c) LiF difference: (total) — (perfect sharing) distribution

Figure 14-18 Electron distributions in hydrogen fluoride and lithium fluoride (electrons per cubic angstrom).

difference maps. The contours show that the dipole moments in the two fluorides are caused by charge movement toward fluorine and away from its lower ionization energy partner, H or Li.

As mentioned in the last section, the dipole moment measures the product of the charge displaced and the distance it is moved. An elementary way of expressing this product is to assume that the experimental dipole moment is caused by movement of a fraction of an electronic charge all the way from one atom to the other. This fractional charge is usually symbolized δ (delta), and it is calculated by dividing μ by the electron charge and the equilibrium bond length, r_0. The values of δ for our three molecules are given in the last column of Table 14-6. These can be taken as a measure of ionic character in these bonds. Symmetrical sharing would cause no charge displacement, so $\delta = 0$, and the ionic character would be zero. The most extreme situation would be that obtained with $\delta = 1$, in which both bonding electrons are centered on one of the atoms. We would call that a bond with 100 percent ionic character. Table 14-6 shows that on this scale HF has 43 percent ionic character, whereas LiF has 84 percent.

(e) ELECTRON SHARING IN AN IONIC BOND

Our consideration of the energetics of bond formation led to the conclusion that valence electrons must be simultaneoulsy near two nuclei. Yet it is often stated that ionic bonds form *because* an electron is removed from an atom that releases its electrons readily and placed on an atom that will hold it tightly. This is not true—even in the skewed electron distributions, the bonding electrons must remain near both nuclei or the bond would not form. This can be seen for LiF, one of the most ionic bonds known, by considering its formation in steps.

$$Li(g) \longrightarrow Li^+(g) + e^- \qquad \Delta H = +124 \text{ kcal} \qquad (14\text{-}11a)$$

$$e^- + F(g) \longrightarrow F^-(g) \qquad \Delta H = -79.5 \text{ kcal} \qquad (14\text{-}11b)$$

step 1
formation of ions

$$Li(g) + F(g) \longrightarrow Li^+(g) + F^-(g) \qquad \Delta H_1 = +44 \text{ kcal} \qquad (14\text{-}12)$$

step 2
bond formation
from ions

$$Li^+(g) + F^-(g) \longrightarrow LiF(g) \qquad \Delta H_2 = ? \qquad (14\text{-}13)$$

over-all process
step 1 + step 2

$$Li(g) + F(g) \longrightarrow LiF(g) \qquad \Delta H = -137 \text{ kcal} \qquad (14\text{-}14)$$

Since the two-step process (14-12) plus (14-13) gives the over-all reaction, the heat effects in (14-12) and (14-13) must sum to the bond energy given in (14-14),

$$\Delta H_1 + \Delta H_2 = \Delta H$$

$$+44 + \Delta H_2 = -137 \text{ kcal}$$

$$\Delta H_2 = -181 \text{ kcal} \qquad (14\text{-}15)$$

Assessing these quantities, we see that the formation of ions (14-12) by no means lowered the energy of the system. Just the opposite! Forty-four kilocalories of energy are *absorbed* to make a mole of gaseous Li^+ and F^- ions. Even though lithium has a relatively low ionization energy, it is well above the electron affinity of the fluorine atom. So in our two-step process, the energy is lowered in the second step as the F^- ion moves its electrons back toward Li^+. In fact, at the equilibrium bond length, the F^- electrons are 1.56 Å away from Li^+, on the average. This is *closer* to the lithium nucleus, on the average, than is the 2s valence electron in an isolated lithium atom. In Table 13-1, the average radius of lithium was calculated to be 2.52 Å. So the effect of skewing electron distribution in ionic bonds definitely does not *remove* valence electrons from the vicinity of either atom. The redistribution tends to center them near the atom that holds them more tightly, while keeping them simultaneoulsy near both nuclei. *All bonds form because electrons are simultaneously near two or more nuclei.*

14-6   On to bigger game

We have considered six molecules: H_2^+, H_2, Li_2, LiH, HF, and LiF. These simple examples display most of the ideas needed to understand all chemical bonding.

(i) When any bond forms, it is because the potential energy drops as the atoms come together.

(ii) Potential energy can drop as atoms come together because the electrons can then be near two or more nuclei simultaneously.

(iii) As in atoms, it is useful to describe the movement of electrons in a molecule in terms of orbitals. However, an electron moving simultaneously near two nuclei occupies an orbital that is molecular in character, that is, a *molecular orbital* (or an M.O.).

(iv) The Pauli Principle applies to M.O. occupancy just as it does to atomic orbitals: At most, two electrons can occupy a given M.O. (and these electrons have opposed spins).

(v) The nature of an M.O. is determined by its nodal surfaces, which can be deduced from the parent atomic orbitals of the separated atoms. These parent orbitals can be considered to contribute to the M.O. either in phase with each other, or out of phase, as governed by their wave nature. The in-phase combination concentrates electrons in the binding region and hence lowers the energy. This is a bonding M.O. The out-of-phase combination has a nodal surface between the nuclei and concentrates electrons in the antibinding region. This is an antibonding M.O.

(vi) Bond order is defined to be the difference between the number of pairs of electrons in bonding M.O.'s minus the number of pairs in antibonding M.O.'s.

(vii) For a given pair of atoms, as the bond order goes up, both the bond energy and the stretching force constant go up, and the bond length goes down.

(viii) When two different kinds of atoms are bonded, the electron distribution will tend to be skewed toward the atom that holds the electrons more strongly, while still remaining close to both nuclei. Such a skewed electron distribution strengthens the bond and it gives rise to an electric molecular dipole. A molecule with an electric dipole is said to have a *dipole moment* and its bond is said to have ionic character.

So far, though, we've been careful to consider only simple atoms in simple molecules. Turning to atoms with more electrons and to molecules with more atoms, we find the chemical bonding problem more difficult to handle. Chemists then use approximate methods based upon the principles we've just reviewed—those established for the simpler molecules. The next chapter shows how.

Problems

1.   A hydrogen atom in the 2p state emits a quantum of light of energy 235.2 kcal/mole and drops to the 1s state.
(a) On a mole basis, calculate for each state the energy $\bar{E}$ (expression (12-11)) and, from the Virial Theorem, the values of $\bar{T}$ and $\bar{V}$.
(b) From the energy change in the light emission, $\Delta \bar{E} = -235.2$ kcal/mole, and the Virial Theorem, deduce the values of $\Delta \bar{T}$ and $\Delta \bar{V}$.
(c) Calculate the values of $\Delta \bar{E}$, $\Delta \bar{T}$, and $\Delta \bar{V}$ from your answers to (a), and verify the values obtained in (b).

2.   The reaction between two oxygen atoms to form an oxygen molecule liberates 118 kcal/mole

$$O(g) + O(g) \longrightarrow O_2(g) \qquad \Delta \bar{E} = -118 \text{ kcal}$$

Use the Virial Theorem to calculate the changes in potential energy, $\Delta \bar{V}$, and in the kinetic energy of the electrons, $\Delta \bar{T}$, that accompany the reaction.

3.   Which of the following true statements can be said to provide an explanation of why a chemical bond can form between two chlorine atoms as they approach.
(a) The positional randomness of two moles of chlorine atoms exceeds that of one mole of Cl_2 molecules.
(b) For the reaction $2\ Cl(g) \longrightarrow Cl_2(g)$, $\Delta S^0 = -25.7$ cal/deg.
(c) For the reaction $2\ Cl(g) \longrightarrow Cl_2(g)$, $\Delta H^0 = -57.9$ kcal/mole.
(d) The nuclear–nuclear repulsion of two chlorine atoms at the Cl_2 internuclear distance is $+48,000$ kcal/mole.
(e) The electron–electron repulsions in Cl_2 exceed those in two separated Cl atoms by more than 40,000 kcal/mole.
(f) The electron kinetic energy in Cl_2 exceeds that in two separated atoms by 58 kcal/mole.
(g) The electron–nuclear potential energy is more negative for Cl_2 than for two separated Cl atoms.

4.   The repulsion between two nuclei A and B gives a potential energy

$$\bar{V}_{AB} = + \frac{(Z_A q)(Z_B q)}{r_{AB}}$$

where q is the charge on the proton, Zq is the charge on the nucleus, and r_{AB} is the internuclear distance. Substituting $q = 4.803 \cdot 10^{-10}$ esu and converting to kcal/mole, this expression becomes

$$\bar{V}_{AB} = +332.0 \frac{Z_A Z_B}{r_{AB}}$$

with r_{AB} in Å.

(a) Calculate $\bar{V}_{AB}$ for two protons ($Z_A = Z_B = 1$) at $r_{AB} = 0.25$, 0.76 (r_0 in H_2), 1.06 (r_0 in H_2^+), 2.50, and 5.00 Å.

(b) Make a plot of these values and draw a smooth curve through the points.

(c) Calculate the ratio of the bond energy to the nuclear repulsion energy at the equilibrium bond length for each molecule, H_2^+ and H_2.

5. Repeat Problem 4 for the gaseous lithium hydride molecule, LiH, at distances 0.75, 1.00 1.61 (r_0 in LiH), 2.00, and 5.00 Å. Add this curve to the plot obtained in Problem 4.

6. Repeat Problem 4 for the gaseous dilithium molecule, Li_2, at distances 2.00, 2.68 (r_0 in Li_2), 4.00, and 5.00 Å.

7. The following table reproduces the data for H_2^+ in Table 14-1 and adds the corresponding quantities for H_2. From the Virial Theorem and the results of Problem 4, fill in all of the blanks for H_2.

	H_2^+ ($r_0 = 1.06$ Å)			H_2 ($r_0 = 0.76$ Å)		
	Separated Atoms $H + H^+$	Mole-cule H_2^+	Net Change	Separated Atoms $H + H$	Mole-cule H_2	Net Change
Contributions to V						
electron–electron repulsions	—none present—			_____	+681	_____
nuclear–nuclear repulsions	0	+313	+313	_____	_____	_____
electron–nuclear attractions	−627	−1069	−442	_____	_____	_____
$\bar{V}$	−627	−756	−129	_____	_____	_____
$\bar{T}$	+313.6	+378	+64.5	_____	_____	_____
$\bar{E}$	−313.6	−378	−64.5	−627	_____	−104

8. If a hydrogen atom and a proton come together with zero translational energy, they "roll downhill" on the potential function shown in Figure 14-9. Unless energy is removed (through simultaneous collision with a third particle) they "roll up" the steep repulsive wall at close approach until all of the energy is again potential energy. Then the particles roll backward and separate again without reaction. According to Figure 14-9, what is the internuclear distance at this "turn-around" point?

9. Redraw the correlation diagram for H_2^+ (see Fig. 14-6) using a linear internuclear distance scale extending from 0 to 5 Å and a compressed energy scale extending from +1500 to −3000 kcal. Take as the zero of energy the system $H^+ + H^+ + e^-$ and use the tabulated data in Exercise 14-4 where useful.

(a) On the right-hand vertical axis, indicate for $H^+ + H$ the total energy, $\bar{E} - \bar{V}_{AB}$, the electron kinetic energy, $\bar{T}$, the electron–nuclear potential energy, $\bar{V} - \bar{V}_{AB}$, and the nuclear–nuclear repulsive energy, $\bar{V}_{AB}$ (which equals zero at $r_{AB} = \infty$).

(b) On the left-hand vertical axis, indicate for He^+ the values of $\bar{E} - \bar{V}_{AB}$, $\bar{T}$, and $\bar{V} - \bar{V}_{AB}$. (Notice that $\bar{V}_{AB} = \infty$ at $r_{AB} = 0$.)

(c) At $r_{AB} = 1.06$ Å, draw a vertical line and indicate for H_2^+ the values of $\bar{E} - \bar{V}_{AB}$, $\bar{T}$, $\bar{V} - \bar{V}_{AB}$, and $\bar{V}_{AB}$.

(d) Add to the graph the values of $\bar{V}_{AB}$ calculated in Problem 4, and draw a smooth curve through these points and another curve through the three $\bar{T}$ points.

(e) Use your estimate of the "turn-around" point from Problem 8 to give another estimate of $\bar{E}$, hence of $\bar{E} - \bar{V}_{AB}$ to add to your graph. The difference between $\bar{E} - \bar{V}_{AB}$ and your graphical estimate of $\bar{T}$ at this "turn-around" point gives you one more point on the steep plot of $\bar{V} - \bar{V}_{AB}$. Add this point and draw smooth curves through all points.

(f) Read off values of $(\bar{E} - \bar{V}_{AB})$ at $r = 0.30, 0.40, 0.60, 0.80, 1.06$, and 1.40 Å. To each value, add the appropriate $\bar{V}_{AB}$ and draw a curve for $\bar{E}$ versus r_{AB}.

10. In the reaction considered in Problem 2, there are eight electrons in each oxygen atom, hence sixteen in O_2. For the separated atoms, there are $2 \cdot 8 = 16$ electron–nuclear attractions and $2 \cdot [\frac{1}{2}(8 \cdot 7)] = 56$ electron–electron repulsions.

(a) For the O_2 molecule, how many electron–nuclear attractions are there?

(b) For the O_2 molecule, how many electron–electron repulsions are there?

(c) Calculate the nuclear–nuclear repulsive energy, $\bar{V}_{AB}$, at the equilibrium bond length in O_2, 1.21 Å.

(d) Combine the answer to (c) with that of Problem 2 to deduce the sum of the energy drop due to increased electron–nuclear attractions and energy rise due to increased electron–electron attractions.

(e) Divide the answer to (d) by the bond energy to demonstrate that the bond energy is a small difference among very large energy contributions.

11. If two neutral helium atoms, $He(1s^2)$, come together, they have an almost negligible bond energy (0.02 kcal/mole). However, if an excited helium atom, $He(1s,2s)$ approaches another helium atom, $He(1s^2)$, a relatively strongly bonded molecule forms, with bond dissociation energy 59 kcal (*Journal of Chemical Physics*, Vol. **49**, p. 4817, 1968).

(a) The $He(1s,2s)$ state is 476 kcal/mole above the $He(1s^2)$ ground state. What is its ionization energy?

(b) Use the one-electron approximation to calculate Z^* and $\bar{r}$ for the 2s electron.

(c) The bond length in this excited He_2 molecule is about 1.05 Å. Discuss the bonding in terms of the orbital occupancy of the H_2^+ molecular orbitals (see Fig. 14-8), and explain why the excited electron seems to occupy neither a bonding nor an antibonding orbital.

12. (a) Calculate the "perfect sharing" bond energies $\bar{D}$ for HF, HCl, HBr, and HI from the bond dissociation energies in Table 7-3.

(b) Calculate the bond energy excesses, $(D_0 - \bar{D})$ for each of these molecules as an indication of the extent of electron skewing, i.e., of ionic character.

(c) Use the experimental dipole moments and bond lengths for these molecules, given in Appendix e, to calculate the charge movement δ needed to account for the molecular dipole moments

$$\delta = \frac{\mu}{r_0 q} = \frac{\mu(D) \cdot 10^{-18} \text{ esu-cm}}{r_0(\text{cm}) \cdot 4.80 \cdot 10^{-10} \text{ esu}}$$

(d) Plot δ versus $(D_0 - \bar{D})$ to see if these two measures of ionic character are consistent for these similar molecules.

13. Repeat parts (a) and (b) of Problem 12 for the *average* CH, CF, CCl, and CBr bonds in, respectively, CH_4, CF_4, CCl_4, and CBr_4.

(a) List the four bonds in order of decreasing ionic character.

(b) Which of these three molecules would you expect to have the largest molecular dipole moment and which the smallest: H_2CBr_2, H_2CCl_2, H_2CF_2?

14. Repeat Problem 12 for the gaseous diatomic molecules LiF, LiCl, LiBr, and LiI. The bond energies are, respectively, 137, 112, 100, and 85 kcal/mole; the dipole moments and bond lengths are given in Appendix e. (While each of Problems 12 and 14 shows a good correlation between δ and $D_0 - \bar{D}$, the two plots superimposed show that only quite similar molecules can be compared.)

fifteen simple molecular orbitals

The Schroedinger Equation can be solved exactly for the electronic energies of the hydrogen molecule ion H_2^+. With high-speed digital computers, chemists have correctly calculated properties for other small molecules— molecules such as H_2, Li_2 and LiH. However, even the last example, LiH, contains only two atoms and four electrons. Meanwhile, experimentalists are busily preparing and working with hundreds of thousands of molecules, most of which contain atoms by the dozen and electrons by the gross. These chemists evidently have some straightforward and quite effective ways of predicting and rationalizing molecular structures. Now, since quantum mechanics has proved itself, it permits us to evaluate and understand the earlier, empirical models of bonding. More important, it guides us as we address the as-yet unsolved problems of chemical bonding and as we improve the existing approximate theories.

The molecular orbital treatment is an approximate theory. To be sure, the M.O. concept is exact in a molecule containing only one electron, as in H_2^+, but in a many-electron molecule, it becomes a useful approximation, in precise analogy to the atomic orbital concept applied to many-electron atoms. The electrons are considered to occupy M.O.'s that are about the same as they would be if electron repulsions could be turned off. The Pauli Principle is applied as the orbitals are filled and, of course, we expect to see effects due to electron–electron repulsions when finer details are considered.

The diatomic molecules across the first row of the Periodic Table provide some more good examples. However, we'll return for a moment to the M.O.'s in H_2 to illustrate the approximate methods needed.

15-1 H_2^+: A source of approximate M.O.'s for H_2

Figure 15-1 is one way to look at the interaction of two 1s orbitals as a bond forms between two hydrogen atoms. The solid line in Figure 15-1(a) indicates how the wave function varies along the line connecting the two protons. The upward direction represents one phase and the downward direction the opposite phase (think of the vibrating guitar string). Figures

559

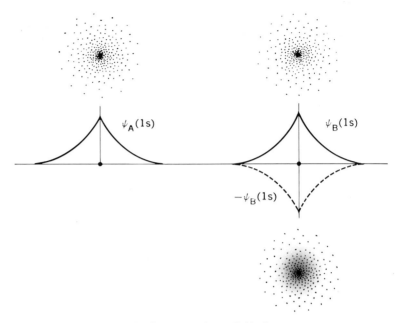

$\psi_A(1s)$

$\psi_B(1s)$

$-\psi_B(1s)$

(a)   Separated Atom Orbitals

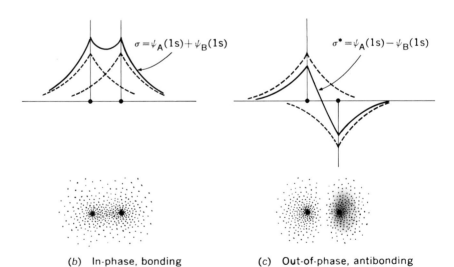

$\sigma = \psi_A(1s) + \psi_B(1s)$

$\sigma^* = \psi_A(1s) - \psi_B(1s)$

(b)   In-phase, bonding
MO combination

(c)   Out-of-phase, antibonding
MO combination

Figure 15-1   *Approximate molecular orbitals in* H_2. *(Phase relationships are shown by shading in the probability plots.)*

15-1(b) and 15-1(c) show how the M.O.'s form as the two possible phase combinations occur. A possible first description of the σ bonding M.O. is $\psi_A(1s) + \psi_B(1s)$; simply the superposition of the two 1s orbitals at the equilibrium bond length, with the in-phase relationship. The σ^* antibonding M.O. can be roughly described as $\psi_A(1s) - \psi_B(1s)$, the out-of-phase superposition. The approximations to σ and σ^* are not sufficiently accurate to furnish quantitative estimates of energy, bond length, and stretching constants, but they certainly do give a correct qualitative picture. The most important qualitative feature they display is the *nodal pattern* of the molecular orbital.

We have already used these nodal patterns to develop the molecular orbitals of H$_2^+$ (in Section 14-3(c)). One of the principal values of the correlation diagram is to provide tie points (at either end of the diagram) at which the nodal pattern is known. Thus, the connection to the He$^+$ 1s orbital indicates that one of the H$_2^+$ molecular orbitals has no nodal surfaces except the one at infinity. The phase relationships then give us the clue to the nodal pattern of the next M.O. It connects to the 2p$_x$ orbital, with a nodal plane perpendicular to the line along which the two H atoms join to form the united He atom. While the two atoms are still spatially separated, this central nodal plane identifies this M.O. as an antibonding orbital.

Having now decided on the nodal patterns of the two lowest-energy M.O.'s of H$_2^+$, we used them to discuss the two-electron molecule H$_2$. In conformity with the Pauli Principle, both of the electrons can occupy the lower orbital, the bonding one. Even though electron repulsions can be expected to affect the molecular properties, to a first approximation the bond in H$_2$ should be about twice as strong as that in H$_2^+$. It should require about twice the energy to break the H$_2$ bond (as it does), the atoms should be closer together (as they are), and the "spring constant" (the vibrational force constant) should be higher (as it is).

Evidently this simple scheme works well—at least qualitatively. As the two atoms H$_A$ and H$_B$ come together, their lowest orbitals 1s$_A$ and 1s$_B$ are considered to offer parentage to two molecular orbitals. What are these two M.O.'s like? That's easy; the scheme says they have the nodal patterns of the in-phase and out-of-phase combinations,

$$\sigma(1s) = 1s_A + 1s_B$$

$$\sigma^*(1s) = 1s_A - 1s_B$$

The nodal patterns, in turn, identify $\sigma(1s)$ as a bonding orbital (it concentrates electron probability in the region between the nuclei) and $\sigma^*(1s)$ as an antibonding orbital (it moves electron probability away from the region between the nuclei). The energy level diagram is qualitatively defined. The number of electrons to be accommodated and the Pauli Principle take it from there. This approach is summarized in Figure 15-2.

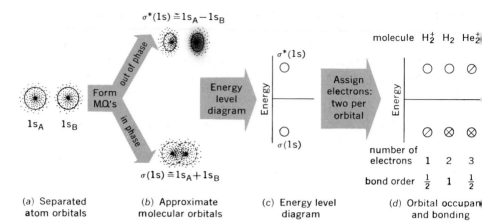

Figure 15-2   The molecular orbital idea.

15-2   Approximate M.O.'s for 2p orbitals

We might exercise these ideas by imagining the bonding questions two H atoms would ask themselves if they approached each other in energetically excited, 2p states. There are two situations to consider, 2p orbitals directed along the molecular axis ($2p_x$ orbitals) and 2p orbitals directed perpendicular to the molecular axis ($2p_y$ and $2p_z$ orbitals). We'll take the axial case first.

(a)   AXIAL $2p_x$ M.O.'s

Figure 15-3(a) shows two separated H atoms, each excited to a $2p_x$ state, with the $2p_x$ orientation along the internuclear axis. These axially directed p orbitals give σ type molecular orbitals.

Figures 15-3(b) and 15-3(c) use the same ideas introduced in Figure 15-1. The "in-phase" combination concentrates electron probability between the nuclei to give a bonding orbital, $\sigma(2p_x) \cong \psi_A(2p_x) + \psi_B(2p_x)$. The out-of-phase combination has a nodal plane between the two nuclei. It is an anti-bonding orbital, $\sigma^*(2p_x) \cong \psi_A(2p_x) - \psi_B(2p_x)$.

(b)   PERPENDICULAR $2p_y$ AND $2p_z$ M.O.'s

Consider now two H atoms, excited to the 2p state, approaching each other but with their 2p orbitals oriented perpendicular to the bond axis. Either the $2p_y$ or the $2p_z$ orientation may be pictured—they are exactly equivalent except for a 90° rotation around the bond axis. Figure 15-4 shows the M.O. formation in this case, as suggested by the in-phase and out-of-phase combinations of the parent orbitals.

There is, however, an obvious difference between these M.O.'s and those pictured thus far in Figures 15-1 and 15-3. In Figure 15-4, no electron probability is found along the bond axis. Instead, the M.O. retains the xz nodal plane possessed by the parent $2p_y$ orbitals. Here is another nodal characteristic with which to classify our M.O.'s. The sigma (σ) orbitals, concentrated along the bond axis, have no nodal surface passing through the bond axis, whether they are formed from s atomic orbitals or axially directed p orbitals. An M.O. formed from the perpendicular p orbitals does have such a surface. Such perpendicular M.O.'s, parented by p orbitals, are called π (pi) orbitals. Because of the difference in spatial distribution, sigma (σ) and pi (π) M.O.'s do not generally have the same energy. On the other hand, there are two bonding M.O.'s that are exactly equivalent; one formed from the $2p_y$ orbitals and one from the $2p_z$ orbitals. Since they are equivalent (except for a rotation through 90° around the bond axis), their energies are identical. The same is true for the two antibonding M.O.'s formed from the $2p_y$ and $2p_z$ orbitals.

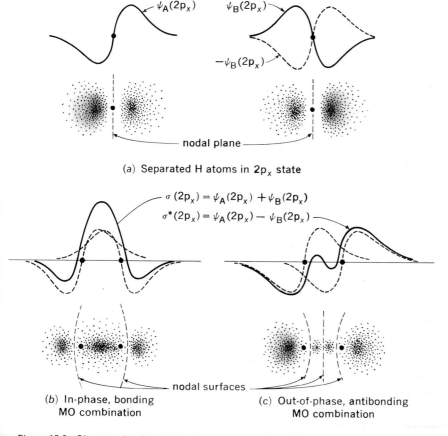

(a) Separated H atoms in $2p_x$ state

$$\sigma\,(2p_x) = \psi_A(2p_x) + \psi_B(2p_x)$$
$$\sigma^*(2p_x) = \psi_A(2p_x) - \psi_B(2p_x)$$

nodal surfaces

(b) In-phase, bonding
MO combination

(c) Out-of-phase, antibonding
MO combination

Figure 15-3   Sigma molecular orbitals from axial p orbitals. (The relative phases of the wave functions are shown by shading.)

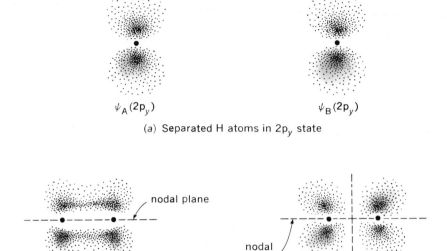

$\psi_A(2p_y)$ $\psi_B(2p_y)$

(a) Separated H atoms in $2p_y$ state

nodal plane

$\pi(2p_y) = \psi_A(2p_y) + \psi_B(2p_y)$

nodal planes

$\pi^*(2p_y) = \psi_A(2p_y) - \psi_B(2p_y)$

(b) In-phase bonding
M.O. combination

(c) Out-of-phase antibonding
M.O. combination

Figure 15-4   Pi molecular orbitals from perpendicular p orbitals. (The relative phases of the wave functions are shown by shading.)

Figure 15-5 summarizes the M.O.'s that can be considered to have their parentage primarily in the 2p orbitals of excited hydrogen atoms. All of the orbitals are placed on an energy scale appropriate to separated 2p orbitals. The bonding M.O.'s are lower and the antibonding orbitals higher in energy than the energies of the isolated 2p orbitals. The σ M.O.'s are shown below the π M.O.'s, although it is not necessarily that way. We can always expect the axial and perpendicular orbitals to have different energies, however, since their geographic distributions are so different.

15-3   M.O.'s for the oxygen molecule

All of this permits us to consider the M.O. energy level diagram for a molecule with quite a few electrons. Oxygen, O_2, is a good place to begin.

Figure 15-6 shows the energy levels of a single oxygen atom. There are large energy separations between the 1s and 2s orbitals, between the 2s and 2p orbitals, and again between the 2p and still higher orbitals. The implication is that the parentage of M.O.'s will be fairly simply related to these isolated atomic orbitals.

Figure 15-7 shows M.O. formation as two oxygen atoms approach. The M.O.'s all look familiar—the bottom part of the diagram looks just like that

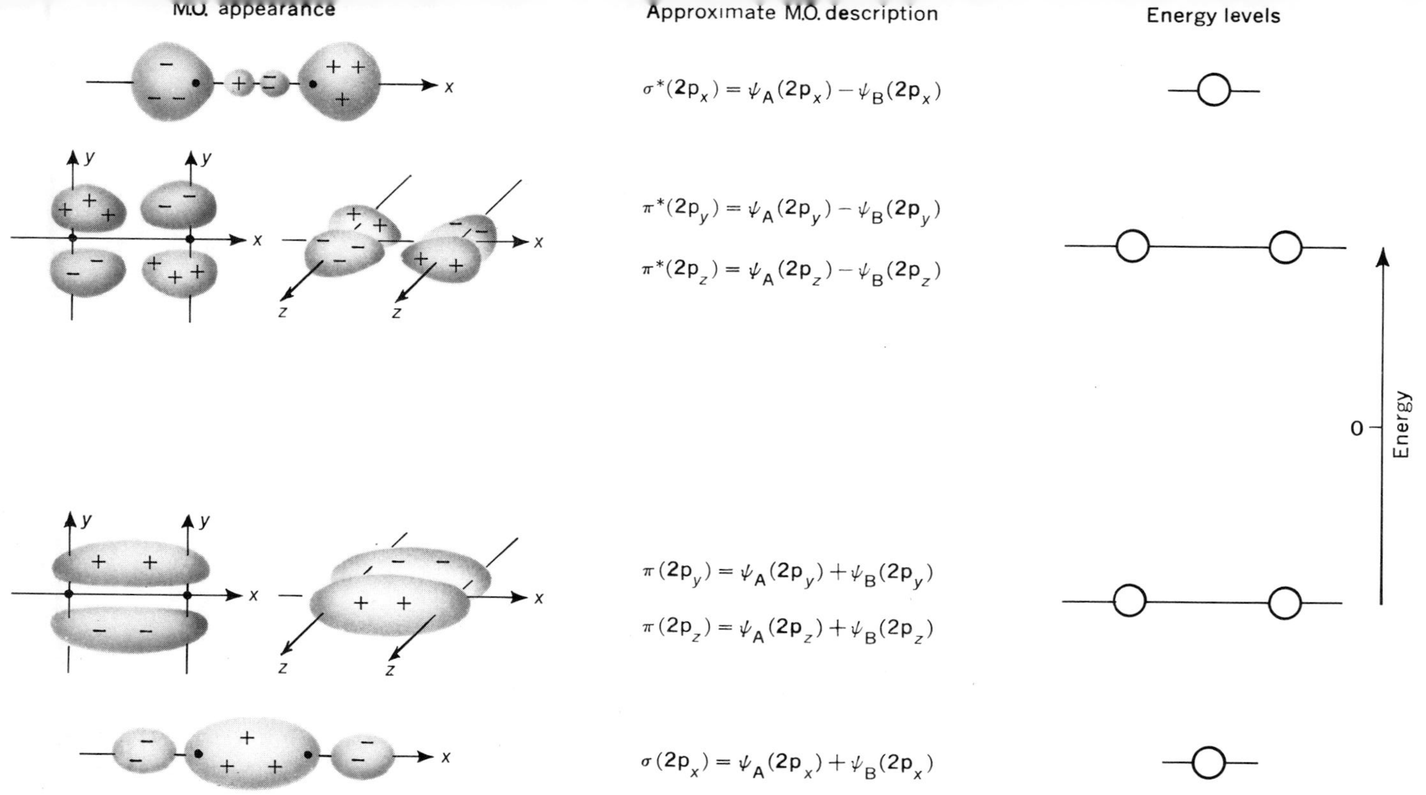

Energy levels

Approximate M.O. description

M.O. appearance

Energy

0

$\sigma^*(2p_x) = \psi_A(2p_x) - \psi_B(2p_x)$

$\pi^*(2p_y) = \psi_A(2p_y) - \psi_B(2p_y)$

$\pi^*(2p_z) = \psi_A(2p_z) - \psi_B(2p_z)$

$\pi(2p_y) = \psi_A(2p_y) + \psi_B(2p_y)$

$\pi(2p_z) = \psi_A(2p_z) + \psi_B(2p_z)$

$\sigma(2p_x) = \psi_A(2p_x) + \psi_B(2p_x)$

Figure 15-5 Molecular orbitals from 2p orbitals. The boundaries represent approximately the 95 percent probability surface. Within, there is a 95 percent chance of finding an electron. The relative phases are shown by plus and minus signs. These signs have nothing whatsoever to do with charge—they simply point out that the wave function that describes the orbitals has phase properties.

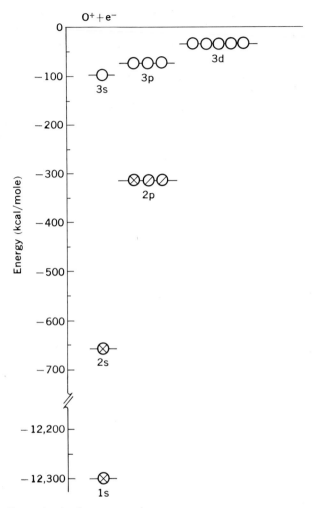

Figure 15-6 Energy levels of an oxygen atom.

of dilithium, Figure 14-11. The upper part is the same as Figure 15-5. All we need do is decide orbital occupancy in accordance with the Pauli Principle.

Each oxygen atom gives us eight electrons. Four of these reside in the $\sigma(1s)$ and $\sigma^*(1s)$ orbitals, which make no net contribution to the bonding. Four more electrons can be placed in the next orbitals, $\sigma(2s)$ and $\sigma^*(2s)$. Again, because both bonding and antibonding orbitals are filled, there is no net bonding. We now have $16 - 4 - 4 = 8$ electrons remaining. The energy level diagram shows three bonding M.O.'s—$\sigma(2p_x)$ and the pair $\pi(2p_y)$, $\pi(2p_z)$. Six electrons into these bonding M.O.'s give three pairs of bonding electrons—sufficient for a bond order of three. However, we have disposed of only 14 of the 16 electrons. The remaining pair of electrons must go into the antibonding orbitals, reducing the bond order, again, to two. Oxygen should have a double bond.

This last pair of electrons has two options. They could both enter one of the π^* orbitals, with opposite spin, of course. This orbital occupancy is shown in Figure 15-8(a). However, another possibility is shown in Figure 15-8(b). As far as the simplified, "turned-off electron" approximation is concerned, these two occupancies have the same energy.

It is not difficult to see how the electron repulsions will differentiate the two occupancies. Two electrons in the same orbital $\pi^*(2p_y)$, give large electron repulsions because the electrons occupy the same region of space.

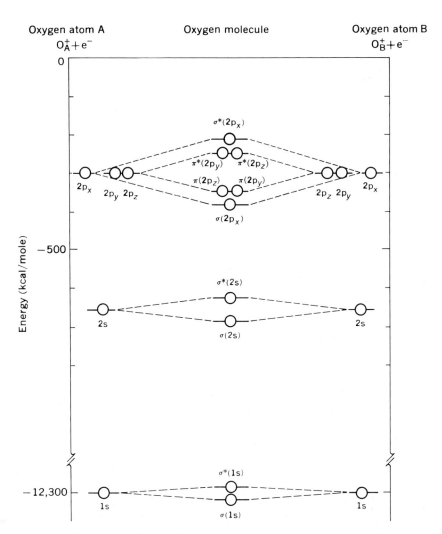

Figure 15-7   Molecular orbitals for O_2.

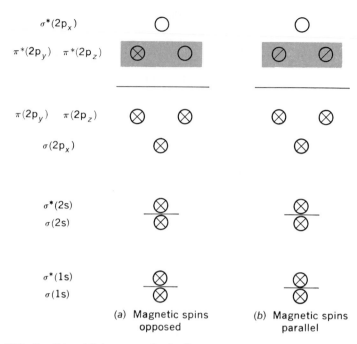

Figure 15-8   Possible orbital occupancies for O_2.

When one electron is moved to the second orbital $\pi^*(2p_z)$, it occupies a different region of space. Then the electrons are farther apart, so electron repulsions are reduced. The right-hand electron configuration has lower energy.

We conclude that there should be two states of O_2 reasonably close in energy. Both states have three pairs of bonding electrons and one pair of antibonding electrons in the uppermost M.O.'s, so both states have double bonds. The lower of the two states—lower because of reduced electron repulsions—can accommodate the last two electrons with parallel magnetic spins. This means the molecule is a tiny magnet and it should respond, somehow, to external magnetic fields. Because of this, certain isolated spectroscopic transitions appear as triplets in an imposed magnetic field. The higher of the two states, the one with the last two electrons in the same orbital $\pi^*(2p_y)$, cannot accept this last pair unless they have opposed spins. There will be no magnetic behavior—isolated spectroscopic transitions will remain as singlet lines even if a magnetic field is imposed.

Table 15-1 shows how well these expectations are realized. The M.O. picture is completely consistent with the known properties of the oxygen molecule in its two lowest energy states. In fact, this was one of the most important successes of the molecular orbital view, early in the development of our quantum mechanical view of bonding.

Table 15-1   Two Lowest-Energy States of Oxygen Molecule

Upper Orbital Occupancy	$\sigma(2p_z)^2\pi(2p_y)^2\pi(2p_z)^2\pi^*(2p_y)\pi^*(2p_z)$	$\sigma(2p_z)^2\pi(2p_y)^2\pi(2p_z)^2\pi^*(2p_y)^2$
name	"triplet"	"singlet"
energy	(0)	+22.5 kcal/mole
bond energy	118 kcal/mole	96 kcal/mole
bond length	1.21 Å	1.22 Å
force constant	11.4 mdyne/Å	10.7 mdyne/Å
apparent bond order	2	2
magnetic properties	magnetic	nonmagnetic

15-4   M.O.'s for other first-row homonuclear diatomic molecules

Figure 15-9 shows the energies of the 2s and 2p electrons for the atoms boron to neon. Fluorine and neon have a large energy gap between these orbitals: over 350 kcal—as does oxygen. These atoms should produce M.O. energy level diagrams qualitatively like that of O_2, like Figure 15-7. Boron, carbon, and nitrogen differ; their 2s and 2p states are only about 100 kcal

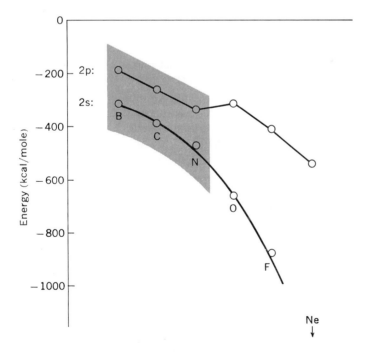

Figure 15-9   Atomic 2s and 2p energy levels for first-row atoms. The 2s and 2p orbitals of B, C and N (shaded) are close enough together to permit a significant interaction of their resultant M.O.'s.

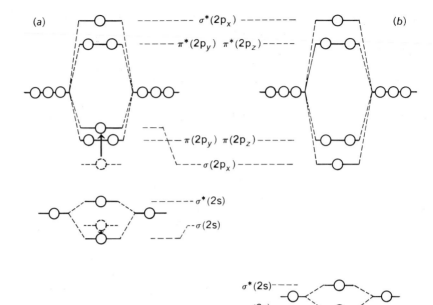

Figure 15-10 Molecular orbitals for first-row homonuclear diatomics: (a) B_2, C_2, N_2; (b) O_2, F_2.

apart. The 2s and 2p states no longer act independently and the energy level diagram is altered. The $\sigma(2p_x)$ M.O. is the one affected most because it has a nodal pattern like that of the nearby $\sigma(2s)$ M.O. Experiment shows that the effect is sufficient to move the $\sigma(2p_x)$ M.O. upward in energy, above the $\pi(2p_y)$, $\pi(2p_z)$ pair.

Figure 15-10 shows the two kinds of energy level diagrams that result. The left-hand diagram, obtained when the 2s–2p energy separation is small, is found applicable to boron, carbon, and nitrogen. The right-hand diagram applies to oxygen, fluorine, and neon.

With this background, we can proceed with the game of counting electrons to determine orbital occupancy and bond order. Figure 15-11 does this and it includes the experimental data about each molecule so that we can test the utility of the M.O. scheme.

The bond orders are predicted to begin at B_2 with a single bond, to increase to a double bond at C_2 and to a triple bond at N_2. Then the bond order decreases progressively from O_2 and F_2 and, finally, to neon, where no bond at all is expected. All of the experimental criteria are in agreement, as shown in Figure 15-12. In a reasonably symmetric way, the bond energies and force constants peak at N_2 and the bond lengths are shortest at N_2. The double bond in O_2 is close to, but weaker than, that in C_2. The difference is attributable to the higher electron–electron repulsions in O_2. The single bond in F_2 is close to, but weaker than, that in B_2, again because F_2 has many more electron repulsions affecting its bonds. No molecular species Ne_2 is predicted, nor is one observed.

These bond energies have direct relationship to the chemistry of these substances. Neon is called an inert gas—it forms no compounds, including no diatomic molecule Ne_2. Fluorine, F_2, has a very weak bond and it is extremely reactive. The double bond of oxygen makes it reasonably stable, but far more reactive than nitrogen. The different roles of atmospheric oxygen and nitrogen in life processes are causally related to this reactivity difference.

Figure 15-11 displays one more opportunity for testing the M.O. description—the magnetic properties (see Fig. 15-13). Of the six molecules, only two have orbital occupancies that permit unpaired magnetic spins; B_2 and O_2. Again this agrees with the observation that these are the only two that have magnetic properties.

This optimistic situation is strengthened even more by the properties of the two gaseous molecule ions N_2^+ and O_2^+. The M.O. diagrams suggest that the removal of an electron from N_2 should have quite a different effect from that caused by the removal of an electron from O_2. The uppermost electron

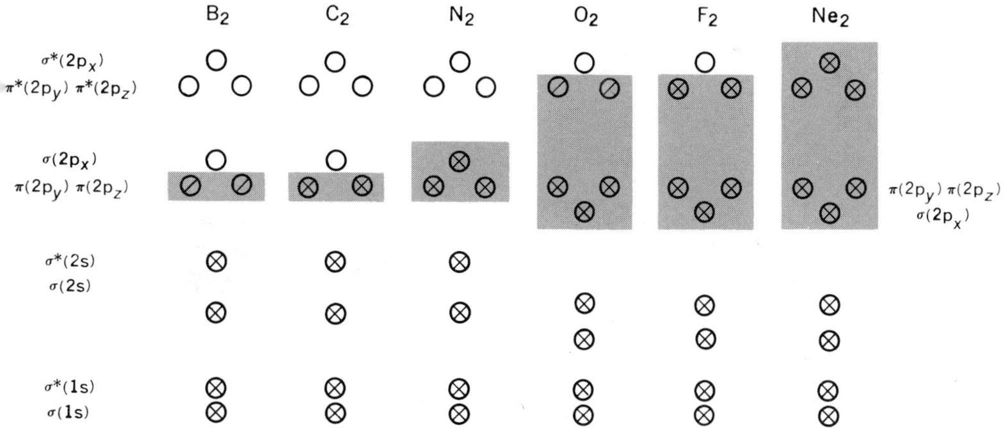

	B_2	C_2	N_2	O_2	F_2	Ne_2	
$\sigma^*(2p_x)$							
$\pi^*(2p_y)\ \pi^*(2p_z)$							
$\sigma(2p_x)$							$\pi(2p_y)\ \pi(2p_z)$
$\pi(2p_y)\ \pi(2p_z)$							$\sigma(2p_x)$
$\sigma^*(2s)$							
$\sigma(2s)$							
$\sigma^*(1s)$							
$\sigma(1s)$							
Predicted: Bond Order	1 (single)	2 (double)	3 (triple)	2 (double)	1 (single)	0 (none)	
Experimental:							
Bond energy (kcal/mole)	69	144	225	118	36	—	
Bond length (Å)	1.59	1.24	1.10	1.21	1.44	—	
Force constant (mdyne/Å)	3.5	9.3	22.4	11.4	4.5	—	
Magnetic	yes	no	no	yes	no	—	

Figure 15-11 *Orbital occupancies and bond properties of first-row homonuclear diatomic molecules.*

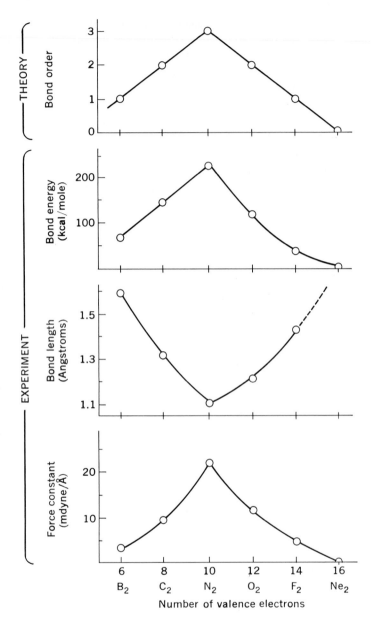

Figure 15-12   Trends in bond properties and predicted bond orders in first-row homonuclear diatomic molecules.

in N_2 is a bonding electron. Its loss should weaken the bond. The uppermost electron in O_2 is an antibonding electron. Its loss should strengthen
the bond! Figure 15-14 shows that this is exactly what is observed. The N_2^+
molecule has properties between C_2 and N_2—the nitrogen–nitrogen bond
is weaker in N_2^+ than in N_2. In contrast, the O_2^+ molecule lies between N_2
and O_2—*the oxygen–oxygen bond is stronger in O_2^+ than in O_2.*

Also shown in Figure 15-14 are the orbital occupancies and stretching
force constants for the gaseous ions N_2^- and O_2^-. (The bond energies are not
yet known, and the bond length for O_2^- refers to the crystal KO_2.) The
former, N_2^-, is the more interesting. Addition of an electron to N_2 requires
occupancy of an antibonding orbital. Hence N_2^- has a bond order of $2\frac{1}{2}$, as
do both N_2^+ and O_2^+. However, N_2^- has the same orbital occupancy as O_2^+, so,
from the molecular orbital point of view, N_2^- should be more like O_2^+ than
like N_2^+. The same electron repulsions are present in N_2^- and O_2^+, so they
should be alike (as they are), and they should both have weaker $2\frac{1}{2}$ order
bonds than the $2\frac{1}{2}$ order bond of N_2^+ (as they do). Even the small difference
between N_2^- and O_2^+ is consistent and understandable. These two ions have
the same electron repulsions, but N_2^- has lower nuclear charges with which
to cope with them. Hence the bond weakening in N_2^- is more severe than
that in O_2^+.

We see that this discussion of the first-row homonuclear diatomic
molecules successfully accounts for their observed properties in terms of
two concepts, the occupancy of a set of molecular orbitals considered in

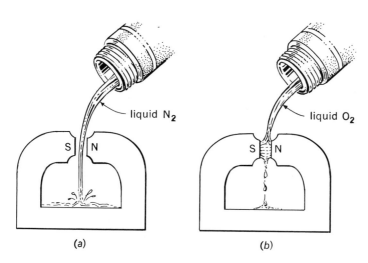

(a)   (b)

*Figure 15-13 The oxygen molecule is magnetic, the nitrogen molecule is not. (a) Liquid
nitrogen (b.p. = 77 degK) can be poured directly between the poles of a strong magnet.
(b) Liquid oxygen (b.p. = 90 degK) is attracted by the magnet and fills up the gap between
the two poles. Oxygen has two unpaired electrons (in antibonding orbitals) and is magnetic.
Nitrogen has all its electrons paired and is not magnetic.*

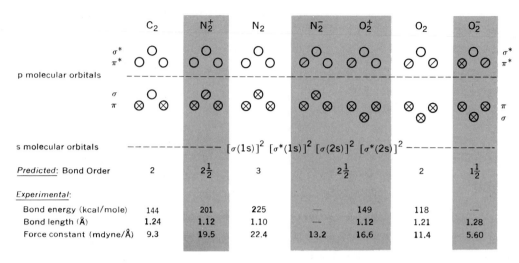

	C₂	N₂⁺	N₂	N₂⁻	O₂⁺	O₂	O₂⁻

Figure 15-14 Orbital occupancy and bond properties of some homonuclear diatomic ions.

the light of electron repulsions. This success gives significant encouragement to the M.O. description. In particular, the fact that the *removal* of an electron from O_2 *strengthens* its bond (as in O_2^+), whereas the *addition* of an electron to N_2 or O_2 *weakens* its bond (as in N_2^- and O_2^-), gives strong credibility to the concept of antibonding orbitals. Further exploration of this bonding model is warranted.

Exercise 15-1.   What would be the expected bond order for F_2^+?

Exercise 15-2.   The ionization energies of N_2 and O_2 are, respectively, 359 and 279 kcal. Discuss this large difference in terms of the orbital occupancies shown in Figure 15-14.

15-5   M.O.'s and first-row heteronuclear diatomic molecules

We have seen that the p orbital M.O. description is sufficient to describe the bonding in the homonuclear diatomics B_2 through the nonexistent Ne_2. The model took the "turned-off electron" approach—electron repulsions were ignored initially. Furthermore, except for noting the s–p energy difference, no attention was paid to the specific atoms involved. The same molecular orbitals are generated by quite different atoms, and the bond properties seem to depend only on the number of electrons available to occupy these orbitals. This behavior suggests that the atoms in a bond need not even be identical. Perhaps we can also predict the bond orders and bond properties of heteronuclear diatomic molecules.

Carbon monoxide, CO, and nitric oxide, NO, are two familiar heteronuclear diatomic molecules. A third example is CN, a well-known molecule, but less available because it is quite reactive. Figure 15-15 shows the orbital occupancies for these three molecules if all valence electrons are treated as community property. (Only the p orbital M.O.'s are shown—in addition, each molecule has eight electrons held in the 1s and 2s M.O.'s.) Carbon monoxide is predicted to resemble nitrogen—it should have a triple bond. Both CN and NO should have $2\frac{1}{2}$ order bonds, respectively intermediate between C_2 and N_2 and N_2 and O_2. The experimental facts bear out these expectations quite well. This agreement can be assessed in Figure 15-16. The broken lines, transcribed from Figure 15-12, show the homonuclear molecular trends. The circles display the experimental facts for the heteronuclear diatomic molecules. Force constants and bond lengths for BN, NF and OF are also included; the bond energies, however, are not yet firmly established. There is noticeable scatter, but the bond properties are generally within about one quarter of a bond order of the expected values based upon the homonuclear trends. NO and CN do have $2\frac{1}{2}$ order bonds, while CO has a triple bond. The OF molecule, with a force constant of 5.41 mdyne/Å, is between F_2 and O_2, consistent with the $1\frac{1}{2}$ bond order prediction. Only NF seems out of line. The presently accepted values for its bond length (1.317 Å) and force constant (6.0 mdyne/Å) suggest that NF has a $1\frac{1}{2}$ order bond rather than a double bond like O_2.

The primary conclusion from Figures 15-12, 15-14, and 15-16 is that the M.O. concept provides a useful description of the bonding in the first-row diatomic molecules. Apparently electron repulsions and skewed charge distributions can be ignored in the first analysis. The method is sufficiently successful and easy to pursue that it permits us to apply quantum mechanical ideas of bonding to more complicated molecules, including polyatomic molecules far beyond the numerical capabilities of the largest

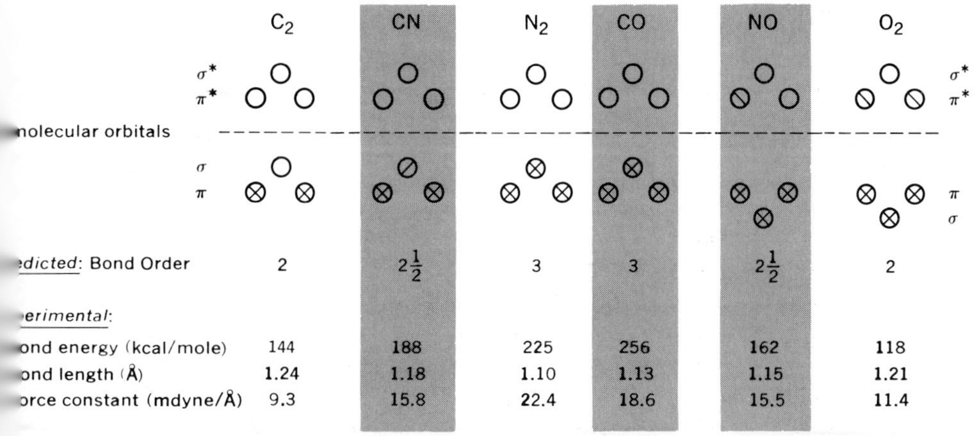

	C₂	CN	N₂	CO	NO	O₂	

molecular orbitals

| | | C₂ | CN | N₂ | CO | NO | O₂ | |

Predicted: Bond Order

| | 2 | $2\frac{1}{2}$ | 3 | 3 | $2\frac{1}{2}$ | 2 |

Experimental:

	C₂	CN	N₂	CO	NO	O₂
Bond energy (kcal/mole)	144	188	225	256	162	118
Bond length (Å)	1.24	1.18	1.10	1.13	1.15	1.21
Force constant (mdyne/Å)	9.3	15.8	22.4	18.6	15.5	11.4

Figure 15-15 Orbital occupancy and bond properties of some heteronuclear diatomic molecules.

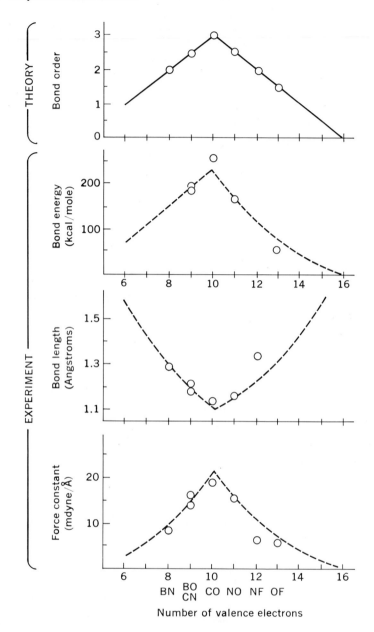

Figure 15-16 Trends in bond properties and predicted bond orders in first-row hetero-nuclear diatomic molecules.

computers. We'll proceed one more step in that direction by considering two simple triatomic molecules, CH$_2$ and CO$_2$.

Exercise 15-3. The cyanide ion, CN$^-$, has a bond length 1.09 Å in solid NaCN. Compare this to the expectation based upon orbital occupancy (see Fig. 15-15).

15-6 Approximate M.O.'s for methylene, CH$_2$

Of all the molecules that have been definitely identified, methylene is one of the most reactive. It can be produced by photoexcitation and fragmentation of a number of more complicated molecules. After birth, it usually needs only two or three collisions with other molecular species to react and end its independent career. Nevertheless, we know its structure, its bond energies, its force constants, and much about its chemistry. It is a good example with which to pursue the consideration of bond and molecular orbitals.

The lowest energy state of methylene is found to be linear, with equal C–H bond lengths of 1.03 Å (see Fig. 15-17). The atoms lie in a line with the carbon at the center of the molecule. The two hydrogen atoms, H$'$ and H$''$, occupy equivalent positions; if the molecule were turned end for end, it would appear the same.

A correlation diagram approach is helpful. Consider the formation of CH$_2$ when a carbon atom is approached simultaneously by two H atoms, one at a distance R on the positive x axis and the other at the same distance R on the negative x axis. As R is decreased, always maintaining the final molecular symmetry, the orbitals begin to interact to form molecular orbitals. Just as in the diatomic cases, the orbitals on the molecular axis interact with each other, but not with those perpendicular to the molecular axis. Each hydrogen atom contributes a 1s valence orbital on the x axis, 1s$'$ and 1s$''$, and the carbon atom contributes two such valence orbitals, 2s and 2p$_x$. The carbon atom also has two perpendicular valence orbitals, 2p$_y$ and 2p$_z$, but they have different nodal properties (π nodal surfaces) so they won't be engaged in the axial (σ) M.O.'s.

As we frame M.O.'s from these four component orbitals, 1s$'$, 1s$''$, 2s and 2p$_x$, it is helpful to consider the M.O.'s that would be formed from the hydrogen atom orbitals alone. If we pretend the carbon atom is not there, the M.O.'s become those of H$_2^+$ again, as pictured in Figure 15-2. For example, the σ^*(1s) orbital has a nodal surface halfway between the atoms. In CH$_2$ that's exactly where the carbon atom is placed and, of course, the 2p$_x$ orbital also has a nodal surface here. Having the same nodal surfaces, 2p$_x$ and σ^*(1s) can join to form M.O.'s. On the other hand, this nodal surface is quite incompatible with the nodal pattern of the centrally located carbon 2s orbital. The 2s orbital does not join in the σ^*(1s) M.O.

The 2p$_x$ and σ^*(1s) orbitals join to form M.O.'s in the now familiar

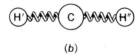

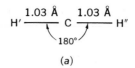

(a)

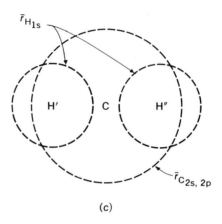

(b)

(c)

Figure 15-17 The structure of methylene, a reactive molecule. The structure of CH_2 is shown in (a) and, on the same scale, in the ball-and-spring model of (b). The average radii of the valence orbitals (see Table 13-1) are shown in (c) on the same scale as (a) and (b). The hydrogen nuclei are essentially "buried" in the carbon 2s and 2p orbitals.

manner. Figure 15-18 shows how $2p_x$ can join with $\sigma^*(1s)$ in an in-phase or an out-of-phase relationship. The in-phase relation concentrates electron probability between the atoms, in the binding region, to give a bonding M.O. The out-of-phase relation moves electron probability out of the binding region and an antibonding M.O. results. These two M.O.'s are both derived from the $2p_x$ and the antibonding $\sigma^*(1s)$ orbitals. The energy effects from the antibonding $\sigma^*(1s)$ parentage are relatively unimportant because in CH_2 the hydrogen atoms are too far apart to interact with each other strongly.

Returning to the H_2^+ M.O.'s, the $\sigma(1s)$ orbital has no nodal surface at the center of the molecule where the carbon atom lies. This nodal pattern is like that of the carbon atom 2s orbital. Hence the 2s and $\sigma(1s)$ orbitals join to form M.O.'s. Figure 15-19 shows the resultant CH_2 M.O.'s—again, one bonding and one antibonding.

Thus the four atomic orbitals, $1s'$, $1s''$, $2s$ and $2p_x$, form four CH$_2$ M.O.'s, two bonding and two antibonding. The makeup of these four M.O.'s is readily deduced from the nodal patterns, taking account of the fact that the carbon atom is exactly halfway between the two hydrogen atoms in a linear arrangement.

Now we can construct the molecular orbital energy level diagram of CH$_2$, adding the $2p_y$ and $2p_z$ perpendicular orbitals of the carbon atom. They interact only weakly with the hydrogen atoms so they furnish essentially nonbonding orbitals (see Fig. 15-20). These nonbonding orbitals are

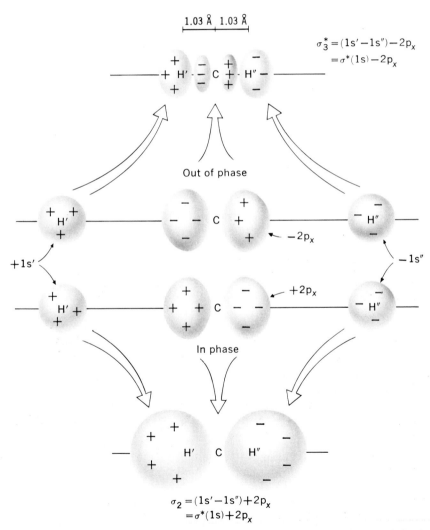

Figure 15-18 *The* CH$_2$ *molecular orbitals from* $\sigma^*(1s)$ *and* $2p_x$ *(the atomic orbitals are shown to scale).*

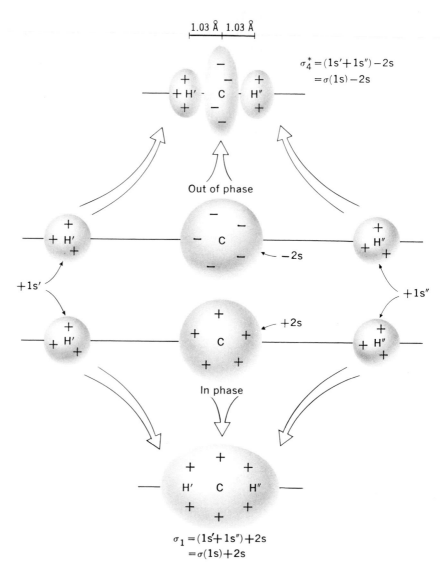

Figure 15-19 *The CH$_2$ molecular orbitals from σ(1s) and 2s (the atomic orbitals are drawn to scale).*

shown at $E = 0$, the energy corresponding to the energy of atoms not interacting at all. They do not change the molecular energy as the atoms approach. The bonding orbitals σ_1 and σ_2 are at negative energies; they lower the molecular energy as the atoms approach. The antibonding orbitals σ_3^* and σ_4^* have the opposite effect; they are at positive energies.

It only remains to add the electrons: two per orbital. The carbon atom furnishes four valence electrons and each hydrogen furnishes one. That

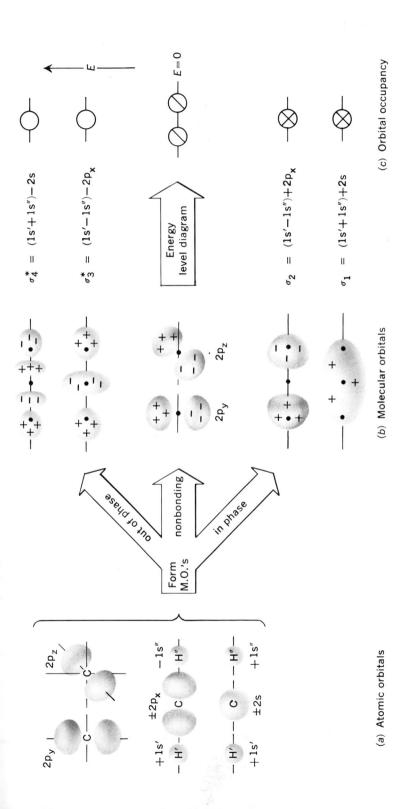

Figure 15-20 The molecular orbitals of linear methylene, CH_2.

is enough to fill σ_1 and σ_2 and leave one electron each for $2p_y$ and $2p_z$. Note that two pairs of electrons, enough to form two single bonds, act to hold three atoms together. Each of these pairs occupies an M.O. that is shared by two C–H bonds, so each M.O. contributes half a bond order to each bond. The net result of adding the bond contributions of σ_1 and σ_2 is that each C–H bond has a bond order of $\frac{1}{2} + \frac{1}{2} = 1$, a single bond. This is in accord with the observed C–H bond length, 1.03 Å, appropriate to a normal, single bond.

The orbital diagram reveals one more expectation for linear CH_2. The last two electrons can have parallel spins, since they occupy different orbitals to minimize electron repulsion. With parallel spins the molecule should display a net magnetic behavior, and spectroscopic studies show that it does.

15-7   Approximate M.O.'s for carbon dioxide, CO_2

Carbon dioxide differs from methylene in the same way that O_2 differs from H_2: Perpendicular 2p orbitals contribute to the bonding in CO_2, as they do in O_2, whereas they do not in CH_2 (because the hydrogen atom 2p orbitals are too high in energy). As for O_2, we must first consider the axial (sigma) M.O.'s and then the perpendicular (pi) M.O.'s.

(a) APPROXIMATE SIGMA M.O.'s FOR CO_2

Carbon dioxide has a linear, symmetric structure like that of CH_2. Hence, framing σ M.O.'s for CO_2 is little different from the process we just went through for CH_2. In fact, the only difference is that the end atoms, oxygen atoms, contribute to the σ M.O.'s with axial $2p_x$ orbitals instead of the 1s orbitals contributed by hydrogen. The oxygen atom 2s orbitals are so low in energy that they do not interact noticeably (see Fig. 15-9). That leaves the oxygen $2p_x$ orbitals with which the carbon atom axial orbitals, 2s and $2p_x$, can form M.O.'s. The results are entirely analogous to those of CH_2, as shown in Figure 15-21. There are two bonding sigma M.O.'s, σ_1 and σ_2, and two antibonding sigma M.O.'s, σ_3^* and σ_4^*.

(b) APPROXIMATE PI M.O.'s FOR CO_2

Since the terminal atoms possess 2p π atomic orbitals like those of the central atom, we must consider the pi M.O.'s as well. As in the case of O_2, it is only necessary to look at the $2p_y$ orbitals and the M.O.'s they form. Whatever the $2p_y$ orbitals do, the $2p_z$ orbitals will do as well, since they are equivalent.

Figure 15-22 shows how the $2p_y$ orbitals of carbon and the two oxygen atoms form M.O.'s as the molecule is formed, always in the linear, symmetric geometry. The carbon atom $2p_y$ orbital, directed along the y axis,

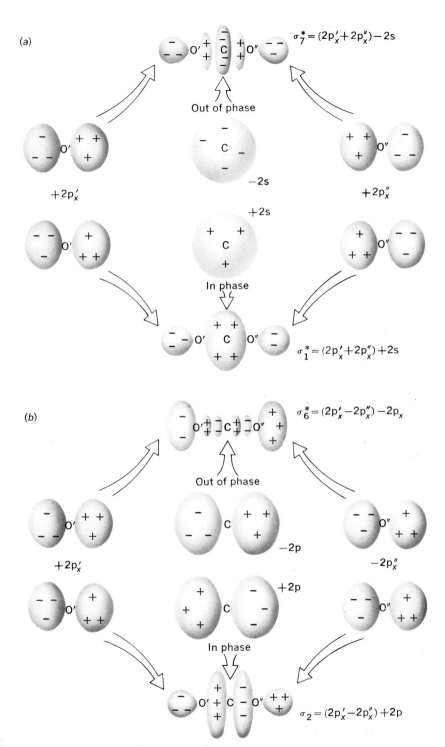

Figure 15-21 Axial (sigma) molecular orbitals for CO_2: (a) sigma molecular orbitals using the carbon 2s orbitals; (b) sigma molecular orbitals using the carbon 2p orbital.

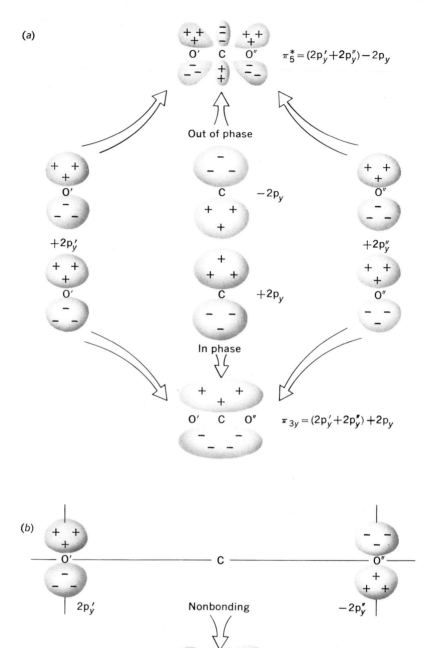

Figure 15-22 Pi (π) molecular orbitals for CO_2 formed from p_y atomic orbitals. (An identical set of orbitals can be drawn for p_z atomic orbitals.) (a) Bonding and antibonding orbitals. (b) Nonbonding orbitals.

forms a bonding M.O. and an antibonding M.O. with the combination of oxygen atom orbitals $(2p_y' + 2p_y'')$. There is, however, a combination of oxygen atom orbitals $(2p_y' - 2p_y'')$ with which it cannot interact. The combination $(2p_y' - 2p_y'')$ is a π combination with a nodal plane perpendicular to the molecular axis. The carbon atom has no π orbitals with such a nodal surface. Hence, the $(2p_y' - 2p_y'')$ combination is itself a molecular orbital, but it only involves atoms that are far apart. In fact, the distance between the oxygen atoms is so great that there is no appreciable interaction or change in electron distribution. Without such interaction the energy is not changed, so this is neither a bonding nor an antibonding orbital. *A molecular orbital that involves no adjacent atoms is called a* **nonbonding** *M.O.* Its energy is the same as that of the separated atoms. We will identify its nonbonding character with a superscript zero, as in π^0.

(c) THE CO$_2$ ENERGY LEVEL DIAGRAM

We can now unite Figures 15-21 and 15-22 to give the CO$_2$ energy level diagram. Figure 15-23 shows the result. We need only plunk in the electrons,

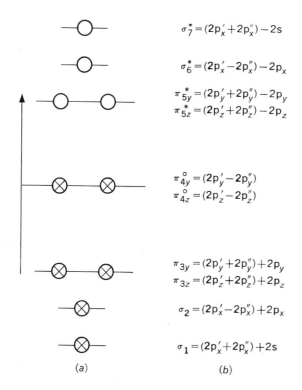

$$\sigma_7^* = (2p_x' + 2p_x'') - 2s$$

$$\sigma_6^* = (2p_x' - 2p_x'') - 2p_x$$

$$\pi_{5y}^* = (2p_y' + 2p_y'') - 2p_y$$
$$\pi_{5z}^* = (2p_z' + 2p_z'') - 2p_z$$

$$\pi_{4y}^0 = (2p_y' - 2p_y'')$$
$$\pi_{4z}^0 = (2p_z' - 2p_z'')$$

$$\pi_{3y} = (2p_y' + 2p_y'') + 2p_y$$
$$\pi_{3z} = (2p_z' + 2p_z'') + 2p_z$$

$$\sigma_2 = (2p_x' - 2p_x'') + 2p_x$$

$$\sigma_1 = (2p_x' + 2p_x'') + 2s$$

(a)   (b)

Figure 15-23 *The energy level diagram and orbital occupancy of CO$_2$. (a) The 12 electrons occupy four bonding and two nonbonding orbitals, thus contributing a net bond order of two per C–O bond. (b) The approximate molecular orbital description of the energy levels.*

two per orbital, with opposite spins, as long as the electrons last. Each oxygen atom gives us six valence electrons, two of which are used by each atom to fill its low energy 2s orbital. The remaining eight electrons (four per oxygen atom) are joined by four carbon atom valence electrons. So 12 electrons placed into the orbitals shown in Figure 15-23 fill all of the four bonding M.O.'s and the two nonbonding M.O.'s, but none are left over to go into antibonding M.O.'s. We find, then, four pairs of electrons in bonding M.O.'s, each M.O. sharing its bonding capacity between two bonds and contributing a one-half bond order to each bond. We conclude that each bond in CO_2 will have a bond order of two and that the molecule should be nonmagnetic.

The experimental data for CO_2 are in rough agreement with this expectation. The average bond energy—one half the energy needed to pull the molecule apart into a carbon and two separated oxygen atoms—is 192 kcal/mole, the carbon–oxygen stretching force constant is 15.5 mdyne/Å, and the C–O bond length is 1.16 Å. These magnitudes compare rather well with the corresponding figures for CN (see Fig. 15-15) and suggest about a $2\frac{1}{2}$ order bond. As in the case of diatomic molecules, that is about the accuracy we can expect for this very simple approach, $\pm$ one half a bond order.

And now we have nonbonding molecular orbitals added to our stable of ideas on bonding.

Exercise 15-4. The carbon–oxygen vibrational force constant of CO_2^+ is almost identical to that of CO_2. Compare this to the expectation based upon orbital occupancy (see Fig. 15-23).

15-8 Approximate M.O.'s for carbonate ion: exit resonance

Carbonate ion, CO_3^{-2}, is one of the molecular species listed in Table 2-4 whose bonding could be explained using electron dots *only* if the resonance argument is invoked. The resonance argument rationalizes the existence of several equally good electron-dot representations by suggesting that all contribute to the bonding. To use carbonate ion as an example, we have three electron-dot formulas that give each atom an inert gas electron population. Each formula predicts one double and two single bonds, one short and two long bonds. Since experiment shows that carbonate ion has three equal bonds, each of bond length between single and double bonds, the resonance rationale is invoked, as shown in Figure 15-24.

The basic idea of resonance, bonds moving about in the molecule, intuitively suggests the molecular orbital approach. After all, molecular orbitals merely admit at the beginning that electrons are mobile and they move in a fashion determined by the whole molecule, not by just a particular pair of atoms (as is suggested by the electron-dot approach given in

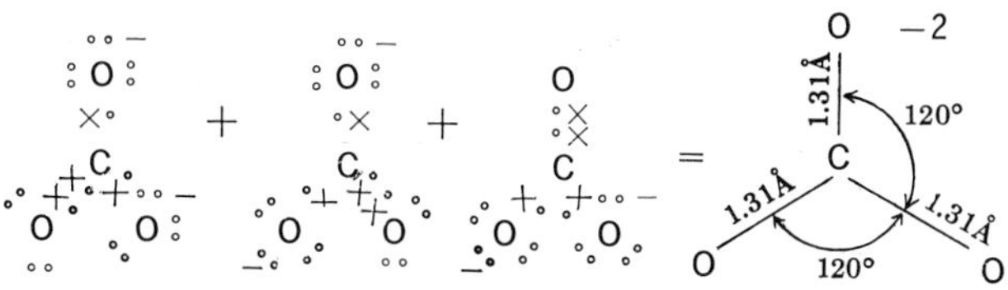

Figure 15-24 How resonance "explains" the carbonate ion structure.

Section 2-4). Hence we should expect that the bonding in molecules like carbonate ion should be more easily described with M.O.'s than with electron dots.

This expectation turns out to be realized, but with a reservation. Drawing an electron-dot formula for a many-atom molecule is quite easy, but drawing molecular orbitals for the same molecule can be a real chore. Fortunately, we find that a mixture of the two approaches almost always gives us the benefits of both systems. Figure 15-24 gives the clue that guides us. The three electron-dot formulas for CO_3^{-2} differ only in two respects: There is one double bond pair of electrons and there are two electron charges that can't decide where to sit. The rest of the electrons are quite happy.

That suggests a simplifying approach that generally works quite well. We assume that *all of the sigma bonds are adequately described by electron-dot formulas, but all pi orbitals must be considered in terms of molecular orbitals.* Let's see how it works out.

(a) THE IN-PLANE BONDING IN CARBONATE ION

The discussion of the bonding is divided into two parts, that involving orbitals confined to the molecular plane and that involving orbitals with nodes in the molecular plane (pi orbitals). If we call the molecular plane the xy plane, each of the four atoms has three valence orbitals in this plane, 2s, $2p_x$ and $2p_y$. Using only these orbitals in an electron-dot representation, we would arrive at the picture shown in Figure 15-25, in which these 12 orbitals accommodate 18 electrons, three pairs of which form sigma

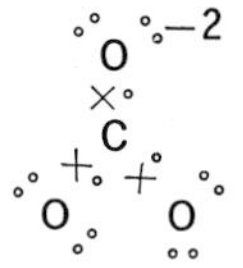

Figure 15-25 The electron-dot representation of the in-plane orbitals of carbonate ion. Each atom "sees" six electrons, enough to occupy the three in-plane orbitals of each.

bonds. Since CO_3^{-2} has 22 valence electrons (six from each oxygen and four from carbon) and two extra (the ion has a -2 charge), there are six electrons unaccounted for in Figure 15-25. These must occupy the pi orbitals. This is where we begin using the molecular orbital description.

(b) THE OUT-OF-PLANE, PI BONDING IN CARBONATE ION

Molecular orbitals must be constructed from the four orbitals that have not yet been considered. These are pictured in Figure 15-26. The carbon atom pi orbital, $p_z(C)$, is directed along the molecular axis (Fig. 15-26(a)) and it has only one nodal surface, the xy plane. The three oxygen atom p_z orbitals $p_z(O_1)$, $p_z(O_2)$, and $p_z(O_3)$ can be combined so that their sum also has only one nodal surface, the xy plane. That is obtained by combining them all in the same phase, as pictured in Figure 15-26(b). With the same nodal properties, $p_z(C)$ and the combination of $p_z(O_1) + p_z(O_2) + p_z(O_3)$ can interact to form two molecular orbitals. If they combine in phase, a bonding M.O. results. If they combine out of phase, an antibonding M.O. is formed (see Fig. 15-27).

Now we have combined four p_z orbitals to form two M.O.'s, one bonding and one antibonding. Four orbitals must give four molecular orbitals, so we have two more to discover. We must seek some other way to combine the three oxygen atom p_z orbitals. Any way we do that, however, involves a phase change between two oxygen atoms to introduce another nodal surface, one that the central carbon atom does not possess. Hence the central carbon atom is not engaged in this molecular orbital at all. But if this is true, the two remaining M.O.'s must be nonbonding since none of

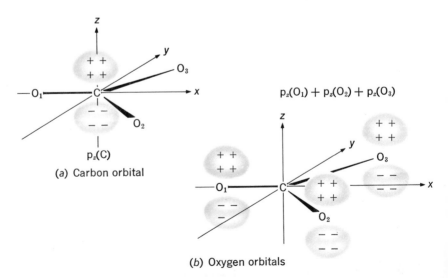

(a) Carbon orbital

$p_z(O_1) + p_z(O_2) + p_z(O_3)$

(b) Oxygen orbitals

Figure 15-26 The p_z (pi) orbitals of carbonate ion.

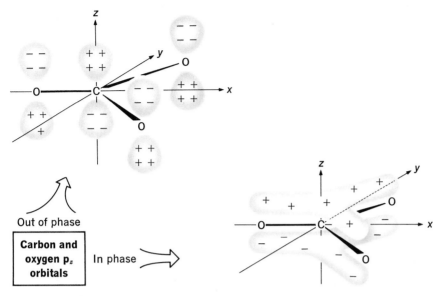

Figure 15-27 The bonding and antibonding pi M.O.'s of carbonate ion.

the atoms involved are close together. Without even trying to draw these two M.O.'s, we conclude that they are nonbonding.

(c) THE PI ORBITAL ENERGY LEVELS OF CARBONATE ION

We have concluded that the four p_z orbitals in CO_3^{-2} give rise to one bonding, one antibonding and two nonbonding M.O.'s. These four M.O.'s must accommodate six electrons. Figure 15-28 shows the energy levels and the orbital occupancies. There is one bonding pair of electrons that distributes its bonding capability equally among three bonds. Therefore each bond receives a bond order of one third from this M.O. Superimposed on the single bonds provided by the in-phase bonding (shown in Fig. 15-25) this pi M.O. leads to a total bond order of $1\frac{1}{3}$ and every bond is alike.

The nonbonding M.O.'s are the last to fill, and they contain the two electrons that give carbonate ion its double negative charge. We conclude that

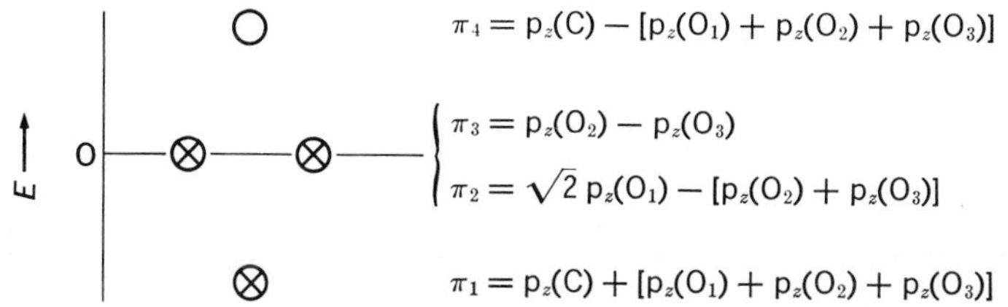

$$\pi_4 = p_z(C) - [p_z(O_1) + p_z(O_2) + p_z(O_3)]$$

$$\begin{cases} \pi_3 = p_z(O_2) - p_z(O_3) \\ \pi_2 = \sqrt{2}\, p_z(O_1) - [p_z(O_2) + p_z(O_3)] \end{cases}$$

$$\pi_1 = p_z(C) + [p_z(O_1) + p_z(O_2) + p_z(O_3)]$$

Figure 15-28 Pi energy levels and orbital occupancy in carbonate ion.

this charge is primarily lodged on the oxygen atoms, equally shared, so that each oxygen atom has a net charge of about two thirds of an electron charge.

The carbonate ion bond lengths are measured to be 1.31 Å, to be compared to the triple bond length of carbon monoxide 1.13 Å, the double bond lengths in C_2 and O_2, respectively, 1.24 Å and 1.21 Å, and the $1\frac{1}{2}$ order bond length in O_2^-, 1.28 Å (see Figs. 15-14 and 15-15). The 1.31 Å bond length is a bit longer than the bond length in O_2^- so the bond is a bit weaker than a $1\frac{1}{2}$ order bond (say, about like $1\frac{1}{3}$?).

(d) RESONANCE AND M.O.'s

We see that the M.O.'s formed from the pi orbitals do rather naturally what the resonance picture does awkwardly. It leads readily to the three conclusions needed to explain the bonding: There is only one pair of electrons engaged in double bond formation; this pair of electrons is equally shared by three bonds; and the excess charge on the ion resides on the peripheral, oxygen atoms, again equally shared.

The pattern of this problem is very common. Among a large proportion of the molecules whose bonding requires resonance (in the electron-dot scheme), the bonding is easily explained by using electron-dots for the sigma bonds (in-plane bonds, for planar molecules) and molecular orbitals for the pi bonds (out-of-plane bonds, for planar molecules).

15-9 Wrap-up on M.O.'s

We have considered the molecular orbital approach first for H_2^+, then H_2, and we continued on to more ornate examples like CH_2 and CO_2. As we proceeded, more and more use was made of the word "approximate." As we continue to consider bonding, the molecules will become so complicated that all of our considerations will be approximate and they will become increasingly subservient to empirical facts. Yet, at this time in the evolution of the understanding of chemical bonding, quantum mechanical ideas finally unify and give coherence to most aspects of bonding. Chemists used to catalog a dozen "bond types," empirically derived, and independently considered. Now we can see the origin of each type within a single framework that considers the parent valence orbitals, how they interact, and the orbital occupancy.

There is one important distinction to be made about the terminology of the molecular orbital approach. In its generic sense, the term molecular orbital indicates that, insofar as it is possible to discuss the properties of a molecule in terms of electrons in orbitals, each orbital is shaped by all of the nuclei in the molecule. Hence, in a molecule, the electrons certainly occupy *molecular orbitals*. In the same way that the electronic structure of an atom is soundly based on a consideration of its atomic orbitals, so also, the properties of a molecule can be expected to be evident and explainable through a consideration of its molecular orbitals. For both atoms and molecules, electron repulsion effects are sufficiently muted to be brought in only as a refining influence.

At another level of consideration we actually attempted to frame the molecular orbitals by taking parent atomic orbitals as our ingredients. Thus we wrote for H_2^+, and also for H_2, that there are two molecular orbitals, $\sigma(1s) = 1s_A + 1s_B$ and $\sigma^*(1s) = 1s_A - 1s_B$. However, the properties of H_2^+ and H_2 are different enough that these two molecules cannot have identical molecular orbitals. These specific representations are, then, only rough approximations. They give the principal atomic orbital parentage of each M.O. and they accurately show its nodal surfaces. But by no means can we assume that the simple superposition of two 1s orbitals at the equilibrium bond length of H_2^+ (or of H_2) accurately gives the electron distribution in the molecule.

With that caveat, let us turn to the prevailing representations of chemical bonds in complex molecules. We will see that they can now be given credentials and meaning with the aid of our quantum mechanical point of view.

Problems

1.   The oxygen molecule has an excited electronic state (103 kcal above the ground state) in which a bonding pi electron has been raised to a pi antibonding orbital. Use the O_2 M.O. diagram to predict the bond order. With the data in Figure 15-11, replot Figure 15-12 to aid in predicting the bond energy, bond length, and force constants of this excited molecule. Assign an uncertainty to each estimate appropriate to $\pm\frac{1}{4}$ bond order. Compare your answers to the best experimental data for this state, which give $r_0 = 1.53$ Å and $k = 2.81$ mdyne/Å. From the discrepancies, do you expect your estimate of bond energy to be high, low, or just right?

2.   Use the orbital occupancy to predict the bond orders of the gaseous ions C_2^+ and C_2^-. Use the plot prepared in Problem 1 to predict the bond energies, bond lengths, and force constants for these two molecule ions. Assume that a positive charge adds and a negative charge detracts by $\frac{1}{4}$ bond order because of electron repulsion effects.

3.   Use the M.O. diagram derived for O_2 to determine the bond orders, bond energies, bond lengths, and force constants for boron oxide, BO; boron carbide, BC; and oxygen fluoride, OF. Assign uncertainty appropriate to $\frac{1}{4}$ bond order. Compare your predictions to the available data in Appendix f.

4.   Repeat Problem 2 for the gaseous molecule ions CN^+ and CN^-.

5.   Repeat Problem 2 for the gaseous molecule ions NO^+ and NO^-.

6.   Which of the following molecules can be expected to display magnetism associated with unpaired electron spins: B_2, BN, CO, CN, NO, NF?

7.   Which of the following molecule ions can be expected to display magnetism associated with unpaired electron spins: BC^+, C_2^+, CN^+, CN^-, NO^+, NO^-?

8. Explain why O_2^- has a longer bond than O_2 (1.21 versus 1.12 Å), whereas CN^- has a shorter bond than CN (1.09 versus 1.18 Å).

9. The electron affinities of some diatomic molecules are given in Appendix c. Explain, in terms of molecular orbital occupancies, why those of O_2 and NO are small ($\sim$10 and 21 kcal, respectively) while those of CN and C_2 are large (88.1 and 71.5, respectively).

10. The ionization energies of some diatomic molecules are given in Appendix c. Explain, in terms of molecular orbital occupancies, why that of NO, 213.3 kcal, is much smaller than those of CN or N_2 (349 and 359 kcal, respectively).

11. Appendix f gives the bond energy and Appendix c gives the ionization energy of nitric oxide, NO. Combine these values with the ionization energy of the oxygen atom to calculate the bond energy of NO^+. Compare to your estimate in Problem 5.

12. Appendix f gives the bond energy and Appendix c gives the electron affinity of gaseous CN. Calculate the bond energy of CN^-, assuming it dissociates to C^- and N. Compare to your estimate in Problem 4.

13. Use the electron affinities of NO and CN (see Appendix c) to predict ΔH for the reaction

$$NO^-(g) + CN(g) \longrightarrow NO(g) + CN^-(g)$$

14. The two molecule ions NO_2^+ and N_3^- have linear, symmetric structures. Discuss their bonding from a molecular orbital point of view, relating them to CO_2. What bond orders will be expected in each of these molecules?

15. Both BF_3 and nitrate ion, NO_3^-, have the same symmetric, planar, triangular structure. Discuss their bonding from a molecular orbital point of view, relating them to carbonate ion, CO_3^{-2}. What bond orders are to be expected?

16. Compare the bonding relationship between BH_3 (which is a transient species) and BF_3 to the bonding relationship between CH_2 (which is an extremely reactive molecule) and CO_2.

17. Deduce the sigma molecular orbitals that would describe the bonding in bifluoride ion, $(FHF)^-$, which has a linear, symmetric structure. Form molecular orbitals from the axial $2p_z$ orbitals on the fluorine atoms and the 1s orbital on the hydrogen atom. Classify each orbital as bonding, nonbonding, or antibonding. Assume all other fluorine atom valence orbitals are filled and nonbonding. What is the bond order in each H–F bond? What orbital accommodates the last electron; i.e., where does the negative charge on the ion reside?

sixteen   molecular geometry

Molecular geometry proves to be almost as important as bond energies in fixing chemical properties. This seems to be particularly so in biological functions. The outcome of competing reactions can be entirely determined by spatial relationships connected with the molecular configurations of the competitors. A molecule with the wrong shape may not react at all, whereas a similar one, differently configured, may react in a trice. Many biological growth and reproductive processes depend upon "lock and key" shape relationships between reactants and products. Furthermore, many of our newer drugs are effective because of their similarity in molecular appearance to some misbehaving biological molecule. The look-alike drug takes the place of the offending molecule and blocks its action.

So chemists are vitally interested in molecular geometries. The experimentalist wants to be able to measure them and all of us want to generalize the results into a guiding theory. This theory should allow us to understand the bond angles and molecular shapes that are known and to predict with reasonable confidence those that are not.

In this chapter we'll examine two explanations of the bond angles observed among the elements in the first three rows of the Periodic Table. One model concentrates on the atomic orbital parentage of the bonds—as does the molecular orbital representation of bonding. This model is called *"orbital hybridization."* The second model ignores the orbital parentage and concentrates only on the number of pairs of valence electrons near each atom—like the electron-pair representation of bonding. This approach is called the *"electron repulsion"* model. We'll examine these two theories and then consider the implications of molecular shape on the molecular dipole moment.

16-1   Orbital hybridization and bond angles

There is, of course, only one question to ask about the molecular geometry of a diatomic molecule: What is its bond length? We have already discovered that, for a given pair of atoms, the higher the bond order, the shorter the bond. This relationship is entirely consistent with our representations of bonding as presented in the dot diagrams (2-20) and (2-17) for H_2O_2 and O_2 (oxygen bond lengths 1.47 and 1.21 Å for single and double bonds, respectively). This bond length–bond order correlation (the higher the bond order, the shorter the bond) applies equally well to polyatomic molecules.

(a) THE MOLECULES H_2O AND NH_3

Molecular structures become more interesting when we pass on from diatomic molecules to polyatomics like H_2O and NH_3. The water molecule could be imagined to be linear or it could be bent. Its two oxygen–hydrogen bonds could be of equal or unequal length. Ammonia could be pyramidal or it could have all its atoms in the same plane; its NH bonds could all be of equal length or they could differ. We need some guiding principles to tell us which of these to expect.

Our orbital picture gives us some guidance and it correlates the known facts. Examine the representation for H_2O in Figure 16-1. We see that each hydrogen is shown sharing electrons with an oxygen 2p orbital. There is no reason to expect one 2p orbital to be more effective than the other, so the picture implies two equivalent OH bonds (as they are). Furthermore, we might expect that the orbital geometry of the 2p orbitals would be represented in the bonding geometry. The 2p orbitals lie at right angles, so the picture implies a bent molecule with a bond angle near 90°. Figure 16-1 compares these expectations with the experimental facts: *The water molecule is a bent molecule with two equivalent bonds and a bond angle of 104.5°.*

Turning to ammonia, we see that each hydrogen is shown sharing electrons with a nitrogen 2p orbital. We expect, then, three equivalent bonds. The perpendicular geometry of p orbitals suggests a pyramidal structure for NH_3. *The ammonia molecule is a pyramid with three equivalent bonds and its bonds have 107° angles between them.* Once again, the qualitative aspects of the molecular shape are represented, even if the bond angle expectations are 15–17° lower than observed.

(b) THE METHYLENE MOLECULE, CH_2

We considered the bonding in the lowest energy state for CH_2 in Section 15-6. This state has been pictured to involve a bond to a 2s orbital and a bond to a $2p_x$ orbital. These are the lowest energy, unfilled orbitals, hence the best for bonding. Now we must deduce the angle between the bonds in accordance with our assumption that the bond angle is fixed by the directional features of the bonding orbitals of the central atom.

One of the central-atom bonding orbitals, the 2s orbital, is spherically symmetric and so it offers no immediate influence on the bond angle. The other carbon atom orbital, $2p_x$, is directed along the x coordinate axis and so it favors formation of bonds along that axis. Thus, the $2p_x$ orbital seems to dictate a linear molecule. Furthermore, it directs as much electron

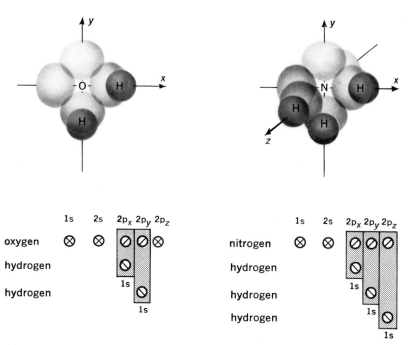

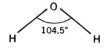

	1s	2s	2p$_x$ 2p$_y$ 2p$_z$
oxygen	⊗	⊗	
hydrogen			
hydrogen			

	1s	2s	2p$_x$ 2p$_y$ 2p$_z$
nitrogen	⊗	⊗	
hydrogen			
hydrogen			
hydrogen			

H_2O: bent, 90° bond angles, equivalent bonds

NH_3: pyramidal, 90° bond angles, equivalent bonds

(a) Orbital prediction

H_2O: bent, 104.5° bond angle, equivalent bonds

NH_3: pyramidal, 107° bond angles, equivalent bonds

(b) Observed structures

Figure 16-1   The molecular shapes of water and ammonia.

probability along the x axis to the right as it does to the left (as does the 2s orbital). Neither bond should be stronger than the other. We see that the carbon atom orbitals ought to lead to two equivalent bonds directed 180° from each other. This view is consistent with our molecular orbital treatment of linear CH_2, presented in Chapter Fifteen (Section 15-6). We learned from this three-center M.O. consideration that a carbon 2s and its axial 2p orbital form two bonding molecular orbitals with two equal bonds in a linear configuration. Thus the two C–H bonds are equivalent. We shall call these bonds *sp hybrids*, the word "hybrid" indicating that each bond is a mixture of the carbon 2s and $2p_x$ orbitals (see Fig. 16-2(a)). To convey this hybridization in an orbital, or "pigeonhole," diagram, we might represent CH_2 as follows:

CH_2 (6 valence electrons)

$$(16\text{-}1)$$

The hybridizing orbitals on carbon are enclosed in a box. The bonds to hydrogen 1s orbitals are represented by a solid line. (In Figure 16-2(b) the bonding orbitals are also enclosed in boxes.) This orbital analysis and hybridization idea prove to have general applicability, and they give us a predictive tool. *Whenever the orbital pigeonhole diagram indicates that the central atom forms two bonds with an s and a p orbital, we can expect linear geometry and two equivalent (sp) bonds.*

(c) THE METHYL MOLECULE, CH_3

A pigeonhole diagram for methyl can be interpreted in a similar way (see Fig. 16-2(c)). If the lowest energy orbitals are used for bonding, the carbon 2s orbital and two of its 2p orbitals are the important ones. A molecular orbital approach like that for CH_2 is more complicated than we can cope with here, but it leads to results that can be derived intuitively in the light of the triplet CH_2 treatment. First, the two 2p orbitals of carbon define a plane in which bonding will take place. As in CH_2, the spherical symmetry of the 2s orbital implies no preferred direction. Hence the directional properties of the bonds are entirely fixed by the two 2p orbitals: *All three CH bonds will be in the same plane.*

Experimentally it is found that CH_3 is planar and that *all three C–H bonds are identical.* This last result proves to be easily accommodated in the molecular orbital development, though it is not readily predicted from theory alone. In a symmetric, planar structure, with three C–H bonds formed at $120°$ angles to each other, each bond will involve a hybrid mixture formed from a 2s and two 2p orbitals. Extending the sp notation used for CH_2, we'll call these bonds sp^2 hybrids. Henceforth we shall expect that *whenever the central atom has at its disposal a half-filled s and two half-filled p orbitals, three equivalent sp^2 hybrid bonds will be formed at $120°$ angles in a plane.*

(d) THE METHANE MOLECULE, CH_4

It is well to be reminded that while CH_2 and CH_3 are well-defined molecules with accurately known structures, they are also extremely reactive because of their unused valence orbital bonding capacity. Methyl has an average

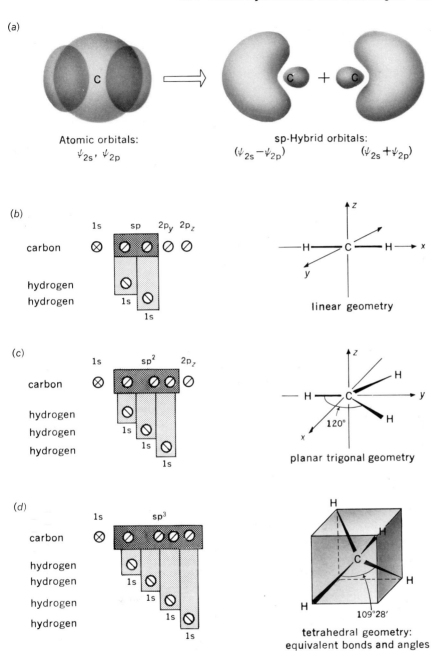

Figure 16-2   Orbital shape (a) and molecular geometry (b), (c), and (d) of hybrid bonds.

lifetime no longer than 10 to 100 microseconds under normal conditions. One of the ways in which it satisfies its unused bonding capacity is by adding a hydrogen to form methane, CH_4.

Again, a molecular orbital consideration of the bonds to be formed from one 2s and three 2p orbitals can be brought into agreement with the known structure of CH_4. Since the three 2p orbitals are mutually perpendicular, we can expect a three-dimensional molecule. As in both CH_2 and CH_3, the s orbital is uniformly divided among four exactly equivalent molecular orbitals. We designate these as sp^3 hybrids. To be equivalent, the bonds must point at the four vertices of a regular tetrahedron; 109°28' bond angles are thus obtained (see Fig. 16-2(d)).

Again we shall expect that *whenever the central atom has at its disposal a half-filled* s *and three half-filled* p *orbitals, four* sp^3 *bonds will be formed at tetrahedral angles.* According to the simple molecular orbital treatment, these angles will be obtained independent of the atom to be bonded. This is true to remarkable accuracy, as shown by the many halogen-substituted methanes whose bond angles are well known. Some of these are listed in Table 16-1. In every case the bond angles are within one or two degrees of the tetrahedral angle 109°28'.

Figure 16-2 summarizes the molecular geometry that accompanies sp, sp^2, and sp^3 hybrid bonds. We must remember, though, that neither the molecular orbital treatment nor the hybridization idea is sufficiently rigorous to prove that the structures of CH_2, CH_3, or CH_4 must be what they are observed to be. Rather, these elementary theories provide a framework that is consistent with the experimental facts, such as those shown in Table 16-1. The value of the theory is that, once tied to firm structures, it provides a basis for predicting the structures of other molecules from an examination of the orbitals available for bonding. Since the molecular structure is important in the chemistry of a molecule, this is a significant step toward understanding chemistry.

Table 16-1   Some Bond Angles in Halogen-Substituted Methanes

	Angle H—C—H	Angle H—C—X	Angle X—C—X
CH_4	109°28'	—	—
CH_3F	110°	108°56'	—
CH_2F_2	111°54'	109°9'	108°18'
CHF_3	—	110°38'	108°48'
CF_4	—	—	109°28'
CH_3Cl	110°30'	108°26'	—
CH_3Br	111°12'	107°44'	—
CH_3I	111°30'	107°26'	—

(e) THE MOLECULES BH₃ AND BF₃

In our earlier discussions, we did not mention the bonding of boron and beryllium. Considering boron first, a pigeonhole representation would initially predict stable molecules BH and BF.

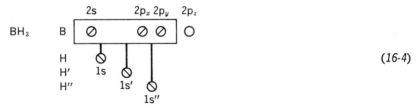

$$\text{(16-2)}$$

$$\text{(16-3)}$$

However, as in CH_2, it is possible to promote an electron to a vacant 2p orbital, at once reducing electron repulsions in the 2s orbital and increasing the number of electrons available for sharing in bonds. Now there are three half-filled orbitals, a 2s orbital and two 2p orbitals. Following the treatment for CH_3, we would expect three equivalent sp^2 hybrid molecular orbitals. The sp^2 hybrid orbitals are all in the same plane at angles of 120°. Hence we would expect planar, symmetric molecules BH_3 and BF_3.

$$\text{(16-4)}$$

$$\text{(16-5)}$$

There is a most significant difference between representations (16-4) for BH_3 and (16-5) for BF_3. The borine (BH_3) representation displays a completely vacant $2p_z$ valence orbital on boron. This orbital, which is oriented perpendicular to the molecular plane, should cause borine to have residual

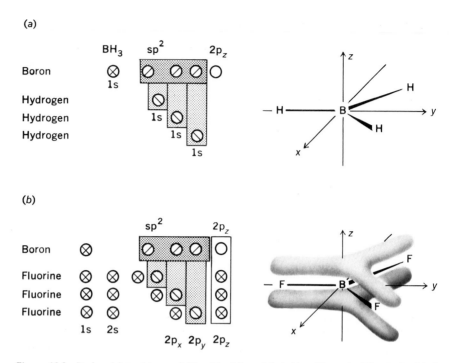

Figure 16-3   Borine (a) and boron trifluoride (b): sp² hybrids with and without pi orbital stabilization.

bonding capacity. The molecule should be reactive. In contrast, the molecule BF_3 places the boron $2p_z$ orbital within a cluster of three fluorine atom $2p_z$ valence orbitals, all perpendicular to the molecular plane (see Fig. 16-3). We can expect the fluorine $2p_z$ electrons (the "pi" electrons) to occupy molecular orbitals that extend over and partially occupy the boron $2p_z$ orbital, just like the pi bonding M.O. deduced for carbonate ion (see Fig. 15-27). Hence additional bonding can result and boron trifluoride, BF_3, should be more stable than borine, BH_3.

These expectations are all in accord with experience. While BF_3 shows some residual bonding capacity that can be attributed to its $2p_z$ orbital, the molecule is an inexpensive stockroom chemical. The molecules of this gaseous substance have a planar symmetrical structure. On the other hand, BH_3 is a transient molecule that is almost surely produced during reactions of a number of boron compounds, but it has never been spectroscopically detected. The search for BH_3 continues (using methods specially developed for such reactive species), and when it is finally found, the most interesting question to be answered will be whether it has the planar structure we have predicted.

(f) THE MOLECULES BeH₂ AND BeF₂

For BeH₂ or BeF₂ the pigeonhole representations are as follows, provided an electron is promoted to a vacant 2p orbital.

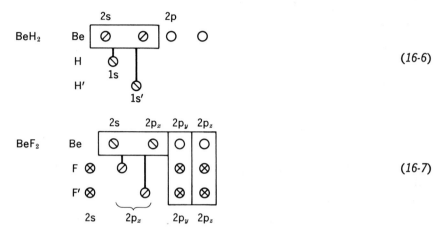

$$(16\text{-}6)$$

$$(16\text{-}7)$$

These two representations, (16-6) and (16-7), both predict linear structures with equivalent sp bonds. As in the case of BH₃, we can expect BeH₂ to be extremely reactive due to the completely vacant $2p_y$ and $2p_z$ valence orbitals of beryllium. This situation is less uncomfortable for BeF₂ because the fluorine $2p_y$ and $2p_z$ orbitals can form molecular orbitals with the boron orbitals and furnish electrons to add to the bonding.

Neither of the triatomic molecules BeH₂ or BeF₂ is well known. The beryllium halides are, of course, easily prepared, but as solids in which the $2p_y$ and $2p_z$ valence orbitals participate in the bonding within the crystal. The closest we can come to testing our structural expectations for these two molecules is to compare the molecular geometries of other alkaline

Table 16-2 *Bond Angles in Gaseous Alkaline Earth Dihalides,* MX₂: *First Value, Electron Diffraction;* Parenthetical Value, Molecular Beam, Electric Deflection†

	F	Cl	Br	I
Mg	180 ± 30°	180 ± 10°	180 ± 10°	
Ca	180°	180 ± 10°	180 ± 10°	180 ± 10°
	(bent)	(180°)	(180°)	
Sr	180°	180 ± 30°	180 ± 10°	180 ± 10°
	(~120°)	(bent)	(180°)	(180°)
Ba	180°	180 ± 40°	180 ± 30°	180 ± 20°
	(~120°)	(~120°)	(bent)	(bent)

* Electron diffraction; P. A. Akishin, V. P. Spiridonov, G. A. Sobolev, and V. A. Naumov, *Zhur. Fiz. Khim.*, **31**, 1871 (1957); **32**, 58 (1958).
† Molecular beam, electric deflection; L. Wharton, R. A. Berg, and W. Klemperer, *J. Chem. Phy.*, **39**, 2023 (1963).

earth dihalides, a number of which have been detected in high temperature reactions. Table 16-2 shows that two experimental techniques have been used and that they are not in complete agreement. The evidence suggests that some alkaline earth dihalides are indeed linear ($CaCl_2$, $CaBr_2$, $SrBr_2$ and SrI_2), but that some are likely to be bent molecules (CaF_2, SrF_2, BaF_2 and $BaCl_2$). Thus, limitations of our simple theory based on sp hybrid molecular orbitals are in evidence here.

16-2   Electron repulsion and bond angles

There is another simple scheme for explaining molecular bond angles that is based upon electron repulsions. The scheme is tuned to the Lewis electron-dot representation of bonding. It embodies the following premises:

(i) Each atom forms bonds until it fills its valence orbitals with electrons to reach the inert gas configuration.

(ii) These electrons are considered to distribute to act as pairs (of opposite spin).

(iii) The molecule takes that geometric structure that keeps these electron pairs as far apart as possible.

(a)   ELECTRON REPULSIONS IN H_2O, NH_3, AND CH_4

This scheme is very easy to apply. For example, for H_2O, NH_3, and CH_4, each central atom has eight electrons in its valence orbitals (some electrons provided by the hydrogen atoms, of course). In each of these molecules there are, then, four pairs of electrons repelling each other and striving to stay apart. The four pairs will tend to be directed towards the vertices of a regular tetrahedron. Figure 16-4 shows the result. In each molecule the bond angle expected is the tetrahedral angle 109°28'. For water, two of the electron pairs link H atoms to the oxygen atom (at the tetrahedral angle) while the other two pairs are merely "unused pairs." The

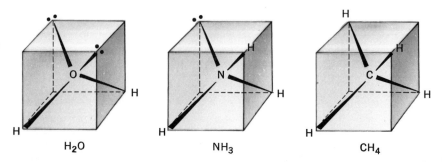

Figure 16-4   Electron repulsions and the structures of H_2O, NH_3 and CH_4.

molecule is expected to be bent. For ammonia, three electron pairs link H atoms and one pair is unused in bonding. The mutually tetrahedral bonding predicts a pyramidal molecule with 109°28′ bond angles. Finally, for methane, four bonds at the tetrahedral angles are expected.

These predictions can be contrasted with those based on the directional characteristics of the occupied valence orbitals. For water, the p^2 bonding predicts a bent molecule with a 90° bond angle, whereas electron repulsion predicts it to be 109°28′. Experimentally we find that the molecule is bent, with a measured angle of 104°30′. For ammonia, both the valence orbitals p^3 and the electron repulsion predict a pyramidal molecule, the former with 90° angles, the latter with 109°28′ angles. The molecule is pyramidal and its bonds have 107.3° angles between them. For methane, the electron repulsion predicts tetrahedral bonds in a three-dimensional molecule, exactly as observed. The same is true for the sp^3 hybrid orbital prediction, but only after a molecular orbital argument to see how one s and three 2p orbitals can form molecular orbitals that correspond to four equivalent, tetrahedral bonds.

The observed deviations from tetrahedral angles in NH_3 and H_2O can also be interpreted within the context of the electron repulsion scheme. Ammonia has one pair of electrons not engaged in bonding. Such a pair is called a "lone pair." In the first approximation, it is considered to be exactly equivalent to the three "bonding pairs" that bind the protons in the molecule. However, we find empirically that the angles between bonds (hence, between bonding pairs) are 107.3°, less than tetrahedral. This can be attributed to a difference between bonding pair–bonding pair repulsions (of which there are three) and the lone pair–bonding pair repulsions (of which there are also three). Since the observed bond angles are less than tetrahedral, we conclude that a lone pair repels a bonding pair more than two bonding pairs repel each other. In shorthand notation, we write

$$\begin{matrix} \ell p\text{–}bp \\ \text{repulsions} \end{matrix} > \begin{matrix} bp\text{–}bp \\ \text{repulsions} \end{matrix} \qquad\qquad (16\text{-}8)$$

Water takes us one more step. This molecule has two lone pairs and two bonding pairs. This gives one bonding pair–bonding pair repulsion, four lone pair–bonding pair repulsions, and, now, one lone pair–lone pair repulsion. It is a reasonable guess that lone pair–lone pair repulsions are even larger than lone pair–bonding pair repulsions, so the bond angle in H_2O should be even smaller than those in NH_3. This is what is observed: H_2O has a bond angle of 104.5°. We can expand (16-8) to include our guess about lone pair–lone pair repulsions.

$$\begin{matrix} \ell p\text{–}\ell p \\ \text{repulsions} \end{matrix} > \begin{matrix} \ell p\text{–}bp \\ \text{repulsions} \end{matrix} > \begin{matrix} bp\text{–}bp \\ \text{repulsions} \end{matrix} \qquad\qquad (16\text{-}9)$$

We shall find these empirical rules useful in more complex situations, but, before discussing them, we should examine the structures of molecules that do not have eight electrons around the central atom.

(b) ELECTRON REPULSIONS IN BH_3, BF_3, BeH_2, AND BeF_2

The pigeonhole representations for the boron compounds BH_3 and BF_3 show three electron pairs shared in each molecule. The fact that the $2p_z$ orbital is partially occupied in BF_3 is ignored. Hence, in each molecule, electron repulsions act to separate the three bonds as much as possible. This criterion dictates that the three bonds lie in a plane at angles of 120°. This is the same prediction as that deduced from the sp^2 orbital hybridization, in agreement with the known structure of BF_3. Beryllium compounds, with only two shared pairs, are even simpler. Clearly two bonds are farthest apart when they point in opposite directions, that is, with a bond angle of 180°. This also agrees with the sp hybridization picture, and again the expectation is only in partial agreement with a cloudy experimental picture (see Table 16-2).

(c) ELECTRON REPULSIONS IN PCl_5 AND SF_6

The compound PCl_5 must dispose ten valence electrons around the central phosphorus atom: five from phosphorus and one each from the five chlorine atoms. According to our electron repulsion scheme, the molecule will take that geometric structure that separates five electron pairs as much as possible. However, there is no way to distribute five bonds so that all of the angles are equal. The best we can do is to arrange them as shown in Figure 16-5(a). The bonds should point to the corners of the structure obtained when two three-sided pyramids are placed base-to-base (*a trigonal*

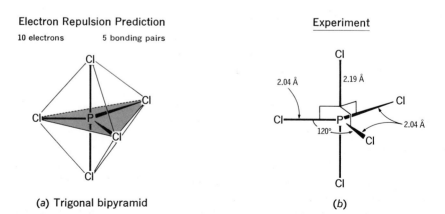

Electron Repulsion Prediction   Experiment

10 electrons   5 bonding pairs

(a) Trigonal bipyramid   (b)

Figure 16-5   Electron repulsions and the structure of PCl_5.

Electron Repulsion Prediction   Experiment

12 electrons   6 bonding pairs

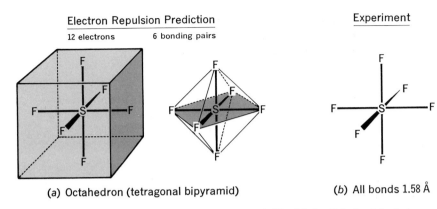

(a) Octahedron (tetragonal bipyramid)   (b) All bonds 1.58 Å

Figure 16-6 Electron repulsions and the structure of SF₆. (a) Predicted octahedral con-figuration showing how the octahedron fits inside a cube. (b) Observed structure.

bipyramid). In this structure there are two kinds of P–Cl bonds, one kind that forms the base of the pyramids and the other kind that is directed axially. Indeed, PCl_5 does have this structure and the experimentally measured bond lengths differ noticeably, as shown in Figure 16-5(b).

Sulfur hexafluoride, SF_6, has twelve valence electrons around the central sulfur atom; six from sulfur and one each from the six fluorine atoms. This time, the bonds should point at the corners of the structure obtained when two four-sided pyramids are placed base-to-base. Such a structure can be called a *tetragonal bipyramid*, but it is usually called an octahedron (it has eight sides). The angle between any pair of bonds is 90°, as pictured in Figure 16-6(a). The experimentally observed structure of SF_6 is in accord— all bond angles are 90° and all bond lengths are equal, 1.58 Å, in a perfect octahedral structure.

(d) ELECTRON REPULSIONS IN SF₄ AND XeF₄

Sulfur tetrafluoride, SF_4, has the same number of valence electrons as PCl_5 (10) so, according to the electron repulsion view, its structure should be similar. The bonds should be directed at the corners of a trigonal bipyramid. Figure 16-7 shows, however, that the molecule still has two options. In Figure 16-7(a), the lone pair is placed in the base of the pyramids, whereas in Figure 16-7(b), it is placed in an axial position. The implication of (a) is that there are two ℓp–bp and four bp–bp interactions at the acute, 90° angle. In (b), these are equally divided, three ℓp–bp and three bp–bp interactions at the acute angle. Since ℓp–bp repulsions exceed bp–bp repulsions (from 16-8), structure 16-6(a) should be preferred. Experiment shows that it is, as displayed in Figure 16-7(c).

The inert gas compound XeF_4 has twelve valence electrons around the xenon atom, hence its structure should be related to SF_6. Again the mol-

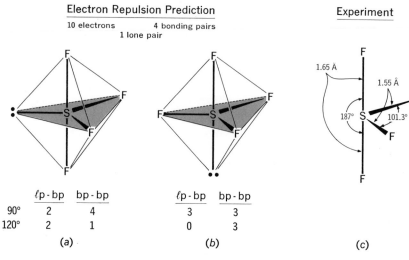

Figure 16-7   *Electron repulsions and the structure of* SF_4.

ecule has two options. In the structure shown in Figure 16-8(a), the bonds are in a square, planar arrangement and there are eight ℓp–bp, but no ℓp–ℓp, repulsions. The structure shown in Figure 16-8(b) (which resembles the SF_4 structure) has one ℓp–ℓp and six ℓp–bp repulsions. To explain the observed square, planar molecular geometry (Fig. 16-8(c)), we must postulate that ℓp–ℓp repulsions exceed ℓp–bp repulsions (as guessed in (16-9)) and by more than ℓp–bp repulsions exceed bp–bp repulsions.

16-3   Contrast of the hybridization and electron repulsion models

We have, then, two bases for explaining bond angles. The bond angles could reflect, and hence be predicted from, the orbital parentage of the

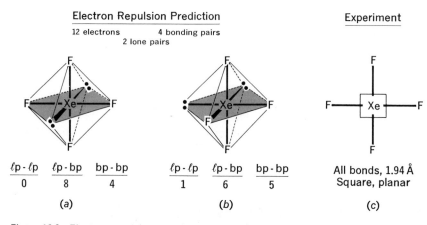

Figure 16-8   *Electron repulsions and the structure of* XeF_4.

Table 16-3 *Contrast of Orbital Hybridization and Electron Repulsion Predictions*

	Orbital Hybridization			Electron Repulsion	
Number of bonds	Hybridization	Bond angles	Number of bonds	Number of electron pairs	Bond angles
2	sp	180°	2	2	180°
3	sp²	120°	3	3	120°
2	p²	90°	2	4	109°28′
3	p³	90°	3	4	109°28′
4	sp³	109°28′	4	4	109°28′
5	dp,sp²	trigonal bipyramid	5	5	trigonal bipyramid
6	d²sp³	octahedron	6	6	octahedron

bonds. On the other hand, the bond angles could be fixed merely by the repulsions between pairs of electrons in valence orbitals (ignoring energy differences between s and p orbitals, as well as their directional properties). The two theories qualitatively agree in all cases (see Table 16-3); they differ quantitatively in their bond angle predictions for H_2O and NH_3. In fact, the bond angles are closer, for these two molecules, to the electron repulsion expectation of 109°28′. There are, however, other opportunities to test these two theories lower in the Periodic Table. The other elements in the sixth column of the Periodic Table should, like oxygen, form dihydrides and difluorides, with the same bond angles predicted. In a similar way, the elements below nitrogen in the fifth column of the Periodic Table should form trihydrides and trihalides with angles like those in ammonia. The known facts are summarized in Table 16-4.

Table 16-4 *Bond Angles in Fifth- and Sixth-Column Hydrides and Fluorides*

H_2O	104.5°	F_2O	102°	Cl_2O	110°				
H_2S	92.2°	—		Cl_2S	102°				
H_2Se	91.0°	—		—					
H_2Te	88.5°	—		—		Br_2Te	98°		
NH_3	107.3°	NF_3	102°	—					
PH_3	93°	PF_3	104°	PCl_3	100°	PBr_3	101.5°	PI_3	98°
AsH_3	91.5°	AsF_3	102°	$AsCl_3$	98°	$AsBr_3$	101°	AsI_3	98.5°
SbH_3	91.3°	SbF_3	88°	$SbCl_3$	99.5°	$SbBr_3$	97°	SbI_3	99°
				$BiCl_3$	100°	$BiBr_3$	100°		

Bond Angle Predictions		
	H_2M, X_2M	H_3M, X_3M
hybrid orbitals	90°	90°
electron repulsion	109°28′	109°28′

Among the hydrides we see that only H_2O and NH_3 are close to the electron repulsion prediction of a tetrahedral angle. All the others are remarkably close to the 90° angle expected for p orbital bonds. For all the halide molecules the bond angles are found to be irritatingly close to half-way between the two predictions. Thus there is little basis for preferring either theory. Nevertheless, proponents of each theory seek rationalizations for the discrepancies. The orbital hybridization advocates speak of repulsions between the electrons in the bonding pairs and repulsions between the terminal atoms, both of which tend to open the bond angles from 90° (as observed in H_2O and NH_3). As the central atom becomes larger (as in the series O, S, Se, Te or N, P, Sb, As), the repulsion becomes less important and the angle approaches the orbital prediction. Thus, the fact that the H_2O and NH_3 angles are larger than predicted, while those analogs lower in the Periodic Table are quite close to the expected values, is comfortably explained. We have already seen that electron repulsion supporters argue that repulsions between nonbonding electron pairs must also be considered. Empirically it is found that nonbonding electrons repel more strongly than bonding electrons. Thus, the nonbonding electron repulsions open the angles involving unused (and unseen) electron pairs and, hence, close the angles between the bonds from the tetrahedral angle. Unfortunately, only a rather strained empirical argument can be mustered to account for the almost 90° angles in H_2S, H_2Se, and H_2Te. The same is true for PH_3, AsH_3, and SbH_3. On the other hand, the natural way in which tetrahedral angles in carbon compounds and the planar structure of BF_3 are rationalized is appealing.

Though there is frequent controversy over the relative merits of these two schemes, they have a generic likeness that is seldom expressed. The electron repulsion theory assumes wrongly that the 2s and 2p orbitals are exactly equivalent, that is, that the s and p orbitals are *completely* hybridized in the first approximation in all cases. The orbital hybridization view probably overemphasizes the energy difference between s and p orbitals and hence, tends to "under-hybridize." Both views, however, embody the concept of "mixing" s and p character to form bonds. If we knew where the valence electrons in H_2O spend most of their time, and if we attempted to describe the distribution in a series expansion based on hydrogen atom orbitals, the expansion would surely involve predominantly 2p orbitals, with some contribution from the 2s orbital. The electron repulsion and orbital hybridization schemes merely offer different first guesses as to the extent of this contribution.

The most important point to remember, though, is that both of these extremely simple theories correlate the bond angles and molecular structures of a multitude of molecules. A detailed quantum mechanical calculation would claim a major accomplishment if it were to predict within a few degrees (without prior knowledge of the experimental result) even one bond angle in a polyatomic molecule. This sort of predictive accuracy can be assumed using either the electron repulsion or orbital hybridization models, provided it is applied within bounds limited by the experimental data.

16-4 Determination of molecular geometry

A variety of reasons for the chemist's great interest in molecular geometries was given in the introduction to this chapter. The geometric influence on molecular charge distribution determines the molecular dipole moment. This dipole moment, together with molecular shape, reflects into such properties as boiling point, melting point, crystal structure, solvent properties, ease of reaction, and a host of other chemically important phenomena. These days chemists feel that a molecular formula or even a molecular structure is not well characterized until the entire three-dimensional perspective is known.

Either the orbital hybridization or the electron repulsion approach allows us to anticipate very many molecular structures. By no means, however, are these methods infallible. Furthermore, they only embody a multitude of experimental facts collected laboriously by quite a variety of methods. Entire fields of physics have been essentially taken over by chemists because of the importance of such data. These fields, which include infrared spectroscopy, nuclear magnetic resonance, X-ray diffraction, electron diffraction, and several others, make up the general subject of molecular spectroscopy. A primary goal of molecular spectroscopy is the determination of molecular geometry.

Problems

1. Two molecules whose structures are not yet measured are SF_2 and NCl_3. Predict the geometry, the bond angles, and the orbital hybridization for each.

2. According to the pigeonhole diagram for a sulfur atom, what is the expected orbital hybridization and bond angle for each sulfur atom in the molecule S_2Cl_2? Propose a likely molecular structure.

3. The ground state configuration of carbon atom is $(1s^2 2s^2 2p_x 2p_z)$, but it has an excited state that permits the 2s orbital to be used in bonding $(1s^2 2s 2p_x 2p_y 2p_z)$. The latter state is involved in the bonding of linear CH_2 (Sections 15-6 and 16-1(b)). According to orbital hybridization, what would be the structure of CH_2 based upon the ground state configuration?

4. Draw a pigeonhole diagram for a nitrogen atom and predict the molecular geometry of the reactive free radical, NH_2.

5. The free radical NH_2 absorbs light with the energy needed to excite an electron from the 2s orbital to one of the half-occupied 2p orbitals (see Problem 4). With the half-occupied 2s orbital now available, what new structure is accessible to the excited NH_2?

6. Methane is often drawn as shown below. What misconceptions are associated with this representation? What does it suggest correctly?

```
      H
      |
 H —  C — H
      |
      H
```

7. Using toothpicks for bonds and gumdrops (or corks) for atoms, make molecular models of NH_3, BF_3, and CH_4.

8. Predict the geometry of the ion BH_4^-.

9. The amide ion NH_2^- is known to exist in alkali metal amide salts and in ammonia solutions. The ammonium ion NH_4^+ forms many stable salts (e.g., ammonium chloride, NH_4Cl) and is a common species in aqueous solutions. The NH_3^+ ion is known only as a transient and reactive ion. Predict the molecular geometries of each of these ions.

10. The molecule ClF_3 can be considered to involve ten electrons around the chlorine atom; seven of them are the Cl valence electrons and three are F atom valence electrons. According to the electron repulsion point of view, these will arrange themselves as five pairs in the geometry of a trigonal bipyramid. Fluorine atoms are attached at three of the positions (bonding pairs) and the other two positions are nonbonding pairs. Draw the three possible geometries and decide which is more stable on the basis of the empirical repulsion rules deduced in Section 16-2(a).

11. Repeat Problem 10 for the inert gas compound XeF_2, in which the xenon atom can be considered to have ten electrons (eight from its own valence electrons and two from the two fluorine atoms). Now how many structures are predicted? Draw them and decide which is most stable and which least stable.

12. Make a toothpick-gumdrop model of ethane (C_2H_6), showing tetrahedral bonding geometry (sp³) around each carbon atom. Now replace a hydrogen atom at each end with a chlorine atom (a green gumdrop). What structural questions about this molecule are *not* answered by the bond angles deduced from orbital hybridization (or electron repulsion)? Deduce with your model the structure that will be most stable if bonds repel each other and chlorine atoms repel each other even more. Draw a picture of this structure.

13. Make a toothpick-gumdrop model of methyl alcohol, H_3COH, and of methyl amine, H_3CNH_2, fixing the bond angles around the carbon, oxygen, and nitrogen atoms in accord with the orbital hybridization and electron repulsion descriptions. In each molecule, decide how many times an equivalent structure is obtained as the O–H group (NH_2 group) is rotated through 360° around the C–O bond (C–N bond).

14.   Repeat Problem 13 for hydroxyl amine, H_2NOH, and for hydrazine, H_2NNH_2.

15.   Make a toothpick-gumdrop model of CH_2FCl (using differently colored gum-drops for different atoms). Now make a second CH_2FCl that is exactly like the *image* of the first when it is viewed with a mirror. Is it possible to superimpose these two models simply by rotating and turning one of them?

16.   Repeat Problem 15 for CHFClBr.

seventeen chemical bond types

Chemists still identify a variety of chemical bond types, despite the unifying influence of quantum mechanics. The value of these bond categories is undeniable—they enable the chemist to predict with reasonable confidence when a bond can form and what bond properties to expect. They are responsible for the ultimate measure of success—new compounds, never before seen on earth, currently being produced at a rate of 300 per day, including Saturdays, Sundays, and the Fourth of July.

17-1 Covalent bonds

Most of the bond situations we have considered thus far involve two identical atoms, each with a valence orbital that contains a single electron. Such *a pair of electrons, equally shared between two atoms (one electron furnished by each atom) forms* a **covalent bond.** The bond forms because the two half-occupied valence orbitals give rise to a bonding and an antibonding M.O. Both electrons can occupy the bonding M.O. They lower the potential energy because both electrons are then simultaneously near two nuclei.

(a) COVALENCY: SHARE AND SHARE ALIKE

A large fraction of the known chemical compounds possess some bonds with the parentage just described. When the atoms bonded are identical, the electron probability distribution is symmetrically disposed, favoring neither atom.

We have represented in a very approximate way the M.O.'s of H_2^+ and H_2 in the form

$$\sigma(1s) = \psi_A(1s) + \psi_B(1s) \tag{17-1}$$

$$\sigma^*(1s) = \psi_A(1s) - \psi_B(1s) \tag{17-2}$$

These approximations correctly describe the qualitative features of the probability distribution—particularly the nodal properties. They also make an explicit statement about where the electron is likely to be found. Hence the bond energy can be calculated for a hypothetical H_2 molecule, with an

612

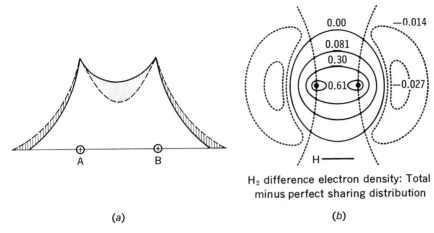

H₂ difference electron density: Total
minus perfect sharing distribution

(a)   (b)

Figure 17-1 (a) The electron probability distribution in H_2 (——) contrasted with the superposition of two 1s probability distributions (- - -). The simple combination of atomic orbitals underestimates the electron density between the nuclei (shaded portion) and overestimates the electron density on the outside of the nuclei in the antibinding regions (crosshatched portions). (b) Difference electron density map of H_2 showing increased electron density in binding region.

electron distribution in which one electron remains in the 1s orbital centered on proton A and the other remains in the 1s orbital centered on proton B. This calculation falls far short of a quantitatively accurate description; it predicts a bond energy only about 10 percent as large as the observed value. It is most revealing how this approximation electron distribution differs from an accurate portrayal. Figure 17-1(a) shows the discrepancy, in terms of the probability density along the line connecting the nuclei. The actual distribution has a far higher density between the nuclei (see the extra, shaded area) than does the superposition of the 1s probabilities. Figure 17-1(b) shows the calculated difference electron density, illustrating the shift of electron density from the antibinding regions into the binding regions. This extra likelihood of finding the electrons between the nuclei (in the binding region) was stolen from the antibinding region. Hence this redistribution accounts for most of the energy-lowering. Crude approximations of bonding M.O.'s like (17-1) always underestimate the extent to which the electrons are placed between the nuclei in the binding region and they underestimate the bond energy, correspondingly.

(b) COVALENT BOND ENERGIES IN HOMONUCLEAR DIATOMIC
     MOLECULES

The energy-lowering that actually results from the equal sharing ("perfect sharing") of electrons in homonuclear diatomic molecules varies quite a bit across the Periodic Table. Figure 17-2 shows the bond energies presently

double bonds
triple bonds

							B₂ / C₂ / N₂ / O₂ columns

IA	IIA			Cu group	Zn group	B group	C group	N group	O group
H_2 104									
Li_2 25	Be_2 17					B_2 69	C_2 144 (72)	N_2 225 (75)	O_2 118 (59)
Na_2 18						Al_2 40	Si_2 76 (38)	P_2 117 (39)	S_2 101 (50)
K_2 12				Cu_2 47	Zn_2 6	Ga_2 < 35	Ge_2 65 (33)	As_2 91 (30)	Se_2 73 (37)
Rb_2 11				Ag_2 39	Cd_2 2.1	In_2 23	Sn_2 46 (23)	Sb_2 69 (23)	Te_2 53 (27)
Cs_2 10				Au_2 52	Hg_2 1.4	Tl_2 15	Pb_2 23 (12)	Bi_2 39 (13)	

Figure 17-2 Bond energies for homonuclear diatomics. Bond energy per bond order in parentheses (where applicable).

known, 32 of them. They range from the very weak interactions of Hg_2, Cd_2 and Zn_2 up to the 225 kcal bond of nitrogen. There are obvious vertical trends. With only two exceptions, covalent bonds become weaker as we move down a particular column in the Periodic Table. One of the exceptions is fluorine, which has a much weaker bond than is consistent with the rest of the halogens. This 37 kcal bond accounts for the extreme reactivity of fluorine.

Figure 17-3 displays these energies graphically. The peaked shapes of the curves plainly reveal the triple bonds throughout the nitrogen family (N_2, P_2, As_2, Sb_2, and Bi_2). They also show that double bonds are obtained throughout the carbon family (C_2, Si_2, Ge_2, and Pb_2) and the oxygen family (O_2, S_2, Se_2, and Te_2). Equally interesting is the Zn column with effectively zero bond energies (Zn_2, Cd_2, and Hg_2). This is appropriate to the orbital occupancies of these elements as gaseous atoms:

Zn · · · · 3d 4s 4p   (17-3)

Cd · · · · 4d 5s 5p   (17-4)

Hg · · · · 5d 6s 6p   (17-5)

In each case there are just enough electrons to fill the valence d and s orbitals, with none left over for the valence p orbitals. Apparently the energy needed to promote an electron from the valence s state to the valence p state exceeds the energy-lowering that would result from the resultant covalent bonding that could occur.

Lastly, note the very low bond energies of the alkali diatomics (Li_2, Na_2, K_2, Rb_2, and Cs_2). These low values reflect another trend, a tendency for bond energies to rise as we move to the right along a row. The effect is masked, at first glance, by the multiple bonds on the right-hand side of the Periodic Table. It can be revealed, however, by dividing each bond energy by its bond order to obtain the "bond energy per bond order." These values are shown parenthetically in Figure 17-2. The most uniform trend is obtained in the second row, as shown in Figure 17-4. The first row is unique in that its multiple bonds are especially stable and the single bond of fluorine is

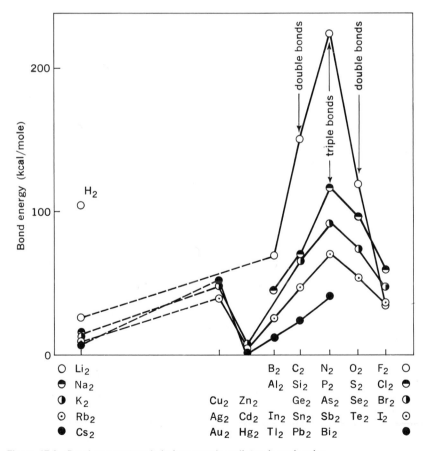

Figure 17-3 Bond energy trends in homonuclear diatomic molecules.

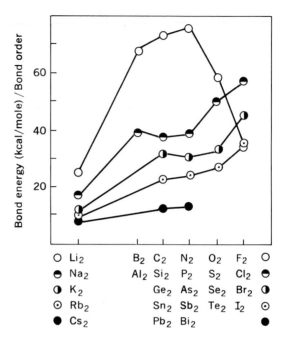

Figure 17-4   Bond energy per bond order trends in homonuclear diatomics.

especially weak. All the other rows show the rising trend appropriate to the tendency for electron affinity to rise as one moves to the right in the Periodic Table (see Fig. 13-10).

(c) COVALENT HOMONUCLEAR BOND ENERGIES IN POLYATOMIC MOLECULES

The oxygen–oxygen bond in the O_2 molecule is a double bond—its bond energy is 118 kcal/mole. However, in hydrogen peroxide, H_2O_2, there are two oxygen atoms bonded to each other, but they share only one pair of electrons. This should be a single bond. Experimentally we find it is so. To break this bond (to give two OH molecules) requires 51 kcal, about half the energy needed to break the O_2 bond. The other measures of bond order— bond length and vibrational force constant (see Section 14-4(d))—are also consistent. Table 17-1 shows the variations of bond properties with bond order for the four elements carbon, nitrogen, oxygen and fluorine. As was found for diatomic molecules in Chapter Fifteen, the three bond order criteria, bond energy, bond length and force constant, correlate and indicate the bond strength in polyatomic molecules. Any one of these measurements can be a useful indicator to the nature of the bonding in a polyatomic molecule.

Of the data in Table 17-1, the energy per bond order is of particular

Table 17-1   *Bond Properties as a Function of Bond Order—Homonuclear Bonds*

Compounds	C	N	O	F
single bond	H_3C-CH_3 ethane	H_2N-NH_2 hydrazine	$HO-OH$ hydrogen peroxide	$F-F$ fluorine
double bond	$H_2C{=}CH_2$ ethylene	$HN{=}NH$ diimide	$O{=}O$ oxygen	—
triple bond	$HC{\equiv}CH$ acetylene	$N{\equiv}N$ nitrogen	—	—
Bond Energies **(kcal)**	**C**	**N**	**O**	**F**
single	83	60	51	37
double	143	?	118	—
triple	194	225	—	—
Bond Energy **Bond Order** **(kcal)**	**C**	**N**	**O**	**F**
single	83	60	51	37
double	72	?	59	—
triple	65	75	—	—
Bond Length **(Å)**	**C**	**N**	**O**	**F**
single	1.54	1.47	1.49	1.45
double	1.34	1.25*	1.21	—
triple	1.20	1.10	—	—
Force Constants **(mdyne/Å)**	**C**	**N**	**O**	**F**
single	4	4	4	4.5
double	11	12	11.4	—
triple	15	22.4	—	—

* For $FN{=}NF$.

interest. Contrast the decreasing energy per bond order for the carbon compounds with the opposite trend for the nitrogen compounds. These trends undoubtedly contribute to the fact that acetylene, $HC{\equiv}CH$, is quite reactive, whereas elemental nitrogen, $N{\equiv}N$, is quite inert.

(d) COVALENT BOND LENGTHS

Table 17-1 lists covalent bond lengths for carbon and nitrogen as measured in compounds with single, double, and triple bonds. These trends are use-

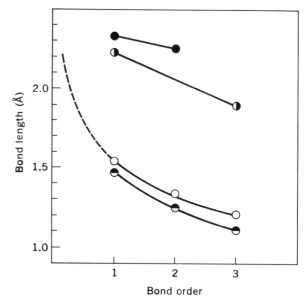

Figure 17-5 Change of bond length with bond order: ⊙, *C—C bonds;* ◓, *N—N bonds;* ●, *Si—Si bonds;* ◑, *P—P bonds.*

ful in the evaluation of the bonding in molecules with unorthodox structures. Figure 17-5 shows a plot of covalent bond length against bond order for carbon, nitrogen, and their second-row counterparts, silicon and phosphorus. The curve for the carbon–carbon bonds is extended to show that the bond length must approach infinity as the bond order approaches zero.

Figure 17-6 shows two important carbon compounds whose bonding is not readily explained with conventional bonding rules. Benzene, for example, can be written with two equally acceptable bond representations. Both structures imply alternate single and double bonds that are alternately long and short. The observed molecular structure shows six *equal* bond lengths of 1.40 Å! This length corresponds to a bond order near $1\frac{2}{3}$. The equal bond lengths force us to invoke the resonance argument that was used to discuss the bonding in sulfur dioxide, nitrate ion, carbonate ion, and so on, in Section 2-4(f). Neither of the individual benzene structures shown in Figure 17-6 accurately describes the molecule—we need a superposition of the two. However, the bond order is about $1\frac{2}{3}$, a stronger bond than would be suggested by the simple average of a single and a double bond. This is generally the case: When the resonance argument is needed, the bonds are somewhat stronger than indicated by averaging. (Note that a molecular orbital description of the pi contribution to the benzene bonds does predict equal bond lengths and bond orders.)

Butadiene, the second molecule in Figure 17-6 has only one conventional representation, but the bond lengths show that the center bond has a bond order of 1.4, much stronger than a normal single bond, while the end bonds have a bond order near 1.9. Again this failing of the line representation is corrected in the molecular orbital description of the pi contribution to the bonding. Figure 17-7 shows the nodal properties of the pi M.O.'s and the resultant energy level diagram. With four electrons in these M.O.'s, the lowest two M.O.'s are occupied. The lowest M.O. contributes to the bonding of the center as well as to the end C–C bonds. A single electron pair divided among three bonds will contribute to each bond about one third of a bond order. The next M.O. strengthens only the end bonds and hence contributes one half of a bond order to each. These pi contributions must be added to the single bond contributions of the sigma orbitals to each bond. We are led to expect the end bond orders to be $1 + \frac{1}{3} + \frac{1}{2} = 1.8$ and the center bond order to be $1 + \frac{1}{3} = 1.3$.

Returning to Figure 17-5, we see that N–N covalent bonds are shorter than C–C bonds. On the other hand, the figure shows that phosphorus and silicon bonds are much longer than their nitrogen and carbon counterparts. We want to understand the factors that determine these trends.

We already have the answer at hand from our consideration of the hydrogen atom. For a one-electron atom, an increase in nuclear charge pulls the electron in closer—the average radius depends on $1/Z$. On the other hand, the radius increases with the square of n, the principal quantum number.

These two factors can be seen to be at work in many-electron atoms. In fact, we have already used a one-electron approximation of atomic size in

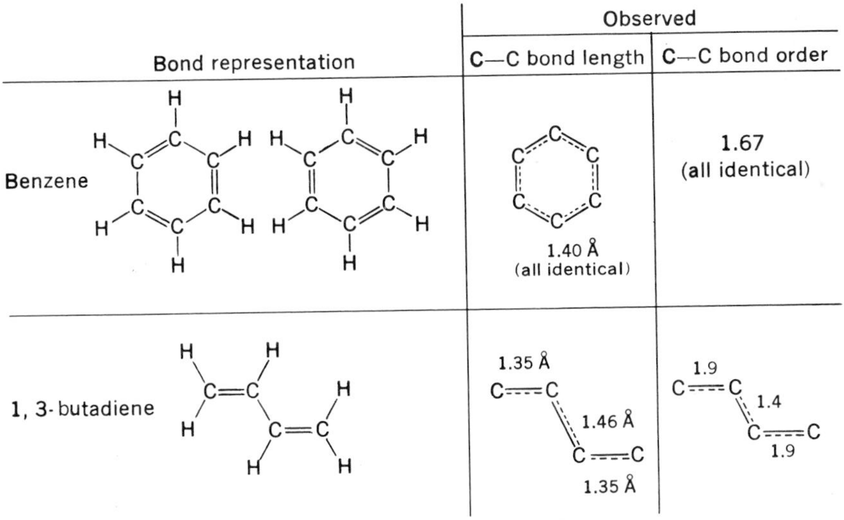

	Bond representation	Observed	
		C—C bond length	C—C bond order
Benzene		1.40 Å (all identical)	1.67 (all identical)
1, 3-butadiene		1.35 Å 1.46 Å 1.35 Å	1.9 1.4 1.9

Figure 17-6 *Molecules with more than one double bond: benzene and 1,3-butadiene. The observed bond orders are obtained from the graph of Figure 17-5 at the observed bond lengths.*

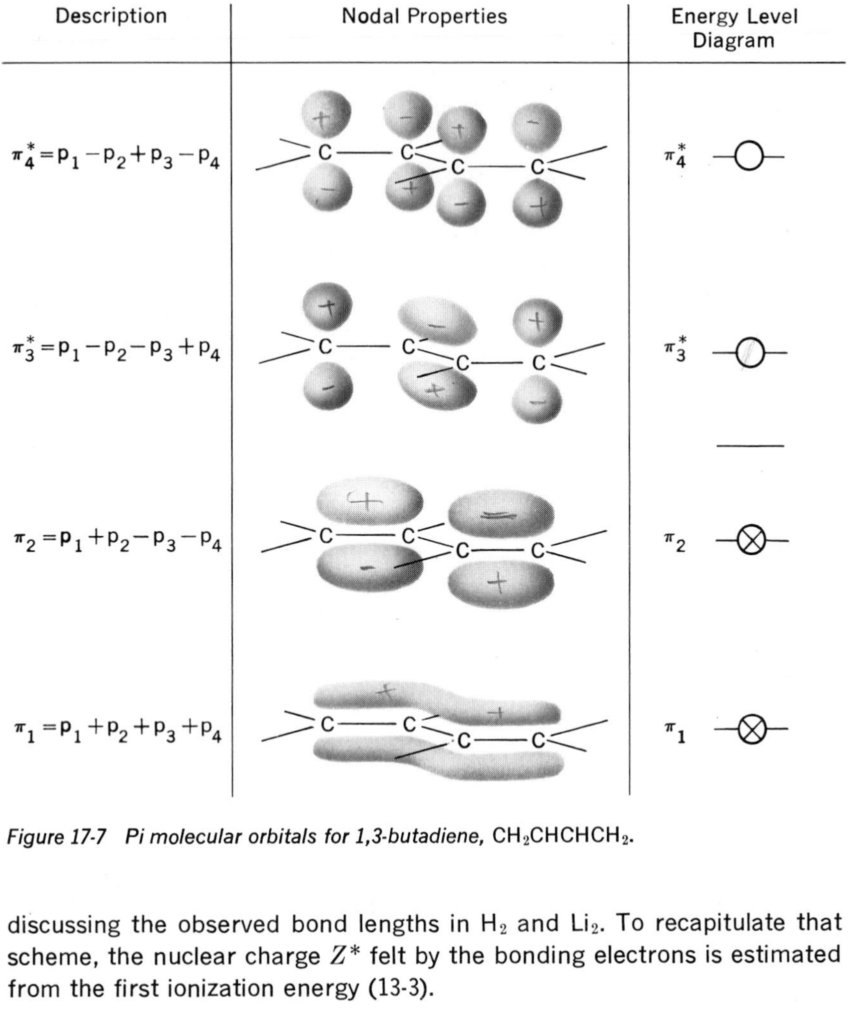

Description	Nodal Properties	Energy Level Diagram
$\pi_4^* = P_1 - P_2 + P_3 - P_4$		π_4^*
$\pi_3^* = P_1 - P_2 - P_3 + P_4$		π_3^*
$\pi_2 = P_1 + P_2 - P_3 - P_4$		π_2
$\pi_1 = P_1 + P_2 + P_3 + P_4$		π_1

Figure 17-7   Pi molecular orbitals for 1,3-butadiene, $CH_2CHCHCH_2$.

discussing the observed bond lengths in H_2 and Li_2. To recapitulate that scheme, the nuclear charge Z^* felt by the bonding electrons is estimated from the first ionization energy (13-3).

$$E_1 = -313.6 \frac{Z^{*2}}{n^2}$$

This value of Z^* fixes the average radius of a hypothetical one-electron atom having the same quantum numbers (n and ℓ) and the same ionization energy as the valence electron (13-4).

$$\bar{r}(\text{Å}) = \frac{0.529n^2}{Z^*} \left\{ \frac{3}{2} - \frac{\ell(\ell + 1)}{2n^2} \right\}$$

Figure 17-8 shows the calculated trends in $\bar{r}$ for the first 18 elements. Also shown, as crosses, are 11 covalent bond lengths (single bonds only).

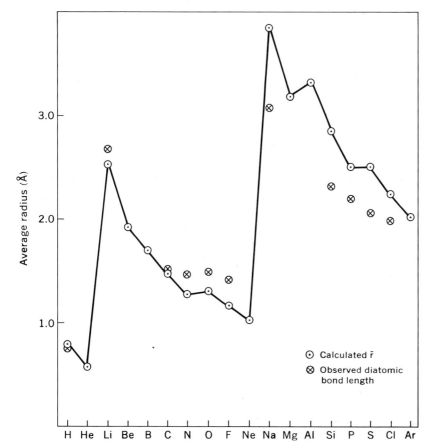

Figure 17-8 Sawtooth systematics in covalent bond lengths; ⊙, calculated $\bar{r}$; ⊗, observed diatomic bond length.

The observed bond lengths clearly reflect the sawtooth systematics attributable to the trend in atomic size, as indicated by $\bar{r}$. The periodic spikes are associated with a change in n, and the general shortening of bonds across a row is caused by an increase in nuclear charge with constant principal quantum number.

17-2 Heteronuclear bonds: ionic bonds

A bond between atoms of different elements is called a heteronuclear bond. We have encountered them in earlier chapters. In Section 15-5, the bonding in first-row heteronuclear diatomic molecules was adequately considered in terms of the molecular orbitals of homonuclear diatomics. Earlier, though, in Section 14-5(d), three heteronuclear molecules were found to differ from homonuclear counterparts. In each of the molecules LiH, HF, and LiF, the electron distribution is skewed toward one of the atoms, producing a dipole moment. The electron movement takes place because the energy is lowered even more than the energy-lowering obtained from a "perfect sharing" (or covalent) distribution. These are important consequences—a dipole moment and a specially strong bond can make a molecule's fortune!

(a) IONIC BOND CHARACTER AND EXCESS BOND ENERGY

A bond in which the electrons concentrate nearer to one of the atoms is said to have *ionic character*. This ionic character is manifested in a dipole moment μ and in bond energy D_0 or, rather, in the excess bond energy above that to be expected from a "perfect sharing" distribution, $\bar{D}_{AB}$. The reference bond energy between two atoms A and B, $\bar{D}_{AB}$, is best taken to be the geometric mean of the bond energies of D_{A_2} and D_{B_2}, the bond energies of the covalent molecules A_2 and B_2. Either μ or $D_0 - \bar{D}_{AB}$ could be adopted as a quantitative measure of ionic character (indeed, both are used). Unfortunately they do not always agree, as we saw in Table 14-6. The molecule HF has a value of $D_0 - \bar{D}$ equal to 73 kcal/mole, midway between those of LiH and LiF, but it has the lowest dipole moment of the three.

Bond energies are generally more influential in determining chemistry than are dipole moments. Hence chemists have placed more emphasis on this feature in their attempts to assess ionic character. The magnitudes of these effects and their trends are shown in Table 17-2 for some molecules that involve only single bonds.

The first molecule listed, LiF, is composed of two elements that form very weak covalent bonds. The diatomic molecules Li_2 and F_2 have bond energies of only 25 and 36 kcal/mole, respectively. Yet the gaseous diatomic molecule LiF has a bond energy of 137 kcal/mole! As was discussed in Section 14-5(e),

Table 17-2 *Excess Bond Energies in Heteronuclear Single Bonds* (kcal/mole)

MX	$D_0(MX)$	$D(M_2)$	$D(X_2)$	$\bar{D}$	$D_0 - \bar{D}$
LiF	137	25	37	30	107
LiCl	115	25	57	38	77
LiBr	101	25	46	34	67
LiI	81	25	36	30	51
NaF	107	17	37	25	82
NaCl	98	17	57	31	67
NaBr	88	17	46	29	59
NaI	71	17	36	25	46
HF	135	103	37	62	73
HCl	102	103	57	77	25
HBr	87	103	46	69	18
HI	71	103	36	61	10

this extra energy must be attributed to an electron distribution skewed toward the fluorine atom, but it cannot be caused by "removal" of an electron from the lithium atom. The formation of gaseous ions Li^+ and F^- from neutral atoms requires an input of energy of 44 kcal/mole. The extra bond energy results from the fact that the movement of the lithium atom's electron over to the fluorine atom localizes that electron and holds it even closer to the lithium nucleus than in the neutral lithium atom. (The bond length in LiF is 1.56 Å, whereas the average radius of the lithium 2s orbital is 2.52 Å.) The importance of this effect is emphasized by the observation that the electron affinity of fluorine, 79.5 kcal, is 60 percent of the entire bond energy. The remaining 56 kcal must be attributed to the interaction of the bonding electrons with the lithium nucleus. This is more than double the Li_2 bond energy!

Turning to the next lithium halide entry, we see that the value of $D_0 - \bar{D}$ decreases to 77 kcal/mole in LiCl, despite the fact that both the bond energy and the electron affinity of chlorine are above those of fluorine. The larger size of the chlorine atom implies that localization of the bonding electrons on the chlorine is less beneficial to the lithium atom. The bond length in LiCl, 2.02 Å, is still less than the average radius of the lithium 2s orbital but almost 0.5 Å larger than the LiF bond length.

The remaining lithium halides continue the trend. As we move down in the Periodic Table, the excess bond energy attributable to ionic character decreases. The same trend is exhibited by the sodium halides and the hydrogen halides. We conclude that in each of these series, the ionic character decreases as the halogen atom becomes larger.

(b) EXCESS BOND ENERGY AND ELECTRONEGATIVITY

Over three decades ago, Linus Pauling attempted to use these excess bond energies to define a scale that encompassed what was then known about ionic character. Chemists had earlier attributed to each element a quality called "electronegativity." Pauling decided to establish a quantitative scale of electronegativity based on the excess bond energy. Each element would be assigned a number x, such that when two elements A and B form a bond, the difference between x_A and x_B would determine the excess bond energy. Well, that's the reverse of what Pauling did. He took the known excess bond energies and hunted around for suitable values of x so that the differences $(x_A - x_B)$ would be consistent with the facts. The functional form he worked with was

$$D_0 = \bar{D}_{AB} + 23(x_A - x_B)^2 \tag{17-6}$$

$$\bar{D}_{AB} = \sqrt{D_{A_2} D_{B_2}} \tag{17-7}$$

The difference $(x_A - x_B)$ appears squared to express the expectation that

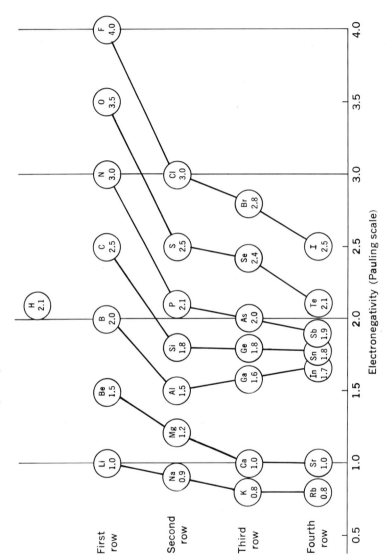

Table 17-3 Pauling's Electronegativity Scale Based upon Excess Bond Energies. Elements in the Same Column of the Periodic Table Are Joined by Heavy Lines

a difference between the x's is what is important, not its sign. The number 23 is an accident of history—it merely adjusts energy units to kilocalories. With the bond energies available to him, Pauling selected the electronegativity values shown in Table 17-3. This scale guides the qualitative thinking of most chemists. A bond between two elements that differ quite a bit in electronegativity can be expected to be strong and to involve a significant charge separation.

Quantitatively, things are not too keen. From expression (17-6), D_0 can be calculated from the electronegativities in Table 17-3. However, Table 17-4 shows that bond energies are not accurately calculated if electronegativity differences are too large.

The other dimension of the electronegativity concept is its relationship to charge separation, or dipole moment. Once again, large charge movement should correlate with a large electronegativity difference. The extent of charge movement is best measured by the quantity $\delta = \mu/re$, in which r is the bond length and e is the charge on an electron. This hypothetical quantity δ is a fraction of an electron charge. It has a magnitude such that if $+\delta$ were to be placed on one atom and $-\delta$ on the other, the molecule would have the observed dipole moment μ. Figure 17-9 shows how δ and $(x_A - x_B)$ correlate for the molecules listed in Table 17-4. The hydrogen halides depend smoothly enough upon $(x_A - x_B)$, but they do not link up well with the alkali halides.

Many alternative electronegativity scales have been drawn up to polish the defects of the Pauling approach. None of these really reaches quantitative predictive value and Pauling's scale remains a useful guide to a chemist's intuition concerning the extent of ionic character.

Table 17-4   Bond Energies Calculated from Pauling's Electronegativities (kcal/mole)

AB	$x_A - x_B$	D_0 (calc.)	D_0 (expt.)	Discrepancy
LiF	3.0	238	137	101
LiCl	2.0	130	115	15
LiBr	1.8	109	101	8
LiI	1.5	82	81	1
NaF	3.1	247	107	140
NaCl	2.1	133	98	35
NaBr	1.9	111	88	23
NaI	1.6	84	71	13
HF	1.9	144	135	9
HCl	0.9	95	102	−7
HBr	0.7	80	87	−7
HI	0.4	64	71	−7

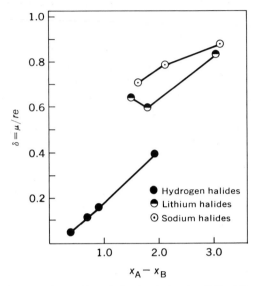

Figure 17-9 Charge separation and its relation to electronegativity difference: ●, *hydrogen halides;* ◓, *lithium halides;* ⊙, *sodium halides.*

17-3 Electron donor–acceptor bonds

Covalent bonds with and without ionic character are bonds resulting from the proximity of two atoms each of which possesses a half-filled valence orbital (an orbital occupied by only one electron). Chemists call such an orbital "half-filled" since two electrons can occupy each orbital, be it an atomic orbital or a molecular orbital.

It is observed that even after all the "half-filled" valence orbitals of an atom have become engaged in bonding, the resultant molecule will still be reactive if it has other, completely vacant valence orbitals. We have already encountered some examples: CH_2, which reacts on almost every collision; BH_3, which is so reactive it has never been detected spectroscopically, and BeF_2, which can be produced in the gas phase only at very high temperatures because of the strong chemical bonds between BeF_2 groups in the solid state. The bonds formed by molecules containing atoms with vacant valence orbitals are very important—we'll begin investigating them with the aid of BH_3 and its fluorine counterpart BF_3.

(a) BF_3 AND BH_3: VACANT VALENCE ORBITALS FOR RENT

The bonding in BF_3 and that expected in BH_3 were described in Section 16-1(e). The pigeonhole representations (16-4) and (16-5) display three sp^2

sigma bonds for each molecule. In each of these bonds, each of the two atoms linked contributes one electron to make up an electron pair. The remaining p_z orbital of boron is vacant.

In each case, the energy can be lowered still further if electrons can occupy this p_z orbital, because such electrons would thus be placed near the positive charge of the boron nucleus. In the case of BF_3, the $2p_z$ electrons of the three fluorine atoms do just that to some extent, accounting for the stability of BF_3 relative to BH_3. Nevertheless, this $2p_z$ orbital is not used to its full effectiveness in bonding. The boron atom would like to form a fourth bond, but it doesn't have the one electron to contribute to a normal bond.

From the molecular orbital point of view, this should not matter. If a boron atom approaches another atom M, there will develop a bonding and an antibonding M.O. A bond can be formed if *either* the boron atom or M furnishes an electron or two to occupy the bonding M.O. Since boron atom is short of electrons, what we need is another molecule with valence electrons to spare. Ammonia is such a molecule. After forming three bonds with hydrogen atoms, the nitrogen has a remaining pair of electrons in a valence orbital. This pair is called a "lone pair" or an "unused pair." The orbital it occupies can interact with the vacant boron valence orbital to give a σ bonding M.O. (as well as a σ^* antibonding M.O.). Then, with nitrogen's permission, its unused pair of electrons can occupy this bonding M.O. to form a boron–nitrogen bond. As is usual in chemical bonding, the two electrons thus contributed by nitrogen are not "lost" to it—rather, they occupy a bonding M.O. that permits them to be simultaneously near both the nitrogen and the boron nuclei. An electron-dot representation shows the situation:

$$
\begin{array}{cc}
\text{H} \quad \text{H} & \text{H} \quad \text{H} \\
\text{H} : \text{B} + \overset{\times}{\underset{\times}{\text{N}}} : \text{H} \longrightarrow \text{H} : \text{B} \overset{\times}{\underset{\times}{\text{N}}} : \text{H} \\
\text{H} \quad \text{H} & \text{H} \quad \text{H}
\end{array}
\qquad (17\text{-}8)
$$

In (17-8) we see that boron, with its vacant valence orbital, acts as an *electron acceptor* and the nitrogen atom, with an unused valence pair of electrons, acts as an *electron donor*. The bond that can be formed is called an *electron donor–acceptor bond* (or, by some chemists, a "dative" bond). We now see that the existence of electron donor–acceptor bonds is entirely in accord with the molecular orbital description. Just as we found for heteronuclear diatomic molecules (see Section 15-5), the *occupancy* of the M.O.'s is the important factor that determines bond order. The electron parentage seems to be a secondary matter.

The properties of some of the donor–acceptor compounds formed by BH_3 are well known, despite the illusory nature of BH_3 itself. For example, the bond energy of BH_3 with the ammonia-like compound trimethylamine $N(CH_3)_3$ has been measured to be 31.5 kcal/mole.

$$BH_3(g) + N(CH_3)_3(g) \longrightarrow H_3B\!-\!N(CH_3)_3(g) \qquad\qquad (17\text{-}9)$$

$$\Delta H = -31.5 \text{ kcal/mole}$$

In this case, the electron donor–acceptor bond energy is similar in magnitude to the weakest covalent bonds known, those of the alkali metal diatomic molecules (in the range 10–25 kcal/mole) and those of F_2 and I_2 (each about 36 kcal/mole).

Despite the extra stability of BF_3 (because of the fluorine $2p_z$ electrons), it forms quite stable donor–acceptor bonds like those of BH_3.

$$BF_3(g) + NH_3(g) \longrightarrow F_3B\!-\!NH_3(g) \qquad\qquad (17\text{-}10)$$

$$BF_3(g) + N(CH_3)_3(g) \longrightarrow F_3B\!-\!N(CH_3)_3(g) \qquad\qquad (17\text{-}11)$$

$$\Delta H = -26.5 \text{ kcal/mole}$$

Again the bond energy is low compared with most chemical bonds. This is generally true; electron donor–acceptor bond energies are generally in the range 10–50 kcal/mole. To designate the weakness of this bond, the molecule is often written with a dot rather than a line—as in $BF_3\!\cdot\!NH_3$ or $BF_3\!\cdot\!N(CH_3)_3$—and called a "molecular complex." These notations suggest that the molecular integrities of the reactants BF_3 and NH_3 are fairly well preserved in the new molecule. This is only partially true, as can be seen in the $F_3B\!-\!NH_3$ compound, whose molecular geometry is reasonably well known.

Figure 17-10 shows the structure of $F_3B\!-\!NH_3$. The hydrogen atoms are not very "visible" in the experimental technique that was used (which was electron diffraction), but all the heavy atoms are well located. Notice the dramatic change inflicted upon BF_3 as the bond forms. The parent molecule BF_3 is planar, with three equal BF bonds at 120° angles to each other. The BF bond lengths are 1.295 Å. In the donor–acceptor compound F_3BNH_3 the

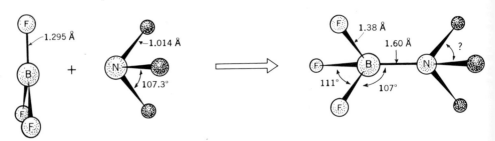

(planar, $\angle FBF = 120°$)

Figure 17-10 $BF_3 + NH_3 \longrightarrow BF_3\!:\!NH_3$. *Planar boron trifluoride folds into a pyramidal shape when it forms a donor–acceptor bond with the lone pair on ammonia. Note that the B–F bond lengthens in the complex since pi bonding is not possible.*

three BF bonds are folded back into a pyramid. The F—B—F bond angles are now 111°, only 1.5° from the perfect tetrahedral angle. The bond lengths, too, are significantly changed. The extension to 1.38 Å shows that the BF bonds are weakened, presumably because of the sacrifice of the extra pi bonding that stabilizes BF_3. The new bond formed, that between boron and nitrogen, is 1.60 Å in length. This is relatively long and suggests a bond order of about $\frac{1}{2}$ to $\frac{3}{4}$. The relative weakness of the bond can be attributed to the necessity to "rehybridize" from planar sp^2-π bonding in planar BF_3 to tetrahedral sp^3 bonding in the complex. This rehybridization raises the energy, but it permits M.O. formation and, hence, bonding.

Oxygen compounds can also act as electron donors. For example, consider the reaction between BF_3 and dimethyl ether, CH_3OCH_3. Both the electron-dot and the pigeonhole representations display the donor–acceptor bond possibility.

$$\begin{array}{ccc}
\text{:F:} & CH_3 & \text{:F: }CH_3 \\
\text{:F: B} & +\text{:O:} \longrightarrow & \text{:F: B :O:} \\
\text{:F} & CH_3 & \text{:F: }CH_3
\end{array}$$ (17-12)

$$\Delta H = -13.9 \text{ kcal/mole}$$

or

(17-13)

In the compound, the boron–oxygen bond length is 1.50 Å, to be compared with a normal single bond length of about 1.38 Å. This length and the low bond energy, 13.9 kcal/mole, indicate a bond order of $\frac{1}{2}$ or less.

Table 17-5 collects the bond energy and bond length data for the presently known BH_3 and BF_3 electron donor–acceptor compounds. Two distinctive features should be noted. First, as carbon monoxide forms the compound H_3B–CO, borine carbonyl, the C–O bond length remains at 1.13 Å, virtually unchanged from the triple bond length of the parent CO molecule. This is not atypical. The geometrical and bond length changes tend to concentrate in the electron acceptor, in which rehybridization is needed. The other feature is that the *nitrogen electron donors form stronger bonds than the oxygen donors.* In fact, pyridine, the last compound in Table 17-5, forms a donor–acceptor bond to boron with a 50.6 kcal/mole bond energy—a husky chemical bond!

Table 17-5   Electron Donor–Acceptor Compounds of BH_3 and BF_3

Electron Acceptor	Electron Donor	Compound	Bond Energy $\left(\dfrac{kcal}{mole}\right)$	Donor–Acceptor Bond Length (Å)
BH_3 borine	NH_3 ammonia	$H_3B—NH_3$	—	—
BH_3	$N(CH_3)_3$ trimethylamine	$H_3B—N(CH_3)_3$	31.5	1.62
BH_3	$C\equiv O$ carbon monoxide	$H_3B—C\equiv O$	—	1.54
BF_3 boron tri-fluoride	NH_3 ammonia	$F_3B—NH_3$	—	1.60
BF_3	$N(CH_3)_3$ trimethylamine	$F_3B—N(CH_3)_3$	26.5	1.58
BF_3	$O(CH_3)_2$ dimethyl ether	$F_3B—O(CH_3)_2$	13.9	1.50
BF_3	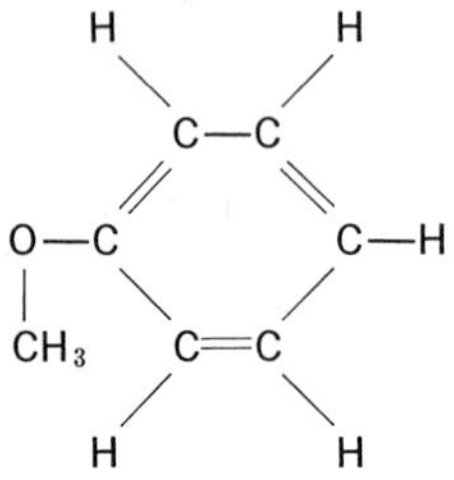 anisole	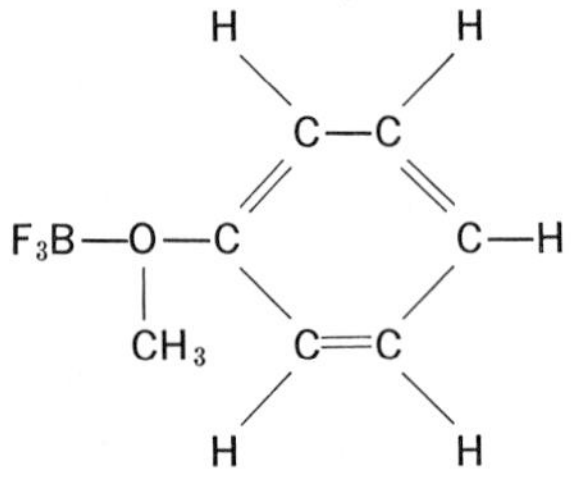	12	—
BF_3	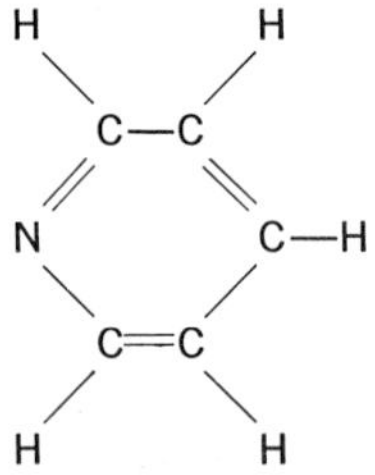 pyridine	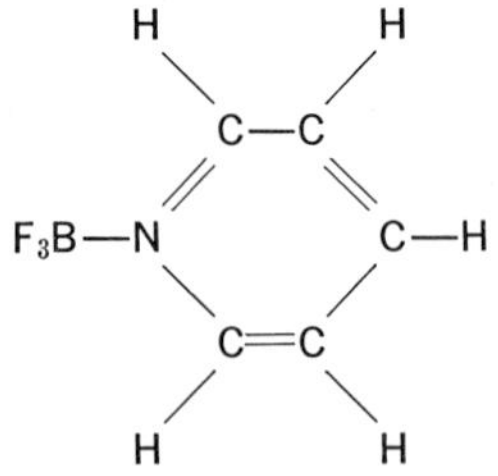	50.6	—

For the anisole row, electron donor structure:

$$
\begin{array}{c}
\text{H} \qquad\qquad \text{H} \\
\diagdown \qquad\qquad \diagup \\
\text{C—C} \\
\diagup\diagup \qquad\qquad \diagdown \\
\text{O—C} \qquad\qquad \text{C—H} \\
| \\
\text{CH}_3 \qquad \text{C}=\text{C} \\
\diagup \qquad\qquad \diagdown \\
\text{H} \qquad\qquad \text{H}
\end{array}
$$

anisole

Compound structure:

$$
\begin{array}{c}
\text{H} \qquad\qquad \text{H} \\
\text{C—C} \\
\text{F}_3\text{B—O—C} \qquad \text{C—H} \\
\text{CH}_3 \qquad \text{C}=\text{C} \\
\text{H} \qquad\qquad \text{H}
\end{array}
$$

For the pyridine row, electron donor structure:

$$
\begin{array}{c}
\text{H} \qquad\qquad \text{H} \\
\text{C—C} \\
\text{N} \qquad\qquad \text{C—H} \\
\text{C}=\text{C} \\
\text{H} \qquad\qquad \text{H}
\end{array}
$$

pyridine

Compound structure:

$$
\begin{array}{c}
\text{H} \qquad\qquad \text{H} \\
\text{C—C} \\
\text{F}_3\text{B—N} \qquad \text{C—H} \\
\text{C}=\text{C} \\
\text{H} \qquad\qquad \text{H}
\end{array}
$$

(b) BERYLLIUM AND ALUMINUM—NOT TO BE OUTDONE

If vacant orbitals have capacity for bonding, beryllium should be twice as able as boron in this respect. A compound like BeF_2 requires that only two of the valence orbitals be employed—there will be two vacant valence orbitals. Hence, we might expect to find BeF_2 (or $BeCl_2$) forming donor–acceptor bonds to two electron donors. Two examples are sufficient to validate this expectation.

Beryllium chloride forms a compound with diethyl ether, $O(C_2H_5)_2$, analogous to the BF_3 compound cited in (17-12). Because of the presence of two vacant orbitals, each $BeCl_2$ molecule combines with two ether molecules.

$$BeCl_2 + 2\ O(C_2H_5)_2 \longrightarrow Cl_2Be\begin{array}{l}\diagup O(C_2H_5)_2 \\ \diagdown O(C_2H_5)_2\end{array} \qquad (17\text{-}14)$$

The second example is a bit different—hence more interesting. The substance $BeCl_2$ readily crystallizes, and its crystal structure arranges the molecules into infinitely long, bridged chains, as pictured in Figure 17-11. Each chlorine atom is equivalently bonded to two beryllium atoms, so that each beryllium is bonded to four chlorines. This can be seen to be the ultimate result of a combination of one bond in which a chlorine electron is shared along with a beryllium electron in a "normal" bond and another bond in which the chlorine donates an electron pair into a vacant beryllium valence orbital. With this parentage, the bonding is better described in terms of molecular orbitals. These M.O.'s give, in now familiar fashion, two equivalent bonds to each chlorine atom instead of a strong one and a weak one. The observed bond lengths, 2.02 Å, are rather long, indicating a bond order less than unity. Thus, with an M.O. point of view, the crystal structure of $BeCl_2$ can be attributed to electron donor–acceptor bonding. The electronegativity difference between beryllium and chlorine is 1.5, so the bonds are only moderately ionic in character. However, we see the interesting feature that chlorine is equivalently bonded to two atoms rather than to only one.

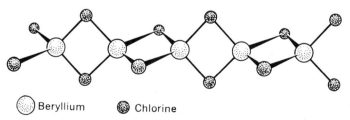

◯ Beryllium   ● Chlorine

Figure 17-11   The structure of solid beryllium chloride, $BeCl_2(s)$. Each beryllium atom is surrounded by four chlorine atoms occupying the corners of a tetrahedron elongated along the chain axis. Every Be–Cl bond length is 2.02 Å.

Aluminum is, of course, a member of the boron family. Hence we should expect donor–acceptor reactions like those shown in Table 17-5. The following is an example:

$$AlCl_3(g) + NH_3(g) \longrightarrow Cl_3Al{-\!}NH_3 \qquad \Delta H = -40 \text{ kcal} \qquad (17\text{-}15)$$

Even more interesting is that aluminum chloride shows a fondness for the bridged structures like those in crystalline $BeCl_2$. However, a single such bridge expends the possibilities, so $AlCl_3$ merely dimerizes, as shown in Figure 17-12, instead of forming an infinite chain. Again we find the central chlorine atoms equivalently bonded to two aluminum atoms. The bond lengths show that these bridge Al–Cl–Al bonds are weaker than the terminal Al–Cl bonds. As for $BeCl_2$, the possibility of electron donor bonding is signalled by orbital occupancy. This bonding capability results in a bridge structure whose equivalent bonds are comfortably rationalized with a molecular orbital description.

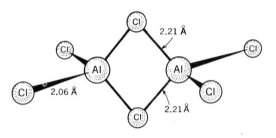

Figure 17-12 The structure of aluminum chloride dimer, $(AlCl_3)_2(g)$. Each aluminum atom is bonded to four chlorine atoms.

(c) POSITIVE IONS (CATIONS) IN SOLUTION

One of the most striking aspects of aqueous chemistry is the appearance, in solution, of ions as separate entities, despite the enormous energies needed to form such ions in the gas phase. Consider, for example, the dissolving of $CuCl_2(s)$ in water to form $Cu^{+2}(aq)$ and 2 $Cl^-(aq)$. Without even worrying about the vaporization of $CuCl_2$ and the breaking of the copper–chlorine bonds to form atoms, we can see that large energies are involved in forming gaseous Cu^{+2} and Cl^- ions.

$$
\begin{array}{ll}
Cu(g) \longrightarrow Cu^+(g) + e^- & \Delta H = +178 \text{ kcal} \\
Cu^+(g) \longrightarrow Cu^{+2}(g) + e^- & \Delta H = +468 \\
2\,e^- + 2\,Cl(g) \longrightarrow 2\,Cl^- & \Delta H = -172 \\
\hline
Cu(g) + 2\,Cl(g) \longrightarrow Cu^{+2}(g) + 2\,Cl^-(g) & \Delta H = +474 \text{ kcal}
\end{array}
$$

Yet solid copper chloride blissfully and readily dissolves in water to form aqueous Cu^{+2} and Cl^- ions.

$$CuCl_2(s) \longrightarrow Cu^{+2}(aq) + 2\,Cl^-(aq) \qquad \Delta H = -12 \text{ kcal} \qquad (17\text{-}16)$$

Nor does the formation of aqueous ions require the absorption of any 474 kcal of energy—exactly the opposite, a modest amount of energy (12 kcal) is *released* as heat. The only conclusion we can reach is that both Cu^{+2} and Cl^- involve strong interactions with the solvent.

In the classical view of this interaction the water molecules are pictured as dipoles, and the favorable orientation of these dipoles around the ions is considered to account for the stability of $Cu^{+2}(aq)$. We can, however, view this aquation process rather differently now, if we choose. The orbital occupancy of Cu^{+2} leaves four vacant valence orbitals:

$$Cu^{+2} \quad \overset{3d}{\text{OOOOO}} \quad \overset{4s}{\text{O}} \quad \overset{4p}{\text{OOO}} \qquad (17\text{-}17)$$

In keeping with our current discussion, Cu^{+2} can act as an electron acceptor, making use of its vacant 4s and 4p orbitals. It should be a particularly effective acceptor because of the net charge on the ion. Any electron donor nearby should cause significant energy-lowering—for example, an oxygen atom from a water molecule. We might picture an electron donor–acceptor interaction involving four water molecules:

$$Cu^{+2} + 4\,H_2O \longrightarrow Cu(OH_2)_4^{+2} \qquad (17\text{-}18)$$

This type of reasoning would also lead us to expect that other electron donors would be able to replace the H_2O molecules. For example, ammonia might do so—nitrogen is supposed to be a better electron donor than oxygen. We might expect a sequence of reactions such as

$$Cu(OH_2)_4^{+2} + NH_3 \longrightarrow CuNH_3(OH_2)_3^{+2} + H_2O \qquad (17\text{-}19a)$$

$$CuNH_3(OH_2)_3^{+2} + NH_3 \longrightarrow Cu(NH_3)_2(OH_2)_2^{+2} + H_2O \qquad (17\text{-}19b)$$

$$Cu(NH_3)_2(OH_2)_2^{+2} + NH_3 \longrightarrow Cu(NH_3)_3(OH_2)^{+2} + H_2O \qquad (17\text{-}19c)$$

$$Cu(NH_3)_3(OH_2)^{+2} + NH_3 \longrightarrow Cu(NH_3)_4^{+2} + H_2O \qquad (17\text{-}19d)$$

The equilibrium constants for all the reactions (17-19a–d) are known, and it is easy to calculate the effect of adding ammonia to an aqueous solution of Cu^{+2}. For example, suppose we add 0.15 mole/liter of ammonia to a 0.010 M Cu^{+2} solution. Calculations show that, at equilibrium, 92 percent of the $Cu^{+2}(aq)$ will have been converted to $Cu(NH_3)_4^{+2}$, and most of the rest will be present as $Cu(NH_3)_3(OH_2)^{+2}$. About one quarter of the ammonia has been consumed in forming the copper–ammonia complexes. Yet, in a 0.15 M NH_3 solution there are 370 times more water molecules competing for the opportunity of donating electrons to the Cu^{+2} electron acceptor! The NH_3 is obviously a better competitor—again the nitrogen compound is a better electron donor than the oxygen compound. This is contrary to expectations based upon dipole moments, since NH_3 has a molecular dipole moment smaller than that of water (1.47 D and 1.80 D, respectively).

(d) OXYGEN ATOM AS AN ELECTRON ACCEPTOR

There is one more important electron acceptor we should consider. The gaseous oxygen atom has, in its ground state, its orbital occupancy represented by

$$\text{O} \quad \overset{1s}{\bigcirc} \quad \overset{2s}{\bigcirc} \quad \overset{2p}{\bigcirc\bigcirc\bigcirc} \quad {}^3\text{P} \tag{17-20}$$

This state, called the "triplet P" state, is lower in energy by 45 kcal than the next higher state because of decreased electron repulsion. This higher state, called the "singlet D" state, has the following pigeonhole diagram:

$$\text{O} \quad \overset{1s}{\bigcirc} \quad \overset{2s}{\bigcirc} \quad \overset{2p}{\bigcirc\bigcirc\bigcirc} \quad {}^1\text{D} \tag{17-21}$$

The representations (17-20) and (17-21) suggest two possible bonding situations for the oxygen atom. The ^{3}P state could bond with any other pair of atoms, each with a half-filled valence orbital, giving such normal compounds as HOH, FOF, H_3COCH_3, HOCl, etc. The ^{1}D state, however, is an electron acceptor, which suggests that oxygen atoms should be able to form donor–acceptor bonds with atoms holding unused pairs. The energy requirement is that the bond energy exceeds the ^{1}D $-$ ^{3}P energy difference, 45 kcal.

A chloride ion, Cl^-, may be taken as the simplest example. This ion has four unused electron pairs and one might expect that one, two, three, or four oxygen atoms could bond to it through donor–acceptor bonding. The species thus predicted are listed in Table 17-6 and they are all well known, both in salts and in aqueous solutions.

Table 17-6 also shows the electron-dot formulas and molecular structures for these ions. Concentrating on electron parentage, we see that the dot formula intrinsically describes the donor–acceptor nature of the bonding. The structures of these reveal a decreasing bond length as oxygen atoms are added. For contrast, the normal Cl–O bond length is 1.69 Å (as it is in Cl_2O). This shortening of the bonds as oxygen atoms are added is generally attributed to increasing ionic character of the Cl–O bonds. Since oxygen has the higher electronegativity, there will tend to be electron charge movement from chlorine to oxygen, leaving the chlorine slightly positive. As successive oxygen atoms bind to the chlorine, each pulls a little more negative charge off the chlorine, so its positive charge increases. This effect, in turn, pulls in the negatively charged oxygen atoms closer and closer.

Similar ions are formed by the other halogens to varying extents. Bromate BrO_3^-, iodate IO_3^-, periodate IO_4^- and, most recently discovered,

Table 17-6 ¹D Oxygen as an Electron Acceptor

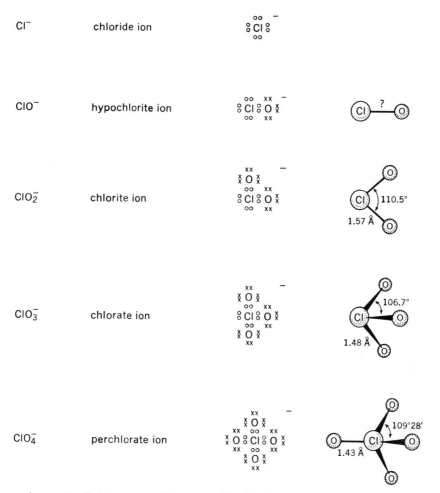

perbromate BrO_4^-, are well known. No fluorine counterparts have been discovered.

From the bonding in these molecules, it is possible to predict some of their chemistry. As we invoke the ¹D description of the oxygen atom to explain the bond formation, it is implied that these compounds should be a potent source of oxygen atoms. These expectations are consistent with the observed chemistry—each of the species is a good oxidizing agent, as shown by the large positive reduction potentials displayed.

$$HClO_2 + 2\,H^+ + 2\,e^- \longrightarrow HClO + H_2O \qquad \mathcal{E}^0 = +1.64 \text{ volts}$$

$$ClO_3^- + 3\,H^+ + 2\,e^- \longrightarrow HClO_2 + H_2O \qquad \mathcal{E}^0 = +1.21 \text{ volts}$$

$$ClO_4^- + 2\,H^+ + 2\,e^- \longrightarrow ClO_3^- + H_2O \qquad \mathcal{E}^0 = +1.19 \text{ volts} \quad (17\text{-}22)$$

Another way of comparing these compounds as oxidizing agents is to examine the exothermicity of some of their reactions. For example, we can compare the heats of reaction with $H_2(g)$ to form liquid water, using the similar reaction with oxygen itself as a reference.

$$H_2(g) + \tfrac{1}{2} O_2(g) = H_2O(\ell)$$
$$\Delta H = -68.3 \text{ kcal/mole} \qquad (17\text{-}23)$$

$$3 H_2(g) + HClO_4(aq) = 3 H_2O(\ell) + HClO(aq)$$
$$\Delta H = 3(-68.3) + 3.6 \text{ kcal/mole} \quad (17\text{-}24)$$

$$2 H_2(g) + HClO_3(aq) = 2 H_2O(\ell) + HClO(aq)$$
$$\Delta H = 2(-68.3) - 4.3 \text{ kcal/mole} \quad (17\text{-}25)$$

$$H_2(g) + HClO_2(aq) = H_2O(\ell) + HClO(aq)$$
$$\Delta H = -68.3 - 14.2 \text{ kcal/mole} \quad (17\text{-}26)$$

These exothermicities show that, from an energy point of view, the molecules $HClO_4$, $HClO_3$, and $HClO_2$ are as good oxygen atom sources as oxygen itself and, in the case of $HClO_2$, substantially better.

A most exciting implication of this type of bonding is connected with the similarity between the halide ions and their adjacent neutral inert gas atoms. Thus, both Cl^- and Ar have eight electrons in their 3s and 3p orbitals. They are called "isoelectronic." Similarly, Br^- and Kr are alike, as are I^- and Xe as well. We are led to expect the existence of molecular analogs, such as ArO_3, KrO_3, and XeO_3, to ions like ClO_3^-, BrO_3^-, and IO_3^-. In fact, one of these, XeO_3, was discovered in 1963. Its structure is extremely close to that of IO_3^-, as shown in Figure 17-13. In some respects, even its chemistry is similar! Like chlorate ion, iodate ion is a good oxidizing agent: It readily accepts electrons by giving up its electron-accepting oxygen atoms. Xenon trioxide is an even better oxidizing agent! It gives up its electron-accepting oxygen atoms so enthusiastically that it detonates when merely touched with oxidizable material such as a piece of tissue.

(e) SUMMARY

Thus, the electron donor–acceptor bonding view provides a basis for discussion of bonding in boron trihalide–ammonia complexes, crystalline

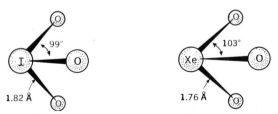

Figure 17-13 The structures of iodate ion, IO_3^-, and xenon trioxide, XeO_3.

$BeCl_2$, aquated and ammoniated cations in water, the oxides of chlorine and even in the oxide of xenon. We see that electron donor–acceptor bonding can be regarded as a link between the bonding in a wide variety of molecular types. Most important, it emphasizes once again that all bonding depends ultimately upon placing electrons simultaneously near two nuclei. A vacant valence orbital signals that possibility just as much as does a half-filled valence orbital. An atom (or molecule) with a vacant valence orbital needs to find a partner with valence electrons to burn—an electron donor. An atom (or molecule) with a half-filled valence orbital needs to find a partner with the same type of orbital occupancy—another half-filled valence orbital. An atom or molecule will continue to display residual bonding capacity until it has used all its valence orbital space as effectively as possible. We recognize this as merely a paraphrase of the classical bonding rule that atoms seek the inert gas electron configuration.

17-4 Electron-excess compounds

There are two more classes of compounds to add to our catalog of bond types. Both classes challenge the classical representation of bonding presented in Section 2-4. Some compounds with laboratory-shelf stability have more electrons than the older bonding rules seem to allow—these we will call *electron-excess* compounds. Other easily stocked substances have too few electrons—these mavericks are called *electron-deficient* compounds. We'll deal with the electron-excess class first.

(a) TRIHALIDE IONS: I_3^- AND ITS FAMILY FRIENDS

The electron-excess compounds are well typified by the trihalide ion I_3^-. This ion forms in aqueous solution by reaction (17-27) and is found in ionic crystals as a linear, symmetric molecule.

$$I^-(aq) + I_2(aq) \longrightarrow I_3^-(aq) \qquad (17\text{-}27)$$

The trihalide ion long mystified theorists of chemical bonding. There is no convenient electron-dot formulation. Furthermore, neither I^- nor I_2 is apparently an electron acceptor, since both I^- and I_2 have achieved filled valence orbitals. From the classical bonding point of view, there are apparently too many electrons.

The molecular orbital picture saves the day for these electron-excess compounds. Following the pattern used in discussing the bonding of CH_2 (Section 15-6), we can examine the nodal surfaces of the molecular orbitals formed from the three axial (sigma) p orbitals. As in the case of CH_2, we must consider the nodal pattern of the central p orbital and link it with orbital combinations of the terminal atoms that possess the same nodal behavior. Figure 17-14 shows the molecular orbitals so formed. One,

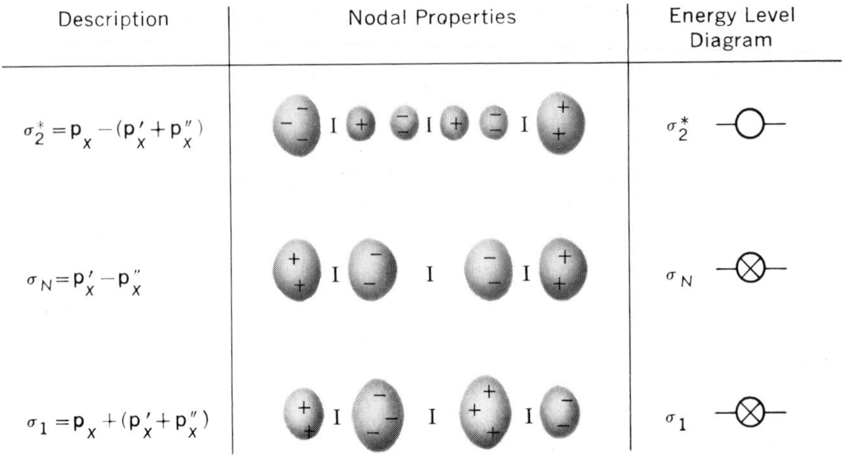

Description	Nodal Properties	Energy Level Diagram
$\sigma_2^* = p_x - (p_x' + p_x'')$		σ_2^*
$\sigma_N = p_x' - p_x''$		σ_N
$\sigma_1 = p_x + (p_x' + p_x'')$		σ_1

Figure 17-14   Axial molecular orbitals for triiodide ion, I_3^-.

$p_x + (p_x' + p_x'')$, is a bonding orbital because it concentrates electrons between the atoms. The other, $p_x - (p_x' + p_x'')$, is antibonding since it does the opposite.

There is a third molecular orbital, $p_x' - p_x''$. If we regard all the iodine atom 5s orbitals to be ineffective in bonding, then there is no axial valence orbital on the central atom that doesn't have a nodal surface perpendicular to the molecular axis. Consequently the $p_x' - p_x''$ molecular orbital does not involve the central atom at all. Since p_x' and p_x'' are geographically distant, this M.O. is a *nonbonding* orbital. It neither helps nor hinders the bond as it forms.

We must now determine the orbital occupancy of the axial molecular orbitals shown in Figure 17-14. There are $3 \times 7 = 21$ valence electrons available from the 5s and 5p electrons of the three iodine atoms. Then there is one extra electron to give the ion its negative charge—so our total is 22. First, we place six electrons in the three low-energy 5s orbitals. Then 12 more go into the three $5p_y$ and three $5p_z$ orbitals, filling them completely. We are left with $22 - 6 - 12 = 4$ electrons with which to occupy the axial molecular orbitals. Two will go into the bonding M.O., $\sigma_1 = p_x + (p_x' + p_x'')$, and two into the nonbonding M.O., $\sigma_N = (p_x' - p_x'')$. The occupied bonding M.O. indicates that there will be a bond, hence there may be a stable species. Of course, one pair of electrons must occupy a three-center molecular orbital, so it will give only $\frac{1}{2}$ order bonds. But, energy-wise, we are about as well off with the I_3^- product as we were with the reactants I_2 and I^-. The I_2 molecule involves one single bond, and I_3^- involves two $\frac{1}{2}$ order bonds.

Turning to the other halogens, we might expect there to be other trihalide ions, such as Br_3^-, Cl_3^-, F_3^-, IBr_2^-, $BrCl_2^-$, and so on. However, our theory predicts very small energy effects as these ions form, so some may be easily prepared while others may be nonexistent. That is the case. For example, no trihalide ion has been discovered in which a fluorine atom occupies the central position. Table 17-7 lists some of the well-known trihalide ions and contrasts the bond lengths with those of a suitable parent halogen. All of them display weakened bonds—one pair of bonding electrons is trying to keep a three-atom molecule's body and soul together. Note that some of the trihalide ions are not symmetric. This will somewhat modify the molecular orbital description needed, but not in a way that contradicts the description. We conclude that there is nothing unique about I_3^- except its common occurrence—in dealing with standardized aqueous I_2 solutions, every freshman chemistry student (knowingly or un-) depends upon I_3^- to reduce I_2 volatility.

(b) OTHER POLYHALOGEN MOLECULES

We have explained the stability of such molecules as I_3^- and ICl_2^- with a molecular orbital description that places two electrons in a bonding M.O. and two in a nonbonding M.O. This raises a question. Why would an iodine atom be limited to one such three-center arrangement? Only one of the valence p orbitals of the central halogen is involved—there are two more such orbitals (both filled) perpendicular to the first one. Either one of these could contribute its pair of electrons to another three-center M.O. system with another pair of lucky halogen atoms. We are led to expect reactions such as:

$$I_3^- + I_2 \longrightarrow I_5^- \qquad (17\text{-}28)$$

$$ICl_2^- + Cl_2 \longrightarrow ICl_4^- \qquad (17\text{-}29)$$

$$ICl_4^- + Cl_2 \longrightarrow ICl_6^- \qquad (17\text{-}30)$$

Table 17-7 *Some of the Known Trihalide Ions*

(X—Y—Z)⁻	Bond Lengths		Reference Bond Lengths		Notes
	X—Y	Y—Z	X—Y	Y—Z	
(I—I—I)⁻	2.90	2.90	2.66	2.66	a
	2.83	3.04	2.66	2.66	b
(Br—Br—Br)⁻	2.53	2.53	2.28	2.28	
(Cl—I—Cl)⁻	2.36	2.36	2.32	2.32	c
(Cl—I—Br)⁻	2.38	2.50	2.32	2.47	
(Cl—Cl—Cl)⁻	—	—	2.01	2.01	d
(Cl—Br—Cl)⁻	—	—	2.14	2.14	e

(a) In crystalline compounds $N(C_2H_5)_4 \cdot 2\ I_2 \cdot I_3$ and $(C_6H_5)_4AsI_3$; (b) in the crystalline compound CsI_3; (c) known also through its vibrational spectrum: ICl force constant in ICl_2^- is 0.46 that of ICl; (d) known through its vibrational spectrum: ClCl force constant in Cl_3^- is 0.3 that of Cl_2; (e) known through its vibrational spectrum: BrCl force constant in $BrCl_2^-$ is 0.3 that of BrCl.

For ICl_4^- it is natural to picture a second Cl–I–Cl bond formation whose orientation is perpendicular to the first Cl–I–Cl group. All the atoms should be in the same plane and the bond lengths should be close to those of ICl_2^-. This is the observed structure as shown in Figure 17-15.

A similar arrangement for I_5^- is not observed. Rather, the I_5^- structure places the common atom in the end position, while preserving the two three-center bonds at about a 90° angle. This difference from ICl_4^- is easy to rationalize. The ICl_4^- structure for I_5^- would have to crowd four large iodine atoms around a central atom of the same size. The actual structure places the terminal iodine atoms in a less crowded region. The M.O. description does not demand that the iodine atom furnishing two electrons be the center atom—it can, as well, be on the end.

The third ion mentioned, ICl_6^-, is not yet known. It will turn up, no doubt, one of these days. Fluorine counterparts, IF_6^- and BrF_6^- have been prepared.

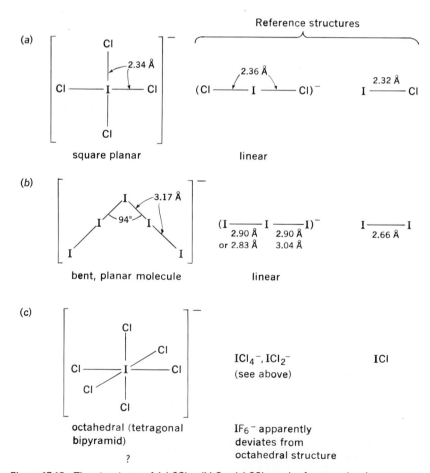

Figure 17-15   The structures of (a) ICl_4^-, (b) I_5^-, (c) ICl_6^- and reference structures.

With this success we can expect another kind of polyhalogen molecule. If one of the filled valence p orbitals in I_3^- or ICl_2^- can form its own three-center M.O. bond, why can't the same happen with I_2 or ICl? This would lead to neutral polyhalogen molecules with structures reminiscent of those for the ions shown in Figure 17-15. Several of these are known and some are displayed in Figure 17-16. We will discuss only ClF_3, but the bond relationships in the others are clearly related.

In ClF_3, the chlorine atom forms a normal covalent bond with one fluorine atom. This is shown by the 1.60 Å bond length, which is quite close to the 1.63 Å bond length in the diatomic ClF. Then using one of its perpendicular and filled p orbitals, the chlorine can form a pair of three-center bonds to two other fluorine atoms that should resemble those of ClF_2^-. The 1.70 Å

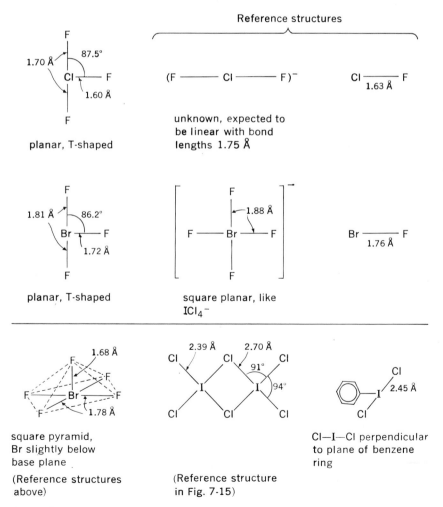

Figure 17-16   *The structures of some neutral polyhalogens.*

bonds, oriented almost perfectly in line and perpendicular to the short bond, fill this bill.

The other molecules shown in Figure 17-16 demonstrate the variety that exists, all understandable with the simple three-center M.O. argument developed for I_3^-. There are a few known polyhalogens with structures that may prove to be more complicated (e.g., IF_7 and I_8). However, even if a few polyhalogen structures prove to be unlike those pictured in Figures 17-15 and 17-16, we must be struck by the similarity of the others and the adequacy of a simple, three-center M.O. description of the bond lengths and bond angles for these electron-excess molecules.

(c) INERT GAS COMPOUNDS

As in the case of the IO_3^-–XeO_3 similarity, there are exciting possible analogs between the interhalogen compounds and their inert gas counterparts. Consider, for example, the reactions (17-31) and (17-32):

$$I^-(aq) + Cl_2 \longrightarrow (Cl\text{—}I\text{—}Cl)^- \qquad (17\text{-}31)$$

$$Xe(g) + Cl_2 \longrightarrow Cl\text{—}Xe\text{—}Cl \qquad (17\text{-}32)$$

Again, the identical orbital occupancies of I^- and Xe imply that as we explain the existence of ICl_2^-, we are predicting the possible existence of $XeCl_2$. In fact, such compounds were predicted on just this basis in 1951, long before they were discovered. The first such inert gas compounds were XeF_2 and XeF_4, prepared ten years after their being predicted from the molecular orbital description of I_3^-. Today, five simple inert gas–halogen compounds are known. Their structures, shown in Figure 17-17, are obviously related to those for the polyhalogens shown in Figures 17-15 and 17-16. It is an interesting sidelight that only two of the isoelectronic polyhalogens are actually known, ICl_2^- and IF_6^-. It seems significant that XeF_2 and XeF_4 are relatively easy to prepare but that no one has yet reported the preparation of either an IF_2^- or an IF_4^- salt.

The bond energies in XeF_2, XeF_4, and XeF_6 are about 30 kcal/mole. These are comparable to the bond energies in F_2 and I_2, and, in fact, all the xenon fluorides are energetically stable with respect to decomposition into the elements. The opposite is true for KrF_2 in which the bond energies are only 12 kcal/mole.

$$XeF_2(g) \longrightarrow Xe(g) + F_2(g) \qquad \Delta H = +25.9 \text{ kcal/mole} \qquad (17\text{-}33a)$$

$$XeF_4(g) \longrightarrow Xe(g) + 2\,F_2(g) \qquad \Delta H = +51.5 \text{ kcal/mole} \qquad (17\text{-}33b)$$

$$XeF_6(g) \longrightarrow Xe(g) + 3\,F_2(g) \qquad \Delta H = +70.4 \text{ kcal/mole} \qquad (17\text{-}33c)$$

$$KrF_2(g) \longrightarrow Kr(g) + F_2(g) \qquad \Delta H = -14.4 \text{ kcal/mole} \qquad (17\text{-}34)$$

On the other hand, these bond energies are sufficiently low that the margin of stability can be expected to be lost in many other inert gas

Inert gas compound	Discovered by (year)	Isoelectronic with
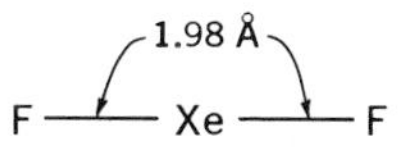 F —— Xe —— F (1.98 Å) linear, symmetric	Weeks, Chernick and Matheson (1962)	IF_2^- (not known)
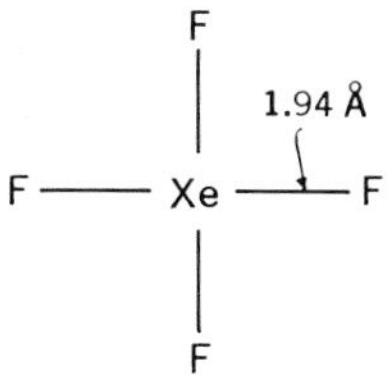 F —— Xe —— F with F above (1.94 Å) and F below; square planar	Claassen, Selig and Malm (1962)	IF_4^- (not known) (BrF_4^-: See Fig. 17-16)
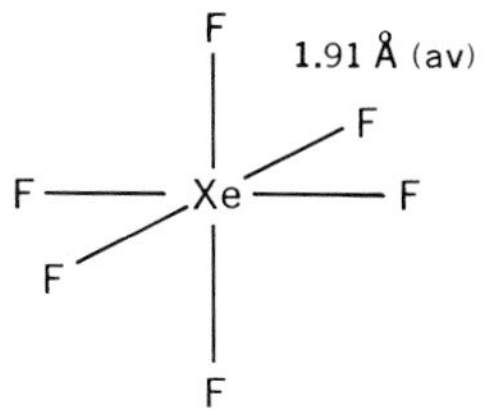 Xe bonded to six F (1.91 Å (av)); Gas phase structure uncertain: not perfectly octahedral	Slivnick (1962)	IF_6^-: structure may not be perfectly octahedral
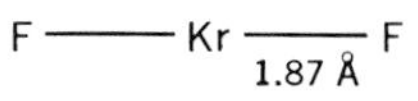 F —— Kr —— F (1.87 Å) linear, symmetric	Turner and Pimentel (1963)	BrF_2^- (not known) (BrF_4^-: See Fig. 17-16) (BrF_3 : See Fig. 17-16)
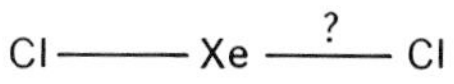 Cl —— Xe —— Cl (?) linear, symmetric	Nelson and Pimentel (1967)	(ICl_2^-: See Fig. 17-15)

Figure 17-17 The structures of some inert gas compounds.

possibilities, particularly those with Cl_2 or Br_2 for which the X_2 bond energy is much higher than that of F_2. For example, concerted efforts to prepare ArF_2, $KrCl_2$, and $XeBr_2$ have as yet been unsuccessful.

(d) HYDROGEN BIFLUORIDE ION, HF_2^-

Perhaps the most important electron-excess compound known is the hydrogen bifluoride ion HF_2^-. As for I_3^-, this ion is formed from two species, F^- and HF, both of which have filled valence orbitals.

$$F^-(aq) + HF(aq) \longrightarrow FHF^-(aq) \qquad (17\text{-}35)$$

(a)

Reference Structure

1.13 Å

(F ——— H ——— F)⁻

0.92 Å

H ——— F

(b)

Description	Nodal Properties	Energy Level Diagram
$\sigma_2^* = s_H - (p'_x + p''_x)$		σ_2^*
$\sigma_N = (p'_x - p''_x)$		σ_N
$\sigma_1 = s_H + (p'_x + p''_x)$		σ_1

Figure 17-18 The structure (a) and molecular orbitals (b) of bifluoride ion, HF_2^-.

Yet, the HF_2^- ion is present in every aqueous HF solution, since there is some F^- formed through dissociation of HF itself. Of course the concentration is raised if a fluoride salt is added (NaF, KF, etc.). Solid bifluoride salts are also known, $NaHF_2$ and KHF_2 being the commonest.

The structure of the HF_2^- ion was first deduced to be linear and symmetric (as shown in Fig. 17-18) through entropy measurements. This conclusion has been firmly corroborated through infrared, neutron diffraction, and nuclear magnetic resonance spectra. The hydrogen atom is centrally placed between the fluorine atoms.

As in the trihalide ions, the molecular orbital point of view readily explains the bonding in HF_2^-. The M.O. energy level diagram is exactly the same as that for I_3^- (Fig. 17-14), although the bonding and antibonding orbitals have no nodal plane through the hydrogen atom since they involve its 1s orbital.

Just as in the trihalides, the last two electrons of HF_2^- are placed in the nonbonding orbital σ_N. Note that in both molecular types, the nonbonding orbital concentrates the extra charge on the terminal atoms. Consequently, these molecules will be most stable if the end atoms have high electronegativity, fluorine being the optimum. This expectation is most strikingly seen in the other possible bihalide ions. Until a few years ago it was felt that *only* fluorine formed such species. Since 1960, however, many bihalide ions have been discovered. Table 17-8 lists a number of these with their bond energies, all much weaker than that of HF_2^-. Furthermore, infrared spectra have shown that HCl_2^- is linear but that its two bond lengths are different. Its hydrogen atom lies between the two chlorine atoms, bonded to both, but not quite equivalently. Again the situation is reminiscent of the polyhalogens. In some cases the three-center bond places the center atom equidistant from the end atoms, but in other cases it is closer to one than the other.

(e) THE HYDROGEN BOND—AN ELECTRON-EXCESS BOND

All the ions listed in Table 17-8 involve a hydrogen atom bonded simultaneously to two other atoms. A hydrogen atom so placed is said to form a *hydrogen bond*. This name is intended to distinguish these situations in Table 17-8 from normal bonds to hydrogen, as in H_2 or HF. The importance of HF_2^- stems from its prototype relationship to hydrogen bonds. Hydrogen bonds crucially influence the structure and chemistry of most biologically active molecules.

Experience tells us that whenever a hydrogen atom that is bonded to an atom A displays acidic properties, that hydrogen atom can form hydrogen bonds.

$$A—H + B \longrightarrow A—H \text{------------} B \qquad\qquad (17\text{-}36)$$

The types of molecules B that can react with A–H are those we identified earlier (in Section 17-3) as good electron donors. For example, HCl forms a reasonably strong hydrogen bond to diethyl ether—ΔH is probably around 6 kcal/mole. Carboxylic acids, alcohols, phenols, and water—all of which are weak acids in aqueous solution—form hydrogen bonds to electron donors such as ethers R–O–R, ketones R–CO–R, ammonia NH_3, or amines RNH_2. The acidic proton acts as the electron acceptor.

Thus we have two explanations for the existence of the hydrogen bond. It can be regarded as an electron-excess compound with the bonding

Table 17-8   *Bond Energies of Some Hydrogen Bihalide Ions*

$HX + Y^- \longrightarrow X—H—Y^-$	
$X—H—Y^-$	ΔH(kcal/mole)
(F ————H————F)⁻	−37
Cl ———H------------- Cl⁻	−14
Br———H------------- Br⁻	−13
I ———H------------- I⁻	−12
Cl———H------------- Br⁻	−9
F ———H------------- Cl⁻	a
F ———H------------- Br⁻	a
F ———H------------- I⁻	a
Cl———H------------- I⁻	a

(a) Bihalide ions known through infrared spectrum; bond lengths and bond energies not yet measured.

described in the molecular orbital framework as done here for the bihalide ions. Because the nonbonding electrons are placed on the terminal atoms, these atoms should have high electronegativity. However, we can also regard the hydrogen atom as an electron acceptor because of charge displacement in its bond toward its highly electronegative partner atom (F, O, or N). The dipole moment of HF, 1.82 D, indicates that charge is moved away from the proton somewhat, leaving its valence orbital region partially vacant. This permits an electron acceptor–donor interaction with an electron donor. The smaller dipole moment of HCl, 1.07 D, suggests that it should form somewhat weaker interactions of this sort. In water, the molecular dipole moment, 1.82 D, implies a bond dipole moment of 1.49 D (see Section 16-4). Its hydrogen bonds should be intermediate to those of HCl and HF. If so, we can expect to find strong hydrogen bonds between water and negative ions like F^-, Cl^-, Br^-, or I^-. Table 17-8 shows the $Cl—H \cdots Cl^-$ bond energy to be 14 kcal and that of $Cl—H \cdots Br^-$ to be 9 kcal. The water–Cl^- bond energy ought to be about 20–25 kcal and that of water–Br^- about 15–20 kcal. Furthermore, in aqueous solution the abundance of water molecules implies that each halide ion will form several such hydrogen bonds. As in the chlorate ion, the four electron pairs of Cl^- might form four donor–acceptor bonds. Hence a gaseous Cl^- ion ought to release a large amount of energy when it dissolves in water—the bond energy of about four $H—O—H \cdots Cl^-$ hydrogen bonds.

$$Cl^-(g) + 4\,H_2O \longrightarrow Cl^- \cdot 4\,H_2O(aq)$$
$$\Delta H \cong 4(-20 \text{ to } -25) \cong -(90) \text{ kcal} \quad (17\text{-}37)$$

$$Br^-(g) + 4\,H_2O \longrightarrow Br^- \cdot 4\,H_2O(aq)$$
$$\Delta H \cong 4(-15 \text{ to } -20) \cong -(70) \text{ kcal} \quad (17\text{-}38)$$

The estimates, based upon a hydrogen bond view of the aquation of anions, compare favorably with the accepted aquation energies of Cl^- and Br^- (based upon heats of solution of HCl and HBr), 87 and 80 kcal/mole, respectively.

(f) HYDROGEN BONDING IN WATER

Now we see why water is such a remarkable solvent. As a substance such as NaCl dissolves, Na^+ ions find that water molecules furnish an excellent environment because of their electron-donor capabilities, and Cl^- ions feel the same comfort because of water's electron-acceptor capabilities.

This schizophrenic personality of water is evident in practically every measurable property. Since H_2O can act both as an electron acceptor and as an electron donor, hydrogen bonds between water molecules interlink most of the molecules in liquid water. The most obvious result is that the boiling point of water is very much higher than those of similar molecules that do not possess the same hydrogen bonding capability. This is evident in Figure 17-19, which shows the distinctive boiling point of water among its

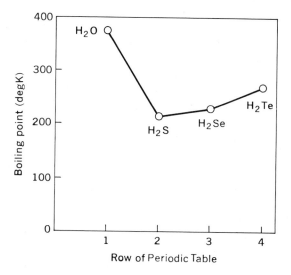

Figure 17-19   *Boiling points of the oxygen family hydrides.*

Periodic Table counterparts. The viscosity of water is also abnormally high—it is five times the viscosity of diethyl ether and three times that of hexane. A most important manifestation of this intermolecular hydrogen bonding is found in the entropy of vaporization of water. The presence of special molecular orientations (to preserve approximately linear hydrogen bonds) implies a high degree of order in the liquid. Hence the disorder created on vaporization is much larger than that of the usual liquid. We find $\Delta S_{vap.} =$ 26 cal/mole degK for water, whereas the norm (as expressed in Trouton's Rule; see Problem 9-21) is 21 cal/mole degK. This entropy shows up in the solvent properties of water. A solute entering an aqueous environment must elbow into the interlinked liquid, breaking hydrogen bonds, which raises both the energy and the entropy. Whether the net effect works in favor of high solubility depends upon the extent to which solute–solvent interactions lower the energy again or create new, highly ordered orientations.

Finally, it is appropriate to note the crystal structure of ice. In this structure, each water molecule finds itself perfectly placed to form the maximum number of hydrogen bonds. It has four, tetrahedrally oriented neighbor oxygen atoms. To two of these the central water molecule forms hydrogen bonds by acting as the electron acceptor. To the other two neighbors it acts as the electron donor, with the neighbors furnishing the hydrogen atoms. The water–water hydrogen bond energy is 5–6 kcal/mole.

(g) MORE HYDROGEN BONDS

As we remarked earlier, there are many substances acidic enough to form hydrogen bonds. Most of these are both hydrogen bonding electron

Table 17-9   Some Hydrogen Bonds and Their Bond Energies, A–H $\cdots$ B

Alcohols	ΔH(kcal/ mole)	Acids	ΔH(kcal/ mole)
CH_3O—H $\cdots$ $O(C_2H_5)_2$	2.5	(cyclic dimer structure) $O \cdots H$—O / H—C ... C—H / O—$H \cdots O$	14.0(= 2 $\cdot$ 7)
CH_3O—H $\cdots$ $N(C_2H_5)_3$	3.0		
Phenols		**Chloroform**	
C_6H_5O—H $\cdots$ $O(C_2H_5)_2$	3.7	Cl_3C—H $\cdots$ $O{=}C(CH_3)_2$	2.5
C_6H_5O—H $\cdots$ $N(CH_3)_3$	5.8	Cl_3C—H $\cdots$ $N(C_2H_5)_3$	4.0
C_6H_5O—H $\cdots$ $O{=}C$—C_2H_5 $\quad\quad\quad$ $\overset{|}{O}CH_3$	5.7	**Inorganics**	
		$N{\equiv}C$—H $\cdots$ $N{\equiv}C$—H	3.3
Amines			
$(C_6H_5)_2N$—H $\cdots$ $O(C_2H_4)_2O$	2.3	O—$H \cdots O$ (with H atoms)	5.0
		F—$H \cdots F$—H	7.0
Amides			
(amide dimer structure) $C_2H_5C{=}O$, N—$H \cdots O$, CH_3, C—C_2H_5, H—N, CH_3	3.6	Cl—$H \cdots Cl^-$	14
		$(F$—H—$F)^-$	37

acceptors as well as donors. As for water molecules, their hydrogen bonding capability reflects into almost every measurable property. For example, it is a firm generalization that any molecule that can form hydrogen bonds to other molecules like itself, will surely crystallize in a lattice linked by such bonds. It is also generally true that hydrogen bonding compounds tend to be soluble in water. Their molecular configurations can be critically influenced by intramolecular hydrogen bonds, as is the case for many biologically important molecules. The protein molecule is held in its helical shape by hydrogen bonds and the double strands of DNA are held together by this same bond type.

Table 17-9 displays some hydrogen bond structures and their associated hydrogen bond energies. Typically they fall in the range 3–7 kcal/mole per hydrogen bond. Whenever an A–H $\cdots$ B hydrogen bond forms, the geometry is close to linear, and the A $\cdots$ B distance is a few tenths of an angstrom shorter than the A $\cdots$ B van der Waals separation expected with no hydrogen between. The A–H stretching frequency is substantially lowered by the hydrogen bond formation. For similar hydrogen bonds, there tends to be an inverse monotonic relationship between the bond energy and the bond length, and a direct monotonic relationship between the bond energy and the downward shift in the A–H stretching frequency.

17-5 Electron-deficient compounds

The last class of compounds we'll consider confounds the classical valence rules as do the electron-excess compounds, but for the opposite reason: They seem to have too few electrons. Boron–hydrogen compounds furnish good examples.

(a) DIBORANE, B_2H_6

As we have said several times, the simplest borohydride, BH_3, has only transient existence. This reactivity arises, we explain, from the completely vacant valence orbital yearning for some electron occupancy. If an electron donor comes by (such as ammonia), a donor–acceptor bond is formed *toute suite*. If no such well-matched reaction partner appears, however, two BH_3 molecules react to form the simplest stable borohydride, diborane, B_2H_6.

The formula B_2H_6 is immediately reminiscent of ethane, C_2H_6. However, the ethane structure requires 14 electrons to form the necessary bonds, shown in (17-39).

$$
\begin{array}{c}
\text{H} \quad \text{H} \\
\text{H} \overset{\circ\circ}{:} \text{C} \overset{\circ\circ}{:} \text{C} \overset{\circ\circ}{:} \text{H} \\
\text{H} \quad \text{H}
\end{array}
\qquad\qquad (17\text{-}39)
$$

If diborane were to adopt the ethane structure, some of the bonds would be short of electrons. Instead, it takes an entirely different structure in which two hydrogen atoms bridge between the two boron atoms (as pictured in Fig. 17-20). One is immediately struck by the similarity to hydrogen bonding—the central hydrogen atoms, in defiance of classical bonding rules, are each equally bonded to two atoms. The bond lengths contrast,

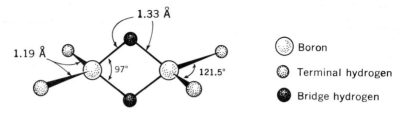

Figure 17-20 The structure of diborane, B_2H_6, an electron-deficient compound.

too, as in hydrogen bonds—the four terminal B–H bonds, which are more normal, have bond lengths of 1.19 Å, whereas the B–H distances in the hydrogen bridges are longer, 1.33 Å.

In fact, the analogy to the hydrogen bond is a close one. Recalling the molecular orbital view of the linear FHF⁻ ion, we find the last two electrons lodged in a nonbonding orbital that localizes them on the highly electronegative fluorine atoms. This last pair of electrons is missing in the B–H–B bridge, as might be considered to be appropriate since the boron atoms have rather low electronegativity. More important, however, is the fact that the orbital so vacated is *nonbonding*, so the bond is not weakened.

In noting the geometry of the four bonds around the boron atom, we see that the boron orbitals are close to sp^3, tetrahedral hybridization. This can account for the nonlinearity of the B–H–B bridge, again in contrast to the linear hydrogen bond. Nevertheless, the simplest description of each arm of the bridge is a three-center molecular orbital description like the hydrogen bond description, except that it involves sp^3 hybrids from the terminal atoms, a bent geometry, and only two electrons instead of four. Thus diborane solves its electron-deficiency problem.

(b) SOME HIGHER BORANES: B_5H_9 AND $B_{10}H_{14}$

Diborane can react with itself to form higher boranes. As it does so, it loses hydrogen.

$$2\tfrac{1}{2}\, B_2H_6 \longrightarrow B_5H_9 + 3\, H_2 \qquad\qquad (17\text{-}40)$$

$$5\, B_2H_6 \longrightarrow B_{10}H_{14} + 8\, H_2 \qquad\qquad (17\text{-}41)$$

Both these structures involve some hydrogen bridges and some terminal B–H bonds like those that hold diborane together. The loss of hydrogen (in reactions (17-40) and (17-41)), however, accentuates the electron-deficiency problem. Consider pentaborane, B_5H_9. Each of the five boron atoms furnishes four valence orbitals (a 2s and three 2p), and the nine hydrogen atoms supply nine more. These 29 valence orbitals must make do with $5 \cdot 3 + 9 = 24$ electrons, only 12 pairs! Such an electron deficiency (or orbital excess) always causes "clustered" structures for which molecular orbitals, extending over many atoms, furnish the bonding. Thus B_5H_9 has a tetragonal pyramid structure, one boron atom "sitting" on top of four borons in a plane. The pyramid base is linked by four single hydrogen bridges, as shown in Figure 17-21(a). If we assign one pair of electrons to each B–H bond and one pair to each B–H–B bridge, 18 electrons are thus consumed. The remaining six electrons must occupy molecular orbitals extending over the whole boron pyramid, binding the apex boron atom to the base. Thus we can say that these last six electrons are "delocalized;" they occupy molecular orbitals that extend over the whole boron cluster.

Decaborane has an even more clustered, boat-like structure, as shown in Figure 17-21(b). There are ten normal B–H bonds, to each of which can be assigned an electron pair, and four B–H–B bridges that consume four more pairs. Twenty-eight electrons are thus used up. The entire molecule has $10 \cdot 3 + 14 = 44$ electrons, so 16 now remain. These eight pairs must link 19 boron–boron bonds (not counting the hydrogen-bridged B–B bonds). Again the electrons must occupy molecular orbitals that extend over the entire boron molecular skeleton.

We will not pursue these electron-deficient, orbital-excess compounds any further at this time. It is sufficient to note that *all metals are made up of electron-deficient, orbital-excess atoms.* Invariably, this type of atom condenses to a clustered or tightly packed solid and the electrons occupy molecular orbitals extending over the whole crystal. The result is that all

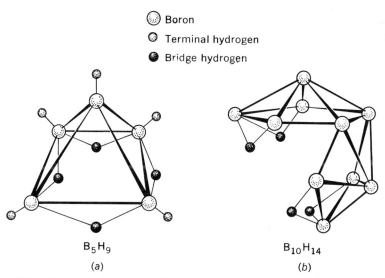

Boron
Terminal hydrogen
Bridge hydrogen

B_5H_9

(a)

$B_{10}H_{14}$

(b)

Figure 17-21 The cluster structures of (a) pentaborane, B_5H_9, and (b) decaborane, $B_{10}H_{14}$. The decaborane structure is drawn without the ten terminal hydrogen atoms—one on each boron. The boron framework is very close to a regular icosohedron with two vacant vertices. (An icosohedron is a figure with 12 vertices and 20 identical triangular faces.)

the atoms share all the valence electrons, thereby occupying as effectively as possible the many vacant orbitals with the few bonding electrons available. Naturally an electron in an orbital that extends over an entire crystal has great mobility, which is the origin of the high electrical conductivity and reflectivity of metals.

17-6 Reprise on chemical bonding

It has only recently become possible to sense any real coherence in the subject of chemical bonding. For many years, chemists had many more or less unrelated classes of chemical bonds: covalent bonds, ionic bonds, dative, metallic, one-electron, three-electron, bridge, donor–acceptor, chelate, coordinate, and hydrogen bonds all were used. Now, with the guidance of quantum mechanics and with the use of the Pauli Principle, all of these types can be interrelated. The older terminology persists and will be useful for some time to come, but gradually the knowledge of one kind of bond is being correlated with that of others. Some useful generalities have emerged.

1. All bonds form because the potential energy is lowered as the bonding atoms approach.

2. The energy is lowered as a bond forms because electrons are then able to move simultaneously near the positive charges of two or more nuclei.

3. Electrons of one atom can move simultaneously near the positive charge of a second atom when that atom has valence orbitals that are either half-filled or vacant. Even when an atom has formed one or more bonds, it retains reactivity as long as there remain additional half-filled or vacant valence orbitals.

4. When a chemical bond forms between two atoms, each with a single, half-filled valence orbital, the bond energy will be in the range 30–130 kcal/mole. The bond energy is higher if one atom attracts electrons more strongly than the other, giving rise to the concepts of effective nuclear charge and electronegativity.

5. When a chemical bond forms between an atom with a vacant valence orbital and an atom with an unused pair, the bond energy will usually be in the range 10–50 kcal/mole.

6. The bonding electrons in a molecule move in the field of all of the nuclei, hence in molecular orbitals. When the number of electron pairs engaged in bonding equals the minimum number of bonds needed to hold the atoms together, localized two-atom bond orbitals usually provide an adequate description (as shown in electron-dot representations or in conventional "valence" theory). However, when there are either electrons in half-filled orbitals, additional vacant orbitals, or too many or too few bonding electrons, the molecular orbital description is needed. Then, new con-

cepts come into play: antibonding orbitals, nonbonding orbitals, multiple bonds, triplet states, promotion, hybridization, and delocalization.

We will find that these generalizations give us a substantial basis for discussing the bonding in condensed phases, the subject of the next chapter.

Problems

1. Elemental phosphorus (white) has a tetrahedral molecular structure, P_4, in which each atom is pyramidally bonded to three other phosphorus atoms. The bond length of each of the six bonds in the pyramid is 2.21 Å, presumably a normal single bond length between phosphorus atoms. The average bond energy is 48 kcal/mole. At high temperatures, gaseous P_4 molecules ($\Delta H_f^0 = +13.1$ kcal) dissociate into gaseous P_2 molecules ($\Delta H_f^0 = +33.8$) with bond length 1.894 Å.
(a) According to molecular orbital occupancy, what is the bond order in P_2?
(b) What energy is needed to dissociate P_4 into four phosphorus atoms?
(c) What is ΔH^0 for the reaction $P_4(g) = 2\ P_2(g)$?
(d) What is the bond energy in P_2?
(e) What is the bond energy per bond order in P_2?
(f) Compare the bond energy per bond order in P_2 to that in N_2 and explain why N_2 is the stable form of elemental nitrogen (N_4 is unknown), but P_4 is the stable form for phosphorus.

2. Elemental sulfur (rhombic) has an eight-membered ring structure, S_8. The bond length of each of the eight bonds is 2.07 Å and the average bond energy is 50.2 kcal. At high temperatures, gaseous S_8 molecules ($\Delta H_f^0 = +24.1$ kcal) dissociate into gaseous S_2 molecules ($\Delta H_f^0 = +29.9$ kcal) with bond length 1.89 Å.
(a) According to molecular orbital occupancy, what is the bond order in S_2?
(b) What energy is needed to dissociate S_8 into eight sulfur atoms?
(c) What is ΔH^0 for the reaction $S_8(g) = 4\ S_2(g)$?
(d) What is the bond energy in S_2?
(e) What is the bond energy per bond order in S_2?
(f) Compare the bond energy per bond order in S_2 to that in O_2 and explain why O_2 is the stable form of elemental oxygen (O_8 is unknown), but S_8 is the stable form for sulfur.

3. Table 17-1 shows that carbon and nitrogen have opposite trends in bond energy per bond order. This difference strongly influences the thermodynamics of the triple-bonded substances nitrogen, $N\equiv N$, and acetylene, $HC\equiv CH$, relative to the analogous single-bonded substances hydrazine, H_2N-NH_2, and ethane, H_3C-CH_3. The thermodynamic properties are given in Appendix d, except for $N_2H_4(g)$ for which $\Delta H_f^0 = +22.8$, $\Delta G_f^0 = +38.1$, and $S^0 = +57.0$.
(a) For each hydrogenation reaction, calculate ΔH^0, ΔG^0, and ΔS^0.

$$2\ H_2(g) + N_2(g) = H_2NNH_2(g)$$
$$2\ H_2(g) + HCCH(g) = H_3CCH_3(g)$$

(b) Explain why ΔS^0 is negative and about the same for these two molecules.
(c) Explain, in terms of bond energy per bond order, why one reaction is exothermic and one endothermic.
(d) To show the importance of this effect in the chemistry of nitrogen and acetylene, calculate the equilibrium constants for these two reactions.

4. (a) The heat of formation of gaseous P_2H_4 is 5.0 kcal/mole. Combine this with the data in Problem 1 and in Problem 3 to calculate and compare the ΔH^0 for the analogous hydrogenation reactions:

$$2 H_2(g) + N_2(g) = H_2NNH_2(g)$$
$$2 H_2(g) + P_2(g) = H_2PPH_2(g)$$

(b) How does the bond energy per bond order for nitrogen–nitrogen bonds and that for phosphorus–phosphorus bonds relate to these ΔH values?
(c) The absolute entropy of $P_2H_4(g)$ is not known. Guess ΔS^0 for the reaction and use it to calculate the equilibrium constant for its formation from H_2 and P_2.

5. The carbon–oxygen bond lengths in carbon monoxide, $C{\equiv}O$, formaldehyde, $H_2C{=}O$, and dimethyl ether, $H_3C{-}O{-}CH_3$, can be taken as prototype triple, double, and single bond lengths—respectively, 1.128, 1.210, and 1.416 Å.
(a) Plot these C–O bond lengths versus bond order and draw a smooth curve through the points.
(b) When a carbon atom is attached to an oxygen atom by a double bond and *also* to another oxygen atom by a single bond, the pi molecular orbital of the double bond includes a small participation by the single-bonded oxygen. This tends to weaken the double bond somewhat and to strengthen the single bond correspondingly. Use the bond lengths in formic acid and in methyl acetate, both shown below, to deduce the bond orders of the carbon–oxygen bonds.
(c) For each compound, add the bond orders of the CO bonds and compare to the bond order sum implied by the traditional formulas:

1.245 Å   O
H—C
1.312 Å   O—H

1.22 Å   O
H₃C—C
1.36 Å   O—CH₃
1.46 Å

6. The pi molecular orbitals pictured in Figure 17-7 for 1,3-butadiene could be considered to be applicable to the molecule biacetyl, shown below. Use the bond lengths (Figure 17-5) and your plot in Problem 5 to deduce the bond orders for all C–C and C–O bonds.

1.47 Å
O   O   1.20 Å
1.20 Å
      C — C
1.54 Å   1.54 Å
H₃C   CH₃

7. The nitrogen–oxygen bonds in nitric oxide, $N{\equiv}O$, nitroxyl, $HN{=}O$, and hydroxylamine, $H_2N{-}OH$, can be taken as prototype $2\frac{1}{2}$ order, double and single bonds. Their stretching force constants are, respectively, 15.5, 10.5, and 3.9 mdyne/Å.
(a) Plot these k's versus bond order and draw a smooth curve through the points (together with the point $k = 0$, bond order $= 0$).
(b) Nitric oxide has two excited states, one 125.7 kcal and the other 131.4 kcal above the ground state. Their vibrational frequencies reveal their stretching force constants, respectively, 24.1 and 4.6 mdyne/Å. With your plot, deduce the bond order of each of these states.
(c) From the bond orders, propose possible molecular orbital occupancies.

8. The molecule FNO seems to be closely related to HNO, but its bonding is quite different. The F–N bond length is 1.52 Å, much longer than the normal single bond

length, 1.37 Å in NF_3. The infrared spectrum shows that the NO stretching force constant is 15.0 mdyne/Å. Use the curve from Problem 7 to deduce the NO bond order in FNO.

9. The bond length and force constant are inversely related. For example, the three states of nitric oxide mentioned in Problem 7 have the following r_0 and k values:

E	k(mdyne/Å)	r_0(Å)
0.0	15.5	1.151
$+125.7$	24.1	1.064
$+131.4$	4.6	1.385

(a) Plot these values and draw a smooth curve through the points.
(b) For most molecules, it is easier to measure the force constant (through the infrared spectrum) than the bond length. For example, the NO force constant for nitroxyl (HNO), $k = 10.5$ mdyne/Å, provided the first evidence about the NO bond length. Use your curve to estimate this bond length in HNO.

10. From the NO force constant given in Problem 8 and the curve obtained in Problem 9, estimate the NO bond length in FNO.

11. The bond length in a heteronuclear single bond, A–B, can be approximated as the geometric mean, $\sqrt{\bar{r}_A \bar{r}_B}$, of the average atomic radii, $\bar{r}$, as approximated by $\bar{r}$ in the one-electron approximation, using the first ionization energy to estimate Z^*. This bond length is a compromise between the optimum bond length for each atom.
(a) Estimate r_0 for the first-row hydride bonds LiH, CH, NH, OH, FH.
(b) From Appendix f, select experimental values for these bond lengths, using the stable molecules LiH, CH_4, NH_3, H_2O, and HF.
(c) Calculate the average percent error to gain a feeling of the validity of this estimate of r_0.

12. Repeat Problem 11 for the second-row hydride bonds, SiH, PH, SH, and ClH. For the stable molecule bond lengths, use SiH_4, PH_3, H_2S, and HCl from Appendix f.

13. Repeat Problem 11 for the first-row single bonds to carbon, C–C, C–N, C–O, and C–F. The experimental single bond lengths are: C–C in C_2H_6, 1.54 Å; C–N in H_3CNH_2, 1.47 Å; C–O in H_3COCH_3, 1.42 Å; C–F in CF_4, 1.32 Å.

14. Repeat Problem 11 for the second-row single bonds to oxygen, OCl, OP, and OS. The experimental single bond lengths are: O–Cl in Cl_2O, 1.70 Å; O–P in P_4O_6, 1.66 Å; O–Si in SiO_2, 1.61 Å.

15. On the basis of electronegativity difference, decide which molecule of each pair is expected to have the larger molecular dipole moment. For *every* molecule, decide which atom (atoms) is expected to carry negative charge.
(a) H_2O or F_2O? (c) NH_3 or NF_3?
(b) H_2O or H_2S? (d) NH_3 or PH_3?
Compare your predictions to the actual molecular dipole moments listed in Appendix e. What is the qualitative batting average for correct predictions?

16. Repeat Problem 15 for the following molecules.
(a) ClF or BrF? (c) PCl_3 or $SbCl_3$?
(b) IBr or ClBr? (d) $SbCl_3$ or $SbBr_3$? ($\mu(SbBr_3) = 3.30$ D)

17. On the basis of electronegativity differences, predict:
(a) the end of a normal C–H bond that will carry negative charge;
(b) the end of a normal C–Cl bond that will carry negative charge;
(c) whether the bond contributions to the molecular dipole moment in methyl chloride, H_3CCl, will be additive or tend to cancel.

18. On the basis of electronegativity differences, predict:
(a) which of the methyl halides H_3CCl, H_3CBr, or H_3CI will have the highest and which will have the lowest molecular dipole moment (see Problem 17);
(b) which will have the largest bond energy enhancement due to ionic character in the bonds (the sum of $D_0 - \bar{D}$ for all bonds $= \Sigma(D_0 - \bar{D})$).

19. From thermodynamic heats of formation, the heats of atomization of the methyl halides can be calculated:

$$H_3CX(g) \longrightarrow 3\ H(g) + C(g) + X(g) \qquad \Delta H^0_{at.}$$

X	Cl	Br	I
$\Delta H^0_{at.}$	376.6	363.2	348.5

(a) Calculate the "perfect sharing" bond energy $\bar{D}$ for C–H and C–X bonds from the average bond energies given in Table 7-3.
(b) Calculate the energy of atomization for each of the methyl halides above if each bond energy were the "perfect sharing" bond energy $\bar{D}$.
(c) Calculate the bond energy enhancement, the difference between the experimental values of $\Delta H^0_{at.}$ and $\Sigma \bar{D}$ for each of the molecules. Compare to your prediction in Problem 18.
(d) Plot the energy enhancements, $\Sigma(D_0 - \bar{D})$ versus the molecular dipole moments for these molecules as listed in Appendix e.

20. The heats of formation of four BF_3–M complexes are listed in Table 17-5, where M is an electron donor. The activation energies for BF_3 exchange in two of these complexes have been measured.

$$B'F_3 + M{-}BF_3 \longrightarrow F_3B'{-}M + BF_3 \qquad E_a$$

M = dimethyl ether, $E_a = 7.7$ kcal
M = pyridine, $E_a = 1.8$ kcal

(a) List the four electron donors in Table 17-5 in decreasing order of electron donor ability (i.e., decreasing "base strength").
(b) Two possible mechanisms for the exchange process are:

A. $BF_3{-}M \xrightarrow{k_1} BF_3 + M$
    $BF'_3 + M \xrightarrow{k_2} F_3B'{-}M$
B. $F_3B{-}M + B'F_3 \xrightarrow{k_1'} (F_3B{-}M{-}B'F_3) \longrightarrow F_3B + M{-}B'F_3$

The second mechanism is the one generally preferred. Why is the first mechanism not acceptable?

21. The equilibrium constant for the replacement of one H_2O molecule from the Cu^{+2} aquation sphere by ammonia is $8 \cdot 10^{+5}$.

$$Cu(OH_2)_4^{+2} + NH_3(aq) = Cu(OH_2)_3NH_3^{+2} + H_2O \qquad K = 8 \cdot 10^{+5}$$

(a) Calculate the ratio of the concentration of $Cu(OH_2)_3NH_3^{+2}$ to $Cu(OH_2)_4^{+2}$ in a $0.00010\ M$ Cu^{+2} solution if NH_3 is added to a concentration of $0.0055\ M$ at which the H_2O/NH_3 concentration ratio is 10,000 to 1 favoring H_2O.

(b) Calculate the concentration of uncomplexed NH_3 if the concentrations in (a) are reversed, i.e., if 0.00010 M NH_3 is added to 0.0055 M Cu^{+2}.

22. (a) Calculate ΔG^0 for the reaction discussed in Problem 21.
(b) The energy of aquation of gaseous Cu^{+2} ions is -508 kcal.

$$Cu^{+2}(g) \longrightarrow Cu(OH_2)_4^{+2} \qquad \Delta H^0 = -508 \text{ kcal}$$

Calculate the energy per bond in the donor–acceptor bond between H_2O and Cu^{+2}.
(c) Assuming ΔS^0 is near zero for the reaction in Problem 21, what is ΔH^0 for the formation of an NH_3–Cu^{+2} donor–acceptor bond?
(d) The dipole moments of H_2O and NH_3 are, respectively, 1.82 and 1.47 D. Are these consistent with the answers to (b) and (c)?

23. Oxygen atom can act as an electron acceptor with bromine and iodine, to form oxygenated counterparts to the chlorine species considered in Section 17-3(d). Their reduction potentials from oxidation number $+5$ to $+1$ are:

$$BrO_3^- + 5\,H^+ + 4\,e^- \longrightarrow HBrO + 2\,H_2O \qquad \mathcal{E}^0 = +1.49$$

$$IO_3^- + 5\,H^+ + 4\,e^- \longrightarrow HIO + 2\,H_2O \qquad \mathcal{E}^0 = +1.14$$

(a) For each of the first two half-reactions in (17-22), calculate ΔG^0 ($\Delta G^0 = -n\mathcal{E}^0\mathcal{F}$; see Section 10-4(b)).
(b) Combine these two half-reactions and calculate ΔG^0 for the reaction

$$ClO_3^- + 5\,H^+ + 4\,e^- \longrightarrow HClO + 2\,H_2O$$

(c) Calculate $\mathcal{E}^0$ for the reaction in (b).
(d) List the three XO_3^- ions in order of decreasing tendency to release oxygen atoms, thus acting as oxidizing agents.

24. A positive ion, such as Cu^{+2}, can simultaneously act as an electron acceptor toward several electron donor atoms (as in $Cu(OH_2)_4^{+2}$). The compound ethylene diamine, $H_2NCH_2CH_2NH_2$, can simultaneously act as an electron donor at two molecular sites, forming a cyclic structure. Such a structure is called a "chelate complex" ("chelate" is pronounced key-late) and ethylene diamine is called a "chelating agent." For example, Cu^{+2} forms chelate complexes $Cu(C_2N_2H_8)^{+2}(aq)$ and $Cu(C_2N_2H_8)_2^{+2}(aq)$.
(a) Draw a picture of ethylene diamine, using normal bonding rules and picturing normal bond angles.
(b) Draw pictures of the two copper chelate complexes mentioned above, assuming Cu^{+2} has its 4s and 4p orbitals available as electron acceptors.

25. The molecular structure of the electron-excess compound ICl_3 has not been accurately measured. However, its thermodynamic properties (in the solid state) are well established (*Journal of Physical Chemistry*, Vol. **73**, p. 755, 1969). $ICl_3(s)$: $\Delta H_f^0 = -21.3$; $\Delta G_f^0 = -5.5$; $S^0 = +40.4$.
(a) Predict the geometry, bond angles, and bond lengths in ICl_3, assuming the bonding is a composite of that in ICl and that in ICl_2^-.
(b) Calculate the average bond energy of the two bonds ruptured in the process

$$ICl_3(g) = ICl(g) + 2\,Cl(g)$$

(Let's guess the heat of sublimation of $ICl_3(s)$ to be the same as that of BrF_3, $+10$ kcal/mole). (Use the ΔH_f^0 values in Appendix d.)
(c) Calculate the ratio of this average bond energy to the bond energy of $ICl(g)$.
(d) Calculate the equilibrium constant for the reaction

$$ICl_3(s) = ICl(g) + Cl_2(g)$$

26.   The bonding in SF_4 was discussed from an electron repulsion point of view in Section 16-2(d) (see Fig. 16-7). Now this molecule can be related to the electron-excess compounds. Assuming the bonding is a composite of normal "perfect sharing" bonding in SF_2 and three-center electron-excess bonding like that in XeF_2, predict the structure and bond lengths in SF_4. (Note: A normal S–F single bond length is 1.54 Å).

27.   In the M.O. view of the bonding in XeF_4, the xenon atom has an unused pair of electrons in its third valence p orbital. Assuming oxygen atom acts as an electron acceptor, predict the structure of the molecule $XeOF_4$, which is known to be a colorless, clear liquid at room temperature.

28.   The inert gas compound XeF_2 can be expected to form oxides (as does XeF_4; see Problem 27) with oxygen acting as electron acceptor. The compound $XeOF_2$ is well known as a crystalline compound and XeO_2F_2 is thought to be an impurity present in $XeOF_4$ preparations. Combine the M.O. description of XeF_2 with the electron acceptor properties of oxygen atom to predict the structure of each of these two compounds, $XeOF_2$ and XeO_2F_2.

29.   From the following thermodynamic data (all relevant to CCl_4 solutions), list the electron acceptors (the "acids") in order of decreasing hydrogen bonding ability. (Note: Chloroform, $HCCl_3$, is one of the few compounds in which a C–H proton displays hydrogen bonding.)

	ΔH
chloroform + triethylamine	
$HCCl_3$ + $N(C_2H_5)_3$ ⟶ $Cl_3CH \cdots N(C_2H_5)_3$	-4.0
chloroform + acetone	
$HCCl_3$ + $O{=}C(CH_3)_2$ ⟶ $Cl_3CH \cdots OC(CH_3)_2$	-2.0
methanol + acetone	
H_3COH + $O{=}C(CH_3)_2$ ⟶ $H_3COH \cdots O{=}C(CH_3)_2$	-2.5
methanol + dioxane	
H_3COH + $O(C_2H_4)_2O$ ⟶ $H_3COH \cdots O(C_2H_4)_2O$	-2.8
t-butanol + dioxane	
$(CH_3)_3COH$ + $O(C_2H_4)_2O$ ⟶ $(CH_3)_3COH \cdots O(C_2H_4)_2O$	-2.9
t-butanol + pyridine	
$(CH_3)_3COH$ + NC_5H_5 ⟶ $(CH_3)_3COH \cdots NC_5H_5$	-4.0
phenol + pyridine	
C_6H_5OH + NC_5H_5 ⟶ $C_6H_5OH \cdots NC_5H_5$	-5.0
phenol + triethylamine	
C_6H_5OH + $N(C_2H_5)_3$ ⟶ $C_6H_5OH \cdots N(C_2H_5)_3$	-6

30.   Use the data in Problem 29 to list the electron acceptors (the "bases") in order of decreasing hydrogen bonding ability.

31.   Methanol forms hydrogen bonds with an electron donor, B, in CCl_4 solution.

$$H_3COH(CCl_4) + B(CCl_4) = H_3COH \cdots B(CCl_4)   \Delta H$$

The most easily measured property that reveals this interaction is in the infrared spectrum where the O–H stretching vibration absorbs. It is found that the force constant $k(OH)$ decreases markedly when a hydrogen bond is formed. The most interesting datum, though, is ΔH, which is difficult to measure.

(a) Make a plot of ΔH versus k(OH) to see how they correlate.

(b) Use your plot to estimate ΔH of hydrogen bond formation between methanol and benzophenone $C_6H_5 - \overset{\displaystyle O}{\overset{\displaystyle \|}{C}} - C_6H_5$, for which $k = 7.01$.

(c) Repeat (b) for methanol and triethylamine, for which $k = 5.84$.

B	ΔH(kcal)	k(mdyne/Å)
CCl$_4$	0.0	7.36
acetone	−2.52	6.91
dioxane	−2.80	6.86
pyridine	−3.88	6.25

32.   The equilibrium constant for the formation of a hydrogen bonded complex between ethanol, CH_3CH_2OH, and pyridine, C_5H_5N, has been measured in carbon tetrachloride solution at 280°K and also at 333°K. Use the expression (10-46) and the equilibrium constants, $K(280°K) = 3.50$ and $K(333°K) = 1.22$ to determine ΔH^0 for hydrogen bond formation.

33.   Repeat Problem 32 for the hydrogen bonded complex between ethanol and the cyclic ether dioxane, OC_4H_8O. $K(280°K) = 1.47$ and $K(333°K) = 0.60$.

34.   When two liquids are mixed, the heat effect is called the heat of mixing. For example, chloroform, Cl_3CH, dissolves readily in carbon tetrachloride, CCl_4; benzene, C_6H_6; acetone, $CH_3\overset{\displaystyle O}{\overset{\displaystyle \|}{C}}CH_3$; and triethylamine, $N(C_2H_5)_3$. When one mole of Cl_3CH is dissolved in one mole of each of these liquids, the heat effects are as follows:

solvent (1 mole)	CCl$_4$	C$_6$H$_6$	CH$_3$COCH$_3$	N(C$_2$H$_5$)$_3$
ΔH(mixing)	+20	−90	−495	−1200 cal

(a) Discuss these values in the light of Problems 29 and 30.

(b) Use the ΔH values in Problem 29 to decide what fraction of the Cl_3CH is present as the hydrogen bonded complex in the acetone and in the triethylamine solutions.

Most of the substances around us are in the liquid or the solid state. Yet we have considered only gaseous molecules thus far. Liquids and solids are more difficult to deal with—a molecule of methane has only five atoms, but a small grain of sand may contain 10^{18} atoms. More difficult, yes, but fortunately, not in proportion to the number of atoms. In fact, the principles of bonding developed for gas molecules serve admirably as we classify the types of liquids and solids and as we explain their properties.

18-1 A general description of liquids and solids

The miracle of condensation is familiar. When night falls, vaporous water molecules join together to sparkle in a drop of dew. At a gust of winter weather, they join hands in the frigid array of a delicate snowflake. But then they scatter to the wind if caught by the warming sun.

Every substance has its own night, its own winter, and its own summer. Every element and every compound will condense to a liquid if cooled sufficiently. If cooled still further, every one of these liquids ultimately becomes a solid.* Conversely, every solid and every liquid can be vaporized if the temperature is raised sufficiently.

Liquids are readily recognized by their fluidity and solids by their rigidity. When a solid melts, its molar volume increases by a few percent (ice is a rare exception). More notable, however, is the fact that both liquids and solids are virtually incompressible compared with the responsiveness of a gas. Raising the pressure on a gas from one to two atmospheres halves its volume. Under the same pressure change, a typical liquid decreases in volume by about 0.01 percent and a solid by even less. The condensed phases are similarly resistant to temperature-induced volume change. Both liquids and solids behave as though their atoms and molecules are held together at spacings near their "natural size." Great effort is needed to crowd them closer together, and it takes work to pull them apart.

---

* Helium is a lone, willful exception. It does not become a solid at any temperature unless the pressure exceeds 25 atmospheres.

(a) MELTING AND BOILING POINTS—HEAT EFFECTS, TOO

Within these generally uniform characteristics, liquids and solids display wide variety and individuality. Some solid substances are lustrous, malleable, and easily classed as *metals*. Others are crystalline, with regular crystal faces and sharp cleavage planes—some of these are classified as *salts* or *ionic crystals* and some as *covalent solids*. Still other solids are soft and retain many properties of the gaseous molecules from which they are condensed—these are *molecular crystals*. Table 18-1 shows the range of boiling points, melting points, and associated heat effects for these four types.

The molecular solids have melting and boiling points extending from almost absolute zero to above room temperature. In contrast, covalent solids, metals, and ionic solids generally melt at much higher temperatures. The energies holding the molecular solids and liquids together must be quite weak relative to those that cause covalent solids, metals, and ionic solids to condense. This conclusion is corroborated by the heats of fusion and vaporization.

The thermodynamic properties are quite informative. First, notice that *every ΔH entry is positive*. Every liquid has higher energy than its solid form at the melting point. Every gas has higher energy than its liquid form at equilibrium. Next, notice that *every ΔS entry is positive*. Every liquid is more random than its solid form at the melting point. Every gas is more random than its liquid form at equilibrium.

The magnitudes of the entropies are also meaningful. In the entire table, $\Delta H_{fus.}$ ranges from 5 to 11,000 calories, while $\Delta S_{fus.}$ ranges only from 2 to 17 cal/mole degK. The entropies of vaporization are even more nearly constant. Though $\Delta H_{vap.}$ ranges from 20 to 122,000 calories, all the values of $\Delta S_{vap.}$ lie between 16 and 30 cal/mole degK (if we omit helium).

These entropies paraphrase in quantitative thermodynamic language what everyone knows about solids and liquids. Crystals involve atoms close together in regular lattice arrangements, whereas liquids have their atoms close together, but packed in random fashion. The regularity of the crystal makes it energetically more stable than the liquid and this energy effect accounts for the rigidity of the solid. Once the temperature is high enough for the randomness of the liquid to dominate the energy stability of the solid, melting occurs. Then the intrinsic disorder of the liquid permits it to flow, to take the shape of its container, and to act as a solvent for other molecules of quite different shape.

In turn, gases involve atoms far apart and even more disordered than in a liquid because of the positional randomness available to the gaseous particles. The liquid–gas entropy change is dominated by the randomness of the gas, hence $\Delta S_{vap.}$ is approximately the same for all gases (near 21.6 cal/mole degK; this is Trouton's Rule) unless there is some special order in

Table 18-1 Melting Points, Boiling Points, and Associated Heat Effects

Substance		Melting Point (°K)	Boiling Point (°K)	ΔH(kcal/mole)		ΔS(cal/mole degK)	
				Fusion	Vapor-ization	Fusion	Vapor-ization
Molecular Solids							
helium	He	1.4 (26 atm)	4.2	0.005	0.020	1.5	4.7
hydrogen	H_2	14	20	0.03	0.22	2.0	15.8
neon	Ne	24	27	0.080	0.431	3.26	17.5
nitrogen	N_2	63	77	0.17	1.33	2.7	17.2
argon	Ar	83	87	0.28	1.56	3.4	17.9
methane	CH_4	89	112	0.23	2.2	2.6	19.7
xenon	Xe	161	166	0.55	3.02	3.4	18.3
chlorine	Cl_2	172	239	1.53	4.88	8.9	20.4
n-nonane	C_9H_{20}	220	424	3.7	9.0	16.8	21.3
carbon tetrachloride	CCl_4	250	350	0.64	7.1	2.6	20.4
water	H_2O	273	373	1.4	11.3	5.3	30.1
benzene	C_6H_6	278	353	2.4	8.3	8.5	23.5
naphthalene	$C_{10}H_8$	353	491	4.6	9.7	12.9	19.7
Covalent Solids							
beryllium chloride	$BeCl_2$	678	760	3.0	25	4.5	30
germanium	Ge	1233	2973	8.3	—	6.7	—
silicon	Si	1683	2560	11	—	6.5	—
silica (quartz)	SiO_2	1883	—	2.0	—	1.1	—

carbon (graphite)	C(gr)	3273	5100	—	—	—	—
carbon (diamond)	C(dia)	dec. to C(gr)		—	—	—	—
Metals							
mercury	Hg	234	630	0.58	15.5	2.5	24.5
sodium	Na	371	1153	0.63	24.6	1.7	21.1
lithium	Li	353	1599	1	—	3	—
lead	Pb	600	2023	1.2	43.0	2.0	21.3
aluminum	Al	933	2600	2.6	67.6	2.8	26.0
silver	Ag	1234	2466	2.7	60.7	2.2	24.6
platinum	Pt	2042	4100	5.2	122	2.5	29.8
Ionic Solids							
potassium nitrate	KNO₃	610	673*	2.8	—	4.6	—
silver bromide	AgBr	703	1806	2.2	37.0	3.1	20.5
silver chloride	AgCl	728	1830	3.2	43.7	4.3	23.9
magnesium chloride	MgCl₂	987	1691	10.3	32.7	10.4	19.3
potassium chloride	KCl	1045	1680	6.1	38.8	5.8	23.1
sodium chloride	NaCl	1081	1738	6.8	40.8	6.3	23.5
barium chloride	BaCl₂	1235	1462	5.4	57	4.4	27

* Decomposes.

the liquid state. Water is the usual example cited—hydrogen bonding is the cause.

(b) THE ELEMENTS AND THE PERIODIC TABLE

An element has but one kind of atom, so there is no basis for charge separation or the formation of ionic solids. We are left with three classes: molecular solids, covalent solids, and metals. Figure 18-1 shows the distribution of these types across the Periodic Table. There are only about 15 elements that are obviously molecular solids—these are crowded into the upper right corner. In contrast, there are about 70 elements that are metallic—those off to the left in the periodic array. Between the metals and the molecular solids there is a zone of elements that includes the covalent solids as well as some solids not readily classified. Some elements, like arsenic and antimony, exhibit both molecular and metallic forms. Phosphorus displays both covalent and molecular solid forms. These borderline elements are particularly important because of their intermediate character, and we'll pay particular attention to them.

Compounds between two different nonmetallic elements always form

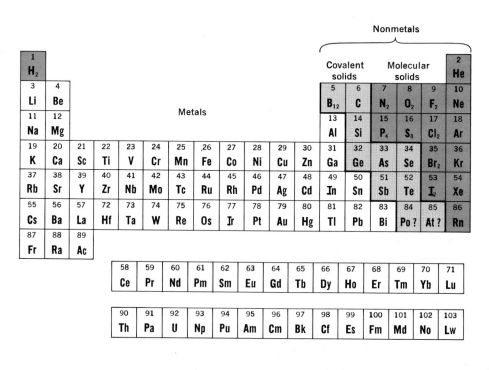

Figure 18-1 The nature of the elemental solids.

molecular or covalent solids. A compound between a metallic and a non-metallic element will usually form an ionic or a covalent solid. Two metals may form one or more metallic compounds or, more often, a range of metallic solutions of one element dissolved in the other.

With this general feeling for the geography of the solid state, we'll tackle the various bonding situations one at a time. The solids will be considered roughly in order of melting points, lowest first. This means we'll begin with the inert gases.

18-2 The inert gases

The elements in the last column of the Periodic Table are distinctively unreactive. Only a few weakly bonded compounds of xenon are known and only a single compound of krypton—no compounds of argon, neon, or helium have ever been prepared. These elements are well described by the family name "inert gases." The reason for this inertness is understandable in terms of the traditional bonding rules and orbital occupancy. All the valence orbitals are filled—there remains no capacity for normal chemical bond formation.

Yet these elements, too, liquify at a sufficiently low temperature and solidify at still lower temperatures. What is the origin of the energy-lowering that causes this condensation? We reply without hesitation: The energy can be lowered relative to the gas phase if electrons of one atom have an opportunity to be attracted to other nuclei in the condensed state. The plot remains the same—simple electric interactions are at the heart of all chemistry.

Since the valence orbitals of each atom are all occupied, the only possible place for electron-sharing is in the high-energy, empty orbitals of its neighbors. This assertion is readily verified with the one-electron, effective-charge model of the neon atom. In Table 13-1, the effective nuclear charge Z^* felt by the most easily removed neon electron was given as 2.52. The average radius of a valence 2p electron near a nucleus with $Z^* = 2.52$ is 1.05 Å. However, the 2p orbitals are all completely occupied; if another atom approaches, it must occupy the space defined by the outer orbitals. The 3s orbital is the best of those available—it has the lowest energy of any of the outer orbitals. How large is the 3s orbital in an atom with this same Z^*? That is easily calculated (using equation (13-4)) to be 2.83 Å. That radius should be approximately equal to the spacing of two (or more) neon atoms clustered together. This estimate is in reasonable agreement with the observed internuclear spacing in solid neon—3.18 Å for the nearest neighbors. Figure 18-2 portrays the situation. The valence electrons of the left atom occupy the vacant 3s orbitals of the right atom, and conversely. We see that it is intuitively natural to picture the two atoms as spheres in contact, each with a radius equal to half the internuclear distance.

Of course, attractive forces at such large separations can be expected to

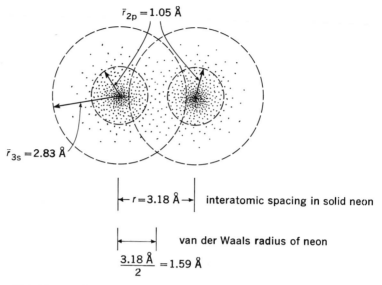

$\bar{r}_{2p} = 1.05\ \text{Å}$

$\bar{r}_{3s} = 2.83\ \text{Å}$

$\leftarrow r = 3.18\ \text{Å} \rightarrow$ interatomic spacing in solid neon

van der Waals radius of neon

$\dfrac{3.18\ \text{Å}}{2} = 1.59\ \text{Å}$

Figure 18-2 The spacing of neon atoms in the solid contrasted to average orbital radii.

be quite weak. That is why the boiling points and the heats of vaporization of the inert gases are so low. These two factors, interaction distance and energy of interaction, are given in Table 18-2, together with comparative data for the adjacent halogen molecule. In every row, the availability of a half-filled valence orbital of a halogen atom permits close approach and a high bond energy. The completely filled valence orbitals of an inert gas permit only outer-orbital approach and very low bond energy. Because large energy differences are readily evident in many phenomena, they are distinguished by name. The valence orbital interactions are, of course, called chemical bonds. The outer-orbital bonding is called *van der Waals bonding*, after the Dutch scientist who studied this type of interaction. The size of the atom, defined to be half the internuclear separation in the solid, will be called the *van der Waals radius*.

Table 18-2 Comparison of Covalent Bonds to van der Waals Bonds: Distance and Energy

Periodic Row	Halogen X_2	Inert Gas M	Distance (Å)		Energy (kcal/mole)	
			r_{XX}	r_{MM}	D_{X_2}	$\Delta H_{\text{sub.}}(\text{M})^{*}$
1	F_2	Ne	1.42	3.18	37	0.511
2	Cl_2	Ar	1.99	3.82	57	1.846
3	Br_2	Kr	2.28	4.02	46	2.557
4	I_2	Xe	2.66	4.40	36	3.578

* $\Delta H_{\text{sub.}}(\text{M})$ = heat of sublimation (solid $\longrightarrow$ gas) for the crystalline inert gas.

(a) INERT GAS CRYSTAL STRUCTURES

Figure 18-2 encourages us to picture inert gas atoms as spheres in contact. This proves to be a useful model, despite the quantum mechanical conclusion that atoms have no boundaries. It is particularly valuable in deducing the crystal structure.

Figure 18-3(a) shows how a third atom would nestle up close to two other like inert gas atoms if getting as close as possible is its aim. The vacant outer orbitals of each atom are so spacious and are occupied so incompletely by a single adjacent atom that there is plenty of room for a third neighbor. For simplicity we'll represent this situation with circles in contact, as in Figure 18-3(b). Now there is one sensible spot for a fourth atom. It can rest comfortably in the natural cradle atop the trio, thereby nudging up close to all three of its neighbors (18-3(c)).

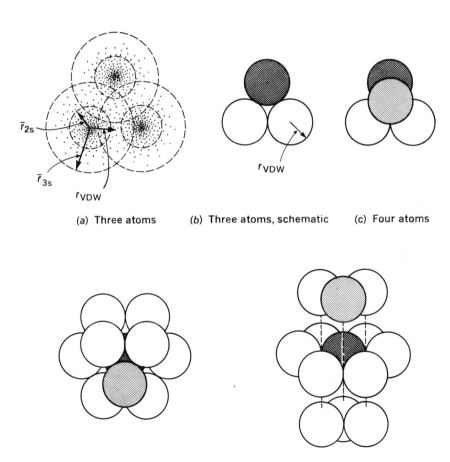

(a) Three atoms   (b) Three atoms, schematic   (c) Four atoms

(d) Ten atoms   (e) Thirteen atoms, "exploded" view

Figure 18-3   *The growth of an inert gas crystal.*

Figure 18-3(d) makes a jump to ten atoms. The cluster of four atoms has been nestled into a larger group of atoms, so that the center atom now has six atoms around it and three cradled above. The shaded atom finds itself well surrounded on the sides and top—and to everyone's advantage, since so many nearest neighbors find themselves at the "in-contact" distance. In fact, there is only one other place left near the shaded atom—three more atoms could be placed below the first plane as shown in Figure 18-3(e). The trio above and the trio below have been moved away to clarify the arrangements. Now the shaded atom has twelve neighbors at the "in-contact" distance. More important, a continuation of this type of growth will ultimately give every atom the same 12 neighbor environment that the shaded atom possesses.

The packing finally developed is seen every day in the grocery store, wherever neatness prevails and oranges are on sale. It is one of the most effective ways of packing spheres. It is called "cubic closest packing," and X-ray studies show that all the inert gas atoms crystallize in this arrangement. The r_{MM} internuclear distances listed in Table 18-2 were measured in these crystal studies. The "in-contact" radius to be assigned to each atom is $\frac{1}{2}r_{MM}$. As we noted earlier, this is a conventional indication of atomic size for an inert gas.

There is much to learn from this closest-packing-of-spheres model of inert gas crystals. We explained the existence of bonding by the desire of nuclei to place themselves near the electrons of other atoms under the circumstances that only vacant outer orbitals are available. The vacant outer orbitals then exert no directive influence on the packing of neighbors. This is contrary to the principles developed in Chapter Sixteen for covalent bond formation. Half-filled valence orbitals form bond angles connected either to the directional properties of the parent orbitals or to the repulsion of electron pairs. The situation for the inert gases is reminiscent of the clustered structures found in the orbital-excess, electron-deficient boron compounds B_5H_9 and $B_{10}H_{14}$ (see Section 17-5(b)). Wherever there are lots of vacant orbitals ineffectively filled, closest-packed arrangements are found.

(b) INERT GAS LIQUIDS

The melting phenomenon is pictured as a change from a well-ordered, closely packed solid array, as displayed in Figure 18-3, to a highly disordered, though still closely spaced, arrangement. Figure 18-4 shows, in two dimensions, how a solid and a liquid might differ if we could see them with submicroscopic vision. Any given atom in the liquid state tends to have about the same number of nearest neighbors as in the solid lattice, but irregularly spaced and with voids here and there. The extent of these voids can be seen in the difference in liquid and solid densities or, more meaningfully, in the molar volumes of the two states. The molar volumes are given in Table 18-3. The liquids expand, on the average, about 15 percent for these

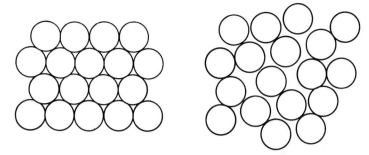

Figure 18-4   Two-dimensional representations of a solid and a liquid.

elements. Of course there are not really voids in the liquid since none of the atoms has a boundary. Rather, the atoms are disarrayed and moved away from the optimum, energy-minimum distance by a few percent. Note that it only takes about a 5 percent increase in the average radius assigned to each atom to increase the molar volume by 15 percent.

One other datum in Table 18-3 catches attention. The molar volume of helium is quite out of line—far bigger than suggested by extrapolation from the other inert gases. This special feature is attributed to the extra mobility of this element due, in turn, to its low mass and weak interatomic forces. These same aspects of helium account for its failure to crystallize except under pressure, for the existence of two "kinds" of liquid helium, and for some unique viscosity properties of one of these liquid states. None of these special phenomena can be interpreted without a quantum mechanical description of the liquid state.

18-3 Molecular solids and liquids

When two chlorine atoms come together, they have the opportunity to form a strong covalent bond through the sharing of two electrons in half-filled valence orbitals. When two chlorine molecules come together, they find themselves in about the same fix as when two argon atoms approach

Table 18-3   Molar Volumes of Solid and Liquid Inert Gases

	Solid V_s(cc/mole)	Liquid V_ℓ(cc/mole)	$\dfrac{V_\ell}{V_s}$
He	—	31.7	—
Ne	13.9	16.8	1.21
Ar	22.6	24.2	1.07
Kr	27.7	32.2	1.16
Xe	36.3	48.6	1.32
Rn	50.4	55.5	1.10

each other. Through the sharing of electrons, each atom in a Cl_2 molecule has already filled all its valence orbitals. The only way it can interact and share electrons with another chlorine molecule is through its outer orbitals —those outside the valence orbitals. Just as for the inert gases, we can expect such interactions to take place at rather large distances and with very weak bonds. Thus, the filled-orbital occupancy achieved through chemical bonding causes most molecules to condense as loosely bound liquids and solids in which the integrity of the gaseous molecule is retained. These are called molecular liquids and solids and, like the inert gas condensed phases, they are held together by weak attractions called van der Waals forces.

(a) THE HALOGENS

Although the forces between chlorine molecules are similar to those between argon atoms, the chlorine crystal must take account of the non-spherical molecular shape of Cl_2. Consequently none of the halogen crystals have the simple grocery store, orange-counter arrangement found in the inert gas crystals. Molecular shape affects the crystal packing in all molecular crystals, so only a few have the simple cubic, closest-packed structure.

The chlorine, bromine, and iodine crystal structures are all known and are quite alike. Figure 18-5 shows the packing in solid chlorine, which consists of interconnected parallel layers of Cl_2 molecules. One such layer is shown in Figure 18-5 as the shaded atoms. There is a layer below (the unshaded atoms) and a layer above (not shown), each displaced halfway between atoms. Within the plane of a single layer, there are two nearest-neighbor distances. Each chlorine atom has one neighbor at a distance of 2.02 Å, very close to the gas phase Cl_2 internuclear distance 2.00 Å. In addition, each atom has two neighbors at 3.34 Å, one almost parallel to the Cl–Cl bond direction and one almost perpendicular. Then, each chlorine atom has neighbors in the layer above and below it at distances of 3.69, 3.73 Å, and longer. These distances are most informative.

The shortest distance, 2.02 Å, shows that molecules of chlorine are still present in the solid, with bond order virtually unchanged from the gas phase. The Cl–Cl distances between layers, 3.69 and 3.73 Å, are quite close to the internuclear distance in solid argon, 3.82 Å. The intermediate distance of 3.34 Å is longer by 60 percent than the Cl_2 distance, but shorter by 13 percent than the argon–argon distance. It seems that within the layers there is a significant component of polyhalogen bonding of the type described in Section 17-4(b). Table 18-4 provides further evidence of this extra bonding. If the two chlorine atoms merely looked like two argon atoms to other chlorine molecules, the heat of sublimation ought to be about double that of solid argon. Instead, it is almost four times as large, as are all the other halogen–inert gas sublimation heat ratios. Furthermore, the bromine and iodine nearest-neighbor distances are also quite short com-

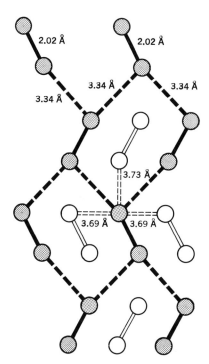

Figure 18-5 Crystalline chlorine: a molecular crystal. The shaded molecules are in one layer and the white molecules are in a lower layer.

pared with the adjacent inert gas distances. The distances between layers for I_2 are almost exactly equal to the xenon–xenon crystal spacing. We can conclude that within layers there is halogen–halogen bonding like that encountered in the polyhalogens, while between the layers the interaction is due to van der Waals forces like those that bind the inert gas crystals together.

Table 18-4 Comparisons Between Solid Halogens and Their Adjacent Inert Gases

$\dfrac{X_2}{M}$	$r_{XX'}$* Å	$\dfrac{r_{XX'}}{r_{MM}}$	$\Delta H_{sub.}(X_2)$ (kcal/mole)	$\dfrac{\Delta H_{sub.}(X_2)}{\Delta H_{sub.}(M)}$
F_2/Ne	—	—	1.9	3.7
Cl_2/Ar	3.34	0.87	6.4	3.5
	3.69	0.97		
	3.73	0.98		
Br_2/Kr	3.30	0.82	9.9	3.9
I_2/Xe	3.56	0.81	14.9	4.2
	4.35	0.99		
	4.40	1.00		

* Nearest nonbonded neighbor distances.

(b) VAN DER WAALS RADII

Crystal structures like that of the halogens give us a basis for assigning size to an atom, though we know it has no edges. Operationally, the nonbonded nearest-neighbor distances indicate how close two atoms will approach each other. For example, the 3.69 Å Cl–Cl distance in solid chlorine seems to define the size of chlorine atoms when they do not bond (the 3.34 Å distances probably involve some bonding). If we divide the 3.69 Å between the two atoms, each would be assigned a radius of about 1.8 Å. The size, so determined, is called the *van der Waals radius.*

The value in this determination of the nonbonded size of a chlorine atom is that it proves to be reasonably applicable to other crystals that also involve nonbonded Cl–Cl contacts (as, for example, in solid CCl_4). Furthermore, nonbonded contact distances between chlorine and other atoms can be estimated as the sum of the van der Waals radius of chlorine plus the van der Waals radius of the other atom, similarly determined.

Figure 18-6 shows commonly accepted van der Waals radii for the elements that form molecular crystals. These radii represent averages of the nonbonded distances found in different solids. They help us understand molecular crystal arrangements, but, even more important, they guide us as we study the molecules themselves. If a molecule is prepared in which two nonbonded atoms are held closer than the sum of their van der Waals radii, that molecule will tend to be unstable. Also, the flexibility of a molecule will be constrained to avoid bent structures that bring nonbonded atoms too close together. These ideas are so useful that most chemists use molec-

1 H 1.2			2 He
7 N 1.5	8 O 1.4	9 F 1.4	10 Ne 1.6
15 P 1.9	16 S 1.9	17 Cl 1.8	18 Ar 1.9
33 As 2.0	34 Se 2.0	35 Br 2.0	36 Kr 2.0
51 Sb 2.2	52 Te 2.2	53 I 2.2	54 Xe 2.2

Figure 18-6   *van der Waals radii of some nonmetals (given in angstroms).*

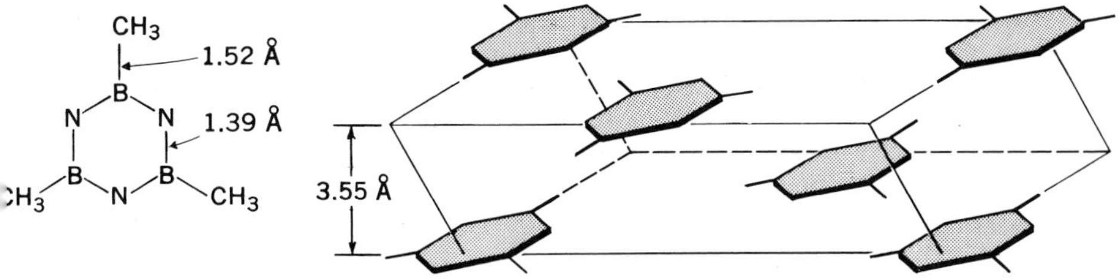

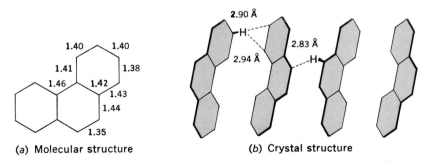

(a) Molecular structure   (b) Crystal structure

Figure 18-7 The crystal and molecular structure of B-trimethylborazole. The B—N bond lengths are remarkably similar to those found in benzene, with which it is isoelectronic. The interlayer spacing in the crystal is slightly greater than that in the graphite modification of boron nitride (see Fig. 18-10), probably as a result of the methyl groups.

ular models embodying these van der Waals radii—they are called "space-filling" models, to distinguish them from ball-and-stick models that show only bonded distances.

(c) SOME MOLECULAR CRYSTALS

Figures 18-7 and 18-8 show examples of two more complicated molecular crystals. The first example, B-trimethylborazole, consists of planar, hexagonal molecules stacked, one above another, in a head-to-tail arrangement. The hexagonal ring has alternating boron and nitrogen atoms, and the six B–N bond lengths are identical, within experimental uncertainty, at 1.39 Å. This is the same as the C–C bond length observed in benzene. The non-bonded distances shown extend from the methyl group carbons to a nitrogen atom of an adjacent molecule, 3.28 and 3.35 Å, respectively. These distances cannot immediately be interpreted in terms of van der Waals radii since the hydrogen atoms on the carbon are not located. However, a useful estimate of the van der Waals radius of a methyl group, hydrogens

(a) Molecular structure   (b) Crystal structure

Figure 18-8 The crystal and molecular structures of phenanthrene, $C_{14}H_{10}$. Hydrogen atoms have been omitted from the drawings.

included, is 2.0 Å. Adding this to the van der Waals radius of 1.5 Å for nitrogen, as listed in Figure 18-6, we are led to expect contact distances of 3.5 Å, only about 0.2 Å larger than actually observed. The vertical distance between two molecules is 3.55 Å.

Figure 18-8 shows the packing of phenanthrene, $C_{14}H_{10}$, in its crystal lattice. In a particular molecule, there are nine different C–C bond lengths determined, ranging from 1.347 to 1.455 Å. These lengths can be rationalized with resonance and with molecular orbital arguments. The molecules are not stacked in parallel planes, but are tilted such that the C–H bonds of one molecule point towards the plane of a neighbor at almost normal incidence. The nearest contacts are hydrogen–carbon distances of 2.90 and 2.83 Å. This awkward, tilted packing of flat molecules is found in other aromatic molecular crystals: e.g., benzene, naphthalene, and anthracene. Substitutions, however, can cause the rings to become coplanar (e,g., in trinitrobenzene).

18-4   Covalent solids

Some solids seem to be held together in exactly the way molecules are bonded, except in an endless network. The conventional bonding principles suffice to explain the observed structures and their properties.

(a) DIAMOND AND GRAPHITE: JEKYLL AND HYDE

Carbon is an element particularly suited to bonding in the solid state. Each carbon atom, as we have seen, is capable of forming four equivalent bonds directed to the corners of a tetrahedron. Methane, CH_4, uses up this bonding capacity by forming four carbon–hydrogen bonds. A more complicated molecule is neopentane, $C(CH_3)_4$, in which the central carbon is now bonded to four other carbons. If we imagine larger and larger molecules built up in this way, we will eventually end up with a three-dimensional network of carbon atoms, each tetrahedrally bonded to four other carbon atoms. This is the form of elemental carbon known as diamond. Strong covalent bonds are formed between each pair of atoms—in fact, diamond might be thought of as one giant molecule C_n. The bond lengths are 1.544 Å, exactly the same as in ethane. Hence the individual bond energies must be similar to that in ethane, about 83 kcal/mole.

A second form of elemental carbon, graphite, has quite a different structure. Carbon atoms are bound to one another to form two-dimensional sheets. The C–C distances are all equal, 1.42 Å, so each carbon is bonded to three others by sp^2 hybrid bonds as in benzene. Figure 18-9 shows the structure of graphite. Notice the six-membered ring that is the basic repeating unit. Fragments of two layers are shown. The spacing between the layers, 3.5 Å, is much greater than the bonded C–C distance. In fact, it is the same as the distance between two of the borazole rings shown in

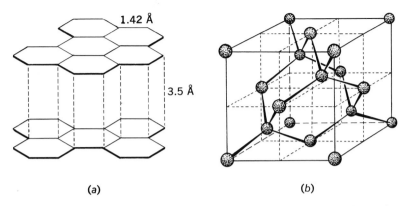

Figure 18-9 Carbon structure: (a) graphite, showing fragments of two adjacent layers; (b) diamond.

Figure 18-7—it is a van der Waals distance. So graphite is a van der Waals crystal in one dimension and a covalent crystal in the other two.

With this understanding of the bonding in diamond and graphite, it is interesting to contrast their properties. Their familiar appearances strike a chord that pervades the comparison. Diamond is clear and its brilliant sparkle casts its role as a precious gem; graphite is a pedestrian grey. Diamond is hard and is used as an abrasive; graphite rubs off the lead of a pencil to make marks on a paper and it is used as a dry lubricant. Diamond is an electrical insulator (though a bit expensive for the purpose); graphite is used as an electrical contact (a "wiper") to the commutators of motors and generators.

Let's begin with stability. In both the sp³ bonding of diamond and the sp²–π bonding of graphite, all the valence orbitals are fully engaged in bonding. The crystals should have comparable stability. They do. Though diamond is unstable with respect to graphite, the energy difference is very small, only about half a kilocalorie.

$$\text{C(diamond)} \longrightarrow \text{C(graphite)} \qquad \Delta H = -0.4532 \text{ kcal/mole} \qquad (18\text{-}1)$$

On the other hand, diamond has a higher density than graphite, 3.51 gm/cc compared with 2.25 gm/cc. This is because of the loose packing between the chickenwire layers in graphite (see Fig. 18-9). This implies that as the pressure is raised, the stability of diamond will improve relative to that of graphite. This is the secret of synthetic diamonds—at pressures above several thousand atmospheres, diamond becomes more stable, and if the temperature is raised to hasten the process, graphite changes into diamond.

Diamond is hard because its strong covalent bonds link each atom to every other in a three-dimensional network. Graphite, on the other hand, has hexagons of carbon atoms tightly bonded in planar sheets, but the sheets are quite loosely held together. Graphite breaks (or cleaves) along these planes quite easily and these sheets slide over each other and over other molecular surfaces so readily that graphite has a slippery character and acts as a lubricant.

Diamond's covalent bonds keep all of the bonding electrons localized between the atoms they are binding together. This localization implies that diamond is a poor electrical conductor. Graphite is quite different. It places a quarter of its electrons into a π molecular orbital that extends along the plane of the hexagons from one end of the crystal to the other. In this two-dimensional M.O., the π electrons are completely delocalized. With their high mobility they are quite responsive to external electric fields, joyfully rolling downhill to the positive end of the field. This mobile charge makes graphite a good electrical conductor in the two dimensions parallel to the bonded sheets. These mobile electrons also account for both the absorption of light and the shiny appearance of graphite. Diamond has such strong bonds that photons of light in the visible spectral region do not have enough energy to reach an antibonding orbital and permit light absorption. When a diamond is colored, it is because there are impurities present.

(b) SILICON, GERMANIUM, AND TIN

These elements, Si, Ge, and Sn, are below carbon in the Periodic Table. They have identical electronic structures and have similar bonding in the solid. All three exist in the diamond structure, although tin also has two other, more complex crystal structures that are metallic in character. None of the three, however, forms a sheet structure like graphite. This peculiarity of carbon is undoubtedly due to the special strength of π bonds in the first row of the Periodic Table. Figure 17-4 showed the unique bond energy per bond order for C–C and N–N multiple bonds that probably accounts for the stability of graphite.

Like diamond, extremely pure silicon and germanium solids are excellent insulators. However, with only minute concentrations of suitable impurities, the electrical properties change dramatically. An element like phosphorus, for example, at a concentration of only one part in 10^7 in silicon, gives the crystal electrical conductivity because the phosphorus atom with its five valence electrons, must take the lattice position of a silicon atom, which has only four electrons. This impurity conductivity characterizes the impure solid as a *semiconductor*—like neither a metal nor a covalent solid—and its special behavior makes transistors possible. We'll speak more of semiconductors in Chapter Twenty following a discussion of the bonding in metals.

(c) BORON NITRIDE—ARTIFICIAL CARBON

Network solids are not limited to elements. Boron nitride, BN, provides a fine example. Two structures have been observed, both with the empirical

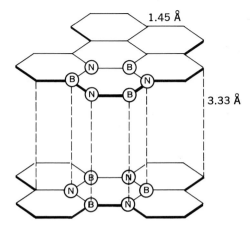

1.45 Å

3.33 Å

Figure 18-10 The graphite modification of boron nitride. The layers are arranged so that unlike atoms are above each other. Notice the remarkable similarity of in-plane and inter-plane distances between BN and graphite (Fig. 18-9). The diamond modification of boron nitride can be prepared from the graphite form at high temperature and pressure.

formula BN. One is just like graphite and the other like diamond, except that boron and nitrogen atoms alternate in the lattice positions.

Let's look at the atomic structures of these atoms to see how the bonding can be explained. For the graphite structure, both boron and nitrogen can easily form the necessary sp² hybrid orbitals by each promoting one s electron into a p state. Now each atom can bond to three neighbors in the planar hexagonal network for which sp² hybrid orbitals are perfectly oriented. Further, the boron $\pi(p_z)$ orbital is vacant; it is an electron acceptor. But it is surrounded by three nitrogen $\pi(p_z)$ orbitals ready to act as electron donors. This situation is just like that in B-trimethylborazole shown in Figure 18-7 in which the bond lengths duplicate those of benzene. With such units interlinked as in Figure 18-10, there are exactly enough electrons in the π molecular orbitals to duplicate the graphite bonding.

Needless to say, boron nitride in the graphite structure has properties similar to those of graphite. When BN is prepared in the diamond structure, it is even harder than diamond itself. Thus the chemist uses his knowledge of bonding to extend his horizons and, sometimes, to improve upon Nature.

(d) QUARTZ—A SILICON–OXYGEN NETWORK SOLID

Network solids in great variety are formed by silicon and oxygen. They are called *silicates*, and they make up 87 percent of the earth's crust. We will discuss at this time only one of these—the prototype, which is called silica, SiO_2. Silica occurs in three different crystal forms, quartz, tridymite, and cristobalite. The most familiar of these, quartz, is found in mineral deposits as crystals or as a crystalline constituent of many rocks, such as

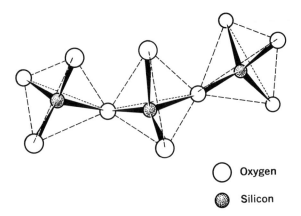

Oxygen

Silicon

Figure 18-11 The crystal structure of quartz, SiO_2. Each silicon atom is surrounded by a tetrahedron of oxygen atoms $(r(Si—O) = 1.61$ Å$)$. Each oxygen atom serves to join two tetrahedra together. In the low-temperature form of quartz, these chains of tetrahedra are helically arranged.

granite. In the quartz crystal (shown in Fig. 18-11), each silicon atom is surrounded by a tetrahedron of covalently bonded oxygen atoms. Each oxygen atom is bonded to two silicons. The structure is linked in three dimensions by an Si–O–Si network of bonds, so a hard, high-melting crystal results.

18-5 Hydrogen bonded solids

In Section 17-4(g) it was noted that when hydrogen bonding can occur as a molecule condenses, the crystal *always* has a structure that takes advantage of that possibility. That is a relatively minor triumph for hydrogen bonds compared with their crucial role in biologically important molecules. It is no exaggeration to claim that life on our planet would have assumed radically different forms—if any at all—were hydrogen bonding not present in water and in the proteins and nucleic acids that compose living cells and that transmit hereditary traits. Thus one of the weakest bonds we know, the hydrogen bond, occupies a preeminent position in the biological scheme of things.

(a) HYDROGEN BONDING IN ICE IS NICE

Water, our most abundant chemical, is ideally suited for the formation of hydrogen bonds. The oxygen atom in gaseous H_2O has two bonds to hydrogen at an angle of 104.5° and two unshared electron pairs. This geometry permits the formation of four hydrogen bonds at tetrahedral angles. The ability to form tetrahedral bonds accounts for the special stability of the diamond lattice. Water forms a similar solid with, of course, bonds con-

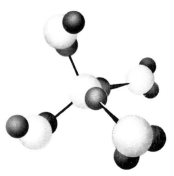

Figure 18-12   The coordination of one water molecule in ice.

siderably weaker than the covalent C–C bonds. Nevertheless, ice is much more strongly held together than such near relatives as F_2O and Cl_2O, both of which form weak van der Waals solids.

The environment of one oxygen atom in ice is illustrated in Figure 18-12, and a larger fragment of the crystal is shown in Figure 18-13. The latter drawing emphasizes the fact that we can't pin down a given pair of hydrogens to one specific oxygen. The energy of the crystal is not affected by an

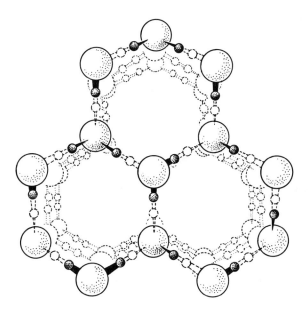

Figure 18-13   Crystal structure of ice. Each hydrogen lies between two oxygen atoms at one of the two positions shown. The arrangement is random and hydrogen atoms change positions, but always maintain around each oxygen atom four tetrahedrally placed hydrogen neighbors, two close (1.0 Å) and two far (1.8 Å). Notice the open channels, characteristic of the ice structure, that account for the low density of ice.

irregular arrangement of the bonds as long as the number remains four per oxygen atom. This irregularity results in a residual entropy or disorder in ice, even at very low temperatures. An interesting aspect of the ice structure is the presence of open "channels" down through the centers of the puckered, six-membered rings (see Fig. 18-13). This explains the density of ice, which is particularly low due to this open structure. Liquid water lacks the regularity of ice (even though considerable hydrogen bonding is still present), so this liquid is more dense than its solid. This very unusual property causes ice to float, accounting for, among other things, the fact that, in cold climates, rivers and lakes don't freeze from the bottom, which would kill off all the water life and wreck the ice-skating.

(b) HYDROGEN BONDING IN SOLID ACIDS AND AMIDES

An organic acid is so named because it contains the carboxyl functional group $-\underset{\underset{OH}{|}}{C}=O$.

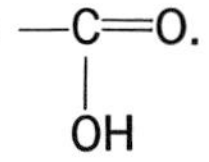

Some simple carboxylic acids are

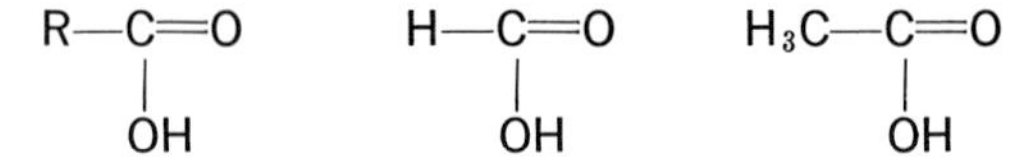

$$R-\underset{\underset{OH}{|}}{C}=O \qquad H-\underset{\underset{OH}{|}}{C}=O \qquad H_3C-\underset{\underset{OH}{|}}{C}=O$$

general acid   formic acid   acetic acid

In aqueous solution each of these molecules can release a proton:

$$RCOOH = H^+(aq) + RCOO^-(aq) \tag{18-2}$$

$$CH_3COOH = H^+(aq) + CH_3COO^-(aq) \tag{18-3}$$

The carboxyl group is capable of acting both as a proton donor and a proton acceptor in hydrogen bond formation. In the gas phase, the formation of dimers results:

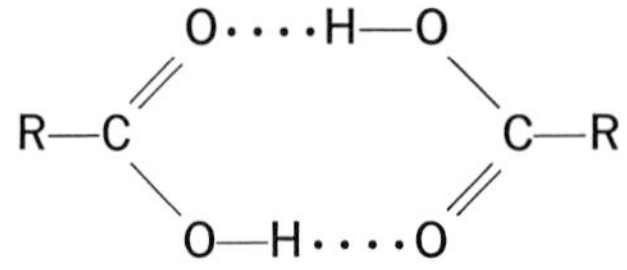

$$R-C\underset{\underset{O-H\cdots\cdots O}{}}{\overset{\overset{O\cdots\cdots H-O}{}}{}}C-R \tag{18-4}$$

In the solid, hydrogen bonded chains are common. Figure 18-14(a) shows a fragment of such a chain in solid formic acid. The zigzag chains are held together by weak van der Waals forces.

Another type of compound can be obtained if the $-OH$ part of the carboxyl group is replaced by an amine group $-NH_2$. Such compounds are

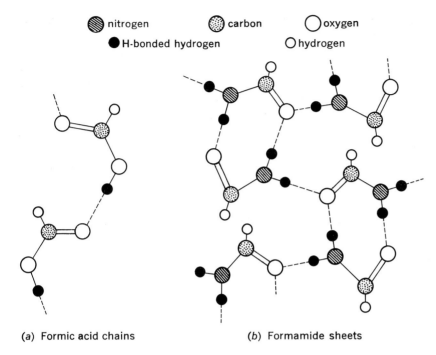

(a) Formic acid chains (b) Formamide sheets

Figure 18-14 *Structure of formic acid,* HCOOH, *and formamide,* HCO(NH₂), *in the solid. Hydrogen bonds are shown by dotted lines.*

called *amides*. The following are examples:

$$
\underset{\text{general }amide}{R-\overset{\overset{\displaystyle O}{\|}}{C}-NH_2}
\qquad
\underset{\text{form}amide}{H-\overset{\overset{\displaystyle O}{\|}}{C}-NH_2}
\qquad
\underset{\text{acet}amide}{H_3C-\overset{\overset{\displaystyle O}{\|}}{C}-NH_2}
$$

The amine group, with two protons, is extremely active in hydrogen bonding circles. Like acids, it can form zigzag chains

$$
\cdots O \quad \overset{\overset{\displaystyle H}{|}}{N}-H\cdots O \quad \overset{\overset{\displaystyle H}{|}}{\underset{\underset{\displaystyle H}{|}}{N}}-H\cdots \tag{18-5}
$$

which, in turn, may be joined to one another through the second hydrogen atom attached to each nitrogen atom. This results in the formation of *sheets*, loosely held together by van der Waals bonds. Figure 18-14(b) shows one of these cross-linked sheets.

18-6   Metals

Most elements are metals. Referring back to Figure 18-1, we see that, roughly speaking, any element to the left of carbon is metallic. Carbon has its 2s and 2p valence orbitals half-filled. The elements to the left of carbon have fewer electrons—they have electron-deficient, orbital-excess occupancies. This orbital-excess situation is the salient property of metals. But before we discuss why a metal is a metal, we should examine the characteristics of the metallic state.

(a) CHARACTERISTIC PROPERTIES OF METALS

The physical attributes of metals are so familiar that classification is easy. The characteristics that identify a metal are

- —a bright, lustrous appearance
- —high electrical conductivity
- —high thermal conductivity
- —malleability

The electrical and thermal conductivities are most readily assessed quantitatively. Table 18-5 lists the electrical conductivities of representative solids. Except for graphite, which is quite unusual, there is an enormous difference between the electrical conductivity of metals and that of any other type of crystal. In fact, the electrical conductivity of a metal is its most distinctive property. An explanation is found in our understanding of the chemical bonding in metals, as exemplified by lithium.

Table 18-5 *Electrical Conductivities of Various Types of Solids*

Substance	Solid Type	Conductivity (ohm cm)$^{-1}$
silver	metal	$6 \cdot 10^5$
zinc	metal	$2 \cdot 10^5$
graphite	covalent	$5 \cdot 10^4$
sodium chloride	ionic	10^{-7}
diamond	covalent	10^{-14}
quartz	covalent	10^{-14}
sulfur	molecular	10^{-17}
paraffin	molecular	$2 \cdot 10^{-19}$

(b) BONDING IN LITHIUM METAL

Lithium has a single valence electron and four valence orbitals. We have already discussed the bonding in Li_2 (Section 14-5(a)) in terms of the molecular orbitals formed from two 2s orbitals. However, each lithium atom has additional bond-forming capacity because of its vacant orbitals. This electron-deficient situation is what causes boron to settle into clustered structures (as in B_5H_9 and $B_{10}H_{14}$, see Section 17-5(b)). By clustering, each atom gets to share its few electrons with several neighbors—and to share theirs. This clustering characterizes the structures of metals.

It is possible to explain the bonding and properties of lithium through a molecular orbital treatment that considers only the 2s orbitals. It should be kept in mind, though, that the vacancy of the 2p orbitals is an essential requirement (otherwise, fluorine and hydrogen would be predicted to be metallic). Consideration of all of the valence orbitals is necessary for all metals other than the alkali metals. Nevertheless, the essential ideas that underlie our understanding of the metallic state are well expressed in this simple 2s molecular orbital argument.

Figure 18-15 shows a sequential development of the molecular orbitals that can be formed from two, three, and four 2s orbitals arranged in a linear array. In 18-15(a) we see the Li_2 problem—one bonding and one antibonding orbital are formed, and only the bonding orbital is occupied. In 18-15(b) a linear Li_3 arrangement is considered. A third M.O. now divides the energy spacing, but it is a nonbonding orbital. With three electrons, the orbital occupancy does not require the use of the antibonding orbital. The linear Li_4, shown in 18-15(c) produces another pair of bonding and antibonding M.O.'s, and the spacing of the energy levels is decreased further. Again we find that half the orbitals are bonding orbitals and, with one electron per atom, only these need be occupied.

Finally, as the linear array is made longer and longer, the M.O.'s come closer and closer together in energy until they give a band of levels that is virtually a continuum. However, the number of levels is always just equal to the number of component orbitals. With one valence electron per atom, only half the M.O.'s in the band of levels need be occupied—the bonding M.O.'s.

There are two characteristics of this band of M.O.'s that are crucially connected to the properties of the metallic state. First, each M.O. extends over the entire array of lithium atoms (in three dimensions, over the entire crystal). This implies electron mobility on a macroscopic scale. Second, there is no energy gap between occupied and vacant M.O.'s in the band. This means that very small perturbations of energy levels can cause occupancy changes that will show up in the physical properties. For example, an electric field can change the relative energies of levels that involve electron motion with or against the field. Figure 18-16 portrays this in a schematic way. If we apply an electric field that favors momentum states that move

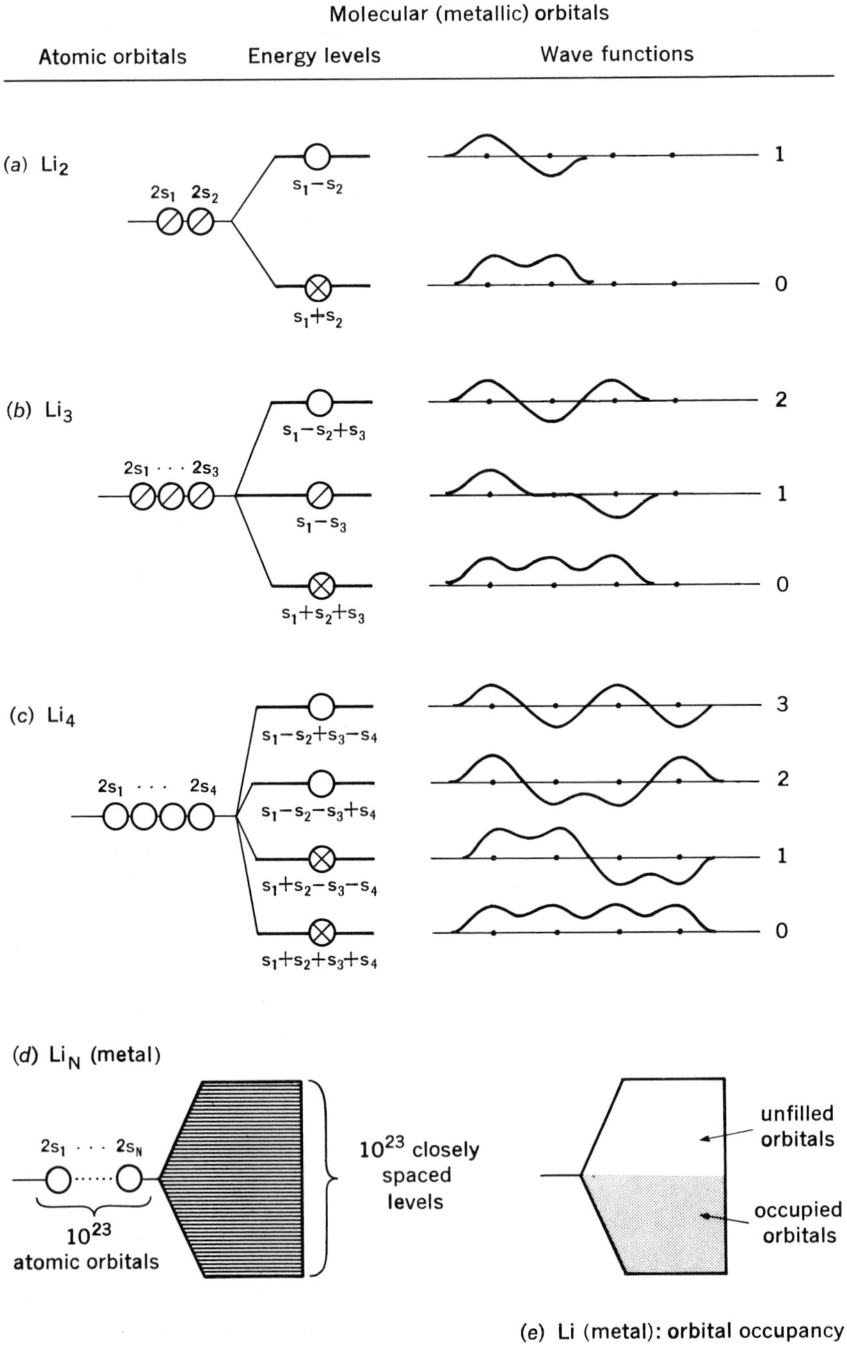

Figure 18-15 *Molecular orbital development of the band theory of metals.*

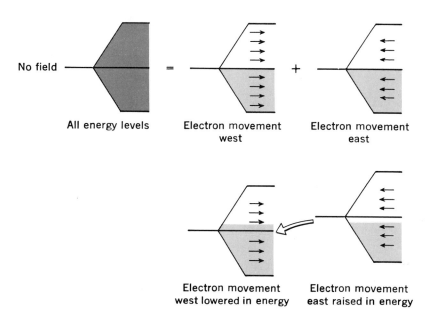

Figure 18-16 Effect of an electric field on electron movement in a metal.

the electrons westward, those states are lowered in energy relative to those that move the electrons to the east. Because the states are so closely spaced, the smallest such displacement brings many vacant westbound levels below filled eastbound levels. The orbital occupancy readjusts, and we find more electrons going west than east—electric current flows.

Whether current continues to flow depends upon whether these westbound electrons are removed from the west end of the crystal (through another electrical conductor) and replaced at the east end (from, for example, an electrochemical cell). The high electrical conductivity of metals depends, then, upon the mobility of electrons in M.O.'s that extend over the whole crystal and upon the absence of an energy gap between occupied and unoccupied orbitals. The absence of an energy gap occurs because the band is only partially filled.

The spatial extent of the orbitals explains the mobility of the electrons as reflected in the electrical conductivity. This same mobility explains the thermal conductivity of metals. If one end of a metal is heated, the electrons at that end acquire higher momentum. This extra momentum is quickly conveyed to the other end as the conducting electrons race back and forth. If the electrical conductivity is very high, the thermal conductivity also will be high. If the electrical conductivity is quite low, then momentum will have to be conducted through the lattice vibrations, which are less effective.

(c) BONDING IN BERYLLIUM METAL

Our discussion of lithium required that the M.O. band be only partially filled in order to obtain metallic conductivity. This is achieved in lithium: The N_0 2s orbitals give N_0 M.O.'s, but only $\frac{1}{2}N_0$ of the orbitals need be occupied because there is one valence electron per atom. Beryllium, with two electrons per atom, has just enough electrons to fill completely the N_0 M.O.'s in the 2s band. Yet, beryllium is a metal.

This example demonstrates the active role of the other vacant valence orbitals. If we had added, in Figure 18-15, the 2p orbitals, they would be placed some 40 kcal higher for the separated atoms. Figure 18-17 shows the more complete M.O. diagram for lithium as well as that for beryllium. In each case, the lowest bonding M.O.'s in the 2p band intermingle with the uppermost antibonding M.O.'s in the 2s band. This is unimportant for

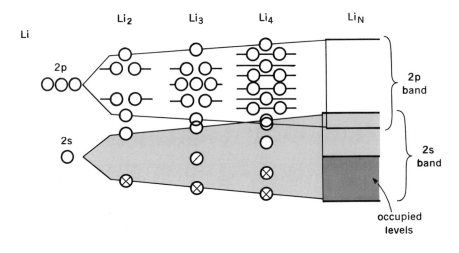

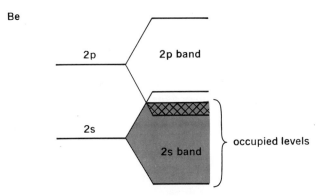

Figure 18-17 *Metallic M.O.'s for lithium and beryllium, including p orbitals.*

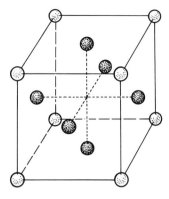

Figure 18-18   Face-centered-cubic closest packing, "f.c.c." For clarity, atoms in the face centers are shaded.

lithium because the lowest band is only half filled. For beryllium, however, the number of electrons would just fill the 2s M.O.'s if they did not inter-mingle in the band generated from the 2p orbitals. Because of the cross-over, however, some of the 2p M.O.'s are occupied, as shown in Figure 18-17, so neither the 2s band nor the 2p band is fully occupied. Metallic properties result.

(d) CRYSTAL STRUCTURES OF METALS

With the electron-deficient, orbital-excess electron configurations that characterize metals, they can be expected to form condensed structures in which each atom seeks as many neighbors as possible. Just as for the inert gas solids, the packing of spheres indicates how atoms can pack to maximize the number of nearest neighbors. Figure 18-3 shows the buildup of one such arrangement. Examination of this array from another angle shows that it contains atoms arranged at the corners of a cube with other atoms centered in each face of the cube (see Fig. 18-18). Consequently this crystal type is called face-centered-cubic closest packing (f.c.c.).

There are other ways, however, to stack oranges and to pack atoms close together. One of them involves only a subtle change from the f.c.c. arrange-ment. The f.c.c. packing was obtained by arranging six atoms around the first, all in a plane. Then three more atoms were cradled above and three more below. No fuss was made about it, but Figure 18-19 shows that there are actually two different ways in which the second trio could be added. It could be cradled directly below the upper trio or it could be cradled equally well after being rotated through 60°. In each arrangement the central atom (i.e., *every* atom) has twelve nearest neighbors. They differ, however, in their next-nearest-neighbor relationships. The f.c.c. arrangement is the one in which the third layer is rotated relative to the first layer (see Figs.

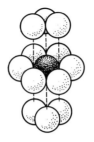
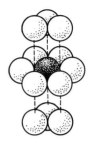

(a) f.cc. (structure III)   (b) h.cp. (structure II)

Figure 18-19   Two closest packing crystal structures: (a) face-centered-cubic closest packed and (b) hexagonal closest packed. Notice that the top and the bottom layers are identical in the h.c.p. structure, but are rotated 60° relative to each other in the f.c.c. structure.

18-19(a) and 18-3(e)). The other arrangement, in which the third layer is placed directly under the first, is called *hexagonal closest packing* (h.c.p.).

There is one more structure that is somewhat less effective. Instead of packing twelve neighbors next to each atom, eight can be spaced around it. They occupy, then, the corners of a cube with the central atom at its center, as shown in Figure 18-20. This arrangement is called the *body-centered-cubic* structure (b.c.c.).

(e) ORBITAL OCCUPANCY AND METALLIC CRYSTAL STRUCTURES

Sodium, magnesium, and aluminum are good examples to study. The first of these, sodium, has one valence electron with which to form bonds to other atoms, and the rest of its valence orbitals are vacant. This orbital-

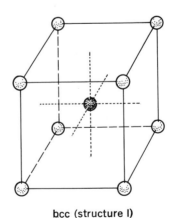

bcc (structure I)

Figure 18-20   Body-centered-cubic packing, "b.c.c." The body center position is shaded for clarity.

Element	Sodium		Magnesium		Aluminum	
	3s	3p	3s	3p	3s	3p
Ground state	···⊘	OOO	···⊗	OOO	···⊗	⊘OO
Bonding state	···⊘	OOO	···⊘	⊘OO	···⊘	⊘⊘O
Promotion energy (kcal/mole)	0		80		83	
Number of s and p electrons for bonding	1		2		3	
Crystal structure	I bcc		II hcp		III fcc	

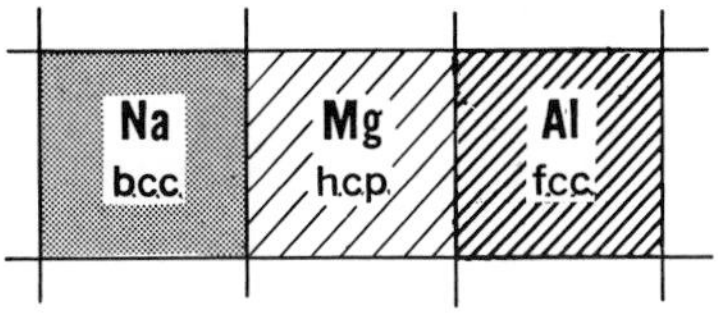

Figure 18-21 The bonding states and crystal structures of sodium, magnesium, and aluminum.

excess situation calls for a closest-packing structure, but with only one s electron to be shared so widely, a sodium atom settles for only eight nearest neighbors, the body-centered-cubic structure.

Magnesium has a ground state orbital occupancy of $1s^2$, $2s^2 2p^6$, $3s^2$. However, by paying a promotional energy price of 80 kcal/mole, both of the 3s electrons become available for bonding (see Fig. 18-21). With a 2s and a 2p electron available for sharing in the many vacant orbitals of its neighbors, magnesium selects one of the twelve-nearest-neighbor arrangements, the hexagonal closest packing.

Aluminum has one more electron and orbital occupancy $1s^2$, $2s^2 2p^6$, $3s^2 3p$ in its ground state. At a cost of 83 kcal, three electrons can be made available for bonding (again, see Fig. 18-21). With three valence electrons to share in its four 3s and 3p orbitals, aluminum also can afford twelve nearest neighbors, but it picks the face-centered-cubic alternative.

(f) ORBITAL OCCUPANCIES AND HEATS OF SUBLIMATION

The heat of sublimation of a metallic crystal is determined by the number of valence electrons participating in the bonding. This is readily apparent in Figure 18-22, which gives the sublimation heats for the first three columns

Li (I)[a]	Be (II)	B (*)
2s	2s2p	$2s2p^2$
32.2[b]	53.5	129
	(2×26.8)[c]	(3×43)
Na (I)	Mg (II)	Al (III)
3s	3s3p	$3s3p^2$
23.1	31.5	67.9
	(2×15.8)	(3×22.6)
K (I)	Ca (III, I)	Sc (II, I)
4s	3d4s	3d4s4p
18.9	36.6	73
	(2×18.3)	(3×24.3)
Rb (I)	Sr (III, I, II)	Y (II, I)
5s	4d5s	4d5s5p
18.1	33.6	94
	(2×16.8)	(3×31)
Cs (I)	Ba (I)	La (II, I, III)
6s	5d6s	5d6s6p
16.3	35.7	96
	(2×17.8)	(3×32)

*Twelve-sided, clustered unit with covalent bonds.
[a] Observed crystal structures: I = bcc; II = hcp; III = fcc.
[b] Heat of sublimination, kcal/mole.
[c] Heat of sublimination, kcal/mole per bonding electron.

Figure 18-22   Heats of sublimation and heats of sublimation per bonding electron for the metals of the first three columns of the Periodic Table.

of the Periodic Table. This table shows also that d orbitals contribute to the bonding, although it is found that they do not exert directive influence on the crystal arrangement.

The trend in heats of sublimation across the Periodic Table can be examined in a number of ways. Figure 18-23 shows two of these. In the upper plot, the heats are plotted against a row in the Periodic Table for part of the way across. In the lower plot, the sublimation energy per bonding electron is shown. Yet another way of examining the data would be to add to the values of $\Delta H_{sub.}$ the promotional energy needed to obtain the bonding state. All these approaches show a rising stability for metallic crystals up to the fifth column, and then a decline. Nevertheless, the data show that metallic stability is enhanced as the number of bonding electrons increases and that the energy per bonding electron is 20–40 kcal per electron, or 40–80 kcal per electron pair. These energies are normal chemical bond

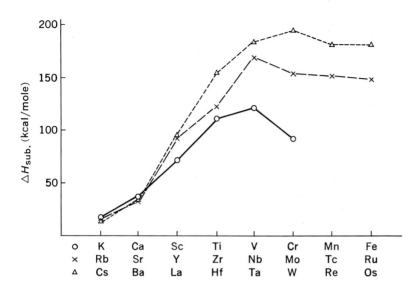

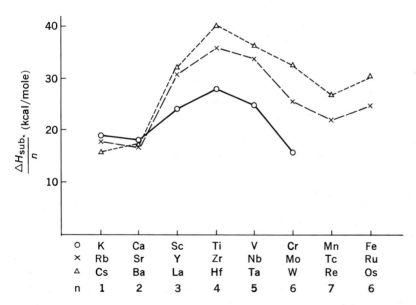

Figure 18-23   Variations of $\Delta H_{\text{sub.}}$ and $\Delta H_{\text{sub.}}$ per electron among metals.

energies, but because the electron pairs must bind many nearest neighbors, individual atom–atom distances are characteristic of quite weak bonds. For example, the bond energy in the diatomic molecule Na_2 is 17 kcal/mole and its bond length is 3.03 Å. The metallic sodium bond energy is 23.1 kcal per electron, or 46.2 kcal per electron pair. This bond energy is distributed over bonds to eight nearest neighbors giving 5.8 kcal per interaction. The bond length should be much longer in the metal–as it is, 3.72 Å. All the alkali metal elements have metallic bonds about 15–20 percent longer than their diatomic bond lengths.

18-7 Ionic solids

The gaseous, diatomic molecules $Cl_2(g)$ and $NaCl(g)$ both form chemical bonds by electron-pair sharing in half-occupied valence orbitals. The bond energies are 57 and 98 kcal/mole, respectively. Chlorine condenses to a colored, molecular crystal, but only if the temperature is lowered to the liquid air range. Figure 18-5 shows that the crystal contains Cl_2 molecules weakly bound together. In contrast, NaCl(g) condenses to a transparent crystal even at temperatures exceeding 1000°K. The heat of sublimation of this crystalline material is high.

$$NaCl(solid) \longrightarrow NaCl(gas) \qquad \Delta H = +48 \text{ kcal} \qquad (18\text{-}6)$$

Evidently NaCl(g) has substantial residual bonding capacity. Furthermore, the properties of solid NaCl are distinctively different from those of solid Cl_2—for example, X-ray studies show no evidence of discrete NaCl molecules in the crystal. In fact, solid NaCl is different from all the solid types considered thus far—molecular, covalent, and metallic crystals. Sodium chloride is typical of an *ionic solid*, the fourth class of solids listed in Table 18-1. Examination of NaCl, first as a gas, then as a solid, serves as a valuable introduction to ionic solids.

(a) SALT—A TYPICAL SALT

The substance sodium chloride has the household name "salt." This very common chemical is found all around us and even within us. Many dried-up lake beds furnish commercial amounts. The endless ocean contains dissolved sodium chloride to the tune of $\frac{1}{2}$ mole per liter, and human blood evolved with an almost identical concentration. This ubiquitous substance has properties that typify most solid compounds formed between an element far to the left in the Periodic Table and an element far to the right. Because NaCl is familiar and typical, its name was gradually transferred to this general class—compounds such as KBr, KCl, $MgCl_2$, CaF_2, and so on, all have been called "salts" since the time of the alchemists. We shall, however, use the name *ionic solids*, a term that more meaningfully represents the bonding that explains their properties.

The gaseous NaCl molecule has a high bond energy that is attributed to ionic bond character. This ionic character gives the gaseous NaCl molecule a high electric dipole moment, 8.97 D. This corresponds to the movement of eight tenths of an electron charge through the bond length, 2.36 Å, from the sodium to the chlorine atom. Thus the molecule is reasonably well characterized as a positive sodium ion embedded in the outer electron distribution of a negative chloride ion. As in the case of LiF, discussed in Section 14-5(e), this skewed electron distribution actually holds the bonding electrons closer, on the average, to sodium than the average radius of a neutral sodium atom.

From an orbital occupancy point of view, the sodium atom in NaCl(g) has considerable residual bonding capacity. It has, after all, three vacant valence orbitals and, to the extent that the electron distribution is skewed away, perhaps four. We can expect that the sodium atom should bond readily to a good electron donor—for example, a chlorine atom in another NaCl(g) molecule. Sodium chloride should readily form molecular dimers in the gas phase. In fact it does, and the bonds are strong: 44.6 kcal is the enthalpy required to break apart the dimer.

$$Na_2Cl_2(g) \longrightarrow 2 NaCl(g) \qquad \Delta H = +44.6 \text{ kcal} \qquad (18\text{-}7)$$

The Na_2Cl_2 structure is not known, but chemists are confident that it has a condensed structure like one of those shown in Figure 18-24. In (a), the dimer is pictured as two rather weakly bound NaCl molecules, each with bond length the same as that of the monomer, 2.36 Å. The representation (b) envisages the same square arrangement, but with four equal bonds, each a bit longer than the 2.36 diatomic bond length. The third version (c) is like (b)—the bond lengths are equal, but a planar, trapezoidal geometry is presumed. Of these, the equal-bond-length models (b) and (c) are generally preferred, and one of them is likely to be correct. (For example, approximate calculations of the bond energy based upon the trapezoidal configuration gave the correct energy with an internal Cl–Na–Cl angle of 105° and uniform bond lengths of 2.59 Å.) If this is so, already in the dimer

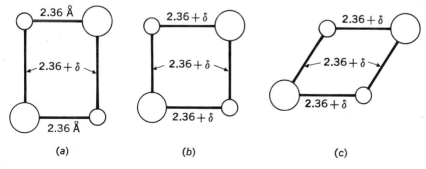

(a)   (b)   (c)

Figure 18-24 Possible structures of NaCl dimer.

we no longer discern individual NaCl molecules. Each sodium atom is equally bonded to two chlorine atoms.

Since the dimerization energy is so large and since each sodium still has a lot of unfilled valence orbital space, we can expect further aggregation. In fact, the situation reminds us of the inert gases with their completely vacant extra-valence orbitals, and of the metals with their almost vacant valence orbitals. Both cases display close-packing arrangements without specific bonding. It is almost obvious that two dimers like (b) could cluster, one above the other, to give a stable tetramer. And certainly the upper square group would be rotated 90° so that each sodium had a negatively charged chlorine above it, rather than a similar, positively charged sodium atom (see Fig. 18-25(a)). This clustering will continue until each sodium atom

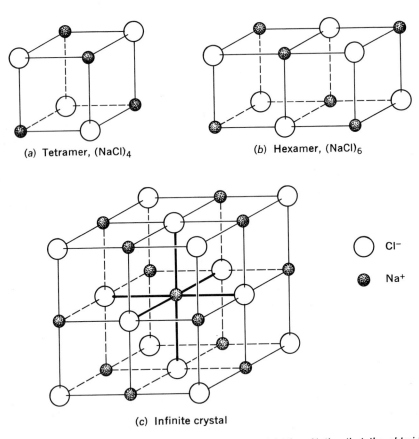

(a) Tetramer, (NaCl)$_4$ (b) Hexamer, (NaCl)$_6$

○ Cl$^-$
● Na$^+$

(c) Infinite crystal

Figure 18-25 The buildup of a sodium chloride crystal lattice. Notice that the chloride ions define a face-centered cube. The sodium ion in the center of the cube is surrounded by six equidistant chloride ions (heavy lines). This position in the f.c.c. lattice is known as an octahedral hole. If the lattice were extended, it would be clear that the sodium ions themselves have a f.c.c. arrangement with chloride ions in octahedral holes. The NaCl structure is, then, two interleaved f.c.c. arrangements of ions.

is surrounded by six chlorine atoms and each chlorine atom by six sodium atoms (see Fig. 18-25(c)). With the skewed charge distribution appropriate to the different electronegativities, this arrangement permits each Na^+ to bask in the comfortable negative environment provided by six nearest Cl^- neighbors. The Na^+–Cl^- distances are longer than in the diatomic molecule (2.814 Å is observed in the crystal, 2.36 Å in the gas), indicating a weaker bond. This is more than compensated, however, by the fact that each sodium atom bonds equally to six neighbors. Thus we are led to the most characteristic aspect of ionic crystals. Their structures are best described without mentioning molecules (despite the stability of the parent molecules such as NaCl(g)). Instead, the crystal is aptly pictured as a closest-packed array of positive and negative ions, alternating in the lattice to give maximum electrostatic attraction between nearest neighbors. Such a lattice is called an *ionic* solid.

(b) OTHER IONIC CRYSTALS

Without being too explicit, Figures 18-24 and 18-25 suggest that Cl^- is much larger than Na^+, the building blocks of the NaCl lattice. This is consistent with our estimates of size based upon the one-electron, effective charge approximation. Since the ionization energy of Cl^- is quite low (the electron affinity of Cl is 83 kcal/mole) and that of Na^+ is very high (the *second* ionization energy of Na is 1091 kcal/mole), Cl^- must be much larger than Na^+. That means our packing problem is more complicated than it was with inert gas or metal atoms all of uniform size. Figure 18-26 shows how the packing in a single layer of the NaCl crystal arrangement is affected by the ion sizes if they differ. The sizes pictured in (a) are labeled "just right." The X^- ions are crowded right up against M^+, but are not repelling each other excessively. In contrast, (b) is a poor fit. If the X^- ions move in close to M^+, they crowd together, repelling each other and raising the energy. If they hold each other apart, then they are not very close to the oppositely charged M^+. Finally, (c) doesn't allow the X^- ions close enough to feel any X^-–X^- repulsion at all.

Figure 18-26 suggests that if M^+ is too large (as in (c)) or if M^+ is too small (as in (b)), a different MX crystal packing might be obtained. In fact when M^+ is quite large, as in CsCl, another arrangement is found in which Cs^+ has *eight* nearest neighbors of Cl^- (instead of six as in NaCl). The CsCl lattice is shown in Figure 18-27(a). On the other hand, if M^+ is too small, only four negative ions pack around the positive ions. Zinc sulfide provides two common examples of this situation. The arrangement shown in Figure 18-27(b) is called *zinc blende*, or cubic ZnS. This lattice is just like diamond except that the atoms of Zn and S alternate. Figure 18-27(c) is the *wurtzite* form of zinc sulfide, or hexagonal ZnS. In each of these lattices, every atom has four nearest neighbors of opposite kind—the result of the large size of sulfur relative to zinc.

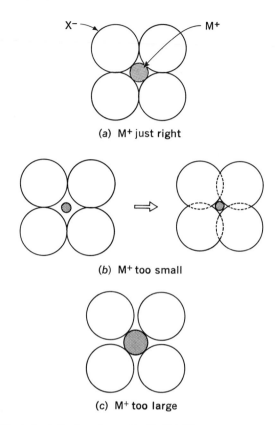

Figure 18-26 Effect of relative ion size on the NaCl lattice.

Figures 18-25 and 18-27 show ways of packing positive and negative ions so that each has 4, 6, or 8 nearest neighbors of opposite charge. The number of nearest neighbors is called the *coordination number*. The observed coordination is determined by the relative sizes of the ions. If the positive ion is much smaller than the negative ion, the coordination numbers of 4 are observed. If the positive ion is quite large, then 8 will probably be observed. For most MX crystals, the NaCl lattice is found, coordination number 6. These rules apply for alkali halides, alkaline earth oxides, alkaline earth sulfides, and a number of other substances, as shown in Table 18-6.

It goes without saying that more complicated crystal structures are needed for MX_2 compounds, such as $CaCl_2$ or Na_2O. However, most of them are understandable on the same simple basis used for the MX ionic crystals. The ions pack so as to surround positive ions with negative ions and vice versa, as dictated by the compound formula and the relative sizes of the two ions and as influenced by the tendency toward ionic or covalent bonds. Figure 18-28 shows two common crystal structures, fluorite (as

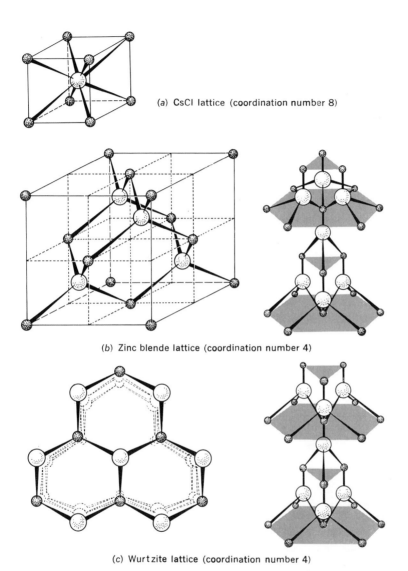

(a) CsCl lattice (coordination number 8)

(b) Zinc blende lattice (coordination number 4)

(c) Wurtzite lattice (coordination number 4)

Figure 18-27 Crystal lattices for MX crystals with coordination numbers eight and four. (a) CsCl lattice: a b.c.c. arrangement (see Fig. 18-20) in which the negative ions (anions) occupy body center positions in the positive ion (cation) lattice. Alternatively, it may be described as two interleaved simple cubic lattices. This lattice is not closest packed, but represents a slightly less efficient packing arrangement. (b) Zinc blende lattice: a f.c.c. lattice of cations with negative ions in alternate tetrahedral sites. The cubic nature of the structure is shown in the first part of the figure and a different view similar to Figure 18-19 is shown in the second. If the atoms in the lattice are identical, it becomes the diamond structure (see Fig. 18-9(b)). (c) Wurtzite lattice: a h.c.p. arrangement (see Fig. 18-19) with alternate tetrahedral sites occupied. This becomes the ice structure (see Fig. 18-13) when the atoms are identical.

Table 18-6 *Crystal Lattices for Ionic MX Crystals*

	Coordination number	8—CsCl lattice			
		6—NaCl lattice			
		4Z—zinc blende lattice			
		4W—wurtzite lattice			
LiF(6)	LiCl(6)	LiBr(6)	LiI(6)	BeO(4W)	BeS(4Z)
NaF(6)	NaCl(6)	NaBr(6)	NaI(6)	MgO(6)	MgS(6)
KF(6)	KCl(6)	KBr(6)	KI(6)	CaO(6)	CaS(6)
RbF(6)	RbCl(6)	RbBr(6)	RbI(6)	SrO(6)	SrS(6)
CsF(6)	CsCl(8)	CsBr(8)	CsI(8)	BaO(6)	BaS(6)
AgF(6)	AgCl(6)	AgBr(6)	AgI(4Z)	ZnO(4Z, W)	ZnS(4Z, W)
				CdO(6)	CdS(4Z, W)
					HgS(4Z)

shown by CaF_2, SrF_2, BaF_2, $SrCl_2$, $BaCl_2$, ZrO_2, ThO_2, UO_2, etc.) and rutile (as shown by TiO_2, MnO_2, GeO_2, SnO_2, MgF_2, MnF_2, NiF_2, ZnF_2, etc.).

(c) MOLTEN SALTS

The comparison of Cl_2 and NaCl is again informative as we consider the liquid state obtained when an ionic solid is melted. Table 18-7 shows the heat effects that accompany fusion and sublimation ($\Delta H_{\text{sub.}} = \Delta H_{\text{fus.}} + \Delta H_{\text{vap.}}$) of these two substances.

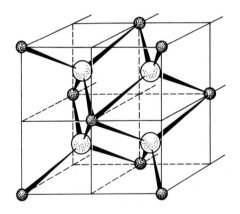

(a) Fluorite (CaF_2) (coordination number 4)   (b) Rutile (TiO_2) (coordination number 6

Figure 18-28 *Crystal lattices for MX$_2$ crystals. (a) Fluorite lattice: the anions occupy every tetrahedral site in a f.c.c. array of cations. For clarity, only half of the face-centered cube is drawn. (b) Rutile lattice: the cations are at the corners and body center positions of a rectangular cell with two equal sides, a, and one side, c, of a different length. Each cation is coordinated by six anions in an octahedral arrangement, while each anion is coordinated by three cations (hence the designation 6(3)).*

Table 18-7 Heats of Fusion and Sublimation of Cl_2 and NaCl

	$\Delta H^0_{fus.}$	$\Delta H^0_{sub.}$	$\dfrac{\Delta H^0_{fus.}}{\Delta H^0_{sub.}}$
Cl_2	1.5	6.4	0.23
NaCl	6.8	47.6	0.14
	$\Delta S^0_{fus.}$	$\Delta S^0_{sub.}$	$\dfrac{\Delta S^0_{fus.}}{\Delta S^0_{sub.}}$
Cl_2	8.9	29.3	0.30
NaCl	6.3	29.8	0.21

The heat of sublimation of solid NaCl is 7.5 times that of solid Cl_2, showing that the forces holding the ionic crystal together are much larger than those binding the van der Waals molecular crystal. This is explained in terms of chemical bonding of the ionic type in the NaCl crystal. Yet the heat of fusion of solid NaCl is a smaller fraction of its heat of sublimation (14 percent) than the corresponding fraction for Cl_2 (23 percent). This suggests that the forces holding the ionic crystal together are fairly intact in the liquid.

The entropies are even more surprising. The entropies of sublimation of solid Cl_2 and NaCl are about the same, because this quantity is dominated by the randomness of the gaseous state for each compound. Since the vapor in each case is made up of diatomic molecules (Cl_2 or NaCl), their entropies are similar. The entropies of fusion show, however, that melting solid sodium chloride, with its regular packing of positive and negative ions, increases randomness even less than melting solid chlorine. This relatively small entropy of fusion shows that the structure of the molten salt must be rather similar to that of the crystal. This implies that the ions are still present and that the sodium ions are still fairly well surrounded by chloride ions (and vice versa). To be sure, the packing is less precise and the ions have the mobility of the liquid state, but they still maintain an electrostatically favorable arrangement.

The presence of these ions in the molten salt is more positively and dramatically shown by electrical conductivity measurements. Whereas solid NaCl is an excellent insulator, molten NaCl is an excellent conductor (not as good as a metal, but very much better than most liquids and comparable to or higher than aqueous salt solutions). We conclude that molten salts contain ions, regularly but not rigidly distributed, and through their mobility, these ions are able to conduct an electrical current.

18-8 Chemical bonds and the future

Quantum mechanics has given structure and coherence to chemistry by providing a basis for understanding the chemical bond. This central concept can be developed logically and convincingly, beginning with the simpler molecules that are fully clarified by the theory. Then the various bond types can be seen to be firmly connected as each leads outward in its own direction toward more complex molecules only recently placed on the chemical shelf.

Despite this optimistic state of our knowledge, we have much to learn and there remains a strong dependence upon empirical tie-points. New experimental advances keep the subject exciting. It is alive with developments, extending from *ab initio* computations on increasingly complex molecules to discoveries of entirely new and unpredicted bond types. Thus the wheel of chemistry turns ever faster, but it rides smoothly on its sturdy hub, the concept of the chemical bond.

Problems

1. Solids are relatively incompressible compared to gases and they expand relatively little on heating. Here are some thermal expansion coefficients, the percent increase in length of a rod of material per degree temperature rise.

Solid Substance	Thermal Expansion Coefficient (%)
aluminum	$26 \cdot 10^{-4}$
calcium fluoride	$20 \cdot 10^{-4}$
diamond	$1 \cdot 10^{-4}$
graphite	$8 \cdot 10^{-4}$
ice	$51 \cdot 10^{-4}$
iodine	$84 \cdot 10^{-4}$
lead sulfide	$20 \cdot 10^{-4}$
platinum	$9 \cdot 10^{-4}$
sodium chloride	$40 \cdot 10^{-4}$
silicon	$8 \cdot 10^{-4}$
silver	$17 \cdot 10^{-4}$

(a) On a piece of graph paper, make a vertical scale from 0.000 to 0.010%. Assign the first column to molecular crystals, the second to ionic solids, the third to metals, and the fourth to covalent crystals. Write the name of each substance at the proper height (determined by the thermal expansion coefficient) and in the correct column (see Table 18-1). List the solid types in order of increasing rigidity, as suggested by the thermal expansion coefficient.

(b) The thermal expansion coefficient of elementary phosphorus (white) is 0.012 and that of the element below it in the Periodic Table, arsenic (gray), is 0.00039. What types of solids are likely for these two chemically similar elements? Add them to your graph, circled to indicate that they represent your expectation.

(c) Repeat (b) for sulfur and selenium, with thermal expansion coefficients, respectively, 0.0064 and 0.0037.

(d) The thermal expansion coefficient of solid paraffin is 0.012. What type of solid is it likely to be?

2. Solids display an enormous range of electrical conductivities. The conductivity can be expressed in dimensions $(\text{ohm-cm})^{-1}$, the reciprocal of a resistivity. A few conductivities are as follows.

Solid Substance	diamond	iodine	platinum	sodium chloride	silver
Conductivity σ(ohm^{-1}-cm^{-1})	$1 \cdot 10^{-14}$	$7 \cdot 10^{-10}$	$1 \cdot 10^5$	$1 \cdot 10^{-7}$	$6 \cdot 10^5$

(a) Make a graph like that in Problem 1 with $\log_{10} \sigma$ on the vertical scale (σ = conductivity, ohm^{-1}·cm^{-1}) extending from $+10$ to -15.
(b) The elements phosphorus (white) and arsenic (gray) have $\sigma = 10^{-11}$ and $3 \cdot 10^{+4}$, respectively. Place these on your graph, using the solid classification you deduced in Problem 1.
(c) Repeat (b) for sulfur and selenium, with $\sigma = 5 \cdot 10^{-18}$ and $8 \cdot 10^{+4}$, respectively.
(d) The conductivity of solid paraffin is $2 \cdot 10^{-16}$. Add it to your plot, using the classification of Problem 1.

3. Because of thermal expansion, railroad tracks have soft dividers between sections of rail. Steel has a thermal expansion coefficient equal to 0.0013%. By how many millimeters would the length of a rail segment increase if its length is 35 feet (11 meters) and if the temperature changed from $-10°C$ in the dead of winter to $+36°C$ (100°F) during the following summer?

4. The conductivity of a wire is proportional to the cross-sectional area and inversely proportional to the length. The proportionality constant is σ. How thick a film of paraffin (a molecular solid with $\sigma = 2 \cdot 10^{-16}$ ohm^{-1}·cm^{-1}) would be needed to decrease the conductivity of a copper wire (a metallic solid with $\sigma = 6 \cdot 10^5$ ohm^{-1}·cm^{-1}) by a factor of one million if that wire extended from Hoover Dam to Los Angeles, a wire length of 200 miles (320 kilometers)?

5. The molar heat capacity of a solid is determined by the vibrational movements of the atoms around their equilibrium positions, each atom being able to move in three dimensions. The molar heat capacities are given in Appendix d for many of the substances listed in Table 18-1.
(a) Look up the molar heat capacities for the six solid metals listed in Table 18-1. Calculate the average heat capacity and the average deviation from the average.
(b) Look up the molar heat capacities for the four ionic solids in Table 18-1 that have the simple formula AB. What is the average heat capacity *per mole of atoms?* Is it distinguishably different from the metals, as indicated by the average deviation from the average?
(c) Repeat (b) for the ionic solids with formula AB$_2$ and add CdCl$_2$, CuCl$_2$, and Cu$_2$S to give more examples.
(d) Look up the molar heat capacities for the ionic solids NaNO$_3$, KNO$_3$, AgNO$_3$, BaSO$_4$, MgSO$_4$, and CdSO$_4$. Explain how these heat capacities reflect the fact that nitrate ion, NO$_3^-$, and sulfate ion, SO$_4^{-2}$, are strongly bound groups that retain their identities in the ionic lattice, rather than vibrating as independent, loosely bound atoms. How much heat capacity is to be assigned to NO$_3^-$ and to SO$_4^{-2}$ if these crystals are regarded as AB crystals?

6. Metals and simple ionic solids all have molar heat capacities near 6.2 cal/deg per mole of atoms at room temperature and above (see Problem 5). This is called the Law of Dulong and Petit. Use Appendix d to see if this rule is applicable at room temperature to the more tightly bound covalent solids listed in Table 18-1 (Ge, Si, SiO$_2$, graphite, and diamond).

7. The heat capacity per gram of rhyolite, one of the most abundant forms of granite, is 0.192 cal/g. Analysis of rhyolite can be represented in the percentages

of various oxides, usually about 70% SiO_2, 15% Al_2O_3, 5% K_2O, 5% Na_2O, and 5% FeO.

(a) How many calories are needed to raise the temperature of 100 grams of granite (rhyolite) by one degree?

(b) How many moles of atoms does 100 grams of such granite contain?

(c) Calculate the heat capacity per mole of atoms for rhyolite. Compare the result to those deduced in Problems 6 and 7 for metals and covalent crystals. Which is granite, a silicate mineral, more like?

8.   Ordinary solder has a one-to-one mole ratio of lead to tin. Assuming a perfect metallic solution, so that the solid follows the Law of Dulong and Petit (see Problem 6), calculate the heat needed to warm one gram of solder from room temperature, 300°K, to the melting point, 500°K. By what percent does the calculated value differ from experiment, 8.2 cal?

9.   Bronze is an alloy of copper and tin. Its heat capacity per gram is 0.086 cal/g.

(a) Assuming that tin dissolves in copper metal so that it behaves according to the Law of Dulong and Petit (see Problem 6), calculate the number of moles of atoms in 100 grams of bronze.

(b) From the atomic weights of Cu and Sn, and the result in (a), calculate the approximate fraction by weight of copper in bronze.

10.   The molar volumes of solid and liquid neon are given in Table 18-3, respectively, 13.9 cc/mole and 16.8 cc/mole.

(a) Calculate the density of solid neon (in g/cc).

(b) What is the ratio of the density of solid neon to the density of ice (0.9168 g/cc at 0°C)?

(c) Calculate the density of liquid neon (in g/cc).

(d) What is the ratio of the density of liquid neon to the density of water (0.9999 g/cc at 0°C)?

(e) Ice floats in water. Would solid neon float in liquid neon?

11.   Solid xenon has a cubic closest packed lattice (see Fig. 18-3) with internuclear spacing 4.40 Å. If this spacing is assigned to spheres of radius a in contact, then for xenon, $a = 2.20$ Å. The molar volume of solid xenon is 36.3 cc/mole, about double that of water.

(a) Calculate from the molar volume the volume per atom of xenon, $V_{at.}$. Express the answer in cubic Angstroms (1 Å $= 10^{-8}$ cm).

(b) Calculate the size of a cube (in Å) that has the volume $V_{at.}$. This is the size we would attribute to xenon atoms if they packed as cubes.

(c) Characterize the cubic closest packed structure by the ratio of the volume occupied per atom, $V_{at.}$, to the volume occupied by a sphere of radius a, $V_0 = \frac{4}{3}\pi a^3$, using the data for solid xenon: $r_{ccp} = V_{at.}/V_0$.

12.   Use the result in Problem 11(c) to calculate the molar volume of solid krypton and the density of solid krypton from the X-ray determination that crystalline krypton has a cubic closest packed structure with internuclear distance 4.02 Å. Compare to the value of V_s given in Table 18-3.

13.   Generally, a liquid has about 15% lower density than its solid form because of irregular packing of the molecules. Yet liquids are quite incompressible. For example, the percent volume reduction per atmosphere pressure for liquid mercury is 0.0004% per atmosphere. If this compressibility remained constant and solidification did not occur, how high a pressure would be needed to raise the density of liquid mercury by 10%? (Note: Pressures in excess of one million atm can be produced in the laboratory.)

14. Molecular liquids tend to be more loosely packed than liquid metals, hence they tend to be more compressible. For example, the percent volume reduction per atmosphere pressure is 0.0049, 0.0112, and 0.0207 for water, ethyl alcohol C_2H_5OH, and diethyl ether $(C_2H_5)_2O$, respectively. Repeat Problem 13 for each of these liquids.

15. The heat of sublimation of solid xenon, 3.6 kcal/mole, can be attributed to the breakage of nearest-neighbor contacts, each at an internuclear distance of 4.40 Å. For this cubic closest packed crystal, how many such contacts are ruptured and what is the energy per contact?

16. Use Figure 18-5 to decide for one *molecule* of Cl_2 in crystalline chlorine how many neighbor contacts it has at internuclear distance 3.34 Å, how many it has at 3.69 Å, and how many at 3.73 Å.

17. The solid I_2 crystal is like that of solid Cl_2 except the internuclear distances are 3.56 Å within the plane and 4.35 Å and 4.40 Å above and below. The heat of sublimation of $I_2(s)$ is 14.9 kcal/mole.
(a) Calculate the energy per nearest-neighbor contact (3.56 Å). Assume that the next-nearest-neighbor contacts (either at 4.35 Å or 4.40 Å) contribute the same energy requirement to the heat of sublimation as xenon–xenon contacts in solid xenon (see Problem 15). Refer to Problem 16 for the number of contacts.
(b) It has been proposed that there is an approximately logarithmic dependence of bond energy on bond length for a given pair of atoms. Make a plot of log (bond energy) versus iodine–iodine distance using four points $r_0 = 4.40$ Å, 3.56 Å, 2.66 Å (as in I_2, with bond energy 36.1 kcal) and $r_0 = 2.90$ Å (as in I_3^-, with bond energy $\approx \frac{1}{2}(36.1) = 18.0$ kcal) to see if a straight line fits the data reasonably well.
(c) In the cesium triiodide crystal, there are two different I–I bond lengths, 2.83 and 3.04 Å. Use the graph derived in (b) to decide on ΔH when two bonds of 2.90 Å are changed into one of 2.83 and one of 3.04 Å (omitting crystal lattice energy effects).

18. The van der Waals radii (derived from solids) can be used to understand the instability of some molecular structures that crowd atoms too close together. To illustrate this, draw on a piece of graph paper the two structures of the planar molecule, 1,2-dichloroethylene, HClC═CHCl. In one structure, the two chlorines are on the same side of the molecule (the *cis* form) and in the other, they are on opposite sides (the *trans* form). Use the bond lengths given in Appendix g and assume sp² hybrid bond angles (120°). To represent the size of H and Cl atoms in nonbonded directions, draw circles with radius equal to the appropriate van der Waals radius (see Fig. 18-6). The *cis* form is less stable than the *trans* form. Is this consistent with your drawing?

19. Figure 18-14(a) shows the arrangement of formic acid molecules in solid formic acid. The two oxygen–hydrogen bond lengths shown have been measured to be 1.04 Å and 1.54 Å.
(a) What is the effect on the O–H bond length when the hydrogen atom is placed near a second oxygen atom? (For a normal O–H bond length, see Appendix g.)
(b) What is the ratio of the longer O—H distance (1.54 Å) to the sum of the van der Waals radii of oxygen and hydrogen atoms?
(c) What is the ratio of the oxygen–oxygen distance to twice the oxygen atom van der Waals radius?
(d) Discuss how the ratios obtained in (b) and (c) show that the hydrogen atom placed between two oxygen atoms exerts an attractive force and lowers the energy.

20. Combine the normal bond lengths (Appendix g) and the van der Waals radii to decide the approximate radius of a methane molecule, CH_4, and of a carbon

tetrachloride molecule, CCl_4. These size estimates are useful for calculations of collision rates in the interpretation of reaction rate data.

21.   The atmosphere of Jupiter contains lots of methane CH_4, ammonia NH_3, hydrogen H_2, and helium He. The atmospheric pressure is unknown, but much higher than that here on earth. The temperature of the atmosphere is quite low, possibly around $100°K$, because Jupiter is so far from the sun (five times the earth–sun distance). Here are some facts about two of Jupiter's atmospheric constituents.

	m.p. (°K)	Normal b.p. (°K)	Molar Volume Liquid (cc/mole)	Density Solid (g/cc)	Molecular Dipole Moment (D)	ΔH_{fusion} (kcal)	$\Delta H_{vap.}$ (kcal)
NH_3	195	240	25.2	0.83	1.47	1.35	5.6
CH_4	89	112	38.6	~0.5	0	0.23	2.2

Here are some questions about what it might be like on Jupiter.
(a) Would an ammonia iceberg float in an ammonia sea?
(b) Would a methane iceberg float in a methane sea?
(c) Would an ammonia iceberg float in a methane sea?
(d) Would an ammonia sea be salty, as are our seas?
(e) Would a methane sea be salty?
(f) What evidence in the data above suggests that solid ammonia involves hydrogen bonds, but that solid methane does not?
(g) What evidence in the data above suggests that liquid ammonia involves hydrogen bonds, but that liquid methane does not?

22.   Solid carbon dioxide and solid silicon dioxide have similar empirical formulas, CO_2 and SiO_2, but their properties are not at all alike. Solid CO_2 (dry ice) sublimes with one atmosphere pressure at a low temperature, $195°K$, while solid SiO_2 (quartz) melts at a very high temperature, $1883°K$. The properties of the solids show that solid CO_2 is a molecular crystal whereas SiO_2 is a three-dimensional covalent crystal, with each silicon atom strongly bound in a tetrahedral geometry to four oxygen atoms.
(a) CO_2 has an energy of atomization equal to 384 kcal

$$CO_2(g) \longrightarrow C(g) + 2 O(g) \qquad \Delta H = +384 \text{ kcal}$$

Using the average bond energy of C–O single bond (see Appendix g), explain why CO_2 forms a stable molecular solid, with each carbon atom forming two double bonds, rather than a quartz-like covalent crystal, with each carbon atom forming four C–O single bonds.
(b) What are the implications about bond energy per bond order for silicon-oxygen bonds in the fact that SiO_2 exists in a variety of high-melting hard, covalent, crystals (quartz, tridymite, and cristobalite)?

23.   The Martian polar cap was shown to be solid carbon dioxide ("dry ice") by the 1969 Mariner Infrared Spectrometer (*Science*, Vol. **166**, p. 496, 1969). Solid carbon dioxide is a molecular solid with a low heat of sublimation, 6.03 kcal/mole. The average solar energy reaching one square centimeter of Martian surface at the equator averages about 50 cal/hour at the equator at noon time. At sixty degrees south, the edge of the polar cap in springtime, the solar energy is only about 25 cal/hour per square centimeter at noontime.
(a) How many moles of solid CO_2 will sublime away between 11:00 AM and 2:00

PM per square centimeter at 60°S on the Martian polar cap? (Assume constant solar heating at 25 cal/hour.)

(b) From the density of solid CO_2, 1.56 g/cc, calculate the thickness in millimeters of the solid layer that will sublime away during this three-hour period.

24.   Of all the metallic elements, the four with the highest electrical conductivity are silver, copper, gold, and aluminum. These elements also have high thermal conductivities.

(a) Draw a valence orbital pigeonhole diagram for each of these elements, showing the orbital occupancies.

(b) Plot their electrical conductivities versus their thermal conductivities to test the hypothesis that the mobile electrons that cause the high electrical conductivity also conduct heat.

	Ag	Cu	Au	Al
electrical conductivities $(ohm\text{-}cm)^{-1}$	$6.3 \cdot 10^5$	$6.0 \cdot 10^5$	$4.6 \cdot 10^5$	$3.8 \cdot 10^5$
thermal conductivities $(cal/cm\ sec\ °K)$	1.01	0.99	0.70	0.50

25.   The Na–Na distance in metallic sodium is 3.72 Å, and in diatomic Na_2 it is 3.08 Å. The NaCl distance in solid NaCl is 2.814 Å, and in diatomic NaCl, it is 2.36 Å. Calculate for each case the ratio of bond length in the solid to that in the gas. The similarity can be attributed to the fact that sodium, in both the metal and in salt, shares electrons with several atoms simultaneously in its vacant valence orbitals.

26.   Copy the picture of the sodium chloride lattice in Figure 18-25. Ignoring the chloride ions completely, draw lines connecting six of the sodium atoms that are arranged in a regular hexagon around the central sodium atoms. Now draw lines connecting the three sodium atoms above the hexagon. Draw lines connecting the three sodium atoms below the hexagon. Compare this drawing to the closest packed structures shown in Figure 18-19 and decide if it matches either of them.

27.   Both carbon and nitrogen react at temperatures near 1000°K with titanium, vanadium, zirconium, niobium, hafnium, and tantalum to form compounds with very high melting points, hardness comparable to diamond, lustre, and high electrical conductivity. Irrespective of the pure metal crystal structure, all of these carbides and nitrides have formula MC or MN and the atoms are arranged in the NaCl lattice.

(a) Draw the orbital occupancies of the six metals listed above. Are they electron-rich or electron-poor insofar as valence orbital occupancy is concerned?

(b) In view of Problem 26, the small carbon (or nitrogen) atom can be regarded to be occupying the "holes" in a metallic face-centered-cubic lattice. What is the evidence that the C (or N) atoms are strongly linked to the neighboring metal atoms to give a three-dimensional covalent character to the lattice?

28.   Here are four solid compounds commonly found lying around the house:

(i) polyethylene toothbrush handle
(ii) table salt
(iii) a penny
(iv) a diamond ring

Which would be the best candidate for improvising to meet each of the following needs?

(a) desire to scratch initials on a plate of glass

(b) need an electrical insulator that would become an electrical conductor if melted

(c) need an electrical insulator that would remain an electrical insulator if melted

(d) need an electrical conductor to substitute in an emergency for a broken switch

29. (a) Classify each of the four following solids by type: molecular, covalent, ionic, or metallic.

(i) dentist's amalgam: 70% Hg, 30% Cu

(ii) naphthalene (mothball): $C_{10}H_8$

(iii) tungsten carbide: WC

(iv) rubidium chloride: RbCl

(b) Assign each of the following properties to the one of these four solids to which the property is most applicable.

A. as hard as diamond

B. as soft as soap

C. melting point above 2000°K

D. high electrical conductivity

E. malleable

F. metallic lustre

G. clear crystal that cleaves readily when struck

H. soluble in water to give a conducting solution

I. electrical insulator that becomes a conductor on melting

J. readily detectable odor

nineteen chemical principles applied: carbon compounds

In these days of specialization it is not surprising to find chemists who do nothing but study the properties and reactions of five or six related elements. Well over a third of ALL chemists, however, are specialists in the chemistry of only one element and its compounds with hydrogen, oxygen, and a scattering of other elements. This element is carbon, and the chemistry of carbon is called *organic chemistry*. Large numbers of chemists are needed to cope with the million-plus known organic compounds, more of which are being invented daily. It is not unusual, for example, for a single preparative organic chemist to prepare several hundred new compounds during his career—after which several hundred *new* chemists are needed if we are to understand the properties and reactions of these compounds.

Interest in the chemistry of carbon is not justified simply because there is so much of it, however. Indeed, a million *useless* compounds would not engender much enthusiasm, even among chemists. We are fascinated with carbon chemistry because it embraces the chemistry of life. Furthermore, we see around us countless applications of organic compounds that alleviate, or sometimes aggravate, the difficulties of man's existence. We have petroleum, pesticides, plastics, and polymers—fertilizers, fabrics, and food additives—detergents and defoliants—serums and smog—medicines and mace. . . . The list can go on and on. There are few moments in our day-to-day lives that do not bring us in contact with some substance that had its origin in the beakers and test tubes of an organic chemist.

The birth of organic chemistry as we know it today can be said to have occurred about one hundred and fifty years ago. That was when a German chemist, Friedrich Wöhler, was surprised to discover that he could synthesize the biologically important substance urea from a common laboratory chemical, ammonium cyanate. Previously, urea had been derived only from living systems and it was generally believed that a "vital force" possessed only by living organisms was needed for the synthesis of all such substances. Wöhler's discovery and others like it during the next two decades finally dispelled the "vital force" superstition and opened the door to the synthesis in the laboratory of a host of biochemical substances we find around and within us.

For the next century thereafter, chemists developed their understanding of carbon compounds and their reactions in a relatively empirical way. Today, however, the practicing organic chemist takes full advantage of the guiding principles that we have already explored in this text: equilibrium thermodynamics, reaction kinetics, bonding, and molecular structure. At the same time, he must have a working knowledge of organic substances—of the reactions they undergo and the reagents that bring these reactions about. We cannot, in this text, present a comprehensive treatment of organic chemistry. We can, however, present enough representative material to indicate what an exciting and fruitful field it has become.

19-1 Introduction to organic chemistry

(a) WHY CARBON?

A carbon atom can form four bonds, at least one of which can be to another carbon atom. The proliferation of carbon compounds is primarily due to the ability of carbon–carbon bonds to form and persist under the normal conditions we find around us. Part of the explanation lies in the C–C bond energy. Figure 19-1 compares the bond energies and reactivities of the dinuclear hydrides of six elements, each of which is capable of

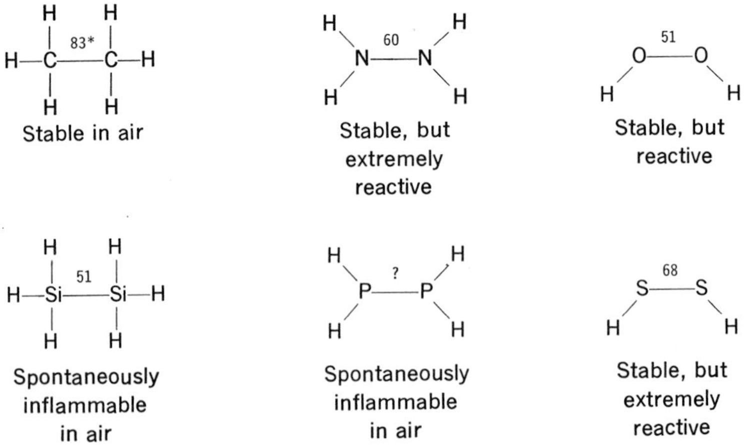

*Bond dissociation energy, kcal/mole

Figure 19-1 Bond energies (kcal/mole) of hydrides: Of the six dinuclear hydrides shown, ethane, C_2H_6, is by far the most stable. Some higher members of the series are known (silanes up to Si_6H_{14}, sulfanes up to H_2S_6), but their stabilities decrease with increasing size. Hydrocarbons, C_nH_{2n+2}, on the other hand, can have chains of any length.

forming quite strong bonds to itself. Of these six compounds, C_2H_6 has the strongest M–M bond and is the most stable under ordinary conditions—that is, under conditions normal on earth. Another cooler planet, such as Mars, might provide a favorable thermal environment for long-chain molecules of some other element, perhaps nitrogen. While many long-chain nitrogen species are known in the lab, they don't play any significant role in the chemistry of living things—a chemistry completely dominated by carbon compounds. A hotter climate, such as that of Venus, might need even stronger bonds than C–C bonds. Silicon–oxygen bonds, for example, are stable at higher temperatures and quite a variety of long-chain compounds (polymers) have been prepared with silicon–oxygen skeletons (the "silicones"). Such compounds would be capable of the information storage and transmission that are essential to reproduction, hence to life. So we see that it is quite possible to imagine life forms based on chemistry other than carbon chemistry that would be more suited to quite different ambient conditions.

However, we are here on earth, and the carbon–carbon bond abounds in our living environment. Let's examine the stability of the simplest compound of this type. Ethane, C_2H_6, is stable in air—yet it is a good fuel. This is fortunate, since it is a major constituent of abundant natural gases. The thermodynamics of the combustion process shows us how ready it is to react with the oxygen in the atmosphere:

$$C_2H_6(g) + \tfrac{7}{2} O_2(g) \rightleftarrows 2\, CO_2(g) + 3\, H_2O(g) \qquad (19\text{-}1)$$

$$\Delta H^0_{comb.} = -372.8 \text{ kcal/mole}$$
$$\Delta G^0_{comb.} = -354.5 \text{ kcal/mole}$$

Clearly, equilibrium favors the products very much. Nevertheless, ethane does not react with oxygen unless the temperature is raised—by a spark or lighted match, for example. The room temperature stability is, then, a kinetic rather than a thermodynamic one. The mechanism of an oxidation reaction such as this is very complicated and has, as yet, defied detailed analysis. Investigators have postulated a free radical chain mechanism beginning with the following step:

$$C_2H_6(g) + O_2(g) \rightarrow C_2H_5(g) + HO_2(g) \qquad (19\text{-}2)$$

The ΔH for reaction (19-2) can be calculated from the appropriate bond energies

$$\Delta H = D_{H-CH_2CH_3} - D_{H-OO}$$
$$= 98.8 - 45$$
$$= +54 \text{ kcal/mole}$$

using the best estimate of the bond energy in HO_2. Thus the first step is quite endothermic and the activation energy will be at least this high. It is not surprising, then, that such reactions do not proceed noticeably at room temperature. Once over the activation energy hump, however, the reaction releases large amounts of heat. This heat elevates the temperature, speeding up the first step and making the reaction self-sustaining.

So we have a combination of strong carbon–carbon bonds in, literally, over a million compounds whose stability is based upon kinetic effects. At the same time, the intrinsic thermodynamic instability of organic compounds makes them useful as fuels—fuels to run engines and heat houses and fuels to supply the energy necessary for the biochemical factories that ensure the maintenance of life.

(b) FUNCTIONAL GROUPS AND SKELETAL INTEGRITY

There is a familiar family of compounds called alcohols. Table 19-1 lists three of them, ethyl alcohol, n-butyl alcohol, and isoamyl alcohol, all produced through fermentation of natural products. The similar names give away the fact that they are alike. Table 19-2 shows how. Four reagents of very different kind are listed, and the three alcohols react identically with them.

—Reaction with an active metal to release hydrogen is a traditional identification of an acid (see Chapter Five). All three alcohols so react with metallic sodium to give a salt and gaseous H_2. In each case, the metal reacts with the OH hydrogen, and the carbon skeleton with its many hydrogens remains intact.

—Reaction with phosphorus trichloride displaces the entire OH group, substituting a chlorine atom. All three alcohols so react, again leaving the carbon skeleton intact.

—Reaction with concentrated sulfuric acid plucks off the OH group and hydrogen from the adjacent carbon, to produce a carbon–carbon double bond. Except for the loss of H_2O, the carbon skeleton remains intact.

Table 19-1   Three Compounds with Similar Chemistry

Common Name	Systematic Name	Molecular Structure	Fermentation Source
ethyl alcohol ("booze")	ethanol	CH_3CH_2OH	grapes
normal butyl alcohol	1-butanol	$CH_3CH_2CH_2CH_2OH$	starch
isoamyl alcohol ("fuel oil")	3-methyl-1-butanol	$CH_3CHCH_2CH_2OH$ $\quad\mid$ $\quad CH_3$	grain

Table 19-2 Some Reactions of Three Alcohols

Acid Reaction with Metallic Sodium

$$CH_3CH_2OH(\ell) + Na(s) \longrightarrow CH_3CH_2ONa(s) + \frac{1}{2} H_2$$

$$CH_3CH_2CH_2CH_2OH(\ell) + Na(s) \longrightarrow CH_3CH_2CH_2CH_2ONa(s) + \frac{1}{2} H_2$$

$$CH_3CHCH_2CH_2OH(\ell) + Na(s) \longrightarrow CH_3CHCH_2CH_2ONa(s) + \frac{1}{2} H_2$$
$$\quad\ \ \ | \qquad\qquad\qquad\qquad\qquad\qquad\qquad\qquad |$$
$$\quad\ \ CH_2 \qquad\qquad\qquad\qquad\qquad\qquad\qquad CH_3$$

$RCH_2OH + Na \longrightarrow RCH_2ONa + \frac{1}{2} H_2$

$(19\text{-}3a)$

Substitution Reaction with Phosphorus Trichloride

$$3\ CH_3CH_2OH(\ell) + PCl_3 \longrightarrow 3\ CH_3CH_2Cl(\ell) + P(OH)_3$$

$$3\ CH_3CH_2CH_2CH_2OH(\ell) + PCl_3 \longrightarrow 3\ CH_3CH_2CH_2CH_2Cl(\ell) + P(OH)_3$$

$$3\ CH_3CHCH_2CH_2OH(\ell) + PCl_3 \longrightarrow 3\ CH_3CHCH_2CH_2Cl(\ell) + P(OH)_3$$
$$\qquad\ \ | \qquad\qquad\qquad\qquad\qquad\qquad\qquad\qquad |$$
$$\qquad\ CH_3 \qquad\qquad\qquad\qquad\qquad\qquad\qquad CH_3$$

$3\ RCH_2OH + PCl_3 \longrightarrow 3\ RCH_2Cl + P(OH)_3$

$(19\text{-}3b)$

Dehydration Reaction with Concentrated Sulfuric Acid

$$CH_3CH_2OH(\ell) \xrightarrow{H_2SO_4} CH_2\!=\!CH_2 + H_2O$$

$$CH_3CH_2CH_2CH_2OH(\ell) \xrightarrow{H_2SO_4} CH_3CH_2CH\!=\!CH_2 + H_2O$$

$$CH_3CHCH_2CH_2OH(\ell) \xrightarrow{H_2SO_4} CH_3CHCH\!=\!CH_2 + H_2O$$
$$\qquad\ | \qquad\qquad\qquad\qquad\qquad\qquad\qquad |$$
$$\qquad CH_3 \qquad\qquad\qquad\qquad\qquad\qquad CH_3$$

$RCH_2CH_2OH \xrightarrow{H_2SO_4} RCH\!=\!CH_2 + H_2O$

$(19\text{-}3c)$

Oxidation Reaction with Dichromate

$$3\ CH_3CH_2OH(\ell) + 2\ Cr_2O_7^{-2} + 16\ H^+ \longrightarrow 3\ CH_3COOH + 4\ Cr^{+3} + 11\ H_2O$$

$$3\ CH_3CH_2CH_2CH_2OH(\ell) + 2\ Cr_2O_7^{-2} + 16\ H^+ \longrightarrow$$
$$3\ CH_3CH_2CH_2COOH + 4\ Cr^{+3} + 11\ H_2O$$

$$3\ CH_3CHCH_2CH_2OH(\ell) + 2\ Cr_2O_7^{-2} + 16\ H^+ \longrightarrow$$
$$\qquad\ | $$
$$\qquad CH_3 \qquad\qquad\qquad\qquad\qquad\qquad 3\ CH_3CHCH_2COOH + 4\ Cr^{+3} + 11\ H_2O$$
$$\qquad\qquad\qquad\qquad\qquad\qquad\qquad\qquad\qquad\qquad |$$
$$\qquad\qquad\qquad\qquad\qquad\qquad\qquad\qquad\qquad CH_3$$

$3\ RCH_2OH + 2\ Cr_2O_7^{-2} + 16\ H^+ \longrightarrow 3\ RCOOH + 4\ Cr^{+3} + 11\ H_2O$

$(19\text{-}3d)$

—Reaction with aqueous dichromate ion oxidizes the carbon atom to which the OH group is attached giving a "carboxylic acid group," –COOH. Once again, the carbon skeleton remains intact.

In each of the four cases, it is plain that the reaction occurs at the terminal carbon with its OH group. In each case, the remainder of the molecule, the carbon skeleton, maintains its integrity—there are no structural rearrangements to change the carbon–carbon linkages. This experience permits us to generalize the reactions of the OH group; we can represent the carbon skeleton ambiguously as R, to indicate that it does not become involved. The functional part of the molecule is what matters— the –CH$_2$OH group. We can infer that once we have learned the chemistry of this functional group, we can expect to find it duplicated in other molecules that also contain the –CH$_2$OH group.

These are the two most important guiding principles in organic chemistry.

Functional Group Chemistry

Certain atomic linkages (such as –CH$_2$OH) *impart characteristic chemistry to an organic molecule* more or less independent of the over-all molecular structure.

Skeletal Integrity

When a functional group (such as –CH$_2$OH) *undergoes its characteristic reactions, the remainder of the molecule generally remains intact.*

These two principles have enormous implications. There are thousands and thousands of alcohols known, yet merely by studying Table 19-2, we can predict with confidence that most of them will react with sodium to give hydrogen, with PCl$_3$ to give the corresponding chloride, with sulfuric acid to give the double-bonded compound, and with dichromate to give the carboxylic acid. Except for the changes in the OH group, the rest of the molecule just goes along for the ride. Only a few experiments on prototype molecules are sufficient to give reliable predictive power.

Here is another example. A carbon atom bonded to an oxygen atom and placed between two carbon atoms has a characteristic chemistry. For example, all such groups react with the reducing agent lithium aluminum hydride to produce an alcohol

(19-4)

Because this and other reactions of the C=O group are so characteristic, this linkage can be classified as a functional group. All compounds containing this functional group are classified together with a family name, *ketones*. We can expect reaction (19-4) to occur equally well if the ketone is a simple one, like acetone, or a complicated one, like progesterone, a human pregnancy hormone.

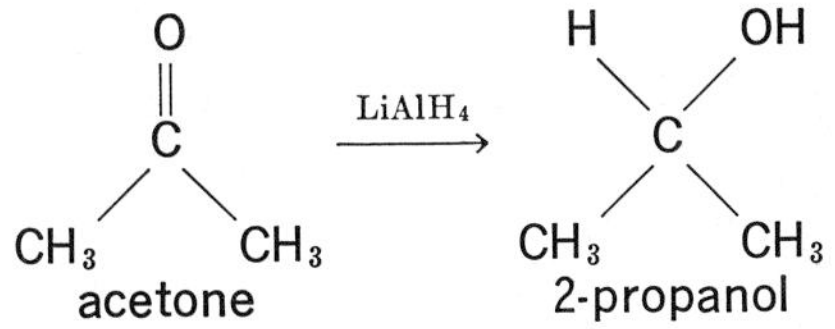

(19-5)

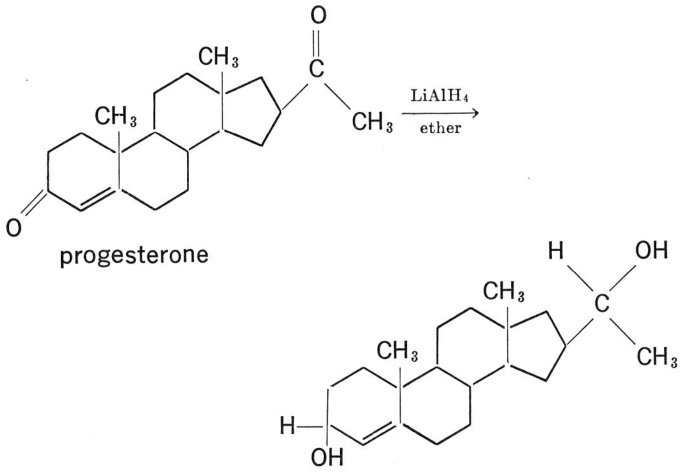

(19-6)

Here we see our two guiding principles in full flower. Progesterone actually contains two ketone groups (it is a diketone), and each reacts exactly like the ketone group in the simple prototype, acetone. Meanwhile, the complicated four-ring skeleton floats along unchanged through the reaction without batting an atomic eyelash. Functional group chemistry and skeletal integrity at work! These simple rules, empirically discovered, permit us to come to grips with the preparation and reactions of a multitude of different compounds merely by understanding the behavior of a couple dozen different functional groups.

Compounds with relatively few carbon atoms can have their physical properties as well as their chemical properties dominated by the functional groups present. For example, there is a family of carbon–hydrogen compounds with the general formula C_nH_{2n+2}. These compounds, called *alkanes*, have very low solubility in water. In any of these hydrocarbons, however, the replacement of one hydrogen by an –OH group changes this property

Table 19-3   Solubilities of Straight-Chain Alkanes and Alcohols in Water (20°C)

Alkane	Solubility (moles/ℓ)	Alcohol	Solubility (moles/ℓ)
methane CH_4	0.0014	methanol CH_3OH	miscible*
ethane CH_3CH_3	0.0017	ethanol CH_3CH_2OH	miscible*
propane $CH_3CH_2CH_3$	0.0015	1-propanol $CH_3CH_2CH_2OH$	miscible*
1-butane $CH_3CH_2CH_2CH_3$	0.0012	1-butanol $CH_3CH_2CH_2CH_2OH$	1.1
. . . .		. . . .	
octane $(CH_3CH_2)_6CH_3$	0.00013	1-octanol $CH_3(CH_2)_6CH_2OH$	0.27

* Miscible means that the two liquids dissolve in all proportions.

entirely. For example, methanol, H_3COH, and ethanol, CH_3CH_2OH, are infinitely soluble in water. The –OH functional group can form hydrogen bonds, so it causes the alcohols to dissolve in water. As the length of the carbon chain increases, the physical properties of the alcohol become more and more like those of the parent alkane. This is illustrated in Table 19-3 for five alkanes and their corresponding alcohols. With three or fewer carbon atoms, the alcohol dissolves in water in all proportions (it is "miscible"), but as the carbon chain is lengthened, the alcohol solubility drops, ultimately approaching that of the corresponding alkane.

In the following section, we will examine the essentials of bonding in carbon compounds. In the process we will become acquainted with most of the important functional groups.

(c) BONDING TO CARBON

At the heart of carbon chemistry is the tetrahedral carbon atom as exemplified by that in methane, CH_4. The bonding in methane was discussed in Chapter Sixteen in terms of sp^3 hybrid orbitals on the carbon atom. Derivatives of methane in which a hydrogen atom has been replaced by another atom are common. For example, on reaction with $Cl_2(g)$, methyl chloride is formed

$$CH_4 \quad + Cl_2 \longrightarrow \quad CH_3Cl \ + HCl \qquad\qquad (19\text{-}7)$$
methane methyl
chloride

Methyl chloride, in turn, reacts with aqueous strong base to produce methanol

$$CH_3Cl \ + NaOH \longrightarrow \ CH_3OH \ + NaCl \qquad\qquad (19\text{-}8)$$
methyl methanol
chloride

or with ammonia to produce the compound methyl amine.

$$CH_3Cl \ + NH_3 \longrightarrow CH_3NH_2 + HCl \qquad\qquad (19\text{-}9)$$
methyl methyl
chloride amine

In the presence of two moles of sodium, two moles of methyl chloride react to form the two-carbon compound, ethane.

$$2\,CH_3Cl + 2\,Na \longrightarrow CH_3CH_3 + 2\,NaCl \qquad\qquad (19\text{-}10)$$
methyl ethane
chloride

The carbon–carbon bond contains two electrons in the sigma bonding orbital formed from the in-phase overlap of two sp^3 hybrid orbitals. This is shown in the form of a pigeonhole diagram in Figure 19-2(a). The geometry around each carbon remains almost exactly tetrahedral, as pictured in Figure 19-2(b) but the possibility of rotation around the C–C bond exists (see Section 19-2(a)).

Reaction of ethane with chlorine yields ethyl chloride, CH_3CH_2Cl, from

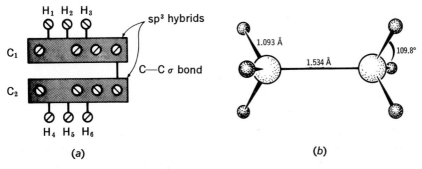

Figure 19-2   The bonding and structure of ethane, C_2H_6.

which HCl can be eliminated by reaction with an alcoholic solution of base to yield the compound ethylene, C_2H_4

$$CH_3CH_2Cl + KOH \text{ (alcohol)} \rightleftharpoons CH_2CH_2 + KCl + H_2O \qquad (19\text{-}11)$$

ethyl
chloride
ethylene

A clue about the bonding in ethylene can be obtained from thermo-dynamic considerations. The heat of formation of ethylene, easily obtained from the heat of combustion, is $+12.4$ kcal/mole. By using the heats of formation of gaseous carbon atoms, 171.3 kcal/mole, and gaseous hydrogen atoms, 52.1 kcal/mole, we can calculate the energy necessary to break all the bonds in this molecule. This will be called the enthalpy of atomiza-tion, ΔH^0_{atoms}.

$$C_2H_4(g) \longrightarrow 2\ C(g) + 4\ H(g) \qquad (19\text{-}12)$$

$$\Delta H^0_{atoms} = 2\Delta H^0_f[C(g)] + 4\Delta H^0_f[H(g)] - \Delta H^0_f[C_2H_4(g)]$$

$$= 2(171.3) + 4(52.1) - 12.4$$

$$= 538.6 \text{ kcal/mole}$$

This energy must be simply related to the bond energies in this molecule

$$\Delta H^0_{atoms} = D_{CC} + 4D_{CH}$$

Using the average CH bond energy from Appendix g, we can find the carbon–carbon bond energy

$$D_{CC} = 538.6 - 4(98.8)$$

$$= 143 \text{ kcal/mole}$$

This is a lot higher than the carbon–carbon single bond energy in ethane, 82.6 kcal/mole. An explanation can be found in the pigeonhole diagram of Figure 19-3(a). Each carbon forms three sigma bonds from sp^2 hybrid orbitals—one to the other carbon and two to hydrogen atoms. This leaves one half-filled p_z orbital on each carbon. As illustrated in part (b), overlap of these two atomic orbitals gives rise to a pi-type molecular orbital between the two carbon atoms. The carbon–carbon bond is thus a double bond, a prediction consistent with the bond energy calculated above and the 1.33 Å bond length shown in Figure 19-3(c), significantly shorter than the 1.54 Å found in ethane. The molecule is planar and there is a slight decrease in the HCH angle from the 120° expected of pure sp^2 hybrid orbitals.

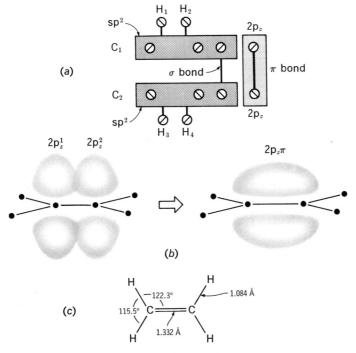

Figure 19-3 The bonding and structure of ethylene, C_2H_4.

Organic compounds containing C=C linkages are known as *alkenes* and, because the doubly bonded carbons are each bonded to only three atoms, they are said to be *unsaturated* molecules.

Carbon forms strong double bonds to other atoms as well. For example, if an alcohol is heated in the presence of a copper catalyst, elimination of H_2 (dehydrogenation) occurs. Dehydrogenation of ethanol and isopropanol yields two compounds with C=O bonds, ethanal and acetone.

$$CH_3CH_2OH \xrightarrow[Cu]{300°C} CH_3 \overset{O}{\underset{}{C}} H + H_2 \qquad (19\text{-}13)$$

ethanol → ethanal (acetaldehyde)

$$CH_3 \underset{H}{\overset{OH}{C}} CH_3 \xrightarrow[Cu]{300°C} CH_3 \overset{O}{\underset{}{C}} CH_3 + H_2 \qquad (19\text{-}14)$$

isopropyl alcohol → acetone

Compounds containing one oxygen in a C=O group, a *carbonyl group*, fall into the two types illustrated above: When the carbonyl carbon is bonded to a hydrogen and to one other carbon atom, the compound is called an *aldehyde*. We have already met the compounds in which the carbonyl carbon is bonded to two other carbons; these are called *ketones*.

The bonding in carbonyl compounds is similar to that in the alkenes, as shown in Figure 19-4(a) and (b). In part (c) the structure of acetone is compared with that of isopropyl alcohol. The C–O bond shortening is consistent with the double bond description. Other strong double bonds are found between carbon and nitrogen and carbon and sulfur.

If ethane, C_2H_6, is reacted with an excess of chlorine, a dichloride, $C_2H_4Cl_2$, is formed instead of the monochloride, C_2H_5Cl. Reaction of this molecule with alcoholic KOH results in the elimination of two moles of HCl

$$CH_2Cl-CH_2Cl \xrightarrow[\text{alcohol}]{\text{KOH}} C_2H_2 + 2\,HCl \tag{19-15}$$

The product, acetylene, contains two carbons, each of which forms only two sigma bonds. This suggests the sp hybrid bond situation illustrated in Figure 19-5(a). If each carbon forms an sp sigma bond to one carbon and one hydrogen, there are two perpendicular p orbitals, each with one electron, left over for pi bonding. The predicted result is similar to that found in the N_2 molecule. The carbon atoms are joined by one sigma and two pi bonds, and a triple bond is formed. Figure 19-5(b) shows two schematic representations of the pi bonds and part (c) gives the structure of this linear molecule. The C≡C bond is only 1.20 Å in length—considerably shortened from the double bond value of 1.33 Å found in ethylene.

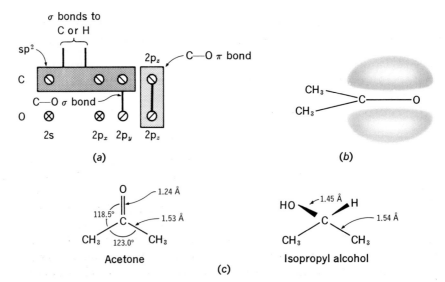

Figure 19-4   *The bonding and structure of acetone,* CH_3COCH_3.

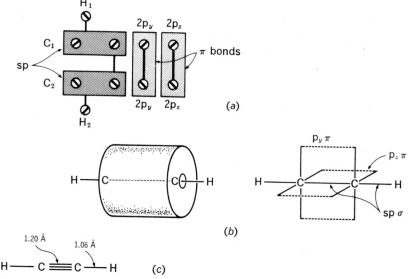

Figure 19-5 The bonding and structure of acetylene, C_2H_2.

Notice, however, that the $C=O$ bond at 1.24 Å is almost as short as the triple bond. This reflects the extra bond strength resulting from the skewing of electron density toward the electronegative oxygen atom. Carbon also forms strong triple bonds to nitrogen.

By way of summary, Table 7-3 and Appendix g list the average energies of the various bonds to carbon. The strength and stability of these bonds accounts for the great diversity of carbon chemistry.

Exercise 19-1. The heat of formation of acetylene is 54.2 kcal/mole. Calculate the carbon–carbon bond energy in this compound using D_{CH} from Appendix g, following the example of (19-12).

Exercise 19-2. Draw a pigeonhole diagram for the bonding in a cyano group, $-CN$.

(d) NOMENCLATURE OF ORGANIC COMPOUNDS

In order to deal logically with the huge number of organic compounds, systematic rules of nomenclature have been developed. The official names follow rules approved by the International Union of Pure and Applied Chemistry (IUPAC). Besides the formal IUPAC names, countless "common" names have been assigned to molecules—sometimes because their discovery predated establishment of the rules and sometimes because the formal names are quite long and complicated. These common names often have no chemical significance, but relate instead to personal, historical, or geographical events associated with the compound's initial synthesis or identification. Although a brief summary of the naming rules will be presented here, their mastery is not necessary in order to read the rest of the chapter and may well be postponed until a first course in organic chemistry.

The IUPAC names of most organic compounds are derived from the common names of those alkane hydrocarbons that have the longest possible carbon–carbon chain. Names and formulas of the first ten unbranched alkanes are given in Table 19-4. Notice that regular Greek prefixes are used beginning at the C_5 compound. The suffix "ane" signifies an alk*ane*.

We find we can prepare two compounds of formula C_4H_{10}, butane and a "branched" compound, isobutane,

$$CH_3CH_2CH_2CH_3 \qquad \begin{array}{c} CH_3 \\ | \\ CH_3CHCH_3 \end{array} \qquad (19\text{-}16)$$
$$\text{n-butane} \qquad\qquad \text{isobutane}$$

Before the formal rules were invented, these were given the names shown. The extended-chain molecule was called "n" for "normal" and the branched-chain molecule "iso." Pentane has three isomers of identical formula, C_5H_{12}.

$$CH_3(CH_2)_3CH_3 \qquad \begin{array}{c} CH_3CHCH_2CH_3 \\ | \\ CH_3 \end{array} \qquad \begin{array}{c} CH_3 \\ | \\ CH_3-C-CH_3 \\ | \\ CH_3 \end{array} \qquad (19\text{-}17)$$
$$\text{n-pentane} \qquad\qquad \text{isopentane} \qquad\qquad \text{neopentane}$$

The third compound in (19-17) was named "neo" (Greek *neos* means new) because it was discovered last among the C_5H_{12} compounds. The hexanes

Table 19-4 *Straight-Chain Alkanes*

No. Carbons	Condensed Formula		Name
1	CH_4	CH_4	methane
2	C_2H_6	CH_3-CH_3	ethane
3	$n\text{-}C_3H_8$	$CH_3-CH_2-CH_3$	propane
4	$n\text{-}C_4H_{10}$	$CH_3-(CH_2)_2-CH_3$	butane
5	$n\text{-}C_5H_{12}$	$CH_3-(CH_2)_3-CH_3$	pentane
6	$n\text{-}C_6H_{14}$	$CH_3-(CH_2)_4-CH_3$	hexane
7	$n\text{-}C_7H_{16}$	$CH_3-(CH_2)_5-CH_3$	heptane
8	$n\text{-}C_8H_{18}$	$CH_3-(CH_2)_6-CH_3$	octane
9	$n\text{-}C_9H_{20}$	$CH_3-(CH_2)_7-CH_3$	nonane
10	$n\text{-}C_{10}H_{22}$	$CH_3-(CH_2)_8-CH_3$	decane

Table 19-5 IUPAC Rules for Alkane Nomenclature

1. *Longest chain:* Alkanes are named as derivatives of the longest continuous carbon chain that can be found in the molecule.

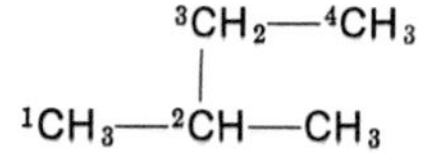

$^1CH_3-^2CH-CH_3$ with $^3CH_2-^4CH_3$
a substituted *butane*

NOT

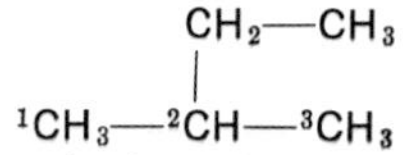

$^1CH_3-^2CH-^3CH_3$ with CH_2-CH_3
a substituted *propane*

2. *Lowest numbers:* The chain is assigned numbers in such a way that the sum of the numbers of the substituent groups is as low as possible.

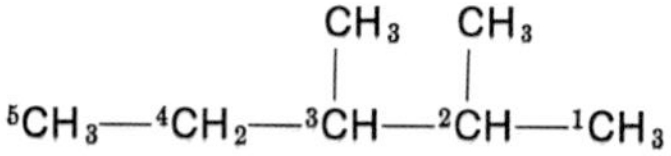

$^5CH_3-^4CH_2-^3CH-^2CH-^1CH_3$ with CH_3 CH_3
2,3-dimethylpentane
(sum = 5)

NOT

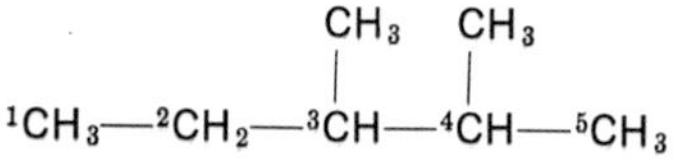

$^1CH_3-^2CH_2-^3CH-^4CH-^5CH_3$ with CH_3 CH_3
3,4-dimethylpentane
(sum = 7)

3. *Two identical substituents* at one position are each supplied with a number.

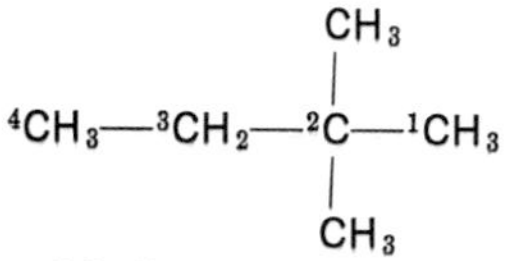

$^4CH_3-^3CH_2-^2C-^1CH_3$ with CH_3 (above) and CH_3 (below)
2,2-dimethylbutane

NOT

2-dimethylbutane

4. *Numbering of substituents* starts at the point of attachment.

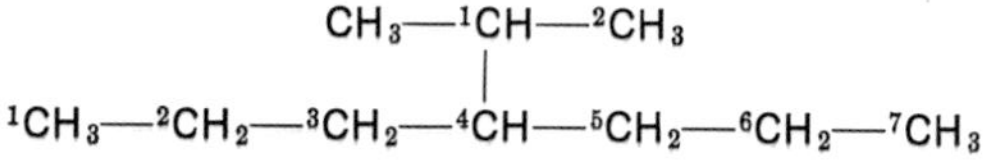

$^1CH_3-^2CH_2-^3CH_2-^4CH-^5CH_2-^6CH_2-^7CH_3$ with $CH_3-^1CH-^2CH_3$
4-(1-methylethyl)-heptane
or
4-isopropylheptane

5. *Two or more substituents* are named in alphabetical order.

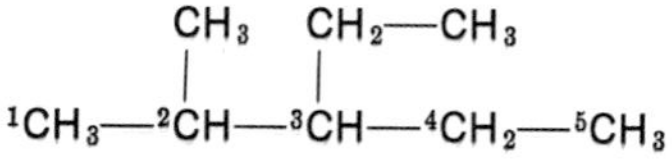

$^1CH_3-^2CH-^3CH-^4CH_2-^5CH_3$ with CH_3 and CH_2-CH_3
3-ethyl-2-methylpentane
or
3-isopropylpentane

6. *Punctuation:* (a) words are run together; (b) dashes separate numbers and words; (c) commas separate numbers and words; (d) parentheses may enclose side chain name.

have five isomers, $C_{10}H_{22}$ has 75, and $C_{30}H_{62}$ has 4,111,846,763. Clearly, assigning names such as normal, iso, and neo would not work well for $C_{30}H_{62}$. The IUPAC system brings order out of this chaos. The formal rules for naming alkanes are summarized in Table 19-5.

The names of the substituent groups are derived from the name of the parent alkane by replacing the -ane suffix with -yl. IUPAC approved names for some familiar substituent groups are listed in Table 19-6. It should be noted that many of these are common names—a result stemming from habit and tradition. Formal IUPAC names also can be given to these groups and be considered equally correct. In Table 19-5 both names are given if there is a choice. By way of summary, IUPAC names for the nine heptane isomers are given in Table 19-7.

Exercise 19-3.   Write down the carbon skeletons and IUPAC names of the five hexane isomers, formula C_6H_{14}.

The names of organic compounds containing other functional groups are usually derived from the name of the parent alkane. Table 19-8 sum-

Table 19-6 *IUPAC Approved Names for Simple Substituent Groups*

C_1	CH_3-	methyl
C_2	CH_3CH_2-	ethyl
C_3	$CH_3CH_2CH_2-$ $CH_3-CH-CH_3$ $\vert$	n-propyl isopropyl
C_4	$CH_3CH_2CH_2CH_2-$ CH_3 $\diagdown$ $\quad CH_2CH_2-$ $\diagup$ CH_3	n-butyl isobutyl
	CH_3CH_2 $\diagdown$ $\quad\quad CH-$ $\diagup$ CH_3	s-butyl (secondary or secbutyl)
	CH_3 $\vert$ CH_3-C- $\vert$ CH_3	t-butyl (tertiary or tertbutyl)

Note: The following distinctions are made between carbon atoms in a chain: primary, joined to one other carbon; secondary, joined to two other carbons; tertiary, joined to three other carbons. The distinction is made to show the nature of the carbon at which the substituent group is attached. Isopropyl might also be known as s-propyl by these rules.

Table 19-7 IUPAC Approved Names for the Nine Heptane Isomers

	Carbon Skeleton	Name
heptanes	C—C—C—C—C—C—C	heptane
hexanes	C | C—C—C—C—C—C	2-methylhexane
	C | C—C—C—C—C—C	3-methylhexane
pentanes	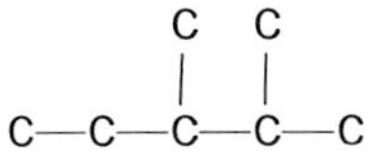 C C | | C—C—C—C—C	2,3-dimethylpentane
	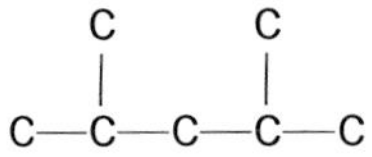 C C | | C—C—C—C—C	2,4-dimethylpentane
	C | C—C—C—C—C | C	2,2-dimethylpentane
	C | C—C—C—C—C | C	3,3-dimethylpentane
	C—C—C—C—C | C—C	3-ethylpentane
butanes	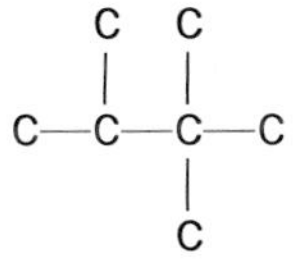 C C | | C—C—C—C | C	2,2,3-trimethylbutane

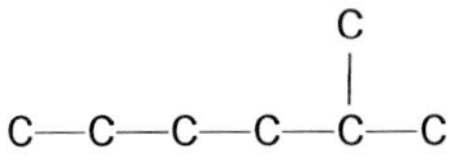

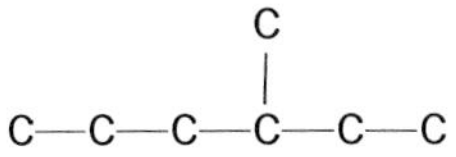

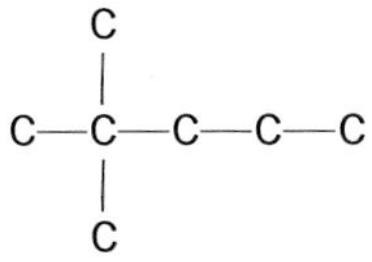

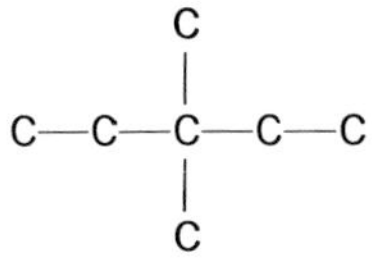

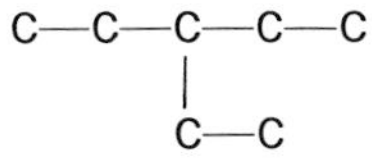

marizes the important functional groups and gives the naming rule for each. This table includes neither all functional groups nor all the information necessary for correct naming. Throughout this chapter we will introduce the name conventions where necessary and also assign common names where they are traditionally used. For example, propanone has already been discussed under its common name, acetone.

Table 19-8   *Common Functional Groups*

Name	Formula	IUPAC Suffix: Replace "-ane" by	Example	
alkane	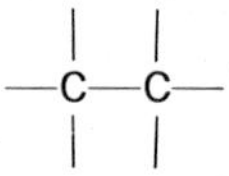 $-\overset{\displaystyle \mid}{\underset{\displaystyle \mid}{C}}-\overset{\displaystyle \mid}{\underset{\displaystyle \mid}{C}}-$	—	CH_3CH_3	ethane
alkene	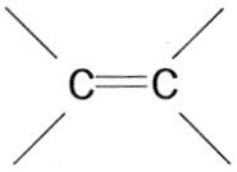 $\overset{\diagdown}{\diagup}C{=}C\overset{\diagup}{\diagdown}$	-ene	$CH_2{=}CH_2$	ethene (ethylene)
alkyne	$-C{\equiv}C-$	-yne	$CH{\equiv}CH$	ethyne (acetylene)
alcohol	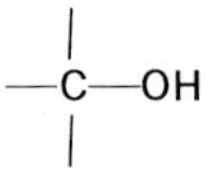 $-\overset{\displaystyle \mid}{\underset{\displaystyle \mid}{C}}{-}OH$	-anol	CH_3CH_2OH	ethanol
aldehyde	$\overset{\displaystyle O}{\overset{\displaystyle \|}{\underset{C\quad H}{C}}}$	-al	CH_3COH	ethanal (acetaldehyde)
ketone	$\overset{\displaystyle O}{\overset{\displaystyle \|}{\underset{C\quad C}{C}}}$	-anone	$\overset{O}{\overset{\|}{CH_3CCH_3}}$	propanone (acetone)
acid	$\overset{\displaystyle O}{\overset{\displaystyle \|}{{-}C{-}OH}}$	-anoic acid	$\overset{O}{\overset{\|}{CH_3C{-}OH}}$	ethanoic acid (acetic acid)
amine	$-NH_2$	substituted amine	$CH_3CH_2NH_2$	ethyl amine
halide	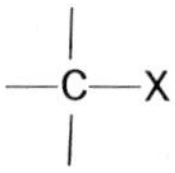 $-\overset{\displaystyle \mid}{\underset{\displaystyle \mid}{C}}{-}X$	halo derivative	CH_3CH_2Cl	chloroethane
ether	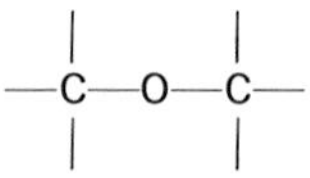 $-\overset{\displaystyle \mid}{\underset{\displaystyle \mid}{C}}{-}O{-}\overset{\displaystyle \mid}{\underset{\displaystyle \mid}{C}}-$	substituted ether	$CH_3CH_2{-}O{-}CH_3$	ethyl-methyl-ether

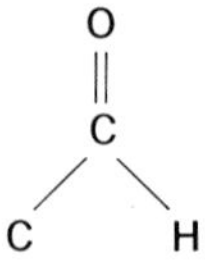

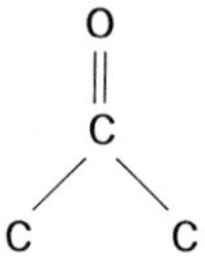

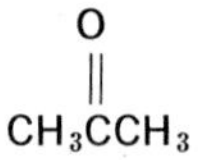

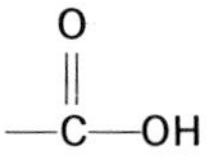

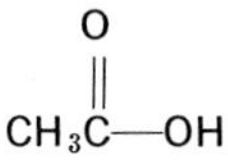

19-2 Hydrocarbons

Millions of years ago, in the primeval jungles, the decomposition of plant life began the process that over the ages has filled the cracks and fissures of the earth's crust with natural gas and the rich black goo known as petroleum.* These substances are the lifeblood of our industrial twentieth century. They provide fuels to heat our houses and to run our engines and raw materials for our plastics, synthetic fabrics, and drugs.

Petroleum consists of a solution of hundreds of hydrocarbons—alkanes, cyclic alkanes, alkenes and benzene derivatives ("aromatics"). It also contains small amounts of sulfur-containing hydrocarbons, which ultimately contribute to air pollution as sulfur dioxide, SO_2. Industrially, this solution (crude oil) is separated by distillation into "fractions" boiling over different temperature ranges. In terms of their familiar names, these are: (1) gas fraction, including natural gas (C_1 to C_5 compounds, boiling range to 40°C); (2) gasoline (C_6 to C_{10} compounds, boiling range 40° to 180°C, over 100 compounds in all); (3) kerosene, for heating and jet fuels (C_{11}, C_{12} compounds, boiling range 180° to 230°C); (4) various furnace oils and diesel fuels (C_{13} to C_{17}, boiling range 230° to 305°C); (5) heavy oils and lubricants (C_{18} to C_{25}, boiling range 305° to 405°C); and finally (6) paraffin waxes and vaseline (C_{26} to C_{38}, boiling range 405° to 515°C). The residues are known as asphalts.

Because petroleum is at the heart of the industrial use and manufacture of organic materials, it is appropriate to begin our study of organic chemistry with a brief look at some important hydrocarbons—their structures and reactions. First, however, we have to come to grips with the artistic problem of how to draw three-dimensional organic molecules on a two-dimensional page. In the process we will learn about the spatial arrangements of atoms in typical hydrocarbons.

(a) REPRESENTATIONS OF ORGANIC STRUCTURES

In discussing organic compounds, there is always a problem of representing three-dimensional structures with two-dimensional drawings. Information about the bonds formed—but not the molecular geometry— is given in the *condensed formulas* we have used so far in this chapter. For example, we write ethylene (ethene), propane, and isobutane (2-methyl propane) as follows:

ethylene: CH_2CH_2 or $CH_2{=}CH_2$
propane: $CH_3CH_2CH_3$

$$\begin{array}{c} CH_3 \\ | \\ \text{isobutane:} \quad CH_3CHCH_3 \quad \text{or} \quad CH_3CH(CH_3)CH_3 \end{array}$$

---

* More recently, petroleum has been found with increasing frequency on the surfaces of our oceans and beaches!

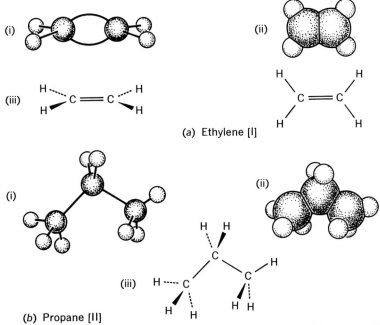

(a) Ethylene [I]

(b) Propane [II]

Figure 19-6 Representations of organic structures of (a) ethylene (b) propane. (i) Ball-and-stick model, (ii) space-filling model, (iii) dotted-line–wedge model: Solid bonds are in plane of paper, dotted bonds behind and tapered bonds come out of plane.

Considerable latitude is allowed in a condensed formula as long as it is perfectly clear which atoms or groups are bonded to which.

Various three-dimensional representations of ethylene and propane are compared in Figure 19-6. Model kits are often used for building three-dimensional models of organic molecules.* Two of the most popular types are the ball-and-stick (i) and space-filling varieties (ii) shown in the figure. The ball-and-stick structure of ethylene illustrates another representation of the double bond. With springs instead of sticks, this gives a practical way to use the tetrahedral holes on a carbon atom model for double and triple bonds as well as single bonds. The dotted-line–wedge representation is the most useful for textual illustration and will be used whenever the three-dimensional arrangement of atoms is essential to the discussion.

So far we have not mentioned the relative positions of the hydrogens on alkane carbons. Figure 19-7(a) shows two extreme possibilities for ethane. A view down the carbon–carbon bond is represented in part (b). The conversion of one into the other is achieved by rotation about the carbon–carbon single bond. Any repulsive forces between hydrogens on opposite ends of the molecule will be lowest in the "staggered" configuration and greatest when the hydrogens are directly in front of one another.

---

* Relatively inexpensive model kits are available and are highly recommended for any student intending to take further courses in organic chemistry.

This latter form is called the "eclipsed" configuration. The presence of repulsive forces in the eclipsed configuration will provide barriers to completely free rotation about the bond. Figure 19-7(c) shows the variation in H–H repulsive energy on rotation of one methyl group relative to the other. The height of the energy barrier, about 3 kcal/mole, is low enough to make the rotation from one staggered configuration to another extremely rapid at room temperature.

The game changes when some of the hydrogens are replaced by bulkier groups. For example, replacement of one hydrogen on each ethane carbon by a methyl group gives us butane. Now we expect a much higher interaction energy when the two methyl groups are eclipsed. Figure 19-8(a) shows these energy barriers. It takes between 5 and 6 kcal/mole to push the methyl groups past each other, but only about 3.3 kcal/mole to push a methyl past a hydrogen. There are also two kinds of minima, the lowest energy corresponding to the methyl groups being as far apart as possible. Once again, we expect this molecule to spend more of its time in the positions of lowest energy. It has never been possible to isolate at room temperature butane species corresponding to one or the other conformation, so rotation about the central C–C bond is still extremely rapid.

(b) ALKANES

Of all the reactions of organic molecules, one of the most important is that of the alkanes with oxygen—combustion. In the presence of excess oxygen, combustion results in complete conversion to carbon dioxide and

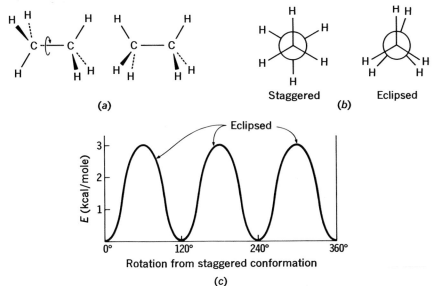

(a)

Staggered (b) Eclipsed

Eclipsed

(c)

Figure 19-7 Conformations of ethane.

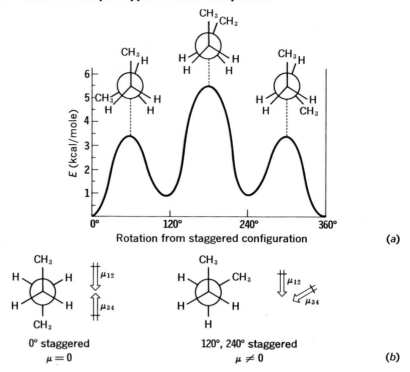

Figure 19-8 *Conformations of butane. (a) Energy barriers to rotation about C_2-C_3 bond. (b) Dipole moments in staggered configurations.*

water. Incomplete oxidation in internal combustion engines is one of the principal sources of hydrocarbon pollutants. The reaction with oxygen is, of course, an oxidation–reduction reaction. Heats of combustion are easily calculated from tabulated heats of formation. For example, for methane

$$\Delta H^0_{\text{comb.}} = \Delta H^0_f(CO_2(g)) + 2\Delta H^0_f(H_2O(g)) - \Delta H^0_f(CH_4(g)) \qquad (19\text{-}18)$$

$$= \quad (-94.1) \quad + \quad 2(-57.8) \quad - \quad (-17.9)$$

$$= -191.8 \text{ kcal/mole}$$

Conversely, measured heats of combustion are routinely used to determine ΔH^0_f's of organic molecules (see Section 7-4(c)), but notice that the heats of combustion quoted in Table 7-6 refer to liquid water as a combustion product).

Estimates of heats of combustion can be based upon average bond energies listed in Table 7-3. For example, let's work out the *extra* heat of combustion that would be found in heptane over hexane—that is, $\Delta H^0_{\text{comb.}}$ due to the extra CH_2-group. When hexane becomes heptane we gain two

Table 19-9 Enthalpies of Combustion of Straight-Chain Alkanes Found in Gasoline

$$C_nH_{2n+2}(\ell) + \frac{(3n+1)}{2} O_2(g) \longrightarrow n\, CO_2(g) + (n+1)H_2O(g)$$

Compound	$\Delta H^0_{comb.}$ (kcal/mole)	Difference
hexane	−928.9	
		147
heptane	−1075.9	
		147
octane	−1222.8	
		147
nonane	−1369.7	
		147
decane	−1516.6	
		147
undecane	−1663.5	

more C–H bonds and another C–C bond as well. We need to consider the combustion of the circled fragment—including the extra C–C bond, $D_{CC} = 82.6$ kcal/mole.

$$CH_3-CH_2-\overset{\frown}{\left(CH_2\right)}CH_2-CH_2-CH_2-CH_3$$

$$-CH_2 + \tfrac{3}{2} O_2(g) \longrightarrow CO_2(g) + H_2O(g)$$

$$\Delta H^0_{comb.} = D_{CC} + 2D_{CH} + \tfrac{3}{2}D_{O=O} - \Delta H_{atom}(CO_2) - \Delta H_{atom}(H_2O)$$

$$= 82.6 + 2(99) + \tfrac{3}{2}(118) - 384.1 - 221.1$$

$$\Delta H^0_{comb.} \cong -148 \text{ kcal/CH}_2 \text{ group} \tag{19-19}$$

We expect, then, that the lengthening of an alkane chain by one CH_2-group should make the combustion reaction about 148 kcal/mole more exothermic. Table 19-9 lists the measured heats of combustion of the straight-chain alkanes from C_6 to C_{11}—these are the principal components of gasoline. The third column shows the differences between these values—in every case essentially constant and in good agreement with our calculated value.

Exercise 19-4. Calculate the heat of combustion of ethane from average bond energies. The experimental value is -341.4 kcal/mole.

The use of halogenated alkanes in the synthesis of alcohols, amines, alkenes, and alkynes was touched on in the last section. Several chlorine compounds are important in their own right: Chloroform ($HCCl_3$) is an anesthetic; carbon tetrachloride (CCl_4) is a cleaning solvent and a fire-extinguishing fluid; trichloroethylene ($CHCl=CCl_2$) is a degreasing solvent.

Fluorine attacks hydrocarbons in violent reactions that produce mixtures of mono- and polyfluorinated alkanes. The reactions are highly exothermic and spontaneous on contact. For example,

$$CH_4(g) + F_2(g) \longrightarrow CH_2F_2(g) + H_2(g) \qquad (19\text{-}20)$$

methane   difluoromethane

$\Delta H^0 = -88.9$ kcal/mole

$\Delta G^0 = -88.1$ kcal/mole

Ease of fluorination is aided by the weak fluorine–fluorine bond and the strong carbon–fluorine bonds that result.

The reaction of chlorine with methane, on the other hand, is not so enthusiastic.

$$CH_4(g) + Cl_2(g) \longrightarrow CH_3Cl(g) + HCl(g) \qquad (19\text{-}21)$$

methane   chloromethane

$\Delta H^0 = -23.5$ kcal/mole

$\Delta G^0 = -24.5$ kcal/mole

$K = 9.3 \times 10^{17}$

Despite the large equilibrium constant, no reaction occurs when methane and chlorine are mixed at room temperature in a dimly lighted room. If irradiated with intense violet or ultraviolet light, however, reaction is immediate and often explosive. Clearly there is an activation energy problem which can be overcome with the addition of some light energy. Two possible first steps can be imagined, the breaking of a carbon–hydrogen bond to produce hydrogen atoms and methyl radicals

$$CH_3\text{--}H \xrightarrow{\text{light}} CH_3 + H$$

$\Delta H^0 = 102$ kcal/mole*

or the breaking of a chlorine–chlorine bond

$$Cl_2(g) \longrightarrow 2\ Cl(g)$$

$\Delta H^0 = 57.1$ kcal/mole

We can easily find out which of the two reactions initiates the reaction by varying the energy of the light used for irradiation. Suppose we use a

---

* While the *average* CH bond energy in methane is 98.8 kcal/mole, the energies needed to remove each successive hydrogen differ markedly. The value quoted refers to the first dissociation, and is usually designated $D(CH_3\text{--}H)$. The other dissociation energies in methane are: $D(CH_2\text{--}H) = 87$ kcal/mole; $D(CH\text{--}H) = 125$ kcal/mole; $D(C\text{--}H) = 81$ kcal/mole.

mercury photolysis lamp enclosed in a pyrex jacket. The pyrex will filter out any light of wavelength shorter than 3200 Å, but will transmit the strong mercury emission line at 3650 Å. This will be the highest energy light available for reaction. When the reaction vessel is illuminated, the reaction occurs. Let's see how much energy we can hope to get, per quantum, at this wavelength. Expression (12-2) will help us here:

$$E = h\nu = hc/\lambda$$

$$= \frac{\left(9.54 \times 10^{-14}\ \frac{kcal}{mole}\ sec\right)(3.0 \times 10^{10}\ cm/sec)}{(3650 \times 10^{-8}\ cm)}$$

$$= 78.4\ kcal/mole$$

So we have enough energy to dissociate chlorine but not enough to disturb the C–H bond in methane. Chlorine atoms must be sufficient to initiate reaction. A possible mechanism would be

$$Cl_2 \xrightarrow{light} 2\ Cl \qquad \Delta H^0 = 57.1\ kcal/mole \qquad (19\text{-}22a)$$

$$Cl + CH_4 \longrightarrow CH_3 + HCl \qquad \Delta H^0 = -1\ kcal/mole \qquad (19\text{-}22b)$$

$$CH_3 + Cl_2 \longrightarrow CH_3Cl + Cl \qquad \Delta H^0 = -23\ kcal/mole \qquad (19\text{-}22c)$$

Notice that one chlorine atom, photochemically produced, results in the formation of one molecule of product *and* another chlorine atom in this three-step process. This is an example of a *chain reaction*, like the H_2–Cl_2 reaction discussed in Section 11-4(e).

Experimental data are available for the chain propagation steps, (19-22b) and (19-22c). For (19-22b),

$$-\frac{\Delta[CH_4]}{\Delta t} = k_1[CH_4][Cl]$$

$$k_1 = (5 \times 10^{10})\ e^{-3900/RT}\ cc/mole\text{-sec} \qquad (19\text{-}23)$$

and for reaction (19-22c)

$$-\frac{\Delta[Cl_2]}{\Delta t} = k_2[CH_3][Cl_2]$$

$$k_2 = (8 \times 10^9)\ e^{-2300/RT}\ cc/mole\text{-sec} \qquad (19\text{-}24)$$

We see that the activation energies, 3.9 and 2.3 kcal/mole, respectively, are both very small and the chain reaction, once started, will proceed rapidly. Chain lengths—product molecules produced per chlorine atom— as large as 10^6 have been observed in practice.

(c) ALKENES: SOURCES AND STRUCTURES

In terms of consumption, ethylene, C_2H_4, is probably the most important petrochemical produced in a highly industrialized country. In 1964, for example, over eight billion pounds were produced in the U.S. alone. It is used for the production of polyethylene (34%), ethylene oxide (26%)— which is used in turn to make ethylene glycol, the principal constituent of antifreeze—and ethanol (16%). Ethylene is produced commercially from gas fraction alkanes by the process known as "cracking." For example, if ethane is heated to 500–600°C, fragmentation results and radical propagated chain reactions occur. A few of the possible reactions occurring in the cracking of ethane might be

$$\text{initiation} \quad CH_3CH_3 \underset{heat}{\overset{\displaystyle CH_3 + CH_3}{\diagup\diagdown}} C_2H_5 + H \tag{19-25}$$

$$\begin{array}{ll}
\text{chain} & CH_3 + C_2H_6 \longrightarrow CH_4 + C_2H_5 & (19\text{-}25a)\\
\text{propagation} & C_2H_5 \longrightarrow C_2H_4 + H & (19\text{-}25b)\\
\text{reactions} & H + C_2H_6 \longrightarrow C_2H_5 + H_2 & (19\text{-}25c)
\end{array}$$

$$\begin{array}{ll}
\text{chain} & H + C_2H_5 \longrightarrow C_2H_6 & (19\text{-}25d)\\
\text{termination} & \\
\text{reactions} & CH_3 + H \longrightarrow CH_4 & (19\text{-}25e)
\end{array}$$

Notice that methane is produced as well as ethylene. Cracking of long-chain petroleum hydrocarbons is one of the principal sources of the lower carbon compounds.

A convenient laboratory synthesis of alkenes was displayed in Table 19-2—through removal of H_2O from an appropriate alcohol. This dehydration reaction can be caused by heating the alcohol with concentrated sulfuric acid. Primary alcohols give only one product. Secondary and tertiary alcohols can give a mixture, as shown in reactions (19-26) and (19-27).

$$CH_3CH_2CH_2CH_2OH \xrightarrow[180°C]{\text{conc. } H_2SO_4} CH_3CH_2CH{=}CH_2 + H_2O \tag{19-26}$$

$$CH_3-\underset{\underset{CH_3}{|}}{\overset{\overset{OH}{|}}{C}}-CH_2CH_3 \xrightarrow[180°C]{\text{conc. } H_2SO_4}
\begin{cases}
CH_3-\underset{\underset{CH_3}{|}}{\overset{\overset{CH_3}{|}}{C}}{=}CHCH_3 + H_2O \ (\sim 85\%) & (19\text{-}27a)\\[1.5em]
CH_2{=}\underset{\underset{CH_3}{|}}{C}-CH_2CH_3 + H_2O \ (\sim 15\%) & (19\text{-}27b)
\end{cases}$$

Alkenes, with two fewer hydrogens than the corresponding alkane, have the general formula C_nH_{2n}. They are named after the longest carbon chain *containing* the double bond, the position of which is indicated by the number of the lower-numbered carbon atom in the double bond. For example, let's consider the five-carbon alkene, C_5H_{10}. There are two possibilities for placement of the double bond in a five-carbon chain. They are

$$CH_2{=}CH{-}CH_2{-}CH_2{-}CH_3 \qquad CH_3{-}CH{=}CH{-}CH_2{-}CH_3$$
1-pentene   2-pentene

These two compounds are examples of "structural isomers." They have identical chemical formulas, but their structures are different. But wait—branching of the carbon skeleton is possible in pentanes—more structural isomers can be drawn. 1-Pentene gives rise to two new isomers and 2-pentene to one,

CH_3 CH_3 CH_3

$CH_2{=}C{-}CH_2{-}CH_3$ $CH_2{=}CH{-}CH{-}CH_3$ $CH_3{-}CH{=}C{-}CH_3$

2-methyl-1-butene   3-methyl-1-butene   2-methyl-2-butene

The story has not ended, however. If 2-pentene is carefully distilled, *two* compounds boiling about 0.6° apart can be separated. This suggests that, unlike the situation in the alkanes, rotation about the carbon–carbon double bond is prohibited. If so, two isomeric 2-pentenes are possible

H   H CH_3 H

C=C   C=C

CH_3 C_2H_5 H C_2H_5

cis-2-pentene   *trans*-2-pentene

b.p. = 36.9°C   b.p. = 36.3°C

Any alkene in which neither doubly bonded carbon has two identical groups attached can be resolved into "geometric isomers" of this sort. The isomers are labeled "*cis*" if the substituent groups are on the same side of the double bond and '*trans*" if they are on the opposite side.

More dramatic than the boiling point differences, however, are dipole moments in halogenated alkenes. The easily measured dipole moment quickly identifies the various isomers. The two geometric isomers of 1,2-difluoroethene are shown in Figure 19-9(a). The observed dipole moments are

$$\mu_{cis} = 2.42 \text{ D}$$
$$\mu_{trans} = 0.00 \text{ D}$$

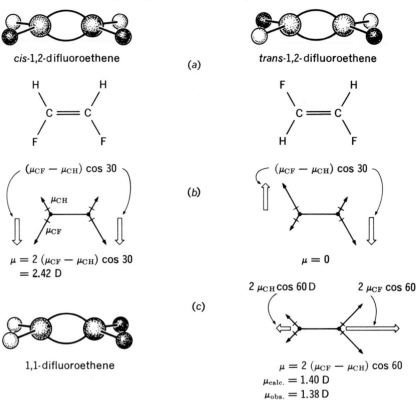

Figure 19-9   Isomers and dipole moments of difluoroethene.

These are consistent with the vector model illustrated in part (*b*) of the figure. The *cis* dipole moment is the resultant of the individual CF and CH bond dipoles resolved along the direction perpendicular to the carbon–carbon bond in the molecular plane. It can be expressed as

$$\mu_{cis} = 2(\mu_{CF} - \mu_{CH}) \cos 30$$

assuming perfect sp² hybridization at the carbon atoms. In the *trans* species the resultant dipoles cancel in all directions. The experimental results confirm this prediction. A zero dipole moment has been measured for this molecule.

There is one more isomer of $C_2H_2F_2$, 1,1-difluoroethene, illustrated in Figure 19-9(*c*). In this molecule the resultant dipole lies along the carbon–carbon bond and may be expressed as

$$\mu = 2(\mu_{CF} - \mu_{CH}) \cos 60$$

We can calculate a value of $(\mu_{CF} - \mu_{CH})$ from the experimental dipole of the *cis* species:

$$(\mu_{CF} - \mu_{CH}) = \mu_{cis}/2 \cos 30$$

$$= 2.42 \text{ D}/2 \cos 30$$

$$= 1.40 \text{ D}$$

Now we can substitute this into the expression for the 1,1-difluoro species

$$\mu_{calc.} = 2(1.40) \cos 60$$

$$= 1.40 \text{ D}$$

This is in excellent agreement with the observed dipole moment, 1.38 D.
    At room temperature, *cis-trans* interconversion of alkenes does not occur at a measurable rate. Isomerization does occur, however, if one isomer is illuminated with light. For example, *trans*-2-butene is converted into a 50–50 mixture of *cis* and *trans* isomers if ultraviolet light is used for illumination.

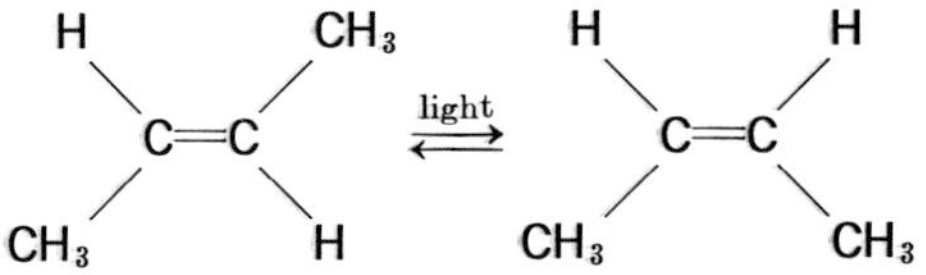

trans-2-butene   *cis*-2-butene

This is easily understood in terms of the molecular orbitals involved in bonding the carbon atoms together. Each doubly bonded carbon forms three sigma bonds using sp^2 hybrid orbitals. Two are to the substituent groups and can be ignored. The third sigma bond is to the other carbon. In addition, the p_z orbitals form the carbon–carbon pi bond. The electron configuration is thus $\sigma^2\pi^2$ and the bond order is two. If one electron is excited by light absorption, that electron must go into the next lowest orbital—the pi antibonding orbital. Thus the excited state obtained from either *cis* or *trans* isomer has the configuration $\sigma^2\pi^1\pi^{*1}$ and a formal *single bond*. Rapid rotation is possible about carbon–carbon single bonds so that the ground state molecule resulting from decay of the singly bonded excited state should be a mixture of isomers—as it is. This process is termed photochemical isomerization and is illustrated in Figure 19-10.

(d) ALKENES: REACTIONS

In the absence of light, alkanes do not react with halogens. Alkenes, on the other hand, react rapidly with bromine in solution to yield dibromo-

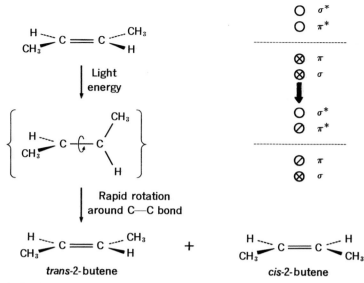

Figure 19-10   *Photochemical isomerization of trans-2-butene.*

alkanes. For example,

$$CH_2{=}CH_2 + Br_2 \rightleftarrows CH_2Br{-}CH_2Br \qquad (19\text{-}28)$$

$$\Delta H^0 = -32 \text{ kcal/mole}$$

$$\Delta G^0 = -21$$

The ease of reaction is no doubt due to the fact that a relatively weak pi bond is being broken while new σ bonds are being formed.

$$D(\text{C–C pi bond}) \approx D_{\text{C=C}} - D_{\text{C–C}}$$

$$= 146 - 83$$

$$= 63 \text{ kcal/mole}$$

Many other reagents add to double bonds. Table 19-10 summarizes some of the important addition reactions of alkenes.

In this section we will look in detail at alkene addition reactions. We will want to know about the mechanisms of the reactions—and this will include a knowledge of the exact geometries of the reactants, products, and hopefully, intermediates. A useful way of accomplishing this is to use a compound that will produce a rigid product. Since the product is always a substituted alkane, this will be the case only if we use cyclic

Table 19-10   Addition Reactions of Double Bonds

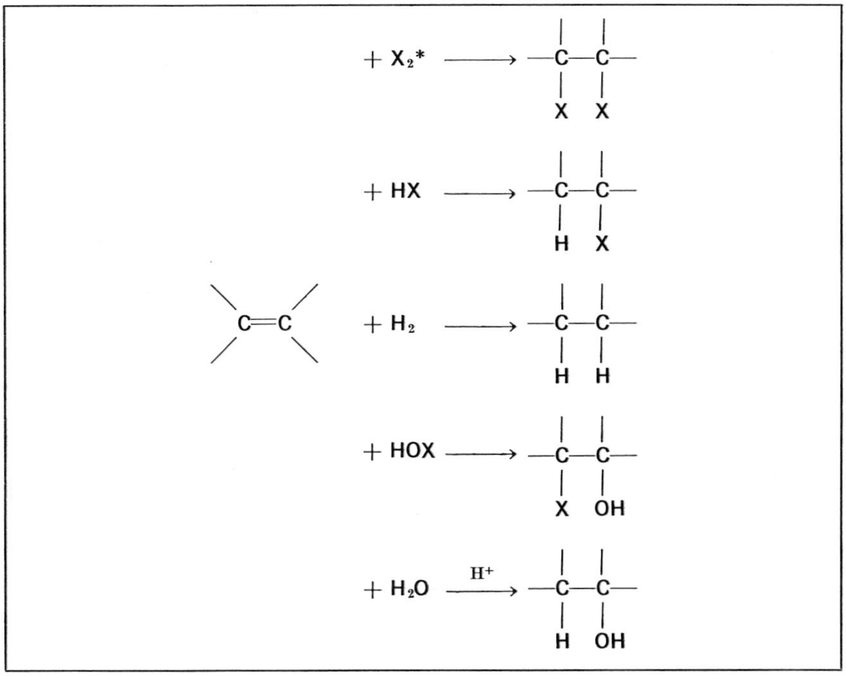

$$\diagdown C = C \diagup$$

$$+ X_2{}^* \longrightarrow -\underset{\underset{X}{|}}{C}-\underset{\underset{X}{|}}{C}-$$

$$+ HX \longrightarrow -\underset{\underset{H}{|}}{C}-\underset{\underset{X}{|}}{C}-$$

$$+ H_2 \longrightarrow -\underset{\underset{H}{|}}{C}-\underset{\underset{H}{|}}{C}-$$

$$+ HOX \longrightarrow -\underset{\underset{X}{|}}{C}-\underset{\underset{OH}{|}}{C}-$$

$$+ H_2O \xrightarrow{\ H^+\ } -\underset{\underset{H}{|}}{C}-\underset{\underset{OH}{|}}{C}-$$

*X = F, Cl, Br, or I.

compounds. The ring structure prevents the free rotation about carbon–carbon bonds that occurs in open-chain alkanes.

We can synthesize a dandy compound using yet another alkene addition reaction—the addition of 1,3-butadiene to ethylene results in the cyclic alkene, cyclohexene

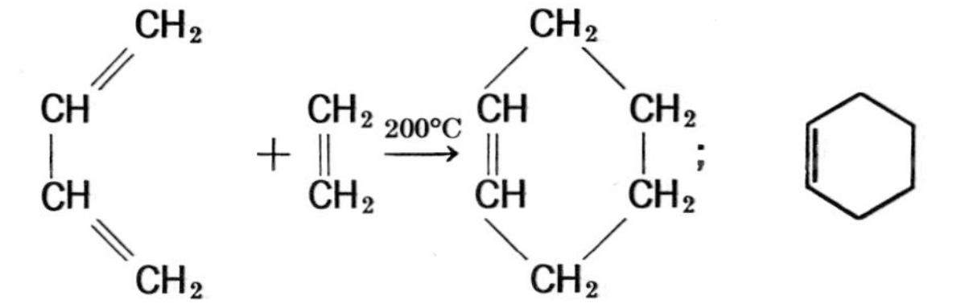

$$(19\text{-}29)$$

1,3-butadiene   cyclohexene   cyclohexene

An abbreviated representation of the ring compound is illustrated at the right in (19-29). Each vertex is a carbon atom and the requisite numbers of hydrogens at each carbon are assumed to be present, though not shown.

On reaction with bromine two geometric isomers of 1,2-dibromohexane

are possible, one with bromine atoms on the same side of the ring (*cis*) and one with bromine atoms on opposite sides (*trans*).

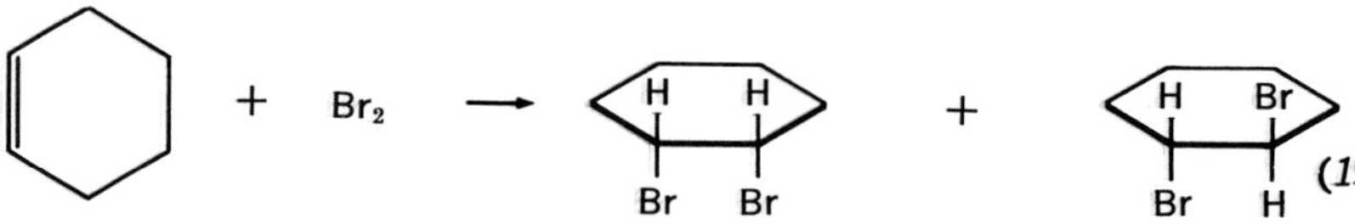

$$\text{(19-30)}$$

cis-1,2-dibromohexane   trans-1,2-dibromohexane

   If a bromine molecule adds across the double bond in a one-step process, the linked bromine atoms must approach on the same side. Then the product would be entirely in the *cis* form. If a two-step mechanism is involved, in which first one and then another bromine adds on, we would expect equal quantities of each isomer. In fact, *only the trans isomer is ever found.* Clearly a two-step mechanism is involved, and the intermediate monobromo species must somehow prohibit attack on the same side by the second bromine.

   Addition of HX to an unsymmetrically substituted double bond also could result in two products. However, under normal conditions, only one is ever found. For example, hydrobromination of propene gives only the 2-bromopropane isomer.

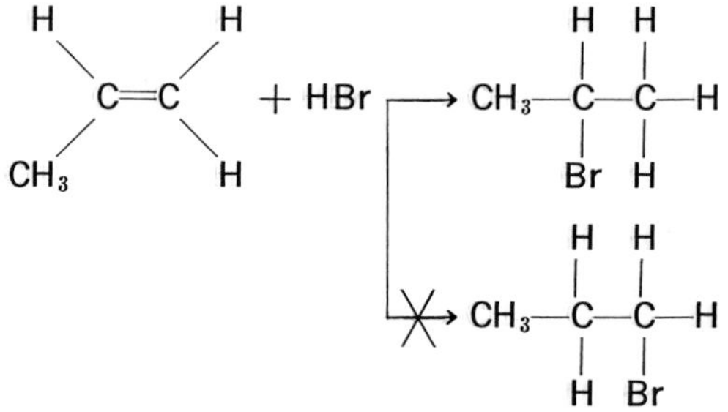

$$\text{(19-31)}$$

This result can be stated as a useful rule-of-thumb, first formulated by Markovnikov in 1870: "In addition of HX to a double bond, the hydrogen goes to the carbon of the double bond that initially had the greater number of hydrogens."

   Unlike the halogens and hydrogen halides, hydrogen does not react with a double bond unless a catalyst is present. Finely divided platinum, nickel, or palladium will do. Once again only one isomer is formed—but this time, the addition is exclusively *cis*.

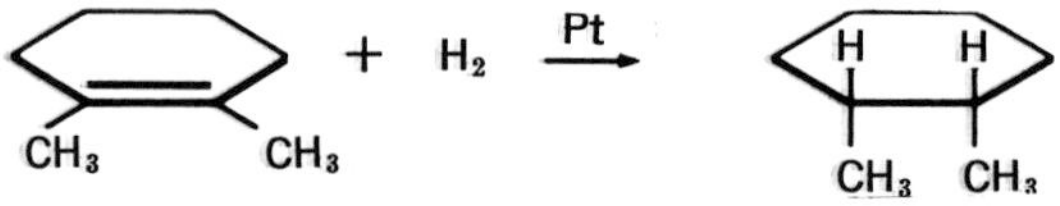

$$\text{(19-32)}$$

To understand this catalyzed reaction we need to consider the interaction of both H_2 and alkenes with transition metals. For a specific example we will use the hydrogenation of ethylene with a platinum catalyst.

$$H_2C{=}CH_2 + H_2 \xrightarrow{\text{Pt}} CH_3{-}CH_3 \qquad (19\text{-}33)$$

Hydrogen is adsorbed on platinum with the release of a fair amount of heat—near 40 kcal/mole. Plainly, chemical bonds between platinum and hydrogen are formed. This is not unreasonable since atoms on the *surface* of the metal have valence orbitals that cannot participate in metallic bonding, leaving them free to bond to other atoms. The evidence is consistent with the formation of platinum hydrides on the surface—that is, with the actual dissociation of hydrogen on the metal.

$$H_2 + 2\ Pt(\text{surface}) \longrightarrow 2\ PtH(\text{surface}) \qquad \Delta H = -40\ \text{kcal} \qquad (19\text{-}34)$$

Obviously a platinum surface covered with hydrogen atoms is a super place for ethylene to pick up some extra H's. But we have to account also for the stereospecificity of the reaction. Only *cis* addition is allowed.

A reasonable explanation is that ethylene is also held firmly on the surface of the metal, leaving only one side open for attack by H atoms. In the absence of hydrogen, ethylene *is* adsorbed onto the platinum surface. The nature of the bond formed is revealed in some interesting monomeric platinum–alkene compounds, typified by $KPtCl_3(C_2H_4){\cdot}H_2O$, Zeise's Salt. The anion of this well-known substance is illustrated in Figure 19-11. The platinum is found bonded in a square planar arrangement to three chlorines and one ethylene molecule, which is perpendicular to the Pt–Cl plane. The interaction presumably involves an empty platinum orbital and the pi electrons in the ethylene double bond.

So now we have the basis for a complete picture of the hydrogenation process, as illustrated in Figure 19-12. An ethylene molecule is bonded to the surface of the platinum, amid a sea of hydrogen atoms. Hydrogenation then could occur only at one side, perhaps at the exposed upper side. The product, ethane, is less strongly adsorbed, so it quickly makes room

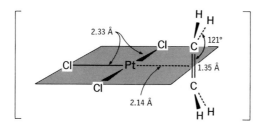

Figure 19-11   The structure of Zeise's Salt. The attached groups are arranged in a square planar array around the platinum atom. The ethylene molecule is perpendicular to the Pt–Cl plane or, possibly, rotating, propeller-wise.

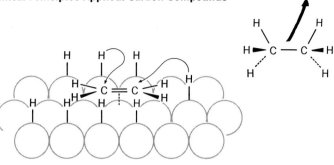

Figure 19-12 Platinum-catalyzed hydrogenation of ethylene.

for another ethylene molecule. Notice that the two hydrogens do not necessarily come from the same H_2 molecule. This is consistent with isotopic studies of the hydrogenation process.

Relative stabilities of isomeric alkenes can easily be deduced from measured enthalpies of hydrogenation. For example, the three butene isomers all give butane on addition of H_2, but the heats of hydrogenation differ.

$$H_2C{=}CH{-}CH_2{-}CH_3 \xrightarrow{\text{H}_2,\ \text{Pt}} CH_3{-}CH_2{-}CH_2{-}CH_3$$

1-butene   n-butane

$$\Delta H^0 = -30.3 \frac{\text{kcal}}{\text{mole}} \qquad (19\text{-}34a)$$

$$\underset{\substack{\diagup \qquad \diagdown \\ \text{CH}_3 \qquad\qquad \text{CH}_3}}{\text{CH}{=}\text{CH}} \xrightarrow{\text{H}_2,\ \text{Pt}} CH_3{-}CH_2{-}CH_2{-}CH_3$$

cis-2-butene   n-butane

$$\Delta H^0 = -28.6 \frac{\text{kcal}}{\text{mole}} \qquad (19\text{-}34b)$$

$$\underset{\substack{\diagup \\ \text{CH}_3}}{\overset{\substack{\text{CH}_3 \\ \diagup}}{\text{CH}{=}\text{CH}}} \xrightarrow{\text{H}_2,\ \text{Pt}} \underset{\text{n-butane}}{CH_3{-}CH_2{-}CH_2{-}CH_3}$$

trans-2-butene

$$\Delta H^0 = -27.6 \frac{\text{kcal}}{\text{mole}} \qquad (19\text{-}34c)$$

This information is put into the enthalpy plot of Figure 19-13. Although we do not know the total enthalpy of each compound, we do know their relative positions. 1-Butene is the least stable and, of the geometric isomers of 2-butene, the *trans* form is more stable than the *cis*. This is reasonable since the *cis* form holds both methyl groups on the same side of the double bond, keeping their mutual repulsions at a maximum.

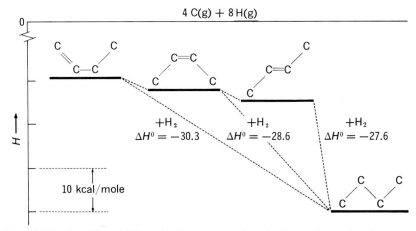

Figure 19-13 *Relative stabilities of butene isomers from hydrogenation enthalpies.*

Exercise 19-5. Calculate ΔH^0 for the isomerization of *cis*-2-butene to *trans*-2-butene to show that the reaction is exothermic by 1.0 kcal.

(e) CYCLIC ALKANES

The *cis-trans* isomers of cyclohexane gave us a useful handle on the stereochemistry of alkene addition reactions. In this section we will look in a little more detail at the structure of these cyclic compounds, of general formula $(CH_2)_n$. Since the bonding capacity of each carbon in such a compound is saturated, the bonding must be derived from sp^3 hybrid orbitals on carbon—which suggests CCC bond angles near 109°. A little thought shows that the smaller rings cannot possibly have bond angles this large. For example, angles of 60° and 90° are required in cyclopropane and (planar) cyclobutane.

cyclopropane

cyclobutane

Any distortion of cyclobutane from planar geometry will decrease the bond angles even further.

This distortion from sp^3 hybridization will raise the energy and should be manifested in lower stability than expected for strain-free rings. We can easily test this using measured heats of combustion. In Table 19-9 the average heat of combustion of a CH_2-group in a straight-chain alkane (all angles tetrahedral) was found to be -147 kcal/mole. Using this value we

can calculate the heat of combustion of a hypothetical strain-free cyclobutane molecule.

$$(CH_2)_4(g) + 6 O_2(g) \longrightarrow 4 CO_2(g) + 4 H_2O(g)$$

$$\Delta H_{calc.} = 4(\Delta H_{comb.} \text{ per } CH_2)$$

$$= 4(-147)$$

$$= -588 \text{ kcal/mole}$$

This can be compared to the observed value

$$\Delta H_{obs.} = -616 \text{ kcal/mole}$$

As illustrated in the enthalpy diagram, Figure 19-14, this corresponds to a
decrease in stability of 28 kcal/mole, which we can blame on strain in the
four-membered ring. Table 19-11 summarizes similar calculations of strain
energy in cyclic alkanes.

The instability of cyclopropane and cyclobutane due to ring strain shows
up in their chemistry. Both substances can be catalytically hydrogenated,
opening the rings, as though they were alkenes (at 120°C for cyclopropane
and 200°C for cyclobutane). Notice that the ring strain falls to zero at
cyclohexane. A few minutes with a model kit will convince you that it is
possible to build a strain-free C_6H_{12} molecule using tetrahedral carbon
atoms.

A model of cyclohexane shows a great deal of flexibility—many different
geometries are possible. Two extreme forms, known as the "boat" and
"chair" forms are drawn in Figure 19-15(a). Examination of the figure—or
better yet of a model—will show that the chair form has six hydrogens
roughly in the average plane defined by the carbon atoms, three more
pointing up, and three pointing down. The six hydrogens pointing out from

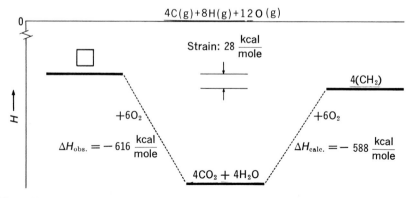

Figure 19-14 Strain destabilization of cyclobutane.

Table 19-11 Strain in Cyclic Alkanes

n	Molecule	$\Delta H^{obs.}_{comb.}$ * kcal/mole	$\Delta H^{obs.}_{comb.}$ per CH_2 kcal/mole	Total Strain† kcal/mole
3	cyclopropane	468	156	27
4	cyclobutane	616	154	28
5	cyclopentane	740	148	5
6	cyclohexane	882	147	0
7	cycloheptane	1036	148	7
8	cyclooctane	1184	148	8
∞	open-chain alkane		147	0

* For gaseous hydrocarbons to gaseous products.
† Strain $= \Delta H^{obs.}_{comb.} - n(147)$.

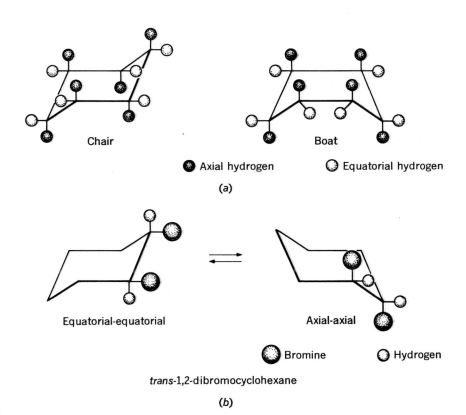

Chair

Boat

● Axial hydrogen   ◐ Equatorial hydrogen

(a)

Equatorial-equatorial   Axial-axial

◐ Bromine   ◐ Hydrogen

trans-1,2-dibromocyclohexane

(b)

Figure 19-15 (a) Cyclohexane: chair and boat conformations. (b) trans-1,2-dibromohexane: two conformations in the chair form (three are possible in the boat form).

the ring are called "equatorial" hydrogens and those pointing up and down are called axial hydrogens. The boat form also has six equatorial hydrogens and six axial, but in this conformation four point down and two up (or vice versa). Nuclear magnetic resonance experiments show that the two forms are in rapid equilibrium at room temperature. At lower temperatures the rate of interconversion is much slower.

Figure 19-15(*b*) shows two of several possible conformations of the *trans*-dibromo compound formed from bromination of cyclohexene. In the chair form, ring flexing switches the two bromine atoms back and forth between a diaxial and a diequatorial arrangement. All three conformations are possible in the boat form—including an equatorial-axial *trans*-1,2-dibromocyclohexane.

Exercise 19-6. Draw the three conformations of *trans*-1,2-dibromocyclohexane in the boat form.

Many important natural products—compounds occurring in living things—contain "fused" cycloalkane rings. These are molecules in which two or more cyclic alkanes share *two* carbon atoms. For example, two cyclohexane rings joined together make the compound decalin, $C_{10}H_{18}$.

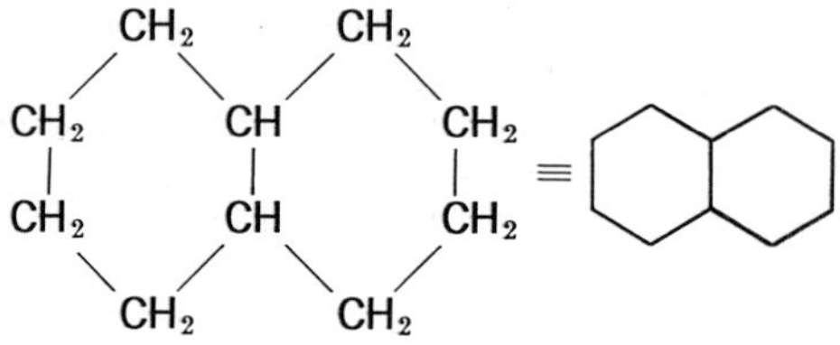

$$
\begin{array}{ccc}
& CH_2 & & CH_2 \\
CH_2 & & CH & & CH_2 \\
CH_2 & & CH & & CH_2 \\
& CH_2 & & CH_2 \\
\end{array}
$$

decalin

Examination of a model shows that the two hydrogens on the bridging carbons may be either *cis* or *trans* to one another as shown in Figure 19-16. The structural difference between the two isomers is profound—and may have great biological importance. For example, cholesterol is a substance that plays an important role in the functioning of the body. There are three

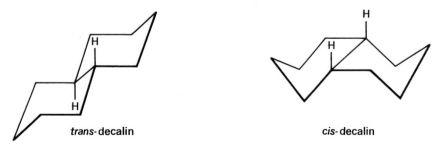

trans-decalin   *cis*-decalin

Figure 19-16 cis- and trans-decalin. Stereochemistry of the ring junction profoundly affects the molecular geometry.

ring junctions, but only the isomer shown below is found in the body. A synthetic substitute for cholesterol would have to have exactly identical geometry at the ring junctions to be biologically active.* (In this representation solid lines to substituents point up and dotted lines back from the plane.)

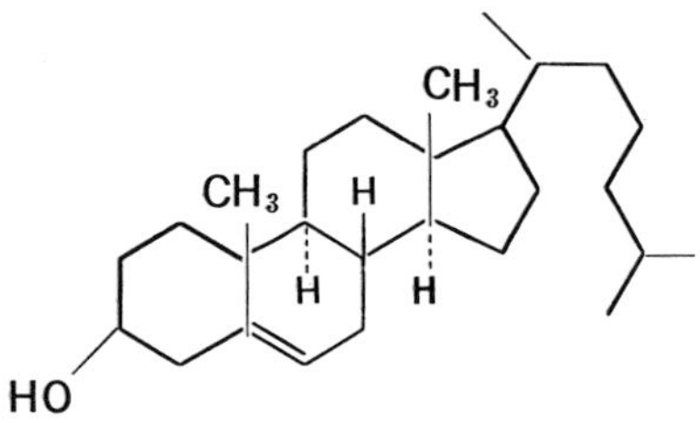

cholesterol

(f) BENZENE AND THE AROMATICS

Derivatives of the molecule benzene, C_6H_6, form an enormous family of compounds. Many of the first members of this family to be studied have strong characteristic odors—the essential components of the volatile oils of cloves, wintergreen, and vanilla, for example. While it is now known that the vast majority of benzene derivatives are neither volatile nor aromatic, the historical family name *aromatic compounds* is still generally used.

As an industrial chemical, benzene is second in importance only to ethylene. In 1966, the U.S. produced over 3.4 billion pounds, mostly for use in the manufacture of polymers. Of the benzene consumed, 40% was converted to styrene (polystyrene), 21% to cyclohexane (nylon and polyurethanes), and 21% to phenol (phenolic resins).

Very early in the game, benzene was found to have a ring structure. The formula is consistent with the compound cyclohexatriene, illustrated in Figure 19-17(a): a distorted hexagon with short alkene-like bonds alternating with longer alkane-like bonds.

This compound should have the chemistry of the alkenes. For example, cyclohexene adds bromine easily to give *trans*-1,2-dibromocyclohexane (19-30). Benzene, on the other hand, reacts reluctantly with bromine and then only in the presence of a metal halide catalyst. What's more, substitution rather than addition occurs.

$$C_6H_6 + Br_2 \xrightarrow{\text{FeCl}_3} C_6H_5Br + HBr \qquad\qquad (19\text{-}35)$$

---

* Actually this is only a small part of the picture. Cholesterol has *256* isomers—resulting from different configurations around the eight carbons that are bonded to four different groups. Only *one* of the 256 is found in the body!

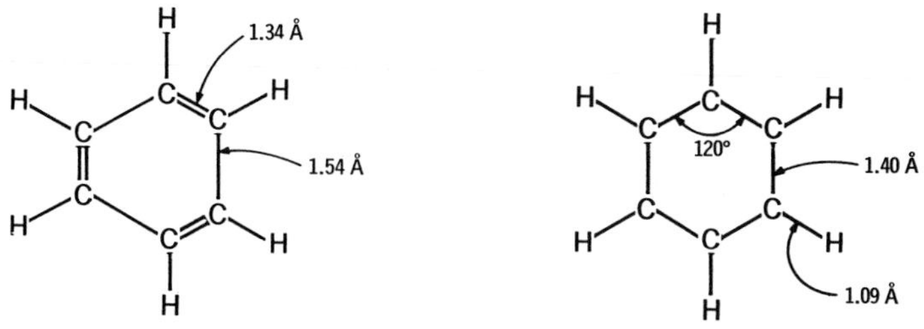

(a) Structure (predicted)
of cyclohexatriene

(b) Observed structure
of benzene

Figure 19-17 Comparison of cyclohexatriene and benzene.

The structure of benzene has been pictured in Figure 17-6. It is obviously not a simple triene, hence its chemical dissimilarity to alkenes. It is a planar molecule with identical carbon–carbon bonds (1.40 Å) and internal angles (120°). The carbon–carbon bond order was deduced to be $1\frac{2}{3}$ from Figure 17-5, a plot of bond order versus bond length for alkanes, alkenes, and alkynes. We must develop a picture of the bonding in this molecule that is consistent with these observations.

Since the bond angles around each carbon are 120°, sp² hybridization must be involved. According to the pigeonhole diagram, Figure 19-18(a), each carbon forms sigma bonds with its sp² orbitals to two other carbons and one hydrogen outside, but in the plane of, the ring. Each carbon now has one, half-filled $2p_z$ orbital sticking up perpendicular to the ring—just what is needed for pi bonding. Since the $2p_z$ orbitals can interact equally with those on either side, we expect the formation of molecular-type pi orbitals that extend over the *entire* benzene ring. A schematic representa-

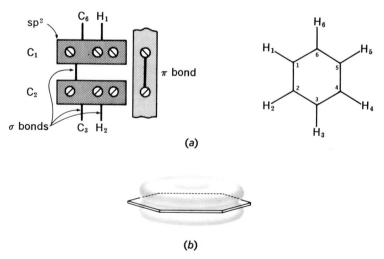

Figure 19-18 Bonding in benzene.

tion of the resulting pi-electron distribution is shown in part (*b*) of the figure. In molecular orbital language, since we start out with six atomic orbitals, we must expect to end up with six molecular orbitals. Detailed consideration shows that three of these are bonding and three antibonding and that all of them are molecular in scope. The six electrons just fill the three bonding M.O.'s, contributing half a bond order to each carbon–carbon bond. Thus the carbons are joined with one sp^2–sp^2 sigma bond and one half (on the average) of a pi bond, giving a resultant bond order of 1.5. This is in good agreement with the bond order predicted on the basis of bond lengths.

Before molecular orbital theory was developed, benzene was described as a resonance hybrid of the two cyclohexatriene structures (see (19-36)). Alternate representations such as (19-37) are sometimes used, but most often, one of the resonance structures is drawn with the understanding that it does not represent reality.

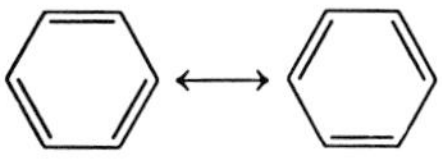

$$(19\text{-}36)$$

$$(19\text{-}37)$$

A favorable energy situation always results when electrons are delocalized over several atoms. On these grounds, benzene should be more stable than cyclohexatriene with its localized double bonds. This can be tested using reactions that result in identical products for both—we will compare enthalpies of hydrogenation. The average enthalpy of hydrogenation of a double bond is found to be -28.6 kcal/mole.

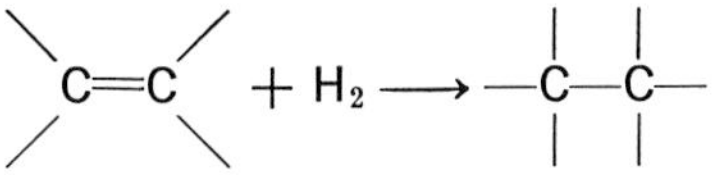

$$(19\text{-}38)$$

$\Delta H = -28.6$ kcal/mole

Cyclohexatriene, with three double bonds, should have a heat of hydrogenation that is three times this value.

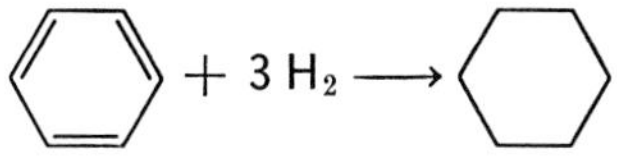

$$(19\text{-}39a)$$

$\Delta H_{\text{calc.}} = 3(-28.6)$

$\qquad = -85.8$ kcal/mole

The measured heat of hydrogenation of benzene is, however, much less than this

$$\langle\bigcirc\rangle + 3\,H_2 \longrightarrow \langle\ \rangle \tag{19-39b}$$

$$\Delta H_{\text{obs.}} = -49.8 \text{ kcal/mole}$$

The difference between these two energies represents the extra stabilization that results when localized pi electrons are permitted to roam over the entire molecule.

$$\Delta H_{\text{stab.}} = \Delta H_{\text{obs.}} - \Delta H_{\text{calc.}}$$

$$= (-49.8) - (-85.8)$$

$$= +36.0 \text{ kcal/mole}$$

This energy difference, illustrated in Figure 19-19, is also called the "resonance" or "delocalization" energy.

Even more significant is the enthalpy change associated with the addition of a single mole of H_2 to benzene. This reaction is actually endothermic!

$$\langle\bigcirc\rangle + H_2 \longrightarrow \langle\ \rangle \tag{19-40}$$

$$\Delta H = +5.6 \text{ kcal/mole}$$

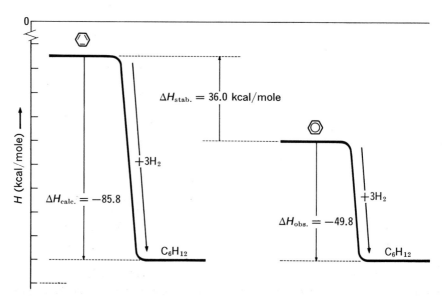

Figure 19-19 Stabilization energy of benzene.

Instead of releasing 28.6 kcal of energy, 5.6 kcal is *absorbed*. This is the reason that benzene and other aromatics tend to be unreactive or to react through attached groups. Here is the explanation of why bromine does not add to the pi system, as it would add to an alkene. Instead, the reagent finds another reaction, displacement of an H atom, that is thermodynamically favorable.

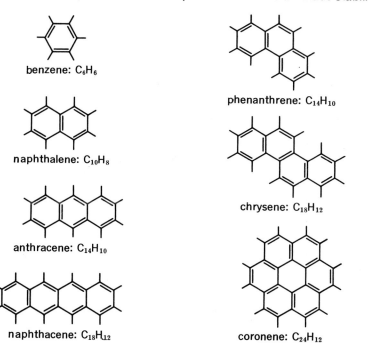

$$\Delta H_{calc.} = D_{CH} + D_{Br_2} - D_{CBr} - D_{HBr}$$
$$= 99 + 46 - 68 - 87$$
$$= -10 \text{ kcal/mole}$$

The substitution reaction is slightly exothermic and, since ΔS is near zero, it has a negative ΔG. Preservation of the conjugated pi bonds makes this substitution thermodynamically more likely than addition.

Figure 19-20 displays some other compounds that have aromatic stability,

benzene: C_6H_6

phenanthrene: $C_{14}H_{10}$

naphthalene: $C_{10}H_8$

chrysene: $C_{18}H_{12}$

anthracene: $C_{14}H_{10}$

naphthacene: $C_{18}H_{12}$

coronene: $C_{24}H_{12}$

Figure 19-20 *Some aromatic compounds as represented through double bonds (one "resonance form" is shown).*

like that of benzene. All of these molecules tend to react through their attached groups, if any, and all are relatively inert when it is realized that each contains an array of multiple bonds.

Exercise 19-7.   Only one of the "resonance structures" of naphthalene is shown in Figure 19-20. Draw the other two and decide which C–C bonds should be the shortest and which should be the longest from the fraction of the structures that attribute double bond character to each bond.

19-3   Oxygen and nitrogen compounds

After carbon and hydrogen, the elements most frequently found in organic molecules are oxygen and nitrogen. In fact, there are few molecules of biological importance that do not include one or both of these elements. In this section, we will take a brief look at some of the more important compounds, their structures, and their reactions. We will consider first their structures, then their oxidation–reduction chemistry, and finally, their acid-base properties.

(a) OXYGEN AND NITROGEN FUNCTIONAL GROUPS

Figure 19-21 lists the names and shows the linkages of the most important oxygen- and nitrogen-containing functional groups, arranged to emphasize analogous structures.

Alcohols and amines can be compared, but notice that the terms primary, secondary, and tertiary have quite different meanings. For alcohols, the notation indicates where in a carbon chain the OH group is located, at a terminal position or at a branched position. For amines, the notation indicates how many of the bonds to nitrogen link it to carbon atoms. Thus primary and secondary amines are chemical analogs to alcohols; they have electron-donating properties (both oxygen and nitrogen have unused electron pairs) and also electron-accepting properties (they have potentially acidic hydrogen atoms: O–H or N–H groups). The tertiary amines are chemical analogs to ethers: Both families have only basic, electron-donating properties. In general, the nitrogen compounds are stronger bases than their oxygen counterparts and weaker acids, as is the case with the prototypes, ammonia and water, and as is appropriate to the lower electronegativity of nitrogen.

From a bonding point of view, aldehydes and ketones have chemical analogs in nitriles and imines, but there is little chemical resemblance. The carboxylic acids can be compared only to nitrous acid (which does not contain carbon because all of the nitrogen bonding capacity is used up in attachments to oxygen). The acid dissociation constant of nitrous acid, $4.5 \cdot 10^{-4}$, is rather close to those of the carboxylic acids (e.g., compare formic acid, $1.8 \cdot 10^{-4}$, and acetic acid, $1.8 \cdot 10^{-5}$).

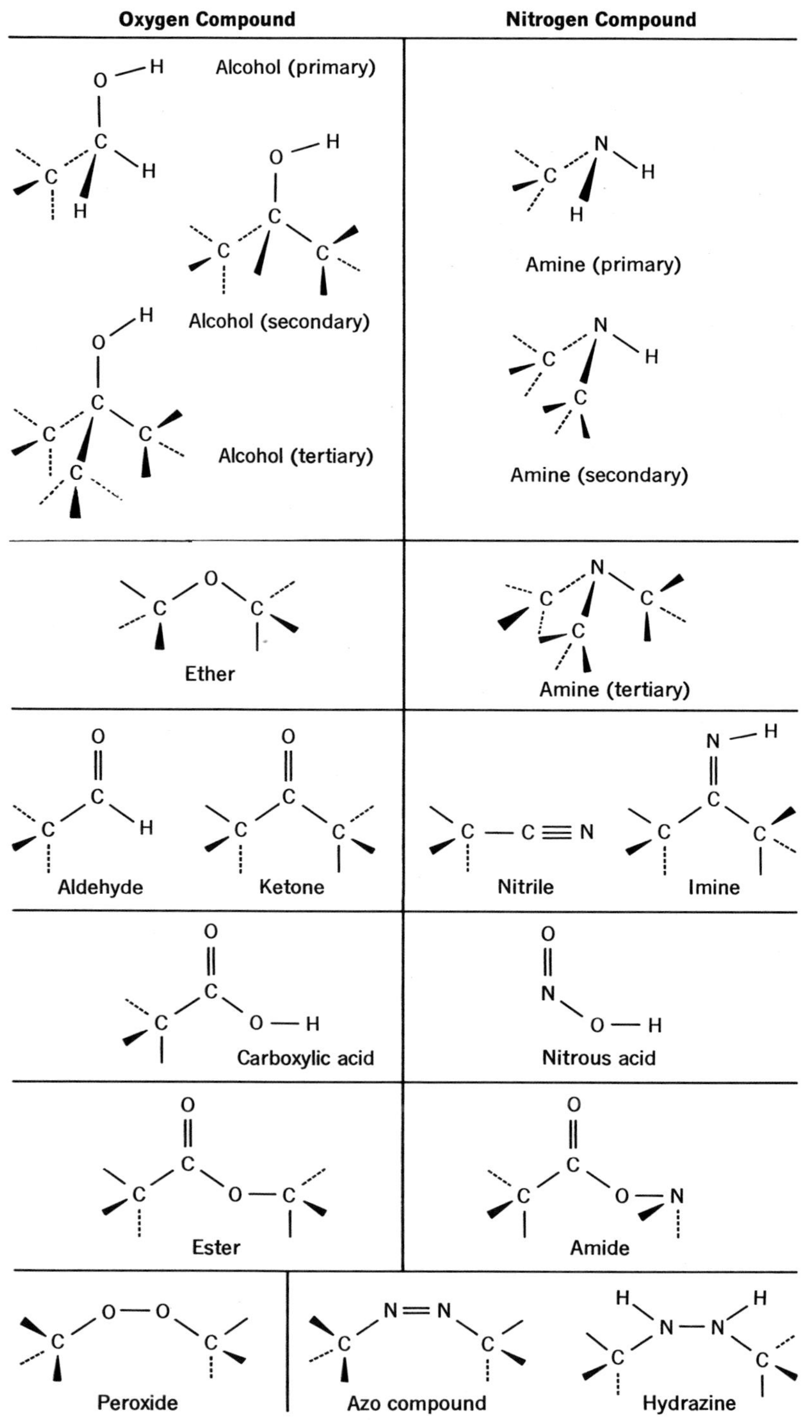

Figure 19-21 *Analogous oxygen and nitrogen compounds.*

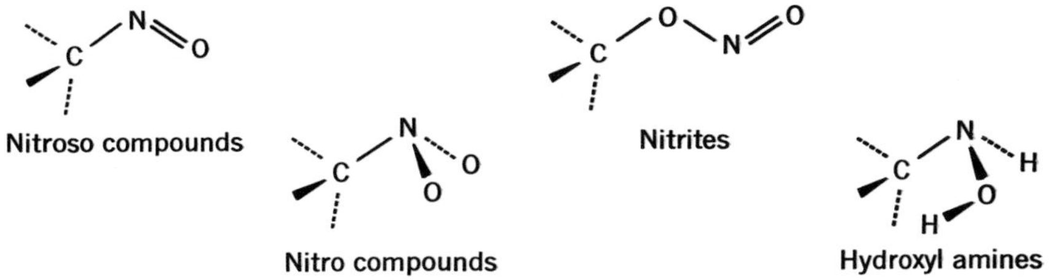

Figure 19-22   *Some nitrogen–oxygen compounds.*

In addition to the compounds shown in Figure 19-21, there are four nitrogen–oxygen compounds that deserve mention: nitroso compounds, nitro compounds, nitrites, and hydroxylamines. These are shown in Figure 19-22 and they complete the catalog we will need to discuss the chemistry of oxygen and nitrogen compounds with carbon.

(b) OXIDATION OF OXYGEN AND NITROGEN COMPOUNDS

In the prototype compounds, methanol CH_3OH, formaldehyde H_2CO, and formic acid $HCOOH$, carbon has oxidation numbers -2, 0, and $+2$. Thus we see that they are related to each other through oxidation.*

The oxidation (or dehydrogenation) of alcohols in the presence of a copper catalyst at 300°C has already been mentioned (see reactions (19-13) and (19-14)). A primary alcohol gives an aldehyde, whereas a secondary alcohol gives a ketone. A tertiary alcohol does not react.

An aldehyde can be oxidized further to a carboxylic acid, so many oxidizing agents will oxidize a primary alcohol all the way to the acid. Dichromate in acid solution is an example.

$$3\ \overset{\displaystyle OH}{\underset{\displaystyle |}{RCH_2}} + 2\ Cr_2O_7^{-2} + 16\ H^+ \longrightarrow 3\ RC\overset{\displaystyle O}{\underset{\displaystyle OH}{\diagup\diagdown}} + 4\ Cr^{+3} + 11\ H_2O \qquad (19\text{-}41)$$

$$3\ \overset{\displaystyle OH}{\underset{\displaystyle |}{RCHR'}} + Cr_2O_7^{-2} + 8\ H^+ \longrightarrow 3\ \overset{\displaystyle O}{\underset{\displaystyle \|}{RCR'}} + 2\ Cr^{+3} + 7\ H_2O \qquad (19\text{-}42)$$

Naturally, dichromate also will oxidize an aldehyde to a carboxylic acid.

$$3\ RC\overset{\displaystyle O}{\underset{\displaystyle H}{\diagup\diagdown}} + Cr_2O_7^{-2} + 8\ H^+ \longrightarrow 3\ RC\overset{\displaystyle O}{\underset{\displaystyle OH}{\diagup\diagdown}} + 2\ Cr^{+3} + 4\ H_2O \qquad (19\text{-}43)$$

---

* Organic chemists have no strong affection for oxidation numbers, though they can be used if desired. In the next chapter, Table 20-1 shows common compounds in which carbon has formal oxidation numbers ranging from $+4$ to -4.

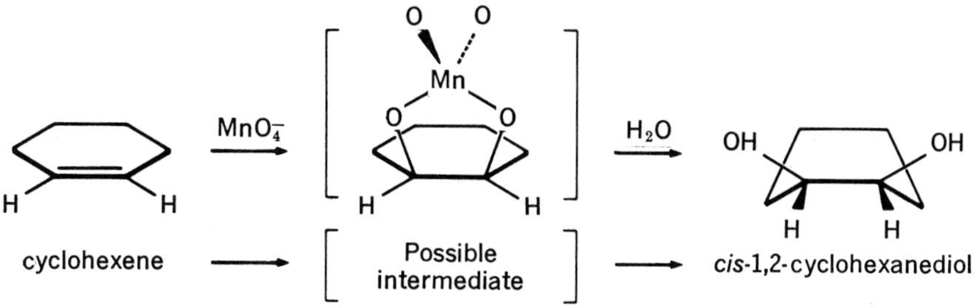

Figure 19-23 *Permanganate oxidation of a cyclic alkene.*

When a molecule contains two functional groups, it is sometimes advanta-geous to be able to oxidize one group while leaving the other intact. Organic chemists have discovered quite a list of such selective reactants. A single example is provided by the oxidation of a molecule containing both a double bond and an alcohol group. With chromic oxide as the oxidiz-ing agent, the alcohol can be oxidized to a ketone, leaving the double bond intact. With permanganate, the double bond is oxidized to a dialcohol.

$$3 \text{ R}-\text{CH}=\text{CH}-\overset{\overset{\displaystyle OH}{|}}{\text{CH}}-\text{R}' + 2 \text{ CrO}_3 + 6 \text{ H}^+ \longrightarrow$$

$$3 \text{ R}-\text{CH}=\text{CH}-\overset{\overset{\displaystyle O}{\|}}{\text{C}}-\text{R}' + 2 \text{ Cr}^{+3} + 6 \text{ H}_2\text{O} \quad (19\text{-}44)$$

$$5 \text{ R}-\text{CH}=\text{CH}-\overset{\overset{\displaystyle OH}{|}}{\text{CH}}-\text{R}' + 2 \text{ MnO}_4^- + 2 \text{ H}_2\text{O} + 6 \text{ H}^+ \longrightarrow$$

$$5 \text{ R}-\overset{\overset{\displaystyle OH}{|}}{\text{CH}}-\overset{\overset{\displaystyle OH}{|}}{\text{CH}}-\overset{\overset{\displaystyle OH}{|}}{\text{CH}}-\text{R}' + 2 \text{ Mn}^{+2} \quad (19\text{-}45)$$

The permanganate reaction provides a convenient route to the prepara-tion of polyhydroxy compounds. When a cyclic alkene is so oxidized, a *cis*-diol results. This suggests the reaction involves a bridged intermediate such as the one illustrated in Figure 19-23.

Ozone is a useful reagent to oxidize alkenes. The ozone bridges across the double bond and then bond cleavage occurs to form two aldehydes. The intermediate in this case is a cyclic ozonide that can be isolated in many cases (though it is usually explosively unstable if isolated).

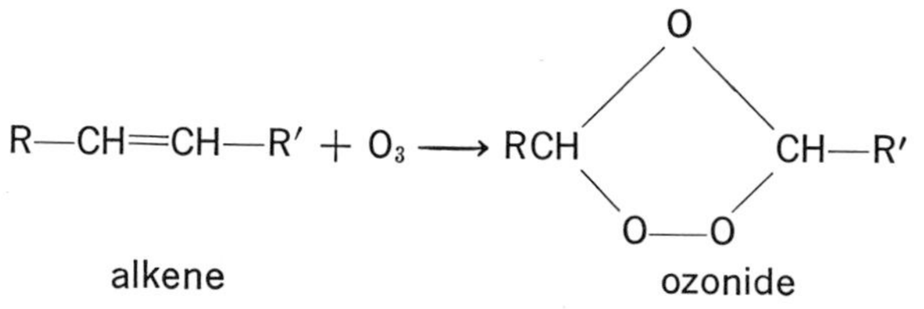

$$\text{R}-\text{CH}=\text{CH}-\text{R}' + \text{O}_3 \longrightarrow \quad (19\text{-}46a)$$

alkene   ozonide

$$RCH \underset{O-O}{\overset{O}{\diagdown}} CH-R' \longrightarrow RC\overset{O}{\underset{H}{\diagup}} + R'C\overset{O}{\underset{H}{\diagup}} + \tfrac{1}{2}O_2 \qquad (19\text{-}46b)$$

Amines are also susceptible to oxidation. A tertiary amine forms an *amine oxide*, a compound that involves a donor–acceptor bond between the nitrogen lone pair and a singlet oxygen atom (see Section 17-3(d)).

$$R_2 \overset{R_1}{\underset{R_3}{\diagdown}} N + H_2O_2 \longrightarrow R_2 \overset{R_1}{\underset{R_3}{\diagdown}} N-O + H_2O \qquad (19\text{-}47)$$

A secondary amine gives a *hydroxylamine* on oxidation

$$R_2 \overset{R_1}{\underset{H}{\diagdown}} N + H_2O_2 \longrightarrow \overset{R_1}{\underset{R_2}{\diagdown}} N-O\underset{H}{\diagdown} + H_2O \qquad (19\text{-}48)$$

A primary amine can be oxidized to a hydroxylamine, but usually further oxidation occurs to give the nitro compound. For example, consider the reaction with the oxidizing agent trifluoroperoxyacetic acid:

$$H \overset{R}{\underset{H}{\diagdown}} N + 3\,CF_3C\overset{O}{\underset{OOH}{\diagup}} \longrightarrow R-N\overset{O}{\underset{O}{\diagup}} + 3\,CF_3C\overset{O}{\underset{OH}{\diagup}} + H_2O \qquad (19\text{-}49)$$

(c) REDUCTION OF OXYGEN COMPOUNDS

Catalytic reduction of alkenes by hydrogen was discussed in Section 19-2(d). Such catalytic reduction can also be used to reduce aldehydes and ketones to alcohols (and nitro compounds to amines). Two other useful reducing agents are lithium aluminum hydride, $LiAlH_4$, and sodium borohydride, $NaBH_4$. Both reagents will reduce aldehydes and ketones to the corresponding alcohols, leaving any $C{=}C$ bonds untouched (see reactions (19-3) and (19-4)). Lithium aluminum hydride—but not sodium borohydride—will also reduce carboxylic acids to alcohols.

$$4\,RC\overset{O}{\underset{OH}{\diagup}} + 2\,LiAlH_4 + 4\,H_2SO_4 \longrightarrow$$

$$4\,RCH_2OH + Li_2SO_4 + Al_2(SO_4)_3 + 4\,H_2O \qquad (19\text{-}50)$$

Exercise 19-8. Balance the equation for the reduction of acetaldehyde, CH_3CHO, to ethanol, CH_3CH_2OH, by sodium borohydride, $NaBH_4$, to give the borate salt, NaH_2BO_3.

Aldehydes and ketones can also be reduced to alcohols by use of an alkyl magnesium halide. These extremely useful compounds—called *Grignard reagents*—are formed in the reaction between organic halides (preferably the iodides) and magnesium metal in ether solution.

$$CH_3\!-\!CH_2I + Mg \xrightarrow{\text{ether}} CH_3CH_2MgI \qquad (19\text{-}51)$$

Reaction of a Grignard reagent with a carbonyl group (ketone or aldehyde) results in the formation of a new carbon–carbon bond. Hence it is a useful way to extend the length of a carbon chain. For example, consider the reaction between acetone and ethyl magnesium iodide.

$$\underset{\substack{\|\\ O}}{CH_3CCH_3} + CH_3CH_2MgI \longrightarrow \begin{bmatrix} OMgI \\ | \\ CH_3\!-\!C\!-\!CH_3 \\ | \\ CH_2 \\ | \\ CH_3 \end{bmatrix} \qquad (19\text{-}52a)$$

$$\begin{bmatrix} OMgI \\ | \\ CH_3\!-\!C\!-\!CH_3 \\ | \\ CH_2 \\ | \\ CH_3 \end{bmatrix} + HCl \xrightarrow{H_2O} \begin{matrix} OH \\ | \\ CH_3\!-\!C\!-\!CH_3 \\ | \\ CH_2 \\ | \\ CH_3 \end{matrix} + MgICl \qquad (19\text{-}52b)$$

Exercise 19-9. Write a sequence of reactions using methyl magnesium iodide by which acetaldehyde, CH_3CHO, could be converted to acetone, CH_3COCH_3, through reduction to an alcohol followed by reoxidation.

Some organic compounds can be reduced in an electrochemical cell. Quinones provide examples—particularly useful ones, in fact. A quinone is a compound with two carbonyls in a cyclic structure that, upon reduction, gives an aromatic diol.

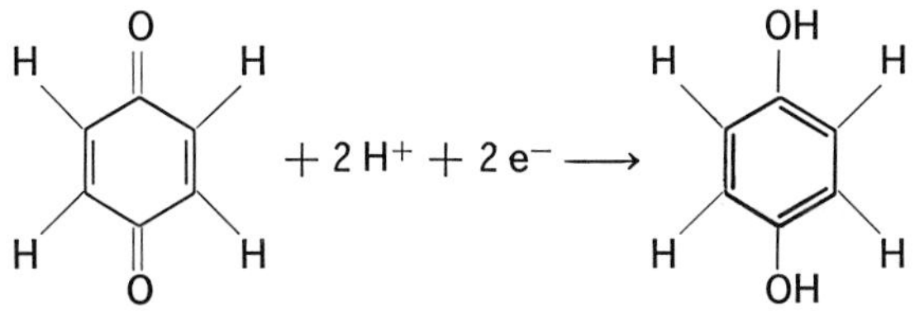

$$+ 2H^+ + 2e^- \longrightarrow \qquad \varepsilon^0 = 0.699 \text{ volt} \qquad (19\text{-}53)$$

quinone   hydroquinone

Table 19-12   Some Oxidizing and Reducing Agents for Organic Functional Groups

Oxidizing Agent	Reacts with	To Produce
O_2 (high T)	$-\overset{\mid}{\underset{\mid}{C}}-$	CO_2, H_2O
O_3	$\underset{/}{\overset{\backslash}{C}}=\underset{\backslash}{\overset{/}{C}}$	$\underset{/}{\overset{\backslash}{C}}=O + O=\underset{\backslash}{\overset{/}{C}}$
dehydrogenation Cu catalyst (300°C)	$-CH_2OH$	$-\overset{O}{\underset{H}{\overset{\parallel}{C}}}$ (aldehyde)
	$-\underset{\mid}{\overset{OH}{\overset{\mid}{CH}}}-$	$-\overset{O}{\underset{\mid}{\overset{\parallel}{C}}}-$
$Cr_2O_7^{-2}$	$-CH_2OH$	$-\overset{O}{\underset{OH}{\overset{\parallel}{C}}}$
	$-\underset{\mid}{\overset{OH}{\overset{\mid}{CH}}}-$	$-\overset{O}{\underset{\mid}{\overset{\parallel}{C}}}-$
MnO_4^-	$\underset{/}{\overset{\backslash}{C}}=\underset{\backslash}{\overset{/}{C}}$	$-\overset{OH}{\underset{\mid}{\overset{\mid}{C}}}-\overset{OH}{\underset{\mid}{\overset{\mid}{C}}}-$

Reducing Agent	Reacts with	To Produce
H_2, Pt or Pd catalyst	$\underset{/}{\overset{\backslash}{C}}=\underset{\backslash}{\overset{/}{C}}$	$-\overset{\mid}{\underset{\mid}{C}}-\overset{\mid}{\underset{\mid}{C}}-$
	$\underset{/}{\overset{\backslash}{C}}=O$	$-\overset{\mid}{\underset{\mid}{C}}-OH$
	$-\underset{\mid}{\overset{\mid}{C}}NO_2$	$-\overset{\mid}{\underset{\mid}{C}}-NH_2$
$LiAlH_4$, $NaBH_4$	$\underset{/}{\overset{\backslash}{C}}=O$	$-\overset{\mid}{\underset{\mid}{C}}-OH$
$LiAlH_4$	$-\overset{O}{\underset{OH}{\overset{\parallel}{C}}}$	$-\overset{\mid}{\underset{\mid}{C}}-OH$
RMgX	$\underset{/}{\overset{\backslash}{C}}=O$	$R-\overset{\mid}{\underset{\mid}{C}}-OH$

This half-reaction is widely used to determine hydrogen ion concentration. Neither quinone nor hydroquinone has a large solubility in water. Hence the simple addition of these two solids to an acidic solution in one compartment of an electrochemical cell gives a voltage that depends upon $\mathcal{E}^0$ and pH through the familiar Nernst Equation (6-40).

Table 19-12 summarizes some of the oxidation and reduction reactions that have been discussed in this and the previous section.

(d) ELIMINATION REACTIONS

Carboxylic acids will react with either alcohols or amines to form a larger organic molecule through the elimination of a mole of water. In the case of alcohols, the products are called *esters*, and in the case of amines, the products are called *amides*. Reactions (19-54) and (19-55) give examples and show how the products are named.

$$CH_3C\overset{O}{\underset{OH}{\big<}} + CH_3CH_2OH \longrightarrow H_2O + CH_3C\overset{O}{\underset{OCH_2CH_3}{\big<}} \qquad (19\text{-}54)$$

acetic acid   ethanol   water   ethyl acetate

$$CH_3C\overset{O}{\underset{OH}{\big<}} + CH_3CH_2NH_2 \longrightarrow H_2O + CH_3C\overset{O}{\underset{\underset{H}{\overset{|}{N}}-CH_2CH_3}{\big<}} \qquad (19\text{-}55)$$

acetic acid   ethyl amine   water   N-ethyl acetamide

For the amide formed in reaction (19-55), notice that the location of the ethyl group, C_2H_5, on the nitrogen atom is specified in the name through the N.

We shall see that the reaction (19-55) is one of the most important reactions in the chemistry of life. The backbones of the molecules of which you and I are made are linked through long chains—polymers—of amide groups. These polymers are called *proteins*. The process by which they are formed, polymerization, is the subject of the next section.

19-4 Polymers: natural and man-made

Mention the chemical industry in the latter half of the twentieth century and polymers spring to mind. Plastics have become an integral part of every aspect of our lives. Much of our clothing is made from synthetic fiber. Man-made rubber may put the rubber tree out of business. Yet all the synthetic polymers known to man cannot approach in importance those produced by nature as the very stuff of living things. The list of natural polymers is a long one, encompassing proteins, cellulose, starch, DNA, and many more. In this section we will examine a few important natural polymers and their man-made imitations. We will see how man's ingenuity plus Nature's guidance pay off in partnership for the synthesis of polymeric materials.

(a) RUBBER

A molecule can form polymers if it has at least two reactive sites—that is, each molecule must be able to react with two others. A molecule with two functional groups is the most obvious example. However, an alkene has but one functional group, the double bond, yet it also fulfills the requirement. Attack at one end of a double bond leaves an unpaired electron at the other end for reaction with a second molecule. If this second molecule is itself an alkene, it can react, in turn, with a third molecule. Continuing this process, many alkene molecules can react in turn to form a very large molecule—a *polymer*. This chain process is called *polymerization*.

To initiate such an alkene polymerization, some double bonds must be "opened." A few stray free radicals will do the trick. For example, suppose ditertiary butyl peroxide, $(CH_3)_3-COOC-(CH_3)_3$ is decomposed by warming or by light. This unstable peroxide decomposes to give two tertiary butoxide radicals, $(CH_3)_3CO$. We will call this free radical RO. Each RO radical has unsatisfied bonding capability and it can attack one end of an alkene double bond. As it does, it forms a new free radical and this initiates the chain polymerization. If the alkene is ethylene, C_2H_4, the familiar plastic polyethylene is formed.

$$ROOR \longrightarrow 2\ RO$$

initiation:

$$RO + H_2C{=}CH_2 \longrightarrow ROCH_2{-}CH_2 \tag{19-56}$$

chain propagation:

$$ROCH_2{-}CH_2 + n(C_2H_4) \longrightarrow ROCH_2CH_2{-}(CH_2CH_2)_{n-1}{-}CH_2CH_2 \tag{19-57}$$

termination:

$$ROCH_2CH_2{-}(CH_2CH_2)_{n-1}{-}CH_2CH_2 + RO \longrightarrow$$
$$ROCH_2CH_2{-}(CH_2CH_2)_{n-1}{-}CH_2CH_2OR \tag{19-58}$$

Commercially, this process is carried out at high temperatures and pressures and yields polymeric chains with 100 to 1000 ethylene units. Polyethylene is widely used for molded articles, electrical insulation, films for packaging, and piping.

We copy Nature in our wide use of 1,3-dienes for the production of a variety of polymers. With these compounds the chain can grow by attachment at different pairs of carbon atoms. For example, consider 1,3-butadiene.

$$\left(\begin{array}{c} {}^{3}CH={}^{4}CH_2 \\ | \\ {}^{2}CH-{}^{1}CH_2 \end{array}\right)_n$$
1,2-addition

(19-59a)

$${}^{1}CH_2={}^{2}CH-{}^{3}CH={}^{4}CH_2 \longrightarrow$$

$$\left[\begin{array}{c} H \qquad\qquad H \\ \backslash \qquad\quad / \\ {}^{2}C={}^{3}C \\ / \qquad\qquad \backslash \\ {}^{1}CH_2 \qquad\qquad {}^{4}CH_2 \end{array}\right]_n$$
cis-1,4-addition

(19-59b)

$$\left[\begin{array}{c} H \qquad\qquad CH_2 \\ \backslash \qquad\quad / \\ C=C \\ / \qquad\qquad \backslash \\ CH_2 \qquad\qquad H \end{array}\right]_n$$
trans-1,4-addition

(19-59c)

Polymerization of 2-methyl-1,3-butadiene (isoprene) in the presence of a catalyst mixture of $Al(C_2H_5)_3$ and $TiCl_4$* gives a 100% yield of the cis-1,4 product, which is a sticky fluid identical to natural rubber.

$$n\ CH_2=\underset{\underset{CH_3}{|}}{C}-CH=CH_2 \longrightarrow$$
isoprene

$$\underset{\underset{CH_3}{/}}{\overset{\overset{CH_3}{\backslash}}{C}}=\underset{\underset{H}{\backslash}}{\overset{\overset{CH_2}{/}}{C}} \left(\begin{array}{c} CH_3 \qquad H \\ \backslash \qquad / \\ C=C \\ / \qquad \backslash \\ CH_2 \qquad CH_2 \end{array}\right)_n \underset{\underset{CH_3}{/}}{\overset{\overset{CH_2}{\backslash}}{C}}=\underset{\underset{H}{\backslash}}{\overset{\overset{CH_3}{/}}{C}}$$

(19-60)

cis-1,4-polyisoprene (natural rubber)

When the long chains of cis-1,4-polyisoprene are cross-linked with sulfur atoms (vulcanization), the firm, elastic substance commonly called "rubber" results. Interestingly, the trans addition product, called gutta percha, is hard and brittle at room temperature

$$\underset{\underset{CH_3}{/}}{\overset{\overset{CH_3}{\backslash}}{C}}=\underset{\underset{CH_2}{\backslash}}{\overset{\overset{H}{/}}{C}} \left(\begin{array}{c} CH_2 \qquad H \\ \backslash \qquad / \\ C=C \\ / \qquad \backslash \\ CH_3 \qquad CH_2 \end{array}\right)_n \underset{\underset{CH_3}{/}}{\overset{\overset{CH_2}{\backslash}}{C}}=\underset{\underset{CH_3}{\backslash}}{\overset{\overset{H}{/}}{C}}$$

(19-61)

trans-1,4-polyisoprene (gutta percha)

* This catalyst is called a Ziegler catalyst after its discoverer, K. Ziegler.

The other type of polymerization is based upon bifunctional molecules. They can be typified by the chemistry used to prepare silicone rubber. A suitable starting material is dichlorodimethyl silane, a substance that reacts so readily with water that it "fumes" when exposed to air at normal humidity.

$$Cl_2Si(CH_3)_2 + 2 H_2O \longrightarrow (HO)_2Si(CH_3)_2 + 2 HCl \qquad (19\text{-}62)$$

Two molecules of the product, dihydroxydimethyl silane, react quite readily to lose a molecule of water and form the Si–O–Si counterpart to an ether linkage

$$
\underset{\overset{|}{CH_3}}{\overset{\overset{CH_3}{|}}{HO-Si-OH}} +
\underset{\overset{|}{CH_3}}{\overset{\overset{CH_3}{|}}{HO-Si-OH}} \longrightarrow
\underset{\overset{|}{CH_3}}{\overset{\overset{CH_3}{|}}{HO-Si-O-}}\underset{\overset{|}{CH_3}}{\overset{\overset{CH_3}{|}}{Si-OH}} + H_2O
$$

$$(19\text{-}63)$$

Reaction (19-63) uses up the reactivity of one OH group on each reactant, but because each molecule contains two OH groups, the product has residual reactivity at either end. Hence the reaction can continue to extend the chain length indefinitely.

$$
\underset{\overset{|}{CH_3}}{\overset{\overset{CH_3}{|}}{HO-Si-O-}}\underset{\overset{|}{CH_3}}{\overset{\overset{CH_3}{|}}{Si-OH}} + n\,
\underset{\overset{|}{CH_3}}{\overset{\overset{CH_3}{|}}{HO-Si-OH}} \longrightarrow
$$

$$
\underset{\overset{|}{CH_3}}{\overset{\overset{CH_3}{|}}{HO-Si-O}}\;
\underset{\overset{|}{CH_3}}{\overset{\overset{CH_3}{|}}{Si-O}}\;
\underset{\overset{|}{n\ CH_3}}{\overset{\overset{CH_3}{|}}{Si-OH}} + n\,H_2O \qquad (19\text{-}64)
$$

If it is desired to limit the chain length, a suitable fraction of the monofunctional silane $ClSi(CH_3)_3$ can be added.

$$ClSi(CH_3)_3 + H_2O \rightarrow HOSi(CH_3)_3 + HCl \qquad (19\text{-}65)$$

$$
\underset{\overset{|}{CH_3}}{\overset{\overset{CH_3}{|}}{HO-Si-O}}\;
\underset{\overset{|}{CH_3}}{\overset{\overset{CH_3}{|}}{Si-O}}_n\;
\underset{\overset{|}{CH_3}}{\overset{\overset{CH_3}{|}}{Si-OH}} +
\underset{\overset{|}{CH_3}}{\overset{\overset{CH_3}{|}}{HO-Si-CH_3}} \longrightarrow
$$

$$
\underset{\overset{|}{CH_3}}{\overset{\overset{CH_3}{|}}{HO-Si-O}}\;
\underset{\overset{|}{CH_3}}{\overset{\overset{CH_3}{|}}{Si-O}}_n\;
\underset{\overset{|}{CH_3}}{\overset{\overset{CH_3}{|}}{Si-O}}\;
\underset{\overset{|}{CH_3}}{\overset{\overset{CH_3}{|}}{Si-CH_3}} + H_2O \qquad (19\text{-}66)
$$

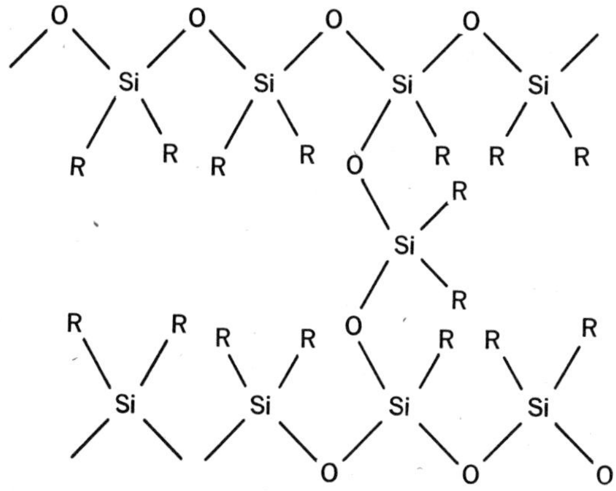

Figure 19-24 A cross-linked silicone polymer (R = –CH₃ or other organic group).

Now the chain is terminated on its right end. Depending upon the amount of the monofunctional additive, statistics will dictate when reaction (19-66) closes off growth at the left end as well, to limit the polymer chain length.

One more possibility remains. Addition of the trifunctional silane, Cl_3SiCH_3, makes chain branching possible. The result is a cross-linked structure that first acquires high viscosity, then elasticity, and finally rigidity, as cross-linking is increased (see Fig. 19-24). Needless to say, the methyl groups can be replaced by a variety of organic groups to tailor the polymer properties to man's needs.

These silicone polymers possess the silicon–oxygen linkages that account for the strength of quartz. Hence silicone rubber can withstand much higher temperatures than natural rubber—witness its use on the heat shields of space capsules. It can also withstand much lower temperatures without cracking. With a high degree of cross-linking, the polymers become even more resistant to heat, oxidation, and chemical attack. Consequently silicone polymers are in wide use as heat-resistant enamels, coatings, and house paints.

(b) AMINO ACIDS AND PROTEIN POLYMERS

Amino acids are bifunctional organic molecules that occupy a central role in the chemistry of life. Each amino acid molecule contains both a carboxylic acid and an amine group. When the amine group is bound immediately adjacent to the carboxyl group, the molecule is called an α-amino acid. The simplest members of the α-amino acids are glycine and alanine:

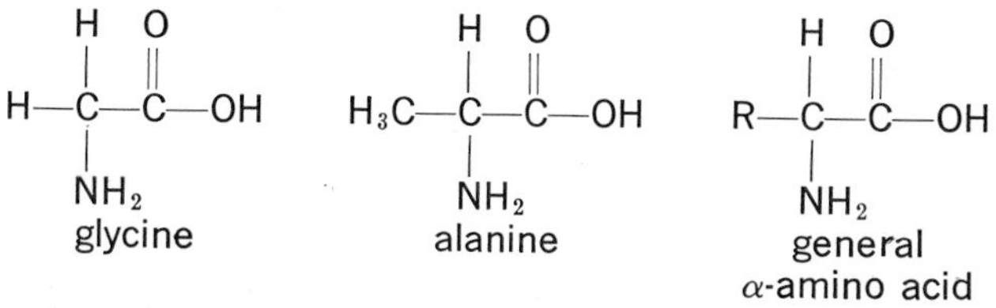

$$
\begin{array}{ccc}
\overset{\displaystyle H}{\underset{\displaystyle NH_2}{H-C}}-\overset{\displaystyle O}{C}-OH
&
\overset{\displaystyle H}{\underset{\displaystyle NH_2}{H_3C-C}}-\overset{\displaystyle O}{C}-OH
&
\overset{\displaystyle H}{\underset{\displaystyle NH_2}{R-C}}-\overset{\displaystyle O}{C}-OH
\\[1em]
\text{glycine} & \text{alanine} & \text{general} \\
& & \alpha\text{-amino acid}
\end{array}
\qquad (19\text{-}67)
$$

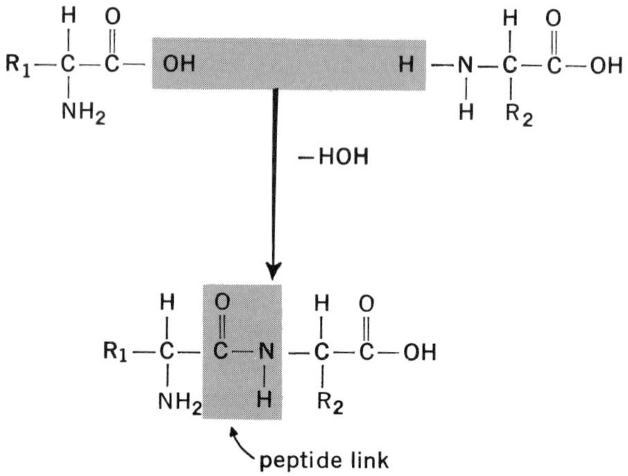

Figure 19-25 Formation of a dipeptide.

We have seen (in Section 19-3(d)) that a carboxylic acid can react with an amine, losing water, to form an amide linkage. When this occurs between groups in the same molecule, a cyclic amide is obtained. These are called *lactams*. However, an α-amino acid cannot do this because the functional groups are too close together. Hence amide formation occurs between different molecules and the bifunctional character of the amino acids can result in polymer formation. For historical reasons, the amide linkage between two α-amino acids is called a *peptide* link and the polyamide polymer chain is called a *polypeptide* chain. In living organisms, these compounds are called proteins. Proteins are, in fact, the building blocks of which you are made. They can be called "macromolecules"—the smallest protein has a molecular weight of about 6000, the largest about 7,000,000. Table 19-13 lists several important types of proteins and summarizes a few of their functions. This list gives an idea of their great importance.

These long polypeptide chains coil themselves into a helix, a structure that may be visualized by imagining the chain being wound in a spiral manner down the length of a small rod. The stability of this structure is

Table 19-13 Biological Functions of Some Proteins

General Type	Biochemical Role
enzymes	catalysis: hydrolysis, oxidation, synthesis
structural	hair, wool, feathers, muscle, silk, what have you
respiratory	oxygen transport and storage (hemoglobin)
antibodies	defend organism against foreign agents: attacks by bacteria and viruses
hormones	regulation of metabolism
nucleoproteins	control of hereditary transmission, protein synthesis (chromosomes)

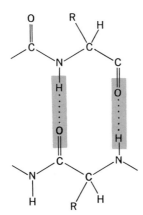

Figure 19-26 Hydrogen bonding between peptide linkages.

due to hydrogen bonds between peptide links, as illustrated in Figure 19-26. These bonds form between a $-\overset{O}{\overset{\|}{C}}-$ group that interacts with an $-\overset{H}{\overset{|}{N}}-$ group located about three amide linkages up the chain, in the next turn of the spiral. Figure 19-27 is a computer-drawn stereo picture of a general helix. It can be viewed in stereo by placing a piece of cardboard between the two views and letting your eyes bring the two images together. The stabilizing hydrogen bonds are shown as light lines, the conventional bonds as dark lines. The hydrogen atoms themselves are omitted to keep the diagram relatively simple. This general structure can accommodate any side chains R to represent any protein. With 3.6 amide groups per turn of the helix, the structure is called an α-helix.

The peptide, or polyamide, linkage that is so important in proteins, is imitated in the man-made fibrous polymer, nylon. There are various nylons, made, not from amino acids, but from diamines and dicarboxylic acids. For example, nylon 66 is made from hexamethylene diamine and adipic acid (both of which contain six carbon atoms, hence the "66" designation).

$$NH_2-(CH_2)_6-NH_2 + HO-\overset{O}{\overset{\|}{C}}-(CH_2)_4-\overset{O}{\overset{\|}{C}}-OH$$

hexamethylene   adipic acid
diamine

$$+ NH_2-(CH_2)_6-NH_2 + \cdots \longrightarrow$$

$$-NH-(CH_2)_6-NH-\overset{O}{\overset{\|}{C}}-(CH_2)_4-\overset{O}{\overset{\|}{C}}-NH-(CH_2)_6\boxed{-NH-\overset{O}{\overset{\|}{C}}-}(CH_2)_4-$$

repeating unit   amide
(peptide)
link

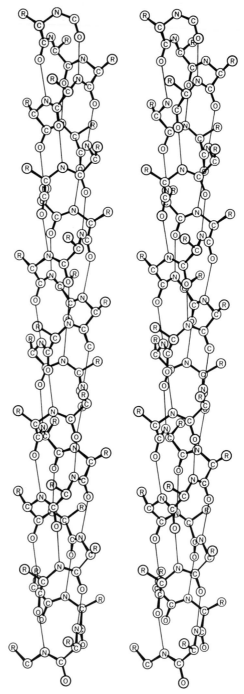

Figure 19-27   Alpha-helix. To view this computer-drawn stereo representation, place a business-size envelope between the two views, put your nose on the edge of the envelope, and allow your eyes to bring the two views into one. For details of the structure, see text. The authors are greatly indebted to Dr. C. K. Johnson of Oak Ridge National Laboratory for providing us with this figure.

Hydrogen bonds between carbonyl oxygens and amine hydrogens on adjacent chains account for the strength of the product, just as they determine the stability of the protein helix. Nylon is a good example of man learning from Nature—and putting the lesson to practical use.

(c) CARBOHYDRATES

Chlorophyll in green plants absorbs sunlight and uses the energy for the synthesis of a sugar, glucose, from water and carbon dioxide.

$$6\ CO_2 + 6\ H_2O \rightarrow C_6H_{12}O_6 + 6\ O_2 \qquad\qquad (19\text{-}68)$$
$$\text{glucose}$$

This is the beginning of the biological processes that are responsible for the manufacture of all living matter.

Glucose is a member of the carbohydrate family whose members have the general formula $(CH_2O)_n$. It exists in two isomeric cyclic forms shown here schematically.

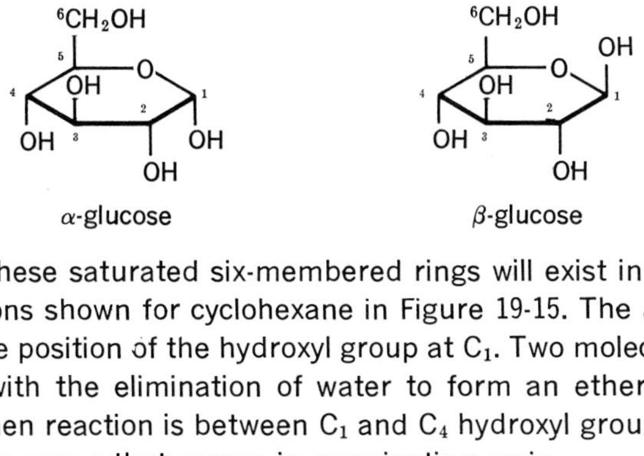

$$\alpha\text{-glucose} \qquad\qquad \beta\text{-glucose} \qquad\qquad (19\text{-}69)$$

Of course these saturated six-membered rings will exist in the chair–boat conformations shown for cyclohexane in Figure 19-15. The α,β-designation refers to the position of the hydroxyl group at C_1. Two molecules of glucose can react with the elimination of water to form an ether-linked dimeric species. When reaction is between C_1 and C_4 hydroxyl groups, the product is maltose, a sugar that occurs in germinating grain

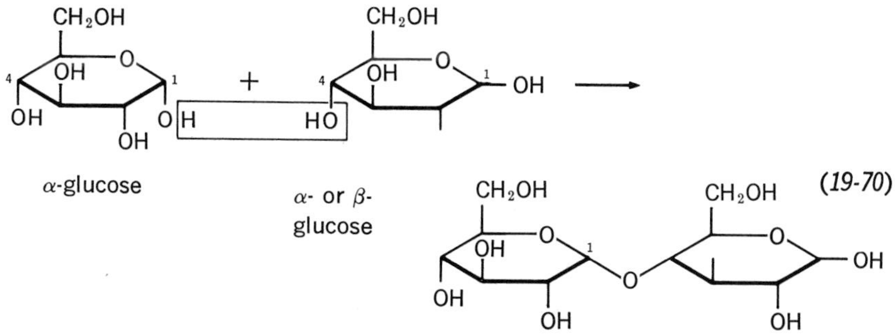

$$(19\text{-}70)$$

$$\alpha\text{- or }\beta\text{-maltose}$$

The familiar table "sugar" extracted from sugar cane or sugar beets is

sucrose, formed from the dimerization of α-glucose and β-fructose by reaction at C_1 on both monomers.

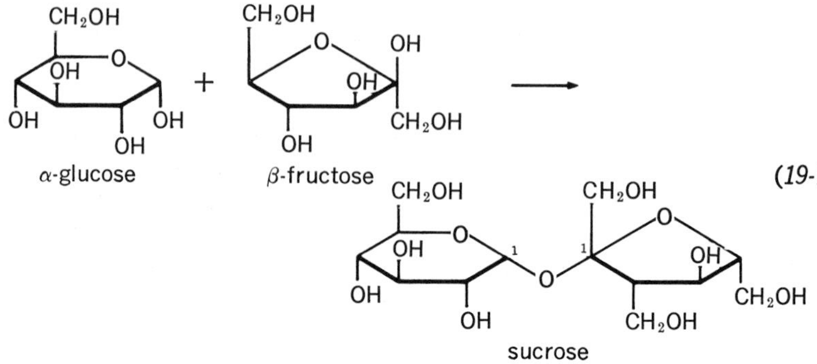

$$(19\text{-}71)$$

Glucose produced in plants is not stored as the monomer; the plant converts it into polymers: cellulose for cell walls and rigid fibers and starch for food. Starch, like maltose, consists of glucose units with α-linkages. Starch from any one source (corn, wheat, rice, potatoes) consists of a particular mixture of linear glucose chains and branched molecules in which linear chains have been joined by 1,6-linkages as shown in Figure 19-28(a).

Animals store glucose in the form of starchlike substances called glyco-

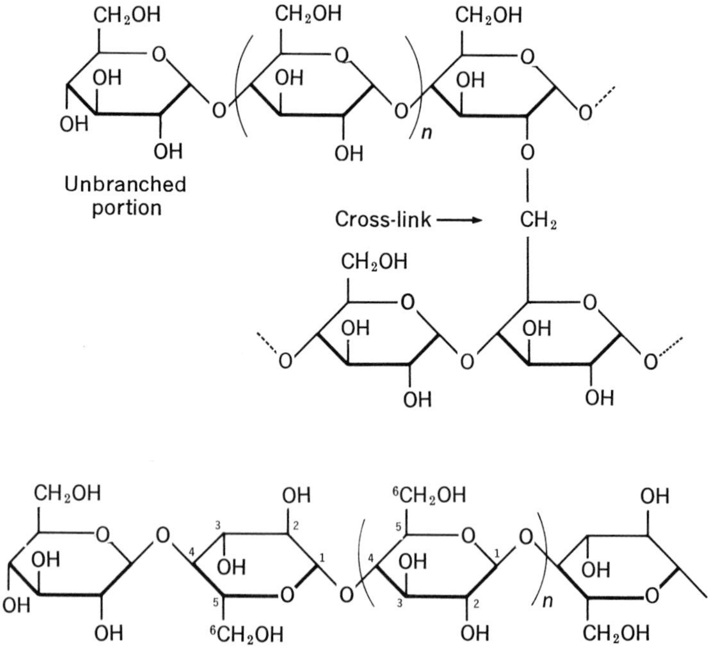

(a) Starch

(b) Cellulose

Figure 19-28 Glucose polymers. Notice that every other glucose unit in cellulose has been "flipped over" from the normal representation.

gens, which differ from starch by the absence of any unbranched molecules. Carbohydrates digested from food are stored in liver and muscle tissue as glycogens. These substances can deliver energy-giving glucose through enzyme-catalyzed hydrolysis as needed between meals.

Cellulose is identical to starch in every way except one—the building block is β-glucose instead of α-glucose. A cellulose fragment is illustrated in Figure 19-28(b). This seemingly simple difference is of vast importance. Starch is an important source of food. Cellulose is completely indigestible by humans. It is, however, the most abundant organic compound on earth!

To choose just one example, cotton fibers are almost 98% cellulose. Cellulose molecules with up to 9000 glucose units are arranged parallel to each other in strands that are given their strength by—as usual—hydrogen bonds between OH groups on adjacent molecules.

(d) DNA AND RNA

Genetic information is stored by the genes, a particular form of a macromolecule commonly known as DNA (deoxyribonucleic acid). The information (e.g., green hair, six toes) carried by the chromosomal DNA molecules is transferred to the appropriate protein-manufacturing cells by another large molecule RNA (ribonucleic acid), which is closely related to DNA. These two types of molecules are at the heart of the reproduction systems

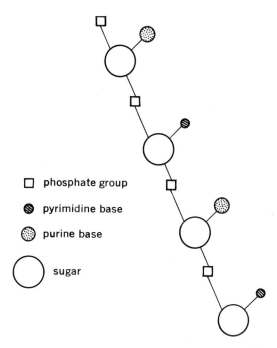

□ phosphate group

◉ pyrimidine base

◉ purine base

◯ sugar

Figure 19-29   General arrangement of nucleic acid.

from amoeba on up (or on down). Because of their enormous molecular size, however, their structures are only gradually coming to light. Once again it is found that hydrogen bonding has a central role in the structure of these molecules. The investigation of the functions and structures of DNA and RNA molecules represents one of the most exciting aspects of contemporary biochemistry—or as it is beginning to be called, *molecular biology.*

The backbones of these species are sugar molecules linked together by phosphate groups, as illustrated schematically in Figure 19-29 and, in more detail, in Figure 19-30. The wide variety of possible molecules results from the different side chains that can be tacked on. These side chains are organic bases—so called because the nitrogen atoms contained in them are good electron donors. Two general types of bases are found, pyrimidine bases and purine bases. The parent compounds are illustrated in Figure 19-31. The replacement of one or more of the hydrogens (marked with arrows) by $-OH$, $-NH_2$ or $-CH_3$ groups accounts for the many different members of these families found in biological systems. Four particular bases are predominantly recurrent in DNA and RNA, two purines and two pyrimidines, although innumerable other bases form a small percentage of the total. The number of base units may be as high as 10^8 in chromosomal DNA and as low as 80 in some RNA molecules. The information-carrying capacity of these species is related to the *order* in which the bases are attached. It only takes a few different bases distributed among one hundred million sites to give the molecule an enormous vocabulary.

The structure of a nucleic acid is a hydrogen bond delight. It consists of a pair of intertwined helices held together by hydrogen bonds between pairs of bases on separate chains. The requirements of the geometry are

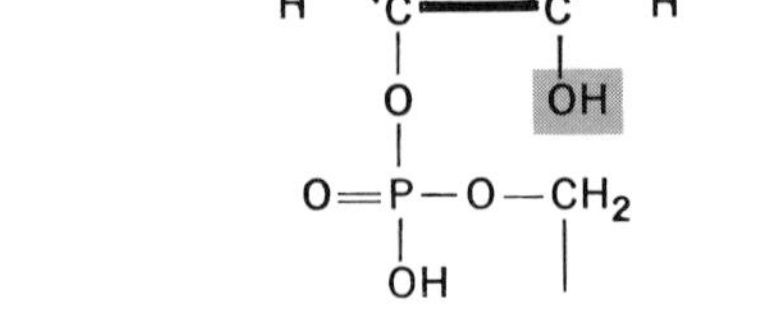

Figure 19-30 Backbone of ribonucleic acid: Five-membered sugar rings are hitched together with phosphate groups. The replacement of the OH group, shown shaded, with a hydrogen atom transforms this to the backbone of DNA, deoxyribonucleic acid. This simple change has a profound effect on the biological functions of this macromolecule.

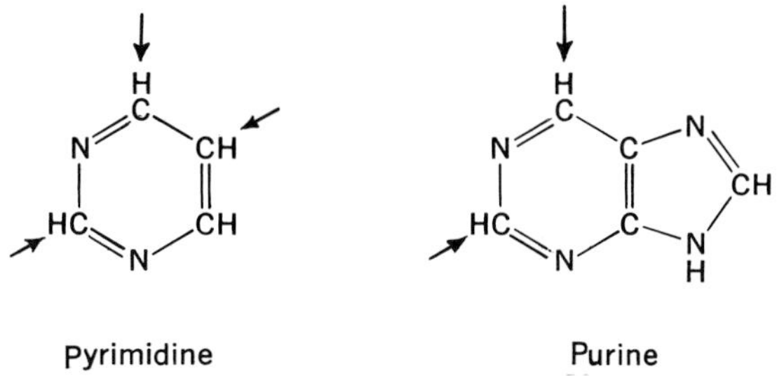

Pyrimidine   Purine

Figure 19-31   *Parent compounds of bases found in DNA and RNA. Different compounds result from the substitution of –NH₂, –OH, and –CH₃ groups for the hydrogens marked by arrows.*

such that just any old pair of bases won't do. The space between helices is properly bridged by a hydrogen bond only when it is from a pyrimidine base to a purine base. Two pyrimidine bases are too small and two purines too large. Figure 19-32 illustrates the pairing arrangement that is most common. This high degree of choosiness about the appropriate partners for hydrogen bonding requires that the two chains be *complementary:* Every base on one chain must be matched by a specific partner base on the other helix. This matching, or lock-and-key relationship, provides the mechanism for molecular reproduction, which is the cornerstone of the chemistry of life.

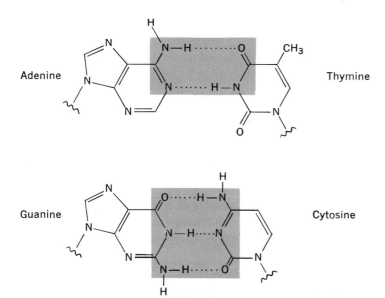

Figure 19-32   *Base pairing by hydrogen bond formation. These four bases predominate in DNA and RNA and the illustrated combinations are the most frequent.*

19-5   Thermodynamics in living systems: bioenergetics

Every spring we witness a seeming thermodynamic miracle. From the brown, raw earth, wet with rain and warmed by the sun, there rises a vast family of identical sprouts. Then as summer comes, the various chemical mixtures in the soil are transformed into fully developed blades of grass. Each blade is like every other, a beautifully organized, highly efficient chemical factory. The whole process takes place spontaneously—in the direction of *increasing* randomness! The starting materials looked random enough—the wet earth with its complex chemical mixture—the final result, the blade of grass, seems far from random! Thermodynamics, wherefore art thou?

The Second Law of Thermodynamics speaks, of course, in terms of the entropy of the universe. True, the processes by which living systems grow and reproduce succeed in creating locally a high degree of order; that is, within the blade of grass, within the cell, within the DNA molecule. But how do these processes drain the surroundings? There we must find the "driving force" to support the local creation of order. There we must find a counterbalancing increase in randomness to make the processes of life occur spontaneously. There we must find our thermodynamic principles at work, making it possible for randomness to *decrease* locally, at the cost of an even larger increase in randomness elsewhere!

Here on earth, the wellspring of life is the sun. As this giant inferno consumes itself, wasting itself for all time, the tiniest fraction of its ever increasing randomness more than suffices to create a garden of order— life—on our lush planet. It does so by sending us light, a form of energy, and we must take it from there.

(a)   ENERGY FROM THE SUN

Man fuels his body with both plant and animal tissue. However, the animals we consume as food, themselves rely mainly on plant life for their nourishment. The plant world is thus the source of all food and energy for the animal world. Ultimately all energy has its origins in the fiery furnace that is the sun. There, extremely high temperatures result from the release of the energies associated with the attractions between nuclear particles. The principal process going on is the "fusion" of hydrogen atoms to form helium, a process that releases enormous quantities of energy in the form of heat.

$$4\,H \longrightarrow He + 2\,e^- + 2\,e^+ \qquad \Delta H = -6.0 \cdot 10^8 \text{ kcal/mole He}$$

This is an endothermic reaction that really tries! Where such a fire burns, temperatures reach millions of degrees (near the center of the sun). Even on the cooler outer edges, the solar gases swirl at 6000°K temperatures.

This is the part of the sun we see; the inner tempest is shrouded from us, fortunately, or the earth would have a brown, cooked surface and life could not exist.

So we see these outer gases at 6000°K, a warm enough climate to break molecules down to atoms and to strip electrons from the atoms. This hot, charged cloud emits light of blinding intensity. The light pours out into space, and far away the planets intercept only a minute fraction of it. The rest goes on to the stars, who see our sun as another tiny, insignificant dot in their heavens, a dot so dim that it could hardly be interesting.

Here on earth, however, the atmosphere, the soil, and the oceans are warmed to produce a reasonably well thermostatted environment with a mean temperature such that molecules can exist, but also undergo chemical transformations at a rate well tuned to the day-night time cycle. In this favorable setting, chlorophyll in green plants also absorbs sunlight and uses the energy for the synthesis of glucose, from water and atmospheric carbon dioxide. (See Section 19-4(c).)

$$6\ CO_2 + 6\ H_2O \longrightarrow C_6H_{12}O_6 + 6\ O_2$$

$$\Delta H^0 = +673\ kcal/mole \tag{19-72}$$

$$\Delta G^0 = +686\ kcal/mole$$

It is clear from ΔH^0 and ΔG^0 that the synthesis of glucose is an uphill fight. Energy has to be supplied. It is estimated that over 35×10^9 tons of carbon per year are made into glucose by photosynthesis, in land and marine plants. This requires an estimated 10^{18} kilocalories per year of energy from the sun. Even this enormous amount of energy is only a drop from the bucket of total energy received on the surface of the earth—in excess of 10^{21} kilocalories per year. Yet the total amount of energy expended in a given year by all man-made machines is guessed to be only about 10^{16} kilocalories. So in using up one one-thousandth of the light energy from the sun, the plant factories channel into useful purposes one hundred times more energy than that used by man with all his machines. It is clear, then, why the study of bioenergetics is of great importance.

(b) BIOLOGICAL ENERGY TRANSFER

Energy originating in hydrogen fusion within the sun is absorbed by green plants and stored as chemical potential energy in the glucose molecule. How is this energy utilized to build cells or contract muscles in living things? At the heart of the biological system of energy transfer is a remarkable pair of compounds, adenosine diphosphate (ADP) and adenosine triphosphate (ATP). These molecules are rather complex organic ring molecules, but we are not concerned with the details of their structure. The key to nearly all movement of energy in biological systems is the following reaction, which represents the hydrolysis of adenosine triphosphate by water to produce adenosine diphosphate and a hydrogen phosphate ion.

$$\bigcirc\!-\!O\!-\!\underset{\underset{O^-}{|}}{\overset{\overset{O}{\parallel}}{P}}\!-\!O\!-\!\underset{\underset{O^-}{|}}{\overset{\overset{O}{\parallel}}{P}}\!-\!O\!-\!\underset{\underset{O^-}{|}}{\overset{\overset{O}{\parallel}}{P}}\!-\!O^- + H_2O \longrightarrow$$

$$(\text{ATP})^{-4}$$

$$\bigcirc\!-\!O\!-\!\underset{\underset{O^-}{|}}{\overset{\overset{O}{\parallel}}{P}}\!-\!O\!-\!\underset{\underset{O^-}{|}}{\overset{\overset{O}{\parallel}}{P}}\!-\!O^- + HPO_4^{-2} + H^+ \qquad (19\text{-}73)$$

$$(\text{ADP})^{-3}$$

The adenosine fragment is represented schematically by a circle. The free energy change associated with this reaction is

$$\Delta G^0(\text{ATP} \longrightarrow \text{ADP}) = -7 \text{ kcal/mole}$$

The free energy change is negative, so the reaction can proceed spontaneously. The hydrolysis of a phosphorus–oxygen bond can be used to perform a maximum of 7 kilocalories per mole of chemical work. This process is responsible for nearly all energy transfer and order-building processes in living things. We shall now see how this is performed.

(c) ATP ENERGY FLOW

A flow diagram for biological energy is shown in Fig. 19-33. Energy stored in glucose is released by the process of oxidation, which is just the reverse of the synthetic process resulting from photosynthesis, (19-72).

$$\text{glucose} + 6\ O_2 \longrightarrow 6\ CO_2 + 6\ H_2O \qquad (19\text{-}74)$$

$$\Delta H^0 = -673 \text{ kcal/mole}$$

$$\Delta G^0 = -686 \text{ kcal/mole}$$

This step is called *respiration*. Energy released during respiration is used to form ATP from ADP (the reverse of (19-73)). The ATP energy is then employed to do useful work in the cells. This work is commonly divided into three categories: biosynthesis (building of new cells), mechanical work

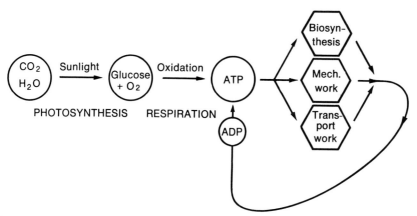

Figure 19-33 Flow of biological energy. Energy absorbed during photosynthesis is transferred to ATP in the respiration step. ATP energy is employed to perform biological work—the resulting ADP then begins the cycle again.

(muscle contraction), and transport work (movement of fluid between regions of differing concentrations). The free energy available on hydrolysis of the P–O bond in ATP is utilized by means of "coupled reactions," some of which we will now describe.

(d) ATP FORMATION

The oxidation of sugars back to carbon dioxide and water takes place in over 70 separate steps, each one catalyzed by a separate enzyme. One such step might be the oxidation of an aldehyde, RCHO, to a carboxylic acid, RCOOH.* In aqueous solution this process proceeds with a decline in free energy

$$\text{RCHO} + \tfrac{1}{2}\,O_2 \longrightarrow \text{RCOO}^-(aq) + H^+(aq) \qquad \Delta G = -7 \text{ kcal/mole}$$

$$\text{aldehyde} \qquad\qquad\qquad \text{acid}$$

$$(19\text{-}75)$$

On the other hand, the production of ATP from ADP and phosphate is uphill on the free energy surface by about this same amount. Clearly, a combination of these two processes would have a ΔG near zero and would result in the formation of ATP without an accompanying increase in free

---

* We will be utterly unspecific about the organic compounds we mention. The symbol R is used to represent any organic fragment.

energy. A sequence of steps that brings about this net result is possible. Such a process is known as a coupled reaction.

$$R-\overset{\overset{\displaystyle O}{\|}}{C}-H + \tfrac{1}{2}O_2 + HPO_4^{-2} \longrightarrow \boxed{RC-O-\overset{\overset{\displaystyle O}{\|}}{\underset{\underset{\displaystyle O^-}{|}}{P}}-O^-} + H_2O \qquad (19\text{-}76a)$$

$$ADP^{-3} + \boxed{R-\overset{\overset{\displaystyle O}{\|}}{C}-O-\overset{\overset{\displaystyle O}{\|}}{\underset{\underset{\displaystyle O^-}{|}}{P}}-O^-} \longrightarrow ATP^{-4} + R-\overset{\displaystyle O}{\underset{\displaystyle O^-}{C}} \qquad (19\text{-}76b)$$

Each of these steps must be catalyzed by an appropriate enzyme. The free energy change associated with the aldehyde oxidation (19-76a) is conserved by the formation of the phosphate ester (box). The free energy of hydrolysis of a phosphate ester is even larger than that of ATP itself. Thus, when the second reaction occurs, the enzymatic transfer of a phosphate group from the ester to ADP, this free energy is used for the formation of ATP. Now the favorable free energy change associated with the oxidation is effectively conserved as ATP free energy, which is used in turn, in further reactions. This is possible only because the pair of reactions (19-76) have a common intermediate, the phosphate ester.

(e) BIOLOGICAL INFORMATION AND ENTROPY

A child playing with a typewriter could eventually fill a page with letters. To others than, possibly, himself, this page would be totally meaningless— a random collection of letters and spaces. On the other hand, consider the decisions made in writing a page of this book.* Information transmitted via the printed word is a culmination of an enormous number of decisions: After each letter, the author has to choose, from among the twenty-six letters of the alphabet, the next one that will create the word that, when assembled with other words, conveys his idea.

We have already discussed energy and positional randomness. Let's consider a new kind, *information randomness*. The two pages described above might contain exactly the same assortment of letters. On the child's page, the information randomness is very high; on the second the randomness has decreased in favor of the information contained in the ordered sets of words. The ordering of the letters in the alphabet into words, and

---

* Any suggestion of similarities between the two situations here described would be considered unkind by the authors. . . .

the words into sentences, represents a very large decrease in informational randomness. In the language of thermodynamics, we would say that the transformation from the child's game to the manuscript page involved a very large decrease in entropy. For this transformation to occur, other processes must occur in which entropy of the surroundings increases even more. As these processes proceed down the free energy hill, a part of the free energy is stored in the informational content of the page. Consideration of information as stored free energy (or stored "order") is a part of the new field "information theory," an extension of both thermodynamics and probability theory. One of the most interesting applications of information theory is to biological systems. We will not go into detail about the methods, but we will describe very briefly one of the ways in which biological information is stored.

Biological instructions for the manufacture of cells in living matter are stored in chemical substances. The most well-known example is DNA (deoxyribonucleic acid), a macromolecule apparently responsible for transmitting hereditary characteristics. The "alphabet" used by DNA to record biological instructions is short, consisting of only four "letters." These letters are really complex molecules, attached in a specific way to the very long-chain DNA molecule. The ordering of these four molecules on DNA present in genes tells the biosynthetic factories whether they should be producing blond or brunette, fingers or toes, four legs or two, and so on.

The amount of biological information stored in even a very simple cell is enormous. When put on a quantitative basis by information theory, it is found that even a very small cell contains more information than a volume of an encyclopedia. From a probability point of view, then, the odds against the processes that store the information are overwhelming. It is only because the entropy decrease in these processes is balanced by the entropy increases associated with the formation of small molecules (CO_2, H_2O, etc.) as by-products of the combustion processes, that life is maintained.

In conclusion, then, we have seen that all biological processes ultimately depend upon energy from the sun to build complex molecules. These complex molecules store both energy and order, the former to activate the biological system and the latter to provide a downhill, free energy path on the way. Finally, the ultimate key to reproduction, chemical information storage, is accomplished through molecular replication. This process is probably the most entropy-expensive step in the chemistry of life—and the most essential.

Problems

1. For each of the following pairs of reactions, predict the structure of the product of the second reaction on the basis of the functional group chemistry shown in the first reaction, assuming skeletal integrity.

(a) 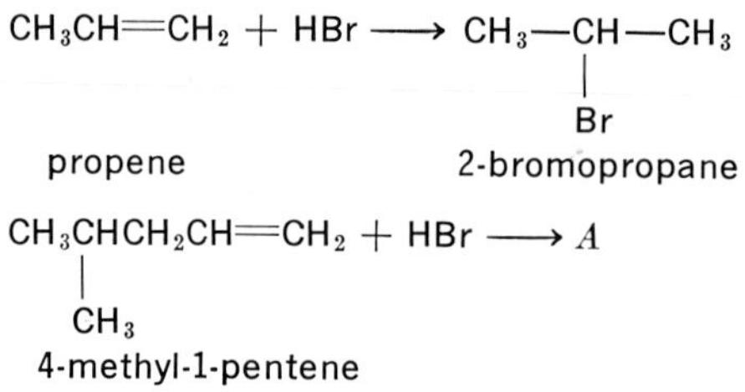 $CH_3CH{=}CH_2 + HBr \longrightarrow CH_3{-}CH{-}CH_3$
$$\underset{Br}{|}$$

        propene   2-bromopropane

$CH_3CHCH_2CH{=}CH_2 + HBr \longrightarrow A$
$$\underset{CH_3}{|}$$
4-methyl-1-pentene

(b) 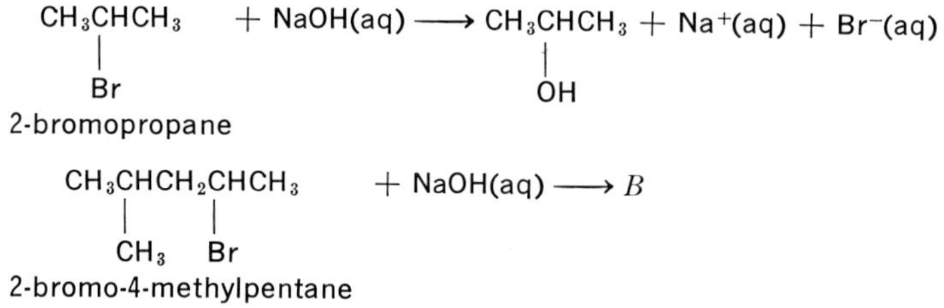 CH_3CHCH_3 $+ NaOH(aq) \longrightarrow CH_3CHCH_3 + Na^+(aq) + Br^-(aq)$
$$\underset{Br}{|} \qquad\qquad\qquad \underset{OH}{|}$$
   2-bromopropane

       $CH_3CHCH_2CHCH_3$ $+ NaOH(aq) \longrightarrow B$
$$\underset{CH_3 \quad Br}{| \qquad |}$$
       2-bromo-4-methylpentane

(c) 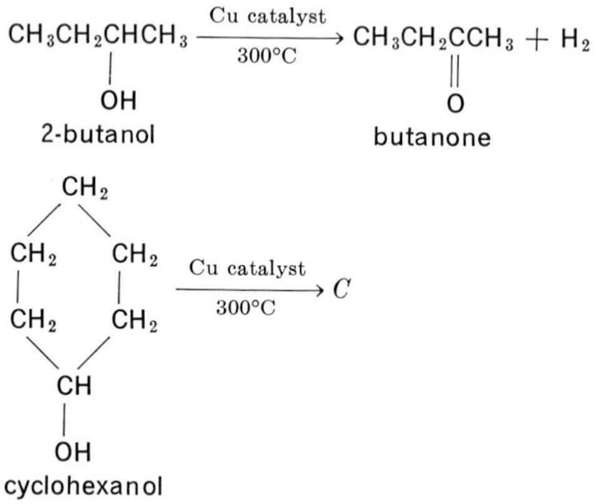 $CH_3CH_2CHCH_3 \xrightarrow[\text{300°C}]{\text{Cu catalyst}} CH_3CH_2CCH_3 + H_2$
$$\underset{OH}{|} \qquad\qquad\qquad\qquad \underset{O}{\|}$$
    2-butanol   butanone

$$
\begin{array}{c}
CH_2 \\
CH_2 \qquad CH_2 \\
CH_2 \qquad CH_2 \\
CH \\
OH
\end{array}
\xrightarrow[\text{300°C}]{\text{Cu catalyst}} C
$$

cyclohexanol

(d) 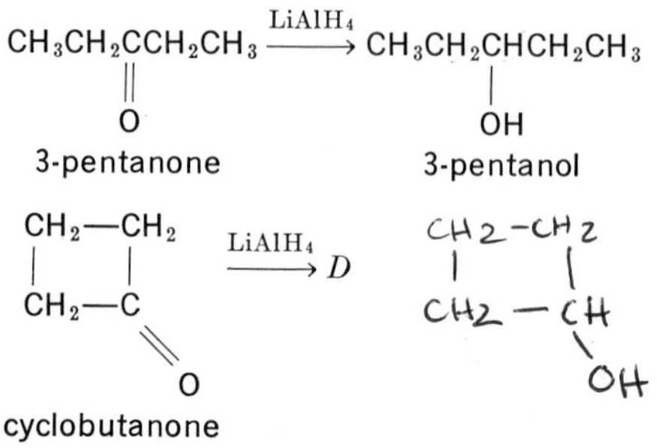 $CH_3CH_2CCH_2CH_3 \xrightarrow{\text{LiAlH}_4} CH_3CH_2CHCH_2CH_3$
$$\underset{O}{\|} \qquad\qquad\qquad\qquad\quad \underset{OH}{|}$$
    3-pentanone   3-pentanol

$$
\begin{array}{c}
CH_2{-}CH_2 \\
CH_2{-}C
\end{array}
\diagdown O
\xrightarrow{\text{LiAlH}_4} D
$$

cyclobutanone

$$
\begin{array}{c}
CH_2{-}CH_2 \\
CH_2{-}CH \\
\diagdown OH
\end{array}
$$

(e) $CH_3CH_2OH \xrightarrow[H_2SO_4]{conc.} CH_2{=}CH_2$

  ethyl alcohol   ethylene

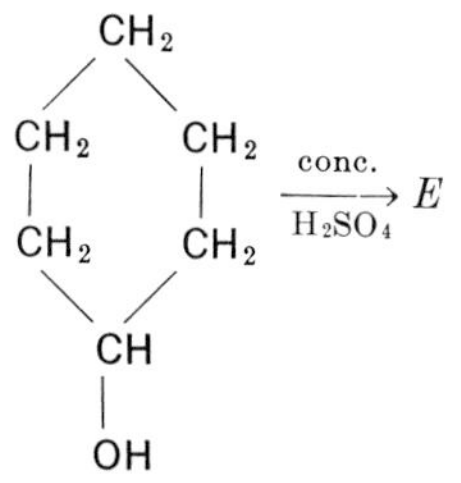

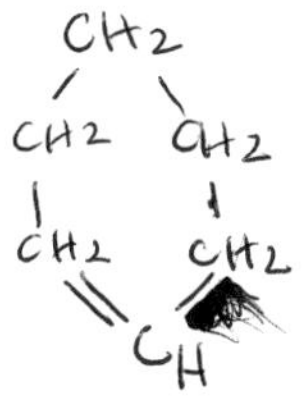

cyclohexanol $\xrightarrow[H_2SO_4]{conc.} E$

2. Formulate a sequence of reactions from Problem 1 by which cyclopentanone could be synthesized from cyclopentene.

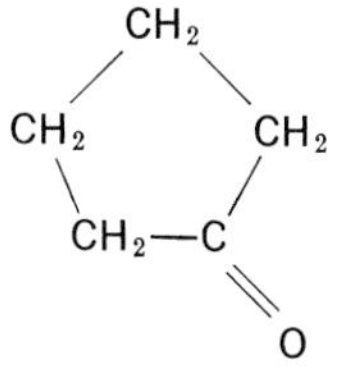

cyclopentanone 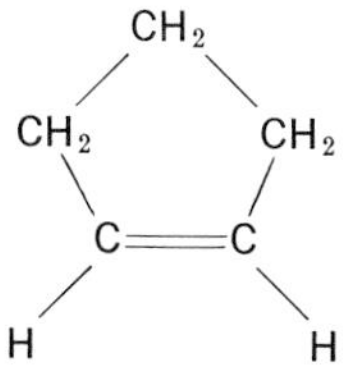   cyclopentene

3. Formulate a sequence of reactions from Problem 1 by which cyclopentene could be synthesized from cyclopentanone.

4. Complete and balance the equations for the reactions that can occur.

(a) $CH_3{-}CH{-}CH_3 + NaOH(aq) \longrightarrow$

  |
  Cl

(b) $CH_3{-}CH{-}CH_3 + KOH(alcohol) \longrightarrow$

  |
  Cl

(c) $CH_3{-}CH{-}CH_3 + Na(metal) \longrightarrow$

  |
  Cl

5. Assign the proper reagent to produce each of the listed products from 1-bromopropane, $CH_3CH_2CH_2Br$. Balance the equation for each reaction.

Reagents	Products
aqueous NaOH	(a) propylamine, $CH_3CH_2CH_2NH_2$
alcoholic KOH	(b) hexane, $CH_3CH_2CH_2CH_2CH_2CH_3$
ammonia, NH_3	(c) 1-propanol, $CH_3CH_2CH_2OH$
sodium, Na	(d) propene, $CH_3CH{=}CH_2$

6.   Irradiation of the cyclic ketone, cyclobutanone (see Problem 1(d)) with ultra-violet light gives a mixture of products that includes carbon monoxide, ethylene, and three more compounds, A, B, and C. Vapor density measurements show that A and B have molecular weights, respectively, 42.1 ± 0.5 and 41.9 ± 0.5. A and B each react on heating with H_2 and a catalyst, adding one mole of H_2 per 42 grams of either A or B. In each case, the product has an infrared spectrum identical to that of propane. A material balance shows that for every mole of cyclobutane there are 0.40 mole of CO, 0.15 mole of A, 0.25 mole of B, and 0.60 mole of ethylene.
(a) Write the structural formulas and names of the products A and B, indicating whether the identification is ambiguous or not.
(b) Deduce the likely molecular formula of C. Propose two conceivable geometries for C that utilize fully the bonding capacity of each atom (four bonds to carbon, two bonds to oxygen). One of these is unknown, presumably because it is un-stable—which of your two structures do you predict is the more stable one?

7.   In separate experiments, the thermodynamics of hydrogenation of A and B in Problem 6 are measured.

$$A + H_2 \longrightarrow CH_3CH_2CH_3 \qquad \Delta H = -38 \text{ kcal} \qquad \Delta S = -24 \text{ cal/deg}$$

$$B + H_2 \longrightarrow CH_3CH_2CH_3 \qquad \Delta H = -30 \text{ kcal} \qquad \Delta S = -31 \text{ cal/deg}$$

(a) What are ΔH and ΔS for the reaction $A \longrightarrow B$?
(b) Identify A and B from the possibilities deduced in Problem 6. Explain why the signs of ΔH and ΔS deduced in (a) identify A and B.

8.   Use the heats of formation of hydrogen cyanide, HC≡N ($\Delta H_f^0 = +31.2$ kcal), and of acetonitrile, $H_3CC≡N$ ($\Delta H_f^0 = +21.0$ kcal) coupled with the average bond energies given in Table 7-3 for the C–N and C–C bonds, the bond energies of H_2 and N_2, and the heat of vaporization of carbon (Appendix d) to provide two esti-mates of the C≡N bond energy.

9.   Of the nine structural isomers of heptane, C_7H_{16}, five are named as sub-stituted pentanes and one as a substituted butane, as given in Table 19-7. Assign one of these names to each of the following carbon skeleton representations. (Two are given twice.)

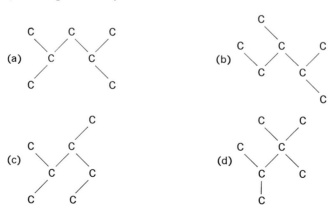

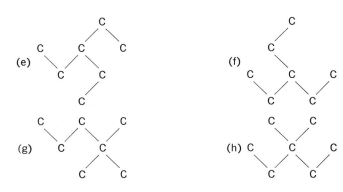

(e)

(f)

(g)

(h)

10. Name each of these five-carbon compounds:

(a) $CH_3CH_2CH_2CH_2CH_2OH$ 1 pentanol

(d) 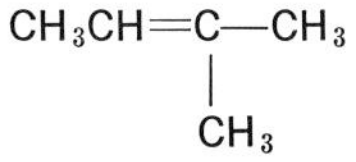 $CH_3CH=C-CH_3$
　　　　　　$|$
　　　　　CH_3

(b)

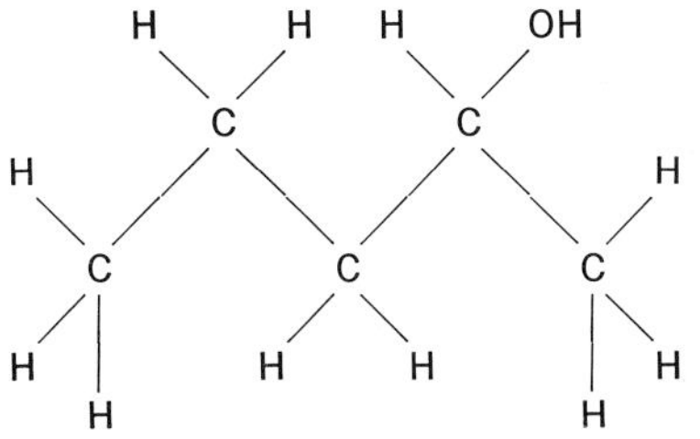

(e) 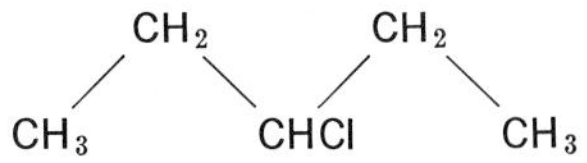
　　CH_2　　　CH_2
　$/$　　$\backslash$　$/$　　$\backslash$
CH_3　　$CHCl$　　CH_3

(c)

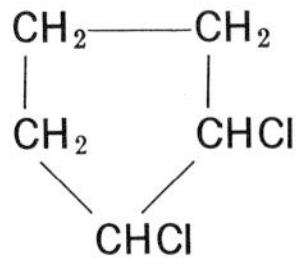

CH_2——CH_2
$|$　　　$|$
CH_2　　$CHCl$
　$\backslash$　$/$
　　$CHCl$

(f)

CH_2——CH_2
$|$　　　$|$
CH_2　　CH
　$\backslash$　$/\!\!/$
　　CH

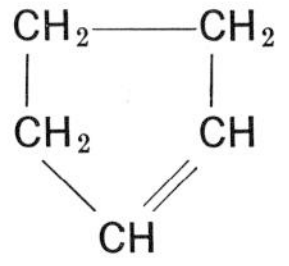

11. Draw and name three compounds, each of which contains three carbon atoms, one oxygen atom, and enough hydrogen atoms so that there are no double bonds.

12. Draw and name three compounds, each of which contains three carbon atoms, one oxygen atom doubly bonded to carbon, and enough hydrogen atoms so that there are no other double bonds.

13. Animal muscles (including yours) are largely protein material. These are very high molecular weight compounds (polymers) with repeating structural units like the one shown in the dashed parentheses

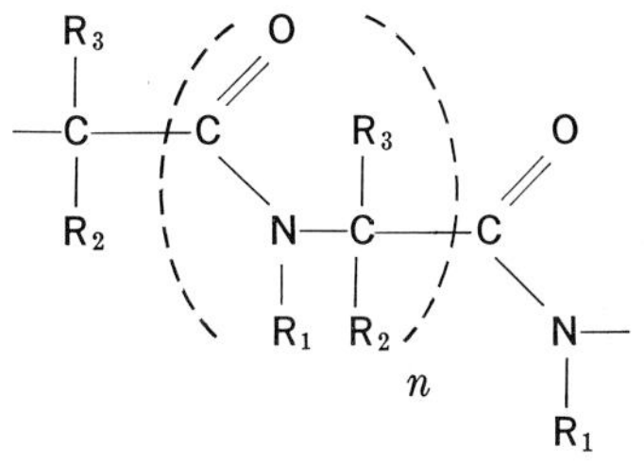

(a) Balance the equation for the combustion of one protein building block (like that in the dashed parentheses) if R_1, R_2, and R_3 are each hydrogen atoms.
(b) Calculate the heat of combustion of this protein building block from the average bond energies (see Appendix g). Count only one of the C–C bonds connecting the unit to the chain and assume that nitrogen ends up as N_2.
(c) Repeat (a) and (b) if R_1, R_2, and R_3 are each methyl groups, $-CH_3$.
(d) Are your muscles inflammable, thermodynamically speaking?

14.   In considering the photochlorination of methane (reaction (19-21)), the text ignores any possible mechanism based upon the first step

$$Cl_2(g) \longrightarrow Cl^+(g) + Cl^-(g)$$

(a) From the bond energy of Cl_2, the ionization energy of Cl, and its electron affinity, calculate ΔH^0 for this reaction.
(b) What wavelength of light will just suffice to initiate this process?

15.   In the photochlorination of methane, the first reaction (19-22b) has the higher activation energy, so it will limit the reaction rate.

$$Cl + CH_4 \longrightarrow CH_3 + HCl \qquad k = 5 \cdot 10^{10}\ e^{-3900/RT}\ cc/mole\text{-}sec$$

Suppose another substance is added to the reaction mixture to remove CH_3 as fast as it is formed (preventing reaction (19-22c)). Calculate the time Δt required for loss by reaction of $\Delta[CH_4] = 0.10$ torr of CH_4 (and of Cl) in an initial mixture of 10 torr CH_4 and 1.0 torr Cl at $298°K$. (Note: 1 torr $= 5.38 \cdot 10^{-8}$ mole/cc at $298°K$.)

16.   Draw the structures of the following compounds, showing both *cis* and *trans* forms when they are possible as structural isomers.
(a) 1,1-dichloroethane, $C_2Cl_2H_4$
(b) 1,1-dichloroethene, $C_2Cl_2H_2$
(c) 1,2-dichloroethane, $C_2Cl_2H_4$
(d) 1,2-dichloroethene, $C_2Cl_2H_2$

17.   In Problem 16, which of the molecules
(a) would have approximately tetrahedral bond angles about carbon?
(b) would have sp^2 hybrid bond angles ($\approx 120°$) about carbon?
(c) would be planar?
(d) would have a zero dipole moment?

18.   A chemist is handed four bulbs, A, B, C, and D, each containing a pure gas labeled "butene, C_4H_8." Their infrared spectra show that no two of these compounds are the same. Catalytic hydrogenation of each sample produces new substances A′, B′, C′, and D′. The infrared spectra of A′, B′, and C′ are identical, whereas that of D′ is distinct. When the original substances are exposed to bromine, new substances A″, B″, C″, and D″ are produced. Now the infrared spectra of A″ and B″ are identical, whereas those of C″ and D″ are each distinct.
(a) Draw the structure and name all compounds (including products) whose identity is positively known.
(b) If there are any compounds (or products) whose identity cannot be specified, give the possibilities.

19.   (a) Propose a series of reactions by which propionaldehyde, CH_3CH_2CHO, can be converted to 1-chloropropane, $CH_3CH_2CH_2Cl$.
(b) Propose a series of reactions by which propionaldehyde can be converted to 2-chloropropane, $CH_3CHClCH_3$.

20. There are three alcohols, three ethers, an aldehyde, and a ketone, all with the same molecular formula, C_3H_6O. Draw their structures.

21. Complete and balance the equations for reactions that can occur.

(a)
$$CH_3-\underset{\underset{CH_2}{\|}}{C}-CH_3 + Cl_2 \longrightarrow$$

(b)
$$CH_3-\underset{\underset{CH_3}{|}}{CH}-CH_3 + Cl_2 \xrightarrow[\text{light}]{\text{UV}}$$

(c) H—⟨benzene ring with H, H, H, H⟩—CH≡CH$_2$ + Cl$_2$ ⟶

(d) H—⟨benzene ring with H, H, H, H⟩—CCl$_3$ + Cl$_2$ $\xrightarrow{\text{heat}}$

22. Instructions for oxidation of propionaldehyde, CH_3CH_2CHO, to propionic acid, CH_3CH_2COOH, read "Add three times the weight of $Cr_2O_7^{-2}$ needed." For one mole of propionaldehyde, how many grams of $K_2Cr_2O_7$ should be weighed out?

23. A chemist is given one cc of a pure substance X, a clear liquid, for identification. He finds that a drop of X dissolves in one cc of benzene and a drop of X dissolves in one cc of water to give a solution with pH = 7. X reacts with aqueous $Cr_2O_7^{-2}$ to produce a liquid substance Y. The product Y is soluble in water and it gives a solution with pH = 4. When Y is dissolved in n-hexane, it reacts with n-pro-panol to produce the ester n-propyl-2-methyl propionate.
(a) Draw the structure of n-propyl-2-methyl propionate.
(b) Draw the structure of Y and name it.
(c) Draw the structures of two compounds, either of which could be X, and name each.

24. A clear liquid mixture of two substances X and Y is given the following series of tests for the purposes of identification.
A. A few drops of the mixture dissolves completely in one cc of water. The pH of the solution is 7.
B. The mixture slowly liberates H_2 when metallic sodium is added.
C. The mixture reacts completely with concentrated sulfuric acid to give a gaseous product P, which, on condensation, shows the sharp boiling point and sharp melting point characteristic of a pure substance. Product P decolorizes dilute Br_2 solution.
D. Product P reacts with HBr to give a product Q that does not decolorize dilute Br_2 solution and that is not soluble in water.

E. Product Q reacts with aqueous NaOH to give a product R that does not de-colorize dilute Br_2 solution and that is soluble in water.

F. Product R reacts with excess $Cr_2O_7^{-2}$ to give a product that is identifiable by odor to be acetone, CH_3COCH_3.

(a) Working backwards from test F, deduce the molecular structures of R, Q, and P.

(b) What are X and Y?

25. Enter the identifying letters of the six compounds shown in Problem 10 wherever applicable below.

(a) It would react catalytically with hydrogen.

(b) It would be oxidized by excess $Cr_2O_7^{-2}$ to a carboxylic acid.

(c) It would have *cis* and *trans* forms.

(d) It would be a suitable starting compound for the preparation in no more than two synthetic steps of a single straight-chain pentene isomer.

(e) It would be a suitable starting compound for the preparation in no more than two synthetic steps of a mixture of pentene isomers.

26. Show a sequence of reactions by which the ester propyl propionate,

$$CH_3CH_2C \overset{O}{\underset{OCH_2CH_2CH_3}{\big\diagup}}$$ could be prepared from 1-chloropropane without the use

of other organic reagents.

27. Which of the following compounds are

(a) amino acids?

(b) α-amino acids?

A.

$$\text{—}\bigcirc\text{—CH}_2\text{NH}_2$$

B.

$$\text{—}\bigcirc\overset{\text{—CH}_2\text{NH}_2}{\underset{\text{COOH}}{}}$$

C.

$$\underset{H_3C}{\overset{O}{\diagdown}}C\text{—}\bigcirc\text{—N}\overset{CH_3}{\underset{CH_3}{}}$$

D.

$$\underset{H}{\overset{H}{\diagdown}}N\overset{CH_3}{\underset{CH_3}{}}^+ \quad Cl^-$$

E.

$$\text{—}\bigcirc\text{—CH}_2\text{—CH—COOH} \quad \overset{NH_2}{\underset{}{}}$$

F.

$$\begin{matrix} CH_2 & O \\ CH_2 & C \\ CH_2\text{—N} \\ H \end{matrix}$$

G.

$$HO\text{—}\bigcirc\text{—C}\overset{O}{\underset{NH_2}{}}$$

H.

$$\underset{H}{\overset{CH_3}{\diagdown}}N\text{—CH}_2\text{—CH}_2\text{—CH}_2\text{—CH}_2\text{—C}\overset{O}{\underset{OH}{}}$$

28. Valene is a five-carbon α-amino acid that is needed in the diet of a normal adult human in order to maintain proper nitrogen equilibrium. Valene has the systematic name 2-amino-3-methylbutanoic acid. Draw its structural formula.

29. Propose a sequence of reactions for the production of the following polyamide polymer entirely from 4-chloro-1-butanol and inorganic reagents.

$$\left[\begin{matrix} \underset{\|}{\overset{O}{C}}-CH_2-CH_2-\underset{\|}{\overset{O}{C}} & & \underset{\|}{\overset{O}{C}}-CH_2-CH_2-\underset{\|}{\overset{O}{C}} & & \underset{\|}{\overset{O}{C}}- \\ -\underset{|}{\overset{}{N}} & \underset{|}{\overset{}{N}}-CH_2-CH_2-CH_2-CH_2-\underset{|}{\overset{}{N}} & & \underset{|}{\overset{}{N}}-CH_2-CH_2-CH_2-CH_2-\underset{|}{\overset{}{N}} \\ H & H & H & H & H \end{matrix}\right]_n$$

30. Propose a sequence of reactions for the production of the following polyester polymer entirely from ethylene, C_2H_4, and inorganic reagents.

$$\left[\begin{matrix} H & \underset{\|}{\overset{O}{C}}-\underset{\|}{\overset{O}{C}} & H\;H & \underset{\|}{\overset{O}{C}}-\underset{\|}{\overset{O}{C}} & H\;H & \underset{\|}{\overset{O}{C}}-\underset{\|}{\overset{O}{C}} & H \\ -\underset{|}{\overset{}{C}}-O & & O-\underset{|}{\overset{}{C}}-\underset{|}{\overset{}{C}}-O & & O-\underset{|}{\overset{}{C}}-\underset{|}{\overset{}{C}}-O & & O-\underset{|}{\overset{}{C}}- \\ H & & H\;H & & H\;H & & H \end{matrix}\right]_n$$

31. A number of sugar-producing plants carry out the reaction

$$\text{glucose} + \text{fructose} \longrightarrow \text{sucrose} + H_2O$$

A set of estimates of the free energies of formation of aqueous solutions of these three sugars are: glucose, -219.2; fructose, -218.8; and sucrose, -364.2 kcal/mole. The ΔG_f^0 for liquid water is -68.3 kcal.
(a) Calculate ΔG^0 for the sucrose formation reaction in aqueous solution.
(b) Does this reaction tend to proceed spontaneously?
(c) Will a catalyst (an enzyme) improve the equilibrium insofar as sucrose formation is concerned?
(d) Suppose this reaction is "driven" by the reaction

$$\text{ATP} \longrightarrow \text{ADP} + \text{phosphate} \qquad \Delta G = -7.0 \text{ kcal}$$

What is the percent efficiency (according to these ΔG_f^0 estimates) of utilizing the ATP $\longrightarrow$ ADP drop in free energy if consumption of one mole of ATP results in formation of one mole of sucrose?

32. Propose a sequence of reactions for the production of the following polyester polymer entirely from 2-pentene and inorganic reagents. Explain the element of randomness that would be likely. (Hint: Ozone splits 2-pentene into a three-carbon and a two-carbon product.)

$$\left[\begin{matrix} H & \underset{\|}{\overset{O}{C}}-\underset{\|}{\overset{O}{C}} & H\;H & \underset{\|}{\overset{O}{C}}-\underset{\|}{\overset{O}{C}} & H\;H & \underset{\|}{\overset{O}{C}}-\underset{\|}{\overset{O}{C}} & H \\ -\underset{|}{\overset{}{C}}-O & & O-\underset{|}{\overset{}{C}}-\underset{|}{\overset{}{C}}-O & & O-\underset{|}{\overset{}{C}}-\underset{|}{\overset{}{C}}-O & & O-\underset{|}{\overset{}{C}}- \\ CH_3 & & CH_3\;H & & CH_3\;H & & H \end{matrix}\right]_n$$

twenty   **chemical principles applied: the p orbital elements**

For the first half of this century most chemists were concerned with the preparation and characterization of chemical compounds. Organic chemists devoted themselves to carbon chemistry, inorganic chemists to the other ninety-odd elements. Physical chemists (a smaller group) were taking the results of the great advances in physics—thermodynamics and quantum mechanics in particular—and applying them to the fruits of the labors of the preparative chemists. Today, however, enormous improvements in relatively inexpensive laboratory instrumentation have made *every* practicing chemist a physical chemist in part.

This change of emphasis has had a dramatic effect on chemistry: The subject can be approached in a logical, rather than a phenomenological, way. Since we have a firm understanding of why atoms and molecules behave as they do, we can logically systematize their behavior with the aid of the guiding principles that have been presented earlier in this book. This will be evident in this chapter as we examine the chemistry of the elements in the right-hand block of the Periodic Table (Fig. 20-1).

The elements lithium through neon head eight columns of the Periodic Table and these columns define eight chemical families, or groups. The Group 3 elements, those under boron, have a single electron in their valence p orbitals. The Group 8 elements under neon have the six p electrons needed to fill these p orbitals. A group between Group 3 and Group 8, say, Group X, is characterized by $(X-2)$ p orbital electrons. This p orbital occupancy proves to be the distinguishing characteristic of the Group 3 to 8 elements, even in the long rows of the Periodic Table. Consequently, we shall call these elements the *p orbital elements*.

20-1 Periodic and group trends: a survey

Many periodic properties have already been discussed: trends in ionization energies (Chapters Two and Thirteen), in diatomic bond properties (Chapter Seventeen), in electronegativities (Chapter Seventeen), and in

784

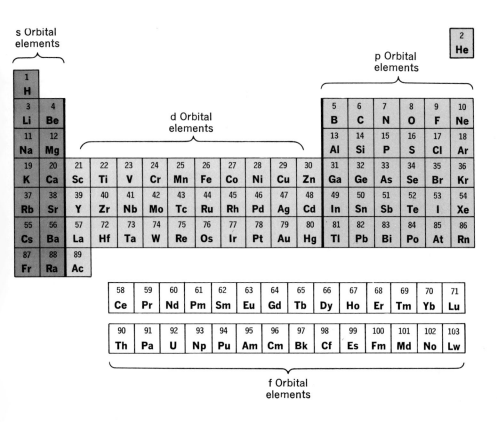

Figure 20-1 Valence orbitals and the Periodic Table.

crystal structures (Chapter Eighteen). Most of these discussions, however, were in terms of variation from left to right across a row. In this section we will concentrate on variations within a group, or family, as we did for the alkaline earth metals in Chapter Two.

(a) VALENCE ELECTRONS

In Chapter Two, we found that the electrons of an atom can be divided into groups by examining the values of E_m/m. Here E_m is the energy needed to remove the mth electron (see Section 2-3). These groups show that every atom has two type-1 electrons and that there can be, at most, eight type-2 electrons. Then, in Chapter Twelve, quantum mechanics revealed the significance of these groupings. The types correspond to the principal quantum numbers, n, of the orbitals from which the electrons are removed. There can be only two 1s electrons and these always have the highest E_m/m because they are closest to the nucleus. There can be at most only eight type-2 electrons because that is the maximum occupancy of the 2s and 2p orbitals. In retrospect, we can perceive in Figure 2-6 the difference between 2s and 2p electrons and even the difference between 2p electrons removed from filled and half-filled orbitals.

With this understanding of E_m/m, we can use it again to investigate the valence orbital energy relationships within each of the p orbital families, including the long-row representatives. Figure 20-2 shows these reduced ionization energies for the Group 3 to Group 8 families. Notice that the np and ns orbitals tend to become closer in energy as p orbital occupancy increases, while the $(n-1)$d orbitals are dropping away more and more. The large energy gap between the ns and $(n-1)$d orbitals explains why there is such a strong resemblance between the first two family members, which have no d electrons, and the latter family members that do. It explains why, for this group of elements, we can draw meaningful electron-dot formulas based on the ns and np orbital occupancy only. *Once filled, the nd orbitals drop so low in energy that they no longer function as valence orbitals.*

Another trend of interest is the slope of E_m/m for increasing n. There is a decreasing binding energy as we drop down in the Periodic Table, particularly between the first and second row. This is another manifestation of the decreasing electronegativity moving downward in a column (see Table 17-3) and it reflects both into bond lengths and bond energies. We shall examine bond lengths—atomic size—first.

(b) SIZE RELATIONSHIPS

The inverse relationship between E_m/m and average atomic size was explored in Chapter Two. Then, in Section 13-1(b), we developed a more sophisticated use of ionization energy to estimate, first, effective nuclear charge, then average radius in a one-electron approximation. With either model, we can expect manifestations of atomic size to be correlated with ionization energy trends.

We have developed two measures of atomic size, the interatomic distances between like, bonded atoms (covalent bond lengths, see Table 17-1) and those between like, but nonbonded atoms (van der Waals distances, see Fig. 18-6). Figure 20-3 shows how the covalent bond lengths change as we drop down in the Periodic Table. The covalent radii refer to half a normal single bond length, measured in appropriate compounds, as exampled in Table 17-1. If no appropriate compound was available, the radius was obtained by subtracting the chlorine or the bromine covalent radius from the bond length in the gaseous halide with the normal bonding

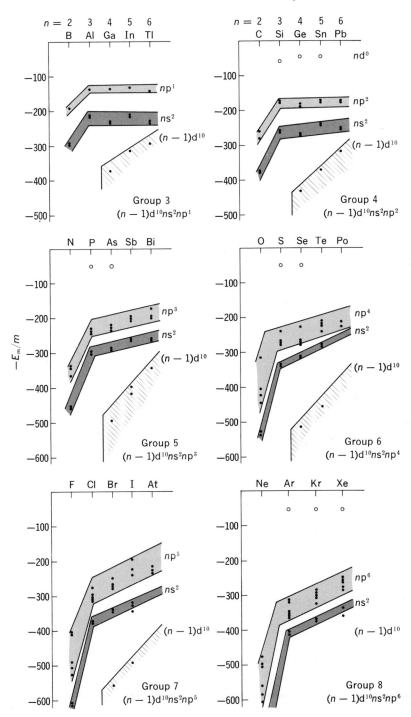

Figure 20-2   *Trends in E_m/m for the p orbital families, Groups 3 to 8.*

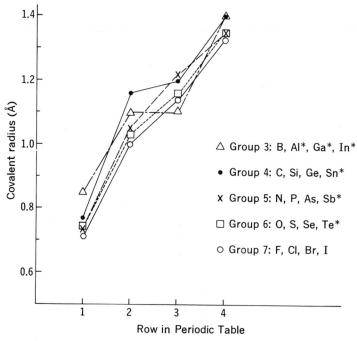

Figure 20-3   *Covalent single bond radius versus row in the Periodic Table.*

capacity. We see that the atoms become larger as we drop down the Periodic Table, as expected from the ionization energy trends. A glance at Figure 18-6 shows that the van der Waals radii also behave as expected. Figure 20-4 combines these two size estimates in a pictorial representation of the Group 6 dihydrides, H_2O, H_2S, H_2Se, H_2Te.

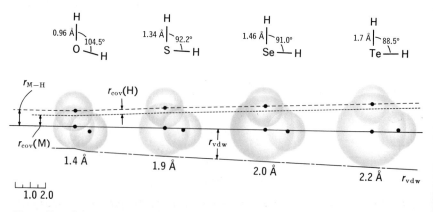

Figure 20-4   *Size trends in the Group 6 hydrides.*

(c) BOND ENERGY RELATIONSHIPS

Figure 20-5 shows some bond energy trends among the p orbital elements. The bond energy per bond order for homonuclear diatomic molecules, as given in Figure 17-2, is plotted *versus* row in the Periodic Table. For each group, the bond energy per bond order decreases steadily from the second to the fifth row, and within a row, it increases from Group 4 to Group 7. This consistent picture is altered in the first row, in which the bond energy per bond order decreases from Group 4 to Group 7. This inversion accentuates the bond energy decrease from C_2 to Si_2, and from N_2 to P_2, it causes the O_2 and S_2 bond energies to be about the same, and it causes the F_2 bond energy to fall well below that of Cl_2.

Figures 20-2 and 20-3 offer a possible explanation for these special behaviors of oxygen and fluorine. Notice that for these two elements, the np^4 and np^5 shaded areas in Figure 20-2 fan out. The spacing of the E_m/m values indicates that electron repulsion in the filled orbitals raises their energy relative to the half-filled, but the effect is significant only in the first row. That is because these atoms are the smallest (see Fig. 20-3), so the electrons are closest together. In oxygen this large electron repulsion arises from one filled p orbital and it detracts from the bond energy of the double bond in O_2. There are two such filled orbitals in fluorine, and the resulting enhancement of electron repulsion detracts even more noticeably from the single bond energy in F_2.

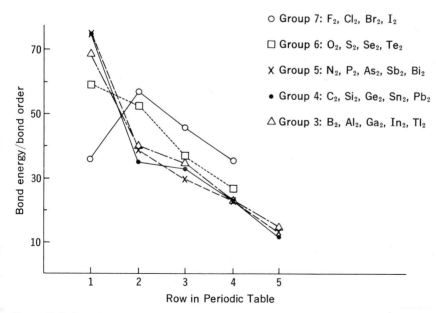

Figure 20-5 *Bond energy per bond order versus row in the Periodic Table. Homonuclear diatomic molecules (data from Fig. 17-2).*

Thus we can expect relatively smooth chemical trends within a family from row 2 to row 5, trends that reflect an increasing size and a decreasing bond energy per bond order. We shall see that these trends are found for bonds between unlike atoms, heteronuclear bonds, as well as for the homonuclear bonds cited above. We can also anticipate rather larger differences for the first row because of electron repulsion, especially for fluorine.

(d) OXIDATION STATES

The oxidation states displayed by an element tell us much about its chemical behavior. Table 20-1 lists compounds formed by three of the p orbital elements, carbon, nitrogen, and chlorine, to illustrate how rich it can be. All three elements exhibit oxidation states all the way from $+x$ to $-(8-x)$, where x is the group number. The relative stability of these states is compactly displayed in the Latimer $\mathcal{E}^0$ reduction potential diagram (see Section 6-5(c)). These are given in (20-1) through (20-4) for nitrogen and chlorine.

These reduction potentials show that nitrate and perchlorate ions are very strong oxidizing agents. Yet they are both regarded as quite inert substances in aqueous solutions because they react very slowly. In concentrated solutions and at elevated temperatures, there is quite a different story. Hot, fuming perchloric acid can explode with great violence in contact with organic material (Table 20-1 shows that almost all organic substances are oxidizable.) Nitric acid, too, becomes a powerful and rapid oxidizing agent in hot, concentrated solution, particularly in company with HCl (a mixture of concentrated HNO_3 and HCl has long been called "aqua regia" because it will even dissolve gold).

The nitrogen reduction potentials show, further, the remarkable stability of elemental nitrogen. This stability is largely due to the quite high bond energy per bond order in the N_2 triple bond (see Figs. 17-4 and 20-5).

Oxygen displays oxidation numbers of -2, -1 (in H_2O_2), and 0 (in O_2 and O_3). That is not, however, because this element leads a drab life, oxidation-wise. It has oxidation number -2 in so many of its compounds because we arbitrarily assign it so. Oxygen compounds are known for every element except the inert gases and even one of these, xenon, forms a series of oxides, such as XeO_3, XeO_4, and $XeOF_2$.

Fluorine has another quirk. This element is so electronegative that its chemistry is dominated by the single oxidation state, -1. The element, F_2, (oxidation state zero) is extremely reactive because of its low bond energy per bond order (Fig. 20-5). The compounds OF_2, HOF, and O_2F_2 are known in which fluorine must be assigned oxidation numbers $+1$ or $+2$ according to our bookkeeping rules. Of course, fluorine has both higher electronegativity and electron affinity than oxygen, so these positive oxidation numbers are again meaningless. In fact, the chemistry of these three compounds reflects fluorine's discomfiture. The first, F_2O, is quite reac-

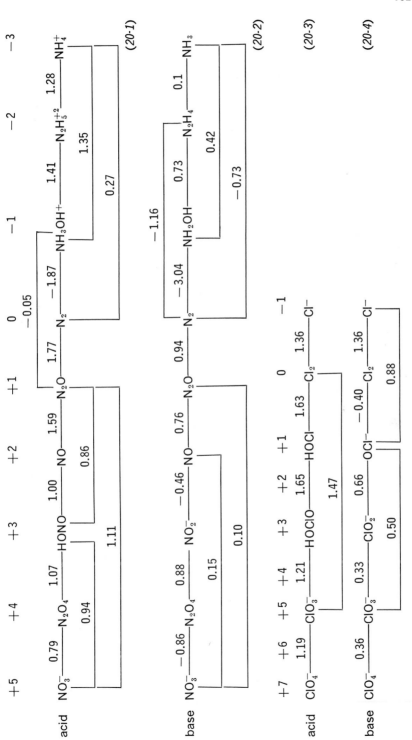

acid

$$+5 \quad\quad\quad +4 \quad\quad\quad +3 \quad\quad\quad +2 \quad\quad\quad +1 \quad\quad\quad 0 \quad\quad\quad -1 \quad\quad\quad -2 \quad\quad\quad -3$$

$$-0.05$$

$$NO_3^- \overset{0.79}{——} N_2O_4 \overset{1.07}{——} HONO \overset{1.00}{——} NO \overset{1.59}{——} N_2O \overset{1.77}{——} N_2 \overset{-1.87}{——} NH_3OH^+ \overset{1.41}{——} N_2H_5^{+2} \overset{1.28}{——} NH_4^+$$

0.94
1.11
0.86
1.35
0.27

(20-1)

base

$$-1.16$$

$$NO_3^- \overset{-0.86}{——} N_2O_4 \overset{0.88}{——} NO_2^- \overset{-0.46}{——} NO \overset{0.76}{——} N_2O \overset{0.94}{——} N_2 \overset{-3.04}{——} NH_2OH \overset{0.73}{——} N_2H_4 \overset{0.1}{——} NH_3$$

0.15
0.10
0.42
−0.73

(20-2)

acid

$$+7 \quad\quad +6 \quad\quad +5 \quad\quad +4 \quad\quad +3 \quad\quad +2 \quad\quad +1 \quad\quad 0 \quad\quad -1$$

$$ClO_4^- \overset{1.19}{——} ClO_3^- \overset{1.21}{——} HOClO \overset{1.65}{——} HOClO \overset{1.63}{——} HOCl \overset{1.63}{——} Cl_2 \overset{1.36}{——} Cl^-$$

1.47

(20-3)

base

$$ClO_4^- \overset{0.36}{——} ClO_3^- \overset{0.33}{——} ClO_2^- \overset{0.66}{——} OCl^- \overset{-0.40}{——} Cl_2 \overset{1.36}{——} Cl^-$$

0.50
0.88

(20-4)

Table 20-1 Oxidation States Displayed by Carbon, Nitrogen, and Chlorine

Carbon

Ox. no.	H + O only	H + F
+4	CO_2, H_2CO_3	CF_4
+3	HOOC–COOH	C_2F_6
+2	CO, H–COOH	CHF_3
+1	OHC–CHO	$C_2H_2F_4$
0	H_2CO	CH_2F_2
−1	$C_2H_4(OH)_2$	$C_2H_4F_2$
−2	CH_3OH	CH_3F
−3	C_2H_6	
−4	CH_4	

Nitrogen

Ox. no.	H + O only	H + F
+5	N_2O_5, $HONO_2$	
+4	NO_2	
+3	N_2O_3, HONO	NF_3
+2	NO	N_2F_4
+1	N_2O	N_2F_2, NHF_2
0	N_2	
−1	NH_2OH	NH_2F
−2	N_2H_4	
−3	NH_3, NH_4^+	

Chlorine

Ox. no.	H + O only
+7	$HClO_3$, ClO_4^-
+6	ClO_3
+5	$HClO_2$, ClO_3^-
+4	ClO_2
+3	HClO, ClO_2^-
+2	ClO
+1	Cl_2O, HOCl
0	Cl_2
−1	Cl^-, HCl

tive; it reacts with water to give O_2 and 2 HF to return to the -1 oxidation state. Hypofluorous acid, HOF, has been detected spectroscopically at very low temperatures, but it decomposes on warming to room temperature, again presumably forming O_2 and 2 HF from 2 HOF. The peroxide, O_2F_2, is dangerously reactive and it has quite unusual bonding. Although the molecule has the same geometry as H_2O_2 (nonplanar, bent) the O–O distance is only 1.22 Å, a double bond distance (compare O_2, $r = 1.21$ Å). The oxygen–fluorine bonds are correspondingly long, 1.58 Å (compare F_2O, $r = 1.42$ Å).

Table 20-2 lists some Latimer reduction potential diagrams for some other p orbital elements for contrast to those discussed earlier.

(e) HYDRIDES

Except for the inert gas elements, all of the p orbital elements form one or more compounds with hydrogen. These compounds tend to be volatile and they form molecular solids (except indium and thallium). Figure 20-6 shows some of these hydrides and their melting and boiling points. It also shows the bonding, via pigeonhole diagrams, for the simplest hydrides. The structures are always in qualitative agreement with the orbital hybridization expectations, with the notable exceptions of the Group 3 electron acceptors, such as BH_3. These give bridged structures (see Fig. 17-20) characteristic of electron-deficient bonding.

To demonstrate group trends, some properties of the Group 5 hydrides are summarized in Table 20-3. With the exception of ammonia, they are all thermodynamically unstable with respect to the elements. Their availability on the chemical shelf indicates kinetic protection—there must be high activation energies needed to rearrange the bonds. We can easily calculate average bond energies from the heats of formation. For example, let's work out the energy, ΔH_{atoms}, required to break up $PH_3(g)$ into atoms.

$$PH_3(g) = P(g) + 3 H(g) \tag{20-5}$$

$$\Delta H_{atoms} = \Delta H_f^0[P(g)] + 3\Delta H_f^0[H(g)] - \Delta H_f^0[PH_3(g)]$$

$$= (75.2) + 3(52.1) - (1.3)$$

$$= 230 \text{ kcal/mole}$$

Since there are three P–H bonds, the average bond energy is

$$D_{PH} = \tfrac{1}{3}\Delta H_{atoms} = \tfrac{1}{3}(230 \text{ kcal})$$

$$= 77 \text{ kcal/bond} \tag{20-6}$$

All the M–H bonds are quite strong—strong enough, apparently, to resist kinetically a negative ΔG^0 for decomposition to the elements. As the

Table 20-2 Oxidation States and Reduction Potentials for p *Orbital Elements*

Group 3

Acid solution

$$Al^{+3} \xrightarrow{-1.66} Al$$

$$Ga^{+3} \xrightarrow{-0.65} Ga_2^{+4} \xrightarrow{-0.45} Ga$$

$$Ga^{+3} \xrightarrow{-0.53} Ga$$

$$In^{+3} \xrightarrow{-0.45} In_2^{+4} \xrightarrow{-0.35} In^{+} \xrightarrow{-0.25} In$$

$$In^{+3} \xrightarrow{-0.342} In$$

$$Tl^{+3} \xrightarrow{+1.25} Tl^{+} \xrightarrow{-0.34} Tl$$

Basic solution

$$Al(OH)_3 \xrightarrow{-2.31} Al$$

$$H_2GaO_3^- \xrightarrow{-1.22} Ga$$

$$In(OH)_3 \xrightarrow{-1.0} In$$

$$Tl(OH)_3 \xrightarrow{-0.05} Tl(OH)$$

Group 4

Acid solution

$$Sn^{+4} \xrightarrow{+0.15} Sn^{+2} \xrightarrow{-0.14} Sn$$

$$PbO_2 \xrightarrow{+1.46} Pb^{+2} \xrightarrow{-0.13} Pb$$

Basic solution

$$Sn(OH)_6^{-2} \xrightarrow{-0.90} HSnO_2^- \xrightarrow{-0.91} Sn$$

$$PbO_2 \xrightarrow{+0.28} PbO \xrightarrow{-0.54} Pb$$

Group 5

Acid solution

$$H_3PO_4 \xrightarrow{-0.94} H_4P_2O_6 \xrightarrow{+0.38} H_3PO_3 \xrightarrow{-0.50} H_3PO_2 \xrightarrow{-0.51} P \xrightarrow{-0.1} P_2H_4 \xrightarrow{0.0} PH_3$$

$$H_3PO_4 \xrightarrow{-0.28} H_3PO_3 \qquad H_3PO_3 \xrightarrow{-0.50} P \qquad P \xrightarrow{-0.065} PH_3$$

$$H_3AsO_4 \xrightarrow{+0.56} H_3AsO_3 \xrightarrow{+0.25} As \xrightarrow{-0.60} AsH_3$$

$$Sb_2O_5 \xrightarrow{+0.48} Sb_2O_4 \xrightarrow{+0.68} SbO^{+} \xrightarrow{+0.21} Sb \xrightarrow{-0.51} SbH_3$$

$$Sb_2O_5 \xrightarrow{+0.58} SbO^{+}$$

$$Bi_2O_5 \xrightarrow{+1.6} BiO^{+} \xrightarrow{+0.32} Bi \xrightarrow{-0.8} BiH_3$$

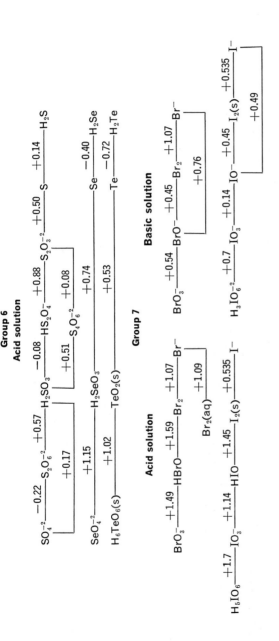

Group 6
Acid solution

SO_4^{-2} —0.22— $S_2O_6^{-2}$ —+0.57— H_2SO_3 —−0.08— $HS_2O_4^-$ —+0.88— $S_2O_3^{-2}$ —+0.50— S —+0.14— H_2S

+0.17

+0.51 —$S_4O_6^{-2}$— +0.08

SeO_4^{-2} —+1.15— H_2SeO_3 —+0.74— Se —−0.40— H_2Se

+1.02

$H_6TeO_6(s)$ —+1.02— $TeO_2(s)$ —+0.53— Te —−0.72— H_2Te

Group 7
Acid solution

BrO_3^- —+1.49— $HBrO$ —+1.59— Br_2 —+1.07— Br^-

$Br_2(aq)$ —+1.09

H_5IO_6 —+1.7— IO_3^- —+1.114— HIO —+1.45— $I_2(s)$ —+0.535— I^-

Basic solution

BrO_3^- —+0.54— BrO^- —+0.45— Br_2 —+1.07— Br^-

+0.76

$H_3IO_6^{-2}$ —+0.7— IO_3^- —+0.14— IO^- —+0.45— $I_2(s)$ —+0.535— I^-

+0.49

3	4	5	6	7
$B_2H_6(g)$ Diborane 108, 181°K	$CH_4(g)$ Methane 89, 112°K	$NH_3(g)$ Ammonia 195, 240°K	$H_2O(l)$ Water 273, 373°K	HF(g) Hydrogen fluoride 189, 293°K
$(AlH_3)_n(s)$ Alane	$SiH_4(g)$ Silane 88, 161°K	$PH_3(g)$ Phosphine 140, 185°K	$H_2S(g)$ Hydrogen sulfide 188, 213°K	HCl(g) Hydrogen chloride 159, 188°K
$Ga_2H_6(l)$ Digallane (?) 252, 412°K	GeH_4 Germane 108, 183°K	$AsH_3(g)$ Arsine 159, 210°K	$H_2Se(g)$ Hydrogen selenide 207, 232°K	HBr(g) Hydrogen bromide 192, 206°K
(InH_3)	$SnH_4(g)$ Stannane 123, 221°K	$SbH_3(g)$ Stibine 182, 256°K	$H_2Te(g)$ Hydrogen telluride 222, 271°K	HI(g) Hydrogen iodide 222, 238°K
	$PbH_4(g)$ Plumbane (unstable)	BiH_3 Bismuthine (unstable)	$H_2Po(g)$ Hydrogen polonide (unstable)	HAt

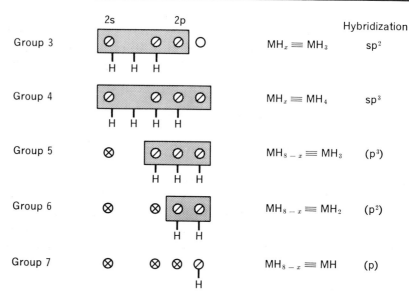

Figure 20-6 Simple hydrides of the p elements (m.p. b.p, degK).

central atoms become larger down the group, the bonds become weaker and longer as expected. The bond angles are interesting in themselves; they were discussed in Section 16-3.

Ammonia is a good electron donor, as mentioned in Section 17-3(a). Since nitrogen is considerably more electronegative than hydrogen, electrons will be attracted toward this atom, accounting for the significant dipole moment. Phosphorus, arsenic, and antimony have electronegativities close to that of hydrogen, implying much less shift of electron densities in PH_3, AsH_3, and SbH_3. This conclusion is in accord with the decreasing dipole moment in the series (see Table 20-3). Since the electronegativity of antimony (1.9) is less than that of hydrogen (2.1), the Sb atom may occupy the positive end of the SbH_3 molecular dipole. This trend also suggests that there should be a lessening of donor properties. In fact, PH_3, AsH_3, and SbH_3 do not function as electron donors at all! All of this shows again how meaningless are the oxidation numbers that seem to attribute a -3 charge to each of these elements in the MH_3 hydrides.

These compounds are really transitional between compounds like HCl in which the hydrogen is definitely positive and the metallic hydrides (e.g., LiH) in which it is undeniably negative. The rules in Chapter Six assign a -1 oxidation number to hydrogen in metallic hydrides. From Table 14-6 we find the fractional charge δ in LiH

$$\delta = \frac{\mu}{r_{LiH} e_0} = 0.76$$

which represents a large shift of electron density towards hydrogen. This suggests that the hydrogen end of a metallic hydride could act as an electron donor. In contrast, hydrides of both of the p orbital elements B and Al are electron deficient and react with metallic hydrides to produce complex hydride ions.

$$2\ LiH + B_2H_6 \xrightarrow{\text{ether}} 2\ LiBH_4 \qquad\qquad (20\text{-}7)$$

$$LiH + AlH_3 \xrightarrow{\text{ether}} LiAlH_4 \qquad\qquad (20\text{-}8)$$

Both the lithium borohydride and lithium aluminum hydride are ionic solids consisting of Li^+ ions and tetrahedral MH_4^- ions. Both BH_4^- and AlH_4^- give up a hydride ion, H^-, quite easily and are excellent reducing agents. The aluminum compound is a common reagent in organic chemistry (see Sections 19-1(b) and 19-3(c)). For example, carboxylic acids are easily

Table 20-3 Properties of Group 5 Hydrides*

	ΔH_f^0 (kcal/mole)	ΔG_f^0 (kcal/mole)	D_{M-H}†	r_{M-H} (Å)	H-M-H (deg)	μ (Debye)
NH_3	-11.0	-3.9	93	1.01	106.8	1.47
PH_3	$+1.3$	$+3.2$	77	1.44	93.5	0.58
AsH_3	$+15.9$	$+16.5$	71	1.52	91.8	0.22
SbH_3	$+35.9$	$+53.1$	61	1.71	91.5	0.12

* Melting and boiling points are given in Figure 20-6.
† Average bond energies calculated using the following enthalpies of formation of gaseous atoms (kcal/mole): N, 113; P, 75.2; As, 72.3; Sb, 62.7.

reduced to alcohols by way of a complex aluminum salt, which is converted to the alcohol by further reduction and acid hydrolysis.

$$LiAlH_4 + CH_3-C\overset{\displaystyle O}{\underset{\displaystyle OH}{\big<}} \quad \xrightarrow{\text{ether}} \quad CH_3-C\overset{\displaystyle O}{\underset{\displaystyle OAlH_3^-Li^+}{\big<}} \quad + H_2$$

acetic acid

$$CH_3COOAlH_3Li + 2\,LiAlH_4 + 3\,HCl \longrightarrow$$
$$CH_3-CH_2-OH + 3\,LiCl + 3\,AlH_3 + H_2O \quad (20\text{-}9)$$
ethanol

20-2 Oxygen compounds: especially oxyacids

Oxygen makes up one fifth by volume of the atmosphere, eight ninths by weight of the water in our oceans, and nearly fifty percent by weight of the earth's crust (mostly as oxides of silicon and aluminum). For this important element, we'll look first at its compounds with the p orbital elements (the oxides) and then with the p orbital elements in combination with hydrogen (the oxy-acids).

(a) OXIDES OF THE p ORBITAL ELEMENTS

Table 20-4 lists the enthalpies and free energies of formation of many of the oxides formed by the p orbital elements. With the exception of the nitrogen oxides and most of the Group 7 (halogen) oxides, these compounds are normally solids and have negative free energies of formation. These negative values of ΔG_f^0 mean that the solid oxides are thermodynamically stable with respect to the elements.

Table 20-5 gives the average bond energies of some gaseous oxides. According to the simple molecular orbital scheme presented in Section 15-5, the Group 4 diatomic oxides have enough electrons to form triple bonds, the Group 3 and 5 oxides would form $2\frac{1}{2}$ order bonds, and the Groups 6, 7, and 8 would have successively lower bond orders, respectively, 2, $1\frac{1}{2}$, and 1. Figure 20-7 shows the data from Table 20-5 and, indeed, each row of the Periodic Table shows the characteristic peak at the 10-valence electron compound, as shown earlier for a variety of first-row compounds in Figures 15-12 and 15-16. Again we can conclude that the bonding in the diatomic oxides of the p orbital elements can be described in molecular orbital language. These elements do form multiple bonds, with decreasing bond energy per bond order as we move down in the Periodic Table, exactly as found for the homonuclear diatomics in Figure 17-4.

It is amusing to extrapolate the curves of Figure 20-7 to the expected

Table 20-4 *Thermodynamic Properties of p Element Oxides: Enthalpies and Free Energies of Formation (298°K) (kcal/mole)*

Oxide	ΔH_f^0	ΔG_f^0	Oxide	ΔH_f^0	ΔG_f^0	Oxide	ΔH_f^0	ΔG_f^0
Group 3			$SnO(s)$	−68.3	−61.4	**Group 6**		
$BO_2(g)$	−71.8		$SnO_2(s)$	−138.8	−124.2	$O_2(g)$	0.0	0.0
$B_2O_2(g)$	−108.7		$PbO(s)$	−51.9	−44.9	$O_3(g)$	34.1	39.0
$B_2O_3(c)$	−304.2		$PbO_2(s)$	−66.3	−51.9	$SO_2(g)$	−70.9	−71.7
$Al_2O(g)$	−31		$Pb_3O_4(s)$	−171.7	−143.7	$SO_3(g)$	−94.6	−88.7
$(AlO)_2(g)$	−94							
$Al_2O_3(s)$	−401	−378	**Group 5**			$SeO_2(s)$	−53.9	
$Ga_2O(s)$	−85					$SeO_3(s)$	39.9	
$Ga_2O_3(s)$	−260.3	−238.6	$NO(g)$	21.6	20.7	$Se_2O_5(s)$	−97.6	
$In_2O_3(s)$	−221.3	−198.6	$NO_2(g)$	7.9	12.3	$TeO_2(s)$	−77.1	−64.6
			$N_2O(g)$	19.6	24.9			
$Tl_2O(s)$	−42.7	−35.2	$N_2O_3(g)$	20.0	33.3	**Group 7**		
$Tl_2O_3(s)$		−74.5	$N_2O_4(g)$	2.2	23.4			
$Tl_2O_4(s)$		−83.0	$N_2O_4(\ell)$	−4.7	23.3	$OF_2(g)$	−5.2	−1.1
			$N_2O_5(s)$	−10.3	27.2	$O_2F_2(g)$	4.3	
Group 4			$N_2O_5(g)$	2.7	27.5	$ClO_2(g)$	24.5	28.8
						$Cl_2O(g)$	19.2	23.4
$CO(g)$	−26.4	−32.8	$P_4O_6(s)$	−392		$Cl_2O_7(\ell)$	56.9	
$CO_2(g)$	−94.1	−94.2	$P_4O_{10}(s)$	−713	−645	$BrO_2(s)$	11.6	
$SiO_2(s)$	−217.7	−204.8	$As_2O_5(s)$	−221	−187	$I_2O_5(s)$	−37.8	
			$As_4O_6(s)$	−314	−274			
$GeO(s)$	−50.7	−56.7	$Sb_2O_4(s)$	−216.9	−190	**Group 8**		
$GeO_2(s)$	−131.7	−118.8	$Sb_2O_5(s)$	−232	−198			
$Ge_2O_2(g)$	−112		$Sb_4O_6(s)$	−344	−303	$XeO_3(s)$	96	
$Ge_2O_3(g)$	−212							

Table 20-5   Average Bond Energies of the p Orbital Gaseous Oxides

Group 3		4		5		6		7		8	
BO	173	**CO**	256	**NO**	162	**O₂**	118	FO		Ne	
BO₂	163	CO₂	192	NO₂	117	O₃	72	F₂O	45		
AlO	120	**SiO**	192	**PO**	145	**SO**	124 or 93.4	**ClO**	64	Ar	
								ClO₂	60		
Al₂O	123					SO₂	119	Cl₂O	49?		
						SO₃	104				
GaO	59.2	**GeO**	157	**AsO**	113	**SeO**	81?	**BrO**	46	Kr	
								Br₂O	>60		
InO	26	**SnO**	132	**SbO**	75	**TeO**	63?	**IO**	43	Xe	
										XeO₃	27
TlO		**PbO**	94?	**BiO**	85?	**PO**		AtO		Rn	

bond energy of the oxides of the inert gases. All of the extensions give bond energies in the range 16 to 30 calories. The inert gas oxide xenon trioxide has an average bond order in this range, 27 kcal per bond.

The oxides of the Group 5 elements typify the similarities and differences within a family. In Table 20-4, we find that all of the elements form $+3$ and $+5$ oxides, M_2O_3 or M_4O_6 and M_2O_5 or M_4O_{10}. In addition, nitrogen forms

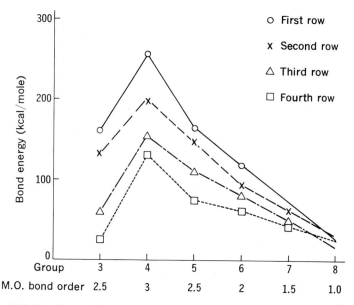

Figure 20-7   Bond energy in gaseous diatomic oxides.

common oxides with oxidation states $+1$, $+2$, and $+4$: nitrous oxide, N_2O; nitric oxide, NO; and nitrogen dioxide, NO_2 and N_2O_4. Figure 20-8 shows the structures of the oxides of nitrogen (as well as of elemental nitrogen). All of the structures involve strong multiple bonds. In some cases, extremely weak bonds are formed, apparently in order to preserve the multiple bonds. For example, N_2O_3 consists of nitric oxide and nitrogen dioxide weakly

Oxidation Number	Name	Structure	Bond Orders		Bond Energies	
			N–O	N–N	N–O	N–N
0	nitrogen	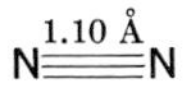 $N\!\equiv\!\equiv\!N$ 1.10 Å		3		225
$+1$	nitrous oxide	$N\!\!\stackrel{1.13\,Å}{=\!=}\!N\!\!\stackrel{1.19\,Å}{=\!=}\!O$	2	2	105	105
			(linear, isoelectronic with CO_2)			
$+2$	nitric oxide	$N\!\!\stackrel{1.15\,Å}{\cdots\cdots}\!O$	2.5		162	
$+3$	dinitrogen trioxide	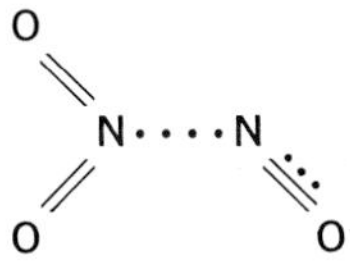 (see structure)	2, 2.5 $<\frac{1}{2}$ (very weak N–N bond, structure not well known)		10	
$+4$	nitrogen dioxide	1.19 Å N 1.19 Å O 134° O	1.5		117	
	dinitrogen tetroxide	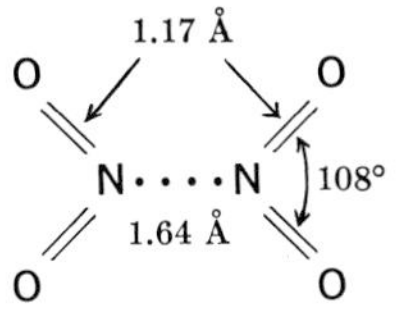 1.17 Å N····N 108° 1.64 Å	2 (crystal, very weak N–N bond)	$<\frac{1}{2}$	13	
$+5$	dinitrogen pentoxide	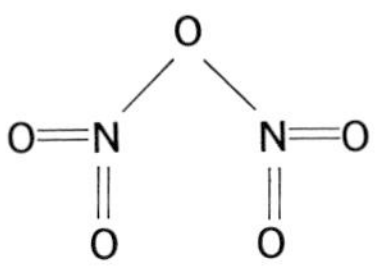 $O\!=\!N\quad N\!=\!O$	2, 1 (presumed gas phase structure: crystalline N_2O_5 is ionic: $[NO_2^+, NO_3^-]$)			
	nitronium ion	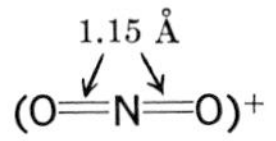 $(O\!=\!N\!=\!O)^+$ 1.15 Å	2 (linear, isoelectronic with CO_2)			
	nitrate ion	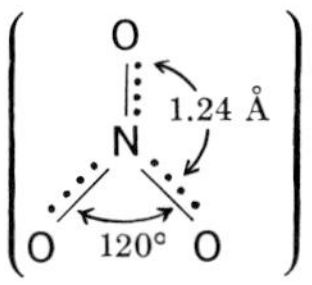 1.24 Å N O 120° O	$1\frac{1}{3}$ (planar)			

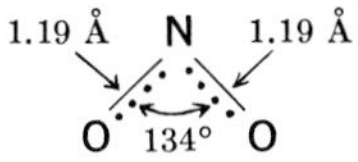

Figure 20-8 The structures of the oxides of nitrogen.

linked by a 10 kcal N–N bond. Such weak linkages appear again in crystalline N_2O_4. Finally, N_2O_5 in the crystalline state rearranges to an ionic crystal in which the cations, NO_2^+, are isoelectronic with CO_2 and the anions are the planar nitrate ions, NO_3^-.

Figure 20-9 shows the two well-characterized oxides of phosphorus, P_4O_6 and P_4O_{10}, as well as the structure of elemental phosphorus (white). In the P_4 molecule, we see that phosphorus forgoes the triple bond structure in favor of the tetrahedral structure in which the three bonds to each atom give a pyramidal geometry. The average bond energy of the six single bonds in P_4 is 48 kcal, whereas that of the triple bond in P_2 is only 116.5 kcal. Hence the formation of P_4 from two P_2 molecules is exothermic by 55 kcal. In contrast, N_4 is unknown, presumably because of the much stronger triple bond in N_2 (225 kcal).

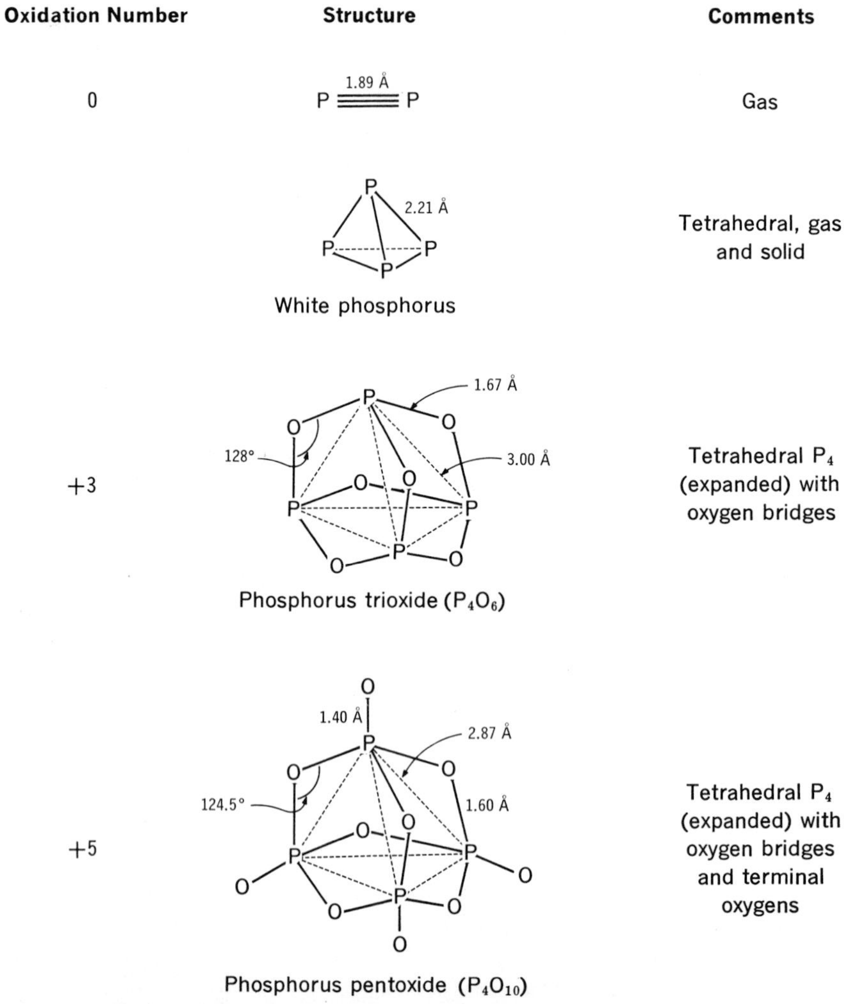

Oxidation Number	Structure	Comments
0	P ≡≡≡ P (1.89 Å)	Gas
	White phosphorus (2.21 Å)	Tetrahedral, gas and solid
+3	Phosphorus trioxide (P_4O_6) (128°, 1.67 Å, 3.00 Å)	Tetrahedral P_4 (expanded) with oxygen bridges
+5	Phosphorus pentoxide (P_4O_{10}) (1.40 Å, 2.87 Å, 124.5°, 1.60 Å)	Tetrahedral P_4 (expanded) with oxygen bridges and terminal oxygens

Figure 20-9   The structures of the oxides of phosphorus.

Surprisingly, this tetrahedral arrangement of four phosphorus atoms is preserved in both P_4O_6 and P_4O_{10}, though with the phosphorus–phosphorus distance lengthened to accommodate the oxygen bridges that link the skeleton together. The P_4O_{10} can be regarded as a P_4O_6 molecule in which each phosphorus has acted as an electron donor to another oxygen atom in the ^1D state acting as an electron acceptor (see Section 17-3(d)).

The structures of the other Group 5 oxides, those of arsenic, antimony, and bismuth, are not so well characterized. Both As_4O_6 and Sb_4O_6 are known to have structures like P_4O_6 (see Fig. 20-9) and it can be expected that the others listed in Table 20-4 will resemble their phosphorus counterparts closely. This pattern, that the first-row oxides are relatively unique, is typical of the p orbital families.

(b) GROUP 7 OXYACIDS

The formation of halogen oxyacids was discussed in Section 17-3(d) and their strong oxidizing powers were noted. Table 20-6 lists the halogen oxyacids that have been detected and, where known, their acid dissociation constants. We see that among the HOX acids, the acidity decreases in the sequence HOCl > HOBr > HOI. This trend can be attributed to changing electronegativity of the halogen atom. The halogen atom in HOX tries to pull electrons from the oxygen atom, tending to give oxygen a positive charge. Since the hydrogen atom also tends to carry a positive charge, these like charges repel and the O–H bond is weakened. The effect is largest with chlorine (electronegativity 3.0) and it decreases successively with bromine and iodine (electronegativity 2.8 and 2.5). Figure 20-10 shows a plot of pK_a $(= -\log K_a)$ versus electronegativities for the hypohalous acids and including the hydrogen counterpart, H_2O. It is interesting that extrapolation in Figure 20-10 to the electronegativity of fluorine suggests that HOF will be a strong acid, with a pK_a below 1.

Table 20-6   Halogen Oxyacids and Their Acid Dissociation Constants

Oxidation Number	Fluorine	Chlorine	Bromine	Iodine
+1	HOF	HOCl $5.6 \cdot 10^{-8}$	HOBr $2 \cdot 10^{-9}$	HOI $1.1 \cdot 10^{-11}$
+3		HOClO $1.1 \cdot 10^{-2}$		
+5		HOClO$_2$ strong	HOBrO$_2$ strong	HOIO$_2$ strong
+7		HOClO$_3$ strong	HOBrO$_3$	HOIO$_3$ H_5IO_6 $K_1 = 5.1 \cdot 10^{-4}$ $K_2 = 2 \cdot 10^{-7}$ $K_3 = 1 \cdot 10^{-15}$

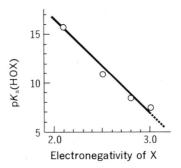

Figure 20-10 *Variation of* $pK_a (= -\log_{10} K_a)$ *with electronegativity of X for the oxyacids* HOX. $(K_a(H_2O) = K_w/[H_2O(\ell)]$

The increasing acid strengths for the higher oxyacids can be regarded in the same light. When an oxygen atom bonds to the halogen atom in HOX to form HOXO, the halogen acts as an electron donor (see Section 17-3(d)). This shifts electrons toward the electron acceptor oxygen atom, leaving the halogen more positive. This positive charge effectively increases the electronegativity of the halogen, hence, increases its acidity. As more oxygen atom electron acceptors become attached to a given halogen, the effect increases, so $HOXO_2$ and $HOXO_3$ are still stronger acids.

We shall find this argument to be very widely applicable. The oxyacids of the higher oxidation states of many elements tend to be strong acids, with acidity increasing for a given element as its oxidation number increases.

Exercise 20-1. Show that in a neutral solution (pH $= 7.0$) containing a hypohalous acid, the concentration ratio $(OX^-)/(HOX)$ changes from about 2 to 500 in the series HOCl, HOBr, HOI because of the lessening acidity of HOX.

(c) GROUP 5 OXYACIDS

When water is added to an oxide of a Group 5 element, an oxyacid is formed without change in oxidation number. Some typical examples are as follows:

$$N_2O_3(s) + H_2O(\ell) \longrightarrow 2\ HONO \qquad \text{nitrous acid} \qquad (20\text{-}10a)$$

$$N_2O_5(s) + H_2O(\ell) \longrightarrow 2\ HONO_2 \qquad \text{nitric acid} \qquad (20\text{-}10b)$$

$$P_4O_6(s) + 6\ H_2O(\ell) \longrightarrow 4\ H_3PO_3 \qquad \text{phosphorous acid} \qquad (20\text{-}11a)$$

$$P_4O_{10}(s) + 6\ H_2O(\ell) \longrightarrow 4\ H_3PO_4 \qquad \text{phosphoric acid} \qquad (20\text{-}11b)$$

$$As_4O_6(s) + 6\ H_2O(\ell) \longrightarrow 4\ H_3AsO_3 \qquad \text{arsenious acid} \qquad (20\text{-}12a)$$

$$As_4O_{10}(s) + 6\ H_2O(\ell) \longrightarrow 4\ H_3AsO_4 \qquad \text{arsenic acid} \qquad (20\text{-}12b)$$

The Group 5 oxyacids can be regarded as derivatives of the trihydrides, MH_3, in which hydrogen atoms are successively replaced by OH groups and in which the unused electron pair of the M atom is donated to an oxygen atom as an electron acceptor. This gives two possible structures for each positive oxidation state. These are shown in Figure 20-11 in the line-diagram counterparts of electron-dot formulas.

The observed nitrogen oxyacids are obviously related to the generalized formulas through loss of a molecule of water. Presumably because it is larger, phosphorus can accommodate two or three OH groups. Table 20-7 gives acid dissociation constants for these oxyacids. We see here the same trends discovered for the halogen oxyacids and explained in terms of the charge movement due to electronegativity and electron donation. Nitric acid can be related to nitrous acid, as shown in Figure 20-11, through addition of an electron-accepting oxygen atom. Hence, nitric acid should be the stronger acid. Looking horizontally in Table 20-7, we see that as the central atom grows in size and decreases in electronegativity, the acidity of a given oxidation state decreases. Again we find the generalizations: For oxyacids, as the central atom oxidation number increases, acidity increases and, as its electronegativity increases, acidity increases.

(d) GROUP 4 OXYACIDS: AMPHOTERISM

If we consider successive replacements of hydrogen by OH groups in the Group 4 hydrides, MH_4, the derived structures relate to the observed oxyacids just as for the Group 5 elements. Figure 20-12 shows the structures and the observed carbon and silicon compounds. Just as observed for nitrogen, the relatively small carbon atom cannot accommodate two or more OH groups and H_2O is eliminated to give the observed acid.

Table 20-7 Group 5 Oxyacids and Their Acid Dissociation Constants

Oxidation Number	Nitrogen	Phosphorus	Arsenic	Antimony
+1	hyponitrous HONNOH $K_1 = 9 \cdot 10^{-8}$ $K_2 = 1 \cdot 10^{-11}$	hypophosphorous H_3PO_2, or $H_2PO(OH)$ $K = 10^{-2}$		
+3	nitrous HNO_2, or HONO $K = 4.5 \cdot 10^{-4}$	phosphorous H_3PO_3, or $HPO(OH)_2$ $K_1 = 1.0 \cdot 10^{-2}$ $K_2 = 2.6 \cdot 10^{-7}$	arsenious H_3AsO_3 $K_1 = 6 \cdot 10^{-10}$ $K_2 = 3 \cdot 10^{-14}$	antimonous H_3SbO_3 $K_1 \cong 10^{-15}$
+5	nitric HNO_3, or $HONO_2$ strong	phosphoric H_3PO_4, or $PO(OH)_3$ $K_1 = 7.5 \cdot 10^{-3}$ $K_2 = 6.2 \cdot 10^{-8}$ $K_3 = 10^{-12}$	arsenic H_3AsO_4, or $AsO(OH)_3$ $K_1 = 2.5 \cdot 10^{-4}$ $K_2 = 5.6 \cdot 10^{-8}$ $K_3 = 3 \cdot 10^{-13}$	

Oxidation Number	Generalized Structure	Nitrogen Compound	Phosphorus Compound
-3	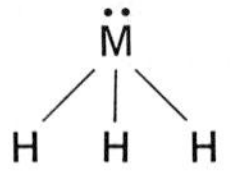	ammonia NH_3	phosphine PH_3
-1	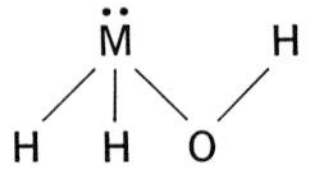	hydroxylamine H_2NOH	—
	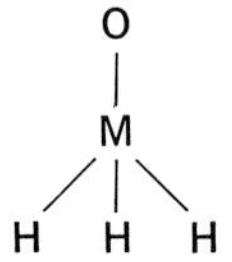	—	—
$+1$	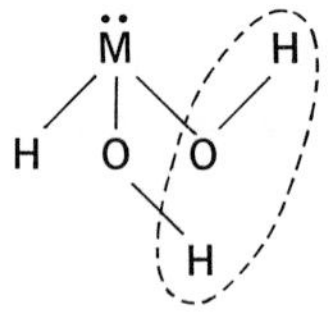	nitroxyl $H—N{=}O$	—
	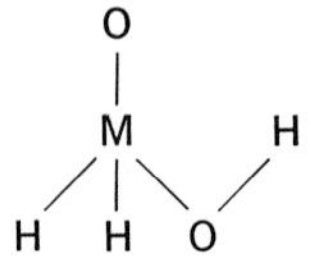	—	hypophosphorous acid H_3PO_2, or $H_2PO(OH)$
$+3$	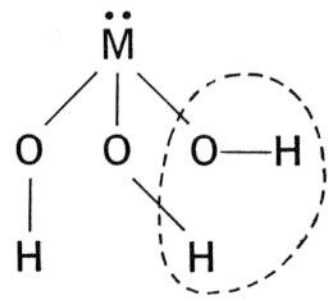	nitrous acid 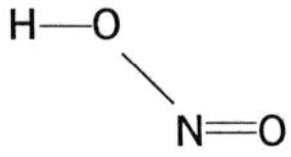 $H—O$ $\diagdown$ $N{=}O$	—
	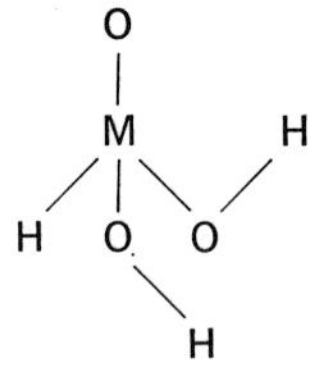	—	phosphorous acid H_3PO_3, or $HPO(OH)_2$
$+5$	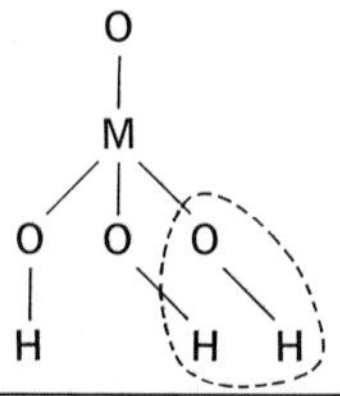	nitric acid 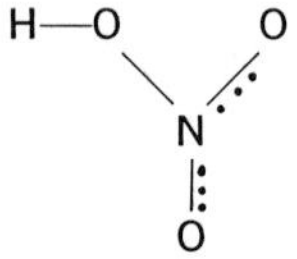 $H—O$ O $\diagdown$ $\diagup$ N $\diagup$ O	phosphoric acid H_3PO_4, or $PO(OH)_3$

Figure 20-11 Group 5 oxyacids as derivatives of the trihydrides.

Oxidation Number	Generalized Structure	Carbon Compound	Silicon Compound
-4	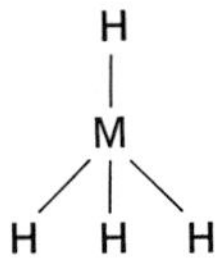	methane CH_4	silane SiH_4
-2	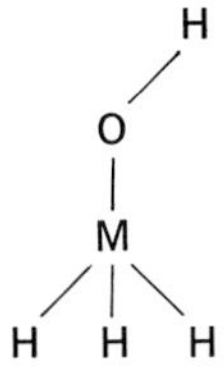	methanol H_3COH $K_a = 1 \times 10^{-17}$	—
0	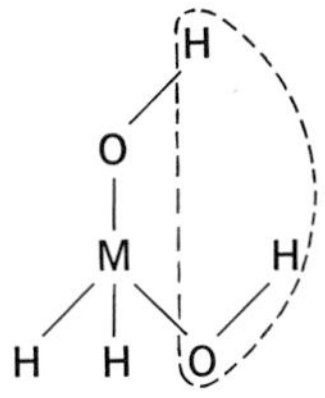	formaldehyde $H_2C{=}O$	—
$+2$	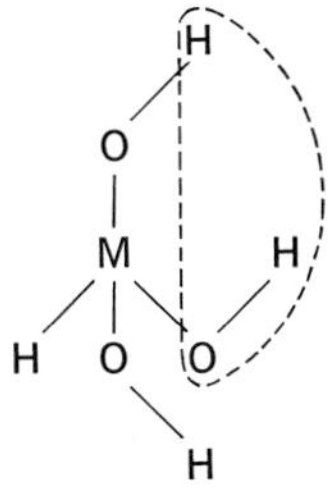	formic acid 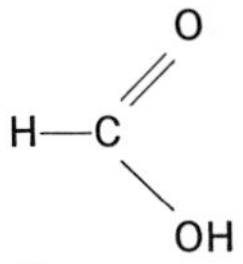 $H{-}C\overset{O}{\underset{OH}{\big\backslash}}$ $K_a = 1.8 \times 10^{-4}$	—
$+4$	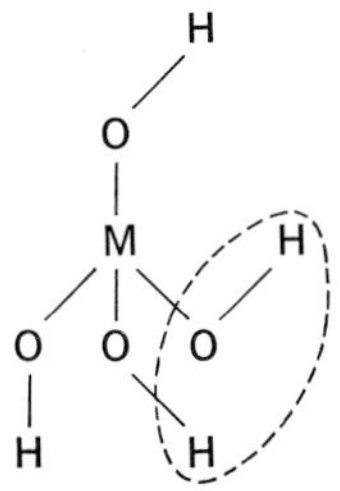	carbonic acid H_2CO_3, or $CO(OH)_2$ $K_1 = 3 \times 10^{-7}$ $K_2 = 6 \times 10^{-11}$	meta-silicic acid H_2SiO_3, or $SiO(OH)_2$ $K_1 = 10^{-10}$ $K_2 = 10^{-12}$

Figure 20-12   Group 4 oxyacids as derivatives of the tetrahydrides.

If we now drop down to lead, a qualitatively different chemistry appears. First, lead in the $+4$ oxidation state, PbO_2, has quite low solubility, so the acidity of this species is not readily determined. The $+2$ state is well known, however, but as $Pb^{+2}(aq)$. By addition of sodium hydroxide to a solution of a soluble lead salt (e.g., lead nitrate, $Pb(NO_3)_2$), a compound $Pb(OH)_2$ precipitates, but no one calls this an oxyacid. This compound is called lead hydroxide. How does this fit with the trends studied earlier in Groups 7 and 5?

Referring back to Tables 20-6 and 20-7, we see that acidity is lowest for an element in its lowest oxidation states, and within a column, it is lowest for the last row of the Periodic Table. We are led to expect $Pb(OH)_2$ to be an extremely weak acid. In fact, that is what is found. Solid $Pb(OH)_2$ dissolves in strong base, acting as an oxyacid with a dissociation constant of $2.1 \cdot 10^{-17}$.

$$Pb(OH)_2(s) = H^+(aq) + HPbO_2^- \qquad K = 2.1 \cdot 10^{-17} \qquad (20\text{-}13)$$

This is as expected. What is new is that solid $Pb(OH)_2$ also dissolves in acid, now with rupture of the lead–oxygen bonds, to release hydroxide ions and form aquated lead ion. Expressed in terms of a solubility product, the equilibrium is:

$$Pb(OH)_2(s) = Pb^{+2}(aq) + 2\,OH^-(aq) \qquad K_{sp} = 2.8 \cdot 10^{-10} \qquad (20\text{-}14)$$

Thus, the compound $Pb(OH)_2$ reacts both with acids and bases. Such a compound is called *amphoteric*.

This behavior can be rationalized with the acidity trends and our earlier explanation of them. In a bonding arrangement M–O–H, a high electronegativity of M implies that electrons are pulled from oxygen, tending to weaken the O–H bond. At the same time, this process tends to strengthen the M–O bond, pulling it in tighter and tighter. This was noted first in Table 17-6, where the higher oxidation states of chlorine show the shortest Cl–O bonds. The reverse trend would accompany an electronegativity shift to lower and lower values. Now the O–H bond would become progressively stronger, and the M–O bond weaker and weaker. If the electronegativity and ionization energy of M become sufficiently low, OH^- can be released into the favorable environment of the ionizing solvent, water. Naturally, there must be some elements and some oxidation states in which both processes can occur—release of H^+ by rupture of the O–H bond and also release of OH^- by rupture of the M–O bond. These cases are the ones we call amphoteric. Furthermore, we can expect elements to become progressively more basic as we move to the left in the Periodic Table, since electronegativity drops in that direction (see Table 17-3). By the time we reach the alkaline earths and the alkali metals, the acid behavior is completely lost. In the middle of the Periodic Table, the lower oxidation states tend to be basic, the higher states tend to be acidic, and those between are often amphoteric.

20-3 Halogen compounds: especially fluorides

All of the halogens are quite electronegative; fluorine is the most electro-negative element in the Periodic Table. As a result, they are extremely reactive and all of the elements except the lighter inert gases form stable halides. Table 20-8 illustrates the stability of fluorides—of the thirty-nine compounds in the table, only three (N_2F_2, O_2F_2, and KrF_2) are unstable with respect to the elements.

Chlorine, bromine, and iodine are easily prepared by the oxidation of the appropriate halide ion. To prepare chlorine in aqueous solution, a strong oxidizing agent is needed: Permanganate ion will do (Cl_2-Cl^-, $\mathcal{E}^0 = +1.36$). Chlorine is made commercially, however, by the electrolysis of NaCl brine or molten NaCl (dissolved in $CaCl_2$ to lower the melting point). Since Br^- and I^- are more readily oxidized (Br_2-Br^-, $\mathcal{E}^0 = +1.09$; I_2-I^-, $\mathcal{E}^0 = +0.54$),

Table 20-8   Fluorides of the p Elements: Enthalpies and Free Energies of Formation (298°K)

	ΔH_f^0	ΔG_f^0		ΔH_f^0	ΔG_f^0
Group 3			**Group 6**		
$BF_3(g)$	−271.8	−267.8	$OF_2(g)$	−5.5	−1.1
$B_2F_4(g)$	−344.2	−337.1	$O_2F_2(g)$	+4.3	
$AlF_3(s)$	−359.5	−450.6	$SF_4(g)$	−185.2	−174.8
$Al_2F_6(g)$	−628		$SF_6(g)$	−289	−264.2
$GaF_3(s)$	−278	−259	$SeF_6(g)$	−267	−243
$InF(g)$	−48.6		$TeF_6(g)$	−315	
$TlF(g)$	−43.6				
			Group 7		
Group 4			$F_2(g)$	0	0
			$ClF(g)$	−13.0	−13.4
CF_4	−221	−210	$ClF_3(g)$	−39	−29
$SiF_2(g)$	−148	−150	$BrF(g)$	−22.4	−26.1
$SiF_4(g)$	−385.9	−375.9	$BrF_3(g)$	−61.1	−54.8
$PbF_2(s)$	−158.7	−147.5	$BrF_5(g)$	−102.5	−83.8
$PbF_4(s)$	−225		$IF(g)$	−22.9	−28.3
			$IF_5(g)$	−196.6	−179.7
Group 5			$IF_7(g)$	−225.6	−195.6
$NF_3(g)$	−29.8	−19.9			
$N_2F_2(cis, g)$	16.6		**Group 8**		
$N_2F_4(g)$	−1.7	19.4	$KrF_2(g)$	∼+14	
$PF_3(g)$	−219.6	−214.5	$XeF_2(g)$	−25.9	
$PF_5(g)$	−381.4		$XeF_4(g)$	−51.5	
$AsF_3(g)$	−220.0	−216.5	$XeF_4(s)$	−62.5	
$SbF_3(s)$	−218.8		$XeF_6(g)$	−70.4	

they are more easily prepared. For example, chlorine itself is a cheap oxidizing agent with ample muscle to oxidize either Br^- to Br_2 or I^- to I_2.

In contrast, even the discovery of elemental fluorine was made extremely difficult by its exceptional reactivity and powerful oxidizing power. While the chemical parallel between HCl and HF was understood by the year 1813, all attempts to prepare F_2 by chemical oxidation of F^- were unsuccessful for over half a century. We now know that F_2 itself is the most powerful oxidizing agent known.

$$F_2 + 2\,e^- = 2\,F^- \qquad \mathcal{E}^0 = 2.87\text{ V} \tag{20-15}$$

so the chemical approaches all were doomed to failure. A great deal of effort was put into electrolytic methods of preparation, but as it was formed, the F_2 always reacted with the electrodes or vessels used. As an added hindrance, both HF and F_2 are extremely toxic, so sickness and premature death were real occupational hazards for the early fluorine chemists. Finally in 1886, a Frenchman, Henri Moissan, succeeded in trapping some of the pale yellow gas. He prepared it by the electrolysis of potassium bifluoride, KHF_2, using a platinum–iridium alloy for the electrodes, which were inserted in stoppers made from CaF_2 itself.

Later it was found that copper containers worked very well for fluorine—not because copper and fluorine don't react—but because the CuF_2 formed on the surface of the metal acts as a protective coating.

Exercise 20-2. Calculate $\Delta \mathcal{E}^0$ and K_{eq} for the reaction between Cu(m) and $F_2(g)$.

The toxicity and corrosive properties of fluorine have hindered the detailed study of its compounds despite their abundance. Today fluorine chemistry is routinely carried out in copper or stainless steel containers and vacuum systems. Glass and quartz are eliminated for careful work because gaseous SiF_4 forms easily and contaminates the sample.

(a) INERT GAS CHEMISTRY

The octet rule postulated by Lewis in 1916 (see Chapter Two) survived the onslaught of quantum mechanics almost intact. We now recognize it as the simple statement that a molecule with all valence orbitals filled or engaged in bonding will be relatively unreactive. Thus, in compounds of the p elements, with ns and np valence orbitals, eight electrons must be held or shared by each atom. There have always been exceptions, however. For many decades, a few compounds have been known in which an atom forms additional bonds after the inert gas population has been reached—compounds like PCl_5, SF_6, HF_2^-, and I_3^-. It is an interesting commentary on science and scientists that the inert gas elements were generally not considered to be susceptible to these frailties.

Of course the ionization energies of the inert gases are quite high, but that of xenon (280 kcal/mole) is not *that* much higher than that of sulfur (239 kcal/mole)! Perhaps the fact that the octet rule was based on the lack of reactivity of the inert gases made the connection between the two just too rigid.

In 1933, a scientist at the California Institute of Technology, D. M. Yost, and a graduate student, A. L. Kaye—apparently acting on a suggestion by Linus Pauling—attempted to prepare some xenon halides. At that time fluorine was not available commercially so it had to be prepared in the lab—no simple task in itself. Xenon was available only in extremely small quantities and was expensive. Nevertheless, the materials were assembled and a Xe/F_2 mixture was subjected to an electric discharge. In true scientific fashion, these scientists cautiously reported on their negative results:

It cannot be said that definite evidence for compound formation was found. It does not follow, of course, that xenon fluoride is incapable of existing. It is known, for example, that nitrogen and fluorine do not combine in an electrical discharge but when prepared directly, NF_3 is a very stable compound.*

This work was not pursued further and it was little noted. Two decades later the possibility that inert gas halides might exist was again considered, this time as a prediction based upon the molecular orbital description of bifluoride and trihalide ions. In 1951, G. C. Pimentel proposed:

It is to be expected that a rare gas could form complexes with halogens. It is interesting to speculate that . . . halogen rare gas compounds . . . might be due to bond formation similar to that in the trihalides.†

In the early 1960's, Neil Bartlett at the University of British Columbia broke the spell by preparing and identifying the first compound involving xenon, $XePtF_6$. He began with no thought of inert gas chemistry. He was attempting to prepare the compound PtF_2 by reacting PtF_4 (then the only known platinum–fluorine compound) with SF_4.

$$PtF_4(s) + SF_4(g) \xrightarrow{\ ?\ } PtF_2(s) + SF_6(g) \qquad (20\text{-}16)$$

However, he first had to remove some bromine impurity from the PtF_4 by reaction with fluorine gas (to form $BrF_5(g)$). The fluorine was introduced in a stream of nitrogen that—enter fate—contained some oxygen gas impurity. Bartlett noticed a red solid forming on the sides of his reaction

---

* D. M. Yost and A. L. Kaye, *Journal of the American Chemical Society*, Vol. **55**, p. 3890, 1933.
† G. C. Pimentel, *Journal of Chemical Physics*, Vol. **19**, p. 446, 1951.

tube—a substance that, after much effort, was identified as $O_2^+ \cdot PtF_6^-$. So far we have one of those typical research paths. A couple of unexpected turns had brought attention to a novel substance—quite different from the originally desired compound. Then came the big step. Bartlett realized that the oxygen molecule and the xenon atom have almost identical first ionization energies, 278.5 and 279.7 kcal/mole, respectively. If PtF_6 could steal an electron from O_2, why not from Xe? The experiment was tried and the results were spectacular! When xenon was allowed to mix with PtF_6, a red solid formed instantly on the walls of the flask. The announcement of these results in 1962 caused great excitement—and provided the psychological release apparently necessary to get chemists working on inert gas chemistry.

Within a very few months, several xenon fluorides were known (XeF_2, XeF_4, XeF_6). Before a year had passed, KrF_2 had been prepared, xenon trioxide had spectacularly demonstrated its explosive properties in several laboratories, and some oxyfluorides ($XeOF_2$, $XeOF_4$) had been found. Since then, several other xenon fluoride species have been identified in the solid state and in aqueous solution. More recently, $XeCl_2$ has been prepared, but so far, all attempts to interest argon, neon, or helium in halogen compound formation have been unsuccessful.

The bonding and structure of krypton difluoride and of the xenon halides were discussed in Chapter Seventeen in conjunction with the inter-halogens. We found that bonds to inert gas atoms and the isoelectronic halide ions always occur in pairs with electronegative fluorine or chlorine atoms on the outside (except in I_3^- and I_5^-). These three-center bonds are long and weak.

Let's take this as a general expectation: Atoms or ions that have achieved an inert gas configuration can form three-center bonds with pairs of electronegative halogen atoms. We will find wide applicability of this "inert gas bonding" to several p orbital elements that form additional electron-excess bonds after satisfying their normal bonding capability.

(b) INERT GAS BONDING IN OTHER p ORBITAL ELEMENTS

The structures BrF_3 and BrF_5 were given in Figure 17-16. They are consistent with bromine achieving its inert gas electron population by forming BrF. Then BrF_3 and BrF_5 result from adding one and two pairs of F's, respectively. These concepts, substantiated by the bond lengths, can be easily verified from bond energy considerations as well. Consider the dissociation energy of BrF.

$$BrF(g) = Br(g) + F(g) \tag{20-17}$$

$$\Delta H_{\text{atoms}} = \Delta H_f^0[Br(g)] + \Delta H_f^0[F(g)] - \Delta H_f^0[BrF(g)]$$

$$= 26.7 + 19 - (-22.4)$$

$$= 68.1 \text{ kcal/mole} = D_{BrF} \tag{20-17a}$$

Now, for BrF_3 we have

$$BrF_3(g) = Br(g) + 3 F(g) \tag{20-18}$$

$$\Delta H_{atoms} = \Delta H_f^0[Br] + 3\Delta H_f^0[F] - \Delta H_f^0[BrF_3]$$

$$= 26.7 + 3(19) - (-61.1)$$

$$= 144.8 \text{ kcal/mole} \tag{20-18a}$$

and the average bond energy, $D_{BrF}[BrF_3]$, would be one third of this: $144.8/3 = 48.3$ kcal/mole. This is much smaller than the 68.1 kcal bond of BrF, confirming our belief that the second and third bonds are weaker.

A more informative approach is to assume that the short BrF bond in BrF_3 has a strength identical to that in BrF itself, as suggested by the bond lengths. With this premise, we can calculate the average energy of the two weak bonds,

$$D'_{Br-F}[BrF_3] = \tfrac{1}{2}[\Delta H_{atoms}(BrF_3) - D_{BrF}]$$

$$= \tfrac{1}{2}[144.7 - 68.1]$$

$$= 38.3 \text{ kcal/mole} \tag{20-18b}$$

Just as we suspected, the two three-center bonds are very much weaker— only 56% as strong as the bond in BrF. This is in accord with the expectation of a bond order near $\tfrac{1}{2}$ for these bonds.

The next member of the series, BrF_5, offers a good test of this approach. The two new bonds should have bond energies near 38.3 kcal/mole as well. Let's use that estimate to calculate the energy required to break BrF_5 apart into atoms.

$$\Delta H_{atoms}[BrF_5, \text{ calc.}] = D_{BrF} + 4D'_{BrF}$$

$$= 68.1 + 4(38.3)$$

$$= 221.3 \text{ kcal/mole} \tag{20-19}$$

This is in excellent agreement with the experimental value, 224.2 kcal/mole.

Sulfur forms several fluorides, including SF_4, SF_6, S_2F_{10}, and S_2F_2 (in two forms, SSF_2 and $FSSF$). Curiously the simplest fluoride, SF_2, is not known. Nevertheless, it is possible to imagine the formation of such a molecule from which SF_4 is formed by adding a pair of fluorines in the usual three-center bond arrangement. This easily accounts for the unusual shape and the bond lengths of this molecule, illustrated in Figure 20-13.

Sulfur tetrafluoride is extremely reactive. For example, it is instantly hydrolyzed by water:

$$SF_4(g) + 2 H_2O(\ell) \longrightarrow SO_2(aq) + 4 HF(aq) \tag{20-20}$$

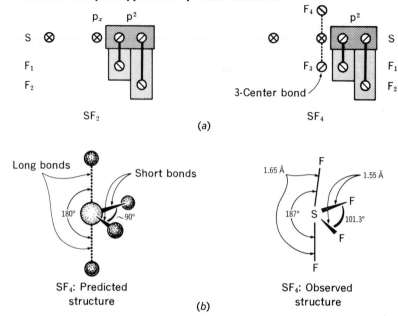

Figure 20-13 (a) Pigeonhole diagrams for SF_2 and SF_4. (b) Structure of SF_4 predicted by inert gas bonding compared to observed structure.

SF_4 is a useful fluorinating agent for organic oxygen compounds—attack is quite selective at the $C{=}O$ group.

$$\begin{array}{c}R_1\\ \diagdown\\ C{=}O + SF_4 \longrightarrow \\ \diagup\\ R_2\end{array}\qquad \begin{array}{cc}R_1 & F\\ \diagdown & \diagup\\ & C\\ \diagup & \diagdown\\ R_2 & F\end{array} + SOF_2 \qquad\qquad (20\text{-}21)$$

$$\begin{array}{c}\qquad O\\ \qquad \parallel\\ R{-}C\\ \qquad \diagdown\\ \qquad OH\end{array} + 2\,SF_4 \longrightarrow R{-}CF_3 + 2\,SOF_4 + HF \qquad\qquad (20\text{-}22)$$

It is easily converted to SF_6:

$$SF_4(g) + F_2(g) \longrightarrow SF_6(g)$$

$$\Delta H^0 = -103.8\ \text{kcal/mole}$$

$$\Delta G^0_{298} = -89.4\ \text{kcal/mole}$$

$(20\text{-}23)$

Sulfur hexafluoride, a regular octahedron, cannot immediately be accommodated within the inert gas bonding scheme. However, a straightforward molecular orbital approach using only sulfur 3s and 3p orbitals does the job. This results in six equal bonds sharing eight electrons—a formal bond order of $\frac{2}{3}$. This prediction is consistent with the observed average bond energy in SF_6, 78 kcal/mole, which is slightly less than that in SF_4, 82 kcal/mole, resulting from two weak bonds and two normal bonds.

Unlike SF_4, SF_6 is exceptionally inert—it will not react with molten KOH, steam at 770°K, or oxygen, even in an electric discharge. (This lack of reactivity was in Bartlett's mind when he attempted to defluorinate PtF_4 with SF_4.) The hydrolysis of SF_6 is favorable thermodynamically

$$SF_6(g) + 3\,H_2O(\ell) \longrightarrow SO_3(g) + 6\,HF(aq)$$

$$\Delta H^0 = -59.7 \text{ kcal/mole} \tag{20-24}$$

$$\Delta G^0_{298} = -79.8 \text{ kcal/mole}$$

so, as usual, kinetic factors must prevent easy reaction.

The chlorination of sulfur yields disulfur dichloride, S_2Cl_2. Using excess chlorine, an equilibrium mixture containing about 85% SCl_2 is obtained. The sulfur dichloride slowly decomposes at room temperature

$$2\,SCl_2 \longrightarrow S_2Cl_2 + Cl_2 \tag{20-25}$$

The dichloride may be converted to crystals of SCl_4, but this compound decomposes at about 240°K. So we have quite different behavior in sulfur fluorides and chlorides. The difluoride, SF_2, is not known—presumably because the tendency to form the higher fluorides is high. This may be due to the high electronegativity of fluorine, a requirement for inert gas bonding. Chlorine is apparently not electronegative enough to be very keen about forming higher chlorides of sulfur.

Many stable chlorides of selenium and tellurium are known, including the octahedral ions $SeCl_6^{-2}$ and $TeCl_6^{-2}$. These ions can be considered to be derived from Se^{-2} and Te^{-2}, ions that have inert gas configurations, through the formation of three three-center bonds. This view is supported by measurements of the electron density at the nuclei by nuclear quadrupole resonance. These measurements point firmly to the use of central atom p orbitals only. This rules out the use of extravalence d orbitals in hybrid bond formation.

Among the Group 5 elements, P, As, and Sb, all form halides of the formula MX_5. All the fluorides and chlorides are known except for $AsCl_5$, the lone holdout. Let's examine PCl_5 to typify these compounds. It is formed from phosphorus trichloride and chlorine

$$PCl_3(g) + Cl_2(g) = PCl_5(g)$$

$$\Delta H^0 = -20.5 \text{ kcal/mole} \tag{20-26}$$

$$\Delta G^0_{298} = -21.0 \text{ kcal/mole}$$

The average bond energy in PCl_5 is 62.1 kcal/mole, much less than that in PCl_3, 77.0 kcal/mole. This is evidence for the existence of a three-center bond. PCl_3 however, is pyramidal, as appropriate to the p^3 nature of the bonding shown in Figure 20-14(a). In order to "free" a phosphorus p orbital for three-center bonding, we can consider the imaginary planar PCl_3 molecule of Figure 20-14(b). This is achieved by the promotion of one electron from the 3s to the $3p_z$ orbital (which requires 170 kcal/mole) to open the door to some sp^2 hybridization. Now two more chlorines can add on, as in part (c) of the figure. The excellence of this prediction is obvious on comparison with the observed structure in Figure 20-14(d).

Now let's suppose that we can transfer D_{P-Cl} from PCl_3 to the equa-

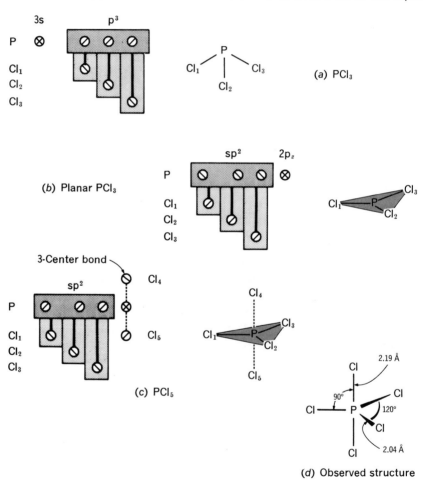

Figure 20-14   The bonding in PCl_3 and PCl_5. (a) p^3 bonding in PCl_3. (b) Hypothetical planar PCl_3. (c) Three-center bonding in PCl_5. (Note that only one 3p orbital is illustrated for each chlorine.)

torial sp^2-bonded atoms in PCl_5. This will give us an estimate of D'_{PCl} for the three-center bonds.

$$D'_{PCl}(PCl_3) = \tfrac{1}{2}[\Delta H_{atoms}(PCl_5) - 3D_{PCl}(PCl_3)]$$

$$= \tfrac{1}{2}[310.3 - 3(77.0)]$$

$$= 39.6 \text{ kcal/mole} \tag{20-27}$$

On this basis, the axial bonds are just 51% as strong as the equatorial ones.

In some solutions PCl_5 retains its trigonal bipyramid structure, but in CCl_4 it appears to be dimeric. In the crystalline state PCl_5 has an ionic structure consisting of discrete PCl_6^- and PCl_4^+ units. In contrast, PBr_5 forms an ionic solid with PBr_4^+ and Br^- units.

The P, As, and Sb hexafluoride ions are all quite stable. As a result, the MF_5 molecules are all good fluoride ion acceptors. For example, in liquid HF, SbF_5 forms the SbF_6^- ion:

$$SbF_5 + HF = SbF_6^- + H^+ \tag{20-28}$$

By accepting a fluoride ion electron pair, SbF_5 is acting as a Lewis acid. PF_5 also is a good Lewis acid, an electron acceptor, toward amines, ethers, and other Lewis bases.

(c) THE NITROGEN HALIDES

Although we have given a lot of attention to the halides of P, As, and Sb, nitrogen has been set aside to be considered alone. Several trihalides of this element are known, but no NX_5 compounds have ever been found. Since NF_5, for example, is predicted to be possible under the inert gas bonding scheme, size limitations must work against its formation. There must not be room to crowd five fluorines around the small nitrogen atom. This is reminiscent of the difficulty nitrogen experiences trying to hold two or three OH groups nearby, a feat that phosphorus handles with ease (see Fig. 20-11).

Nitrogen trifluoride, like SF_6, is remarkably stable and inert. It can be formed from the reaction of ammonia and excess fluorine with a copper catalyst.

$$2 NH_3(g) + 3 F_2(g) \rightleftharpoons 2 NF_3 + 3 H_2 \tag{20-29}$$

$$\Delta H^0 = -37.6 \text{ kcal/mole}$$

$$\Delta G^0_{298} = -32.0 \text{ kcal/mole}$$

If ammonia is in excess, N_2F_2 and N_2F_4 are formed as well.

The chloride derivatives NF_2Cl, $NFCl_2$, and NCl_3 are known, but are reactive and often explosive. Nitrogen trichloride is a dangerously explosive, oily liquid. To understand the difference between NF_3 and NCl_3, we can use average bond energies to derive ΔH for the decomposition processes.

$$2\,NX_3(g) \longrightarrow N_2(g) + 3\,X_2(g) \tag{20-30}$$

$$\Delta H^0 = 6D_{N-X} - D_{N_2} - 3D_{X_2}$$

$$NF_3\text{:} \quad \Delta H^0 = 6(65) - 225 - 3(36) = +57\ \text{kcal/mole}$$

$$NCl_3\text{:} \quad \Delta H^0 = 6(46) - 225 - 3(58) = -123\ \text{kcal/mole}$$

Thus NF_3 decomposes endothermically while the same process is exothermic for NCl_3. The relative stability of NF_3 has two sources: The strong N–F bonds and the weak F_2 bond work hand-in-hand to protect NF_3 from decomposition.

The reaction of iodine with concentrated ammonia produces the explosive crystals popularly known as nitrogen triiodide

$$\tfrac{3}{2}\,I_2(s) + 2\,NH_3(aq) = NH_3{\cdot}NI_3(s) + \tfrac{3}{2}\,H_2(g) \tag{20-31}$$

This solid is a shock- and light-sensitive compound that decomposes explosively into nitrogen, ammonia, and iodine. Because of its sensitivity, this compound is quite dangerous to handle in larger than milligram amounts.

$$NH_3{\cdot}NI_3(s) \longrightarrow \tfrac{1}{2}\,N_2 + NH_3(g) + \tfrac{3}{2}\,I_2(s) \tag{20-32}$$

$$\Delta H^0 = \Delta H_f^0(NH_3) - \Delta H_f^0(NH_3{\cdot}NI_3)$$

$$= (-11.02) - (36.9)$$

$$= -47.9\ \text{kcal/mole} \tag{20-33}$$

20-4 Silicon and germanium semiconductors

The p orbital elements on the right tend to form molecular crystals. Since these crystals are held by weak van der Waals forces, these elements have low melting and boiling points. At the left, the Group 3 elements (excepting boron) form metallic solids. The many vacant valence orbitals give the characteristic mobile electron-sharing that furnishes our explanation of metallic properties. Between these limits are the covalent solids, linked in strong three-dimensional arrays by normal chemical bonds. There is nothing startling about these solids from a bond energy or a molecular architecture point of view (see Section 18-4). Among these covalent solids, however, we find quite novel and very important electrical properties, as we shall see in the properties of silicon and germanium.

(a) THE PREPARATION OF SILICON AND GERMANIUM

Silicon is, of course, extremely abundant in the earth's crust—it is second only to oxygen. It is seen all around in rocks, which are silicates, and in the beach sands, which are silica, SiO_2. Germanium, however, is quite rare. An ore that contains 1% germanium as GeO_2 is relatively rich.

Whereas any Boy Scout can quickly char up a batch of elemental carbon, silicon and germanium are rather difficult to prepare. A typical process for manufacture of pure silicon involves the following steps.

$$SiO_2(s) + C(s) \xrightarrow{\text{red heat}} Si(s) + 2\,CO(g) \qquad (20\text{-}34)$$

$$Si(s) + 2\,Cl_2(g) \longrightarrow SiCl_4(g) \qquad (20\text{-}35)$$

$$SiCl_4(g) + 2\,H_2(g) \longrightarrow Si(s) + 4\,HCl(g) \qquad (20\text{-}36)$$

Any silicon left as silicon carbide, SiC, in the first reaction is also converted to $SiCl_4$ by the chlorine. Then the silicon tetrachloride intermediate compound facilitates purification because it is a volatile liquid (m.p.: $203°K$; b.p.: $331°K$). After distillation, the pure $SiCl_4$ can be reduced with hydrogen to give silicon containing only minute amounts of impurities, below the parts per million level. Germanium in high purity can be prepared in similar fashion.

Believe it or not, these substances can be purified even further, and need to be if the electrical properties are to be exploited. The process used is called *zone-melting*. A rod of the silicon or germanium is heated by induction heating in a high vacuum. Only a narrow zone less than a millimeter in thickness is allowed to melt. Then this molten zone is slowly moved through the length of the rod. Because impurities tend to be rejected by a regular crystal lattice, they concentrate in the liquid film and are swept down to the end of the rod. This end is discarded and the rest of the rod can be left with impurity levels below one part in 10^{10}.

(b) ELECTRICAL PROPERTIES OF SEMICONDUCTORS

As suggested by the term "semiconductor," germanium and silicon display electrical conductivity that is demonstrably different from metallic conductivity. An obvious difference is that Ge and Si are much poorer conductors than metals. Secondly, their electrical conductivities depend sensitively upon minute concentrations of impurities. Most distinctive, though, is their temperature dependence. Figure 20-15(a) shows that the electrical conductivity of a typical metal rises smoothly as the temperature is lowered, apparently heading for infinity at absolute zero. The curve shown in Figure 20-15(b) for a typical semiconductor does just the opposite. The conductivity falls as temperature drops and it precipitously heads for the floor over a rather narrow temperature range. A reasonable interpreta-

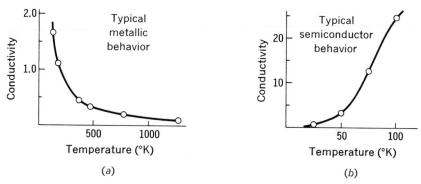

Figure 20-15 *Temperature dependences of the conductivity of metals and semiconductors.*

tion of the latter behavior is that a semiconductor transmits electrical charge through some thermally activated process. The nature of this process and its dependence upon impurities can be understood through the molecular orbital explanation of metallic bonding presented in Section 18-6.

(c) INTRINSIC SEMICONDUCTORS: MOLECULAR ORBITAL DESCRIPTION

Both silicon and germanium crystallize in the diamond lattice pictured in Figure 18-9. All of the valence orbitals (3s and 3p for silicon, 4s and 4p for germanium) are fully used in covalent bonds. A pure crystal can be regarded as a giant molecule linked by conventional chemical bonds. A molecular orbital treatment would be a bit too much to write down explicitly, but it is not difficult to anticipate how the energy level diagram will come out. Our experience with metallic M.O.'s, as pictured in Figure 18-17 is applicable. We can expect bands of very closely spaced energy levels. Figure 20-16 shows how the energy bands develop for a diamond lattice, be it carbon, silicon, or germanium. As Avogadro's Number of atoms are brought together, the 4 valence orbitals give $4N_0$ molecular orbitals. Because of the perfect tetrahedral bonding of the diamond lattice, the $4N_0$ energy levels split into two distinct bands, a low-energy band consist-

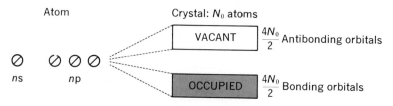

Figure 20-16 *Energy bands for a perfect diamond lattice.*

ing of bonding molecular orbitals and a high-energy band of antibonding molecular orbitals. The lower band contains $4N_0/2$ M.O.'s, so it can accommodate double this number of electrons (two per M.O., by the Pauli Principle). This exactly equals the number of valence electrons, so the lower band will be completely filled and the upper band will be totally empty. All of the valence electrons occupy bonding M.O.'s, so four bonds will be formed to each atom. That describes the diamond lattice.

The energy gap shown in Figure 20-16 is due to the absence of nonbonding M.O.'s in the diamond crystal geometry. The magnitude of the gap, ϵ, is determined by the strength of the covalent bonds formed. These are normal single bonds, so the bond energies will be comparable to those of the M–M bonds in the compounds $H_3M–MH_3$. These bond energies are given in Table 20-9, together with the measured energy gaps for the three elements. With this information, we can understand the electrical properties of these solids.

In Figure 18-16 we attributed metallic conductivity to vacant orbitals immediately adjacent to occupied orbitals. Then a small energy displacement by an external electrical field can cause net charge movement. In the filled band situation shown in Figure 20-16, there will be no vacant orbitals into which electrons can spill under the influence of an electric field. The crystal should be a perfect insulator. Well, not absolutely perfect. There will be some electrons in the upper band, where there are vacant orbitals, just because of thermal excitation. The fraction of the electrons that are so excited is just given by the Boltzmann factor $e^{-\epsilon/RT}$ as discussed in Section 8-2. The last column in Table 20-9 shows that only a small number of electrons are excited at room temperature. These electrons account for the conductivity of the pure crystals. The exponential temperature dependence explains why conductivity rises rapidly in a semiconductor as temperature rises. This conductivity is a property of the perfect crystals, so they are called *intrinsic semiconductors*.

Table 20-9 *Bond Energies in the M_2H_6 Molecules and Semiconductor Band Gaps*

Compound	M–M Bond Energy (kcal/mole)	Element	Band Gap, ϵ (kcal/mole)	$4N_0\, e^{-\epsilon/RT}$ at 298°K
ethane H_3CCH_3	83	carbon (diamond)	122.9	$2.00 \cdot 10^{-66}$
silane H_3SiSiH_3	51	silicon	26.3	$1.27 \cdot 10^{+5}$
germane H_3GeGeH_3	34	germanium	15.5	$1.05 \cdot 10^{+13}$

(d) IMPURITY SEMICONDUCTORS

Although silicon and germanium have a low conductivity due to intrinsic semiconductivity, their electrical conductivities can be raised very much merely by adding as little as one part in 10^7 of an impurity such as aluminum or phosphorus. This added conductivity also rises with rising temperature, but it is obviously connected with a much smaller energy gap than the one shown in Figure 20-16.

An explanation intuitively rises out of the molecular orbital description that leads to the energy bands. Each molecular orbital extends over the whole crystal, but if there is a stranger atom at a particular spot in the lattice, that impurity will not contribute as effectively to the bonding or antibonding character of the M.O.'s. Locally, at the impurity site, some energy levels will split out of the bands. The impurity will contribute some bonding energy levels, but not as effectively bonding as the perfect crystal M.O.'s. It will also contribute some antibonding energy levels, but not as antibonding as the perfect crystal M.O.'s. The energy level diagram will be modified as shown in Figure 20-17.

Given the new energy level picture of Figure 20-17, we must decide the level occupancy. A phosphorus atom, of course, brings one more electron to the lattice than would the silicon atom it replaces. Hence its localized bonding molecular orbital will be filled and there will be an electron left over to occupy its localized antibonding molecular orbital. To get this electron up into the vacant antibonding orbitals depends again upon thermal excitation, but with a much smaller energy gap, ϵ_P, to excite. It is possible to transfer a large fraction of these electrons into the conduction band, so only a small concentration of impurity atoms will suffice to cause impurity conductivity to dominate the intrinsic conductivity.

An aluminum impurity site has quite different orbital occupancy. Here the impurity has one less electron than the silicon atom it replaces. Hence

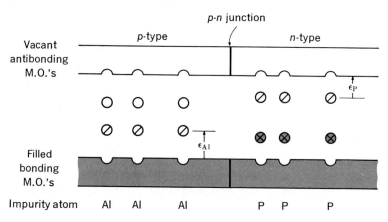

Figure 20-17   Energy levels of p-type and n-type semiconductors.

its bonding orbital will be only half filled. Now an electron can be thermally excited into this orbital, with energy gap ϵ_{Al} determining the extent of excitation. When an electron does get up into this orbital, it must come out of the filled M.O. band. This gives the mirror image of an electron in the vacant M.O. band. An electron occupying one of the many orbitals in the upper band can move its negative charge over the full extent of the crystal. An electron vacancy in the lower band looks like an absence of negative charge—a positive charge—and this charge deficiency can move over the full extent of the crystal. In both cases, impurity electrical conductivity results, but for phosphorus the charge carriers are negative electrons— this is called an n-type semiconductor (n = negative). For aluminum, the charge carriers are electron "holes" in the lower, filled band. This looks like charge transfer by positive electrons, so this is called a p-type semi-conductor (p = positive).

It isn't difficult to see that these impurity semiconductors open up quite new electrical properties. The energy gaps can be varied by proper selection of the impurity—nitrogen, phosphorus, arsenic, and antimony all will give n-type semiconductors, but their different sizes and electro-negativities imply different impurity energy gaps. The amount of electrical conductivity depends upon impurity concentration, again a matter of

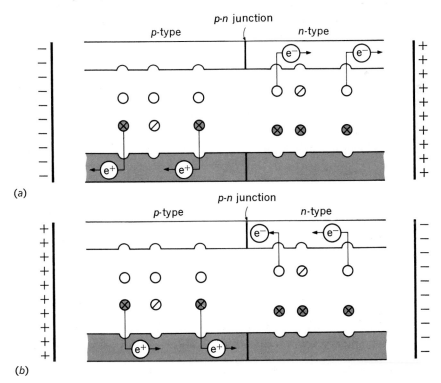

Figure 20-18   Charge movement (a) away from and (b) toward a p-n junction.

laboratory control. Finally, the combination of n- and p-type semiconductors in contact give interesting junction effects. To see just one possibility, consider the different consequences of the imposed electrical fields pictured in Figure 20-18. In the upper case, the external field pulls the negative electrical carriers to the right, away from the junction, and it pulls the positive carriers to the left, again away from the junction. Hence no charge moves through the junction and a potential difference builds up across the junction. Current won't flow. In the lower picture, the opposite electrical field causes both negative and positive charges ("holes") to move toward the junction. They meet at the junction and neutralize each other. Hence current flow can occur in this direction. This describes the electrical function of a familiar radio tube—a "rectifier." It blocks current flow in one direction and passes current in the opposite direction.

The miracle of such an electrical device is that its function can be carried out with a device small enough to fit into a fly's footprint. The result has revolutionized electronic devices. A most dramatic application is the construction of miniaturized computers for our spacecraft that are flying to Venus and Mars. A sophisticated computer can be built that fits into the weight and volume constraints usually associated with a small portable radio. Most of these electrical devices, called *transistors*, are based upon germanium impurity semiconductors.

Problems

1. The covalent radii for some of the row 2 elements can be read from Figure 20-3, as follows: Si, 1.16 Å; P, 1.10 Å; S, 1.04 Å; Cl, 1.00 Å.
(a) Why do the covalent radii tend to become smaller among the p orbital elements as we move to the right across the row?
(b) The bond lengths of the gaseous monoxides of some of these elements have been measured at very high temperatures: SiO(g), 1.51 Å; PO(g), 1.47 Å; SO(g), 1.49 Å. Explain in terms of molecular orbitals why these bond lengths are rather constant, rather than decreasing, as the covalent bond radii dictate.

2. The bond energies of the gaseous monoxides mentioned in Problem 1 are as follows: SiO, 192; PO, 144; SO, 124; ClO, 64 kcal/mole.
(a) Plot bond energy versus column in the Periodic Table.
(b) Plot bond energy/bond order versus column in the Periodic Table (use the same graph as for (a)).
(c) How does the ionic character of these oxide bonds account for the difference between the slope obtained in (b) and that displayed for Si_2, P_2, S_2, and Cl_2 in Figure 17-4?

3. Figure 17-19 shows the anomalous boiling point of water among the oxygen family dihydrides. This is attributable to hydrogen bonding in liquid water. Make a similar plot for the Group 5 hydride boiling points (°K) and decide whether ammonia, NH_3, displays evidence of hydrogen bonding.

normal boiling points:
NH_3, −33°C; PH_3, −88°C; AsH_3, −63°C; SbH_3, −17°C

4. (a) Add to the plot of Problem 3 the normal boiling points of the hydrogen halides and of the Group 6 hydrides, using dotted and dashed lines to differentiate the families.

normal boiling points:
HF, 20°C;　　HCl, −85°C;　　HBr, −67°C;　　HI, −35°C;　　H_2O, 100°C;
H_2S, −60°C;　　H_2Se, −41°C;　　H_2Te, −2°C

(b) From this evidence alone, which species forms the strongest hydrogen bonds to itself and which the weakest: HF, H_2O, or NH_3?

(c) Other evidence shows that HF is the best electron acceptor (best acid) and NH_3 is the best electron donor (best base) of these three molecules. Rationalize these facts with your answer to (b).

5. Make a plot like that of Figure 17-19 for the trifluorides and trihalides to see if the discontinuities shown by H_2O and by NH_3 (Problem 3) could be only a "first-row" effect.

normal boiling points:
NF_3, −129°C;　　PF_3, −101°C;　　AsF_3, 19°C;　　SbF_3, 376°C;
NCl_3, 71°C;　　PCl_3, 76°C;　　$AsCl_3$, 130°C;　　$SbCl_3$, 221°C

6. An electrochemical cell is made up with a silver electrode in 0.20 M Ag^+ in the left cell (Ag^+–Ag, $\varepsilon^0 = 0.80$) and, in the right cell, an inert platinum electrode in a solution of selenate, SeO_4^{-2}, at 0.40 M, selenous acid, H_2SeO_3, at 0.10 M, and H^+ buffered at $1.0 \cdot 10^{-3}$ M. The cell voltage is measured to be 0.050 volt and current flows so that silver metal dissolves.

(a) Is the platinum electrode the anode or cathode?

(b) Balance the equation for the SeO_4^{-2}–H_2SeO_3 half-reaction.

(c) Balance the equation for the over-all cell reaction.

(d) Use the Nernst Equation to calculate ε^0 for the SeO_4^{-2}–H_2SeO_3 half-reaction. Do these data give an answer in agreement with the value given in Table 20-2?

(e) Which of the following reagents can oxidize selenous acid, H_2SeO_3, to selenate, SeO_4^{-2}: Ag^+, Br_2, Cl_2, $Cr_2O_7^{-2}$, Fe^{+3}, H_2O_2, I_3^-? (See Table 6-5.)

(f) Which of the reagents listed in (e) can oxidize sulfurous acid, H_2SO_3, to sulfate, SO_4^{-2}?

7. An effective reagent for the conversion of NO into N_2O is triphenylphosphine, $(C_6H_5)_3P$:

$$2\ NO(g) + (C_6H_5)_3P(s) \longrightarrow N_2O(g) + (C_6H_5)_3PO(s)$$

Nitric oxide is passed into a 100-cc flask until the pressure is 1.00 atm at 27°C. Then 0.500 gram of triphenylphosphine is added.

(a) Which reactant is in excess?

(b) If reaction is complete, what will be the final pressure in the flask (at 27°C)?

(c) Repeat (a) and (b) if only 0.250 gram of triphenylphosphine is added.

8. Hydrazine, N_2H_4, like ammonia, is a good electron donor (a good base). In acid, it forms hydrazinium ion, $N_2H_5^+$, the counterpart to ammonium ion, NH_4^+. The acid dissociation constants of $N_2H_5^+$ and NH_4^+ are, respectively:

$N_2H_5^+(aq) \longrightarrow N_2H_4(aq) + H^+(aq)$　　　$K_a = 1.2 \cdot 10^{-8}$

$NH_4^+(aq) \longrightarrow NH_3(aq) + H^+(aq)$　　　$K_a = 5.5 \cdot 10^{-8}$

(a) What is the equilibrium constant for the acid-base reaction between the Brønsted acid, NH_4^+, and the Brønsted base, N_2H_4?
(b) What is the OH^- concentration in a 0.10 M solution of NH_3? of N_2H_4?
(c) Which is the stronger base, NH_3 or N_2H_4?
(d) Is hydrazine likely to be soluble in methanol? Explain.
(e) Copper(II), $Cu^{+2}(aq)$ forms an ammonia complex $Cu(NH_3)_4^{+2}$. Is $Cu^{+2}(aq)$ liable to form a corresponding hydrazine complex and, if so, will it be more tightly or more loosely bound in the complex? Explain.

9. Until recently, hydrazine N_2H_4 (a good rocket fuel) was prepared in quantity by the Raschig synthesis. This is an aqueous solution reaction that is difficult to carry out because unwanted side reactions are catalyzed by minute concentrations of metal ion impurities. The desired steps are:

$$NH_3(aq) + OCl^-(aq) \longrightarrow NH_2Cl(aq) + OH^-(aq) \qquad \text{fast}$$

$$NH_3(aq) + NH_2Cl(aq) + OH^-(aq) \longrightarrow N_2H_4(aq) + Cl^-(aq) + H_2O \qquad \text{slow}$$

(a) What is the balanced equation for the over-all reaction?
(b) Use the ε^0 values included in (20-2) and (20-4) to calculate ΔG^0 and the equilibrium constant for the reaction.
(c) If there is no side reaction interference or catalysis, what would be the rate law for the reaction, $\Delta[N_2H_4]/\Delta t$?

10. Hydrazine, N_2H_4, can be titrated in aqueous solution with iodate solution, IO_3^-, to give a quantitative yield of nitrogen gas and iodide ion.
(a) Balance the equation for the reaction.
(b) What volume of dry nitrogen gas would be collected if 0.65 g of solid hydrazinium sulfate, $N_2H_5^+HSO_4^-$, is titrated and the gas is collected at one atmosphere and at 27°C?

11. The thermodynamics of the dissociation of N_2O_4 in carbon tetrachloride solution, CCl_4, has been measured. They can be compared to the gas phase values.

	$\Delta H^0(298°K)$	$\Delta S^0(298°K)$
$N_2O_4(CCl_4) = 2\ NO_2(CCl_4)$	+14.6 kcal	+31.8 cal/deg
$N_2O_4(g) = 2\ NO_2(g)$	+13.6	+42.8

(a) Why is ΔS^0 for the reaction in CCl_4 positive?
(b) The reaction $N_2O_4(CCl_4) = 2\ NO_2(CCl_4)$ can be considered to take place in a three-step process, with step I being vaporization of N_2O_4,

$$N_2O_4(CCl_4) \xrightarrow{\Delta S_I} N_2O_4(g)$$

and step III being the reverse of the vaporization of NO_2,

$$2\ NO_2(CCl_4) \xrightarrow{\Delta S_{III}} 2\ NO_2(g)$$

Formulate this cyclic process, calculate $\Delta S_I - \Delta S_{III}$, and discuss the significance of its sign.
(c) Calculate the NO_2 concentration in a 25°C CCl_4 solution prepared by dissolving 1.0 mole of N_2O_4 in 1.0 liter of CCl_4.

12. (a) Estimate, from Table 17-3, the electronegativity of astatine, element 85.
(b) Replot, as in Figure 17-9, the electronegativity difference versus the quantity δ for the hydrogen halides (dipole moments are given in Appendix e). Draw the best smooth curve through the four points that passes through the origin.

(c) From your answer to (a) and using this plot, predict the dipole moment of hydrogen astanide, HAt. (Assume the H—At bond length is 1.8 Å.)
(d) Will HAt have ionic character in its bond and/or substantial hydrogen bonding capability?

13. (a) Replot, with an electronegativity scale extending from 2.0 to 4.0, the data plotted in Figure 20-10, $-\log K_a$ versus electronegativity for the hypohalide acids, HOCl, HOBr, and HOI. Draw the best straight line through the three points.
(b) By extrapolation, estimate the value of K_a for HOF.
(c) Using your electronegativity estimate from Problem 12, estimate the value of K_a for the astatine counterpart, hydrogen hypoastatide, HOAt.

14. Calculate to ten percent accuracy the H^+ concentrations in 0.10 M solutions of HOCl, HOBr, and HOI (see Table 20-6), and, using the results of Problem 13, of HOF and HOAt. State and test any assumptions you make, including the neglect of the dissociation of the solvent, H_2O.

15. Calculate to ten percent accuracy the H^+ concentrations in 0.10 M solutions of HOCl, HOClO, $HOClO_2$, and $HOClO_3$ (see Table 20-6). State and test any assumptions you make.

16. Figure 20-11 shows among the Group 5 oxyacids the formula H_3MO. For nitrogen, this would be called amine oxide. This compound is unknown for M = nitrogen, phosphorus, or any other member of the nitrogen family, but hydroxylamine, NH_2OH, with the same empirical formula, is known. The fluorine counterparts, F_3MO, are known, however, while the fluorine counterpart of hydroxylamine, NF_2OF, is not. For example, F_3NO (trifluoroamine oxide) and F_3PO (phosphorus oxyfluoride) are both known with the structural parameters given below. Compare the M—O and M—F bond lengths to those in the prototype molecules NO, HNO, NF_3, PO, and PF_3 and estimate the bond orders in both F_3NO and F_3PO.

17. In an early thermochemical study of XeF_6 (1962), 0.245 g of solid XeF_6 was allowed to vaporize in a calorimeter. Excess hydrogen was added and reaction occurred at room temperature to produce gaseous HF and xenon. The temperature rise indicated that 306 calories of heat were released.
(a) Calculate the heat of formation of $XeF_6(g)$.
(b) The heat of sublimation of $XeF_6(s)$ at room temperature is 9.0 kcal/mole. What is the heat of formation of $XeF_6(s)$?
(c) Calculate from your result in (a) the heat of atomization of $XeF_6(g)$ and the average Xe—F bond energy in $XeF_6(g)$.

$$XeF_6(g) \longrightarrow Xe(g) + 6 F(g) \qquad \Delta H(\text{atomization})$$

(d) What percent changes have there been between these early values of ΔH_f^0 and Xe—F bond energies and the presently accepted values quoted in Section 17-4(c)?

18.   Xenon trioxide, XeO_3, reacts with aqueous bromide ion, Br^-, to produce xenon gas and aqueous tribromide ion.
(a) Balance the equation for the reaction.
(b) What can be said about ε^0 for the XeO_3–Xe half-reaction in view of the observation that this reaction proceeds?
(c) What do the following rate measurements indicate about the rate law?

$[XeO_3]$	$[H^+]$	$[Br^-]$	$\Delta([Xe]/\Delta t)$
$1.0 \cdot 10^{-3}\ M$	$0.100\ M$	$0.100\ M$	$4.0 \cdot 10^{-3}$
$3.0 \cdot 10^{-3}$	0.100	0.100	$12 \cdot 10^{-3}$
$3.0 \cdot 10^{-3}$	0.050	0.100	$3.0 \cdot 10^{-3}$
$1.0 \cdot 10^{-3}$	0.050	0.200	$4.0 \cdot 10^{-3}$

(d) Deduce the rate law for each of the following reaction mechanisms to decide if any of them is consistent with the observed rate law.

A. $XeO_3 + Br^- \longrightarrow XeO_2 + BrO^-$ fast
    $H^+ + OBr^- \longrightarrow HOBr$ fast
    $XeO_2 + HOBr \longrightarrow XeO + HOBrO$ slow
    $XeO + HOBr \longrightarrow Xe + HOBrO$ fast
    $HOBrO + Br^- \longrightarrow HOBr + BrO^-$ fast
    $HOBr + Br^- \longrightarrow Br_2 + OH^-$ very slow
    $Br_2 + Br^- \longrightarrow Br_3^-$ fast

B.   Same as A except third reaction is fast and fourth is slow.
C.   Same as A except third reaction is fast and fifth is slow.
D. $XeO_3 + 2\ Br^- + 2\ H^+ \longrightarrow Xe + HOBr + HOBrO$ slow
    $HOBrO + Br^- \longrightarrow HOBr + OBr^-$ fast
    $HOBr + Br^- \longrightarrow Br_2 + OH^-$ fast

19.   Liquid ammonia, like water, is weakly conducting (it "autoionizes").

$$2\ NH_3 = NH_4^+ + NH_2^-$$

On the basis of orbital occupancy and hybridization rules, what is the expected geometry of NH_4^+ and NH_2^- if these two compounds are considered to be the hydrides of N^+ and of N^-? With what neutral hydrides are they isoelectronic?

20.   Phosphorus pentachloride, PCl_5, crystallizes in a lattice that is most readily described as an ionic lattice containing PCl_4^+ and PCl_6^- ions.
(a) Describe the bonding and expected structure for PCl_4^+.
(b) The bonding in PCl_5 was described in Section 20-3(b) as three sp_2 hybrid bonds coupled with an inert gas, three-center bond. What geometry would be expected if PCl_6^- were described as two sp hybrid bonds coupled with *two* inert gas, three-center bonds?

21.   Consider the ε^0 values given in Table 20-2 for the H_2SO_3–$S_4O_6^{-2}$–$S_2O_3^{-2}$ half-reactions.
(a) Balance the half-reactions implied by

$$H_2SO_3 \xrightarrow{\ +0.51\ } S_4O_6^{-2}$$

$$S_4O_6^{-2} \xrightarrow{\ +0.08\ } S_2O_3^{-2}$$

(b) Calculate ΔG^0 for each of the half-reactions in (a).

(c) Calculate ΔG^0 and from it ε^0 for the half-reaction

$$2 H_2SO_3 + 2 H^+ + 4 e^- \longrightarrow S_2O_3^{-2} + 3 H_2O$$

22. The sulfur counterpart to hypochlorous acid, HOCl, would be $S(OH)_2$, but only a single compound of this acid is known, the salt formed by Co^{+2}, cobaltous sulfoxylate, $Co(OSO)$.

(a) What are the sulfur counterparts to chlorous acid, $HOClO$, and to chloric acid, $HOClO_2$?

(b) Draw electron-dot drawings for these two counterparts.

(c) Predict the geometries of the ions of these two acids (sulfite ion and sulfate ion) by analogy to the chlorine oxyacids, pictured in Table 17-6.

23. Thiosulfate ion, $S_2O_3^{-2}$, is an important analytical reagent because of its quantitative reaction with iodine to give I^- and tetrathionate ion, $S_4O_6^{-2}$. What would be the structure of $S_2O_3^{-2}$ if it is considered to be derived from SO_3^{-2} acting as an electron donor and sulfur atom acting as an electron acceptor in the manner attributed to oxygen atom in the ^1D state (see Section 17-3(d))? Draw a picture of the ion and its electron-dot drawing.

24. One of the possible errors in iodometric methods is that the reagent thiosulfate, $S_2O_3^{-2}$, can react with oxygen from the air, changing the $S_2O_3^{-2}$ concentration.

(a) Balance the equation for reaction between $S_2O_3^{-2}$ and O_2 to give H_2SO_3 and H_2O. Call this reaction A.

(b) Calculate $\Delta\varepsilon^0$ and ΔG^0 for the reaction considered in (a). (See Problem 21 and Table 6-5.)

(c) Calculate $\Delta\varepsilon^0$ and ΔG^0 for the reaction

$$2 H_2SO_3 + O_2 \longrightarrow 2 HSO_4^- + 2 H^+$$

(See Table 20-2.) Call this reaction B.

25. (a) Suppose the air oxidation of thiosulfate, $S_2O_3^{-2}$, proceeded via reaction A followed by reaction B in Problem 24 with reaction A slow and B fast. What would be the rate law?

$$\frac{\Delta(S_2O_3^{-2})}{\Delta t} = ?$$

(b) Repeat question (a) with reaction A fast and B slow.

(c) According to one report, air oxidation of $S_2O_3^{-2}$ proceeds via the following reaction mechanism:

$$S_2O_3^{-2} + H^+ \xrightarrow{k_1} HSO_3^- + S(s) \qquad \text{slow}$$

$$2 HSO_3^- + O_2 \xrightarrow{k_2} 2 SO_4^{-2} + 2 H^+ \qquad \text{fast}$$

What is the rate law, $\Delta(S_2O_3^{-2})/\Delta t = ?$

(d) What is the over-all reaction for the mechanism in (c)?

(e) Does the mechanism in (c) involve any species that can be properly called a catalyst? If so, which?

(f) By what factor would the air oxidation rate change under the mechanism in (c) if the pH were changed from 5 (a pH at which air oxidation is not too troublesome) to 1?

26. Chlorine dioxide, ClO_2, is used in large commercial quantities because its high oxidizing power makes it useful as a bleaching agent. Thus it is used to bleach paper pulp because legibility is higher on white paper and to bleach flour because the natives like white bread. Two commercial processes for the preparation of ClO_2 involve the acidic reduction of chlorate, ClO_3^-, by oxalic acid, HOOCCOOH, or by sulfur dioxide, SO_2. (It is to be noted that both SO_2 and ClO_2 are noxious gases and ClO_2 is dangerously explosive, as well.)
(a) Balance the equations for the reactions used in these two processes.

$$ClO_3^- + HOOCCOOH \longrightarrow ClO_2(g) + CO_2(g)$$
$$ClO_3^- + SO_2(g) \longrightarrow ClO_2(g) + HSO_4^-(aq)$$

(b) How many kilograms of oxalic acid would be required to produce one kilogram of ClO_2?
(c) How many liters of SO_2 (at one atm and 300°K) would be required to produce one kilogram of ClO_2?

27. Whereas pure diamonds are crystal-clear, the purest silicon and germanium samples look metallic. Yet both silicon and germanium are useful as infrared-transmitting windows. (Note: The infrared region begins at wavelengths longer than about 7000 Å or 0.7 microns; $1 Å = 10^{-8}$ cm, 1 micron $= 10^{-4}$ cm.)
(a) At what wavelength do the quanta have just the right energy to excite electrons across the intrinsic energy gap in pure silicon? (See Table 20-9.) Express the wavelength both in Å and in microns.
(b) Repeat (a) for pure germanium.
(c) Repeat (a) for pure diamond.
(d) What infrared spectral range would be isolated if solar light were first passed through a pure silicon filter and then reflected from a pure germanium mirror?

28. Suppose an n-type impurity is dissolved in silicon at a mole fraction of 10^{-9} and the impurity energy levels lie 2.86 kcal/mole below the conduction band.
(a) How many impurity sites are introduced per mole of silicon?
(b) What fraction of the extra electrons (one per impurity atom) are excited into the conduction band at room temperature?
(c) Calculate the ratio of impurity conduction electrons to intrinsic conduction electrons (see Table 20-9).

29. Repeat Problem 28 except for germanium instead of silicon as the host lattice.

30. Repeat Problem 28 except for germanium at dry ice temperature, $-78°C = 195°K$.

31. The most sensitive infrared light detectors now available are like the "doped" germanium sample described in Problem 29. The detector's electrical conductivity changes when infrared light is absorbed, exciting electrons to the conduction band.
(a) Suppose 10^9 quanta with energy 2.86 kcal/mole are suddenly absorbed by a mole of the "doped" germanium at the dry ice temperature, 195°K. By what factor is the number of conduction electrons increased?
(b) Repeat (a) for a detector cooled with liquid nitrogen (77°K) instead of dry ice.

twenty-one chemical principles applied: the d and f orbital elements

In Figure 2-4 the ionization energies show that the third and fourth rows of the Periodic Table each contains 18 elements, not 8. Then, in Chapter Thirteen, the existence of these long rows was explained. Because of orbital penetration, electron repulsions are more effective in shielding d orbitals than s and p orbitals. Hence the 3d orbitals are higher in the energy than the 3s and 3p—in fact, they are raised to about the energy of the 4s and 4p orbitals. The third-row elements have 4s, 4p, and 3d orbitals as valence orbitals. The fourth-row elements base their chemistry on 5s, 5p, and 4d orbitals. Then in the fifth row, the 4f orbitals join the 6s, 6p, and 5d orbitals as valence orbitals. We shall examine first the elements with partially filled d orbitals (and completely vacant or completely filled f orbitals). These d orbital elements are traditionally called *transition metals* and they are located in the center of the Periodic Table (see Fig. 21-1). Literally and figuratively, they are the most colorful elements in the Periodic Table. The shapes of the orbitals and the spacing of the energy levels show why.

21-1 The d orbitals

The d orbitals correspond to quantum number $\ell = 2$. Harking back to Section 12-4(a), we recall that there are ℓ angular nodal surfaces and $n - \ell$ radial nodal surfaces (including that one at infinity). There is no unique set of pictures with which to represent the d orbitals.* Figure 21-2 shows an acceptable set, however ("acceptable" means that Mr. Schroedinger and his equation would approve). Each picture has two angular nodal surfaces and only one radial nodal surface, at infinity, as appropriate for 3d orbitals ($\ell = 2$, $n - \ell = 3 - 2 = 1$). The nodal surfaces through the nucleus account for the poor penetration that causes these orbitals to be shielded more effectively than s and p orbitals with the same principal quantum

---

* Indeed, though it wasn't mentioned earlier, there is no unique set of pictures of p orbitals, either. The ones shown in Figure 12-14 are one acceptable set, but there are other sets, any of which can be shown to be suitable combinations of the set we have chosen to display.

d Orbital elements

21	22	23	24	25	26	27	28	29	30
Sc	Ti	V	Cr	Mn	Fe	Co	Ni	Cu	Zn
39	40	41	42	43	44	45	46	47	48
Y	Zr	Nb	Mo	Tc	Ru	Rh	Pd	Ag	Cd
57	72	73	74	75	76	77	78	79	80
La	Hf	Ta	W	Re	Os	Ir	Pt	Au	Hg
89									
Ac									

f Orbital elements

58	59	60	61	62	63	64	65	66	67	68	69	70	71
Ce	Pr	Nd	Pm	Sm	Eu	Gd	Tb	Dy	Ho	Er	Tm	Yb	Lu
90	91	92	93	94	95	96	97	98	99	100	101	102	103
Th	Pa	U	Np	Pu	Am	Cm	Bk	Cf	Es	Fm	Md	No	Lw

Figure 21-1   The d and f orbital elements.

number. Recalling Chapter Sixteen, we can also expect these orbital shapes to encourage more complex geometrical structures than those commonly encountered when only s and p orbitals are available.

21-2   Energy levels across the third row

The first ionization energy tells only the ease of removal of the first electron. To appreciate the interesting chemistry of transition metals, we need to see the energies of all the valence orbitals. Fortunately, the line spectra of the neutral atoms reveal these energies. Figure 21-3 shows the energy placement of the valence orbitals for the third row. Here is much food for thought!

For the first two elements, K and Ca, the most easily removed electron comes from a 4s orbital. Then for the next five elements, Sc to Mn, it is a 3d electron. After Mn, the 3d orbitals fall below the 4s orbitals. For Fe, it takes more energy to remove a 3d than a 4s electron to form Fe^+, *even though the neutral atom has lowest energy with orbital occupancy $3d^6 4s^2$*. This apparent contradiction shows that electron repulsions in Fe^0 and Fe^+ differ and that the entire orbital occupancies of both Fe^0 and Fe^+ determine E_1 from a given orbital. Beyond Zn, the 4p orbitals must be occupied.

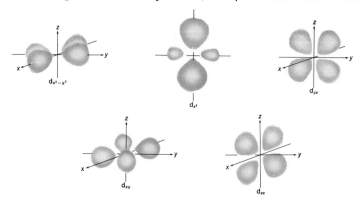

Figure 21-2   A representation of the 3d orbitals.

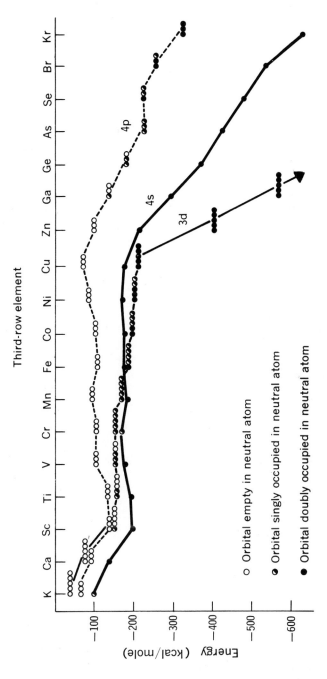

Figure 21-3 Neutral atom orbital energies across the third row.

The 3d orbitals plummet downward, leaving the last six elements, Ga to Kr, with the impression they are completing a row of eight, just like elements B to Ne and Al to Ar. That is why we have called them the p orbital elements.

The trends displayed in Figure 21-3 can be discussed in terms of effective charge, Z^*, and average radius, $\bar{r}$, again on the basis of the one-electron, hydrogen atom model. Applying equations (13-3) and (13-4) once more, we obtain the neutral atom Z^* values and one-electron average radii listed in Table 21-1. All the way from scandium to copper, the effective nuclear charge felt by the 4s electrons is just about constant. The reason is seen in the $\bar{r}$ trends across the row, pictured in Figure 21-4. As the first two electrons enter the 4s orbital, Z^* increases more for the 3d orbitals than for either the 4s or 4p. That occurs because the lower principal quantum number of the 3d orbitals initially gives them a smaller radius, so they lie "inside" much of the 4s distribution. Then, as 3d orbitals begin to fill, these small orbitals are quite effective in shielding, despite the penetration of the 4s orbitals. In contrast, the outermost 4p orbitals, with their nodal surface at the nucleus, are more affected by added electron repulsions than by increasing Z, so Z^* actually decreases as d orbitals fill. From Fe to Zn, as the 3d orbitals are accepting second electrons, the 3d orbitals begin to shrink, increasing their shielding capability for the larger 4s and the much larger 4p orbitals.

Table 21-1 Effective Nuclear Charge and One-Electron Radius for the Third-Row, Neutral Atoms in Their Lowest Energy States

			Z^*			$\bar{r}$		
			4s	3d	4p	4s	3d	4p
K	4s		2.26	1.06	1.79	5.62	5.25	6.79
Ca	$4s^2$		2.67	1.52	2.13	4.75	3.67	5.71
Sc	$4s^23d$		3.18	2.08	2.67	4.00	2.67	4.55
Ti	$4s^23d^2$	d orbitals half-filling	3.15	2.13	2.60	4.04	2.61	4.67
V	$4s^23d^3$		3.03	2.11	2.31	4.19	2.63	5.26
Cr	$4s3d^5$		2.97	2.12	2.33	4.27	2.63	5.23
Mn	$4s^23d^5$		3.04	2.22	2.24	4.18	2.51	5.44
Fe	$3d^64s^2$		3.05	2.34	2.37	4.17	2.38	5.14
Co	$3d^74s^2$	d orbitals filling	3.04	2.38	2.31	4.18	2.33	5.26
Ni	$3d^84s^2$		3.00	2.41	2.14	4.24	2.31	5.68
Cu	$3d^{10}4s$		3.01	2.47	1.93	4.21	2.25	6.31
Zn	$3d^{10}4s^2$		3.33	3.40	2.28	3.82	1.64	5.34
Ga	$3d^{10}4s^24p$		3.87	4.04	2.65	3.28	1.38	4.59
Ge	$3d^{10}4s^24p^2$		4.36	4.96	3.05	2.92	1.12	3.99
As	$3d^{10}4s^24p^3$		4.66	6.11	3.40	2.72	0.91	3.58
Se	$3d^{10}4s^24p^4$		4.95	6.67	3.39	2.57	0.83	3.59
Br	$3d^{10}4s^24p^5$		5.21	7.27	3.73	2.44	0.76	3.26
Kr	$3d^{10}4s^24p^6$		5.66	8.17	4.06	2.24	0.70	3.00

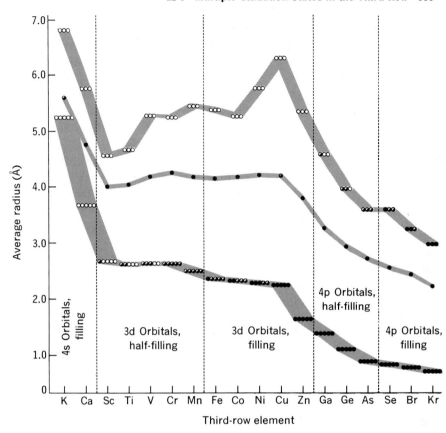

Figure 21-4 Average radius—one-electron approximation.

o Orbital empty in neutral atom   • Orbital singly occupied in neutral atom
• Orbital doubly occupied in neutral atom

Once the 3d orbitals are filled, Z^* increases more rapidly for all orbitals because of the poor shielding provided by the expanded 4p orbitals, as they are occupied. This accounts for the plummetting energies of both 3d and 4s orbitals at the right side of the Periodic Table.

Also evident in Figure 21-3 is the cause of the color of many transition metal compounds. The relatively close spacing of orbital energies implies that there is a good likelihood that two energy levels will have such a small energy difference that a quantum of visible light can be absorbed. That's all it takes to give a substance color.

21-3 Multiple oxidation states in the third row

The alkali metals display only $+1$ oxidation states. The alkaline earth elements, in the second column of the Periodic Table, display only $+2$ oxidation states. The third-column elements display only $+3$ states. Beginning at the fourth column, however, the d orbitals endow the transition elements with multiple oxidation states and a rich oxidation-reduction chemistry. We will look at the trend in oxidation states across the row and then examine three important elements in detail.

(a) OXIDATION STATES

Figure 21-5 shows the oxidation states observed across the third row of the Periodic Table. For each element, the most common and readily prepared oxidation states are shaded. We see that the number of oxidation states increases across the row up to the element manganese, for which the d orbitals are just half occupied. Then, from iron to zinc, the $+2$ state predominates. Beyond zinc, the chemistry is that of the nonmetals.

The three elements chromium, manganese, and iron will be considered in detail to typify transition metal chemistry.

(b) CHROMIUM

Chromium is a hard, lustrous metal familiar because of its common use as an electroplated, protective coat on more readily corroded metals. Its corrosion resistance is kinetic in nature, for chromium has significant

Element	Orbital Occupancy		−2	−1	0	+1	+2	+3	+4	+5	+6	+7
K	$4s$	s orbital elements			0	+1						
Ca	$4s^2$				0		+2					
Sc	$4s^23d$	d orbitals half-filling			0			+3				
Ti	$4s^23d^2$				0		+2	+3	+4			
V	$4s^23d^3$				0		+2	+3	+4	+5		
Cr	$4s3d^5$				0		+2	+3	+4	+5	+6	
Mn	$3d^54s^2$				0		+2	+3	+4		+6	+7
Fe	$3d^64s^2$	d orbitals filling			0		+2	+3			+6	
Co	$3d^74s^2$				0		+2	+3				
Ni	$3d^84s^2$				0		+2		+4			
Cu	$3d^{10}4s$				0	+1	+2					
Zn	$3d^{10}4s^2$				0		+2					
Ga	$3d^{10}4s^24p$	p orbital elements			0		+2	+3				
Ge	$3d^{10}4s^24p^2$				0		+2		+4			
As	$3d^{10}4s^24p^3$				0			+3		+5		
Se	$3d^{10}4s^24p^4$		−2		0				+4		+6	
Br	$3d^{10}4s^24p^5$			−1	0	+1				+5		+7
Kr	$3d^{10}4s^24p^6$				0							

Figure 21-5 Oxidation states in the third row (most common states shaded).

tendency to be oxidized, as revealed by its highly negative reduction potentials.

$$Cr^{+2} + 2\,e^- = Cr(s) \qquad \mathcal{E}^0 = -0.86$$
$$Cr^{+3} + 3\,e^- = Cr(s) \qquad \mathcal{E}^0 = -0.71$$

$(21\text{-}1)$

It is generally accepted that metallic chromium immediately oxidizes when exposed to air, but only a thin, invisible layer of the oxide is sufficient to form a protective coat that prevents further oxidation. Most metals form such oxide coats, but some are effective in preventing corrosion and some are not. The adhesion of the coating to the metal and the continuity of the coverage (absence of minute cracks and fissures) determine the effectiveness.

Chromium displays common oxidation states $+2$, $+3$, and $+6$, though $+4$ and $+5$ states can also be prepared. The oxidation-reduction chemistry among these states is compactly displayed in the Latimer $\mathcal{E}^0$ diagram:

$$\begin{array}{cccc} +6 & +3 & +2 & 0 \end{array}$$

acidic solution
$$Cr_2O_7^{-2} \xrightarrow{+1.33} Cr^{+3} \xrightarrow{-0.41} Cr^{+2} \xrightarrow{-0.86} Cr \qquad (21\text{-}2)$$
$$\underline{\qquad -0.71 \qquad}$$

basic solution
$$CrO_4^{-2} \xrightarrow{-0.13} Cr(OH)_3 \xrightarrow{-1.3} Cr \qquad (21\text{-}3)$$

Chromous solutions, Cr^{+2}, can be prepared by the reaction between $Cr^{+3}(aq)$ and zinc metal, but because $Cr^{+2}(aq)$ reacts rapidly with atmospheric oxygen, the solution must be protected from exposure to air.

Exercise 21-1. From the $\mathcal{E}^0$ values $Zn^{+2} + 2\,e^- \longrightarrow Zn$, $\mathcal{E}^0 = -0.76$, and $Cr^{+3} + e^- \longrightarrow Cr^{+2}$, $\mathcal{E}^0 = -0.41$, calculate ΔG^0 for the reaction between Cr^{+3} and Zn, and show that the equilibrium constant exceeds 10^{+11}.

Dichromate ion, $Cr_2O_7^{-2}$, is one of the most useful and powerful oxidizing agents available in acid solution. The Latimer diagram (21-2) shows us that the value $\mathcal{E}^0 = +1.33$ V is associated with the balanced half-reaction involving $Cr_2O_7^{-2}$ and Cr^{+3}:

$$Cr_2O_7^{-2} + 14\,H^+ + 6\,e^- \longrightarrow 2\,Cr^{+3} + 7\,H_2O \qquad \mathcal{E}^0 = +1.33\ \text{V} \quad (21\text{-}4)$$

The potassium salt, $K_2Cr_2O_7$, is easily prepared in high purity, it has high solubility in water, and its aqueous solutions are stable indefinitely. Despite its oxidizing power, $Cr_2O_7^{-2}$ reacts very slowly at room temperature with chloride ion. This adds to the practical value of dichromate in quantitative applications because many analytical applications must be conducted in the presence of chloride ion.

If base is added to an acidic solution of dichromate ion, the orange dichromate ion color changes to yellow. This color is due to chromate ion, CrO_4^{-2}, which also places chromium in the $+6$ oxidation state. The equilibrium is rapid and it reverses when the pH is lowered again.

$$Cr_2O_7^{-2}(aq) + H_2O \rightleftarrows 2\ CrO_4^{-2}(aq) + 2\ H^+(aq) \qquad K = 2.4 \cdot 10^{-15}$$
$$(21\text{-}5)$$

Exercise 21-2. Sodium hydroxide is slowly added to a 0.15 M solution of $K_2Cr_2O_7$. Show that when the concentrations of $Cr_2O_7^{-2}$ and CrO_4^{-2} become equal, at 0.10 M, the pH $= 6.8$.

Exercise 21-3. Calculate the CrO_4^{-2} concentration at pH $= 3$ with $Cr_2O_7^{-2} = 0.15\ M$.

The structure of $Cr_2O_7^{-2}$ is well known through X-ray studies of crystalline $(NH_4)_2Cr_2O_7$. The $Cr_2O_7^{-2}$ geometry is shown in Figure 21-6. There is one oxygen atom bridged between two chromium atoms, so that there are four oxygen atoms around each Cr. Such oxygen bridges are common in transition metal compounds.

Chromate ion is the salt (Brønsted conjugate base) of a weak acid, $HCrO_4^-$. The acid H_2CrO_4 is a strong acid.

$$H_2CrO_4 = HCrO_4^- + H^+ \qquad \text{strong} \qquad (21\text{-}6)$$

$$HCrO_4^- = CrO_4^{-2} + H^+ \qquad K_2 = 3.2 \cdot 10^{-7} \qquad (21\text{-}7)$$

Chromate ion does not have the strong oxidizing power of $Cr_2O_7^{-2}$ ion. According to (21-3), the equation for the basic solution half-reaction for chromate reduction is:

$$CrO_4^{-2} + 4\ H_2O + 3\ e^- = Cr(OH)_3(s) + 5\ OH^- \qquad \mathcal{E}^0 = -0.13 \quad (21\text{-}8)$$

Chromic hydroxide has a relatively low solubility; it precipitates from a solution of chromic ion, Cr^{+3}, when ammonia solution is added. However,

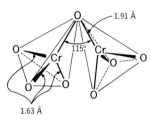

Figure 21-6 The structure of dichromate ion, $Cr_2O_7^{-2}$.

it dissolves again in concentrated sodium hydroxide. This is an example of the amphoteric behavior we identified in Section 20-2(d). Superficially, the sequence of events can be represented by the equations for two simple reactions:

$$Cr^{+3}(aq) + 3\ OH^-(aq) = Cr(OH)_3(s) \quad K = \frac{1}{K_{sp}} = \frac{1}{6.7 \cdot 10^{-31}}$$

$$(21\text{-}9)$$

$$Cr(OH)_3(s) + OH^-(aq) = CrO_2^-(aq) + 2\ H_2O \quad K = 9 \cdot 10^{-3} \quad (21\text{-}10)$$

It is more informative, however, to recognize that this amphoteric behavior can be properly classified as a proton-release by water of aquation.

Aquated chromic ion probably involves six water molecules. The oxygen atom of a water molecule is an electron donor and chromium ion has vacant valence orbitals, so it can act as an electron acceptor. The resulting chromium–oxygen bonds are sufficiently strong to permit release of protons, through breakage of oxygen–hydrogen bonds. This can occur in a sequence of equilibrium steps.

$$Cr(OH_2)_6^{+3} + OH^- \longrightarrow Cr(OH)(OH_2)_5^{+2} + H_2O \qquad (21\text{-}11a)$$

$$Cr(OH)(OH_2)_5^{+2} + OH^- \longrightarrow Cr(OH)_2(OH_2)_4^+ + H_2O \qquad (21\text{-}11b)$$

$$Cr(OH)_2(OH_2)_4^+ + OH^- \longrightarrow Cr(OH)_3(OH_2)_3 + H_2O \qquad (21\text{-}11c)$$

$$Cr(OH)_3(OH_2)_3 + OH^- \longrightarrow Cr(OH)_4(OH_2)_2^- + H_2O$$
$$K = 9 \cdot 10^{-3} \quad (21\text{-}11d)$$

$$Cr(OH)_4(OH_2)_2^- + OH^- \longrightarrow Cr(OH)_5(OH_2)^{-2} + H_2O \qquad (21\text{-}11e)$$

$$Cr(OH)_5(OH_2)^{-2} + OH^- \longrightarrow Cr(OH)_6^{-3} + H_2O \qquad (21\text{-}11f)$$

In the sequence, all of the species formed that have ionic charge are soluble in the ionizing solvent, water. The lone neutral species, $Cr(OH)_3(OH_2)_3$, having no ionic charge, has low solubility and it precipitates within that pH range in which (21-11c) is important. Because the solid incorporates a number of water molecules, it tends to carry with it variable amounts of excess solvent, depending upon precipitation conditions. This excess water content affects both the physical character and the composition of this and many other hydroxide solids. Freshly prepared $Cr(OH)_3$ precipitates as a flocculent, slimy solid and only on prolonged heating does its composition become uniform.

(c) MANGANESE

The most common oxidation states of manganese are the $+2$, $+4$, and $+7$ states as represented by the compounds $MnCl_2$, manganous chloride,

MnO_2, manganese dioxide, and $KMnO_4$, potassium permanganate. However, in both acidic and basic solution, it is possible to prepare the $+3$ and $+6$ states, manganic and manganate. The reduction potentials are as follows:

| +7 | +6 | +4 | +3 | +2 | 0 |

acidic solution

$$MnO_4^- \xrightarrow{+0.54} MnO_4^{-2} \xrightarrow{+2.23} MnO_2 \xrightarrow{+1.1} Mn^{+3} \xrightarrow{+1.5} Mn^{+2} \xrightarrow{-1.18} Mn$$

with intermediate steps $+1.67$, $+1.28$, and overall $+1.51$

$$\hspace{2cm} (21\text{-}12)$$

basic solution

$$MnO_4^- \xrightarrow{+0.54} MnO_4^{-2} \xrightarrow{+0.58} MnO_2 \xrightarrow{+0.5} Mn(OH)_3 \xrightarrow{-0.4} Mn(OH)_2 \xrightarrow{-1.47} Mn$$

with intermediate step $+0.57$

$$\hspace{2cm} (21\text{-}13)$$

In acidic solution, permanganate ion, MnO_4^-, is a more powerful oxidizing agent than dichromate, preferentially reacting to give $Mn^{+2}(aq)$ with most reducing agents. In basic solution, permanganate reacts with reducing agents to give solid MnO_2. Permanganate in acid solution is particularly useful as a quantitative titrant because of the high oxidizing power ($\mathcal{E}^0 = +1.51$ volts), because it reacts quantitatively to give a single product, Mn^{+2}, and because its intense violet color indicates precisely the end-point in the titration (Mn^{+2} is a light pink color in water). This usefulness depends, however, upon the extremely slow reaction of some possible interfering reactions. The oxidizing power of MnO_4^- permits it to react with the solvent, water, either in acid or base solution.

acidic solution
$$4\,MnO_4^- + 4\,H^+ = 4\,MnO_2(s) + 3\,O_2(g) + 2\,H_2O \quad (21\text{-}14)$$

basic solution
$$4\,MnO_4^- + 2\,H_2O = 4\,MnO_2(s) + 3\,O_2(g) + 4\,OH^- \quad (21\text{-}15)$$

Fortunately, these reactions are negligibly slow at room temperature if the acidity is not too high and if the solution is pure. At high H^+ concentration or at elevated temperature, the reduction can become rapid enough to interfere. Once decomposition begins, its rate increases because the product, solid MnO_2, acts as a heterogeneous catalyst and further speeds things up. Such a reaction, in which a product acts as a catalyst for its own production, is called *autocatalytic*.

Exercise 21-4. From the ε^0 values for the $MnO_4^--MnO_2$ and for the H_2O-O_2 half-reactions, calculate the equilibrium constants for reactions (21-14) and (21-15).

Exercise 21-5. Another reaction that is slow in the absence of catalysts is the reaction of Mn^{+2} with MnO_4^- to give $MnO_2(s)$. Balance the equation for this reaction and show that its equilibrium constant is $10^{+23.5}$ (for the reaction involving one mole of MnO_4^-).

Manganous ion, Mn^{+2}, can be precipitated with a number of reagents. Table 21-2 lists some of these with their solubility products.

Table 21-2 Solubility Products for Manganous Salts of Low Solubility

Cation	Formula	K_{sp}
hydroxide	$Mn(OH)_2$	$2 \cdot 10^{-13}$
sulfide	MnS	$7 \cdot 10^{-16}$
phosphate	$Mn_3(PO_4)_2$	$1 \cdot 10^{-22}$
carbonate	$MnCO_3$	$8.8 \cdot 10^{-11}$
oxalate	MnC_2O_4	$1.1 \cdot 10^{-15}$

Exercise 21-6. From the solubility product and the equilibrium constants for the dissociation $H_2S = 2H^+ + S^{-2}$, $K = 1.1 \cdot 10^{-21}$, show that if the pH is slowly raised in a solution containing $0.10\ M$ H_2S and $0.02\ M$ Mn^{+2}, precipitation will begin when the pH $= 3.9$.

As for chromium, the high oxidation states of manganese are quite acidic: Both $HMnO_4$ and H_2MnO_4 are completely dissociated. Furthermore, the low oxidation states are basic. Manganous hydroxide is only weakly amphoteric:

$$Mn(OH)_2(s) + OH^-(aq) = Mn(OH)_3^-(aq) \qquad (21\text{-}16)$$

Metallic manganese has more tendency to oxidize than metallic chromium ($Mn^{+2}-Mn$, $\varepsilon^0 = -1.18$ volts) and its oxide does not form a protective coating. Because of this reactivity, manganese is added to the molten iron during manufacture of steel to remove the last amounts of oxygen and sulfur. These undesired impurities react to form MnO_2 and MnS, which are removed with the slag. In larger concentrations, manganese serves to toughen steel.

Manganese is found in Nature in a variety of oxide and carbonate ores, but the most important mineral is pyrolusite, MnO_2.

(d) IRON

Iron is the fourth most abundant element (by weight) in the earth's crust. On the average, about two out of every 100 atoms in the crust are iron atoms. The core of the earth is thought to be largely metallic iron, and some meteorites are mainly elemental iron. The principal iron ores are hematite, Fe_2O_3, and magnetite, Fe_3O_4. The mineral FeS_2 is called iron pyrites; because of its golden lustre, it is also dubbed "fool's gold." From the oxide ores, metallic iron is produced in enormous quantities, and it shapes our twentieth century technology.

Iron has only two important oxidation states, $+2$ and $+3$.

acidic solution

$$Fe^{+3} \xrightarrow{\ +0.77\ } Fe^{+2} \xrightarrow{\ -0.44\ } Fe \qquad (21\text{-}17)$$

basic solution

$$Fe(OH)_3 \xrightarrow{\ -0.56\ } Fe(OH)_2 \xrightarrow{\ -0.88\ } Fe \qquad (21\text{-}18)$$

Exercise 21-7. Calculate the equilibrium constant for the reaction between Fe^{+3} and Fe to form Fe^{+2} and show that the concentration of Fe^{+3} in equilibrium with $0.1\ M$ Fe^{+2} and metallic iron is immeasurably small.

Exercise 21-8. Calculate the equilibrium constants for the acidic reaction between Fe^{+2} and O_2 to form Fe^{+3} and H_2O_2 (O_2–H_2O_2, $\mathcal{E}^0 = +0.68$). Repeat the calculation for basic solution (O_2–HO_2^-, $\mathcal{E}^0 = -0.08$).

Table 21-3 gives the solubility products for some ferrous and ferric substances of low solubility. Both $Fe(OH)_2$ and $Fe(OH)_3$ are somewhat amphoteric; ferrous hydroxide dissolves in concentrated sodium hydroxide. Both $Fe^{+2}(aq)$ and $Fe^{+3}(aq)$ are considered to be bonded strongly to six water molecules in an octahedral arrangement. The acidic properties of the ferric ion are known in quantitative detail. The first two acid dissociation constants have been measured (compare to the chromium reactions (21-11)).

$$Fe(OH_2)_6^{+3} = Fe(OH)(OH_2)_5^{+2} + H^+ \qquad K = 9 \cdot 10^{-4} \qquad (21\text{-}19)$$

$$Fe(OH)(OH_2)_5^{+2} = Fe(OH)_2(OH_2)_4^+ + H^+ \qquad K = 5.5 \cdot 10^{-4} \qquad (21\text{-}20)$$

If sodium hydroxide is slowly added to a ferric ion solution, a brown, gelatinous mass of hydrous ferric hydroxide appears. Like the gelatinous chromic hydroxide, this slimy precipitate has indefinite composition and it is difficult to handle. Because of its high water content and poorly

Table 21-3   Solubility Products for Ferrous and Ferric Salts of Low Solubility

Ferrous, Fe^{+2}			Ferric, Fe^{+3}		
Cation	Formula	K_{sp}	Cation	Formula	K_{sp}
carbonate	$FeCO_3$	$2.1 \cdot 10^{-11}$	phosphate	$FePO_4$	$1.5 \cdot 10^{-18}$
hydroxide	$Fe(OH)_2$	$1.8 \cdot 10^{-15}$	hydroxide	$Fe(OH)_3$	$6 \cdot 10^{-38}$
sulfide	FeS	$4 \cdot 10^{-19}$	sulfide	Fe_2S_3	10^{-88}

defined crystal structure, the precipitate tends to trap other metal ions that would not precipitate alone at the same pH.

Exercise 21-9. Show that the H^+ concentration in a $0.2\,M$ solution of ferric nitrate exceeds $0.01\,M$.

21-4 Complex ions in the third row

The stability of positive ions (cations) in water solutions can be attributed to electron donor–acceptor bonding (see Section 17-3(c)). The oxygen atoms in water molecules act as electron donors and the cation inevitably has vacant valence orbitals, so it can act as an electron acceptor. This interaction is often displayed explicitly by substituting for the enigmatic symbol (aq) the number of water molecules considered to be attached to a given cation, as in $K(OH_2)_4^+$, $Ca(OH_2)_4^{+2}$, $Sc(OH_2)_6^{+3}$.

In Section 4-3(e), we saw that other electron donors can act as ionizing solvents: Ammonia and sulfur dioxide furnished examples. In fact, ammonia can substitute for water molecules in the aqueous solvation sphere, despite the preponderant excess of H_2O over NH_3. Copper ion was cited in Section 17-3(c) as an attractive example: The azure blue color of $Cu(OH_2)_4^{+2}$ changes to the inky blue of $Cu(NH_3)_4^{+2}$ when ammonia is added. The ions $Cu(OH_2)_4^{+2}$, $Cu(NH_3)_4^{+2}$, and intermediates, $Cu(OH_2)_3(NH_3)^{+2}$, $Cu(OH_2)_2(NH_3)_2^{+2}$, and $Cu(OH_2)(NH_3)_3^{+2}$, are called *complex ions*. Such complex ions add dimension to the chemistry of the transition elements that is as rich and interesting as that provided by their multiple oxidation states.

(a) ORBITALS AND GEOMETRIES

The ionic charge is one of the factors that determines the number of electron donor groups that bind to a cation. For example, the stable ammonia complex of Cu^+ is $Cu(NH_3)_2^+$, that of Cu^{+2} is $Cu(NH_3)_4^{+2}$, and that of Cr^{+3} is $Cr(NH_3)_6^{+3}$. The geometries of these complex ions can be associated in many cases with the available vacant valence orbitals. For the examples cited, the orbital occupancies are as follows:

Cu^+ sp hybrid: linear   (21-21)

Cu^{+2} sp³ hybrid: tetrahedral   (21-22a)

dsp² hybrid: square, planar   (21-22b)

Cr^{+3} d²sp³ hybrid: octahedral   (21-23)

The observed geometry can be discussed in terms of orbital hybridization, as in Chapter Sixteen. For Cu^+, the lowest-energy, vacant orbitals are s and p orbitals, so two bonded groups are attached in the geometry of sp hybrid orbitals. According to Table 16-3, this gives a linear geometry and, indeed, the $Cu(NH_3)_2^+$ ion is linear (see Fig. 21-7(a)). For Cu^{+2}, four groups attach strongly because of the higher nuclear charge. With the orbital occupancy (21-22a), the tetrahedral sp^3 arrangement can be expected, and that is what is most often observed in crystals (Fig. 21-7(b)) (for example, $CuCl_4^{-2}$ in crystalline Cs_2CuCl_4). The orbital occupancy (21-22b) is somewhat higher in energy because of the promotion of a d electron to the 4p state. Such an occupancy would give dsp^2 hybridization, with the groups arranged at the corners of a square and with the copper atom in the middle, as shown in Figure 21-7(c). This arrangement is observed in the crystal $Cu(NH_3)_4SO_4 \cdot H_2O$. The four NH_3 molecules are placed at the corners of a square with a Cu–N distance of 2.05 Å. Perpendicular to this square are two symmetrically placed water molecules, one above and one below, at a much longer distance, 2.59 Å. This situation is not uncommon—other electron donors form both tetrahedral and square planar complexes with

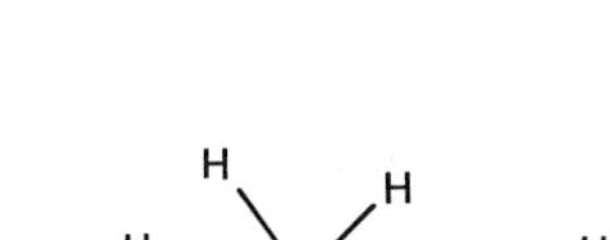

Linear (sp)

(a) Cu^+: $Cu(I)(NH_3)_2^+$

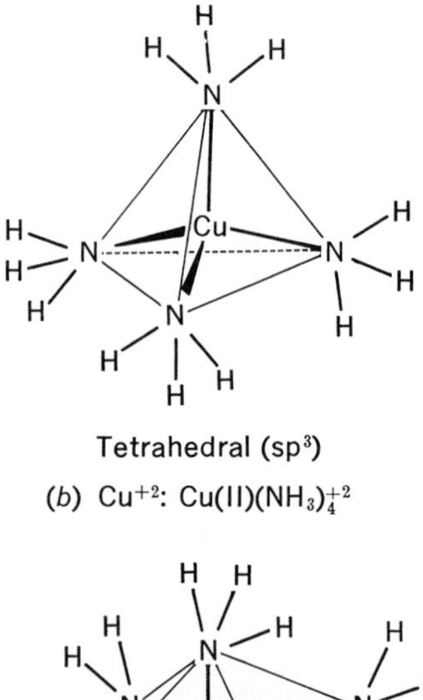

Tetrahedral (sp^3)

(b) Cu^{+2}: $Cu(II)(NH_3)_4^{+2}$

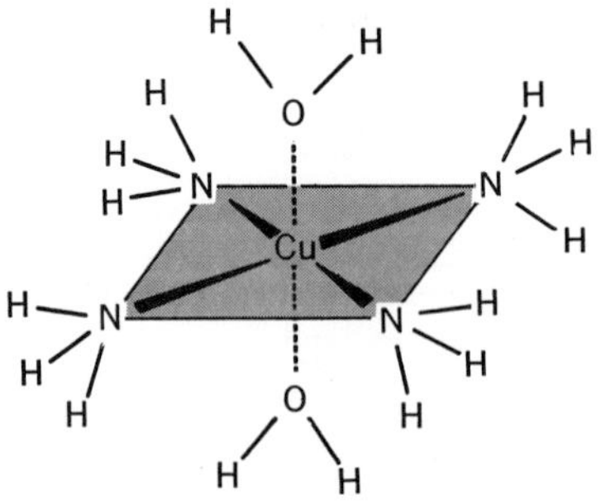

Square planar (dsp^2)

(c) Cu^{+2}: $Cu(II)(NH_3)_4^{+2}$

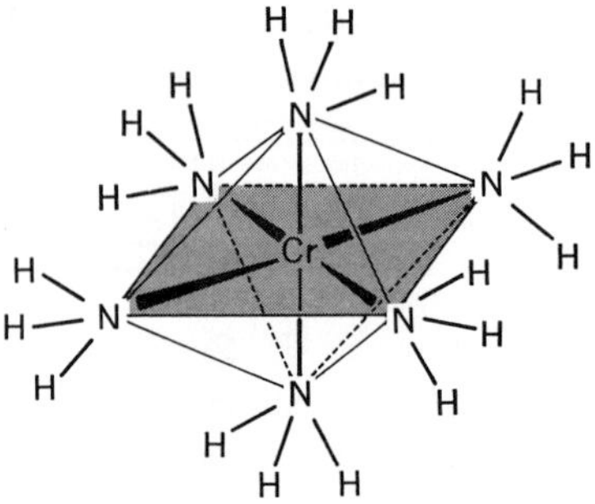

Octahedral (d^2sp^3)

(d) Cr^{+3}: $Cr(III)(NH_3)_6^{+3}$

Figure 21-7 Geometries of some complex ions.

copper(II). The dsp^2 hybrids would be based upon the p_x and p_y orbitals together with the $d_{x^2-y^2}$ orbital (see Fig. 21-2), all of which are directed along coordinate axes in the xy plane.

The third example, Cr^{+3}, with orbital occupancy (21-23), bonds six electron donor groups in most of its complex ions. The symmetry based upon orbital hybridization arguments should be octahedral (d^{2}sp^3 hybridization, see Table 16-3). The d orbitals with suitable geometry would be $d_{x^2-y^2}$ and d_{z^2} (see Fig. 21-2). Figure 21-7(d) shows the resultant geometry.

Almost all transition metal complex ions can be related to these four geometries, though in many cases the tetrahedral or octahedral arrangement is found to be a bit distorted.

(b) ENERGIES AND FREE ENERGIES

The effectiveness of a cation as an electron acceptor increases, of course, as the positive charge goes up. This must be so, for the energy needed to form the ion goes up rapidly with each successive electron removal. This is a direct measure of the energy lowering that can be realized as these electrons are "replaced" by the electron donor.

There is a considerable amount of information available concerning the energies of forming the aquo complexes of cations in water. We have already discussed the aquation energy of $Na^+(g)$ (Section 4-2(d)), of $H^+(g)$ and $K^+(g)$ (Section 5-1(e)), and of $Cu^{+2}(g)$ (Section 17-3(c)). Table 21-4 lists some of these aquation energies for $+1$, $+2$, and $+3$ ions in the third row. The fourth column shows the aquation energy divided by the sum of the ionization energies involved in forming the ion. We see that the aquation energy of the gaseous ion averages $83 \pm 10\%$ of the ionization energy

Table 21-4 Heats of Aquation of Third-Row Gaseous Ions: $M^{+n}(g) \longrightarrow M^{+n}(aq)$, ΔH_{aq}

Cation	$\Delta H_{aq}*$ (kcal)	$\sum_{i=1}^{n} E_i$ (kcal)	$\dfrac{\Delta H_{aq}}{\Sigma E_i}$	$N\dagger$	$\Delta H_{aq}/N$ (kcal)
K$^+$	-84	100	0.84	4	-21
Cu$^+$	-145	178	0.82	4	-36
Ca^{+2}	-389	415	0.94	4	-97
Cr^{+2}	-449	536	0.73	4(6)	-112 (-75)
Fe^{+2}	-467	555	0.84	4(6)	-117 (-78)
Cu^{+2}	-508	646	0.70	4	-127
Sc^{+3}	-960	1019	0.94	6	-160
Cr^{+3}	-1054	1250	0.84	6	-176
Fe^{+3}	-1055	1262	0.84	6	-176

* Calculations based upon assumed heat of aquation for Cl$^-$(g) $\longrightarrow$ Cl$^-$(aq), $\Delta H_{aq} = -84$ kcal.
$\dagger$ $N =$ probable number of water molecules strongly attached.

Table 21-5 Free Energy Changes Accompanying Complex Formation (kcal)

		$A = NH_3$	$A = Cl^-$*
$Cu(OH_2)_2^+ + 2\,A = CuA_2^+ + 2\,H_2O$	$\Delta G^0 =$	-20	-12
$Cu(OH_2)_4^{+2} + 4\,A = CuA_4^{+2} + 4\,H_2O$	$\Delta G^0 =$	-27	-3
$Zn(OH_2)_4^{+2} + 4\,A = ZnA_4^{+2} + 4\,H_2O$	$\Delta G^0 =$	-22	-8

* Notice that the charge on the complex ion changes when A itself carries charge, as for Cl^-.

sum. This shows that the electron donors place their electrons in the spatial region vacated by the valence electrons. The resulting interaction gives a bond energy that increases with charge, as shown in the last column. For $+1$ ions, the energy per bond is very low, about 25 kcal, but it is near 100 kcal per bond for $+2$ ions. For $+3$ ions, the ion–water bond energy is of the order of 175 kcal per bond!

Such strong bonds must be present also in the complexes that form between cations and other electron donors in aqueous solutions. Otherwise, the complexes could not form in competition with 55 moles per liter of solvent molecules. This is evident in the free energy changes that accompany formation of the ammonia and chloride complexes of Cu^+, Cu^{+2}, and Zn^{+2}. Tables 21-5 and 21-6 list the free energy changes associated with the replacement of the water of aquation by NH_3 or Cl^-. All of the entries are negative, showing that equilibrium favors the replacement. Both ammonia and chloride ion are better electron donors (Lewis bases) than is water. (Notice that these free energies apply to the equilibrium constant in which the water concentration, 55 M, is *not* incorporated in the constant.)

(c) RATE EFFECTS IN COMPLEX ION FORMATION

With the strong interactions involved in the transition metal complex ions, it can be expected that some of them are quite slow to react. Generally this is not so for the replacement of a water molecule; this usually can occur in a small fraction of a second. Other electron donors can be quite

Table 21-6 Free Energy Changes and Equilibrium Constants for Complex Formation of Cupric Ion *

	$A = NH_3$		$A = Cl^-$ †	
	ΔG^0 (kcal)	K	ΔG^0 (kcal)	K
$Cu(OH_2)_4^{+2} + A = Cu(OH_2)_3A^{+2} + H_2O$	-8.0	$8\cdot10^5$	-2.1	30
$Cu(OH_2)_3A^{+2} + A = Cu(OH_2)_2A_2^{+2} + H_2O$	-7.1	$2\cdot10^5$	-1.4	11
$Cu(OH_2)_2A_2^{+2} + A = Cu(OH_2)A_3^{+2} + H_2O$	-6.3	$4\cdot10^4$	-0.3	1.7
$Cu(OH_2)A_3^{+2} + A = CuA_4^{+2} + H_2O$	-5.3	$8\cdot10^3$	$+0.6$	0.35

* The equilibrium constants do *not* incorporate the concentration of water into the equilibrium constant.
† See footnote, Table 21-5.

Table 21-7   Mean Lifetimes, $\bar{t}_{\frac{1}{2}}$, of the Aquation Sphere of Third-Row Ions

M^{+2}	$\bar{t}_{\frac{1}{2}}$ (sec)	M^{+3}	$\bar{t}_{\frac{1}{2}}$ (sec)
		Ti^{+3}	$1 \cdot 10^{-5}$
Cr^{+2}	$<1.3 \cdot 10^{-10}$		
Mn^{+2}	$3.2 \cdot 10^{-8}$		
Fe^{+2}	$3.1 \cdot 10^{-7}$	Fe^{+3}	$3 \cdot 10^{-3}$
Co^{+2}	$7.4 \cdot 10^{-7}$		
Ni^{+2}	$3.7 \cdot 10^{-5}$		
Cu^{+2}	$1.2 \cdot 10^{-10}$		
		Ga^{+3}	$5.5 \cdot 10^{-4}$

reluctant to give up the favored position near a cation, particularly negative ions such as halide ions or cyanide ion, CN^-. Some chloride ion complexes may take several seconds, minutes, or even hours to be displaced,
despite an unfavorable equilibrium situation. This is manifested in many
cases by slow changes in the solution color.

The replacement of the immediate aquation sphere is so important that
many chemists have tried to measure the rate at which this process occurs.
Recently, nuclear magnetic resonance studies have made it possible to do
so. A small amount of isotopically labeled water, $H_2^{17}O$, is added to the
solvent. The nucleus of oxygen-17 has magnetic properties that cause it to
absorb light in the microwave region when the solution is placed in a strong
magnetic field. The wavelength absorbed is changed if the oxygen-17
nucleus becomes immobilized near the magnetic nucleus of a transition
metal ion. Hence, a simple frequency measurement reveals the average
time an $H_2^{17}O$ molecule spends in such a position before it is displaced
by a competing, garden-variety $H_2^{16}O$ molecule. Some of these mean
lifetimes are listed in Table 21-7. The $+3$ ions have longer lifetimes than
$+2$ ions, as is reasonable. The longest lifetime given, that of Fe^{+3}, is only
3 milliseconds, not a very long honeymoon. Yet this is an extremely long
time compared to the mobility of the liquid environment and it gives the
Fe^{+3} aquation complex a real identity as a molecular species. At the other
extreme, lifetimes of 10^{-10} second suggest a mobile and fluid situation
in which well-defined aquation structures do not persist long enough to
be identifiable in the chemistry of the ion.

(d) CHROMIUM

Chromium(III), Cr^{+3}, forms literally thousands of complexes like those
pictured in Figure 21-7. It is generally considered to be hexacoordinated
(bound to six electron donors) both in solutions and in crystals. This is
appropriate to the orbital occupancy, (21-23).

d²sp³ hybrid: octahedral   (21-23)

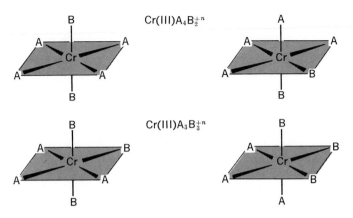

Figure 21-8 Geometrical isomerism in octahedral complexes.

In aqueous solution, one to six of the electron donors can be water molecules. In fact, even in crystals of Cr(III), mixed ammonia–water complexes are known, $Cr(NH_3)_{6-n}(OH_2)_n^{+3}$, for $n = 0, 1, 2, 3, 4,$ and 6, all except 5 (for unknown reasons). Of course, mixed complexes also occur involving electron donor combinations without water. The mixed ammonia–chloride complexes, $Cr(NH_3)_{6-n}Cl_n^{+(3-n)}$, are known for $n = 0, 1, 2, 3, 4,$ and 6. These mixed complexes introduce geometrical isomerism to the molecular structures (analogous to that so commonly found in organic compounds). Figure 21-8 shows the possible structures, in octahedral symmetry, of complex ions with formulas $CrA_4B_2^{+3}$ and $CrA_3B_3^{+3}$. For each ion there are two distinct geometrical isomers. Which will be most stable can often be explained in terms of the size and charge of A and B, but it can be difficult to predict in advance. Sometimes both isomers can be prepared, even when there is a significant energy difference between the two. In fact, chromic ion, Cr^{+3}, furnishes some notable examples of chloride complexes that persist at room temperature for many minutes, sometimes days, after equilibrium conditions have been changed to cause the complex to react.

(e) MANGANESE

Manganese can be prepared in aqueous solution in oxidation states $+2$ and $+3$. The orbital occupancies are:

	3d	4s	4p	
Mn^{+2}	⊘ ⊘ ⊘ ⊘ ⊘	◯	◯ ◯ ◯	(21-24a)
	sp^3: tetrahedral			
or	⊘ ⊘ ⊘ ⊘ ◯	◯	◯ ◯ ⊘	(21-24b)
	dsp^2: square, planar			
or	⊗ ⊗ ⊘ ◯ ◯	◯	◯ ◯ ◯	(21-24c)
	d^2sp^3: octahedral			

Mn^{+3} ⊘ ⊘ ⊘ ⊘ :O | O   O O :O | (21-25a)

sp^3: tetrahedral or dsp^2: square, planar

or   ⊗ ⊘ ⊘ | O O   O   O O O | (21-25b)

d^{2}sp^3: octahedral

Manganous ion, Mn^{+2}, has relatively weak complexing properties. With its higher charge, manganic ion, Mn^{+3}, is more versatile in its complex ion formation. Nevertheless, the aquo complexes cause Mn^{+2}(aq) to be the more important species, presumably because the half-filled d orbital occupancy of Mn^{+2} (21-24a) is specially stable relative to the alternatives. This dominance of Mn^{+2}(aq) is revealed in the oxidation potentials involving Mn^{+3}(aq). The $\mathcal{E}^0$ values show that Mn^{+3} is unstable in acidic solution with respect to self-oxidation-reduction (*disproportionation*).

$$\text{Mn}^{+3} + \text{e}^- = \text{Mn}^{+2} \qquad\qquad \mathcal{E}^0 = +1.5$$
$$(21\text{-}26)$$

$$\text{MnO}_2(\text{s}) + 4\,\text{H}^+ + \text{e}^- = \text{Mn}^{+3} + 2\,\text{H}_2\text{O} \qquad\qquad \mathcal{E}^0 = +1.1$$
$$\overline{2\,\text{Mn}^{+3} + 2\,\text{H}_2\text{O} \longrightarrow \text{Mn}^{+2} + \text{MnO}_2(\text{s}) + 4\,\text{H}^+ \quad \Delta\mathcal{E}^0 = +0.4}$$
$$(21\text{-}27)$$

The reaction (21-27) has a standard potential of +0.4 volt, which corresponds to a free energy change $\Delta G^0 = -9.2$ kcal and an equilibrium constant $K = 5.7 \cdot 10^{+6}$. Consequently Mn^{+3} has only transitory existence as the aquated ion.

This situation can be altered significantly by suitable complexing agents, those that form the more stable Mn^{+3} complex ions. The stability of these ions lowers the reduction potential of manganese(III) with respect to manganese(II). For example, the chloride and fluoride complexes of manganic ion change its reduction potential noticeably and the cyanide complex changes it dramatically.

$$\text{Mn}^{+3} + \text{e}^- = \text{Mn}^{+2} \qquad\qquad \mathcal{E}^0 = +1.5 \qquad\qquad (21\text{-}28)$$
$$\text{MnCl}^{+2} + \text{e}^- = \text{Mn}^{+2} + \text{Cl}^- \qquad \mathcal{E}^0 = +1.44 \qquad\qquad (21\text{-}29)$$
$$\text{MnF}^{+2} + \text{H}^+ + \text{e}^- = \text{Mn}^{+2} + \text{HF} \qquad \mathcal{E}^0 = +1.35 \qquad\qquad (21\text{-}30)$$
$$\text{Mn(CN)}_6^{-3} + \text{e}^- = \text{Mn(CN)}_6^{-4} \qquad \mathcal{E}^0 = -0.22 \qquad\qquad (21\text{-}31)$$

Manganic ion also forms a series of complexes with oxalate ion, $\text{C}_2\text{O}_4^{-2}$. The equilibrium constants for successive additions are as follows:

$$\text{Mn}^{+3} + \text{C}_2\text{O}_4^{-2} = \text{Mn(C}_2\text{O}_4)^+ \qquad\qquad K_1 = 10^{10} \qquad (21\text{-}32)$$
$$\text{Mn(C}_2\text{O}_4)^+ + \text{C}_2\text{O}_4^{-2} = \text{Mn(C}_2\text{O}_4)_2^- \text{ (yellow)} \quad K_2 = 4.0 \cdot 10^6 \quad (21\text{-}33)$$
$$\text{Mn(C}_2\text{O}_4)_2^- + \text{C}_2\text{O}_4^{-2} = \text{Mn(C}_2\text{O}_4)_3^{-3} \text{ (red)} \qquad K_3 = 3.2 \cdot 10^2 \quad (21\text{-}34)$$

Only three oxalate ions can be bonded to manganese, and the structures of the complexes explain why. The oxalate ion can act as an electron donor at either end or, in fact, at both ends simultaneously. In the oxalate complexes this is exactly what happens. Figure 21-9(a) shows, first, the structure of oxalate ion, $C_2O_4^{-2}$, the salt of the dibasic acid, oxalic acid, HOOCCOOH. The negatively charged oxygen atoms at each end of the molecule can act independently as electron donors. Hence one $C_2O_4^{-2}$ molecule can displace two water molecules to form a bridged complex, as in Figure 21-9(c). Then two more can be displaced to give one or both of the geometrical isomers, Figure 21-9(d). Finally, all of the electron donor capacity of Mn^{+3} can be consumed by the bridged oxalate complex bonds, as in Figure 21-9(e).

A complexing agent that can bond at two places, such as oxalate ion, is

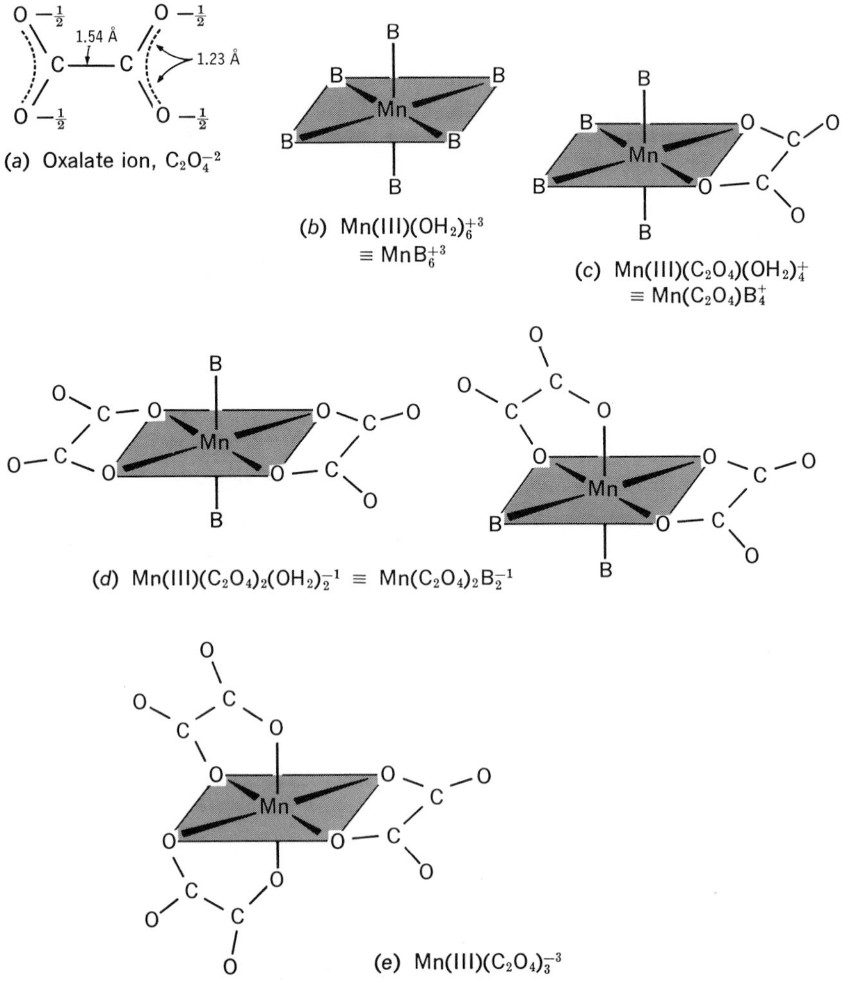

(a) Oxalate ion, $C_2O_4^{-2}$

(b) $Mn(III)(OH_2)_6^{+3}$
$\equiv MnB_6^{+3}$

(c) $Mn(III)(C_2O_4)(OH_2)_4^+$
$\equiv Mn(C_2O_4)B_4^+$

(d) $Mn(III)(C_2O_4)_2(OH_2)_2^{-1} \equiv Mn(C_2O_4)_2B_2^{-1}$

(e) $Mn(III)(C_2O_4)_3^{-3}$

Figure 21-9 The oxalate complexes of manganic ion.

called *difunctional* or *bidentate*. There are quite a number of difunctional electron donors that form bridged complex ions like those shown in Figure 21-9.

(f) IRON

The orbital occupancies of the common oxidation states of iron, Fe^{+2} and Fe^{+3}, differ from those of manganese by one added electron. This gives the special stability afforded by half-filled d orbitals, as enjoyed by Mn^{+2}, to Fe^{+3}. That presumably accounts for the stability of $Fe^{+3}(aq)$ relative to $Fe^{+2}(aq)$, as displayed in the much lower reduction potential ($Fe^{+3}-Fe^{+2}$, $\mathcal{E}^0 = +0.77$; $Mn^{+3}-Mn^{+2}$, $\mathcal{E}^0 = +1.5$).

	3d	4s	4p	
Fe^{+2}	⊗ ⊘ ⊘ ⊘ ⊘	○	○ ○ ○	(21-35a)
	sp³: tetrahedral			
or	⊗ ⊗ ⊗ ○ ○	○	○ ○ ○	(21-35b)
	d²sp³: octahedral			
Fe^{+3}	⊘ ⊘ ⊘ ⊘ ⊘	○	○ ○ ○	(21-36a)
	sp³: tetrahedral			
	⊗ ⊗ ⊘ ○ ○	○	○ ○ ○	(21-36b)
	d²sp³: octahedral			

As for manganese, the Fe^{+2} complexes tend to be weaker and less important in the chemistry of ferrous ion than are the Fe^{+3} complexes. The effect of complexing on the relative stability of Fe(II) and Fe(III) is shown in Table 21-8; the complexes all increase the stability of Fe(III) relative to the Fe(II) species present. The cyanides produce a large change, but by no means as great as that shown by manganese (21-31). Iron forms an interesting series of pentacyano complexes also, as typified in Table

Table 21-8   Reduction Potentials for Some Fe(II)–Fe(III) Complex Ions

	$\mathcal{E}^0$
$Fe^{+3} + e^- = Fe^{+2}$	+0.77
$FeBr^{+2} + e^- = Fe^{+2} + Br^-$	+0.73
$FeCl^{+2} + e^- = Fe^{+2} + Cl^-$	+0.68
$Fe(CN)_5(NO_2)^{-2} + e^- = Fe(CN)_5(NO_2)^{-3}$	+0.52
$Fe(CN)_5(OH_2)^{-2} + e^- = Fe(CN)_5(OH_2)^{-3}$	+0.49
$Fe(CN)_5(NH_3)^{-2} + e^- = Fe(CN)_5(NH_3)^{-3}$	+0.37
$Fe(CN)_6^{-3} + e^- = Fe(CN)_6^{-4}$	+0.36
$Fe(C_2O_4)_3^{-3} + e^- = Fe(C_2O_4)_2^{-2} + C_2O_4^{-2}$	+0.02

Table 21-9 Equilibrium Constants for Some Ferric Complexes

$$Fe^{+3} + 3\, C_2O_4^{-2} = Fe(C_2O_4)_3^{-3} \qquad 10^{20}$$
$$Fe^{+3} + PO_4^{-3} = Fe(PO_4) \qquad 6.7 \cdot 10^{17}$$
$$Fe^{+3} + 5\, F^- = FeF_5^{-2} \qquad 2 \cdot 10^{15}$$

$$Fe^{+3} + NCS^- = FeNCS^{+2} \qquad 9.6 \cdot 10^2$$
$$Fe^{+3} + 3\, NCS^- = Fe(NCS)_3 \qquad 3.8 \cdot 10^4$$
$$Fe^{+3} + 6\, NCS^- = Fe(NCS)_6^{-3} \qquad 1.3 \cdot 10^9$$

$$Fe^{+3} + Cl^- = FeCl^{+2} \qquad 33$$
$$FeCl^{+2} + Cl^- = FeCl_2^+ \qquad 4.5$$
$$FeCl_2^+ + Cl^- = FeCl_3(aq) \qquad 0.1$$

$$Fe^{+3} + Br^- = FeBr^{+2} \qquad 4$$

21-8. The last reduction potential refers to oxalato complexes with the bridged-type structures portrayed in Figure 21-9.

Successive equilibrium constants are known for many ferric ion complexes: Table 21-9 lists some for ferric salts. We see that the electron donors can be listed in order of effectiveness in replacing water in the aquation sphere as follows:

$$\text{oxalate} \approx \text{phosphate} > \text{fluoride} \gg \text{thiocyanate} > \text{chloride} > \text{bromide}$$
$$C_2O_4^{-2} \approx \quad PO_4^{-3} \quad > \quad F^- \quad \gg \quad NCS^- \quad > \quad Cl^- \quad > \quad Br^-$$
$$(21\text{-}37)$$

Figure 21-10 shows the structures of some ferrous and ferric chloride

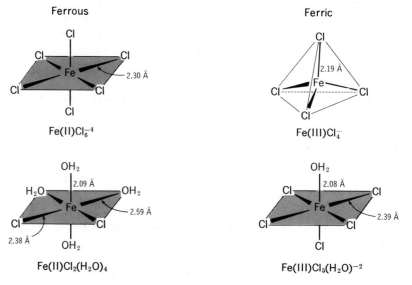

Figure 21-10 The structures of some chloride complexes of iron.

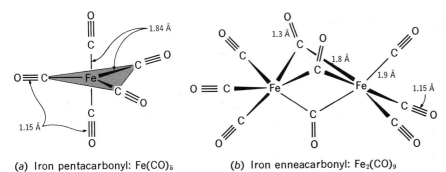

(a) Iron pentacarbonyl: Fe(CO)₅   (b) Iron enneacarbonyl: Fe₂(CO)₉

Figure 21-11   The structures of two iron carbonyls.

complexes as they are found in crystals. The ferrous chloride complex is octahedral, appropriate to orbital occupancy (21-35b), whereas some ferric crystals contain the $FeCl_4^-$ unit in the tetrahedral arrangement appropriate to orbital occupancy (21-36a). The shorter Fe–Cl bond lengths in $FeCl_4^-$ bespeak stronger bonding than found in $FeCl_6^{-4}$, as is reasonable.

Notice also in Figure 21-10 that the ferrous and ferric mixed aqua-chloride complexes have remarkably similar bond lengths. The ferrous dichloride tetraaquato complex is surprising in its choice of geometrical isomers. The chlorine atoms occupy adjacent positions rather than opposed positions (see Fig. 21-8). The latter would be expected due to repulsions if the Cl atoms each carried a full negative charge, as in Cl⁻. Notice also that two of the water molecules have much longer Fe–O distances, hence weaker bonds, than the other two.

Iron forms complex compounds with carbon monoxide that are of particular interest and that typify similar compounds formed by many transition metals. Figure 21-11 shows the structures of iron pentacarbonyl, Fe(CO)₅, and iron enneacarbonyl, Fe₂(CO)₉. The pentacarbonyl is a pale yellow, nonmagnetic molecular liquid at room temperature (m.p. 253°K; b.p. 376°K). The enneacarbonyl is a yellow, molecular crystal that converts on heating to a mixture of Fe(CO)₅ and a higher carbonyl, Fe₃(CO)₁₂.

In Fe(CO)₅, the C–O bond lengths are quite close to those in carbon monoxide (1.13 Å), showing that the molecule is acting as an electron donor (as in H₃B—C≡O, see Section 17-3(a)). The absence of magnetic properties (diamagnetism) requires the orbital occupancy (21-38). There are, then, five orbitals left vacant and dsp³ hybrid orbitals are expected. As indicated in Table 16-3, the dsp³ hybrids may be considered to be the superposition of three sp² hybrids (sp_xp_y) in the xy plane and two dp hybrids (d_{z^2}, p_z) along the z axis. These symmetries lead to a trigonal bipyramid prediction, just as observed, rather than alternative possibilities, such as a tetragonal pyramid.

Fe(CO)₅   ⊗ ⊗ ⊗ ⊗ | ○   ○   ○ ○ ○ | (21-38)

dsp³ = dp,sp²: trigonal bipyramid

The enneacarbonyl structure is much more complicated. The bridged bond lengths reveal C–O bonds of $1\frac{1}{2}$ bond order (compare: formaldehyde, $H_2C{=}O$, 1.21 Å; methanol, H_3C–OH, 1.43 Å). The terminal CO groups, on the other hand, are like those of $Fe(CO)_5$. It is plain that the end groups are acting as electron donors and the bridged groups participate in a delightful molecular orbital bonding situation.

Although a molecular orbital description would be too complicated to develop here, the bonding is susceptible to rationalization, based on the bond-order clue provided by the bridge carbonyls. First, the geometry around each iron can be looked on as a distorted octahedron. That leads us to consider the bonding that is possible with the d^2sp^3 orbitals. Six of the eight iron electrons occupy the three nonbonding 3d orbitals, leaving two electrons for bond formation. Each of the three terminal CO groups donates two electrons, so we now have a total of eight. How many electrons are contributed by the bridge carbonyls?

If a carbon atom forms a covalent, $1\frac{1}{2}$ order bond, it must contribute half of the three electrons needed. Hence, of the four valence electrons, $1\frac{1}{2}$ are used to bond to oxygen and $2\frac{1}{2}$ are thrown into the molecular orbital pot that binds the bridging structure together. Since there are three such carbon atoms, they provide about $3(2\frac{1}{2}) = 7\frac{1}{2}$ electrons. If we round off $7\frac{1}{2}$ to 8, then the carbonyls provide a total of 8 electrons to bind to the two iron atoms. That is 4 electrons per iron atom. Now we find a total of $2 + 6 + 4 = 12$ electrons available for bonding around each iron, just the number needed to form the six Fe–C bonds displayed in Figure 21-11(b).

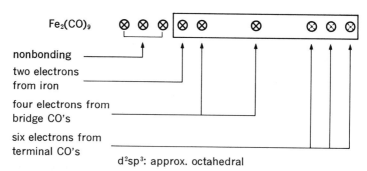

d^2sp^3: approx. octahedral

Figure 21-12 shows two other complexes of iron that display the rich chemical possibilities provided by complex formation. The first compound, ferrocene, is a laboratory curiosity. Two planar pentadienyl groups, C_5H_5, lie parallel to each other and the iron nestles inside. The bonding is explainable in the molecular orbital language and several transition metals have been made up into such sandwich compounds.

The second compound is generated in every human factory. It is called hemin, and it is a part of hemoglobin, the red pigment in the red corpuscles

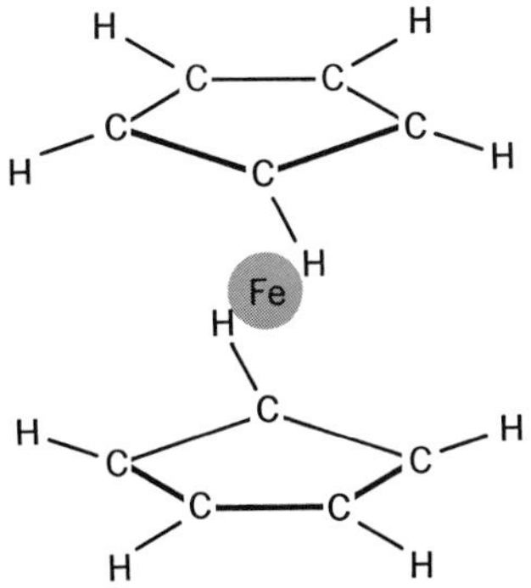

(a) Ferrocene: Fe(C₅H₅)₂

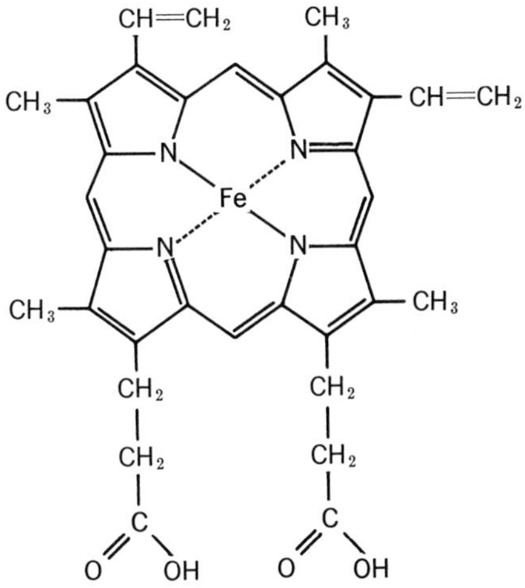

(b) Hemin

Figure 21-12 Two complicated iron complexes.

of the blood. There is, apparently, residual complex-forming capacity for iron placed in the square-planar environment of the hemin molecule. With this bonding capability the iron atoms in hemin can bind to oxygen molecules. Thus, the hemin can act as a bus, carrying oxygen through the blood stream from the lungs down to the socks and up to the hat where oxygen molecules are needed. The poisonous action of carbon monoxide derives from its stronger complexing power, as an electron donor. If CO is available, it binds so strongly to the iron atoms in the hemoglobin that oxygen displaces it very slowly. Suffocation results. That is why carbon monoxide formation by auto exhaust gases is a particularly serious contribution to urban air pollution.

(g) COBALT: MAGNETS AND M.O.'S

Many of the complexes described in the preceding sections are paramagnetic—the metal ion has one or more unpaired electrons, even after complex formation. Figure 15-13 showed how the paramagnetism of O_2 can be demonstrated for the liquid state using a strong magnet. Since transition metal complexes can be studied as either solid salts or in solution, a quantitative measure of their magnetism is easy to obtain. Figure 21-13 shows a simplified drawing of a Gouy balance, a device widely used for the study of magnetic compounds. The sample, placed in a small tube, is weighed in and then again out of the magnetic field. The apparent mass change, Δm, can be related to the molecular magnetic moment if the

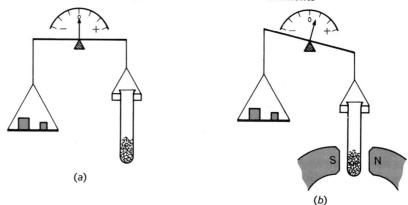

Figure 21-13 Measurement of molecular paramagnetism using a Gouy balance. (a) Sample is weighed out of magnetic field. (b) Sample is attracted into magnetic field.

apparatus is calibrated with a reference compound for which the magnetic moment is accurately known.

According to quantum mechanics, the magnetic moment due to the spin of n unpaired electrons, μ_S, is

$$\mu_S = \sqrt{n(n+2)} \text{ B.M.} \qquad (21\text{-}39)$$

The abbreviation B.M. stands for the "Bohr Magneton," a convenient unit. Thus a simple measurement of the molecular magnetic moment should lead directly to the number of unpaired electrons. In practice, however, there is usually a contribution to the observed moment, $\mu_{obs.}$, from motion of the electrons other than spin. As a result, μ_S is a lower limit on the observed magnetic moment. That is,

$$\mu_{obs.} \geq \mu_S = \sqrt{n(n+2)} \text{ B.M.} \qquad (21\text{-}40)$$

If the chemical composition is known, the geometry of a new complex often can be established from a magnetic measurement alone. For example, cobalt(II) forms both tetrahedral (sp^3) and square planar (dsp^2) four-coordinated complexes:

$$\text{Co}^{+2} \qquad \otimes \ \otimes \ \oslash \ \oslash \ \oslash \qquad \boxed{\bigcirc \qquad \bigcirc \ \bigcirc \ \bigcirc} \qquad (21\text{-}41)$$
$$sp^3\text{: } n = 3 \text{ (high spin)}$$

$$\text{Co}^{+2} \qquad \otimes \ \otimes \ \otimes \ \oslash \boxed{\bigcirc} \qquad \bigcirc \qquad \bigcirc \ \bigcirc \boxed{\bigcirc} \qquad (21\text{-}42)$$
$$dsp^2\text{: } n = 1 \text{ (low spin)}$$

Experimentally, tetrahedral Co(II) complexes have $\mu_{obs.}$ in the range $4.4 - 4.8$ B.M. (compare to $\mu_S = \sqrt{3(3+2)} = 3.9$ B.M.) while the less common square planar complexes have $\mu_{obs.}$ in the range 2.3 — 2.8 B.M. ($\mu_s = \sqrt{1(1+2)} = 1.7$ B.M.).

Exercise 21-10. From the observed magnetic moments, predict the electronic configuration and geometry of the following Co(II) complexes:

$[Co(PEt_3)_2(NCS)_2]$ $\mu_{obs.} = 2.3$ B.M. (Et $= C_2H_5$)

$[Co(NCS)_4]^{-2}$ $\mu_{obs.} = 4.4$ B.M.

Equally straightforward magnetic measurements can be used to distinguish between four- and six-coordinated complexes. For example, Mn(II) forms low-spin octahedral complexes (21-24c), but high-spin four-coordinated complexes with five unpaired electrons (21-24a,b). Notice also that the square planar Mn(II) configuration (21-24b) suggests there should be a second, high-spin form of square planar Co(II)

Co^{+2} ⊗ ⊗ ⊘ ⊘ | O   O   O O | ⊘   (21-43)

dsp^2: $n = 3$ (high spin)

but no such form has ever been observed. This is presumably because the energy needed to promote a d electron to the 4p orbital is not offset by the accompanying decrease in electron–electron repulsion. There are, however, many cases in which both high- and low-spin complexes exist for the same geometry. We will now examine one such case and develop an explanation.

Cobalt(III) forms many complexes—*all* of which are octahedral. In terms of a pigeonhole diagram, the electronic configuration is

Co^{+3} ⊗ ⊗ ⊗ | O O   O   O O O | (21-44)

d^{2}sp^3: $n = 0$ (low spin, diamagnetic)

Thus *every* Co(III) octahedral complex is expected to be diamagnetic. This is not, however, observed to be the case. At least one well-known complex, $[CoF_6]^{-3}$, is paramagnetic with $\mu_{obs.} = 5.3$ B.M., suggesting four unpaired electrons. This is reminiscent of the O_2 problem. A simple pigeonhole approach, such as that below, predicts a doubly bonded, diamagnetic molecule.

p^2: $n = 0$ (21-45)

But oxygen isn't diamagnetic—it has a magnetic moment appropriate to two unpaired electrons. In Chapter Fifteen we found that a molecular orbital description resolved this problem. Once antibonding orbitals entered the picture, the paramagnetism of O_2 was easily explained. The simple pigeonhole approach fails here because it considers only bonding orbitals.

Our present problem—the existence of both low- and high-spin Co(III) complexes can also be resolved in the molecular orbital framework. In a great majority of cases, an M.O. treatment is not necessary for transition metal complexes, but occasionally, as in $[CoF_6]^{-3}$, the full treatment is necessary.

Figure 21-14 shows the beginnings of an energy level diagram for an octahedral complex. The ordering of levels shown is based on experiment and detailed calculation. The M.O.'s formed from metal ion 4s and 4p orbitals are similar to those described for CO_2 in Section 15-7. Each 4p orbital forms a bonding and antibonding M.O. with a pair of donor atom orbitals—like the CO_2 sigma orbitals shown in Figure 15-21. All six donor atoms contribute to the M.O.'s formed with the metal 4s orbital.

In our pigeonhole description of the bonding in octahedral complexes, we found that only two 3d orbitals could participate in the formation of a d^2sp^3 hybrid: the $3d_{z^2}$ and $3d_{x^2-y^2}$. In the M.O. picture, these two atomic orbitals are capable of forming bonding and antibonding M.O.'s with the appropriate donor atom orbitals. Bonding combinations of orbitals are shown in Figure 21-15(a) and (b), in which "L" represents the donor group or atom and "M" the transition metal ion. The other 3d orbitals, the $3d_{xy}$, $3d_{xz}$, and $3d_{yz}$, do not point toward any of the six ligands and, in the

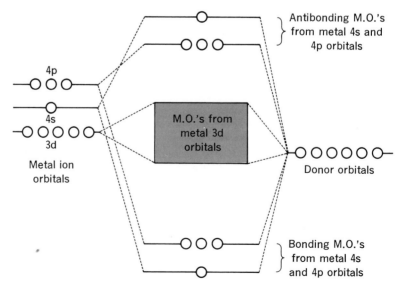

Figure 21-14   Octahedral complex: M.O. energy levels from 4s and 4p metal orbitals.

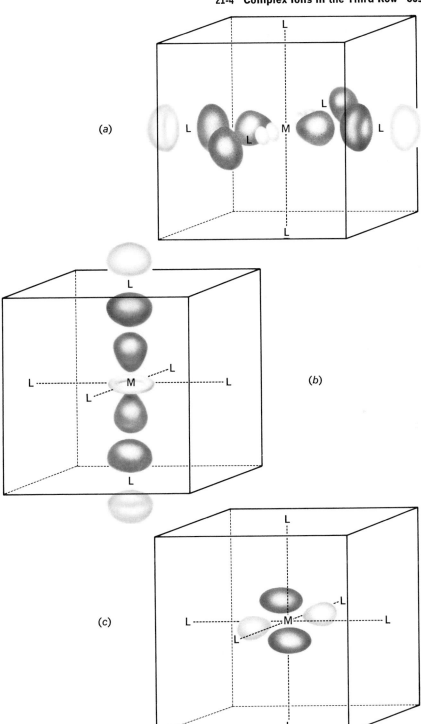

Figure 21-15   *Bonding and nonbonding combinations of orbitals in an octahedral complex. (a) $3d_{x^2-y^2}$ orbital can bond to four L orbitals; (b) $3d_{z^2}$ orbital can bond to two L orbitals; (c) $3d_{xy}$ orbital does not point toward any L orbitals and is nonbonding. (Shading distinguishes the phase of the atomic wave functions.)*

simple M.O. approximation, are nonbonding orbitals. One of these is illustrated in Figure 21-15(c).

A complete energy level diagram is shown in Figure 21-16 with spacing appropriate to the complex $[Co^{III}(NH_3)_6]^{+3}$. We can now proceed to put electrons into the M.O.'s. We have six Co^{+3} d electrons plus twelve electrons donated by the six ammonia molecules to account for. Using the familiar rules we arrive at the configuration shown in the figure—all electrons are paired. The complex is thus predicted to be diamagnetic, in agreement with both experiment and the pigeonhole prediction.

New in this picture, however, is the more complete explanation for the color of transition metal complexes. The antibonding orbital derived from atomic 3d orbitals lies very near the three 3d nonbonding orbitals. In $Co(NH_3)_6^{+3}$ spectroscopic measurements place this energy difference at $\Delta E = 69$ kcal/mole, in the blue part of the spectrum. In general, the presence of low-lying antibonding orbitals accounts for the electronic transitions in the visible region that are so characteristic of transition metal complexes.

Now back to the hexafluoride complex. Experimentally it is found that the 2p orbitals of F^- are much lower in energy than the lone-pair orbital on the ammonia nitrogen. This energy difference affects the relative spacings of the M.O. energy levels as shown in Figure 21-17. Of particular importance is the decrease in the energy difference, ΔE, between the nonbonding orbitals and the lowest antibonding orbitals. In CoF_6^{-3} spectroscopic measurements show this to be $\Delta E = 37$ kcal/mole. Different donors result in different ΔE's, but there is always some critical value

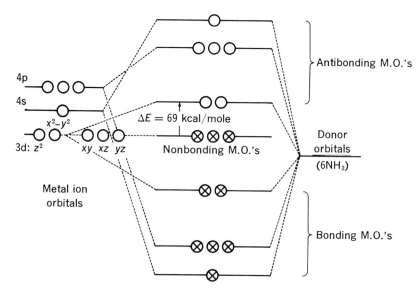

Figure 21-16 Octahedral complex: levels and electron occupancy appropriate to $[Co^{III}(NH_3)_6]^{+3}$.

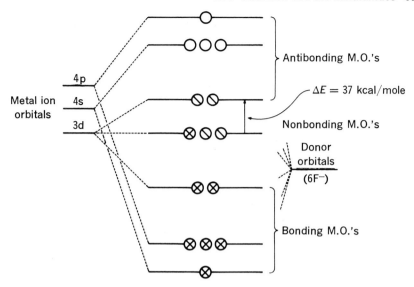

Figure 21-17 *Octahedral complex: high-spin case. Orbital occupancy appropriate to* CoF_6^{-3}.

below which there can be a net energy gain if electrons are left unpaired. For example, in CoF_6^{-3} itself the electron configuration illustrated is consistent with the observed magnetic moment. Clearly the extra energy needed to put electrons into the higher, antibonding orbitals is more than offset by the decreased electron–electron repulsion. Hence the CoF_6^{-3} complex can have unpaired electron spins.

The existence of both high- and low-spin complexes for the same geometric arrangement of donor groups is quite common in transition metal chemistry. Once again simple M.O. theory provides a straightforward explanation.

21-5 Fifth row and the lanthanides

The fourth row of the Periodic Table also contains 18 elements, rubidium to xenon. The valence electrons occupy the 4d, 5s, and 5p orbitals and the familial resemblances to the row above are close and useful. The next row, the fifth row, opens a new door, as is revealed immediately by the presence of 32 elements in the row.

(a) FIFTH-ROW ORBITAL OCCUPANCIES

The explanation for the length of the fifth row is obvious: The 4f orbitals are near in energy to the 5d, 6s, and 6p orbitals that must serve as valence orbitals. Because of their lower principal quantum number, these 4f orbitals are relatively small in average radius. Proceeding across the row,

Table 21-10   Orbital Occupancies of Gaseous Fifth- and Sixth-Row Elements

	4f	5d	6s			5f	6d	7s
Cs			$6s$	87	Fr			$7s$
Ba			$6s^2$	88	Ra			$7s^2$
La*		$5d^1$	$6s^2$	89	Ac†		$6d^1$	$7s^2$
Ce	$4f^1$	$5d^1$	$6s^2$	90	Th		$6d^2$	$7s^2$
Pr	$4f^3$		$6s^2$	91	Pa	$5f^2$	$6d^1$	$7s^2$
Nd	$4f^4$		$6s^2$	92	U	$5f^3$	$6d^1$	$7s^2$
Pm	$4f^5$		$6s^2$	93	Np	$5f^4$	$6d^1$	$7s^2$
Sm	$4f^6$		$6s^2$	94	Pu	$5f^6$		$7s^2$
Eu	$4f^7$		$6s^2$	95	Am	$5f^7$		$7s^2$
Gd	$4f^7$	$5d^1$	$6s^2$	96	Cm	$5f^7$	$6d^1$	$7s^2$
Tb	$4f^9$		$6s^2$	97	Bk	$5f^9$		$7s^2$
Dy	$4f^{10}$		$6s^2$	98	Cf	$5f^{10}$		$7s^2$
Ho	$4f^{11}$		$6s^2$	99	Es	$5f^{11}$		$7s^2$
Er	$4f^{12}$		$6s^2$	100	Fm	$5f^{12}$		$7s^2$
Tm	$4f^{13}$		$6s^2$	101	Md	$5f^{13}$		$7s^2$
Yb	$4f^{14}$		$6s^2$	102	No	$5f^{14}$		$7s^2$
Lu	$4f^{14}$	$5d^1$	$6s^2$	103	Lw	$5f^{14}$	$6d$	$7s^2$
Hf	$4f^{14}$	$5d^2$	$6s^2$	104				
Ta	$4f^{14}$	$5d^3$	$6s^2$	105				
W	$4f^{14}$	$5d^4$	$6s^2$					
Re	$4f^{14}$	$5d^5$	$6s^2$					
Os	$4f^{14}$	$5d^6$	$6s^2$					
Ir	$4f^{14}$	$5d^7$	$6s^2$					
Pt	$4f^{14}$	$5d^9$	$6s^1$					

* Lanthanides.
† Actinides.

they behave in an analogous manner to the 3d orbitals in the third row (see Figure 21-4), shrinking in size as Z increases and falling below the energy of the other valence orbitals. Experimentally, we find that the 4f orbitals fall below the 5d, 6s, and 6p orbitals at cerium. Thereafter, they shrink faster than the larger valence orbitals and, consequently, the next thirteen elements continue with the filling of these inner 4f orbitals. Only after element lutecium does the orbital occupancy of the 5d and 6s orbitals change. Only then does the chemistry return to that of transition elements. The accepted orbital occupancies of the elements of this row are given in the first two columns of Table 21-10.

The elements lanthanum to lutecium are all chemical lookalikes. Consequently they are classed together as a chemical family, named after the first of the tribe: These elements are called the *lanthanides.**

(b) THE CHEMISTRY OF THE LANTHANIDES

Lanthanum has only one important oxidation state in aqueous solution, the +3 state. With few exceptions, that statement tells the whole boring

* In the older literature, these elements are called rare earths.

Table 21-11 Oxidation-Reduction Chemistry of the Lanthanides

Ele-ment	Z	E_1 (kcal)	Oxidation States	$\mathcal{E}^0$			K_{sp} M(OH)$_3$
				M^{+3}/M	M^{+3}/M^{+2}	M^{+4}/M^{+3}	
La	57	129	+3	−2.52			$1.1 \cdot 10^{-19}$
Ce	58	150	+3, +4	−2.48		+1.61	$1.5 \cdot 10^{-20}$
Pr	59	131	+3, +4	−2.47		+2.86	$2.7 \cdot 10^{-20}$
Nd	60	131	+3	−2.44			$1.9 \cdot 10^{-21}$
Pm	61	—	+3	−2.42			$1 \cdot 10^{-21}$
Sm	62	130	+2, +3	−2.41	< −0.9		$6.8 \cdot 10^{-21}$
Eu	63	131	+2, +3	−2.41	−0.43		$3.4 \cdot 10^{-22}$
Gd	64	142	+3	−2.40			$2.1 \cdot 10^{-22}$
Tb	65	155	+3, +4	−2.39			$2.0 \cdot 10^{-22}$
Dy	66	157	+3	−2.35			$1.4 \cdot 10^{-22}$
Ho	67	—	+3	−2.32			$5 \cdot 10^{-23}$
Er	68	—	+3	−2.30			$1.3 \cdot 10^{-23}$
Tm	69	—	+3	−2.28			$3.3 \cdot 10^{-24}$
Yb	70	144	+2, +3	−2.27	−0.578		$2.9 \cdot 10^{-24}$
Lu	71	141	+3	−2.25			$2.5 \cdot 10^{-24}$

story about the other fourteen lanthanides. Table 21-11 shows first ionization energies, E_1, the oxidation states, and some $\mathcal{E}^0$ values. The $\mathcal{E}^0$'s for reduction to the metals are all extremely negative, (compare sodium, $Na^+ + e^- = Na$, $\mathcal{E}^0 = -2.7$) showing that the lanthanide elements are readily oxidized to the +3 oxidation state.

Cerium and praseodymium form +4 states, but both are extremely powerful oxidizing agents. Praseodymium(IV) oxidizes water in acid solution, evolving oxygen gas.

$$2\ Pr^{+4} + H_2O \longrightarrow 2\ Pr^{+3} + \tfrac{1}{2}\ O_2(g) + 2\ H^+$$
$$\Delta\mathcal{E}^0 = +1.53 \text{ volts} \tag{21-46}$$

Ceric ion, Ce^{+4}, on the other hand, reacts sufficiently slowly with water to make it a valuable oxidizing agent in aqueous solutions.

Table 21-11 shows that three elements form +2 states in water, samarium, europium and ytterbium. All three are very strong reducing agents, though neither Eu^{+2}(aq) or Yb^{+2}(aq) reduce water to H_2 rapidly at low hydrogen ion concentrations. The appearance of the +2 oxidation states is attributed to special stability of the half-filled set of 4f orbitals (Eu^{+2}) and of the filled set of 4f orbitals (Yb^{+2}).

The last column of Table 21-11 typifies the remarkable similarity in the chemistry of the lanthanides. The solubility products for the hydroxides, M(OH)$_3$, change on the average by a factor of only 2 per element. The lanthanides all form fluorides of low solubility and chlorides of high solubility. Thus these fifteen elements form a family whose chemistry is more uniform than is that of any other chemical family in the Periodic Table.

The copy-cat chemistry of these elements is attributable to the very small radius of the 4f orbitals relative to the other valence orbitals. The shielding by the 4f electrons, as these orbitals fill, is so effective that the ionization energies are virtually constant (see Table 21-11). This signifies a sameness about the 5d, 6s, and 6p orbitals, whose properties determine the chemistry of the ions and their salts.

21-6 Sixth row and the actinides

Our experience with the 4f orbitals leads to the expectation that the sixth row should again consist of 32 elements. The 5f orbitals must dip below the 6d, 7s, and 7p orbitals somewhere across this row, though it is not easy to predict where. Table 21-10 lists the accepted electron configurations of the gaseous elements 87 to 103. According to these occupancies, the first occupancy of 5f orbitals occurs at protactinium and the 6d orbital retains occupancy through element 93, neptunium. Apparently the 5f and 6d orbitals are quite close in energy, an important matter in the chemistry of these elements.

The elements actinium, Ac, to lawrencium, Lw, are called *actinides*, a name suggesting that these elements should parallel the chemistry of their fifth row counterparts, the lanthanides. The expectation leads to some surprises.

(a) OXIDATION STATES OF THE ACTINIDES

Table 21-12 contrasts the known oxidation states of the sixth-row actinides to the oxidation states of the lanthanides and the subsequent fifth-row transition metals. The chemistry of the lanthanides is dominated by the $+3$ oxidation state and when oxidation states $+2$ or $+4$ appear, they are much less stable than the $+3$ states. The subsequent elements, hafnium through mercury, display the multiple oxidation states and rich chemistry that characterize the transition metals in the third and fourth rows. The sixth-row elements, thorium through americium, can't seem to make up their minds. Up to plutonium, the large number of stable oxidation states and their sequencing make the elements look like transition metals. Only beginning at curium does the chemistry settle down to the lanthanide-like drabness. The $+3$ state of protactinium is not known while both uranium and neptunium are readily oxidized to the $+4$ state and higher. The chemistry suggests that the 5f orbitals do not fall markedly below the energies of the other valence orbitals until americium. The orbital occupancies of the gaseous atoms (Table 21-12) tend to confirm this conclusion. Thorium has no 5f occupancy (nor a $+3$ oxidation state), and the 6d orbital retains occupancy through neptunium.

Mendelevium and nobelium contrast interestingly with their lanthanide

Table 21-12 Positive Oxidation Numbers of Fifth- and Sixth-Row Elements

Lanthanides		Fifth-Row Elements Following Lutetium		Sixth-Row Elements		$\mathcal{E}^0$ M^{+3}/M^{+2}	$\mathcal{E}^0$ M^{+4}/M^{+3}
		Cs	1	Fr	1		
		Ba	2	Ra	2		
La	3	Lu	3	Ac	3		
Ce	3 4	Hf	4	Th	4		
Pr	3 4	Ta	5	Pa	4 5		
Nd	3	W	2 3 4 5 6	U	3 4 5 6		−0.63
Pm	3	Re	3 4 5 6 7	Np	3 4 5 6 7		+0.147
Sm	2 3	Os	2 3 4 6 8	Pu	3 4 5 6 7		+0.98
Eu	2 3	Ir	3 4 6	Am	2 3 4 5 6		+2.6
Gd	3	Pt	2 4	Cm	3		
Tb	3 4	Au	1 3	Bk	3 4		~+1.6
Dy	3	Hg	1 2	Cf	2 3		
Ho	3	Tl	1 3	Es	2 3		
Er	3	Pb	2 4	Fm	2 3		
Tm	3	Bi	3 5	Md	2 3	−0.2	
Yb	2 3	Po	2 4	No	2 3	+1.4	
Lu	3	At	(−1)	Lw	3		

counterparts, thulium and ytterbium. The +2 state of Md is quite stable and that of nobelium so stable that it is difficult to oxidize to +3.

(b) COMPOUNDS OF URANIUM, NEPTUNIUM, PLUTONIUM, AND AMERICIUM

Many similarities are found among these elements, superimposed upon systematic trends. There are many *isostructural* compounds, that is, compounds with the same chemical formulas and also the same crystal structure. Table 21-13 lists some of these.

There are, of course, some differences too. Uranium forms a trihydride, UH_3, whereas plutonium and americium are like thorium in forming dihydrides, PuH_2 and AmH_2, as well as intermediate (nonstoichiometric) hydrides up to $MH_{2.7}$. Nevertheless, Np, Pu, and Am bear such striking resemblances to U that they were called "uranides" by some until it became vogue to refer to the elements from actinium on as *actinides.*

Table 21-13 Some Isostructural Solid Oxides and Fluorides of U, Np, Pu, *and* Am

dioxide	UO_2	NpO_2	PuO_2	AmO_2
sesquioxide	U_3O_8	Np_3O_8		
trifluoride	UF_3	NpF_3	PuF_3	AmF_3
tetrafluoride	UF_4	NpF_4	PuF_4	AmF_4
hexafluoride	UF_6	NpF_6	PuF_6	

(c) AQUEOUS CHEMISTRY OF URANIUM, NEPTUNIUM, PLUTONIUM, AND AMERICIUM

The similarities among these four elements are also displayed in their solution chemistries. All four elements form $+6$ ions analogous to uranyl(VI) ion, UO_2^{+2}. One of the interesting properties of the nitrate salt, uranyl nitrate, is that it is quite soluble in numerous organic ethers, alcohols, ketones, and esters. Very few other metal nitrates have this property, but neptunyl nitrate, plutonyl nitrate, and americyl nitrate also dissolve in these organic solvents. Consequently, these elements in the $+6$ oxidation state can be removed from aqueous solution by extraction with one of these organic liquids. For example, the dissolved metal taken from a uranium reactor will contain uranium, plutonium, and half the Periodic Table from fission processes. To remove the host of radioactive fission products, the solution can be treated with dichromate to oxidize uranium and plutonium to UO_2^{+2} and PuO_2^{+2}, then extracted with methyl isobutyl ketone. The organic layer will contain most of the uranyl nitrate and plutonyl nitrate, but almost all of the fission products will remain in the aqueous layer.

The lower oxidation states, $+3$ and $+4$, can be precipitated as fluorides of low solubility, MF_3 and MF_4. The higher oxidation states are soluble in aqueous fluoride solutions.

The Latimer $\mathcal{E}^0$ diagrams for these four elements show a systematic trend of increasing stability for the lower oxidation states (see Table 21-14).

Table 21-14 Reduction Potentials for Uranium, Neptunium, Plutonium, and Americium in 1 M HClO$_4$ at 298°K

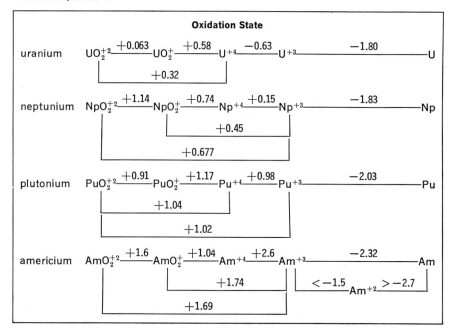

These trends can be exploited to separate these elements when present together in solution. For example, ferrous will reduce PuO_2^{+2} and NpO_2^{+2}, but not UO_2^{+2}. Then an extraction with methyl isobutyl ketone will remove uranyl nitrate, but it will leave Pu^{+4} and Np^{+4} behind in the aqueous layer. Or, alternately, fluoride ion can be added to the reaction mixture to precipitate PuF_3 and NpF_4, leaving UO_2^{+2} behind in the aqueous solution. This sort of chemistry is used to separate and to purify plutonium in quantities of 10 tons per year (as announced by the AEC). That is not because plutonium has interesting chemical application—it is the nucleus of plutonium that we are after. To explain why, we need to consider the subject of nuclear decomposition and radioactivity.

21-7 Nuclear processes

Our attention, throughout this text, has been focussed upon the valence electrons, the orbitals they occupy, and how these factors determine the chemistry of an atom. Isotopes of the same element have different nuclear masses, but almost identical chemical properties. Yet the nuclei have a lively world of their own, one that cannot be ignored by anyone living in this century.

(a) NATURAL RADIOACTIVITY

Twenty-two elements found in nature contain nuclei that spontaneously decompose into other nuclei. Such unstable nuclei are said to be *radioactive*. Two of the processes by which these nuclear decompositions occur are called alpha-decay (α-decay) and beta-decay (β-decay). In the first, there is ejected from the unstable nucleus a particle that decreases the nuclear charge by two units and the nuclear mass by four units. Thus the ejected particle, the α-particle, has the mass and charge of a helium nucleus. In β-decay, an electron is ejected, leaving the nuclear charge increased by one, but nominally leaving the nuclear mass the same. Examples of these two decay types are provided by the thorium isotope $^{232}_{90}\text{Th}$ and its radioactive offspring, $^{228}_{88}\text{Ra}$.

$$^{232}_{90}\text{Th} \longrightarrow \,^{228}_{88}\text{Ra} + \,^{4}_{2}\text{He} \qquad \tau_{\frac{1}{2}} = 1.4 \cdot 10^{10} \text{ years} \qquad\qquad (21\text{-}47)$$

$$^{228}_{88}\text{Ra} \longrightarrow \,^{228}_{89}\text{Ac} + e^{-} \qquad \tau_{\frac{1}{2}} = 6.7 \text{ years} \qquad\qquad (21\text{-}48)$$

These decays are characterized by two readily measured properties, the rate at which parent nuclei decompose and the energy of the ejected particle. The decay rate is specified in terms of the "half-life," $\tau_{\frac{1}{2}}$, which is the time needed for half of a collection of nuclei to decompose. The examples given show the enormous range of the half-lives—that of thorium-232 is longer than the 5-billion-year age of the earth. That is why this isotope is still present on earth, even though it is continually decomposing. The radium isotope, however, has a short half-life, even compared to a man's half-life. The only place we find $^{228}_{88}$Ra is in a deposit of thorium, where it is being formed continuously in reaction (21-47).

Table 21-15 lists all of the radioactive elements found in nature and the decay properties of some of their radioactive isotopes. For all elements beyond bismuth, all isotopes are radioactive. Thorium has four natural nuclear isotopes that decay by α emission (227, 228, 230, and 232) and two that decay by β emission (231 and 234). The table also shows that the energies involved in nuclear changes are 5 to 6 orders of magnitude larger than those observed in chemical changes. These large energies are usually given in millions of electron volts per nuclear decay (Mev). In these terms, the energy of a chemical reaction that releases 23.0 kcal/mole would be only one ev per molecule reacted.

There is one more type of natural radioactive decay that must be mentioned, gamma-decay (γ-decay). A "gamma-ray" is merely a photon of light, and it differs from other photons only in the frequency ν or, since $E = h\nu$, in its energy. Whereas a photon of visible light has an energy of about 3 ev, and an X-ray photon might carry 100 ev, a typical gamma photon energy might be 100,000 ev or 0.1 Mev. These photons do not change the nuclear composition, and they reveal the existence of quantized nuclear energy levels analogous to those observed in atoms and molecules. Gamma emission often accompanies α or β emission, or follows it, showing that the product nucleus is "born" in an excited state.

Before leaving the subject of natural radioactivity, we must note that two expected elements were not found in nature, technetium ($Z = 43$) and promethium ($Z = 61$). A presumptive explanation suggests itself: These elements must have no stable isotopes, all of the isotopic half-lives must be short compared to the age of the earth, and there are no natural parents of long life that decay to form either element. Through artificial synthesis, using nuclear transformations, we now know that all of these implications are correct.

(b) NUCLEAR TRANSFORMATIONS

We see that nuclear decay changes one element into another. That is a science with a thousand-year history ("transmutation"). To change lead into gold was the fondest dream of the early alchemists. They knew nothing of nuclei, of course, and they were foredoomed to failure. With only chemical energies at their disposal, they were trying to pry open coconuts with toothpicks. It was only after the invention of particle accelerators that bombarding projectiles with a chance of cracking the nucleus came under control. A big breakthrough came in 1932 at Berkeley, when E. O. Lawrence's cyclotron began spinning out protons and deuterons (deute-

Table 21-15 *Radioactive Elements Found in Nature*

Element	Z	A	Percent Abundance	Half-Life $\tau_{\frac{1}{2}}$	Decay Type and Energy (Mev)*		
					α	β	γ
carbon	6	14		5700 yr		0.155	
potassium	19	40	0.011	$1.4 \cdot 10^9$ yr		1.4	1.5
rubidium	37	87	27.9	$6 \cdot 10^9$ yr		0.28	
indium	49	115	95.8	$6 \cdot 10^{14}$ yr		0.63	
tin	50	124	6.0	$>1.7 \cdot 10^{17}$ yr		†	
lanthanum	57	138	0.09	$1.2 \cdot 10^{12}$ yr		1.0	
samarium	62	147	14.9	$6.7 \cdot 10^{11}$ yr	2.1		
lutetium	71	176	2.5	$7.2 \cdot 10^{10}$ yr		0.4	0.27, 0.18
rhenium	75	187	62.9	$4 \cdot 10^{12}$ yr		0.043	
thallium	81	206, -7, 8, 10		4, 5, 3, 1 min		†	†
lead	82	210, 11, 12, 14				†	†
bismuth	83	210, 11, 12, 14			†	†	†
polonium	84	210, 11, 12, 14, 15, 16, 18			†	†	†
astatine	85	215, 218		10^{-4}, 2 sec	8.0, 6.7		
radon	86	219, 20, 22			†		
francium	87	223		21 min		1.2	0.05, 0.10
radium	88	223, 24, 26, 28			†		
actinium	89	227, 228		28 yr, 6 hr	†	†	†
thorium	90	227, 28, 30, 31, 32, 34			†	†	
protactinium	91	231		34, 300 yr	4.66, 5.04	†	
uranium	92	234, 35, 38	0.006, 0.71, 99.3	10^5, 10^8, 19^9 yr	4.8, 4.4, 4.2	†	0.10, 0.29

* Daggers indicate decay occurs, but energies are not known or there are several.

rium nuclei) with energies in the Mev range. Not long thereafter, artificial radioactive nuclei were being produced by the dozen.

Nowadays, we have a complete array of bombarding particles with which to carry out nuclear syntheses: electrons, neutrons, protons, deuterons, tritons, α-particles, and heavier nuclei up to and exceeding 12 and constantly moving upward. It is usual for a nuclear reaction induced by a bombarding particle to be accompanied by ejection of another particle. Thus a deuteron might enter a nucleus and a neutron leave. This would be designated a (d,n) reaction, (deuteron in, neutron out). Here are some important examples, including some nuclear reactions that are useful in the synthesis of the missing elements technetium and promethium.

$$_{4}^{9}\text{Be} + {}_{1}^{1}\text{H} \longrightarrow {}_{4}^{8}\text{Be} + {}_{1}^{2}\text{H} \qquad \text{or} \qquad {}_{4}^{9}\text{Be}(p,d){}_{4}^{8}\text{Be} \qquad (21\text{-}49)$$

$$_{7}^{14}\text{N} + {}_{0}^{1}\text{n} \longrightarrow {}_{6}^{14}\text{C} + {}_{1}^{1}\text{H} \qquad \text{or} \qquad {}_{7}^{14}\text{N}(n,p){}_{6}^{14}\text{C} \qquad (21\text{-}50)$$

$$_{41}^{93}\text{Nb} + {}_{2}^{4}\text{He} \longrightarrow {}_{43}^{96}\text{Tc} + {}_{0}^{1}\text{n} \qquad \text{or} \qquad {}_{41}^{93}\text{Nb}(\alpha,n){}_{43}^{96}\text{Tc} \qquad (21\text{-}51)$$

$$_{60}^{144}\text{Nd} + {}_{1}^{2}\text{H} \longrightarrow {}_{61}^{144}\text{Pm} + 2{}_{0}^{1}\text{n} \qquad \text{or} \qquad {}_{60}^{144}\text{Nd}(d,2n){}_{61}^{144}\text{Pm} \qquad (21\text{-}52)$$

By reactions like those above, artificial radioactive nuclei have been prepared for every element in the Periodic Table. Furthermore, a dozen elements beyond the last naturally occurring element, uranium, have been synthesized and positively identified. We have been calmly considering their chemistry, neptunium through mendelevium, without taking note of the miracle of their existence. Only thirty years after the first of these were discovered (neptunium and plutonium, at the University of California, Berkeley), plutonium is being produced at the rate of 10 tons per year. As of 1970, there existed on earth at least five kilograms of curium, an element first viewed by man through a high-power microscope only 25 years earlier.

There are two more types of nuclear transformations that must be mentioned, neither of which depends upon a particle accelerator. These are nuclear fission and nuclear fusion. Nuclear fission is the splitting of a nucleus into two nuclear fragments of comparable mass. This occurs only in the very heaviest of the elements and then, only for certain nuclei. Fusion is the opposite process, the union of two nuclei into one. This occurs only for the very lightest nuclei.

Fission is generally induced by the capture of a neutron by a very heavy nucleus that is already rich in neutrons. Plutonium-239 is an example. Its neutron to proton ratio, N/Z, is 1.54. After a neutron enters the nucleus, its capture liberates in the nucleus all of its binding energy, some 10 Mev. Just as when two atoms react, the resulting product is liable to break up again because it contains this excitation (see Fig. 7-20). The nucleus oscillates wildly, and like a drop of water, it can fly into two smaller droplets if it is sufficiently unstable. The consequences are most important in the history of man. First, two nuclear fragments are formed with the neutron

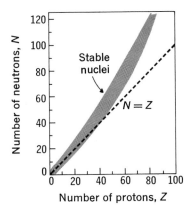

Figure 21-18   *Relationship between number of neutrons and number of protons in stable nuclei. For a given Z, stable nuclei result from any N within the shaded bond.*

to proton ratio of $^{240}_{94}$Pu. However, these two fragments are far down on the stable nucleus curve shown in Figure 21-18, perhaps in the Z range 30 to 64, where the nuclei have stable N/Z ratios of 1.35 to 1.45. The daughter nuclei produced in fission are far, far out of the nuclear stability range. There are two implications: First, lots of energy is liberated as the products decay through beta and gamma emission to move toward a stable nuclear composition—about 200 Mev per fission—and second, a new process that lowers the N/Z ratio becomes important, neutron emission. The energy release is enormous; it corresponds to $4.6 \cdot 10^9$ kcal/mole of fission reactions—the energy obtained in the combustion of 440 tons of gasoline. This frighteningly exothermic fission process is self-propagating; a neutron capture results in fission, and fission produces one or more neutrons to initiate a second capture. This type of process was called, in Chapter Eleven, a chain reaction.

(c) KINETICS OF NUCLEAR TRANSFORMATIONS

When a radioactive nucleus decides to decay, either by fission or by spitting out an α-particle, a β-particle, or a γ-ray, it is surrounded by its electrons, but they seem to exert no influence on nuclear decay. Other nuclei are very far away. From a nucleus-eye point of view, the nucleus disintegrates alone in the world, in its own time and according to its own personality. This transformation is like the chemical decay of N_2O_5 (see Section 11-4(b)) in which each molecule decomposes individually, without aid and comfort from a reaction partner. The rate at which such a process occurs depends upon the nature of the instability that permits an individual to react and upon the number of individuals trying to make up their minds to do so. This can be paraphrased by saying that the number of nuclei decaying per unit time is proportional to the number of nuclei.

$$\frac{\Delta n}{\Delta t} = -kn \qquad\qquad (21\text{-}53)$$

In (21-53) the proportionality constant k describes the probability that one nucleus will decay in a fixed time. We have already labeled a rate law like (21-53); it is called *first order*. We have also seen (Section 11-4(b)) that this rate law implies that if we start with n_0 unstable nuclei at time $t = 0$, the number n that remain later at time t is given by a simple exponential relation:

$$n = n_0\,e^{-kt} \qquad\qquad (21\text{-}54)$$

or

$$\ell n\,\frac{n}{n_0} = -kt \qquad\qquad (21\text{-}55)$$

Expression (21-55) permits us to connect the decay constant k to a readily measured quantity, the time in which n is decreased by a factor of two. This time is what we tabulated in Table 21-15 and called the "half-life," $\tau_{\frac{1}{2}}$. It is related to k, as follows:

at $\quad n = n_0/2,$

$$k\tau_{\frac{1}{2}} = -\ell n\,\frac{n_0/2}{n_0} = -\ell n\,\tfrac{1}{2} = +\ell n\,2 = 0.693$$

or $\quad k = \dfrac{0.693}{\tau_{\frac{1}{2}}} \qquad\qquad (21\text{-}56)$

Experiment shows that nuclear half-lives vary all the way from too long to measure (say, long compared to the age of the earth) to too short to measure (much shorter than a millisecond). Of course, the natural radioactive nuclei with half-lives short compared to the age of the earth must be generated continuously. In Table 21-15, all of the nuclei with short half-lives except ^{14}C are "daughter" nuclei ultimately from parent nuclei with a sufficiently long life to have some representatives still on earth, perhaps ten billion years after they left whatever solar caldron in which they were formed. Carbon-fourteen is continuously formed through cosmic ray bombardment. The half-lives of synthetic radioactive nuclei also vary over this limitless range—many are known with half-lives less than one second.

Fission gives us another opportunity to apply reaction kinetics in a new context. As remarked earlier, the fission process is initiated by neutron capture, but the fission fragments are so neutron-rich that one or more neutrons may "boil off" as the fragments are formed. A typical reaction would be

$$^{235}_{92}U + ^1_0n \longrightarrow ^{236}_{92}U \tag{21-57}$$

$$^{236}_{92}U \xrightarrow{\text{fission}} {}_{56}Ba + {}_{36}Kr + \alpha^1_0n \tag{21-58}$$

If α is unity, we have a chain reaction like that observed in the hydrogen–chlorine explosion. Whenever a neutron is absorbed in (21-57), it is replaced by a neutron in (21-58). Thus the reaction is self-sustaining if no neutrons are lost from the reaction mixture. If α exceeds unity, a more dramatic possibility occurs. If, for example $\alpha = 3$, then one neutron absorbed in reaction (21-57) releases the energy of fission in (21-58) and provides neutron "starters" for *three* new reactions. These three, if absorbed, provide neutrons to induce nine new reactions. Thus the chain is more than self-propagating—it grows. Such "branching chain" reactions have been known for three decades in chemical explosions (the familiar H_2-O_2 reaction is one). They are at work in our nuclear power plants, which run on nuclear fission. The branching chain process is allowed to build up to the desired power level and then neutron monitoring devices regulate operation by inserting neutron absorbers as needed to maintain a constant neutron concentration ("flux").

Nuclear fusion is still another kinetic matter. Two nuclei can fuse only if they get close enough together to feel the special forces that bind the nucleus. These forces seem to reach out no more than 10^{-12} centimeters. Nuclei further apart than that feel only the familiar internuclear electrostatic repulsion. It is useful to calculate how much energy we would need to place two protons at 10^{-12} cm separation assuming only electrostatic repulsion to that point.

Consider the reaction between a tritium and a deuterium nucleus to give a helium nucleus and a neutron.

$$^3_1H + ^2_1H \longrightarrow ^4_2He + ^1_0n \tag{21-59}$$

The electrostatic energy needed to bring these two nuclei to a distance of 10^{-12} cm is:

$$E = \frac{q_1 q_2}{r} = \frac{(4.8 \cdot 10^{-10})^2}{10^{-12}} = 2.3 \cdot 10^{-7} \text{ erg} = 3.3 \cdot 10^6 \frac{\text{kcal}}{\text{mole}} \tag{21-60}$$

At closer approach, the nuclear attractions dominate the electrostatic repulsion, so then the energy drops. Thus (21-60) can be taken as a crude estimate of the activation energy in the bimolecular reaction (21-59). The average kinetic energy of a particle at temperature T is $\frac{3}{2}kT$ (see equation (8-4)) or, per mole, $\frac{3}{2}RT$. If we must surmount a 3.3 million kcal activation energy with an average thermal collision, we would need a temperature such that

$$\tfrac{3}{2}RT = 3.3 \cdot 10^9 \text{ calories}$$

$$T \cong 10^9 \text{ °K} \tag{21-61}$$

That is the sort of temperature that might be found in the center of the sun or in the fireball of a nuclear bomb explosion. In the laboratory, a billion degrees, or even a hundred million degrees, is a pretty warm oven—warmer, in fact, than has been available. Much effort has gone into heating one up, for the stakes in reaction (21-59) are high. If a fusion fire could be ignited, it would sustain itself by its own exothermicity. Thus it would make fusion energy available for controlled and constructive use.

(d) POWER GENERATION

It is informative to examine the weight relations in the fusion reaction (21-59). The nuclear masses are known with very high precision.

$$\begin{matrix} & {}^3_1\text{H} & + & {}^2_1\text{H} & \longrightarrow & {}^4_2\text{He} & + & {}^1_0\text{n} \\ \text{gm/mole} & 3.0170 & & 2.01471 & & 4.00390 & & 1.00893 \end{matrix} \tag{21-62}$$

The products weigh 0.01895 gram less than the reactants! This mass must appear as energy, according to Einstein's relation, $E = mc^2$ (see Section 12-2).

$$E = (0.01895)(2.998 \cdot 10^{10})^2 = 1.70 \cdot 10^{19} \text{ ergs/mole}$$
or
$$E = 4.1 \cdot 10^8 \text{ kcal/mole} \tag{21-63}$$

This energy release is almost too big to contemplate. In nuclear fission, we found that a mole of uranium (238 grams) could release the energy obtained by burning 440 tons of gasoline. Now, with nuclear fusion, this same energy would be released by only 21 grams of the hydrogen reactants. These staggering energies imply that we have a fuel that can replace conventional fuels as they become scarce, or sooner if they make our air unfit to breathe. This energy can be used to run factories and build homes. It can be used to desalt water, to pump it over mountains, to irrigate the great deserts of this and other continents. These new gardens could stave off the famine that seems so sure to come while we learn to stabilize the earth's burgeoning population. Unfortunately, this energy can also be used for destruction in war. The potentiality now exists to kill tens of millions of human beings in a few hours of concerted attack, through the use of nuclear bombs. It is difficult to conceive of a major nuclear war that would not amount to suicidal self-destruction for both protagonists. Thus nuclear energy offers, at once, the chance for man to survive in comfort or to die in a nuclear holocaust. In this country, the decision lies in the hands of our voters as they tell our politicians which fate they prefer.

Problems

1. The effective nuclear charges and one-electron radii listed in Table 21-1 are based upon the actual energy level spacings, as measured spectroscopically (see Figure 21-3). These spacings show that the 4s and 3d levels are quite close so that, in a first approximation, one might assume that their ionization energies are equal. This proves to be useful when the actual spacings are in doubt.
(a) Calculate Z^* and $\bar{r}$ for the 4s electron of chromium atom, assuming that the first ionization energy (156 kcal) refers to removal of the 4s electron instead of a 3d electron. Calculate the percent discrepancy with the values listed in Table 21-1.
(b) Repeat (a) for the 3d electron removal from an iron atom, assuming the first ionization energy is applicable (182 kcal, which refers to removal of the 4s electron).

2. Use the approximation tested in Problem 1, that the 4s and 3d levels have identical ionization energies, to calculate $\bar{r}$ for these orbitals for the iron atom and for each of the ions, Fe^+, Fe^{+2}, and Fe^{+3}. (See Appendix b for the successive ionization energies.)

3. (a) Calculate ΔG^0 and the equilibrium constant for the reaction between chromic ion Cr^{+3} and metallic chromium to produce chromous ion, Cr^{+2}.
(b) If a 0.10 M solution of Cr^{+3} is placed in contact with excess chromium metal (acid solution) until equilibrium is reached in this reaction, what would be the concentrations of Cr^{+2} and of Cr^{+3}?

4. An electrochemical cell is prepared with platinum electrodes with, in the left cell, 0.10 M $Cr_2O_7^{-2}$, 0.10 M Cr^{+3}, and 0.30 M H^+ and, in the right cell, 1.00 M I^-, 0.050 M I_3^-, and 0.10 M H^+.
(a) What will be the voltage of the cell (see Table 6-5), which electrode will be the anode (where oxidation takes place), what will be the over-all reaction in the cell, and what will be the equilibrium constant for that reaction?
(b) Repeat (a), but after solid sodium hydroxide has been added in each half-cell to change their acidities to 0.10 M OH^-.

5. Suppose a solution containing 0.0100 M Cr^{+3} and 0.200 M Fe^{+3} is stirred with an iron stirring rod in dilute acid solution until any reactions between the iron and the two ions are in equilibrium. What will be the composition of the solution?

6. Repeat Problem 5, but stir the solution with a chromium stirring rod (instead of the iron one).

7. Suppose a solution containing 0.0100 M Sn^{+4} and 0.200 M Fe^{+3} is stirred with a tin stirring rod in a dilute acid solution until any reactions between the tin and the two ions are in equilibrium. What will be the concentrations of each of the ions, Fe^{+3}, Fe^{+2}, Sn^{+4}, and Sn^{+2}? The Latimer reduction potential diagram for tin is

$$Sn^{+4} \xrightarrow{+0.150} Sn^{+2} \xrightarrow{-0.136} Sn$$
$$\underset{+0.007}{\rule{3cm}{0.4pt}}$$

8.   Repeat Problem 7, but stir the solution with an iron stirring rod.

9.   Two of our most rapidly growing national resources are aluminum beer cans and tin-coated iron soup cans. You'd think we might figure out a way to use such chemistry as that in Problems 7 and 8 to advantage. (Assume no reaction rate difficulties, though there might be some.) Let's begin by dissolving a soup can entirely in acid to give a 0.200 M Fe^{+2} and 0.010 M Sn^{+2} solution.
(a) Draw an electrolytic cell in which an iron electrode is immersed in 0.190 M Fe^{+2} in the left half-cell (dilute H^+) and a tin electrode is immersed in the soup can solution in the right half-cell (dilute H^+). What direction will current flow, what is the over-all cell reaction, which electrode will be the anode, what will be the initial cell voltage, and at what concentrations in the two cells will current stop flowing?
(b) Draw an electrolytic cell in which a beer can electrode is immersed in 0.10 M Al^{+3} solution (dilute acid) in the left half-cell and an iron electrode is immersed in the soup can solution after step (a) has been completed. (Note: $Al^{+3}\xrightarrow{\quad -1.66\quad}Al$)
Answer all of the questions in (a) (ignoring any concentrations below 10^{-10} M). (Notice that the current flow in each of the cells in (a) and (b) could be used to light a neon sign that says "Recycle").

10.   Chromic hydroxide has substantial amphoteric behavior, as expressed in the equilibrium constants (21-9) and (21-10). Use these two K's to calculate the concentration of Cr(III) remaining in solution if a 0.0100 M Cr^{+3} solution is made basic
(a) with 1.00 M NH_4^+ and 0.10 M NH_3 (for K_b, see Table 5-3);
(b) with 0.50 M NaOH;
(c) with 2.0 M NaOH.

11.   None of the ions, Fe^{+2}, Fe^{+3}, or Mn^{+2} shows appreciable amphoteric solubility in 2.0 M NaOH, in contrast to Cr^{+3} (see Problem 10). Use the equilibrium constants given in Tables 21-2 and 21-3 together with (21-9) and (21-10) to clarify the following separation scheme. (The slimy hydroxide precipitates often entrap significant amounts of impurity ions, but let's assume equilibrium can be reached with the aid of slow precipitation and lots of stirring.)
    A 100-ml mixture of 0.010 M Cr^{+3}, 0.010 M Fe^{+2}, 0.010 M Fe^{+3}, and 0.010 M Mn^{+2} is buffered with 1.0 M NH_4^+ and 0.10 M NH^3 (see Table 5-3). A slimy precipitate P_1 appears and is removed by filtration from the filtrate F_1. Precipitate P_1 is treated with 100 ml of 2.0 M NaOH, but a slimy precipitate P_2 remains, and it is removed by filtration from this filtrate, F_2.
    To F_1 is added enough NH_3 to raise the NH_3 concentration to 2.0 M. A slimy precipitate P_3 appears and is removed by filtration from the filtrate F_3.
(a) At equilibrium, what is the concentration of each metal ion in F_1?
(b) What is the composition of P_1?
(c) Repeat (a) for F_2.
(d) Repeat (a) for F_3.
(e) What is the composition of P_2?
(f) What is the composition of P_3?

12.   What would change in Problem 11 if the original solution is treated with a Br_3^-–Br^- solution before beginning?

13.   What would change in Problem 11 if the original solution is treated with an I_3^-–I^- solution before beginning?

14. Hydrazine N_2H_4 forms a complex with zinc, $Zn(N_2H_4)_2Cl_2$ and bridged com-plexes with transition metals, M, have been found in certain crystals.

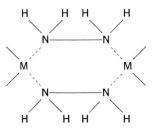

Ammonia forms an analogous zinc complex, $Zn(NH_3)_2Cl_2$, but no bridged com-plexes. Explain the nature of the zinc–hydrazine complex and why hydrazine can form bridged complexes while ammonia cannot.

15. Entropies have been tabulated for quite a few ions in the standard state, $1\,M$, in water, relative to the arbitrary reference that places H^+ (aq, $1\,M$) at zero. Here are a few solution entropies for third-row, d orbital ions.

M^{+n}	Cu^+	Cu^{+2}	Mn^{+2}	Fe^{+2}	Fe^{+3}
$S^0 \left(\dfrac{\text{cal}}{\text{deg}}\right)$	-6.3	-23.6	-20.0	-27.1	-70.1

(a) Explain the trend, proceeding from $n = +1$ to $n = +3$.
(b) Calculate the entropy change for the reaction (under standard conditions):

$$Fe^{+3}(aq) + Cu^+(aq) \longrightarrow Fe^{+2}(aq) + Cu^{+2}(aq)$$

Does this entropy change increase or decrease the tendency for spontaneous reaction between Fe^{+3} and Cu^+?
(c) Calculate $\Delta\mathcal{E}^0$ for the reaction (see Table 6-5) and calculate ΔG^0. Does the reaction proceed spontaneously under standard conditions?

16. Plot the solution entropies given in Problem 15 for $Cu^+(aq)$, $Cu^{+2}(aq)$, $Fe^{+2}(aq)$, and $Fe^{+3}(aq)$ against the total heat of aquation for these same ions, ΔH_{aq}, as listed in Table 21-4. What is the significance of the correlation?

17. (a) Repeat Problem 2 for copper, calculating $\bar{r}$ for Cu^+ and Cu^{+2}.
(b) The positive charge of an ion tends to strengthen its electron donor–acceptor bonds (see $\Delta H/N$ in the last column of Table 21-4). An electrostatic energy of attraction felt by a donor charge, q_d, would be proportional to the ionic charge of the acceptor, q_a, and inversely proportional to the distance between, $\bar{r}_{3d}$. Accord-ingly, the energy per bond should be roughly proportional to the quotient $(q_a/\bar{r}_{3d})$. Using the results of (a) and of Problem 2, calculate for each of the ions Cu^+, Cu^{+2}, Fe^{+2}, and Fe^{+3} the quantity $(q_a/\bar{r}_{3d})$ and the ratio of $(\Delta H_{aq}/N)$ to $(q_a/\bar{r}_{3d})$. For Fe^{+2} and Cu^{+2}, find the ratio for both $N = 4$ and $N = 6$. According to this criterion, which choice seems more consistent for the $+2$ ions, four or six water molecules bound in the closest aquation sphere?

18. The relative electron donor ability of NH_3 and Cl^- is implicit in the free energies given in Table 21-5.
(a) From the ΔG^0 values in Table 21-5, calculate the free energy change ΔG^0 and the equilibrium constant for the replacement of Cl^- in the cuprous chloride com-

plex, $CuCl_2^-$, by NH_3

$$CuCl_2^-(aq) + 2 NH_3 = Cu(NH_3)_2^+ + 2 Cl^-$$

(b) Calculate the ratio of $[Cu(NH_3)_2^+]/[CuCl_2^-]$ if NH_3 is added to a 0.0050 M $CuCl_2^-$ solution in 2.0 M Cl^- until the NH_3 concentration is 0.050 M.

19. The difference in complexing ability (electron donor ability) between Cl^- and NH_3 tends to be magnified as the positive ion charge increases. Repeat Problem 18 except for the cupric complexes, $CuCl_4^{-2}$ and $Cu(NH_3)_4^{+2}$.

20. Draw the geometrical isomers that are possible for a square planar copper(II) mixed ammonia-chloride complex ion, $CuCl_2(NH_3)_2^{+2}$.

21. The chromium(III) mixed chloride–water complexes interconvert very slowly, which permits the following observations. There are two crystalline salts, a violet-colored one, V, and a green-colored one, G, with the following properties.
 A. Chemical analysis shows that V and G each contain 19.52% by weight chromium, 39.92% chloride, and the rest is water.
 B. Prolonged drying under vacuum has no effect on the composition of V, but G loses 13.52% of its weight as water vapor.
 C. When V is dissolved in water and excess $AgNO_3$ is added, silver chloride precipitates and analysis shows that all of the chloride from V is in the AgCl. In the same experiment with G, only one third of the chloride from G is found in the AgCl.
 D. Electrical conductivity measurements show that a 0.10 M solution of V has about 80% of the ionic conductivity of a 0.10 M $LaCl_3$ solution whereas a 0.10 M solution of G has only about one quarter that conductivity. However, the conductivity of G rises to that of V in about six hours.
(a) Calculate the number of moles of Cl and of H_2O per mole of Cr(III) for each of V and G.
(b) What is the formula of G after drying?
(c) Postulate possible structures for the complex ions present in V and G, explaining how each of the observations is explained.
(d) Are there isomeric alternatives for either V or G?

22. (a) From the $\mathcal{E}^0$ values given in (21-28) to (21-30), calculate the equilibrium constants for the Mn(III) complex ion formation equilibria

$$Mn^{+3}(aq) + Cl^-(aq) = MnCl^{+2}(aq)$$

$$Mn^{+3}(aq) + HF(aq) = MnF^{+2}(aq) + H^+(aq)$$

(b) Calculate the equilibrium constant for the replacement of chloride by fluoride in the $MnCl^{+2}$ complex

$$MnCl^{+2}(aq) + HF(aq) = MnF^{+2}(aq) + H^+(aq) + Cl^-(aq)$$

(c) Which is the better electron donor, Cl^- or F^-?

23. Recently, magnetic resonance techniques have made it possible to measure the equilibrium constants for attachment of an ion in the *second* aquation sphere, just outside the nearest-neighbor aquation sphere. This can be called an "ion-pair." For example, the manganous ion, $Mn(OH_2)_n^{+2}$, can form a loose, next-nearest neighbor complex with a halide ion:

$$Mn(OH_2)_n^{+2} + Cl^- = Mn(OH_2)_n^{+2} \cdot Cl^- \qquad K = 0.09$$

$$Mn(OH_2)_n^{+2} + F^- = Mn(OH_2)_n^{+2} \cdot F^- \qquad K = 1.5$$

(a) Calculate to $\pm 15\%$ the concentration of $Mn(OH_2)_n^{+2} \cdot Cl^-$ ion-pair complexes in 0.010 M Mn(II) and 0.20 M Cl$^-$.

(b) Repeat (a) for 0.010 M Mn(II) and 0.20 M F$^-$.

(c) What percentage of the $Mn(OH_2)_n^{+2}$ is *not* engaged in ion-pair formation in each of (a) and (b)?

24.   Recently two Swiss chemists with the unlikely names L'Eplattinier and Calderazzo made a solid compound of osmium by the reaction between H_2 and solid osmium pentacarbonyl, $Os(CO)_5$, at high pressure. They placed 15.2 g of their new compound in a 1.0-ℓ vessel and completely decomposed it to osmium metal, H_2 gas, and CO gas. The pressure in the vessel at 27°C was measured to be 6.15 atm. When the temperature was reduced to just below $-190°C$ (the normal boiling point of CO(ℓ)) the pressure dropped to 1.34 atm. When the temperature was further reduced to just below $-253°C$ (the normal boiling point of $H_2(\ell)$), the pressure dropped to 0.082 atm. The freezing point of CO is $-207°C$.

(a) How many moles of gas mixture were formed when the sample was decomposed?

(b) How many moles of H_2 were in the sample?

(c) How many moles of CO were in the sample?

(d) What is the formula of the compound?

(e) Explain the pressure that was recorded at $-190°C$.

25.   Ceric ion, Ce^{+4}, is a useful and very powerful oxidizing agent in aqueous solution. Calculate $\Delta\mathcal{E}^0$, ΔG^0, and the equilibrium constant for the oxidation of water by Ce^{+4}, the reaction analogous to that given for Pr^{+4} in equation (21-46).

26.   (a) Calculate $\Delta\mathcal{E}^0$, ΔG^0, and the equilibrium constant for the reduction of H^+ by europium, Eu^{+2}.

(b) What would be the equilibrium concentration (to $\pm 10\%$) of Eu^{+2} in equilibrium with 0.10 M Eu^{+3} at $[H^+] = 4 \cdot 10^{-6}$ M and under one atmosphere pressure of H_2?

27.   The small differences in hydroxide solubilities between successive lanthanides are attributed to a small reduction in size from element to element (the "lanthanide contraction"). Test this idea by comparing the ratios of solubility products for each of the successive alkaline earth hydroxides, $Mg(OH)_2/Be(OH)_2$, $Ca(OH)_2/Mg(OH)_2$, $Sr(OH)_2/Ca(OH)_2$, and $Ba(OH)_2/Sr(OH)_2$, (see Table 2-8) to the size ratios as indicated by the molar volumes of the metals (see the last column in Table 2-6).

28.   The 4f, 5d, and 6s orbitals of cerium have about the same energy.

(a) Use the approximation tested in Problem 1, that these orbitals have identical ionization energies, to calculate $\bar{r}$ for each of these orbitals.

(b) Calculate the orbital radius ratios $\bar{r}(6s)/\bar{r}(5d)$ and $\bar{r}(5d)/\bar{r}(4f)$. (The lack of participation of the 4f orbitals in the chemistry of the lanthanides is attributed to their small size.)

29.   Repeat Problem 28 for the 5f, 6d, and 7s orbitals of uranium. Compare the radius ratio obtained for $\bar{r}(6d)/\bar{r}(5f)$ to the ratios obtained in Problem 28. (There is substantial evidence that the 5f orbitals *do* participate in the chemistry of uranium and subsequent elements. This difference from the lanthanides may be due to the larger size of the 5f orbitals relative to the other valence orbitals.)

30.   Radioactivity was discovered in 1896 by Henri Becquerel, and within a few years, Pierre and Marie Curie had proved by chemical separations that several chemical elements were involved. In absence of direct knowledge of the chemical identities, the "daughter" elements were named after the parent. For example, a long sequence of radioactive "decays" were found to originate in thorium samples

and the sequence of products was named as follows:

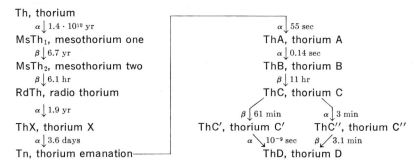

Th, thorium
$\alpha \downarrow 1.4 \cdot 10^{10}$ yr
MsTh$_1$, mesothorium one
$\beta \downarrow 6.7$ yr
MsTh$_2$, mesothorium two
$\beta \downarrow 6.1$ hr
RdTh, radio thorium
$\alpha \downarrow 1.9$ yr
ThX, thorium X
$\alpha \downarrow 3.6$ days
Tn, thorium emanation

$\alpha \downarrow 55$ sec
ThA, thorium A
$\alpha \downarrow 0.14$ sec
ThB, thorium B
$\beta \downarrow 11$ hr
ThC, thorium C
$\beta \downarrow 61$ min $\alpha \downarrow 3$ min
ThC', thorium C'   ThC'', thorium C''
$\alpha \searrow 10^{-9}$ sec $\beta \nearrow 3.1$ min
ThD, thorium D

It was a number of years before it became clear that the sequence originated in the thorium isotope of mass 232, $^{232}_{90}$Th. That, however, was sufficient to identify every daughter nucleus.

(a) List the historical names and, for each, the chemical name and the values of Z and the isotopic mass. (Remember, in α-decay a $^4_2$He nucleus is ejected from the nucleus and in β-decay an electron is ejected.)

(b) Which of the thorium daughters would be gaseous?

(c) Which of the thorium daughters would have the chemistry of an alkaline earth?

(d) Suppose a natural thorium sample were entirely dissolved, lead nitrate added, and then the lead were extracted and carefully purified from all other elements. What radioactivity would be displayed immediately after purification? 24 hours later?

31. The artificially produced, radioactive cobalt isotope $^{60}_{27}$Co has almost completely replaced radium in cancer therapy. This cobalt isotope is a beta emitter with a seven-year half-life and the emitted electron carries away 1.45 Mev of energy. It can be made by three reactions initiated by deuteron, $^2_1$H, or neutron, $^1_0$n, bombardment. For each of the designated nuclear reactions, indicate the appropriate target nucleus.

(a) (d,p) (deuteron in, proton out)

(b) (n,p)

(c) (n,γ)

32. The artificially produced, radioactive iodine isotope $^{131}_{53}$I is commonly used in the treatment of thyroid disorders. When iodide is ingested, it concentrates in the thyroid gland, where the radioactivity can be effective. This iodine isotope $^{131}_{53}$I, is a beta emitter with an eight-day half-life and the emitted electron carries away 0.69 Mev of energy. It can be made by three reactions initiated by deuteron, $^2_1$H, or neutron, $^1_0$n, bombardment. In two of these syntheses, an intermediate radioactive isotope is produced, $^{131}_{52}$Te, which decays by beta emission with a 25-minute half-life. For each of the designated nuclear reactions, indicate the appropriate target nucleus.

(a) (d,n) $^{131}_{53}$I

(b) (d,p) $^{131}_{52}$Te $\xrightarrow{\beta^-}$ $^{131}_{53}$I

(c) (n,γ) $^{131}_{52}$Te $\xrightarrow{\beta^-}$ $^{131}_{53}$I

33. Suppose an artificially radioactive cobalt sample is prepared for medical therapy by nuclear bombardment until 10^{15} radioactive $^{60}_{27}$Co nuclei have been formed.

(a) Use the half-life of this isotope (given in Problem 31) to calculate the decay constant k in expression (21-53). Express k in dimensions sec^{-1}.

(b) With expression (21-53), calculate how many decays will occur in one second within the cobalt sample.

(c) If the cobalt sample is held in proximity to a cancerous area for one hour, how many calories of energy are transferred to the cancerous material? (1 Mev = one million electron volts; 1 ev = $3.83 \cdot 10^{-20}$ calorie.)

34. Repeat Problem 33 except for an iodine sample containing 10^{15} radioactive $^{131}_{53}$I nuclei. (See Problem 32.)

35. How many radioactive nuclei will remain and what will be the rate of decay (decays per second) of the cobalt sample described in Problem 33 after the hospital has used the sample for three months? (30 days = $2.59 \cdot 10^6$ sec)

36. Repeat Problem 35 for the iodine sample described in Problem 34.

37. Radioactive carbon, $^{14}_6$C is generated continuously in the atmosphere by cosmic ray-produced neutrons. These neutrons are absorbed by $^{14}_7$N(n,p)$^{14}_6$C. The carbon-14 decays with β emission and a 5668-year half-life.

The atmospheric $^{14}_6$C becomes oxidized to CO_2, so all atmospheric carbon dioxide has a small steady state concentration of radioactive $^{14}CO_2$. If radioactive $^{14}CO_2$ is incorporated into plant material, through photosynthesis, then it decays without opportunity for replenishment. Hence the level of radioactivity observed in plant material (e.g., wood) is a measure of the time elapsed since the photosynthetic process formed the material.

(a) What is the rate constant k for the decay of $^{14}_6$C, expressed in (years)$^{-1}$?

(b) By what fraction would the decay rate have decreased in an ax handle that was made at the time of the birth of Christ?

38. A wooden carving of the sun god Ra found in the tomb of King Tut showed a carbon-decay rate that had decreased by a factor of 0.666 from the atmospheric value. How old is the carving? (See Problem 37.)

39. If a nuclear power plant consumes one kilogram of uranium-235 per day in a fission process that releases 200 Mev per fission, and if this energy could be converted to electrical energy at 30% efficiency, the power output would be 300,000 kilowatts. This output is about that of a large power plant that furnishes power to a fair-sized city and that consumes about 2500 tons of coal a day.

(a) The density of uranium is 18.5 g/cc. How many cc does one kilogram of uranium occupy?

(b) The density of coal is about 2 g/cc. How many cubic feet do 2500 tons of coal occupy? (1 ton = 2000 pounds = $9.1 \cdot 10^5$ grams; 1 cu ft = 28.3 ℓ).

(c) How many cubic feet of CO_2 are released into the atmosphere per day by the complete combustion of 2500 tons of coal (answer for a temperature of 300°K and one atmosphere pressure).

40. A sequence of nuclear reactions postulated to account for a part of the energy generation in stars is the "carbon-nitrogen" cycle. It involves two steps that evolve a "positron," a positively charged electron e$^+$.

$$^{12}C + {}^1H \longrightarrow {}^{13}N + \gamma$$
$$^{13}N \longrightarrow {}^{13}C + e^+$$
$$^{13}C + {}^1H \longrightarrow {}^{14}N + \gamma$$
$$^{14}N + {}^1H \longrightarrow {}^{15}O + \gamma$$
$$^{15}O \longrightarrow {}^{15}N + \beta^+$$
$$^{15}N + {}^1H \longrightarrow {}^{12}C + {}^4He$$

(a) What is the over-all reaction?

(b) What are the catalysts for this over-all reaction?

(c) The ultimate fate of the positrons is that they combine with ordinary, negative electrons in an "annihilation" process to liberate their energy in the form of a gamma-ray. Hence the net process is consumption of four hydrogen atoms (each with mass 1.00812 g/mole) to produce one helium atom (each with mass 4.00388 g/mole). Calculate the net mass converted into other forms of energy per mole of helium produced. How many kcal of energy are released per 4 grams of hydrogen consumed?

epilogue

So the last chapter ended on a sobering note—and fitting it is. The aims of man's investigation of his environment are to permit him to live at ease in it, to control it so as to prolong his own existence and the existence of his kind. At this, we have been eminently successful over the centuries of modern time.

Yet we now see new threats appearing. The population of the earth gets larger and larger, doubling twice in a man's lifetime, while the earth remains the same size. We can no longer regard our planet as a boundless source of food, shelter, and the things man needs to live. Our environment can no longer diffuse all of the wastes we produce, the auto exhaust, the sewage, the industrial discards. Even our conscious efforts to increase food production and to decrease disease can backfire as excessive and careless use of fertilizers, medicines, and insecticides might poison us and damage ecologies we wish to preserve. Finally, our ability to destroy has reached a fearsome dimension, seeming to make the Apocalypse a matter of choice.

How shall we fare in such a precarious time? That depends upon how man directs his creativity. There is no smog problem that cannot be conquered as soon as we bend to the task. We can find new ways to grow food, we can care more tenderly for the nature around us by understanding it well. Knowledge, creatively directed, has always before enabled man to triumph over threats to his existence and it will enable us to deal with new threats as they arise. Even human relations must be better known—we must find new ways to live that guarantee peace among men.

We cannot afford to fear man's knowledge and his creativity. They are our most powerful possessions. They furnish a sturdy basis for confidence that our problems are soluble and for hope that we will solve them. These precious attributes earn the envy of the limitless Universe for man in his tiny spacecraft, Earth.

appendices

Appendix a: *Physical Constants and Conversion Factors*

Avogadro's number	N_0	6.023×10^{23} mole^{-1}
Planck constant	h	6.626×10^{-27} erg sec
		$9.537 \times 10^{-14} \left(\dfrac{\text{kcal}}{\text{mole}}\right)$ sec
Speed of light (vacuum)	c	2.998×10^{10} cm/sec
Elementary charge	e	1.602×10^{-19} coulomb
		4.803×10^{-10} e.s.u.
Electron rest mass	m_e	9.109×10^{-28} g
Proton rest mass	m_p	1.673×10^{-24} g
Bohr radius	a_0	0.5292×10^{-8} cm
H ionization energy	E_H	313.6 kcal/mole
Gas constant	R	1.987 cal/mole degK
		8.314 joule/mole degK
		62.36 liter torr/mole degK
		8.205×10^{-2} liter atm/mole degK
$2.303\ RT$ (25°C)		1.364 kcal/mole
Volume 1 mole perfect gas		
at STP (1 atm, 0°C)		22.41 liter
at 1 atm, 25°C		24.47 liter
Faraday constant	$\mathcal{F}$	96,487 coulombs per mole of electrons
		23,061 calories per volt per mole of electrons
Standard atmosphere	atm	760 torr = 760 mm of mercury
		1.013×10^6 dyne/cm^2
$\pi = 3.1416 \qquad e = 2.71828$		$\log_e 10 = 2.30259 = 1/0.43429$
		$\log_{10} x = 0.43429 \log_e x$

These values are those recommended by the National Bureau of Standards as listed in the October, 1963, NBS Technical News Bulletin and reprinted in the *Journal of Chemical Education*, **40**, 642 (1963). All values are based on the Unified Atomic Weight Scale in which $C^{12} = 12$ exactly.

	TO CONVERT		MULTIPLY BY ↓
	FROM ⟶	TO	
6.947×10^{-14}	erg/molecule	kcal/mole	1.439×10^{13}
4.336×10^{-2}	$\dfrac{\text{electron volts}}{\text{molecule}}$	kcal/mole	23.061
3.498×10^2	$\dfrac{\text{cm}^{-1}}{\text{molecule}}$	kcal/mole	2.859×10^{-3}
5.034×10^{15}	$\dfrac{\text{cm}^{-1}}{\text{molecule}}$	erg/molecule	1.986×10^{-16}
2.390×10^{-4}	kcal/mole	joules/mole	4.1840×10^3
↑ MULTIPLY BY	TO ⟵	FROM	
	TO CONVERT		

Appendix b: Ionization Energies of the Elements (kcal/mole)

$$A^{+(n-1)} \longrightarrow A^{+n} + e^- \quad \Delta H = E_n$$

At. No.	Elem.	E_0*	E_1	E_2	E_3	E_4	E_5	E_6	E_7	E_8
1	H	17.4	313.6	—	—	—	—	—	—	—
2	He	—	566.8	1254	—	—	—	—	—	—
3	Li	14	124.3	1744	2823	—	—	—	—	—
4	Be	—	214.9	419.9	3548	5020	—	—	—	—
5	B	(7)	191.3	580.0	874.5	5980	7843	—	—	—
6	C	29	259.6	562.2	1104	1487	9034	11300	—	—
7	N	—	335.1	682.8	1094	1786	2257	12727	15377	—
8	O	34	314.0	810.6	1267	1785	2624	3184	17044	20088
9	F	79.5	401.8	806.7	1445	2012	2634	3623	4268	21990
10	Ne	—	497.2	947.2	1500	2241	2913	3641	—	—
11	Na	(12.5)	118.5	1091	1652	2280	3192	3969	4806	6093
12	Mg	—	176.3	346.6	1848	2521	3256	4301	5186	6134
13	Al	(11)	138.0	434.1	655.9	2767	3593	4391	5567	6562
14	Si	(32)	187.9	376.8	771.7	1041	3844	4730	5682	6990
15	P	(18)	241.8	453.2	695.5	1184	1499	5083	6072	7132
16	S	48	238.9	540	807	1091	1672	2030	6711	7582
17	Cl	83.3	300.0	548.9	920.2	1230	1564	2230	2635	8032
18	Ar	—	363.4	637.0	943.3	1379	1730	2105	2860	3308
19	K	(10), (20)	100.1	733.6	1100	1405	1905	2299	2721	3574
20	Ca	—	140.9	273.8	1181	1550	1946	2514	2952	3305
21	Sc	—	151.3	297.3	570.8	1700	2120	2560	3205	3667
22	Ti	(9)	158	314.3	649.0	997.2	2301	2767	3252	3966
23	V	(22)	155	328	685	1100	1499	2975	3482	4013
24	Cr	(23)	156.0	380.3	713.8	1140	1683	2099	3713	4266
25	Mn	—	171.4	360.7	777.0	1200	1753	2306	2744	4520
26	Fe	(13)	182	373.2	706.7	1310	—	2375	2998	3482
27	Co	(22)	181	393.2	772.4	—	—	1916	3067	3759
28	Ni	(29)	176.0	418.6	810.9	—	—	—	—	3874
29	Cu	(41.5)	178.1	467.9	849.4	—	—	—	—	—
30	Zn	—	216.6	414.2	915.6	—	—	—	—	—
31	Ga	—	138	473.0	708.0	1480	—	—	—	—
32	Ge	—	182	367.4	789.0	1050	2153	—	—	—
33	As	—	226	466	653	1160	1444	2940	—	—
34	Se	—	225	496	738	989	1568	1891	3574	—
35	Br	77.6	273.0	498	828	—	1377	2043	2375	4451
36	Kr	—	322.8	566.4	851	—	1492	1810	2560	2906
37	Rb	>5	96.31	634	920	1213	1637	1946	2288	3136
38	Sr	—	131.3	254.3	1005	1300	1651	2094	2444	2820
39	Y	—	147	282.1	473	1425	1776	2145	2675	2975
40	Zr	—	158	302.8	530.0	791.8	1898	2280	2675	3205
41	Nb	—	158.7	330.3	579.8	883	1153	2375	2883	3252
42	Mo	23	164	372.5	625.7	1070	1411	1570	2906	3528
43	Tc	—	168	351.9	681.2	—	—	—	—	—
44	Ru	—	169.8	386.5	656.4	—	—	—	—	—
45	Rh	—	172	416.7	716.1	—	—	—	—	—
46	Pd	—	192	447.9	759.2	—	—	—	—	—
47	Ag	—	174.7	495.4	803.1	—	—	—	—	—
48	Cd	—	207.4	389.9	864.2	—	—	—	—	—
49	In	—	133.4	435.0	646.5	1250	—	—	—	—
50	Sn	—	169.3	337.4	703.2	939.1	1829	—	—	—
51	Sb	—	199.2	380	583	1020	1291	2491	2744	—

Appendix b: Ionization Energies of the Elements (kcal/mole) *(Continued)*

$$A^{+(n-1)} \longrightarrow A^{+n} + e^- \quad \Delta H = E_n$$

At. No.	Elem.	E_0*	E_1	E_2	E_3	E_4	E_5	E_6	E_7	E_8
52	Te	—	208	429	720	880	1383	1683	3159	—
53	I	70.9	241	440.3	—	761	1637	1914	2398	3920
54	Xe	—	279.7	489	740	1015	1384	1914	2352	2906
55	Cs	>4	89.8	579	807	—	—	—	2491	2813
56	Ba	—	120.2	230.7	818.7	—	—	—	—	2929
57	La	—	129	263.6	442.1	—	—	—	—	—
58	Ce	—	149.9	284	461	768	—	—	—	—
59	Pr	—	131	—	235	—	—	—	—	—
60	Nd	—	131	—	—	—	—	—	—	—
61	Pm	—	—	—	—	—	—	—	—	—
62	Sm	—	130	258	—	—	—	—	—	—
63	Eu	—	131	259	—	—	—	—	—	—
64	Gd	—	142	277	—	—	—	—	—	—
65	Tb	—	155	—	—	—	—	—	—	—
66	Dy	—	157	—	—	—	—	—	—	—
67	Ho	—	—	—	—	—	—	—	—	—
68	Er	—	—	—	—	—	—	—	—	—
69	Tm	—	—	—	—	—	—	—	—	—
70	Yb	—	144	279	—	—	—	—	—	—
71	Lu	—	141	339	—	—	—	—	—	—
72	Hf	—	161	344	—	—	—	—	—	—
73	Ta	—	182	374	—	—	—	—	—	—
74	W	12	184	408	—	—	—	—	—	—
75	Re	4	181	383	—	—	—	—	—	—
76	Os	—	201	392	—	—	—	—	—	—
77	Ir	—	207	392	—	—	—	—	—	—
78	Pt	—	207	428	—	—	—	—	—	—
79	Au	—	213	473	—	—	—	—	—	—
80	Hg	—	240.5	432.4	788	1660	1891	—	—	—
81	Tl	—	140.8	470.9	687	1169	—	—	—	—
82	Pb	—	170.9	346.6	736	976	1587	—	—	—
83	Bi	—	168	384.7	589.4	1045	1291	2036	—	—
84	Po	—	194	447	629.6	—	—	—	—	—
85	At	—	219	464	675.7	—	—	—	—	—
86	Rn	—	247.8	493	678	—	—	—	—	—
87	Fr	—	88.3	519	773	—	—	—	—	—
88	Ra	—	121.7	233.9	—	—	—	—	—	—
89	Ac	—	159	279	461	—	—	—	—	—
90	Th	—	—	265	461	677.5	—	—	—	—
91	Pa	—	—	—	—	—	—	—	—	—
92	U	—	92	—	—	—	—	—	—	—
93	Np	—	—	—	—	—	—	—	—	—
94	Pu	—	118	—	—	—	—	—	—	—
95	Am	—	138	—	—	—	—	—	—	—

* E_0 = electron affinity; parenthetical values are not experimental, calculated only.

Values from W. Finkelnburg and W. Humbach, *Naturwiss.*, **42**, 35 (1955); C. E. Moore, Circular of the National Bureau of Standards 467, *Atomic Energy Levels*, Vol. III, 1958; R. W. Kiser, *Tables of Ionization Potentials*, U.S. Atomic Energy Commission TID 6142 (1960).

Appendix c: Ionization Energies and Electron Affinities of Molecules

Ionization Energies of Molecules* (kcal/mole)

Diatomics

H_2	355.8	CO	323.1	F_2	362	HF	363.7
N_2	359.3	CN	348.9	Cl_2	264.8	HCl	293.8
O_2	278.5	NO	213.3	Br_2	243.3	HBr	268.0
OH	304.0	CH	256	I_2	214.0	HI	239.4

Triatomics

CO_2	318.0	N_2O	297.5	H_2O	290.4
OCS	257.6	NO_2	225.5	H_2S	241.2
CS_2	232.5	O_3	295.2	H_2Se	227.8
SO_2	284.6	HCN	320.8	H_2Te	210.7

Polyatomics

methyl chloride	CH_3Cl	260.1	methane	CH_4	299.3
methyl bromide	CH_3Br	242.8	ethane	C_2H_6	268.6
methyl iodide	CH_3I	220.0	propane	C_3H_8	255.3
methanol	CH_3OH	250.2	n-butane	C_4H_{10}	245.1
methanthiol	CH_3SH	217.7	i-butane	C_4H_{10}	243.8
methyl amine	CH_3NH_2	206.9	n-pentane	C_5H_{12}	238.7
nitromethane	CH_3NO_2	255.5	cyclopentane	C_5H_{10}	242.8
acetonitrile	CH_3CN	281.8	ethylene	C_2H_4	242.6
formaldehyde	H_2CO	250.6	propylene	C_3H_6	224.4
acetaldehyde	CH_3CHO	235.5	1-butene	C_4H_8	220.9
formic acid	HCOOH	254.8	trans-2-butene	C_4H_8	210.6
acetic acid	CH_3COOH	239.2	cis-2-butene	C_4H_8	210.6
methyl formate	$HCOOCH_3$	249.4	1-pentene	C_5H_{10}	219.0
methyl formamide	$HCOONH_2$	236.4	cyclopentene	C_5H_8	207.8
toluene	$C_6H_5CH_3$	203.5	acetylene	C_2H_2	263.1
naphthalene	$C_{10}H_8$	187.3	benzene	C_6H_6	213.3

* K. Watanabe, T. Nakayama, and J. Mottl, *J. Quant. Spectroscopy and Energy Transfer,* **2**, 369–382 (1962).

Electron Affinities of Molecules

$$M^- \rightarrow M + e^- \qquad \Delta H = E_0 \text{ (kcal/mole)}$$

Diatomics

O_2	3.5, 10	CH	37	SH	53
CN	88.1	OH	42	SiH	(34)†
NO	21	NH	5.1	PH	(21)†
C_2	71.5				

Triatomics		Polyatomics	
C_3	42	SF_5	84.4
N_3	72–81	SF_6	34.4
SCN	50	C_6H_5	50.7
NH_2	33.3	$C_6H_5CH_2$	20.8

† Parenthetical values are calculated only; no experimental data.

Appendix d: Thermodynamic Properties

Formula	State	ΔH_f^0 (kcal/mole) at 298.16°K (25°C)	ΔG_f^0 (kcal/mole) at 298.16°K (25°C)	S^0 (cal/mole degK)	C_p^0 (cal/mole degK)
Aluminum					
Al	s	0.0	0.0	6.8	5.8
Al_2O_3	s (α, corundum)	−400.5	−378.2	12.2	18.9
AlF_3	s	−359.5	−340.6	15.9	18.0
$AlCl_3$	s	−168.3	−150.3	26.5	22.0
$AlBr_3$	s	−126.0	−122.1	49.0	24.5
AlI_3	s	−75.0	−71.9	38.0	23.6
$Al_2(SO_4)_3$	s	−822.4	−741.0	57.2	62.0
AlN	s	−76.0	−68.6	4.82	7.20
Antimony					
Sb	s	0.0	0.0	10.9	6.0
Sb_2O_3	s	−168.4	−149.0	29.4	24.2
Sb_2O_5	s	−232.3	−198.2	29.9	28.1
$SbCl_3$	g	−75.0	−72.0	80.7	18.3
	s	−91.3	−77.4	44.0	25.8
Sb_2S_3	s (black)	−41.8	−41.5	43.5	28.6
Argon					
Ar	g	0.0	0.0	37.0	5.0
Arsenic					
As	s (α, grey)	0.0	0.0	8.4	5.9
	g	72.3	62.4	41.6	5.0
As_4	g	34.4	22.1	75.0	—
As_4O_6	s (octahedral)	−314.0	−275.5	51.2	45.7
As_2O_5	s	−221.0	−187.0	25.2	27.8
AsH_3	g	15.9	16.5	53.2	9.1
As_2S_3	s	−40.4	−40.3	39.1	27.8
Barium					
Ba	s	0.0	0.0	16.0	6.3
BaO	s	−133.4	−126.3	16.8	11.3
BaO_2	s	−150.5	—	—	—
$BaCl_2$	s	−205.6	−193.8	30.0	18.0
$BaCl_2 \cdot H_2O$	s	−278.4	−253.1	40.0	28.2
$BaCl_2 \cdot 2 H_2O$	s	−349.3	−309.7	48.5	37.1
$BaSO_4$	s	−350.2	−323.4	31.6	24.3
$Ba(NO_3)_2$	s	−237.1	−190.1	51.1	36.1
$BaCO_3$	s (witherite)	−291.3	−272.2	26.8	20.4
Beryllium					
Be	s	0.0	0.0	2.3	4.3
BeO	s	−146.0	−139.0	3.4	6.1
$BeCl_2$	s	−112.0	−101.9	21.5	—
Bismuth					
Bi	s	0.0	0.0	13.6	6.1
Bi_2O_3	s	−137.9	−118.7	36.2	27.2

Appendix d: Thermodynamic Properties (Continued)

Formula	State	ΔH_f^0 (kcal/mole) at 298.16°K (25°C)	ΔG_f^0 (kcal/mole) at 298.16°K (25°C)	S^0 (cal/mole degK)	C_p^0 (cal/mole degK)
Bismuth—cont.					
$BiCl_3$	g	−64.7	−62.2	85.3	19.0
	s	−90.6	−76.2	45.3	—
$BiOCl$	s	−87.3	−77.0	20.6	—
Bi_2S_3	s	−43.8	−39.4	35.3	30.7
Boron					
B	s	0.0	0.0	1.4	2.6
B_2O_3	s	−304.2	−285.3	12.90	15.0
B_2H_6	g	7.5	19.8	55.7	13.5
BF_3	g	−265.4	−261.3	60.7	12.1
BCl_3	g	−96.5	−92.9	69.3	15.0
	ℓ	−102.1	−92.6	49.3	25.5
BBr_3	g	−44.6	−51.0	77.5	16.3
Bromine					
Br_2	ℓ	0.0	0.0	36.4	18.1
	g	7.4	0.7	58.6	8.6
Br	g	26.7	19.7	41.8	5.0
Cadmium					
Cd	s (α)	−0.1	−0.1	12.4	6.2
CdO	s	−61.7	−54.6	13.1	10.4
$CdCl_2$	s	−93.6	−82.2	27.5	17.9
$CdCl_2 \cdot H_2O$	s	−164.5	−140.3	40.1	—
$CdCl_2 \cdot 2.5\ H_2O$	s	−270.5	−225.6	54.3	—
CdS	s	−38.7	−37.4	15.5	13.2
$CdSO_4$	s	−223.1	−196.6	29.4	23.8
Cesium					
Cs	s	0.0	0.0	19.8	7.4
Cs_2O	s	−75.9	—	—	—
CsH	g	29.0	24.3	51.2	7.1
CsF	s	−126.9	−119.7	19.8	12.1
CsCl	s	−103.5	−96.8	23.9	12.6
CsBr	s	−94.3	−91.6	29.0	12.4
CsI	s	−80.5	−79.7	31.0	12.4
Calcium					
Ca	s	0.0	0.0	9.9	6.3
CaO	s	−151.9	−144.4	9.5	10.2
$Ca(OH)_2$	s	−235.8	−214.3	18.2	20.2
$CaCl_2$	s	−190.0	−179.3	27.2	17.4
$CaSO_4$	s (anhydrite)	−342.4	−315.6	25.5	23.8
$CaSO_4 \cdot 0.5\ H_2O$	s (α)	−376.5	−343.0	31.2	28.6
	s (β)	−376.0	−342.8	32.1	29.6
$CaSO_4 \cdot 2\ H_2O$	s	−483.1	−429.2	46.4	44.5
CaC_2	s	−15.0	−16.2	16.8	14.9
$CaCO_3$	s (calcite)	−288.4	−269.8	22.2	19.6

Appendix d: Thermodynamic Properties (Continued)

Formula	State	ΔH_f^0 (kcal/mole) at 298.16°K (25°C)	ΔG_f^0 (kcal/mole) at 298.16°K (25°C)	S^0 (cal/mole degK)	C_p^0 (cal/mole degK)
Carbon					
C	s (graphite)	0.0	0.0	1.37	2.0
	s (diamond)	0.45	0.69	0.57	1.5
	g	171.3	160.4	37.8	5.0
CO	g	−26.4	−32.8	47.2	7.0
CO_2	g	−94.1	−94.3	51.1	8.9
CH_4	g	−17.9	−12.1	44.5	8.4
C_2H_6	g	−20.2	−7.9	54.8	12.6
C_2H_4	g	12.5	16.3	52.5	10.4
C_2H_2	g	54.2	50.0	48.0	10.5
C_3H_8	g	−24.8	−5.6	64.5	17.6
C_6H_6	g	19.8	31.0	64.3	19.5
	ℓ	11.7	29.8	41.3	—
HCHO	g	−27.7	−27.0	52.3	8.5
HCOOH	g	−90.5	−80.2	60.0	—
CH_3OH	g	−48.0	38.7	57.3	10.5
	ℓ	−57.0	−39.8	30.3	19.5
CCl_4	g	−24.6	14.5	74.0	19.9
	ℓ	−32.4	−15.6	51.7	31.5
$COCl_2$	g	−52.3	−48.9	67.7	13.8
CH_3Cl	g	−19.3	−13.7	56.0	9.7
CH_2Cl_2	g	−22.1	−15.8	64.6	12.2
$CHCl_3$	g	−24.6	−16.8	70.6	15.7
	ℓ	−32.1	−17.6	48.2	27.2
CS_2	g	28.0	16.0	56.8	10.8
	ℓ	21.4	15.6	36.2	18.1
Chlorine					
Cl_2	g	0.0	0.0	53.3	8.1
Cl	g	29.1	25.3	39.5	5.2
Chromium					
Cr	s	0.0	0.0	5.7	5.6
Cr_2O_3	s	−269.7	−250.2	19.4	28.4
CrO_3	s	−138.4	−119.9	17.2	—
$CrCl_2$	s	−94.6	−85.1	27.4	16.9
$CrCl_3$	s	−134.6	−118.0	30.0	21.5
Cobalt					
Co	s	0.0	0.0	7.2	6.1
CoO	s	−57.2	−51.5	12.6	—
Co_3O_4	s	−216.3	−188.0	24.5	—
$CoCl_2$	s	−77.8	−67.5	25.4	18.8
Copper					
Cu	s	0.0	0.0	8.0	5.8
Cu_2O	s	−39.8	−34.4	22.4	16.7
CuO	s	−37.1	−30.4	10.2	10.6

Appendix d: Thermodynamic Properties (Continued)

Formula	State	ΔH_f^0 (kcal/mole)	ΔG_f^0 (kcal/mole)	S^0 (cal/mole degK)	C_p^0 (cal/mole degK)
		at 298.16°K (25°C)			
Copper—cont.					
CuCl	s	−32.2	−28.4	21.9	11.6
CuCl$_2$	s	−49.2	−39.0	27.0	19.0
Cu$_2$S	s	−19.0	−20.6	28.9	18.2
CuS	s	−11.6	−11.7	15.9	11.4
CuSO$_4$	s	−184.0	−158.2	27.1	24.1
CuSO$_4$·H$_2$O	s	−259.0	−219.2	35.8	31.3
CuSO$_4$·3 H$_2$O	s	−402.3	−334.6	53.8	49.0
CuSO$_4$·5 H$_2$O	s	−544.5	−449.3	73.0	67.2
CuCO$_3$	s	−142.2	−123.8	21.0	—
Fluorine					
F$_2$	g	0.0	0.0	48.4	7.5
F	g	18.9	14.8	37.9	5.4
F$_2$O	g	5.5	9.7	59.0	—
Germanium					
Ge	s	0.0	0.0	7.4	5.6
GeO$_2$	amorphous	−128.4	−114.4	12.5	—
GeCl$_4$	ℓ	−127.1	−110.6	58.7	—
Gold					
Au	s	0.0	0.0	11.4	6.0
Au$_2$O$_3$	s	19.3	39.0	30.0	—
Au(OH)$_3$	s	−100.0	−69.4	29.0	—
AuCl$_3$	s	−28.3	−11.6	35.4	—
Helium					
He	g	0.0	0.0	30.1	5.0
Hydrogen					
H$_2$	g	0.0	0.0	31.2	6.9
H$_2$O	g	−57.8	−54.6	45.1	8.0
	ℓ	−68.3	−56.7	16.7	18.0
H$_2$O$_2$	g	−32.6	−25.2	55.6	10.3
HF	g	−64.8	−65.3	41.5	7.0
HCl	g	−22.1	−22.8	44.6	7.0
HBr	g	−8.7	−12.8	47.5	7.0
HI	g	6.3	0.4	49.3	7.0
H$_2$S	g	−4.9	−8.0	49.2	8.2
H$_2$Se	g	7.1	3.8	52.3	8.3
H$_2$Te	g	36.9	33.1	56.0	—
Iodine					
I$_2$	s	0.0	0.0	27.8	13.0
	g	14.9	4.6	62.3	8.8
I	g	25.5	16.8	43.2	5.0
ICl	g	4.2	−1.3	59.1	8.5
ICl$_3$	s	−21.1	−5.4	41.1	—

Appendix d: Thermodynamic Properties (Continued)

Formula	State	ΔH_f^0 (kcal/mole) at 298.16°K (25°C)	ΔG_f^0 (kcal/mole) at 298.16°K (25°C)	S^0 (cal/mole degK) at 298.16°K (25°C)	C_p^0 (cal/mole degK) at 298.16°K (25°C)
Iron					
Fe	s	0.0	0.0	6.5	6.0
$Fe_{0.95}O$	s (wustite)	−63.2	−57.5	14.0	11.5
Fe_2O_3	s (hematite)	−196.5	−177.1	21.5	25.0
Fe_3O_4	s (magnetite)	−267.0	−242.8	36.2	36.3
$FeCl_2$	s	−81.5	−72.2	28.6	18.2
$FeCl_3$	s	−96.8	−80.6	32.2	22.8
FeS	s (α)	−22.7	−23.3	16.1	13.1
FeS_2	s (pyrites)	−42.5	−39.8	12.7	14.8
$FeSO_4$	s	−220.5	−194.8	25.7	—
Fe_3C	s (cementite)	5.0	3.5	25.7	25.3
$FeCO_3$	s (siderite)	−178.7	−161.1	22.2	19.6
Krypton					
Kr	g	0.0	0.0	39.2	5.0
Lead					
Pb	s	0.0	0.0	15.5	6.3
PbO	s (red)	−52.3	−45.2	15.9	11.0
	s (yellow)	−51.9	−44.9	16.4	10.9
PbO_2	s	−66.3	−51.9	16.4	15.4
Pb_3O_4	s	−171.7	−143.7	50.5	35.1
PbF_2	s	−158.7	−147.5	26.4	—
$PbCl_2$	s	−85.9	−75.1	32.5	—
$PbBr_2$	s	−66.6	−62.6	38.6	19.2
PbI_2	s	−41.9	−41.5	41.8	18.5
PbS	s	−24.0	−23.6	21.8	11.8
$PbSO_4$	s	−219.9	−194.4	35.5	24.7
Lithium					
Li	s	0.0	0.0	6.7	5.6
	g	37.1	29.2	33.1	5.0
Li_2O	s	−142.4	−133.8	9.1	13.0
LiH	g	30.7	25.2	40.8	7.1
LiOH	s	−116.4	−105.5	10.2	11.9
LiF	s	−146.3	−139.6	8.6	10.0
LiCl	s	−97.7	−91.9	13.9	12.0
Li_2SO_4	s	−342.8	−316.0	29.0	—
Li_2CO_3	s	−290.5	−270.7	21.6	23.3
Magnesium					
Mg	s	0.0	0.0	7.8	5.7
MgO	s	−143.8	−136.1	6.4	8.9
$Mg(OH)_2$	s	−221.0	−199.3	15.1	18.4
MgF_2	s	−263.5	−250.8	13.7	14.7
$MgCl_2$	s	−153.4	−141.6	21.4	17.0
$MgCl_2 \cdot H_2O$	s	−231.1	−206.1	32.8	27.5
$MgCl_2 \cdot 2\,H_2O$	s	−306.0	−267.3	43.0	38.0

Appendix d: Thermodynamic Properties (Continued)

Formula	State	ΔH_f^0 (kcal/mole) at 298.16°K (25°C)	ΔG_f^0 (kcal/mole) at 298.16°K (25°C)	S^0 (cal/mole degK)	C_p^0 (cal/mole degK)
Magnesium—cont.					
$MgCl_2 \cdot 4\ H_2O$	s	−454.0	−390.5	63.1	57.7
$MgCl_2 \cdot 6\ H_2O$	s	−597.4	−505.6	87.5	75.5
$MgSO_4$	s	−305.5	−280.5	21.9	23.0
$Mg(NO_3)_2$	s	−188.7	−140.6	39.2	33.9
$MgCO_3$	s	−266.0	−246.0	15.7	18.0
Manganese					
Mn	s (α)	0.0	0.0	7.6	6.3
MnO	s	−92.0	−86.8	14.4	10.3
MnO_2	s	−124.5	−111.4	12.7	12.9
Mn_2O_3	s	−229.4	−210.8	26.4	25.8
Mn_3O_4	s	−331.4	−306.0	35.5	33.3
$Mn(OH)_2$	amorphous	−165.8	−145.9	21.1	—
MnF_2	s	−189.0	−179.0	22.2	16.2
$MnCl_2$	s	−112.0	−102.2	28.0	17.4
$MnBr_2$	s	−90.7	−87.4	33.0	—
MnI_2	s	−59.3	−59.5	36.0	—
$MnSO_4$	s	−254.2	−228.5	26.8	23.9
$MnCO_3$	s	−213.9	−195.4	20.5	19.5
Mercury					
Hg	ℓ	0.0	0.0	18.2	6.6
	g	14.5	7.6	41.8	5.0
HgO	s (red)	−21.7	−14.0	16.8	10.9
	s (yellow)	−21.6	−14.1	17.5	—
Hg_2Cl_2	s	−63.3	−50.3	46.8	24.3
$HgCl_2$	s	−55.0	−44.0	34.5	18.3
HgS	s (red)	−13.9	−11.7	18.6	11.1
	s (black)	−12.9	−11.1	19.9	11.1
Hg_2SO_4	s	−177.3	−149.1	48.0	31.5
Molybdenum					
Mo	s	0.0	0.0	6.8	5.6
MoO_2	s	−139.5	−126.9	13.6	—
MoO_3	s	−180.3	−161.9	18.7	17.6
MoS_2	s	−55.5	−53.8	15.1	15.2
Mo_2S_3	s	−102.0	−99.5	28.0	—
Neon					
Ne	g	0.0	0.0	34.9	5.0
Nickel					
Ni	s	0.0	0.0	7.2	6.2
NiO	s	−58.4	−51.7	9.2	10.6
$NiCl_2$	s	−73.0	−61.9	23.3	18.6
$NiCl_2 \cdot 6\ H_2O$	s	−505.8	−410.5	75.2	—
$NiSO_4$	s	−213.0	−184.9	18.6	33.4

Appendix d: Thermodynamic Properties (Continued)

Formula	State	ΔH_f^0 (kcal/mole) at 298.16°K (25°C)	ΔG_f^0 (kcal/mole) at 298.16°K (25°C)	S^0 (cal/mole degK)	C_p^0 (cal/mole degK)
Nickel—cont.					
$NiSO_4 \cdot 6\ H_2O$	s (blue)	−642.5	−531.0	73.1	82.0
$Ni(CO)_4$	g	−144.7	−140.3	96.0	—
Nitrogen					
N_2	g	0.0	0.0	45.8	7.0
NO	g	21.6	20.7	50.3	7.1
NO_2	g	7.9	12.3	57.3	8.9
N_2O	g	19.6	24.9	52.5	9.2
N_2O_4	g	2.2	23.4	72.7	18.5
N_2O_5	g	2.7	27.5	85.0	20.2
NH_3	g	−11.0	−3.9	46.0	8.4
HNO_3	ℓ	−41.6	−19.3	37.2	26.3
NH_4NO_3	s	−87.4	−44.0	36.1	33.3
NH_4Cl	s	−75.2	−48.5	22.6	20.1
$(NH_4)_2SO_4$	s	−282.2	−215.6	52.6	44.8
Oxygen					
O_2	g	0.0	0.0	49.0	7.0
O_3	g	34.1	39.0	57.1	9.4
Phosphorus					
P	s (white)	0.0	0.0	9.8	5.7
	s (red)	−4.2	−2.9	5.4	5.1
	g	75.2	66.5	39.0	5.0
P_4	g	14.1	5.8	66.9	16.0
P_4O_{10}	s	−713.2	−644.8	54.7	50.6
PH_3	g	1.3	3.2	50.2	8.9
H_3PO_4	s	−305.7	−267.5	26.4	25.4
PCl_3	g	−68.6	−64.0	74.5	17.2
PCl_5	g	−89.6	−72.9	87.1	27.0
$POCl_3$	g	−133.5	−122.6	77.8	20.3
PBr_3	g	−33.3	−38.9	83.2	18.2
Platinum					
Pt	s	0.0	0.0	10.0	6.3
$Pt(OH)_2$	s	−87.2	−68.2	26.5	—
Potassium					
K	s	0.0	0.0	15.2	7.0
K_2O	s	−86.4	−77.0	23.5	—
KH	g	30.0	25.1	47.3	—
KF	s	−134.5	−127.4	15.9	11.7
KCl	s	−104.2	−97.6	19.8	12.3
$KClO_3$	s	−93.5	−69.3	34.2	24.0
KBr	s	−93.7	−90.6	23.0	12.8
$KBrO_3$	s	−79.4	−58.2	35.6	25.1
KI	s	−78.3	−77.0	24.9	13.2

Appendix d: Thermodynamic Properties (Continued)

Formula	State	ΔH_f^0 (kcal/mole)	ΔG_f^0 at 298.16°K (25°C)	S^0 (cal/mole degK)	C_p^0
Potassium—cont.					
KIO_3	s	-121.5	-101.7	36.2	25.4
K_2SO_4	s	-342.7	-314.6	42.0	31.1
KNO_3	s	-117.8	-94.0	31.8	23.0
$KMnO_4$	s	-194.4	-170.6	41.0	28.5
KOH	s	-101.8	—	—	—
Radon					
Rn	g	0.0	0.0	42.1	5.0
Rubidium					
Rb	s	0.0	0.0	18.3	7.3
RbF	s	-131.3	-124.0	18.0	12.6
$RbCl$	s	-102.9	-96.0	21.9	12.3
$RbBr$	s	-93.0	-90.4	25.9	12.7
RbI	s	-78.5	-77.8	28.2	12.5
Selenium					
Se	s (grey)	0.0	0.0	10.1	6.1
	g	54.3	44.7	42.2	5.0
Se_2	g	34.9	23.0	60.2	8.5
SeF_6	g	-267.0	-243.0	75.0	26.4
Silicon					
Si	s	0.0	0.0	4.5	4.8
SiO_2	s (quartz)	-217.7	-204.8	10.0	10.6
SiH_4	g	8.2	13.6	48.9	10.2
SiF_4	g	-386.0	-375.9	67.5	17.6
$SiCl_4$	g	-157.0	-147.5	79.0	21.6
SiC	s (β, cubic)	-15.6	-15.0	4.0	6.4
	s (α, hexagonal)	-15.0	-14.4	3.9	6.4
Silver					
Ag	s	0.0	0.0	10.2	6.1
Ag_2O	s	-7.3	-2.6	29.1	15.7
AgF	s	-48.5	-44.2	20.0	—
$AgCl$	s	-30.4	-26.2	23.0	12.1
$AgBr$	s	-23.8	-22.9	25.6	12.5
AgI	s	-14.9	-15.8	27.3	13.0
Ag_2S	s (α, rhombic)	-7.6	-9.6	34.8	18.0
Ag_2SO_4	s	-170.5	-147.2	47.8	31.4
$AgNO_3$	s	-29.4	-7.7	33.7	22.2
Ag_2CO_3	s	-120.9	-104.5	40.0	26.8
Sodium					
Na	s	0.0	0.0	12.2	6.8
Na_2O	s	-99.4	-90.0	17.4	16.3
Na_2O_2	s	-123.0	-107.8	22.6	—

Appendix d: Thermodynamic Properties (Continued)

Formula	State	ΔH_f^0	ΔG_f^0	S^0	C_p^0
		(kcal/mole)		(cal/mole degK)	
		at 298.16°K (25°C)			
Sodium—cont.					
NaH	g	29.9	24.8	44.9	7.0
NaOH	s	−102.0	−91.0	15.3	14.3
NaF	s	−136.5	−129.3	12.3	11.0
NaCl	s	−98.6	−92.2	17.4	11.9
NaBr	s	−86.0	−82.9	20.0	12.5
NaI	s	−68.8	−67.5	21.8	—
Na_2SO_4	s	−330.9	−302.8	35.7	30.5
$Na_2SO_4 \cdot 10\ H_2O$	s	−1033.5	−870.9	141.7	140.4
$NaNO_3$	s	−111.5	−87.5	27.8	22.2
Na_2CO_3	s	−270.3	−250.4	32.5	26.4
$Na_2CO_3 \cdot 10\ H_2O$	s	−975.6	—	—	128.0
$NaHCO_3$	s	−226.5	−203.6	24.4	20.9
Strontium					
Sr	s	0.0	0.0	13.0	6.0
SrO	s	−141.1	−133.8	13.0	10.8
$SrCl_2$	s	−198.0	−186.7	28.0	18.9
$SrSO_4$	s	−345.3	−318.9	29.1	—
$SrCO_3$	s	−291.2	−271.9	23.2	19.5
$Sr(OH)_2$	s	−229.3	—	—	—
SrF_2	s	−171.1	—	—	19.0
SrI_2	s	−135.5	—	—	19.5
Sulfur					
S	s (rhombic)	0.0	0.0	7.6	5.4
	s (monoclinic)	0.08	—	—	—
	g	66.6	56.9	40.1	5.7
S_8	g	24.4	11.9	103.0	37.4
SO_2	g	−70.9	−71.7	59.3	9.5
SO_3	g	−94.6	−88.7	61.3	12.1
H_2SO_4	ℓ	−194.5	−164.9	37.5	33.2
SF_6	g	−289.0	−264.2	69.7	23.2
Tellurium					
Te	s	0.0	0.0	11.9	6.1
TeO_2	s	−77.1	−64.6	19.0	—
TeF_6	g	−315.0	−292.0	80.7	—
Tin					
Sn	s (white)	0.0	0.0	12.3	6.5
	s (grey)	−0.5	0.0	10.5	6.2
SnO	s	−68.3	−61.4	13.5	10.6
SnO_2	s	−138.8	−124.2	12.5	12.6
$SnCl_4$	ℓ	−122.2	−105.2	61.8	39.5
SnS	s	−24.0	−23.5	18.4	11.8
SnS_2	s	−40.0	−38.0	20.9	16.8

Appendix d: Thermodynamic Properties (Continued)

Formula	State	ΔH_f^0 (kcal/mole)	ΔG_f^0 (kcal/mole)	S^0 (cal/mole degK)	C_p^0 (cal/mole degK)
		at 298.16°K (25°C)			
Titanium					
Ti	s	0.0	0.0	7.2	6.0
TiO_2	s (rutile)	−218.0	−203.8	12.0	13.2
$TiCl_4$	ℓ	−191.5	−175.3	59.6	37.5
TiN	s	−73.0	−66.1	7.2	8.9
TiC	s	−54.0	−53.0	5.8	8.0
Tungsten					
W	s	0.0	0.0	8.0	6.0
WO_3	s (yellow)	−200.8	−182.5	19.9	19.5
Uranium					
U	s	0.0	0.0	12.0	6.6
UO_2	s	−270.0	−257.0	18.6	15.3
UF_6	g	−505.0	−485.0	90.8	—
Vanadium					
V	s	0.0	0.0	7.0	5.8
V_2O_3	s	−290.0	−271.0	23.6	24.8
VO_2	s	−172.0	−159.0	12.3	14.2
V_2O_5	s	−373.0	−344.0	31.3	31.0
VCl_2	s	−108.0	−97.0	23.2	17.3
VCl_3	s	−137.0	−120.0	31.3	22.3
Xenon					
Xe	g	0.0	0.0	40.5	5.0
Zinc					
Zn	s	0.0	0.0	10.0	6.1
ZnO	s	−83.2	−76.1	10.4	9.6
$Zn(OH)_2$	s	−153.5	—	—	17.3
$ZnCl_2$	s	−99.2	−88.3	26.6	17.0
$ZnBr_2$	s	−78.5	−74.6	33.1	—
ZnI_2	s	−49.7	−49.9	38.5	—
ZnS	s (zinc blende)	−49.2	−48.1	13.8	11.0
	s (wurtzite)	−46.0	—	—	—
$ZnSO_4$	s	−234.9	−209.0	28.6	28.0
$ZnSO_4 \cdot H_2O$	s	−311.8	−270.6	33.1	34.7
$ZnSO_4 \cdot 6\ H_2O$	s	−663.8	−555.6	86.9	85.5
$ZnSO_4 \cdot 7\ H_2O$	s	−735.6	−612.6	92.9	91.6
$ZnCO_3$	s	−194.3	−174.9	19.7	19.1
Zirconium					
Zr	s	0.0	0.0	9.2	6.2
ZrO_2	s	−258.2	−244.4	12.0	13.4
$ZrCl_4$	s	−234.7	−213.4	44.5	28.7

Appendix e: *Molecular Dipole Moments and Bond Lengths (Gas Phase)*

$$(D = \text{Debye Units} = 10^{-18}\ \text{esu-cm})$$

Alkali Halides	μ (D)	r_e (Å)
LiF	6.28	1.56
LiCl	7.09	2.02
LiBr	7.23	2.17
LiI	7.39	2.39
NaF	8.12	1.93
NaCl	8.97	2.36
NaBr	9.09	2.50
NaI	9.21	2.71
KF	8.55	2.17
KCl	10.24	2.67
KBr	10.60	2.82
KI	10.86	3.05
RbF	8.51	2.27
RbCl	10.48	2.79
RbBr	10.78	2.94
RbI	11.26	3.18
CsF	7.85	2.35
CsCl	10.36	2.91
CsBr	10.85	3.07
CsI	11.50	3.32

Other Diatomics	μ (D)	r (Å)
LiH	5.88	1.61
OH	1.66	0.98
HF	1.91	0.92
HCl	1.07	1.27
HBr	0.79	1.42
HI	0.38	1.62
ClF	0.88	1.63
BrF	1.29	1.76
IF	1.6	1.99
BrCl	0.57	2.14
ICl	0.6	2.32
IBr	1.21	
TlF	4.2	2.08
TlCl	4.4	2.48
NO	0.15	1.15
CO	0.13	1.13
ClO	1.70	1.55
SrO	8.90	1.92
BaO	7.93	1.94

Polyatomics	μ (D)	r (Å)	θ
H_2O	1.85	0.96	104.5°
H_2S	0.95	1.33	93.3°
H_2Se	0.4	1.46	91.0°
H_2Te	<0.2	1.7	89.5°
F_2O	0.2	1.42	103.2°
NH_3	1.47	1.01	107.3°
NF_3	0.23	1.37	102.1°
NCl_3	0.6	~1.76	
PH_3	0.55	1.42	93.1°
PF_3	1.0	1.54	104°
PCl_3	0.8	2.04	100.5°
$SbCl_3$	3.9	2.36	95.2
O_3	0.53	1.28	116.8°
NO_2	0.3	1.19	134.1°
N_2O	0.18	1.13/1.19	180°
SO_2	1.61	1.43	119.5°
ClO_2	0.78	1.49	119°
ClF_3	0.6	1.60/1.70	87.5°/180°
BrF_3	1.0	1.72/1.81	86.2°/180°
BrF_5	1.51	1.68/1.80	85°/90°

Polyatomics	μ (D)
CH_3F	1.82
CH_3Cl	1.94
CH_3Br	1.79
CH_3I	1.64
CH_2F_2	1.96
CH_2Cl_2	1.60
CH_2Br_2	1.5
CH_2I_2	1.11
CHF_3	1.60
$CHCl_3$	1.00
$CHBr_3$	1.00
CHI_3	0.8
H_2CO	2.30
Cl_2CO	1.19
Cl_2CS	0.28
HCOOH	1.52
CH_3COOH	1.75
CH_3NO_2	3.46
HCN	2.95

Appendix f: Average Bond Energies and Bond Lengths in Molecules and Ions with Formula BA$_n$**

$(Z_A)_n-Z_B$	Formula	D_0(av.) (kcal/mole)	r_0 (Å)
$n = 1$: Diatomics, AB			
1—1	H$_2$	104	0.76
	H$_2^+$	61	1.06
1—3	LiH	58	1.61
1—4	BeH	53	1.35
1—5	BH	79	1.24
1—6	CH	81	1.12
	CH$^+$	83	1.13
1—7	NH	74, 85	1.05
1—8	OH	102	0.97
	OH$^+$	109	1.03
	OH$^-$	110	0.98
1—9	HF	135	0.92
1—11	NaH	48	1.89
1—12	MgH	47	—
1—13	AlH	68	1.65
1—14	SiH	75	1.52
1—16	SH	82	1.35
1—17	HCl	103	1.27
	HCl$^+$	104	1.32
1—35	HBr	87	1.42
1—53	HI	71	1.62
2—2	He$_2^+$	71	1.08
3—3	Li$_2$	25	2.68
3—9	LiF	137	1.56
3—17	LiCl	112	—
3—35	LiBr	100	—
3—53	LiI	85	—
4—4	Be$_2$	17	—
4—8	BeO	100	1.33
4—9	BeF	124	1.36
4—17	BeCl	137	—
5—5	B$_2$	69	1.59
5—7	BN	152	1.28
5—8	BO	173	1.21
5—9	BF	186	1.27
5—16	BS	119	1.61
5—17	BCl	128	1.72
5—35	BBr	128	1.88
5—53	BI	104	—
6—6	C$_2$	144	1.24
6—7	CN	188	1.18
6—8	CO	256	1.13

Appendix f: Average Bond Energies and Bond Lengths in Molecules and Ions with Formula BA_n (Continued)*

$(Z_A)_n - Z_B$	Formula	D_0(av.) (kcal/mole)	r_0 (Å)
	CO^+	228	1.12
6—9	CF	127	1.27
6—14	CSi	104	—
6—15	CP	140	1.56
6—16	CS	175	1.53
6—35	CBr	96	
7—7	N_2	225	1.10
	N_2^+	146 or 201	1.12
7—8	NO	162	1.15
	NO^+	252	1.06
7—9	NF	63	1.32
7—14	SiN	105	1.57
7—15	NP	139	1.49
7—16	NS	116	1.50
7—35	NBr	67	—
8—8	O_2	118	1.21
	O_2^+	149	1.12
8—12	MgO	94	—
8—13	AlO	120	1.62
8—14	SiO	192	1.51
8—15	PO	144	1.47
8—16	SO	124	1.49
8—17	ClO	64	1.55
8—35	BrO	46	2.29
8—53	IO	43	—
9—9	F_2	37	1.44
9—11	NaF	114	—
9—12	MgF	107	—
9—13	AlF	159	1.65
9—14	SiF	130	1.60
9—17	ClF	61	1.63
9—35	BrF	57	1.76
11—11	Na_2	18	3.08
11—17	NaCl	98	2.36
11—35	NaBr	87	2.50
11—53	NaI	73	2.71
12—17	MgCl	63	1.8
12—35	MgBr	59	—
13—13	Al_2	40	—
13—17	AlCl	118	2.14
13—35	AlBr	107	2.30
13—53	AlI	88	—

Appendix f: Average Bond Energies and Bond Lengths in Molecules and Ions with Formula BA_n*(Continued)*

$(Z_A)_n - Z_B$	Formula	D_0(av.) (kcal/mole)	r_0 (Å)
14—14	Si_2	76	2.25
14—16	SiS	151	1.93
14—17	SiCl	77	—
14—35	SiBr	70	—
15—15	P_2	117	1.90
15—16	PS	70	—
16—16	S_2	101	1.89
17—17	Cl_2	57	2.00
	Cl_2^+	101	1.89
17—35	BrCl	53	2.14
17—53	ICl	51	2.32
33—33	As_2	92	—
34—34	Se_2	66	2.15
35—35	Br_2	46	2.29
35—53	BrI	43	—
50—50	Sn_2	47	—
51—51	Sb_2	72	—
53—53	I_2	36	2.66

$n = 2$: Triatomics, BA_2

$(Z_A)_n - Z_B$	Formula	D_0(av.) (kcal/mole)	r_0 (Å)
$(1)_2$—6	CH_2	91	1.03
$(1)_2$—7	NH_2	88	1.02
$(1)_2$—8	H_2O	111	0.96
$(1)_2$—16	H_2S	88	1.33
$(1)_2$—34	H_2Se	76	1.46
$(1)_2$—52	H_2Te	64	1.7
$(8)_2$—6	CO_2	192	1.16
$(8)_2$—7	NO_2	73	1.19
$(8)_2$—8	O_3	72	1.28
$(8)_2$—16	SO_2	127	1.43
$(8)_2$—17	ClO_2	60	1.48
$(9)_2$—6	CF_2	123	1.30
$(9)_2$—7	NF_2	70	—
$(9)_2$—8	OF_2	51	1.42
$(9)_2$—14	SiF_2	147	—
$(9)_2$—36	KrF_2	12	1.87
$(9)_2$—54	XeF_2	30	1.98
$(16)_2$—6	CS_2	128	1.55
$(17)_2$—8	OCl_2	49	1.70
$(17)_2$—16	SCl_2	65	1.99
$(17)_2$—34	$SeCl_2$	60	—

Appendix f: Average Bond Energies and Bond Lengths in Molecules and Ions with Formula BA$_n$(Continued)*

$(Z_A)n$—Z_B	Formula	D_0(av.) (kcal/mole)	r_0 (Å)
$(35)_2$—34	$SeBr_2$	56	—
$n = 3$: Tetratomics, BA$_3$			
$(1)_3$—6	CH_3	98	1.08
$(1)_3$—7	NH_3	93	1.01
$(1)_3$—15	PH_3	77	1.44
$(1)_3$—33	SbH_3	71	1.52
$(1)_3$—51	AsH_3	61	1.71
$(8)_3$—54	XeO_3	21	1.76
$(9)_3$—5	BF_3	161	1.30
$(9)_3$—7	NF_3	67	1.37
$(9)_3$—13	AlF_3	147	—
$(9)_3$—15	PF_3	117	1.54
$(9)_3$—17	ClF_3	41	1.60, 1.70
$(9)_3$—33	AsF_3	116	1.71
$(9)_3$—35	BrF_3	48	1.72, 1.81
$(9)_3$—51	SbF_3	75	2.36
$(17)_3$—5	BCl_3	116	1.75
$(17)_3$—13	$AlCl_3$	111	—
$(17)_3$—15	PCl_3	76	2.04
$(17)_3$—31	$GaCl_3$	67	—
$(17)_3$—33	$AsCl_3$	74	2.16
$(17)_3$—51	$SbCl_3$	75	2.36
$(35)_3$—5	BBr_3	88	1.87
$(35)_3$—13	$AlBr_3$	87	—
$(35)_3$—15	PBr_3	63	2.18
$(35)_3$—31	$GaBr_3$	72	—
$(35)_3$—33	$AsBr_3$	61	2.31
$(35)_3$—49	$InBr_3$	69	—
$(35)_3$—51	$SbBr_3$	63	2.52
$n = 4$: Pentatomics, BA$_4$			
$(1)_4$—6	CH_4	99	1.09
$(1)_4$—14	SiH_4	77	1.48
$(1)_4$—32	GeH_4	69	1.53
$(1)_4$—50	SnH_4	60	1.70
$(8)_4$—76	OsO_4	123	1.66
$(9)_4$—6	CF_4	117	1.32
$(9)_4$—14	SiF_4	140	1.55
$(9)_4$—16	SF_4	82	1.55, 1.65
$(9)_4$—54	XeF_4	31	1.94
$(17)_4$—6	CCl_4	78	1.77
$(17)_4$—14	$SiCl_4$	96	2.01

Appendix f: Average Bond Energies and Bond Lengths in Molecules and Ions with Formula BA$_n$ (Continued)*

$(Z_A)_n$—Z_B	Formula	D_0(av.) (kcal/mole)	r_0 (Å)
$(17)_4$—32	$GeCl_4$	81	2.08
$(17)_4$—50	$SnCl_4$	75	2.30
$(35)_4$—6	CBr_4	65	1.94
$(35)_4$—14	$SiBr_4$	79	2.15
$(35)_4$—32	$GeBr_4$	67	2.29
$(35)_4$—50	$SnBr_4$	65	2.44
$n = 5$: BA$_5$			
$(9)_5$—15	PF_5	91	1.53, 1.57
$(9)_5$—35	BrF_5	45	1.68, 1.78
$(9)_5$—53	IF_5	63	
$(17)_5$—15	PCl_5	62	2.04, 2.19
$(17)_5$—51	$SbCl_5$	61	2.29, 2.34
$n = 6$: BA$_6$			
$(9)_6$—16	SF_6	78	1.58
$(9)_6$—34	SeF_6	72	1.70
$(9)_6$—52	TeF_6	80	1.84
$(9)_6$—54	XeF_6	32	1.91
$n = 7$: BA$_7$			
$(9)_7$—53	IF_7	55	—

* In order of increasing Z_A.

Appendix g: Average Bond Energies (kcal/mole) *and Bond Lengths* (Å) *in Polyatomic Molecules*

Bond	Average Energy	Average Length
H—C	98.8	1.10 (monosubstituted alkanes)
H—N	93.4	1.00
H—O	110.6	0.97
C—C	82.6	1.54 ⎱ aromatic, 1.39
C=C	145.8	1.34 ⎰
C≡C	199.6	1.20
C—N	72.8	1.47
C=N	147	
C≡N	212.6	1.16
C—O	85.5	1.43 (alcohols, ethers)
		1.36 (carboxylic acids)
C=O	178	1.22 (aldehydes and ketones)
	192.1	1.16 (CO_2)
C—F	116	1.38 (alkanes) 1.33 (others)
C—Cl	81	1.77 (alkanes) 1.72 (alkenes)
		1.70 (aromatics)
C—Br	68	1.94 (alkanes) 1.89 (alkenes)
		1.85 (aromatics)
C—I	51	2.21
N—O	53	1.36
N=O	145	1.21
N—F	65	1.36
N—Cl	46	1.75
XeO	27	1.76
XeF	32	1.97

Appendix h: Integration

In scientific problems it often becomes necessary to measure the area under a curve. We first met this in Section 7-2(f) when discussing pressure–volume work. We found that the work done by an expanding gas can easily be calculated if the pressure remains constant

$$w(P \text{ constant}) = P\Delta V$$

Graphically, the work is numerically equivalent to the area under a plot of P against V between the initial volume V_1 and the final volume V_2 as illustrated in Figure 7-5(a). We want to examine now what happens if, as in Figure 7-5(b) and Figure h-1, the gas pressure does not remain constant, but changes in some regular way during the expansion. Is the area under the curve numerically equal to the work done in this case also? We can show that it is by the following reasoning.

Let's begin by drawing a rectangular path from the initial state P_0V_0 to the final state P_fV_f that will let us calculate the work done. This path, shown on Figure h-2, involves an increase in pressure at constant volume during which $w = 0$ since $\Delta V = 0$. The second step is a constant pressure expansion with the work given by

$$w = P_f(V_f - V_0) = P_f\Delta V$$

Clearly this is not the correct answer. We have done our calculation on a path much different from the desired one. We can come closer to the real path by the process illustrated in Figure h-3. Here we have divided the volume change into two equal parts, $\Delta V'$, and drawn constant pressure lines to the final pressure points in both halves. Now the work done is given by a sum of that done in each of the two steps

$$w = P_1\Delta V' + P_f\Delta V'$$

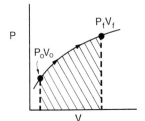

P

P_fV_f

P_0V_0

V

Figure h-1

P

P_f

V

Figure h-2

We still do not have a path that agrees with the desired one, but obviously we have come closer.

If we continue this process, taking smaller and smaller volume intervals, we will come closer and closer to the correct result while still allowing the work done in each expansion to be calculated. Figure h-4 shows divisions into six equal volume increments, $\Delta V''$ with a resulting total work

$$w = P_1 \Delta V'' + P_2 \Delta V'' + \cdots$$

$$= \sum_{i=1}^{6} P_i \Delta V''$$

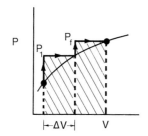

Figure h-3   Figure h-4

The Greek letter sigma, Σ, indicates a summation of the products $P_i \Delta V''$ for each value of i. In our particular case, i can have the values 1, 2, 3, $\cdots$ 6 as shown above and below the sigma sign.

Our aim is now clear. If we continue reducing the size of ΔV, in the limit that ΔV becomes infinitesimally small, we will have a path that is indistinguishable from the desired one while allowing us to calculate the work done during each tiny expansion. This problem is dealt with directly in integral calculus. An infinitesimal change in volume is given the notation dV and the sum of the areas of each narrow rectangle which will give us the work done is indicated by

$$w = \int_{V_0}^{V_f} P dV$$

called the "integral of PdV between the limits V_0 and V_f." The integral sign is related to the letter "S," implying a summation of all the little areas PdV.

Two cases of particular interest in thermodynamics permit a simpler solution of this equation.

(1) Pressure constant: If the pressure is constant it can be removed outside the integral sign

$$w = P \int_{V_0}^{V_f} dV$$

with the simple result, already familiar to us, that

$$w = P(V_f - V_0) = P\Delta V$$

(2) Pressure not constant; ideal gas expanded at equilibrium pressure: If the gas used is ideal (all gases are ideal at low pressures), we can carry out the expansion so that the pressure is inversely related to the volume at every instant,

$$P = \frac{nRT}{V}$$

Substitution in the integral gives us

$$w = nRT \int_{V_0}^{V_f} \frac{dV}{V}$$

The result, stated here without proof, is derived in every introductory calculus class to be

$$w = nRT(\log_e V_f - \log_e V_0)$$

$$w = nRT \log_e \left(\frac{V_f}{V_0}\right)$$

We may conclude by generalizing these two results to any function y of a variable x

$$y = c \text{ (constant)} \quad \int_{x_1}^{x_2} c\,dx = c \int_{x_1}^{x_2} dx = c(x_2 - x_1) = c\Delta x$$

$$y = \frac{1}{x} \quad \int_{x_1}^{x_2} \frac{dx}{x} = \log_e x_2 - \log_e x_1 = \log_e \left(\frac{x_2}{x_1}\right)$$

The exponential e

We just noted that the integral of a commonly occurring function $1/x$ is given by the logarithm to the base "e" of x. What is "e?" This number arises from further consideration of the integral calculus. Any number y can be expressed as some other number a raised to a power, x.

$$y = a^x$$

There is one particular value of a that has great importance because it has the unique property of making y the integral of itself. This particular value of a is called e.

$$\int y\,dx = \int e^x dx = e^x = y$$

in which

$$e = 2.7182 \ldots$$

In general, if

$$y = e^x$$

then the number x is called the logarithm of y to the base e:

$$x = \log_e y$$

or, in a usual notation

$$x = \ell n\ y$$

The symbol ℓn is used to represent the natural logarithm, or the logarithm to the base e. This logarithm is simply related to the more common logarithm to the base 10:

$$\log_{10} y = 0.4343 \log_e y$$

$$\log_e y = 2.303 \log_{10} y$$

When the function $y = e^x$ is plotted against x, the solid curve shown in Figure h-5 results. "Exponential" curves such as these are frequently encountered in chemistry and physics, especially when some form of growth is being considered. This is so because the slope, or rate of increase of y, is always proportional to the value of y itself. Thus the curve is a regularly increasing one. As an example we might expect population growth to follow an exponential rise since the number of births probably increases in the same proportion as the population itself increases.

Of course, the function e^{-x} is simply related to the e^x curve. If $y = e^x$,

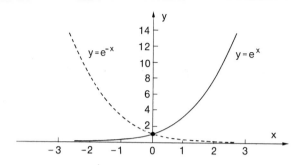

Figure h-5

then $1/y = e^{-x}$. So the value of e^{-x} for any particular value of x is just the reciprocal of e^{+x}. The dashed curve in Figure h-5 shows a plot of $y' = e^{-x}$.

answers to odd-numbered problems

Chapter 1

1. (a) A girl's blush is caused by increased blood flow in the skin. This could be applied to the problem at hand by coloring the animal's skin blue, placing veins or arteries in the skin surface that become more obvious when the animal is active, and by placing a yellow (or green) dye in his blood.

3. $+22°C$, $-83°C$, $195°K$

5. Elements: $N_2(g)$, $N_2(\ell)$, C(graphite), C(diamond)
Compounds: $NH_3(g)$, $CO_2(s)$

7. (a) 1, (b) 2, (c) 2, (d) 1, (e) 1, (f) 2, (g) 4

9. 17.03, 32.05, 28.01, 47.01, 63.01, 60.06, 194.19

11. $4.069 \cdot 10^{23}$

13. (b) 3.60 gms H_2O

15. (a) $Fe_2O_3(s) + \frac{1}{3}CO(g) \longrightarrow \frac{2}{3}Fe_3O_4(s) + \frac{1}{3}CO_2(g)$; (b) 4.17 moles

17. (b) 19.8 g

19. 0.112 g

21. (b) 11.1 pounds

23. C_2H_2 (acetylene)

25. (a) 22.414 ℓ-atm

27. (a) B; (b) can't tell; (c) A; (d) can't tell; (e) 2.1 (two sig. figs.)

Chapter 2

1. 13.60 ev

3. (a) H atom has only one electron to remove; (b) Li and Na each have only one electron at large radius. Hence E_2 must refer to removal of an electron of lower quantum number, which is more tightly bound.

5. (b) radii: (1.00), 0.66, 0.66, 0.13, 0.12

7. r_m/r_H: (1.00), 0.50, 0.33, 0.25, 0.20

9. r/r_{Li}: 0.40, 0.22, (1.00), 0.58, 0.65, etc.

11. Fe^{+2}, 555 kcal/mole; Fe^{+3}, 1262 kcal/mole

13.   :S̈:H   :S̈:S̈:   S—H   S—S

           H   H H | | |
                                     H   H H

       H_2S H_2S_2

15. The eight sulfur atoms are arranged in a ring.

17.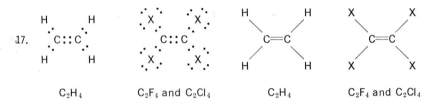

C_2H_4 C_2F_4 and C_2Cl_4 C_2H_4 C_2F_4 and C_2Cl_4

19. $MgCl_2(s) \longrightarrow Mg^{+2}(aq) + 2\ Cl^-(aq)$; $[Mg^{+2}] = 0.25\ M$, $[Cl^-] = 0.50\ M$

21. $[Ca^{+2}] = 0.050\ M$; $[Sr^{+2}] = 0.10\ M$; $[Cl^-] = 0.30\ M$

23. $Ba^{+2}(aq) + SO_4^{-2}(aq) \longrightarrow BaSO_4(s)$
$[Ba^{+2}] = 1.0 \cdot 10^{-5}\ M$; $[SO_4^{-2}] = 1.0 \cdot 10^{-5}\ M$; $[Mg^{+2}] = 0.10\ M$, $[Cl^-] = 0.20\ M$

Chapter 3

1. Yes. That the glass is open to the atmosphere is not important as long as the tea is consumed within a short time compared to the rate of evaporation and of warming of the tea.

3. No. It is a steady state temperature, balanced between heat generation in the cylinders and heat removal via the water cooling system.

5. As humidity increases, the perspiration approaches equilibrium with water vapor in air. At equilibrium, condensation occurs as rapidly as evaporation. With no net evaporation, there is no cooling.

7. $K = \dfrac{[N_2]^3[H_2O]^4}{[N_2H_4]^2[NO_2]^2}$

9. $K = \dfrac{[SO_3]}{[SO_2][O_2]^{1/2}}$ or, equally well, $K' = \dfrac{[SO_3]^2}{[SO_2]^2[O_2]}$ where $K' = K^2$

11. $Cu^{+2}(aq) + 4\ NH_3(aq) = Cu(NH_3)_4^{+2}(aq)$; $K = \dfrac{[Cu(NH_3)_4^{+2}]}{[Cu^{+2}][NH_3]^4}$

Chapter 4

1. $K_c = 2.7 \cdot 10^{-4}$ mole/liter

3. (a) 0.61 torr; (b) A would be empty, B would contain $\frac{1}{2}$ mole $CaCl_2$, $\frac{1}{2}$ mole

$CaCl_2 \cdot 2\,H_2O$, C would be unaffected, the water vapor pressure under the bell jar would be 0.61 torr.

5. 163 ml

7. $K_{sp} = 1.76 \cdot 10^{-6}$

9. $[Ag^+] = [IO_3^-] = 1.7 \cdot 10^{-4}\,M$

11. $[Ba^{+2}] = 6.9 \cdot 10^{-4}\,M$, $[IO_3^-] = 1.4 \cdot 10^{-3}\,M$

13. $[Ag^+] = 3.0 \cdot 10^{-7}\,M$, $[IO_3^-] = 0.10\,M$

15. $[Ag^+] = 1.0 \cdot 10^{-4}\,M$, $[IO_3^-] = 3.0 \cdot 10^{-4}\,M$

17. $[Ca^{+2}] = 2.5 \cdot 10^{-4}\,M$, $[SO_4^{-2}] = 0.10\,M$, $[Cl^-] = 0.20\,M$, $[Na^+] = 0.40\,M$, wt = 13.6 g $CaSO_4$

19. $[Ca^{+2}] = 1.2 \cdot 10^{-8}\,M$, $[SO_4^{-2}] = 0.20\,M$, $[Cl^-] = 0.20\,M$, $[Na^+] = 1.40\,M$, $[CO_3^{-2}] = 0.40\,M$, ppt = 0.10 mole $CaCO_3$

21. NH_3, 39.8 M; SO_2, 22.9 M

23. (a) $5.17 \cdot 10^{-4}$, (b) $5.51 \cdot 10^{-4}$, (c) 6%

25. 1.4 ml

27. $S = 1.78 \cdot 10^{-5}$

Chapter 5

1. $[H^+] = 2 \cdot 10^{-3}\,M$, $[OH^-] = 5 \cdot 10^{-12}\,M$

3. $[H^+] = 2 \cdot 10^{-3}\,M$, $[OH^-] = 5.7 \cdot 10^{-13}\,M$

5. Basic

7. $[H^+] = 0.40\,M$, $[OH^-] = 2.5 \cdot 10^{-14}\,M$

9. pH = 2.7, pOH = 11.3

11. pH = pOH = 5.52

13. 2.64

15. pH = 3.7, pOH = 10.3

17. $1.5 \cdot 10^{-10}$

19. pH = 8.02, pOH = 3.02

21. $[H^+] = 0.023$

23. (b) $K_a = 1.23 \cdot 10^{-7}$; (c) $[OH^-] = 1.10 \cdot 10^{-4}$

25. HF – F^-; benzoic acid–benzoate; hypobromous acid–hypobromite

27. 13.7 ml

29. $[HP] = 0.034\,M$; $[HP \cdot hy] = 0.051\,M$; $[P^-] = 0.014\,M$; $[P \cdot hy^-] = 8.0 \cdot 10^{-4}\,M$

Chapter 6

1. Ag is the cathode, Cd the anode; electrons flow from Cd to Ag.

3. 1.12 g Cd; 2.16 g Ag

5. Left electrode is cathode; cell voltage $= 0.32$ volt

7. Both Cd and Ni electrodes are anodes; $\Delta\varepsilon^0 = 2.25$ volts

9. 0.466 volt; reaction must be slow

11. I^-: IO_3^-, 0.66 volt; $Cr_2O_7^{-2}$; Cl_2; BrO_3^-; Ce^{+4}; MnO_4^-; H_2O_2, 1.23 volts
    I_3^-: Sn^{+2}, 0.39 volt; Cr^{+2}; Zn, 1.30 volts

13. $2.5 \cdot 10^{+47}$

15. (a) $+0.85$; (c) $+0.06$; (d) $K = 1.1 \cdot 10^{+2}$

17. $+1$, $+5$, $+7$, $+7$

19. 0, $+3$, -2

21. (a) 8, 7, 6, 4, 3, 2, 0; (b) 7, 6, 2; (c) H_2O_2, MnO_4^-

23. $VO_2^+ + 2\,H^+ + e^- \longrightarrow VO^{+2} + H_2O$
    $Ti^{+3} + 2\,H_2O \longrightarrow TiO_2^{+2} + 4\,H^+ + 3\,e^-$

25. $Ti^{+3} + RuCl_5^{-2} + 2\,H_2O \longrightarrow Ru + TiO_2^{+2} + 5\,Cl^- + 4\,H^+$

27. (a) $4\,MnO_4^- + 4\,H^+ \longrightarrow 4\,MnO_2 + 3\,O_2 + 2\,H_2O$
    (b) $2\,Cl_2 + ClO_3^- + 2\,H_2O \longrightarrow 5\,ClO^- + 4\,H^+$

29. (a) $IO_3^- + 8\,I^- + 6\,H^+ \longrightarrow 3\,I_3^- + 3\,H_2O$
    (b) $2\,ClO_2 + 2\,OH^- \longrightarrow ClO_2^- + ClO_3^- + H_2O$

31. (a) and (b): $2\,CuFeS_2 + \frac{9}{2}\,O_2 \longrightarrow Cu_2S + Fe_2O_3 + 3\,SO_2$

Chapter 7

1. (a) $t + r$; (b) v; (c) $t + v$; (d) $t + r$

3. (a) 4.91 cal/mole-degree; (b) 1.93 cal/mole-degree

5. (a) exothermic; (b) endothermic; (c) exothermic; (d) thermoneutral

7. (a) (i) heat energy $\longrightarrow$ light energy; (ii) light $\longrightarrow$ chemical (potential) energy; (iii) chemical $\longrightarrow$ chemical energy; (iv) chemical $\longrightarrow$ chemical energy; (v) chemical $\longrightarrow$ gravitational (potential) energy; (vi) chemical $\longrightarrow$ heat and light energy; (b) All

9. (a) -0.100 kcal; (b) -87.0 kcal/mole

11. (a) 592 cal; (b) -592 cal; (c), (d), and (e) 0; (f) $+680$ cal

13. $+9.31$ kcal

15. -155.7 kcal

17. (a) 76.8 kcal; (b) 5.6

19. -204.0 kcal

21. $H_2(g)$, $Hg(\ell)$

23. -39.6 kcal; -41.7 kcal; 5%

25. $\Delta E = +76.2$, $q_p = \Delta H = +76.8$; $w = +0.59$ kcal

27. (c) -13.1 ± 0.2 kcal/mole

Chapter 8

1. C most favors player; D most favors house.

3. 386

5. (b) 4 ϵ, 6.7%; 0 ϵ, 33.3%; (c) $1\frac{1}{3}$ ϵ

7. (b) 35.5%; (c) $K = 0.143$

9. (b) 23.1%; (c) $K = 0.083$

11. (a) 18; (c) 60; (e) A_2, 1.17; A, 1.66; (f) 0.425. Yes.

13. (a) 0.068; (b) 0.0047; (c) 0.261, 0.068

Chapter 9

1. $\Delta E = 0$, $\Delta H = 0$, $w = q = 1 \cdot 10^5$ ℓ-atm $= 2420$ kcal

3. $\Delta E = 0$, $\Delta H = 0$, $w = q = 6.93 \cdot 10^4$ ℓ-atm $= 1.68 \cdot 10^3$ kcal

5. (b) 8.73 · 10³ kcal

7. $w_{elect.} = 57.6$; $w_q = 14.4$; $q - w_q = -29.4$ calories

9. Examples: Process 1, $q' = +22$, $w' = -65$, $\Delta E' = +87$; Process 1', $q' = +194$, $w' = +281$; $\Delta E' = -87$ cal

11. $\Delta S_\infty = 1.40 \dfrac{PV}{T} = 1.40 \cdot nR$; $\Delta S_2 = \Delta S_\infty$

13. $\Delta S_\infty = -0.050$ cal/deg; $\Delta S_2 = \Delta S_\infty$

15. System: (a), (d) increase and (b), (c) decrease; universe: all increase

17. $\Delta H = -87$ kcal/mole; $\Delta S = -50$ cal/deg mole; $\Delta G = -72$ kcal/mole

19. (b) $q = -7.6$ kcal; $w_{elect.} = +38.2$ kcal; $t = 5.20$ hrs; $\Delta S = +38.6$ cal/deg; $\Delta G = -57.3$ kcal

21. $\Delta S_{vap.}$: Cl_2, 20.4 cal/deg; C_6H_6, 20.8 cal/deg; Na, 20.7 cal/deg. The constancy is called "Trouton's Rule."

23. Most positive, (b); most negative, (d)

25. (a) ΔS is positive; (b) ΔH is negative; (c) ΔG is negative and both ΔH and ΔS make ΔG negative

Chapter 10

1. −75.66 kcal

3. −77.98 kcal

5. e.g., (10-1), $\Delta S^0 = -6.61$ cal/deg, negative as predicted

7. $\Delta H = -13.6$ kcal; $\Delta S = -51.1$ cal/deg

9. $\Delta H = -0.51$ kcal; $\Delta S = +0.11$ cal/deg

11. $\Delta G^0 = +0.10$; $\Delta G = -2.63$ kcal; toward products

13. $K = 0.84$; P(HONO) $= 1.1 \cdot 10^{-3}$ atm

15. $\Delta H^0 = +30.00$ kcal; $\Delta G^0 = +21.01$ kcal; $\Delta S^0 = +30.20$ cal/deg; $K = 4.0 \cdot 10^{-16}$

17. $K = 1.11$

19. $K = 0.0307$; P(HONO) $= 5.7 \cdot 10^{-3}$ atm

21. $298°$K, 28.6%; $250°$K, 25.3%

23. (b) CaCO$_3$, $\Delta H = 7.88$ kcal, $\Delta S = 16.74$ cal/deg; (c) $\Delta G(700°$K$) = +15.62$ kcal

25. K(520) $= 2.92 \cdot 10^{-10}$ atm^2; $\Delta G^0(520) = +22.7$ kcal; $\Delta H^0 = +57.6$ kcal; $\Delta S^0 = 67.1$ cal/deg

27. 2.52 volts

29. (e) ΔH^0(gas) $= -239$, $+256$, and -222 kcal; (f) ΔS is approximately zero since all species are monatomic

31. (a) $\Delta H > 0$, $\Delta S > 0$; (b) $\Delta H < 0$, $\Delta S < 0$

33. (b) 3.07 kgm; (c) $\Delta H = -1.73$ kcal/kgm fuel; (e) $\Delta G = -501$ kcal

35. It is a very good approximation.

Chapter 11

1. (b) $2.9 \cdot 10^{-11}$ mole/cc-sec; (c) $-1.45 \cdot 10^{-11}$ mole/cc-sec

3. $k = 1.86 \cdot 10^{+7}$ cc^2/mole^2-sec

5. ninefold increase

7. $8.3 \cdot 10^8$ min (1600 years)

9. 11 collisions/sec

11. 4.5 kcal/mole

13. $Z = 6.0 \cdot 10^{-9}$ cc/molecule-sec $= 3.6 \cdot 10^{15}$ cc/mole-sec; $p = 0.011$

15. $A = 3.9 \cdot 10^{10}$ cc^2/mole^2-sec

17. $E_{a,r} = 42.3$ kcal/mole

19. (a) $\Delta H_I = +22$ to $+24$ kcal; (b) 59 kcal is unreasonable; 35 kcal is best

21. $\Delta(O_3)/\Delta t = -k[O_3]^2/[O_2]$

23. $\tau_{\frac{1}{2}} = 0.693/k$; for ClCOOH at $298°$K, $\tau_{\frac{1}{2}} = 256$ μsec

25. C and E

27. $\Delta[Br_2]/\Delta t = -\dfrac{k_{-1}k_{-2}}{k_2}\dfrac{[H_2O][Br_2][NO]}{[HBr]}$

29. D

31. Chain terminating reactions: Cl + Cl + M $\longrightarrow$ Cl$_2$ + M; Cl + COOH $\longrightarrow$ ClCOOH; COOH + COOH $\longrightarrow$ HOOCCOOH

Chapter 12

1. All do.

3. 23.4 cycles per second

5. 50,000 tons

7. About seven million gallons

9. 5540 Å

11. (b) 3.3 Å

13. 2.7 cm/sec

15. 19.6 kcal/mole

17. (a), (b), (f), (g)

19. $\bar{r}(He^+,2s) = 2\,\bar{r}(H,1s)$

Chapter 13

1. (a) 313.6 kcal; (c) $E_1 = 209$, $E_2 = 314$ kcal

3. $Z^* = 2.93$

5. $\bar{r} = 2.26$ Å

7. $1s^2\,2s^2 2p^6\,3s^2 3p^6$; Cl^- is identical in orbital occupancy to Ar

11. (d) $E_2(Al)$ removes 3s electron, while $E_2(Si)$ removes 3p electron; both $E_1(Al)$ and $E_1(Si)$ remove 3p electrons from half-occupied orbitals, but Si has higher Z.

13. For the state at 170.0 kcal/mole (a) $1s^2 2s^1 3p^1$; (b) $E_1 = 44.9$ kcal/mole; (d) $Z^* = 1.14$; (e) $\bar{r} = 5.80$ Å

15. (a) e.g., $Z^*(Na^+) = 3.73$; (b) $Z^*(F^-) \cong 0.96$; (c) $E_0(F^-) \cong 72$ kcal (experimental value = 79.5)

17. (b) $Z^*(N^-) = 0.3 - 0.5$; (c) $E_0(N^-) = 13 \pm 6$

19. $\bar{r} = 3.9$ Å

21. $\bar{r}$: F^-, 2.63; Ne, 1.05; Na, 3.88 Å

Chapter 14

1. $\Delta\bar{T} = +235.2$; $\Delta\bar{V} = -470.4$ kcal/mole

3. (c) and (g)

5. $r = 2.00$ Å, $\bar{V}_{AB} = +498$ kcal; $D_0/\bar{V}_{AB}$ at $r = r_0 = 0.094$

7. e.g., electron–proton attractions, $H + H$, -1254; H_2, -2580; change $= -1326$ kcal/mole

9. (e) At the "turn-around" point, $\bar{E} = -313.6$, hence $\bar{E}-\bar{V}_{AB} = -866$, $\bar{T} \cong 600$, so $\bar{V}-\bar{V}_{AB} \cong -1466$ kcal.

11. (a) 91 kcal; (b) $\bar{r} = 2.95$ Å

13. (b) $D_0 - \bar{D}$ for CF = 60.7, for CH = 3.8 kcal; (c) CF > CCl > CBr > CH; (d) the experimental dipole moments for H_2CF_2, H_2CCl_2, and H_2CBr_2 are, respectively, 1.96, 1.57, and 1.43 D

Chapter 15

1. $D_0 = 36 \pm 15$; $r_0 = 1.44 \pm 0.05$; $k = 4.5 \pm 1.4$

3. BO: $D_0 = 190 \pm 30$ kcal; $r_0 = 1.16 \pm 0.04$ Å; $k = 14 \pm 3$ (the experimental force constant for BO is $k = 13.3$)

5. NO^+: $D_0 = 250 \pm 25$; $r_0 = 1.07 \pm 0.03$; $k = 26 \pm 4$
   NO^-: $D_0 = 93 \pm 20$; $r_0 = 1.25 \pm 0.05$; $k = 9.5 \pm 2$

7. BC^+, C_2^+, NO^+

9. The available orbitals in O_2 and NO for electron-capture are antibonding, hence high in energy, giving low electron affinities.

11. NO^+: $D_0 = 263$ kcal

13. $\Delta H = -57$ kcal

15. BF_3, NO_3^- and CO_3^{-2} are isoelectronic; they have the same $1\frac{1}{3}$ bond orders.

17. Last electron occupies the nonbonding molecular orbital ($p_x - p_x'$). Each fluorine atom carries half the electron charge.

Chapter 16

1. SF_2: p^2 bonding, bent, bond angle $\sim 100°$
   NCl_3: p^3 bonding, pyramidal, bond angles $\sim 100°$

3. p^2 bonding would give a bent CH_2 with bond angle near that of water. This state is known; it has slightly higher energy than linear CH_2, possibly 10–20 kcal, and the bond angle is 102.4°.

5. Linear NH_2, with sp hybridization. This excited state of NH_2 is known.

9. NH_2^-: p^2, bent, 104° in $KNH_2(s)$; NH_4^+: sp^3, tetrahedral; NH_3^+: sp^2, planar, 120°

11. Three structures in order of decreasing stability: linear; bent, 90° angle; bent, 120° angle. Linear structure has 3 ℓp–ℓp, 6 ℓp–bp, and 0 bp–bp repulsions.

13. Three equivalent structures, repeated every 120°

15. Yes.

Chapter 17

1. (a) triple bond, as in N_2; (b) $\Delta H = +288$ kcal; (e) 38.9 kcal/bond order

3. (d) $K(N_2) = 1.2 \cdot 10^{-28}$; $K(C_2H_2) = 2.9 \cdot 10^{+42}$

5. (c) Formic acid, bond orders $1.74 + 1.36 = 3.10$ compared to formula, $2 + 1 = 3$

7. (b) and (c) 125.7 kcal state: bond order $\cong 3.0$, M.O. occupancy, antibonding electron excited to higher, nonbonding orbital; 131.4 kcal state: bond order $\cong 1.2$, M.O. occupancy, bonding electron excited to antibonding orbital, giving $1\frac{1}{2}$ order bond.

9. (b) 1.24 Å; the best current experimental value is 1.21 Å

11. (a) LiH, r_0(est.) = 1.42; CH, r_0(est.) = 1.07; av. % error = 5%

13. (a) C–C, r_0(est.) = 1.45; C–F, r_0(est.) = 1.32; av. % error = 6%

15. O–H has X_O–X_H = 1.4; O–F has X_O–X_F = −0.5; H_2O should have higher μ, with O carrying negative charge. In F_2O, F atoms carry negative charge. Over-all batting average, $\frac{3}{4}$.

17. (a) C–H bond, C will carry negative charge; (b) C–Cl bond, Cl will carry negative charge; (c) bond dipoles are additively oriented

19. (a) $\bar{D}$(C–H) = 92.8; $\bar{D}$(C–Cl) = 69.2; (b) $\Sigma\bar{D}(H_3CCl)$ = 347.5; (c) $\Sigma(D_0 - \bar{D})$ = 29.1 kcal for H_3CCl.

21. (a) 4400; (b) $2.3 \cdot 10^{-8}\ M$

23. (b) −131.4 kcal; (c) +1.43 volts; (d) $BrO_3^- > ClO_3^- > IO_3^-$

25. (a) Geometry same as BrF_3; (c) $D_0/D_{0(ICl)}$ = 0.74; (d) $K = 1.0 \cdot 10^{-3}$

27. Regular, 4-sided pyramid, oxygen at the apex, Xe in the plane defined by the 4 F atoms

29. Phenol > t-butanol > methanol > chloroform

31. Benzophenone, ΔH(expt.) = −2.2 kcal

33. ΔH = −3.1 kcal

Chapter 18

1. (a) molecular < ionic ≈ metallic < covalent. (b) White phosphorus forms a molecular crystal of P_4 molecules. Arsenic has a similar molecular crystal (yellow arsenic), but it is unstable with respect to a covalent, two-dimensional sheet-like lattice (gray arsenic) that has a bright lustre and electrical conductivity comparable to graphite. (c) Sulfur forms a molecular crystal containing S_8 molecules; selenium forms zig-zag infinite chains that lie parallei to each other in the crystal; selenium has moderate electrical conductivity.

3. 6.6 mm (about $\frac{1}{4}$ inch)

5. (a) 6.5 ± 0.30 cal/deg; (b) 6.05 cal/deg, no; (d) NO_3^-, 10.4 cal/deg

7. (c) 4.0 cal/deg per mole of atoms

9. (b) calc., 76% Cu in bronze (actually, 80%)

11. (a) 60.27 Å³; (b) 3.921 Å; (c) r_{cep} = 1.351

13. 25,500 atm

15. 12 contacts, 300 cal per bond

17. (a) 2.83 kcal; (c) ΔH = +1.8 kcal

19. (a) O–H bond is lengthened by 0.07 Å; (b) 0.59; (c) 0.92; interatomic distances are much less than van der Waals radius sum, revealing bonding.

21. (a), (b), and (c) no; (d) yes; (e) no; (f) high ΔH_{fusion}, high solid density, high m.p. for NH_3; (g) high entropy of vaporization = $\Delta H/T$, low molar volume, high boiling point for ammonia

23.   (a) 0.012 mole/sq. cm.; (b) 3.5 mm

25.   1.21, 1.19

27.   (b) hardness, very high melting point

29.   (a) (i) metallic; (ii) molecular; (iii) covalent; (iv) ionic
(b) A. (iii); B. (ii); C. (iii); D. (i); E. (i); F. (i); G. (iv); H. (iv); I. (iv); J. (ii).

Chapter 19

1.   (a) 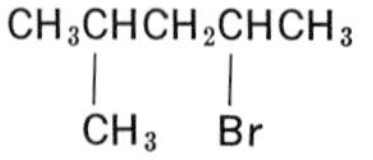 CH$_3$CHCH$_2$CHCH$_3$

        | |
      CH$_3$ Br

    4-methyl-2-bromopentane

(d)   CH$_2$—CH$_2$

      | |
     CH$_2$—CH

            |
           OH

    cyclobutanol

(b) 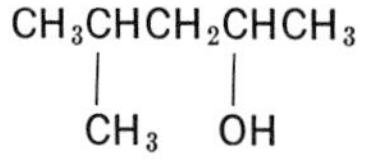   CH$_3$CHCH$_2$CHCH$_3$

        | |
      CH$_3$ OH

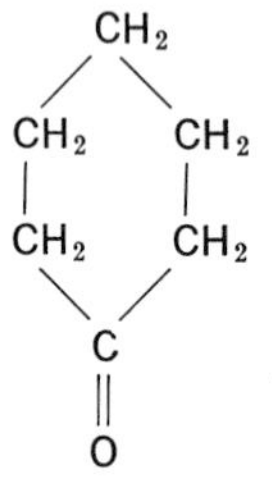

    4-methyl-2-pentanol

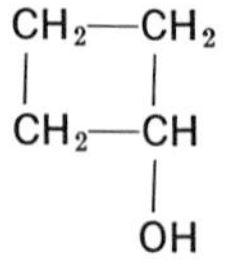

(e) 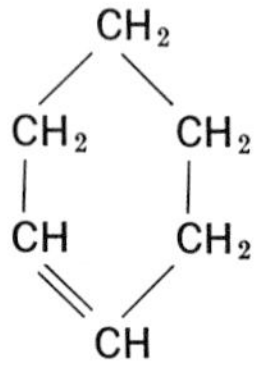   CH$_2$

  CH$_2$ CH$_2$

  CH   CH$_2$

     CH

  cyclohexene

(c)   CH$_2$

  CH$_2$ CH$_2$

  CH$_2$ CH$_2$

     C

     ‖
     O

  cyclohexanone

3. 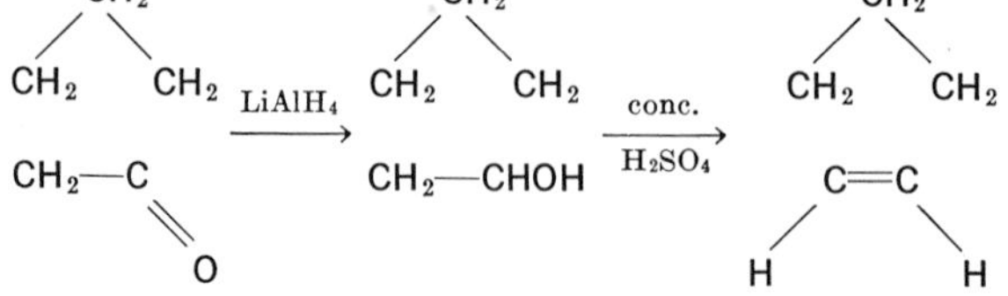   CH$_2$

CH$_2$ CH$_2$ $\xrightarrow{\text{LiAlH}_4}$

CH$_2$—C

      O

CH$_2$

CH$_2$ CH$_2$ $\xrightarrow[\text{H}_2\text{SO}_4]{\text{conc.}}$

CH$_2$—CHOH

CH$_2$

CH$_2$ CH$_2$

  C=C

  H   H

5.   (a) NH$_3$; (b) Na; (c) aqueous NaOH; (d) alcoholic KOH

7.   (a) $\Delta H = -8$; $\Delta S = +7$; (b) Cyclopropane will be less stable than propene because of ring strain. Hence the negative ΔH in the reaction A ⟶ B identifies A as cyclopropane. Because of its symmetry, cyclopropane is more highly ordered than propene, so the positive ΔS (randomness increasing) also shows that A is cyclopropane.

9.   (a) 2,4-dimethylpentane
(b) and (c) 2,3-dimethylpentane
(d) 2,2,3-trimethylbutane

(e) and (f) 3-ethylpentane
(g) 2,2-dimethylpentane
(h) 3,3-dimethylpentane

11. $CH_3CH_2CH_2OH$, 1-propanol;
     CH_3CHCH_3, 2-propanol;

     $\overset{|}{OH}$

     $CH_3CH_2OCH_3$, methyl ethyl ether

(Two other possibilities are cyclopropanol $\overset{CH_2}{\underset{CH_2-CHOH}{\diagup\diagdown}}$ and the cyclic ether,

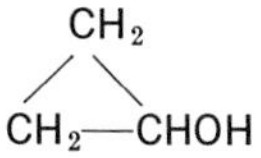

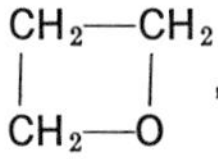

$\underset{CH_2-O}{\overset{CH_2-CH_2}{\mid\qquad\mid}}$, called trimethylene oxide.)

13. (a) $-CONHCH_2- + \tfrac{9}{4}O_2 \longrightarrow 2\ CO_2 + \tfrac{3}{2}H_2O + \tfrac{1}{2}N_2$; (b) $\Delta H_{comb.} = -249$ kcal

15. 2.6 milliseconds

17. (a) a and c; (b) b and d (both *cis* and *trans*); (c) b and d (both *cis* and *trans*);
     (d) d (*trans* only)

19. (a) $CH_3CH_2CHO \xrightarrow{\text{LiAlH}_4} CH_3CH_2CH_2OH \xrightarrow{\text{PCl}_3} CH_3CH_2CH_2Cl$
     (b) $CH_3CH_2CHO \xrightarrow{\text{LiAlH}_4} CH_3CH_2CH_2OH \xrightarrow[\text{H}_2\text{SO}_4]{\text{conc.}} CH_3CH=CH_2 \xrightarrow{\text{HCl}} CH_3CHClCH_3$
     (Markovnikov's rule)

21. (a) CH_3CClCH_3;   (b) CH_3CClCH_3;
         $\overset{|}{CH_2Cl}$ $\overset{|}{CH_3}$

(c) 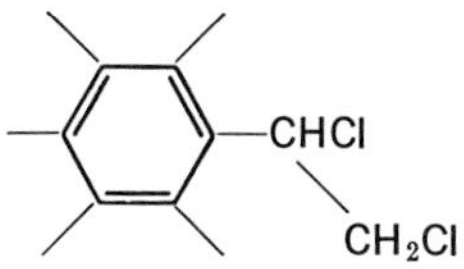   (d)

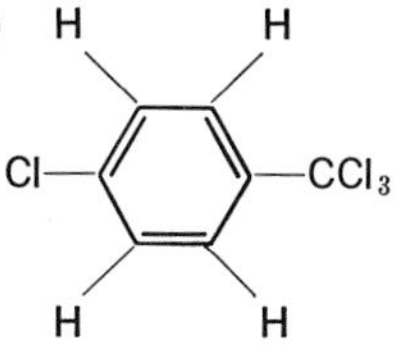

23. (b) 2-methyl propionic acid; (c) 2-methyl-1-propanol; 2-methyl-1-propanal

25. (a) d, f; (b) a; (c) c; (d) a; (e) b, e

27. (a) B, E, H; (b) E

29. 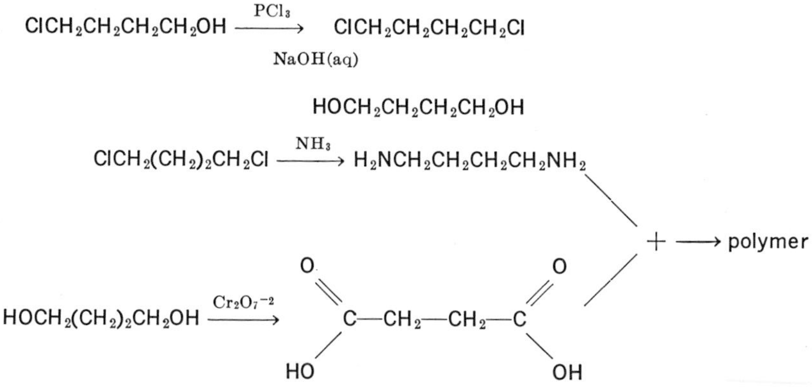 $ClCH_2CH_2CH_2CH_2OH \xrightarrow{\text{PCl}_3} ClCH_2CH_2CH_2CH_2Cl$
     $\xrightarrow{\text{NaOH(aq)}}$
     $HOCH_2CH_2CH_2CH_2OH$

     $ClCH_2(CH_2)_2CH_2Cl \xrightarrow{\text{NH}_3} H_2NCH_2CH_2CH_2CH_2NH_2$

     $+ \longrightarrow$ polymer

     $HOCH_2(CH_2)_2CH_2OH \xrightarrow{\text{Cr}_2\text{O}_7^{-2}} \underset{HO}{\overset{O}{\|}}C-CH_2-CH_2-\underset{OH}{\overset{O}{\|}}C$

31. (b) No; (c) no; (d) $+5.5/7.0 \times 100 = 79\%$

Chapter 20

1. (a) Increasing nuclear charge; (b) multiple bond orders give opposite trend to single bond radius trend: SiO has a triple bond, PO has a $2\frac{1}{2}$ order bond, and SO has a double bond.

3. Liquid ammonia does show evidence of hydrogen bonding.

5. No first-row effect is displayed by the trihalides.

7. (a) Triphenylphosphine; (c) 0.60 atm

9. (b) $K = 3 \cdot 10^{11}$, $\Delta G^0 = -36$ kcal

11. (a) More positional randomness with two particles; (b) $\Delta S_I - \Delta S_{III} = -11$ kcal/deg; negative sign shows less additional randomness for vaporization of a mole of particles than for vaporization of two moles of particles; (c) 0.014 M

13. (b) $K_{HOF} = 1.0$

15. HOCl, $5.6 \cdot 10^{-8}$ M; HOClO, $2.8 \cdot 10^{-2}$ M; HOClO$_2$, 0.10 M; HOClO$_3$, 0.10 M

17. (b) 32.3 kcal; (d) ΔH_f^0, 18%; D_0, 7%

19. Tetrahedral, sp^2; bent, p^2; CH$_4$ and H$_2$O

21. (b) -70.6, -3.7 kcal; (c) -37.1 kcal, $\mathcal{E}^0 = +0.40$ volt

23. Tetrahedral (as it is)

25. (a) $\Delta(S_2O_3^{-2})/\Delta t = -k[S_2O_3^{-2}][O_2][H^+]^2$; (f) faster by 10^4

27. (c) 2326 Å (this is in the ultraviolet); (d) the infrared wavelength range from 1.09 to 1.84 microns

29. (c) 0.46

31. (a) By a factor 1.027, or by 2.7%

Chapter 21

1. (a) 5.3%; (b) 2.2%

3. $[Cr^{+2}] = 0.15$ M; $[Cr^{+3}] = 1.4 \cdot 10^{-9}$ M

5. $[Fe^{+2}] = 0.304$ M; $[Cr^{+2}] = 0.0085$ M; $[Cr^{+3}] = 0.0015$ M; $[Fe^{+3}] =$ negligible

7. $[Fe^{+2}] = 0.200$ M; $[Sn^{+2}] = 0.120$ M; $[Sn^{+4}] = 3.1 \cdot 10^{-12}$ M; $[Fe^{+3}] = 3.5 \cdot 10^{-17}$ M

9. (a) Iron electrode is the anode; left cell final conc., $[Fe^{+2}] = 0.200$ M; right cell final concs., $[Fe^{+2}] = 0.200$ M, $[Sn^{+2}] = 1.1 \cdot 10^{-11}$ M
(b) aluminum electrode is the anode; left cell final conc., $[Al^{+3}] = 0.233$ M

11. (a) $[Fe^{+2}] = [Mn^{+2}] = 0.010$ M, $[CrO_2^-] = 1.6 \cdot 10^{-8}$ M, $[Fe^{+3}] = 1 \cdot 10^{-20}$ M; (b) Fe(OH)$_3$(s) + Cr(OH)$_3$(s); (c) $[CrO_2^-] = 0.010$ M; (d) $[Mn^{+2}] = 0.010$ M, $[Fe^{+2}] = 1.4 \cdot 10^{-4}$ M, $[CrO_2^-] = 1.6 \cdot 10^{-8}$ M; (e) Fe(OH)$_3$(s); (f) Fe(OH)$_2$(s)

13. (a) $[Fe^{+2}] = 0.020$ M, $[Mn^{+2}] = 0.010$ M, $[CrO_2^-] \doteq 1.6 \cdot 10^{-8}$; (b) Cr(OH)$_3$(s); (c) $[CrO_2^-] = 0.010$ M; (d) $[Mn^{+2}] = 0.010$ M, $[Fe^{+2}] = 1.4 \cdot 10^{-4}$ M, $[CrO_2^-] = 1.6 \cdot 10^{-8}$ M; (e) no ppt.; (f) Fe(OH)$_2$

15. (a) Highly charged ions bind H_2O rigidly, creating order; (b) $\Delta S^0 = +79.9 \frac{cal}{deg}$, positive entropy increases reaction tendency; (c) $\Delta \mathcal{E}^0 = +0.618$, $\Delta G^0 = -14.25$ kcal; yes

17. (a) Cu^+, 1.52 Å; Cu^{+2}, 1.13 Å; (b) 4 water molecules

19. (a) $\Delta G^0 = -24$ kcal; $K = 3.9 \cdot 10^{+17}$; (b) $2 \cdot 10^{+10}$

21. (a) $CrCl_3(H_2O)_6$; (d) yes: G could have two structures; see Figure 21-8

23. (c) Cl^-, 98%; F^-, 77%

25. $\Delta \mathcal{E}^0 = +0.381$ volt; $\Delta G^0 = -17.6$ kcal; $K = 7.6 \cdot 10^{+12}$

27. K_{sp} $Mg(OH)_2/K_{sp}$ $Be(OH)_2 = 62.5$ $r_{Mg}/r_{Be} = 1.42$
    K_{sp} $Ba(OH)_2/K_{sp}$ $Sr(OH)_2 = 3.3$ $r_{Ba}/r_{Sr} = 1.05$

29. (b) $\bar{r}_{6d}/\bar{r}_{5f} = 1.35$; $\bar{r}_{7s}/\bar{r}_{6d} = 1.24$

31. (a) and (c) $^{59}_{27}Co$; (b) $^{60}_{28}Ni$

33. (a) $k = 3.14 \cdot 10^{-9}$ sec^{-1}; (b) $3.14 \cdot 10^6$ decays/sec; (c) $6.3 \cdot 10^{-4}$ calories

35. $9.76 \cdot 10^{14}$ radioactive nuclei remain; $3.06 \cdot 10^6$ decays per second

37. (a) $k = 1.223 \cdot 10^{-4}$ year^{-1}; (b) 0.786

39. (a) 54 cc; (b) 40,000 cu ft; (c) 380 million cubic feet (That's a lot of smog!)

formula index

subject index

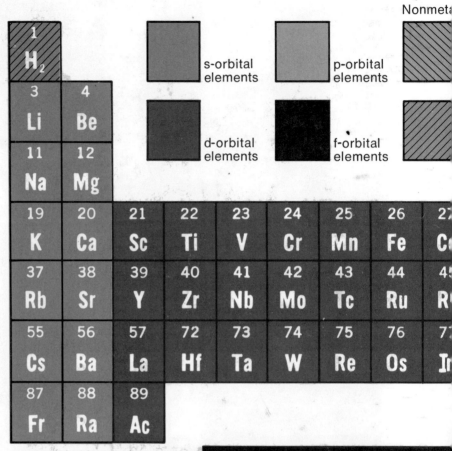